Yu. A. Mitropol'skii

Problems of the Asymptotic Theory of Nonstationary Vibrations

Translated from Russian by
Ch. Gutfreund

Israel Program for Scientific Translations
Jerusalem 1965

This book is a translation of
PROBLEMY ASIMPTOTICHESKOI TEORII
NESTATSIONARNYKH KOLEBANII

Izdatel'stvo "Nauka"
Moskva 1964

IPST Cat. No. 2167

Printed in Jerusalem by S. Monson
Binding: K. Wiener

Table of Contents

ANNOTATION

This book gives approximate asymptotic methods for solving problems in the theory of nonstationary vibrational processes.

The author's methods can be applied to a wide range of nonstationary vibrational processes in nonlinear systems. They help in the solution of problems on the transition through resonance, investigations into nonstationary processes in gyroscopic systems, accelerating devices, automatic regulation systems, and other important problems of physics and engineering.

The book will be useful to a wide class of engineers and scientists interested in the various problems that arise in the theory of vibrations and differential equations containing a small parameter.

Foreword

Research into vibrational processes is of unquestionable importance to modern physics and engineering.

Nonstationary processes that occur because of variations in frequency, mass, and other parameters are among the basic problems in the theory of vibrations in nonlinear vibrating systems. The author has developed a very effective method of investigation, applicable to many cases of vibrational processes. The method only requires that the system parameters vary slowly compared to the "normal periods" of vibrations; this requirement is often satisfied.

Mathematically, the problem reduces to studying nonlinear differential equations with "slowly varying" coefficients, or, which is the same, equations with a small parameter preceding the highest derivative.

In 1955 the author published several articles and a monograph, "Nestatsionarnye protsessy v nelineinykh kolebatel'nykh sistemakh" (Nonstationary Processes in Nonlinear Vibrating Systems). The original book, now difficult to obtain, has been translated into English, Chinese, and Japanese. The methods given have been developed and generalized by the author as well as by several other scientists.

The main purpose of our book is to systemize the method used to study nonstationary vibrations in nonlinear systems with one or many degrees of freedom. A rigorously justified method of studying monofrequency vibrational processes in systems with many degrees of freedom is given.

The investigations into transition through resonance in nonlinear systems with one or many degrees of freedom deserve careful study. Until now such problems were only more or less fully solved in the simpelr cases of linear systems. The author's results are thus of much present and future value.

An effective algorithm is rendered through which approximate solutions, convenient for their practical application to solving widely diverse problems, can be constructed. A series of problems in the theory of differential equations containing a small parameter are considered. Problems of the asymptotic convergence of the approximate solutions are discussed. Special attention is paid to proving intricate theorems on the existence and stability of two-parametric families of solutions; these theorems justify the author's monofrequency method of investigating vibrating systems.

The appearance of this book is thus most welcome. It is a valuable contribution to the important yet insufficiently studied field of nonstationary vibrations and the associated nonlinear differential equations containing a small parameter.

Academician N.N. Bogolyubov

Introduction

It may be claimed without exaggeration that the problems of nonlinear vibrations constitute at present one of the dominating fields in the widely diverse branches of physics and engineering.

The methods of asymptotic expansions in powers of a small parameter, developed by N. M. Krylov and N. N. Bogolyubov, are a most effective tool for analyzing nonlinear vibrations. They allow us to obtain relatively simple schemes of calculation and to explain the detailed features of the oscillatory process in many cases of practical importance. These methods have recently been widely extended and applied to the solution of various problems connected with the study of nonlinear oscillatory phenomena.

Many actual problems of physics and engineering, e. g., the construction of high-speed power and industrial units, the necessity of simplifying their structure, the study of complicated phenomena in the design of regulation systems, in the design of gyroscopic systems, the investigation of oscillatory phenomena in constructing modern powerful accelerating devices, the calculation of rocket trajectories and satellite orbits, the study of transient wave processes, etc., involve the basic problem of the theory of nonlinear vibrations, i. e., the study of nonstationary phenomena arising because of variations in frequencies, masses, and other parameters of a nonlinear vibrating system.

The need to investigate nonstationary vibrating processes arises, for example, in the widely extended problem of transition through resonance (e. g., in calculating the vibrations of turboengines, turbine blades, centrifuges, various gyroscopic systems, etc.). It also arises in the study of vibrations in systems with variable mass and rigidity, in problems associated with vibrations of bridges and cranes subject to mobile loads and pulsating forces, in numerous problems of electrical and radio engineering connected with modulation problems, in the calculation of rocket trajectories in the active section, in the calculation of resonance phenomena in the process of particle acceleration in a synchrophasotron, etc.

Some of these problems have already been investigated, but only in a linear formulation and with many essential simplifications. Thus, for example, the nonstationary processes in vibrating systems under the action of external forces of a variable frequency (the problem of transition through resonance, etc.) have been considered mainly for a system with a single degree of freedom and in a linear formulation with a few simplifications. At the same time, the nonstationary vibrational processes, the study of which is necessary for solving the aforementioned actual problems of modern physics and engineering, are in most cases described by nonlinear differential equations with varying coefficients.

If these equations are close to linear ones and possess some special properties and if their coefficients vary slowly compared to a "natural" time unit (a time unit of the order of the vibration period), then it is possible to work out for such systems with one, as well as with many degrees of freedom (including the case of infinitely many degrees of freedom), a quite general method which allows the problem to be solved completely, not only qualitatively but also quantitatively. The method is based on the ideas of the aforementioned asymptotic methods of nonlinear mechanics.

The development of such a theory was begun in the year 1948 by the author of the present monograph. In the year 1955 he published the book "Nonstationary Processes in Nonlinear Vibrating Systems" /104/. It gave an effective method of investigation applicable to an extensive class of oscillatory processes described by nonlinear differential equations and systems of nonlinear differential equations with variable coefficients. The method can be applied if there is a "slow" variation of the system parameters in comparison with the "normal periods" of vibrations; this is indeed always satisfied in the problems that are generally considered.

After the publication of this book the author obtained a series of new results. In addition, the methods presented in the book found extensive application and were further developed by other authors.

The main purpose of this monograph is to present the method of asymptotic expansions in powers of a small parameter. This method is used when investigating nonstationary modes in nonlinear vibrating systems with slowly varying parameters. An exhaustive study is made of nonlinear equations with slowly varying parameters close to linear ones.

The examples discussed are predominantly illustrative, and there is no claim to cover all the problems encountered in the theory of nonstationary vibrations.

The book has been written for engineers and scientists, and also for postgraduate students and students of higher courses, interested in various problems in the theory of vibrations and differential equations containing a small parameter. In order to make the book accessible to people dealing with applied problems, the results of theoretical investigations are given as formulas, and clear rules are given for solving without too much difficulty many complicated problems in the theory of nonlinear vibrations. In addition, attention is paid to practical methods of solving a series of problems so that the complicated phenomena observed in nonlinear vibrating systems in a nonstationary mode can be analyzed completely. All the methods presented in the book are illustrated by actual problems which are also of independent interest.

It should be mentioned that we have not attempted to develop and present completely the asymptotic methods of solving equations with slowly varying parameters. We have also not given an account of the wide literature, which appeared recently, devoted to theoretical questions as well as to an investigation of numerous actual problems and an explanation of new physical phenomena. This is difficult to do in a single book. We assume that the reader who is interested in the problems touched upon in this book, will, after getting acquainted with the methods and problems presented here, turn to a more detailed study of the mathematical problems that occur in the last chapter, and of the numerous physical phenomena. He will be guided in this by the list of references of the basic literature on the theory of nonlinear vibrations and on problems of nonstationary oscillatory processes that appears at the end of this book.

In conclusion the author expresses deep gratitude to his teacher Academician Nikolai Nikolaevich Bogolyubov. Discussions with him contributed greatly to an understanding of many problems considered during the elaboration of the given theory.

This author also wishes to thank V.M. Volsov, Doctor of physical-mathematical sciences, and O.B. Lykov, Candidate of physical-mathematical sciences, for reading the manuscript and making valuable remarks; also A.Ya. Gadionenko for assistance in preparing the manuscript for press.

Yu.A. Mitropol'skii

Chapter I

EXAMPLES OF DIFFERENTIAL EQUATIONS ENCOUNTERED WHEN INVESTIGATING NONLINEAR VIBRATING SYSTEMS WITH SLOWLY VARYING PARAMETERS

§ 1. Typical differential equations, basic definitions and restrictions

The investigation of nonstationary processes in vibrating systems with one degree of freedom may in many cases be reduced to the solution of a nonlinear second order differential equation with variable coefficients of the type

$$\frac{d}{dt}\left[m(\tau)\frac{dx}{dt}\right]+c(\tau)x=\varepsilon F\left(\tau, \theta, x, \frac{dx}{dt}\right). \tag{1.1}$$

Here x may be taken as a linear displacement, current intensity, angular displacement, or torsion angle; $m(\tau)$, asthe mass of the system, inductance coefficient, or moment of inertia; $c(\tau)$, as the rigidity of the system, or inverse capacitance, etc.; θ is the phase of the perturbing force $\left(\frac{d\theta}{dt}=\nu(\tau)\right.$ is the instantaneous frequency of the perturbing force); $m(\tau)$, $c(\tau)$, $\nu(\tau)$ are some functions of slowing time $\tau=\varepsilon t$, where ε is a small positive parameter. This parameter indicates that the system is close to a linear conservative one, and that the coefficients in (1.1) vary slowly, i. e., that their derivatives with respect to the independent variable t are proportional to a small parameter ε, for example,

$$\frac{dm(\tau)}{dt}=\frac{dm(\tau)}{d\tau}\cdot\frac{d\tau}{dt}=\varepsilon\frac{dm(\tau)}{d\tau}. \tag{1.2}$$

It will be shown below when analyzing actual examples that this condition is not very strict, since the coefficients of (1.1) should vary slowly compared to a "natural time unit" (a time unit of the order of a period of normal vibrations of the system in question). Thus, it is possible to consider in high-frequency systems quite rapid (in the ordinary sense) variations of the system parameters.

The study of differential equations with slowly varying coefficients is equivalent to the study of differential equations in which a small parameter precedes the highest derivative.

Actually, an equation of the type

$$\varepsilon\frac{dx}{dy}+p(y)x=0 \tag{1.3}$$

assumes after introducing the independent variable

$$t=\frac{y}{\varepsilon} \tag{1.4}$$

the form

$$\frac{dx}{dt}+p(\varepsilon t)\,x=0, \qquad (1.5)$$

where the coefficient $p(\varepsilon t)$ is a slowly varying quantity in the previously mentioned sense.

Equations with a small parameter before the highest derivative were considered in numerous works: A. N. Tikhonov /156/, L. S. Pontryagin /135/, I. S. Gradshtein /36/, E. F. Mishchenko /118/, A. B. Vasil'eva /24/, V. M. Volosov /26/, and others. Their results will, however, not be discussed in what follows. The first form of equations — with slowly varying parameters — is retained throughout the present book.

Returning to equation (1.1), we assume that the expression $\varepsilon F\left(\tau, \theta, x, \frac{dx}{dt}\right)$ on the right-hand side is a periodic function of θ with period 2π. Since in most cases it is sufficient to investigate the nonstationary behavior in a finite time interval $[0, T]$, it is adequate to assume that the functions $m(\tau)$, $c(\tau)$, $F\left(\tau, \theta, x, \frac{dx}{dt}\right)$ and $\frac{d\theta}{dt}=\nu(\tau)$ have the desired number of bounded derivatives for any finite values of their arguments for any τ within the interval $0\leqslant\tau\leqslant L$, where $L=\varepsilon T$. It is assumed in addition to this, that the functions $m(\tau)$ and $c(\tau)$ are strictly positive for any τ. Consequently, the solutions of equation (1.1) oscillate in the interval $0\leqslant\tau\leqslant L$.

All these restrictions are compatible with the physical nature of actual examples to be considered in the sequel, and are required in order that asymptotic methods may be applied to the construction of approximate solutions for equations like (1.1).

It should be noticed that the interval $[0, T]$ under consideration, although finite, has a length of the order of $\frac{1}{\varepsilon}$. Therefore, when $\varepsilon\to 0$, the solution of equation (1.1) is approximated in the method suggested over a finite, but as long as desired time interval.

The function $F\left(\tau, \theta, x, \frac{dx}{dt}\right)$ on the right-hand side of (1.1) may also depend on the parameter ε, i. e., have the form $F\left(\tau, \theta, x, \frac{dx}{dt}, \varepsilon\right)$. Since this does not introduce additional essential difficulties, we shall keep for the sake of simplicity to the notation of equation (1.1).

One characteristic property of vibrating systems described by equations of the type (1.1) is that since the coefficients $m(\tau)$ and $c(\tau)$ of this equation, and also the right-hand side depend on the slowing time τ, the vibrations will, generally speaking, not display all the characteristic properties of oscillating processes in linear and nonlinear vibrating systems with constant parameters. Since the "frequency" of the external periodic force $\frac{d\theta}{dt}=\nu(\tau)$, the "normal frequency" of the system $\omega(\tau)=\sqrt{\frac{c(\tau)}{m(\tau)}}$, etc., depend on the slowing time τ, the usual concepts of a constant normal frequency characterizing the vibrating system, of harmonic oscillations, of an external sinusoidal force, etc., become meaningless. However, these terms will be used for simplicity in what follows, keeping in mind that $\omega(\tau)$, $\nu(\tau)$ are some known functions of slowing time.

The investigation of nonstationary processes in vibrating systems with many degrees of freedom leads in many cases to a system of differential equations of the type

$$\frac{d}{dt}\left\{\sum_{i=1}^{N} a_{ij}(\tau)\dot{q}_i\right\}+\sum_{i=1}^{N} b_{ij}(\tau)q_i=$$

$$=\varepsilon Q_j(\tau,\ \theta,\ q_1,\ \ldots,\ q_N,\ \dot{q}_1,\ \ldots,\ \dot{q}_N,\ \varepsilon)\quad (j=1,\ 2,\ \ldots,\ N), \tag{1.6}$$

where q_i $(i = 1, 2, \ldots, N)$ are generalized coordinates; $\varepsilon Q_j(\tau, \theta, q_1, \ldots, q_N, \dot{q}_1, \ldots, \dot{q}_N, \varepsilon)$ $(j = 1, 2, \ldots, N)$ are external perturbing forces, periodic in θ with period 2π; $\tau = \varepsilon t$ is the slowing time; ε is as before, a small positive parameter indicating the closeness of the system to a linear conservative one; $a_{ij}(\tau)$ and $b_{ij}(\tau)$ $(i, j = 1, 2, \ldots, N)$ are inertial and quasi-elastic coefficients, slowly varying in time, and satisfying for any τ, $a_{ij}(\tau) = a_{ji}(\tau)$, $b_{ij}(\tau) = b_{ji}(\tau)$.

Denoting the kinetic and potential energies of the vibrating system under investigation by

$$T=\frac{1}{2}\sum_{i,\ j=1}^{N} a_{ij}(\tau)\dot{q}_i\dot{q}_j,\quad V=\frac{1}{2}\sum_{i,\ j=1}^{N} b_{ij}(\tau)q_iq_j, \tag{1.7}$$

we may express equations (1.6) in the form

$$\frac{d}{dt}\left[\frac{\partial T}{\partial \dot{q}_j}\right]+\frac{\partial V}{\partial q_j}=\varepsilon Q_j(\tau,\ \theta,\ q_1,\ \ldots,\ q_N,\ \dot{q}_1,\ \ldots,\ \dot{q}_N,\ \varepsilon)\ (j=1,2,\ldots,N). \tag{1.8}$$

Nonstationary vibrations of a system with many degrees of freedom are in many cases characterized by a Lagrangian function of the form

$$\mathcal{L}=\frac{1}{2}\left\{\sum_{i,\ j=1}^{N} a_{ij}(\tau)\dot{q}_i\dot{q}_j+2\sum_{i,\ j=1}^{N} g_{ij}(\tau)q_i\dot{q}_j-\sum_{i,\ j=1}^{N} b_{ij}(\tau)q_iq_j\right\}, \tag{1.9}$$

where $a_{ij}(\tau)=a_{ji}(\tau),\ b_{ij}(\tau)=b_{ji}(\tau),\ g_{ij}(\tau)\neq g_{ji}(\tau)\ (i,\ j=1,\ 2,\ \ldots,\ N)$.

Supposing that the system is also subjected to external perturbing forces $\varepsilon Q_j(\tau, \theta, q_1, \ldots, q_N, \dot{q}_1, \ldots, \dot{q}_N, \varepsilon)$ $(j = 1, 2, \ldots, N)$ we obtain the following system of N second order differential equations with slowly varying parameters

$$\frac{d}{dt}\left[\frac{\partial \mathcal{L}}{\partial \dot{q}_j}\right]-\frac{\partial \mathcal{L}}{\partial q_j}=\varepsilon Q_j(\tau,\ \theta,\ q_1,\ \ldots,\ q_N,\ \dot{q}_1,\ \ldots,\ \dot{q}_N,\ \varepsilon)\qquad (j=1,\ 2,\ \ldots,\ N) \tag{1.10}$$

or, more explicitly,

$$\frac{d}{dt}\left\{\sum_{i=1}^{N} a_{ij}(\tau)\dot{q}_i\right\}+\sum_{i=1}^{N} c_{ij}(\tau)\dot{q}_i+\sum_{i=1}^{N} b_{ij}(\tau)q_j=$$

$$=\varepsilon Q_j(\tau,\ \theta,\ q_1,\ \ldots,\ q_N,\ \dot{q}_1,\ \ldots,\ \dot{q}_N,\ \varepsilon)-\sum_{i=1}^{N} \dot{g}_{ij}(\tau)q_i\qquad (j=1,\ 2,\ \ldots,\ N), \tag{1.11}$$

where $c_{ij}(\tau)=g_{ij}(\tau)-g_{ji}(\tau)$ and, hence, $c_{ij}(\tau)=-c_{ji}(\tau)$.

The system of equations (1.11), just like the system (1.6), may easily be replaced by a system of $2N$ first order differential equations. However, as shown in what follows, it is in many cases convenient to keep the equations in the form (1.6) or (1.11), since it is then possible to express the final formulas with the help of the expressions of kinetic and potential energies or by the Lagrangian when constructing approximate solutions. This simplifies the calculations and also allows the approximate solutions to be obtained without first constructing the exact differential equations of the problem.

As in the case of equation (1.1), we impose several restrictions on the coefficients and on the right-hand sides of the systems of differential equations (1.6) and (1.11). The quadratic forms T and V, defined by the expressions (1.7), are assumed to be positive definite in any finite interval $0 \leqslant \tau \leqslant L$, and the functions $a_{ij}(\tau)$, $b_{ij}(\tau)$, $c_{ij}(\tau)$, $\nu(\tau) = \frac{d\theta}{dt}$, $Q_j(\tau, \theta, q_1, \ldots, q_N; \dot{q}_1, \ldots, \dot{q}_N, \varepsilon)$ $(i, j = 1, 2, \ldots, N)$ are assumed to possess a sufficient number of derivatives for any finite values of their arguments $\tau, q_1, \ldots, q_N, \dot{q}_1, \ldots, \dot{q}_N$, and sufficiently small ε.

It has been assumed until now that the parameters characterizing the given vibrating system are known slowly varying functions of time and, consequently, that their variation is completely independent of the motion of the vibrating system according to a certain law.

A much more general case may occur in many interesting problems, when the variation of some parameters of the system is essentially affected by the motion of the given vibrating system, and is directly coupled to it. This coupling is often weak, and it may then be assumed under several restrictions that the parameters vary independently of the vibrating system (which is, for example, obtained in certain problems in the first approximation). However, there are many cases where it is necessary to take into account the mutual relation between the vibrating system and its characteristic parameters, varying in time, even in the first approximation.

A typical example is the problem of interaction between a vibrating system performing forced oscillations and the energy source, which is subject to the action of the vibrating system, — the variation of the vibration parameters is accompanied by a change in the operation behavior of the energy source /67/. Such an interaction becomes especially pronounced when the energy source has a small power reserve.

Such a treatment considerably complicates the study of a vibrational process, and instead of equations (1.1) and (1.6), more complicated systems of differential equations arise. The problem reduces in the most general case, as formulated by V. M. Volosov /32/, to a study of the system of differential equations of the type

$$\left.\begin{aligned} \frac{dx}{dt} &= X(t, x, y, \varepsilon) \equiv X_0(t, x, y) + \varepsilon X_1(t, x, y) + \ldots, \\ \frac{dy}{dt} &= \varepsilon Y(t, x, y, \varepsilon), \end{aligned}\right\} \tag{1.12}$$

where x, y are k- and m-dimensional vectors, respectively $(k + m = n)$, in an n-dimensional Euclidean space E_n; $X(t, x, y, \varepsilon)$ and $Y(t, x, y, \varepsilon)$ are k- and m-dimensional vector functions, respectively. It is assumed that the system (1.12) for $\varepsilon = 0$, i. e., the system

$$\left.\begin{aligned} \frac{dx}{dt} &= X_0(t, x, y), \\ y &= \text{const}, \end{aligned}\right\} \tag{1.13}$$

describes an oscillatory process.

The system (1.12) describes in this case the action of small perturbations, given by

$$\varepsilon X_1(t, x, y) + \varepsilon^2 X_2(t, x, y) + \ldots, \tag{1.14}$$

on the unperturbed system (1.13). Equation (1.12) also shows the existence of slowly varying parameters y, where

$$\frac{dy}{dt} = \varepsilon Y(t, x, y, \varepsilon). \tag{1.15}$$

The right-hand sides of these equations are proportional to a small parameter ε and satisfy some conditions which guarantee a slow variation of y; for example, $Y(t, x, y, \varepsilon)$ should possess a mean with respect to t.

In the case of the previously considered vibrating system with "one" degree of freedom and slowly varying parameters, the variation of which depends on the coordinates and velocities of the vibrating system, we have to deal with the following system of equations

$$\left.\begin{aligned} \frac{d}{dt}\left[m(y)\frac{dx}{dt}\right] + c(y)x = \varepsilon F\left(y, \theta, x, \frac{dx}{dt}\right), \\ \frac{dy}{dt} = \varepsilon f\left(t, x, \frac{dx}{dt}, y, \varepsilon\right). \end{aligned}\right\} \tag{1.16}$$

Assuming that the right-hand side of the second equation of the system (1.16) does not depend on x and $\frac{dx}{dt}$, and integrating it for the initial values $t = t_0$, $y = y_0$, we find

$$y = y(y_0, t_0, t, \varepsilon) = y_0 + \varepsilon\varphi_1(y_0, t_0, t) + \varepsilon^2\varphi_2(y_0, t_0, t) + \ldots \tag{1.17}$$

Substituting the value of y from (1.17) in the first equation of the system (1.16), we obtain the original formulation of the problem. In particular, setting $f \equiv 1$ in the system (1.16), we obtain $y = \varepsilon t = \tau$ $(t_0 = 0,\ y_0 = 0)$, and (1.16) transforms to equation (1.1).

A whole series of actual problems frequently leads to a study of systems of differential equations with slowly varying parameters of the form

$$\left.\begin{aligned} \frac{d}{dt}\left[\frac{\partial \mathscr{L}}{\partial \dot{q}_j}\right] - \frac{\partial \mathscr{L}}{\partial q_j} = \varepsilon Q_j(y, \dot{y}, q_1, \ldots, q_N, \dot{q}_1, \ldots, \dot{q}_N, \varepsilon), \\ \frac{d^2y}{dt^2} = \varepsilon \mathscr{Y}(y, \dot{y}, q_1, \ldots, q_N, \dot{q}_1, \ldots, \dot{q}_N, \varepsilon) \\ (j = 1, 2, \ldots, N), \end{aligned}\right\} \tag{1.18}$$

where

$$\mathscr{L} = \frac{1}{2}\left\{\sum_{i,j=1}^{N} a_{ij}(y)\dot{q}_i\dot{q}_j + 2\sum_{i,j=1}^{N} g_{ij}(y)q_i\dot{q}_j - \sum_{i,j=1}^{N} b_{ij}(y)q_iq_j\right\}, \tag{1.19}$$

and the functions $Q_j(y, \dot{y}, q_1, \ldots, q_N, \dot{q}_1, \ldots, \dot{q}_N, \varepsilon)$ $(j = 1, 2, \ldots, N)$, $\mathscr{Y}(y, \dot{y}, q_1, \ldots, q_N, \dot{q}_1, \ldots, \dot{q}_N, \varepsilon)$ are periodic in y with period 2π, and satisfy conditions analogous to those mentioned before in connection with the system (1.11). The Lagrangian function (1.19) has for $y = \tau$ the same properties as the function (1.9).

The system (1.18) is a particular case of (1.12) and has several advantages which facilitate the construction of approximate solutions. When $\mathscr{Y}(y, \dot{y}, q_1, \ldots, q_N, \dot{q}_1, \ldots, \dot{q}_N, \varepsilon) \equiv 1$, the system (1.18) obviously reduces to (1.10).

We have considered in /104/ actual vibrating systems and derived for them differential equations of the type (1.18) and (1.12).

The investigation of nonstationary oscillatory processes frequently leads to the study of the following differential equation, containing slowly varying parameters and "periodic" coefficients

$$\frac{d}{dt}\left[m(\tau, \theta)\frac{dx}{dt}\right] + c(\tau, \theta)x = \varepsilon F\left(\tau, \theta, x, \frac{dx}{dt}\right), \tag{1.20}$$

where ε is a small positive parameter; $\tau = \varepsilon t$; $\frac{d\theta}{dt} = \nu(\tau)$; and the functions $m(\tau, \theta)$, $c(\tau, \theta)$, $F\left(\tau, \theta, x, \frac{dx}{dt}\right)$ contain the slowing time τ, are periodic in θ with period 2π, and have a sufficient number of derivatives for any finite values of their arguments $\tau,\ x,\ \frac{dx}{dt}$. The study of nonstationary vibrating processes can also lead to the study of a system of differential equations of the form

$$\sum_{j=1}^{n}\left[p_{ij}(\tau,\ \theta)\frac{dx_j}{dt}+q_{ij}(\tau,\ \theta)x_j\right]=\varepsilon f_i(\tau,\ \theta,\ x_1,\ \ldots,\ x_n)\qquad(i=1,\ 2,\ \ldots,\ n),\ \ (1.21)$$

where $\tau=\varepsilon t,\ \frac{d\theta}{dt}=\nu(\tau)$, and the functions $p_{ij}(\tau,\ \theta),\ q_{ij}(\tau,\ \theta),\ f_i(\tau,\ \theta,\ x_1,\ \ldots,\ x_n)$ $(i,\ j=1,\ 2,\ \ldots,\ n)$ are subject to the same conditions as the functions $m(\tau,\ \theta)$, $c(\tau,\ \theta)$ and $F\left(\tau,\ \theta,\ x,\ \frac{dx}{dt}\right)$.

Some particular cases of equations of the type (1.20) and (1.21) were discussed in /65, 83, 84/.

Many interesting problems in the theory of nonstationary vibrating processes reduce in the general case to considering systems of differential equations with slowly varying parameters of the type

$$\frac{dx}{dt}=X(\tau,\ x)+\varepsilon X^*(\tau,\ \theta,\ x,\ \varepsilon),\tag{1.22}$$

where x is an n-dimensional vector of the Euclidean space E_n; X and X^* are n-dimensional vector functions; t is the time; ε is a small positive parameter; $\tau=\varepsilon t$; $\frac{d\theta}{dt}=\nu(\tau)$; and the functions $X^*(\tau,\ \theta,\ x,\ \varepsilon)$ are periodic in θ with period 2π.

It is assumed herewith that for the unperturbed system, corresponding to the system (1.22),

$$\frac{dx}{dt}=X(\tau,\ x),\tag{1.23}$$

(where τ is considered as a constant parameter) we know for any constant τ a set of stable periodic solutions

$$x=x_0(\tau,\ a,\ \omega t+\varphi)\tag{1.24}$$

with period 2π with respect to $\psi(\psi=\omega t+\varphi)$, depending on $k+1$ $(k+1\leqslant n)$ arbitrary constants $a_1,\ a_2,\ \ldots,\ a_k$, and φ, and on τ as a parameter. Generally

$$\omega=\omega(\tau,\ a_1,\ a_2,\ \ldots,\ a_k).\tag{1.25}$$

The problem of the action of "periodic" forces with slowly varying frequencies on vibrating relaxation systems, problems connected with the investigation of oscillatory phenomena in calculating satellite orbits and the perturbation of satellite orbits, problems associated with accelerators, etc., are all reduced to equations of the type (1.22). Equations of this type are considered in /82, 85/.

We note that one may formulate the problem for equation (1.20), and also for the systems of equations (1.21), (1.22), so that the slowly varying parameters are also determined by a system of differential equations, in analogy with what was done before. This however will not be considered here.

§ 2. Examples of vibrating systems described by differential equations of the type (1.1)

As a first example of a vibrating system described by a differential equation of the form (1.1) we quote the well-known problem of vibrations of a pendulum of variable length.

Let m denote the mass of the mathematical pendulum, g, the acceleration due to gravity, ϑ, the angle of deviation from the vertical, $l = l(t)$, the time varying pendulum length. Ignoring friction we obtain the equation

$$\frac{d}{dt}\left[ml^2(t)\frac{d\vartheta}{dt}\right] + mgl(t)\sin\vartheta = 0, \tag{1.26}$$

which for a slow variation (compared to the period of normal vibrations) of the length $l(t)$ is a nonlinear differential equation with slowly varying coefficients, and belongs for small deviation angles ($-45° < \vartheta < 45°$) to equations of the type (1.1).

Many problems of radio engineering reduce to the study of differential equations of the type (1.1). For example, while investigating the operation of a reflex klystron in the regime of super-regenerative amplification we obtain the equation /5/

$$\frac{d^2U}{dt^2} + \varrho(G_\varepsilon - G_e)\frac{dU}{dt} + U = E(\tau)\sin[1 + \mathscr{E}(\tau)]\,t, \tag{1.27}$$

where $t = \omega_0\tau$; $\varrho = \omega_0 L$; $\omega_0 = \frac{1}{\sqrt{LC}}$; $G_\varepsilon = G_R + G_H$ is the total conductance; G_R, the conductance characterizing the resonator losses; G_H, the conductance characterizing the load; G_e, the active part of the electron conductivity; $E(\tau)$, the signal emf; $\mathscr{E}(\tau)$, the variable voltage of superization. Thus, the oscillating system considered represents a linear vibrator under the action of a "sinusoidal" force with varying amplitude and frequency.

Equations of the type (1.1) are also obtained in solving problems of frequency modulation. Thus, for example, we obtain for the characteristic corresponding to the "soft" regime of self-excitation of an ordinary Meissner circuit, after simplifications, the equation (neglecting the grid current and anode reaction)

$$\frac{d^2I}{dt^2} + \frac{\omega_0\varphi}{g(t)}\frac{dI}{dt} + \frac{\omega_0^2 + 2h\dot{g}(t)}{g(t)}I = -\frac{\gamma}{g(t)}I^3, \tag{1.28}$$

where $g(t)$ is some periodic function of time, and ω_0, φ, h, γ are constants.

The examination of transient processes in oscillating circuits with a positive feedback by means of an electronic tube leads after a few transformations to the equation /88/

$$\frac{d^2x}{dt^2} + x = \varepsilon(\tau)f\left(\tau, x, \frac{dx}{dt}\right) + \lambda_0 \sin n(\tau)\,t, \tag{1.29}$$

which by a substitution of variables may be reduced to an equation of the type (1.1).

The problem of longitudinal vibrations of a thread of variable length loaded at its end, which is important, for example, in the theory of elevating devices, may in certain cases be reduced to the differential equation /55, 56, 148/

$$l\left(m + \frac{1}{3}\varrho Fl\right)\frac{d^2\varphi}{dt^2} + \left[\frac{dl}{dt}\left(m + \frac{1}{2}\varrho Fl\right) + \eta\right]\frac{d\varphi}{dt} + EF\varphi = \left(g - \frac{dv_c}{dt}\right)\left(m + \frac{1}{2}\varrho Fl\right), \tag{1.30}$$

where l is determined by the expression

$$l = l(0) + \int_0^t \frac{v_c}{1+\varphi}\,dt\,; \tag{1.31}$$

ϱ is the linear density of the thread; m, the mass of the terminal load; l, the length in undeformed state; v_c, the linear velocity of the thread; E, the modulus of elasticity; η, the friction coefficient; F, the area of the thread cross section.

Equation (1.30) may in many cases (for example, in the case of a not too fast variation of the thread length, or in the case of a very long thread) be also reduced to an equation of the type (1.1).

Equations of this type are also encountered in the study of oscillatory effects in accelerators. Thus, for example, in examining various cases of resonance with slow phase vibrations in magnetic gap accelerators, particularly in calculating the transition through resonance (taking into account the nonlinearity of the vibrations), the well-known phase equation is by means of several transformations reduced to the form /145/

$$\frac{d^2\alpha}{dt^2} + \alpha = -\varepsilon\left(\frac{\alpha^2}{2}\operatorname{ctg}\varphi_0 - \frac{\alpha^3}{6} - D\sin\eta\right) + \varepsilon^2\left(\frac{\alpha^4}{24}\operatorname{ctg}\varphi_0 - \frac{\alpha^5}{120} - \delta\frac{d\alpha}{dt}\right), \tag{1.32}$$

where ε is a small parameter; $\frac{d\eta}{dt} = \xi_0 + \varepsilon\xi_1 t$; D, δ, φ_0 are slowly varying functions of time.

§ 3. Vibrations of systems of a variable mass

In the classical theory of vibrations the mass of the vibrating body is usually considered as a constant quantity. It is however possible to give numerous examples (vibrations of certain systems of centrifuges, etc.) where the mass of the vibrating system varies with time.

We shall consider the case when the mass of the body varies continuously in the course of vibrations. It is then assumed /90/ that: 1) the form and volume, and hence also the position of the center of inertia and the moments and products of inertia vary continuously; 2) the velocities of the variable points (i. e., material points, attaching to the body or detaching from it) are either constant, or continuously vary with time; 3) the projections of the total vector and the total moment of forces acting on the body are at any time expressed by one and the same functions of time, position of the body, and its translational and angular velocities.

If the mass variation proceeds without impacts, i. e., if the variable points do not change their velocity at the moment of attachment to the body or detachment from it, then the differential equations of motion, referred to a body coordinate system with its origin at some point which belongs to the body during the entire course of motion, have the same form as for a body of constant mass. Thus, for example, in the case of a system with one degree of freedom we have the equation

$$\frac{d}{dt}\left[m(t)\frac{dx}{dt}\right] = X, \tag{1.33}$$

x is a coordinate; t, the time; $m(t)$, the variable body mass; X, n of external forces and reactions applied to the body.

In the general case of mass variation the body is subject to impacts of the variable points, and their action may be replaced by a certain system of forces, usually called "additional forces".

The equation of motion then assumes the form

$$\frac{d}{dt}\left[m(t)\frac{dx}{dt}\right] = X + \alpha\frac{dm(t)}{dt}, \tag{1.34}$$

where α denotes the velocity of the center of inertia of the variable points at time t. If the variable mass depends not only on t, but also on the position of the body, i.e., on x ($m = m(t, x)$), then in equation (1.34)

$$\frac{dm}{dt} = \frac{\partial m}{\partial t} + \frac{\partial m}{\partial x}\frac{dx}{dt}. \tag{1.35}$$

If

$$X = -c(\tau)x + \varepsilon F\left(\tau, \theta, x, \frac{dx}{dt}\right), \tag{1.36}$$

where the functions $c(\tau)$ and $F\left(\tau, \theta, x, \frac{dx}{dt}\right)$ satisfy the conditions of the first section, and the mass of the vibrating system m depends on the slowing time τ and through a small parameter also on the displacement x, i.e., if

$$\frac{dm}{dt} = \varepsilon\frac{\partial m}{\partial \tau} + \varepsilon\frac{\partial m}{\partial x}\cdot\frac{dx}{d\tau}, \tag{1.37}$$

then (1.34) is an equation of the type (1.1). If however the mass of the vibrating system depends on the parameter y ($m = m(y)$), whose law of variation is known to us and is given by the equation

$$\frac{dy}{dt} = \varepsilon f\left(t, x, \frac{dx}{dt}, y\right), \tag{1.38}$$

then equation (1.34) together with equation (1.38) belong to the system (1.16).

§ 4. The action of "periodic" forces of variable frequencies and amplitudes on a vibrating system. Transition through resonance

The nonstationary transient modes of vibrating systems involve a series of phenomena whose analysis is of theoretical, as well as of practical interest. Prior to the establishment of a stationary mode in a vibrating system, one frequently observes a series of effects in which the amplitude of vibrations increases considerably. This may be caused by the appearance of beating, a transition of the external frequency through the values of the system's normal frequencies (resonance values) or a variation of other parameters of the system (for example, mass, rigidity coefficient, friction coefficient, etc.). There is a significant increase in the amplitude in a transition through resonance. Hence the investigation of effects accompanying such a transition play the same role in the study of nonstationary modes of vibrating systems as the investigation of the effects associated with an ordinary resonance.

Before discussing the phenomenon of transition through resonance, we consider the generally used concept of resonance as a phenomenon observed in vibrating systems.

External periodic forces acting on a vibrating system in time with its normal vibrations are known to induce large deviations in the latter.

By a resonance in vibrating systems described by a linear differential equation of the type

$$\frac{d^2x}{dt^2}+2h\frac{dx}{dt}+k^2x=H\sin(\nu t+\gamma), \tag{1.39}$$

where h, k, H, ν, γ are constants, is meant the effect that the amplitude of the vibrations induced by the external perturbing force $H\sin(\nu t+\gamma)$ begin at a definite period T or frequency $\nu=\frac{2\pi}{T}$ (usually close to the period of normal vibrations or to a normal frequency) and increase sharply reaching a certain maximum.

We note that in a linear vibrating system the system parameters (normal and external frequencies, amplitude of the perturbing force) remain constant during the setting up of the resonance (during the increase of the vibration amplitude).

The essence of the resonance phenomenon is more complicated in vibrating systems described by more general differential equations, for example, by an equation with periodic coefficients (parametric resonances) or by a nonlinear equation (overtone, fractional, combination resonances, etc.)*. Thus, for example, a normal frequency in nonlinear systems may depend on the vibration amplitude and varies during the setting up of the resonance so long as a mode with constant amplitude is not established.

An example of resonance in general complicated systems is the phenomenon observed in nonautonomous vibrating systems (subject to the action of external periodic forces, explicitly depending on time) consisting in a sharp increase of the vibration amplitude (the amplitude increases while the vibrations set in, after which it takes on some constant value corresponding to the external frequency). A general characteristic property of resonance phenomena in linear, as well as in nonlinear vibrating systems is that a small perturbing force usually leads to a very large vibration amplitude.

A characteristic of the resonance phenomena mentioned above is that in the course of time a stationary vibration mode is established in the system, for which all the parameters of the vibrating system are constant (until this mode is established, some parameters of a nonlinear system may, as indicated above, vary with time).

In vibrating systems under the action of external perturbing "periodic" forces with variable frequencies one observes, unlike the case of a stationary resonance discussed above, a sharp increase of the amplitude during a variation of the external frequency proceeding in such way that it passes through the values of the normal frequencies of the vibrating system. Such a phenomenon is usually called a transition (passage) through resonance.

The transition through resonance differs from an ordinary resonance (i. e., from the case when the vibrating system is in a resonant state).

* This problem is considered in several classical works: N. M. Krylov and I. N. Bogolyubov /72/; L. I. Mandel'shtam and N. D. Papaleksi /88/.

st, the intensive vibrating process is observed during a finite time interval depending on the velocity of transition through resonance, i. e., on the variation velocity of certain parameters of the system, whose change brings the system to a resonance state. A transition through resonance is possible not only when the external variable frequency passes through the values of the normal frequencies of the system (transition through the principal resonance), but also when the mass or rigidity of the system changes.

Second, the vibration amplitudes in a transition through resonances cannot attain their maximum values for a stationary resonance, since this phenomenon occurs during some finite time interval. The remaining characteristic properties and peculiarities of the transition through resonance will be considered in detail in the second chapter while studying numerous examples.

Here we note that it is very difficult to distinguish strictly between resonance phenomena observed in a transition through resonance and those encountered in vibrating systems during the establishment of a stationary resonant state (especially in nonlinear systems). The transition through resonance, just as the processes of a stationary resonance and all the accompanying effects, belongs to the field of resonance phenomena arising when the perturbing frequencies are close to the normal frequencies (or, as will be discussed in detail in the sequel, when the normal frequency ω is close to $\frac{p}{q}\nu$, where p and q are mutually prime small integers, and ν is the frequency of the perturbation). They are characterized by a pronounced increase of the amplitude caused by the action of small perturbing forces.

Let us now write down a few equations describing the vibrations during a transition through resonance for a nonlinear system with one degree of freedom. Let a nonlinear vibrator of mass m be subject to the action of a "sinusoidal" perturbing force $E(\tau)\sin\theta$, whose instantaneous frequency $\frac{d\theta}{dt}=\nu(\tau)$ is a function of the slowing time τ.

The vibrations of the system are then described by the differential equation

$$m\frac{d^2x}{dt^2}+F\left(x,\ \frac{dx}{dt}\right)=E(\tau)\sin\theta. \tag{1.40}$$

Assuming that the vibrating system under consideration is, in the absence of an external "sinusoidal" perturbation, close to a linear conservative one, and that the amplitude of the external force is small, we obtain an equation of the type (1.1).

The transition through resonance may in many cases occur not only in connection with a varying frequency $\nu(\tau)$ of the external perturbing force, but also as a result of a slow variation of the mass of the system, its rigidity, and other parameters. In that case we are led to the equation

$$\frac{d}{dt}\left[m(\tau)\frac{dx}{dt}\right]+c(\tau)x=\varepsilon F\left(\tau,\ x,\ \frac{dx}{dt}\right)+E(\tau)\sin\theta, \tag{1.41}$$

which in the case of a small $E(\tau)$ also belongs to the type (1.1).

Conversely, it is of interest to study cases when a transition through resonance causes a slow change of some parameters of the vibrating system. In that case we have to consider the system of equations

$$\left.\begin{array}{r}\frac{d}{dt}\left[m(y)\frac{dx}{dt}\right]+c(y)x=\varepsilon F\left(y, x, \frac{dx}{dt}\right)+E(y)\sin\theta,\\ \frac{dy}{dt}=\varepsilon f\left(y, t, x, \frac{dx}{dt}\right) \quad \left(\frac{d\theta}{dt}=\nu(y)\right),\end{array}\right\}. \tag{1.42}$$

belonging to the type (1.16).

While investigating the transition through a parametric resonance in a nonlinear circuit, we obtain an equation of the type

$$\frac{d^2x}{dt^2}+2\delta\frac{dx}{dt}+\gamma x^3+\omega^2(1+h\cos\theta)x=0, \tag{1.43}$$

where $\frac{d\theta}{dt}=\nu(\tau)$, and δ, γ, ω, h are constants. Herewith, δ, γ, h are small.

In the case of an oscillating circuit with nonlinear friction (for example, in the case of a parametric excitation of a circuit with an electronic tube feedback), we obtain the equation of motion

$$\frac{d^2x}{dt^2}+2(\delta_0+\delta_2x^2)\frac{dx}{dt}+\omega^2(1+h\cos\theta)x=0, \tag{1.44}$$

where $\frac{d\theta}{dt}=\nu(\tau)$, and δ_0, δ_2, h are small constants.

The last two equations are also of the type (1.1).

§ 5. Examples of vibrating systems described by differential equations of the type (1.6), (1.18), and (1.21)

A system consisting of a shaft rotating together with some masses is a typical mechanical system for which the examination of the nonstationary vibration mode leads to equations of the type (1.6). Such systems are, for example, a motor with a fly-wheel rotating a dynamo, a motor moving a screw propeller, the motor-crankshaft-propeller system, etc.

Let $I_1, I_2, \ldots, I_n$ be the moments of inertia of the rotating masses; $\varphi_1, \varphi_2, \ldots, \varphi_n$, the rotation angles of the shaft sections at which the masses are attached; $c_1, c_2, \ldots, c_{n-1}$, torsional rigidity of the corresponding shaft segments; $M_1, M_2, \ldots, M_n$, external torsional moments acting on the rotating masses. The mass of the shaft will be neglected.

The kinetic and potential energies of the system are given by the expressions

$$\left.\begin{array}{l}T=\frac{1}{2}\{I_1\dot{\varphi}_1^2+I_2\dot{\varphi}_2^2+\ldots+I_n\dot{\varphi}_n^2\},\\ V=\frac{1}{2}\{c_1(\varphi_2-\varphi_1)^2+c_2(\varphi_3-\varphi_2)^2+\ldots+c_{n-1}(\varphi_n-\varphi_{n-1})^2\}.\end{array}\right\} \tag{1.45}$$

Constructing the equations of motion, we obtain

$$\left.\begin{array}{l}I_1\ddot{\varphi}_1+c_1(\varphi_1-\varphi_2)=M_1,\\ I_2\ddot{\varphi}_2+c_1(\varphi_2-\varphi_1)+c_2(\varphi_2-\varphi_3)=M_2,\\ \cdots\cdots\cdots\cdots\cdots\cdots\cdots\cdots\\ I_{n-1}\ddot{\varphi}_{n-1}+c_{n-2}(\varphi_{n-1}-\varphi_{n-2})+c_{n-1}(\varphi_{n-1}-\varphi_n)=M_{n-1},\\ I_n\ddot{\varphi}_n+c_{n-1}(\varphi_n-\varphi_{n-1})=M_n.\end{array}\right\} \tag{1.46}$$

Eliminating one degree of freedom — the rotation, we obtain a system of $n-1$ second order equations. The moments acting on the rotating masses will vary in transient modes (in starting or stopping the motion of the shaft, in transition from one mode of operation to another one, etc.). If these moments are periodic, then their amplitudes and frequencies may vary. In order to prevent resonant vibrations with large amplitudes the shaft may in some cases be furnished with elastic muffs with nonlinear elasticity characteristics.

Taking all these additional conditions into account in constructing the system (1.46), we are led to a system of differential equations of the type (1.6).

In the case when the vibrating system exerts during its motion an influence on the energy source providing the external torsional moments M_i, we obtain a system of equations of the type (1.18).

In solving the problems of the effect of foundations on the critical number of revolutions of rotating shafts one has to consider the system of equations /79/

$$\left.\begin{aligned} M\ddot{x}_0 + c_1x_0 - c(x - x_0 - e\cos\varphi) &= 0,\\ m\ddot{x} + c(x - x_0 - e\cos\varphi) &= 0,\\ m\ddot{y} + c(y - e\sin\varphi) &= 0,\\ I\ddot{\varphi} + cef\sin(\varphi - \psi) &= 0, \end{aligned}\right\} \tag{1.47}$$

where M is the mass of the foundation plate; m, the mass of the rotor; c_1, the rigidity of the foundation stands; c, the rigidity of the shaft for bending; f, the flexure of the shaft in the middle; e, the eccentricity of the rotor on the shaft; x_0, the displacement of the foundation; x, y, coordinates of the rotor mass center; φ, the rotor rotation angle.

The last of equations (1.47) is integrated in an elementary way. Substituting the found value of $\varphi = \varphi(t)$, we may integrate the third equation, which is of the form (1.1). The first two equations belong to a system of the type (1.6).

Equations of the type (1.6) are also found in the problem of torsional vibrations in systems with balance antivibrators. Restricting the expression of the kinetic energy for such a system to terms of the second order with respect to the deviation angles, ψ_1 and ψ_2, from a uniform rotation φ_k, we obtain the equations of motion in the form

$$\left.\begin{aligned} &(\theta_k - \theta_{11})\ddot{\varphi}_k + \theta_{12}\ddot{\psi}_1 + \theta_{13}\ddot{\psi}_2 = L_n\sin\omega t,\\ &\theta_{12}\ddot{\varphi}_k + \theta_{22}\ddot{\psi}_1 + \theta_{23}\ddot{\psi}_2 - \Omega^2 m_{22}\psi_1 - \Omega^2 m_{23}\psi_2 = 0,\\ &\theta_{13}\ddot{\varphi}_k + \theta_{23}\ddot{\psi}_1 + \theta_{33}\ddot{\psi}_2 - \Omega^2 m_{23}\psi_1 - \Omega^2 m_{33}\psi_2 = 0, \end{aligned}\right\} \tag{1.48}$$

where Ω is the angular velocity of rotation of the whole system*. The angular velocity varies in starting the system, in stopping it, and in a transition from one mode of operation to another, and the equations (1.48) reduce to a system of the type (1.6).

The problem of vibrations of a shaft with an unbalanced disk leads to a system of equations of the type (1.18). In the simplest case we obtain the

* The derivation of the equations (1.48) and an investigation of the stationary mode are given in the work of A. I. Chekmarev /168/.

system of equations

$$\left.\begin{aligned} m\ddot{x}+cx&=\varepsilon c\cos\varphi,\\ m\ddot{y}+cy&=\varepsilon c\sin\varphi,\\ mr^2\ddot{\varphi}&=\varepsilon c\,(y\cos\varphi-x\sin\varphi)+M(\dot{\varphi}),\end{aligned}\right\}\tag{1.49}$$

where m is the mass of the disk; r, its radius of gyration; c, a constant, characteristic of the elastic properties of the shaft; $M(\dot{\varphi})$, a moment exerted on the shaft by the drive; $\dot{\varphi}=\omega$, the angular rotation velocity of the rotor.

The problem of the motion of a charged particle in the magnetic field of a cyclotron with a space variation of the magnetic intensity reduces to the following system of equations, derived and investigated by V. P. Dzhelepov, V. P. Dmitrievskii, V. V. Kol'g, et al. /22/,

$$\left.\begin{aligned} \ddot{r}-r\dot{\varphi}^2&=\frac{e}{mc}(r\dot{\varphi}H_z-\dot{z}H_\varphi),\\ r\ddot{\varphi}+2\dot{r}\dot{\varphi}&=\frac{e}{mc}(\dot{z}H_r-\dot{r}H_z),\\ \ddot{z}&=\frac{e}{mc}(r\dot{\varphi}H_r-\dot{r}H_\varphi),\end{aligned}\right\}\tag{1.50}$$

where r, φ, z are cylindrical coordinates; m and e, the mass and charge of the particle; H_z, H_r, H_φ, components of the magnetic field intensity.

Since for the system (1.50) $\dot{r}^2+\dot{r}^2\varphi^2+\dot{z}^2=v^2=\text{const}$, we may pass to the new independent variable φ and obtain the system

$$\left.\begin{aligned} r''-\frac{2r'^2}{r}-r&=-\frac{e}{mc}\frac{\sqrt{r'^2+r^2+z'^2}}{v}\left(rH_z-z'H_\varphi-\frac{z'r'}{r}H_r+\frac{r'^2}{r}H_z\right);\\ z''-\frac{2r'z'}{r}&=-\frac{e}{mc}\frac{\sqrt{r'^2+r^2+z'^2}}{v}\left(r'H_\varphi-rH_r-\frac{z'^2}{r}H_r+\frac{z'r'}{r}H_z\right),\end{aligned}\right\}\tag{1.51}$$

where r and z primed denote differentiation with respect to φ.

Let the form of the magnetic field be such that the total magnetic field vector in the meridial plane be directed along the z-axis. The function characterizing the magnetic field in this plane completely determines the dynamics of the particles' motion and may be represented in the form

$$H_z=H(r)\,[1+\varepsilon f(r,\varphi)],\tag{1.52}$$

where ε is the variation depth of the magnetic field, and $f(r,\varphi)$, a periodic function of both variables with a mean value equal to zero, $H(r)\,r\Big|_{r=R}=\frac{mvc}{e}$.

Substituting (1.52) in (1.51), we obtain, up to terms of the second order, the following system of equations

$$\left.\begin{aligned} &\varrho''+\left[1+n+\varepsilon R\frac{\partial f}{\partial r}+(2+n)\,\varepsilon f\right]\varrho+\left[\frac{d}{R}+\frac{1}{R}+\frac{2n}{R}+\right.\\ &\left.+\frac{\varepsilon}{R}(1+2n+d)\,f+\varepsilon\,(2+n)\frac{\partial f}{\partial r}+\frac{\varepsilon R}{2}\frac{\partial^2 f}{\partial r^2}\right]\varrho^2-\frac{1}{2R}[1-3\varepsilon f]\,\varrho'^2-\\ &-\frac{1}{2}\left[\frac{2d}{R}+\frac{2d}{R}\varepsilon f+2n\varepsilon\frac{\partial f}{\partial r}+\varepsilon\frac{\partial f}{\partial r}+\frac{n}{R}(1+\varepsilon f)+\frac{\varepsilon}{R}\frac{\partial^2 f}{\partial\varphi^2}+\varepsilon R\frac{\partial^2 f}{\partial r^2}\right]z^2-\\ &\qquad\qquad-\frac{\varepsilon}{R}\frac{\partial f}{\partial\varphi}zz'+\frac{1}{2R}(1+\varepsilon f)\,z'^2=-\varepsilon Rf,\\ &z''-\left[n+\varepsilon nf+\varepsilon R\frac{\partial f}{\partial r}\right]z-\left[\frac{2}{R}(n+d)+\frac{2\varepsilon}{R}(n+d)\,f+\right.\\ &\left.+2\varepsilon\,(1+n)\frac{\partial f}{\partial r}+\varepsilon R\frac{\partial^2 f}{\partial r^2}\right]z\varrho+\frac{\varepsilon}{R}\frac{\partial f}{\partial\varphi}z\varrho'-\frac{1}{R}[1-\varepsilon f]\,z'\varrho'=0,\end{aligned}\right\}\tag{1.53}$$

where

$$\varrho = r - R, \quad n = \frac{R}{H(R)} \frac{dH(r)}{dr}\bigg|_{r=R}, \quad d = \frac{1}{2} \frac{R^2}{H(R)} \frac{d^2H(r)}{dr^2}\bigg|_{r=R},$$

and the values of the function f and of its derivatives are taken at $r = R$.

Equations (1.53), obviously belong to a system of the type (1.21).

§ 6. Vibrations about a "quasi-stationary" state of motion

We describe another wide class of problems, whose solution requires the consideration of systems of nonlinear differential equations with slowly varying parameters of the type (1.11) or (1.18).

To this end we introduce the concept of "quasi-stationary" motion. By this we mean such a motion of a system with cyclic or "quasi-cyclic" coordinates, for which the noncyclic coordinates (and consequently, also the velocities corresponding to the cyclic coordinates, as shown below) are slowly varying functions as compared with the frequency of small vibrations, which may appear in the considered state of "slow" motion. By "quasi-cyclic" coordinates we mean coordinates which are cyclic only in the first approximation; this is stated below more precisely.

"Quasi-stationary" motion is, for example, the motion of centrifuges, transmissions, various gyroscopic systems, etc., during a slow transition from one mode to another.

If the constant mode of rotation were retained in the aforementioned mechanical systems, we would have a stationary motion — rotation with a constant angular velocity. However, during a slow transition from one mode to another, the angular velocity of rotation varies slowly and we get a "quasi-stationary" motion.

Let us consider small vibrations about a "quasi-stationary" state and construct their differential equations.

Let $p_1, p_2, \ldots, p_k$ be the cyclic, and $q_1, q_2, \ldots, q_N$, the noncyclic generalized coordinates of the mechanical system under consideration. Then the following k integrals of motion correspond to the k cyclic coordinates

$$\frac{\partial \mathcal{L}}{\partial \dot{p}_r} = \beta_r \qquad (r = 1, 2, \ldots, k), \tag{1.54}$$

where $\beta_1, \beta_2, \ldots, \beta_k$ are integration constants; $\mathcal{L} = \mathcal{L}(q_1, q_2, \ldots, q_N, \dot{q}_1, \dot{q}_2, \ldots, \dot{q}_N, \dot{p}_1, \dot{p}_2, \ldots, \dot{p}_k)$ is the Lagrangian function of the mechanical system under consideration.

If $p_1, p_2, \ldots, p_k$ are "quasi-cyclic" coordinates, then according to the above we have

$$\frac{d}{dt}\left[\frac{\partial \mathcal{L}}{\partial \dot{p}_r}\right] = \varepsilon \ldots \qquad (r = 1, 2, \ldots, k) \tag{1.55}$$

and consequently, ignoring terms of the first order with respect to ε, we also obtain for them k integrals of motion of the type (1.54).

We assume that in vibrational motion, perturbed about the basic motion, the constants $\beta_1, \beta_2, \ldots, \beta_k$ have the same values as in the unperturbed state. We also assume that the system is conservative and that the constraints do not depend on time. The kinetic energy of the system may then be represented in the form

$$T=\frac{1}{2}\sum_{i,\,j=1}^{N}A_{ij}\dot{q}_i\dot{q}_j+\sum_{i,\,\sigma=1}^{N,\,k}B_{i\sigma}\dot{q}_i\dot{p}_\sigma+\frac{1}{2}\sum_{\sigma,\,r=1}^{k}C_{\sigma r}\dot{p}_\sigma\dot{p}_r, \tag{1.56}$$

where the coefficients A_{ij}, $B_{i\sigma}$, $C_{\sigma r}$ ($i, j=1, 2, \ldots, N$; $\sigma, r=1, 2, \ldots, k$) are certain functions of the noncyclic coordinates $q_1, q_2, \ldots, q_N$.

Let the potential energy of the system be of the form

$$V=\frac{1}{2}\sum_{i,\,j=1}^{N}D_{ij}q_iq_j, \tag{1.57}$$

where D_{ij} are functions of $q_1, q_2, \ldots, q_N$; then the integrals of motion corresponding to the cyclic coordinates are the expressions

$$\sum_{\sigma=1}^{k}C_{\sigma r}\dot{p}_\sigma+\sum_{i=1}^{N}B_{ir}\dot{q}_i=\beta_r \qquad (r=1, 2, \ldots, k). \tag{1.58}$$

We denote by $S_{\sigma r}$ the ratio of the minor corresponding to the element $C_{\sigma r}$, and the determinant of the elements $C_{\sigma r}$. Solving the system of equations (1.58) for $\dot{p}_r$, we find

$$\dot{p}_\sigma=\sum_{r=1}^{k}S_{\sigma r}\left[\beta_r-\sum_{i=1}^{N}B_{ir}\dot{q}_i\right] \qquad (\sigma=1, 2, \ldots, k). \tag{1.59}$$

Setting these values in expression (1.56), we obtain

$$T=\frac{1}{2}\cdot\sum_{i,\,j=1}^{N}\left[A_{ij}-\sum_{\sigma,\,r=1}^{k}S_{\sigma r}B_{i\sigma}B_{jr}\right]\dot{q}_i\dot{q}_j+\frac{1}{2}\sum_{\sigma,\,r=1}^{k}S_{\sigma r}\beta_\sigma\beta_r. \tag{1.60}$$

It is now necessary to remove the cyclic coordinates and derive an expression for the modified kinetic potential. We have

$$\overline{\mathcal{L}}=T-V-\sum_{\sigma=1}^{k}\dot{p}_\sigma\frac{\partial\mathcal{L}}{\partial\dot{p}_\sigma}=T-V-\sum_{\sigma=1}^{k}\dot{p}_\sigma\beta_\sigma \tag{1.61}$$

or

$$\overline{\mathcal{L}}=\frac{1}{2}\sum_{i,\,j=1}^{N}\left[A_{ij}-\sum_{\sigma,\,r=1}^{k}S_{\sigma r}B_{i\sigma}B_{jr}\right]\dot{q}_i\dot{q}_j+\sum_{s=1}^{N}\sum_{\sigma,\,r=1}^{k}S_{r\sigma}\beta_rB_{s\sigma}\dot{q}_s-$$

$$-\frac{1}{2}\sum_{\sigma,\,r=1}^{k}S_{\sigma r}\beta_\sigma\beta_r-\frac{1}{2}\sum_{i,\,j=1}^{N}D_{ij}q_iq_j. \tag{1.62}$$

Substituting the value of $\overline{\mathcal{L}}$ into the Lagrange equation, we obtain

$$\frac{d}{dt}\left[\frac{\partial\overline{\mathcal{L}}}{\partial\dot{q}_i}\right]-\frac{\partial\overline{\mathcal{L}}}{\partial q_i}=0 \qquad (i=1, 2, \ldots, N). \tag{1.63}$$

It is now assumed that the basic motion is "quasi-stationary" and is, therefore, by the above definition characterized by the following values of the noncyclic coordinates

$$q_1=f_1(\tau), \quad q_2=f_2(\tau), \ldots, q_N=f_N(\tau) \quad (\tau=\varepsilon t). \tag{1.64}$$

Their time derivatives will then be proportional to a small parameter ε. The velocities $\dot{p}_j$ ($j=1, 2, \ldots, k$), determined from the system of equations (1.58), will also be slowly varying functions of time, since they will depend on the noncyclic coordinates $q_1, q_2, \ldots, q_N$ (1.64) (through the coefficients C_{ij}) and besides, they will "weakly" depend on the velocities $\dot{q}_1, \dot{q}_2, \ldots, \dot{q}_N$ (the dependence being of the order of ε). Thus, determining $\dot{p}_j$ ($j=1, 2, \ldots, k$) from the system of equations (1.58) and taking into account (1.64), we find

$$\dot{p}_1 = \dot{f}_{N+1}(\tau, \varepsilon), \quad \dot{p}_2 = \dot{f}_{N+2}(\tau, \varepsilon), \quad \ldots, \quad \dot{p}_k = \dot{f}_{N+k}(\tau, \varepsilon). \tag{1.65}$$

Substituting now in equations (1.63)

$$q_i = f_i(\tau) + \xi_i, \quad \dot{q}_i = \varepsilon \dot{f}_i(\tau) + \dot{\xi}_i \qquad (i = 1, 2, \ldots, N), \tag{1.66}$$

where ξ_i, $\dot{\xi}_i$ are new variables (presumably small deviations of the coordinates of the given system and the velocities of these deviations from a "quasi-stationary" state), we obtain the system of equations

$$\frac{d}{dt}\left[\frac{\partial \mathcal{L}}{\partial \dot{\xi}_i}\right] - \frac{\partial \mathcal{L}}{\partial \xi_i} = 0 \qquad (i = 1, 2, \ldots, N), \tag{1.67}$$

where

$$\mathcal{L} = \frac{1}{2}\left\{\sum_{i,\,j=1}^{N} a_{ij}(\tau)\dot{\xi}_i\dot{\xi}_j + 2\sum_{i,\,j=1}^{N} g_{ij}(\tau)\xi_i\dot{\xi}_j - \sum_{i,\,j=1}^{N} b_{ij}(\tau)\xi_i\xi_j\right\} + \varepsilon\mathcal{L}_1(\tau, \xi_i, \dot{\xi}_i) \tag{1.68}$$

It is easily shown that $\mathcal{L}_1(\tau, \xi_i, \dot{\xi}_i)$ is, for any values of τ for which $a_{ij}(\tau) = a_{ji}(\tau)$, $b_{ij}(\tau) = b_{ji}(\tau)$, but $g_{ij}(\tau) \neq g_{ji}(\tau)$ ($i, j = 1, 2, \ldots, N$), some holomorphic function expandable in a power series of ξ_i, $\dot{\xi}_i$ beginning at least with the third power. The coefficients of this series are some functions of τ. The functions $a_{ij}(\tau)$, $g_{ij}(\tau)$, $b_{ij}(\tau)$ ($i, j = 1, 2, \ldots, N$) are obviously differentiable to any order for any value of τ in the interval $0 \leqslant \tau \leqslant L$, if the coefficients $A_{ij}, B_{i\sigma}, C_{\sigma r}$ and D_{ij} ($i, j = 1, 2, \ldots, N$; $\sigma, r = 1, 2, \ldots, k$), entering in the expressions of the kinetic and potential energies (1.56) and (1.57), are unlimitedly differentiable functions for all finite values of the variables $q_1, q_2, \ldots, q_N$ and if $q_1, q_2, \ldots, q_N$ in the original "quasi-stationary" motion are unlimitedly differentiable functions for any τ in the interval $0 \leqslant \tau \leqslant L$.

Terms of the order of ε in equations (1.67), appearing on account of the function $\varepsilon\mathcal{L}_1(\tau, \xi_i, \dot{\xi}_i)$, may always be referred to external perturbing forces. Assuming that the system is under the action of external perturbing forces depending on the slowing time τ and periodic in $\theta\left(\frac{d\theta}{dt} = \nu(\tau)\right)$, we are led to equations of the type (1.11).

Thus, the problem of vibrations about a "quasi-stationary" state of motion reduces to the integration of the Lagrange equations in which the Lagrangian function is a homogeneous quadratic form of velocities and coordinates with slowly varying coefficients. The right-hand sides of these equations contain small perturbing forces, nonlinear in the coordinates and velocities, and depending (periodically) on θ and on the slowing time τ.

In problems of vibrations about a "quasi-stationary" state of motion or even about a stationary state of motion, but with "quasi-cyclic" coordinates, we are led to a system of equations of the type (1.18).

§ 7. Examples of vibrations about a state of "quasi-stationary" motion

A characteristic mechanical vibrating system described by differential equations of the type (1.11) is the system consisting of a flexible shaft with a disk pinned on it.

The coordinate system is chosen as follows: the origin is fixed at the left bearing of the shaft, the x-axis is directed along the axis of the undeformed shaft, the z-axis is directed vertically, and the y-axis — perpendicularly to the zx plane. Let the center of gravity of the disk lie on the rotation axis, which in the absence of deformation coincides with the x-axis, and in the presence of deformation assumes the form of a curve of double curvature. We denote the coordinates of the disk's center of gravity by y and z, the angle between the tangent to the elastic line and the yx plane by α, the angle between the projection of this tangent on the yx plane and the x-axis by β. Then, given the angular rotation velocity of the shaft $\varphi = \varphi(t)$, we can determine the disk's position at any moment (with an accuracy of terms of the first order) if we know the four variables: y and z, the coordinates of the disk's center of gravity, and α and β, the angles characterizing the axis orientation (Figure 1).

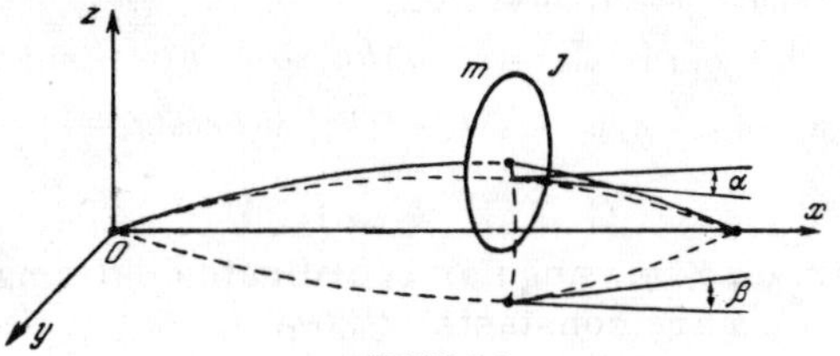

FIGURE 1

After a series of calculations we obtain the Lagrangian function describing the vibrations of the disk in the form (neglecting the mass of the shaft)

$$\mathcal{L} = \frac{1}{2} m(\dot{y}^2 + \dot{z}^2) + \frac{1}{2} I_x \dot{\varphi}^2 + \frac{1}{2} I(\dot{\alpha}^2 + \dot{\beta}^2) + I_x \dot{\varphi} \dot{\beta} \alpha -$$
$$- \frac{1}{2}[s_1(z^2 + y^2) + s_2(\alpha^2 + \beta^2) - 2s_3(\alpha z + \beta y)] + \text{higher order terms}, \quad (1.69)$$

where m is the mass of the disk; I, its moment of inertia with respect to a diametral axis passing through the center of gravity and perpendicular to the symmetry axis; I_x, the moment of inertia of the disk with respect to the symmetry axis; s_1, s_2, s_3, some constants allowing for the conditions at the fastened ends of the shaft, for the disk's position, and also for the material and geometry of the shaft.

Substituting the Lagrangian function (1.69) into equations (1.67), we obtain the following system of differential equations with gyroscopic terms and slowly varying coefficients (assuming that $\dot{\varphi}(t)$ is a slowly varying function)

$$\left. \begin{array}{l} m \frac{d^2 y}{dt^2} + s_1 y - s_3 \beta = \varepsilon Q_1(\tau, y, z, \alpha, \beta, \dot{y}, \dot{z}, \dot{\alpha}, \dot{\beta}, \varepsilon), \\ m \frac{d^2 z}{dt^2} + s_1 z - s_3 \alpha = \varepsilon Q_2(\tau, y, z, \alpha, \beta, \dot{y}, \dot{z}, \dot{\alpha}, \dot{\beta}, \varepsilon), \\ I \frac{d^2 \beta}{dt^2} + I_x \dot{\varphi}(\tau) \frac{d\alpha}{dt} + s_2 \beta - s_3 y = \varepsilon Q_3(\tau, y, z, \alpha, \beta, \dot{y}, \dot{z}, \dot{\alpha}, \dot{\beta}, \varepsilon), \\ I \frac{d^2 \alpha}{dt^2} - I_x \dot{\varphi}(\tau) \frac{d\beta}{dt} + s_2 \alpha - s_3 z = \varepsilon Q_4(\tau, y, z, \alpha, \beta, \dot{y}, \dot{z}, \dot{\alpha}, \dot{\beta}, \varepsilon), \end{array} \right\} \quad (1.70)$$

where the functions on the right-hand sides satisfy all the conditions indicated in the preceding section.

If the mass of the disk is also slowly varying in the course of vibrations

$(m = m(\tau))$, then instead of equations (1.70) we obtain the system

$$\left.\begin{aligned}
&\frac{d}{dt}\left[m(\tau)\frac{dy}{dt}\right] + s_1 y - s_3\beta = \varepsilon\bar{Q}_1(\tau, y, z, \alpha, \beta, \dot{y}, \dot{z}, \dot{\alpha}, \dot{\beta}, \varepsilon),\\
&\frac{d}{dt}\left[m(\tau)\frac{dz}{dt}\right] + s_1 z - s_3\alpha = \varepsilon\bar{Q}_2(\tau, y, z, \alpha, \beta, \dot{y}, \dot{z}, \dot{\alpha}, \dot{\beta}, \varepsilon),\\
&\frac{d}{dt}\left[I(\tau)\frac{d\beta}{dt}\right] + I_x(\tau)\dot{\varphi}(\tau)\frac{d\alpha}{dt} + s_2\beta - s_3 y = \varepsilon\bar{Q}_3(\tau, y, z, \alpha, \beta, \dot{y}, \dot{z}, \dot{\alpha}, \dot{\beta}, \varepsilon),\\
&\frac{d}{dt}\left[I(\tau)\frac{d\alpha}{dt}\right] - I_x(\tau)\dot{\varphi}(\tau)\frac{d\beta}{dt} + s_2\alpha - s_3 z = \varepsilon\bar{Q}_4(\tau, y, z, \alpha, \beta, \dot{y}, \dot{z}, \dot{\alpha}, \dot{\beta}, \varepsilon).
\end{aligned}\right\} \quad (1.71)$$

If the disk is set on the shaft with linear (e) and angular (δ) eccentricities, then after a series of operations /104/, which will be given in detail in Chapter V, § 2, we obtain the following system of differential equations

$$\left.\begin{aligned}
&m\ddot{\eta} + a_{02}\eta + a_{11}\alpha_1 = meh\dot{\varphi}_1^2\cos\varphi_1 + Q_\eta,\\
&m\ddot{\zeta} + a_{02}\zeta + a_{11}\beta_1 = meh\,\dot{\varphi}_1^2\sin\varphi_1 + Q_\zeta,\\
&I\ddot{\alpha}_1 + I_z h\,\dot{\varphi}_1\dot{\beta}_1 + a_{11}\eta + a_{20}\alpha_1 = (I_z - I)\,h\,\dot{\varphi}_1^2\sin\delta\sin(\varphi_1 - \varkappa) + Q_{\alpha_1},\\
&I\ddot{\beta}_1 - I_z h\dot{\varphi}_1\dot{\alpha}_1 + a_{11}\zeta + a_{20}\beta_1 = -(I_z - I)\,h\dot{\varphi}_1^2\sin\delta\cos(\varphi_1 - \varkappa) + Q_{\beta_1},\\
&I_z h^2\ddot{\varphi}_1 = Q_{\varphi_1},
\end{aligned}\right\} \quad (1.72)$$

where η, ζ are linear, α_1, β_1, φ_1, angular coordinates defining the disk's position; m, I, I_z, a_{11}, a_{02}, a_{20}, h are constants; Q_η, Q_ζ, Q_{α_1}, Q_{β_1}, Q_{φ_1} are small perturbing nonlinear functions satisfying the conditions of the preceding section; e and δ are small linear and angular eccentricities; $\dot{\varphi}_1$ is, according to the last equation of the derived system, a slowly varying quantity.

It is easily seen that the system of equations (1.72) belongs to the system of differential equations (1.18).

§ 8. Vibrations of systems with variable constraints

Systems of differential equations of the type (1.67) appear also in certain problems of vibrations in systems with variable constraints. If the constraints of a system depend on time, then the kinetic energy may, as known, contain not only terms of zero or second order in velocities, but also linear terms. Consequently, the equations which describe the vibrations in such systems may contain gyroscopic terms even when the vibrations about a relative equilibrium position are considered.

For the coefficients of the resulting equations to be slowly varying, it is necessary that the functions of time characterizing the constraints vary slowly. If the constraint, for example, rotates about a certain point or a certain axis, then the angular rotation velocity should vary slowly.

As an example one may take the problem of vibrations of a heavy material point about its equilibrium position coinciding with the lowest point of some surface rotating with a variable angular velocity $\dot{\varphi} = \nu(\tau)$ $(\tau = \varepsilon t)$ about a vertical axis through this point /159/.

Let x, y, z be the coordinates of the vibrating point in a coordinate system connected with the surface, the z-axis being directed vertically upwards, and the x- and y-axes — along the tangents to the lines of curvature at the lowest surface point.

The equation of the surface is

$$z=\frac{x^2}{2\varrho_1}+\frac{y^2}{2\varrho_2}+\text{higher order terms}, \tag{1.73}$$

where ϱ_1 and ϱ_2 are radii of curvature.

The kinetic and potential energies of the system have the form

$$\left.\begin{aligned} T&=\frac{1}{2}m\{(\dot{x}-y\dot{\varphi})^2+(\dot{y}+x\dot{\varphi})^2+\dot{z}^2\},\\ V&=mgz, \end{aligned}\right\} \tag{1.74}$$

where m is the mass of the vibrating point.

For the kinetic potential describing the vibrations of the material point we find the expression

$$\mathscr{L}=\frac{1}{2}m\{\dot{x}^2+\dot{y}^2+2\dot{\varphi}(x\dot{y}-y\dot{x})+\dot{\varphi}^2(x^2+y^2)\}-mg\left\{\frac{x^2}{2\varrho_1}+\frac{y^2}{2\varrho_2}\right\}+ \\ +\text{higher order terms}. \tag{1.75}$$

Substituting (1.75) into equations (1.67), we obtain a system of second order differential equations with gyroscopic terms and slowly varying coefficients

$$\left.\begin{aligned} &\frac{d^2x}{dt^2}-2\nu(\tau)\frac{dy}{dt}+x\left(\frac{g}{\varrho_1}-\nu^2(\tau)\right)=\varepsilon f_1(\tau, x, y, \dot{x}, \dot{y}),\\ &\frac{d^2y}{dt^2}+2\nu(\tau)\frac{dx}{dt}+y\left(\frac{g}{\varrho_1}-\nu^2(\tau)\right)=\varepsilon f_2(\tau, x, y, \dot{x}, \dot{y}). \end{aligned}\right\} \tag{1.76}$$

§ 9. The idea and basic assumptions of asymptotic methods in nonlinear mechanics

The method of constructing asymptotic solutions for nonlinear differential equations with slowly varying coefficients, given in the present monograph, is a result of a further development and refinement of the asymptotic methods of nonlinear mechanics worked out by N. M. Krylov and N. N. Bogolyubov /9, 72/. Because of this we consider it necessary to give the contents of some well-known ideas of asymptotic methods of nonlinear mechanics, since they will constantly be applied in deriving asymptotic expansions.

Let us discuss these methods briefly in connection with the simplest nonlinear differential equation of the form

$$\frac{d^2x}{dt^2}+\omega^2x=\varepsilon F\left(x, \frac{dx}{dt}\right), \tag{1.77}$$

where ε is a small parameter.

We arrive at the correct formulation of the method of constructing asymptotic approximations for equation (1.77) by starting from physical concepts on the character of the oscillatory process.

Thus, when there is no perturbation, i. e., when $\varepsilon = 0$, the vibrations will, evidently, be purely harmonic

$$x=a\cos\psi, \tag{1.78}$$

with a constant amplitude and a uniformly rotating phase angle

$$\frac{da}{dt}=0, \quad \frac{d\psi}{dt}=\omega \quad (\psi=\omega t+\theta) \tag{1.79}$$

(the amplitude a and the phase θ of the vibrations are constant in time, depending on the initial conditions).

The existence of nonlinear perturbation ($\varepsilon \neq 0$) results in the appearance of overtones in the solution of equation (1.77), establishes a dependence between the instantaneous frequency $\frac{d\psi}{dt}$ and the amplitude, and finally gives rise to a systematic increase or decrease in the amplitude of the vibrations, depending upon whether the energy is expelled or absorbed by the perturbing forces. All these effects are obviously absent in the limiting case ($\varepsilon = 0$).

Taking all this into account, we seek a general solution of equation (1.77) in the form of an expansion

$$x = a\cos\psi + \varepsilon u_1(a, \psi) + \varepsilon^2 u_2(a, \psi) + \varepsilon^3 u_3(a, \psi) + \ldots, \tag{1.80}$$

where $u_1(a, \psi)$, $u_2(a, \psi)$, ... are periodic functions of the angle ψ with period 2π, and the quantities a, ψ, as functions of time, are defined by the differential equations

$$\left.\begin{aligned} \frac{da}{dt} &= \varepsilon A_1(a) + \varepsilon^2 A_2(a) + \ldots, \\ \frac{d\psi}{dt} &= \omega + \varepsilon B_1(a) + \varepsilon^2 B_2(a) + \ldots \end{aligned}\right\} \tag{1.81}$$

Thus, the problem of constructing asymptotic approximations for equation (1.77) reduces to the choice of suitable expressions for the functions $u_1(a, \psi)$, $u_2(a, \psi)$, ..., $A_1(a)$, $A_2(a)$, ..., $B_1(a)$, $B_2(a)$, .. in such a way that (1.80) would, after replacing a and ψ by the functions of time defined by equations (1.81), serve as a solution of the initial equation (1.77).

As soon as this problem is solved and the explicit expressions for the expansion coefficients occurring in the right-hand sides of (1.80), (1.81) are found, the problem of integrating equation (1.77) is reduced to the simpler problem of integrating equations (1.81) with separable variables (if we are considering equation (1.77)), thus making the investigation by means of well-known elementary methods possible.

The determination of the coefficients in the aforementioned expansions presents no theoretical difficulties. However, in view of the rapidly growing complexity of the formulas, in practice only the first two or three terms may usually be derived effectively.

Confining ourselves to these terms in our expansions, i. e., assuming

$$x = a\cos\psi + \varepsilon u_1(a, \psi) + \varepsilon^2 u_2(a, \psi) + \ldots + \varepsilon^m u_m(a, \psi) \quad (m = 1, 2, \ldots) \tag{1.82}$$

and

$$\left.\begin{aligned} \frac{da}{dt} &= \varepsilon A_1(a) + \varepsilon^2 A_2(a) + \ldots + \varepsilon^m A_m(a), \\ \frac{d\psi}{dt} &= \omega + \varepsilon B_1(a) + \varepsilon^2 B_2(a) + \ldots + \varepsilon^m B_m(a), \end{aligned}\right\} \tag{1.83}$$

we may obtain approximations of the first, second, etc. (in general of a low) order. Hence the practical applicability of the asymptotic method is known to be determined not by the convergence properties of the series (1.80), (1.81) when $m \to \infty$, but by their asymptotic properties for a given fixed value of m when $\varepsilon \to 0$. It is only required that the expression (1.80) should for a small ε give a sufficiently accurate form of the solution of equation (1.77) for a sufficiently long time interval. Hence the expansions (1.80), (1.81) are usually treated as formal expansions necessary for working out the asymptotic approximations (1.82). In other words, the question of constructing asymptotic approximations is usually more carefully stated as the problem of deducing such functions $u_1(a, \psi)$, $u_2(a, \psi)$, ..., $A_1(a)$, $A_2(a)$, ..., $B_1(a)$, $B_2(a)$, ..., that the expression (1.82), in which the time functions a, ψ are defined by "equations of the m-th approximation" (1.83), satisfies equation (1.77) with an accuracy of the order of ε^{m+1}.

In solving numerous practical problems it is sufficient for the construction of approximate solutions to confine oneself to the first approximation, which, as may be shown easily (cf. for example /9/, §1), may for equation (1.77) be represented in the form

$$x = a \cos \psi, \tag{1.84}$$

where a and ψ are defined by the equations

$$\frac{da}{dt} = \varepsilon A_1(a), \quad \frac{d\psi}{dt} = \omega + \varepsilon B_1(a), \tag{1.85}$$

with

$$\left.\begin{aligned} \varepsilon A_1(a) &= -\frac{\varepsilon}{2\pi\omega} \int_0^{2\pi} F(a \cos \psi, \ -a\omega \sin \psi) \sin \psi \, d\psi, \\ \varepsilon B_1(a) &= -\frac{\varepsilon}{2\pi\omega a} \int_0^{2\pi} F(a \cos \psi, \ -a\omega \sin \psi) \cos \psi \, d\psi. \end{aligned}\right\} \tag{1.86}$$

The equations of the first approximation may also be obtained by somewhat different physical arguments leading to the widely known method of averaging.

Since the method of averaging will be applied to the solution of some problems connected with the elaboration of a method of investigating differential equations with slowly varying parameters, we shall now discuss this method and also its relation to the so called u-method of constructing asymptotic approximations, presented above.

First of all, we note that when $\varepsilon = 0$, equation (1.77) has the solution

$$\left.\begin{aligned} x &= a \cos \psi, \\ \frac{dx}{dt} &= -a\omega \sin \psi, \end{aligned}\right\} \tag{1.87}$$

where $\psi = \omega t + \theta$, the amplitude a and the phase of vibration θ appearing as constant quantities. It is, however, not difficult to show that formulas (1.87) may also be retained in the case $\varepsilon \neq 0$, provided that the quantities a and θ are regarded not as constants but as certain functions of time.

We shall now consider (1.87) as representing a change of variables, a and θ— the amplitude and phase of vibration — taken as new unknown functions of time. By determining these functions from (1.87), we can find the required expression for the originally unknown x.

After a series of elementary operations we obtain instead of the single differential second order equation (1.77) with respect to the variable x, two differential first order equations with respect to the variables a and θ:

$$\left.\begin{aligned} \frac{da}{dt} &= -\frac{\varepsilon}{\omega} F(a \cos \psi, \ -a\omega \sin \psi) \sin \psi, \\ \frac{d\theta}{dt} &= -\frac{\varepsilon}{a\omega} F(a \cos \psi, \ -a\omega \sin \psi) \cos \psi \\ &\qquad (\theta = \psi - \omega t). \end{aligned}\right\} \tag{1.88}$$

We note that the right-hand sides of equations (1.88) have a period $\frac{2\pi}{\omega}$ with respect to the independent variable t, and besides, that $\frac{da}{dt}$ and $\frac{d\theta}{dt}$ are proportional to the small parameter ε, so that a and θ are slowly varying functions of time.

Differential equations, when reduced to this form, will be called **equations in the standard form**.

It is easily verified that the right-hand sides of (1.88) may be represented in the form of sums as

$$\left.\begin{aligned} -\frac{\varepsilon}{\omega} F(a\cos\psi, -a\omega\sin\psi)\sin\psi &= \varepsilon\sum_{\nu}\left[F_{\nu 1}^{(1)}(a)\cos\nu\psi + F_{\nu 2}^{(1)}(a)\sin\nu\psi\right], \\ -\frac{\varepsilon}{a\omega} F(a\cos\psi, -a\omega\sin\psi)\cos\psi &= \varepsilon\sum_{\nu}\left[F_{\nu 1}^{(2)}(a)\cos\nu\psi + F_{\nu 2}^{(2)}(a)\sin\nu\psi\right] \end{aligned}\right\} \quad (1.89)$$

According to the above, the form of an approximate solution of the system of equations (1.88) may be determined from the following considerations. As a and θ are changing slowly, we represent them as a superposition of smoothly varying terms $\bar{a}$ and $\bar{\theta}$ and sums of oscillating terms. In the first approximation we put

$$a = \bar{a}, \qquad \theta = \bar{\theta} \qquad (\bar{\psi} = \omega t + \bar{\theta}).$$

Hence, in view of (1.89), we have

$$\left.\begin{aligned} \frac{d\bar{a}}{dt} &= \varepsilon\sum_{\nu}\left[F_{\nu 1}^{(1)}(\bar{a})\cos\nu\bar{\psi} + F_{\nu 2}^{(1)}(\bar{a})\sin\nu\bar{\psi}\right], \\ \frac{d\bar{\theta}}{dt} &= \varepsilon\sum_{\nu}\left[F_{\nu 1}^{(2)}(\bar{a})\cos\nu\bar{\psi} + F_{\nu 2}^{(2)}(\bar{a})\sin\nu\bar{\psi}\right], \end{aligned}\right\} \quad (1.90)$$

or

$$\left.\begin{aligned} \frac{d\bar{a}}{dt} &= \varepsilon F_{01}^{(1)}(\bar{a}) + \text{small sinusoidal oscillating terms}, \\ \frac{d\bar{\theta}}{dt} &= \varepsilon F_{01}^{(2)}(\bar{a}) + \text{small sinusoidal oscillating terms}. \end{aligned}\right\} \quad (1.91)$$

Assuming that these sinusoidal oscillating terms give rise only to small oscillations a and θ about their first approximation $\bar{a}$ and $\bar{\theta}$, and do not affect systematic changes of a and θ, we can neglect them and hence arrive at the equations of the first approximation in the form

$$\left.\begin{aligned} \frac{d\bar{a}}{dt} &= \varepsilon F_{01}^{(1)}(\bar{a}) = \underset{t}{M}\left\{-\frac{\varepsilon}{\omega} F(\bar{a}\cos\bar{\psi}, -\bar{a}\omega\sin\bar{\psi})\sin\bar{\psi}\right\}, \\ \frac{d\bar{\theta}}{dt} &= \varepsilon F_{01}^{(2)}(\bar{a}) = \underset{t}{M}\left\{-\frac{\varepsilon}{\bar{a}\omega} F(\bar{a}\cos\bar{\psi}, -\bar{a}\omega\sin\bar{\psi})\cos\bar{\psi}\right\}, \end{aligned}\right\} \quad (1.92)$$

where $\underset{t}{M}$ is the operator of averaging over the explicit time, with $\bar{a}$ and $\bar{\theta}$ as constants.

It is easily seen that the equations (1.92) for $\bar{a}$ and $\bar{\theta}$ obtained here coincide with those derived before with the help of the equations of the first approximation (1.85) by the u-method. Indeed, averaging the right-hand sides of equations (1.92) and introducing in place of θ the full phase of vibration ψ, we obtain

$$\left.\begin{aligned} \frac{da}{dt} &= -\frac{\varepsilon}{2\pi\omega}\int_0^{2\pi} F(a\cos\psi, -a\omega\sin\psi)\sin\psi\, d\psi, \\ \frac{d\psi}{dt} &= \omega - \frac{\varepsilon}{2\pi a\omega}\int_0^{2\pi} F(a\cos\psi, -a\omega\sin\psi)\cos\psi\, d\psi. \end{aligned}\right\} \quad (1.93)$$

The method of averaging just given will also be applied to constructing higher approximations, but then it will already be necessary to take into account the vibrational terms in the expressions for a and θ.

Generally speaking, the structure of the solutions in the phase plane of the averaged equations (obtained in the asymptotic method) and the structure of the solutions of the exact equations in an infinite time interval are considerably different (in a finite time interval they should, unconditionally, be close — the trajectories should be close to each other). At the same time, while considering the stationary case (steady vibrations), which is of considerable interest in the theory of vibrations, it is possible to study in an infinite interval the averaged instead of the exact equations, obviously after performing an appropriate analysis first.

The idea of asymptotic methods, shown above, may without considerable modifications be applied to the construction of approximate asymptotic solutions for various vibrating systems containing a small parameter with one, as well as with many degrees of freedom. It may also be applied to vibrating systems with so called distributed parameters, described by partial differential equations.

We note, among other things, that the method of averaging differential equations reduced to the standard form, in general simplifies considerably the application of the asymptotic methods of nonlinear mechanics to the construction of approximate solutions for systems of nonlinear differential equations. However, the application of asymptotic methods to the construction of such solutions requires the preliminary solution of a set of ordinary linear differential equations with a number of unknowns proportional to the number of degrees of freedom. This causes considerable difficulties in practical application of these methods. In order to remove these difficulties N. N. Bogolyubov /7/ has suggested a method of asymptotic integration, which the author developed in detail and verified in a series of articles /98, 104/. This method serves for the investigation of monofrequency vibrations in nonlinear systems with many degrees of freedom.

Let us now discuss the idea of the monofrequency method, since it is applied in the present monograph and is developed and verified in application to the study of monofrequency vibrational processes in nonlinear systems with slowly varying parameters.

We first of all remark that the presence of unavoidable internal and external friction in many cases leads in vibrating systems with a finite or infinite number of degrees of freedom to a rapid attenuation of higher harmonics and to the establishment of the basic tone of vibrations. The presence of external perturbing forces may also give rise to intensive vibrations of only one definite tone. It is natural in such cases to consider a vibrating system with many degrees of freedom as approximately a system with all its coordinates performing vibrations at one and the same frequency.

The essence of Bogolyubov's method, suitable for studying monofrequency vibrations, consists in the following. We assume that the vibrations of a system with many degrees of freedom are described by differential equations of the form

$$\frac{dx_k}{dt} - \sum_{q=1}^{n} c_{kq}x_q = \varepsilon f_k(x_1, \ldots, x_n, \varepsilon) \quad (k = 1, 2, \ldots, n), \tag{1.94}$$

where c_{kq} are constants.

We assume that in the unperturbed system (i. e., the system (1.94) for $\varepsilon = 0$) there may occur undamped harmonic vibrations with a certain frequency ω_1 of the form

$$x_k = a\varphi_k e^{i(\omega_1 t+\theta)} + a\varphi_k^* e^{-i(\omega_1 t+\theta)} \quad (k = 1, 2, \ldots, n) \tag{1.95}$$

depending only on two arbitrary constants a and θ. Imposing certain restrictions (the absence of an internal resonance and static solutions, different from the trivial ones $x_k = 0$, in the unperturbed system, etc.), we are looking for the solution of the perturbed equations (1.94), close to the particular solution (1.95) of the unperturbed system corresponding to a monofrequency mode, in the form of series

$$x_k = a\varphi_k e^{i\psi} + a\varphi_k^* e^{-i\psi} + \varepsilon u_k^{(1)}(a, \psi) + \varepsilon^2 u_k^{(2)}(a, \psi) + \ldots \qquad (k = 1, 2, \ldots, n), \quad (1.96)$$

where a and ψ should be determined from the equations

$$\left.\begin{aligned} \frac{da}{dt} &= \varepsilon A_1(a) + \varepsilon^2 A_2(a) + \ldots, \\ \frac{d\psi}{dt} &= \omega_1 + \varepsilon B_1(a) + \varepsilon^2 B_2(a) + \ldots \end{aligned}\right\} \quad (1.97)$$

Thus, the problem of deducing approximate solutions for the system (1.94) is reduced, as above, to determining the functions $u_k^{(1)}(a, \psi)$, $u_k^{(2)}(a, \psi)$, ... $(k = 1, 2, \ldots, n)$, $A_1(a)$, $A_2(a)$, ..., $B_1(a)$, $B_2(a)$, .. and the subsequent integration of a system of two equations.

It should be noted that the expression (1.96) represents, after the insertion of a and ψ determined from equations (1.97), a two parametric family of particular solutions. As the superposition principle is not valid for nonlinear systems, we cannot construct the general solution of the system of equations (1.94) starting from the particular solutions of the type (1.96). However, the two parametric family of particular solutions (1.96) possesses in a series of important cases a special property of stability consisting in the fact that any solution of equations (1.94), whose initial values are sufficiently close to the initial values of the two parameteric family (1.96), approaches the latter with increasing time t. In these cases the study of solutions of the type (1.96) is of physical interest.

The asymptotic method of integration through which approximate solutions for the aforementioned types of differential equations with slowly varying parameters can be constructed effectively will be developed in the following chapters. We will develop the asymptotic method mainly in the form of the u-method, which is, in our opinion, more clear physically for many problems. In applications it almost automatically eliminates the possibility of errors in calculations, since only elementary mathematical operations are involved. However, in order to gain a more complete scope of the possibilities offered by asymptotic methods, we also develop a modified method of averaging which allows us to consider effectively nonstationary processes in nonlinear systems with many degrees of freedom and slowly varying parameters.

Chapter II

"NORMAL" OSCILLATIONS IN NONLINEAR SYSTEMS WITH SLOWLY VARYING PARAMETERS

§ 1. Method of constructing asymptotic solutions for an equation close to a linear one

Let us first consider a nonlinear vibrating system with one degree of freedom for which some of the parameters, e. g., the mass of the system, its rigidity, the friction coefficient, etc., vary slowly with time ("slowly" with respect to a natural time unit — period of normal oscillations), but without any external forces which depend explicitly on time t. In such a case we have to deal with the following nonlinear differential equation with slowly varying coefficients

$$\frac{d}{dt}\left[m(\tau)\frac{dx}{dt}\right]+c(\tau)x=\varepsilon F\left(\tau, x, \frac{dx}{dt}\right), \tag{2.1}$$

which is a particular case of the equation (1.1). Here the right-hand side does not depend on θ. As before, ε is a small positive parameter, $\tau = \varepsilon t$, the "slowing" time.

To construct asymptotic approximate solutions of equation (2.1), we assume that the coefficients $m(\tau)$, $c(\tau)$ in this equation, and also the function $F\left(\tau, x, \frac{dx}{dt}\right)$ have the desired number of derivatives with respect to τ, x, $\frac{dx}{dt}$ for all their finite values, and that $m(\tau) > 0$, $c(\tau) > 0$ for any τ within the interval $0 \leqslant \tau \leqslant L$*.

A correct formulation of the method of constructing asymptotic solutions is attained, as is usually done in nonlinear mechanics, by starting with physical notions about the nature of the oscillatory process described by equation (2.1). By our assumption, the coefficients of equation (2.1) vary slowly, i. e., not too rapidly during a period of vibration.

Together with equation (2.1) we will consider the equation

$$m(\tau)\frac{d^2x}{dt^2}+c(\tau)x=0, \tag{2.2}$$

where the parameter τ is considered as a constant and, consequently, $m(\tau)$ and $c(\tau)$ are constants. When $\varepsilon = 0$ is set in equation (2.1) and τ is taken as some constant parameter, we obtain equation (2.2). This equation will be called the unperturbed equation corresponding to the equation (2.1).

* Even if $F\left(\tau, x, \frac{dx}{dt}\right)$ is not differentiable with respect to x, $\frac{dx}{dt}$, it is still possible in some cases to construct asymptotic approximations (for example, when the elastic restoring force of the vibrating system has a nonlinear characteristic, consisting of rectilinear segments).

Vibrations described by (2.2) are obviously purely harmonic

$$x = a\cos\psi, \tag{2.3}$$

with a constant amplitude and uniformly rotating phase angle

$$\frac{da}{dt} = 0, \quad \frac{d\psi}{dt} = \omega, \tag{2.4}$$

where $\psi = \omega t + \theta$, $\omega = \sqrt{\frac{c(\tau)}{m(\tau)}}$, $\tau = \text{const}$.

The presence of a nonlinear perturbation ($\varepsilon \neq 0$), and also the slow variation of some of the parameters, introduces several new effects in the solution of (2.1) in comparison with those obtained according to (2.3). Thus, the solution may contain overtones, the instantaneous frequency $\frac{d\psi}{dt}$ may cease to be constant and may depend on the amplitude of vibration and on the slowing time τ, a systematic increase or decrease of the amplitude of vibration may occur, etc.

Taking all these into account, it is natural to look for a solution of equation (2.1) in the form of an expansion

$$x = a\cos\psi + \varepsilon u_1(\tau, a, \psi) + \varepsilon^2 u_2(\tau, a, \psi) + \ldots, \tag{2.5}$$

where $u_1(\tau, a, \psi)$, $u_2(\tau, a, \psi)$, ... are periodic functions of the angle ψ with period 2π, and the quantities a and ψ, as functions of time, are defined by the differential equations

$$\left.\begin{aligned} \frac{da}{dt} &= \varepsilon A_1(\tau, a) + \varepsilon^2 A_2(\tau, a) + \ldots, \\ \frac{d\psi}{dt} &= \omega(\tau) + \varepsilon B_1(\tau, a) + \varepsilon^2 B_2(\tau, a) + \ldots, \end{aligned}\right\} \tag{2.6}$$

where $\omega(\tau) = \sqrt{\frac{c(\tau)}{m(\tau)}}$ is the "normal" frequency of the given vibrating system, and $\tau = \varepsilon t$.

The expansions (2.5) and (2.6) are, as indicated in the preceding section, considered as formal expansions, necessary for the determination of explicit expressions of the functions

$$u_1(\tau, a, \psi),\ u_2(\tau, a, \psi),\ \ldots,\ A_1(\tau, a),\ A_2(\tau, a),\ \ldots,\ B_1(\tau, a),\ B_2(\tau, a),\ \ldots, \tag{2.7}$$

which are required for the construction of approximate solutions.

We shall not discuss now problems that arise when estimating the derived approximations. These problems are partly illustrated in /104/, and will partly be considered in the last chapter of this book.

We pass immediately to the problem of constructing approximate solutions, remarking that the determination of the expressions (2.7) involves, as known, a certain arbitrariness.

In fact, suppose that certain expressions have been derived for the functions (2.7). Taking the arbitrary functions

$$\alpha_1(a, \tau),\ \alpha_2(a, \tau),\ \ldots,\ \beta_1(a\ \tau),\ \beta_2(a, \tau),\ \ldots$$

and performing in equations (2.6) and (2.5) a change of variables

$$\begin{aligned} a &= b + \varepsilon\alpha_1(b, \tau) + \varepsilon^2\alpha_2(b, \tau) + \ldots, \\ \psi &= \varphi + \varepsilon\beta_1(b, \tau) + \varepsilon^2\beta_2(b, \tau) + \ldots, \end{aligned}$$

we again obtain an expression of the type (2.5) and a system of equations of the type (2.6), only with somewhat modified expressions of the coefficients

(2.7). We shall therefore proceed as usually in nonlinear mechanics, and ensure a unique definition of these coefficients by imposing on them additional conditions; this, as known, may be done with some arbitrariness. These additional conditions are chosen to be the requirement that the expressions of $u_1(\tau, a, \psi)$, $u_2(\tau, a, \psi)$, .. do not contain the first harmonic of the angle ψ. In other words, these periodic functions of the phase angle ψ are defined so as to satisfy the equalities

$$\left.\begin{aligned} &\int_0^{2\pi} u_1(\tau, a, \psi)\cos\psi\, d\psi = 0, \quad \int_0^{2\pi} u_2(\tau, a, \psi)\cos\psi\, d\psi = 0, \ldots \\ &\int_0^{2\pi} u_1(\tau, a, \psi)\sin\psi\, d\psi = 0, \quad \int_0^{2\pi} u_2(\tau, a, \psi)\sin\psi\, d\psi = 0, \ldots \end{aligned}\right\} \tag{2.8}$$

These conditions are from the physical point of view equivalent to associating the quantity a with the full amplitude of the first fundamental harmonic of vibration.

We shall now deduce appropriate expressions for the functions (2.7) under the conditions (2.8).

Differentiating the right-hand side of (2.5), we obtain

$$\left.\begin{aligned} x &= a\cos\psi + \varepsilon u_1(\tau, a, \psi) + \varepsilon^2 u_2(\tau, a, \psi) + \ldots, \\ \frac{dx}{dt} &= \frac{da}{dt}\left\{\cos\psi + \varepsilon\frac{\partial u_1}{\partial a} + \varepsilon^2\frac{\partial u_2}{\partial a} + \ldots\right\} + \\ &+ \frac{d\psi}{dt}\left\{-a\sin\psi + \varepsilon\frac{\partial u_1}{\partial \psi} + \varepsilon^2\frac{\partial u_2}{\partial \psi} + \ldots\right\} + \\ &+ \varepsilon^2\frac{\partial u_1}{\partial \tau} + \varepsilon^3\frac{\partial u_2}{\partial \tau} + \ldots, \\ \frac{d^2x}{dt^2} &= \frac{d^2a}{dt^2}\left\{\cos\psi + \varepsilon\frac{\partial u_1}{\partial a} + \varepsilon^2\frac{\partial u_2}{\partial a} + \ldots\right\} + \\ &+ \frac{d^2\psi}{dt^2}\left\{-a\sin\psi + \varepsilon\frac{\partial u_1}{\partial \psi} + \varepsilon^2\frac{\partial u_2}{\partial \psi} + \ldots\right\} + \\ &+ \left(\frac{da}{dt}\right)^2\left\{\varepsilon\frac{\partial^2 u_1}{\partial a^2} + \varepsilon^2\frac{\partial^2 u_2}{\partial a^2} + \ldots\right\} + \\ &+ 2\frac{da}{dt}\frac{d\psi}{dt}\left\{-\sin\psi + \varepsilon\frac{\partial^2 u_1}{\partial a\,\partial\psi} + \varepsilon^2\frac{\partial^2 u_2}{\partial a\,\partial\psi} + \ldots\right\} + \\ &+ \left(\frac{d\psi}{dt}\right)^2\left\{-a\cos\psi + \varepsilon\frac{\partial^2 u_1}{\partial \psi^2} + \varepsilon^2\frac{\partial^2 u_2}{\partial \psi^2} + \ldots\right\} + \\ &+ 2\frac{da}{dt}\left\{\varepsilon^2\frac{\partial^2 u_1}{\partial\tau\,\partial a} + \varepsilon^3\frac{\partial^2 u_2}{\partial\tau\,\partial a} + \ldots\right\} + \\ &+ 2\frac{d\psi}{dt}\left\{\varepsilon^2\frac{\partial^2 u_1}{\partial\tau\,\partial\psi} + \varepsilon^3\frac{\partial^2 u_2}{\partial\tau\,\partial\psi} + \ldots\right\} + \\ &+ \varepsilon^3\frac{\partial^2 u_1}{\partial\tau^2} + \varepsilon^4\frac{\partial^2 u_2}{\partial\tau^2} + \ldots \end{aligned}\right\} \tag{2.9}$$

Taking equations (2.6) into account, we also find the following expressions

$$\left.\begin{aligned} \frac{d^2a}{dt^2} &= \varepsilon\frac{\partial A_1}{\partial a}\frac{da}{dt} + \varepsilon^2\frac{\partial A_1}{\partial\tau} + \ldots = \varepsilon^2\left(\frac{\partial A_1}{\partial a}A_1 + \frac{\partial A_1}{\partial\tau}\right) + \varepsilon^3\ldots, \\ \frac{d^2\psi}{dt^2} &= \varepsilon\frac{d\omega(\tau)}{d\tau} + \varepsilon\frac{\partial B_1}{\partial a}\frac{da}{dt} + \varepsilon^2\frac{\partial B_1}{\partial\tau} + \ldots = \\ &= \varepsilon\frac{d\omega(\tau)}{d\tau} + \varepsilon^2\left(\frac{\partial B_1}{\partial a}A_1 + \frac{\partial B_1}{\partial\tau}\right) + \varepsilon^3\ldots, \\ \left(\frac{da}{dt}\right)^2 &= \varepsilon^2 A_1^2 + \varepsilon^3\ldots, \\ \frac{da}{dt}\frac{d\psi}{dt} &= \varepsilon A_1\omega(\tau) + \varepsilon^2[A_2\omega(\tau) + A_1B_1] + \varepsilon^3\ldots, \\ \left(\frac{d\psi}{dt}\right)^2 &= \omega^2(\tau) + 2\varepsilon\omega(\tau)B_1 + \varepsilon^2[B_1^2 + 2\omega(\tau)B_2] + \varepsilon^3\ldots \end{aligned}\right\} \tag{2.10}$$

Inserting the expressions (2.9) in the left-hand side of equation (2.1), taking into account (2.6), (2.10), and expanding the result in powers of ε, we obtain

$$\frac{d}{dt}\left\{m(\tau)\frac{dx}{dt}\right\}+c(\tau)x=\varepsilon\left\{-\left[2\omega(\tau)m(\tau)A_1+a\frac{d[m(\tau)\omega(\tau)]}{d\tau}\right]\sin\psi-\right.$$
$$\left.-2m(\tau)\omega(\tau)aB_1\cos\psi+\omega^2(\tau)m(\tau)\frac{\partial^2 u_1}{\partial\psi^2}+c(\tau)u_1\right\}+$$
$$+\varepsilon^2\left\{-\left[2m(\tau)\omega(\tau)A_2+2m(\tau)A_1B_1+am(\tau)\frac{\partial B_1}{\partial a}A_1+\right.\right.$$
$$\left.+\frac{dm(\tau)}{d\tau}aB_1+am(\tau)\frac{\partial B_1}{\partial\tau}\right]\sin\psi+\left[-2m(\tau)\omega(\tau)aB_2-m(\tau)aB_1^2+\right.$$
$$\left.+m(\tau)\frac{\partial A_1}{\partial a}A_1+\frac{dm(\tau)}{d\tau}A_1+m(\tau)\frac{\partial A_1}{\partial\tau}\right]\cos\psi+m(\tau)\omega^2(\tau)\frac{\partial^2 u_2}{\partial\psi^2}+$$
$$+c(\tau)u_2+2m(\tau)\frac{\partial^2 u_1}{\partial\tau\,\partial\psi}\omega(\tau)+2m(\tau)\omega(\tau)B_1\frac{\partial^2 u_1}{\partial\psi^2}+$$
$$\left.+2m(\tau)\omega(\tau)A_1\frac{\partial^2 u_1}{\partial\psi\,\partial a}+\frac{\partial u_1}{\partial\psi}\frac{d}{d\tau}[m(\tau)\omega(\tau)]\right\}+\varepsilon^3\ldots \tag{2.11}$$

We expand the right-hand side of equation (2.1) in a Taylor series, keeping in mind (2.5) and (2.9), and, expanding in powers of the small parameter ε, we write it in the form

$$\varepsilon F\left(\tau,\ x,\ \frac{dx}{dt}\right)=\varepsilon F(\tau,a\cos\psi,\ -a\omega\sin\psi)+\varepsilon^2\left\{F'_x(a\cos\psi,\ -a\omega\sin\psi)u_1+\right.$$
$$\left.+\left(A_1\cos\psi-aB_1\sin\psi+\omega(\tau)\frac{\partial u_1}{\partial\psi}\right)F'_{x'}(\tau,\ a\cos\psi,\ -a\omega\sin\psi)\right\}+\varepsilon^3\ldots \tag{2.12}$$

In order that the original equation (2.1) would be satisfied by (2.5) up to quantities of the order of ε^{m+1} (evidently, the quantities a and ψ should be determined from the system of equations (2.6) with corresponding accuracy), it is necessary to equate the coefficients of equal powers of ε in the right-hand sides of (2.11) and (2.12) up to and including terms of the m-th order. This leads to:

$$\omega^2(\tau)m(\tau)\left[\frac{\partial^2 u_1}{\partial\psi^2}+u_1\right]=F_0(\tau,\ a,\ \psi)+2m(\tau)\omega(\tau)A_1\sin\psi+$$
$$+2m(\tau)\omega(\tau)aB_1\cos\psi+\frac{d[m(\tau)\omega(\tau)]}{d\tau}a\sin\psi, \tag{2.13}$$

$$\omega^2(\tau)m(\tau)\left[\frac{\partial^2 u_2}{\partial\psi^2}+u_2\right]=F_1(\tau,\ a,\ \psi)+\left\{m(\tau)\left[2\omega(\tau)aB_2-\right.\right.$$
$$\left.\left.-\frac{\partial A_1}{\partial a}A_1+aB_1^2\right]-\frac{d[m(\tau)A_1]}{d\tau}\right\}\cos\psi+$$
$$+\left\{m(\tau)\left[2\omega(\tau)A_2+2A_1B_1+a\frac{\partial B_1}{\partial a}A_1\right]+a\frac{d[m(\tau)B_1]}{d\tau}\right\}\sin\psi, \tag{2.14}$$

. . . . ,

where the following notation has been introduced:

$$F_0(\tau,\ a,\ \psi)=F(\tau,\ a\cos\psi,\ -a\omega\sin\psi), \tag{2.15}$$

$$F_1(\tau,\ a,\ \psi)=u_1F'_x(\tau,\ a\cos\psi,\ -a\omega\sin\psi)+$$
$$+\left[A_1\cos\psi-aB_1\sin\psi+\frac{\partial u_1}{\partial\psi}\omega(\tau)\right]F'_{x'}(\tau,\ a\cos\psi,\ -a\omega\sin\psi)-$$
$$-2m(\tau)\omega(\tau)\left[\frac{\partial^2 u_1}{\partial\tau\,\partial\psi}+\frac{\partial^2 u_1}{\partial a\,\partial\psi}A_1+\frac{\partial^2 u_1}{\partial\psi^2}B_1\right]-\frac{\partial u_1}{\partial\psi}\frac{d[m(\tau)\omega(\tau)]}{d\tau}, \tag{2.16}$$

. . . .

We shall determine the functions $A_1(\tau,a)$, $B_1(\tau,a)$, and $u_1(\tau,a,\psi)$ from equation (2.13). To do this we represent the function $F_0(\tau,\ a,\ \psi)$ (noting that it is periodic in ψ with period 2π) and also the sought for periodic function

$u_1(\tau, a, \psi)$ in the form of series

$$F_0(\tau, a, \psi) = \sum_{n=-\infty}^{\infty} F_{0n}(\tau, a) e^{in\psi}, \tag{2.17}$$

$$u_1(\tau, a, \psi) = \sum_{n=-\infty}^{\infty} g_n(\tau, a) e^{in\psi}, \tag{2.18}$$

where

$$F_{0n}(\tau, a) = \frac{1}{2\pi} \int_0^{2\pi} F_0(\tau, a, \psi) e^{-in\psi} d\psi. \tag{2.19}$$

Since $u_1(\tau, a, \psi)$ should, by (2.8), not contain first harmonics of the angle ψ, it has to be kept in mind that $n \neq \pm 1$ in the summation of (2.18).

Substituting the expression (2.18) of $u_1(\tau, a, \psi)$ and the expression (2.17) of $F_0(\tau, a, \psi)$ in equation (2.13), we obtain

$$\omega^2(\tau) m(\tau) \sum_{\substack{n=-\infty \\ (n\neq\pm1)}}^{\infty} [1-n^2] g_n(\tau, a) e^{in\psi} =$$

$$= \sum_{n=-\infty}^{\infty} e^{in\psi} \frac{1}{2\pi} \int_0^{2\pi} F_0(\tau, a, \psi) e^{-in\psi} d\psi + 2m(\tau)\omega(\tau) a B_1 \cos\psi +$$

$$+ \left\{2m(\tau)\omega(\tau) A_1 + \frac{d[m(\tau)\omega(\tau)]}{d\tau} a\right\} \sin\psi. \tag{2.20}$$

Equating the coefficients of the same harmonics, we obtain

$$g_n(\tau, a) = \frac{1}{2\pi m(\tau)\omega^2(\tau)[1-n^2]} \int_0^{2\pi} F_0(\tau, a, \psi) e^{-in\psi} d\psi \quad (n \neq \pm 1). \tag{2.21}$$

Inserting $g_n(\tau, a)$ from (2.21) into the right-hand side of (2.18), we finally obtain

$$u_1(\tau, a, \psi) = \frac{1}{2\pi c(\tau)} \sum_{n\neq\pm1} \frac{e^{in\psi}}{1-n^2} \int_0^{2\pi} F_0(\tau, a, \psi) e^{-in\psi} d\psi. \tag{2.22}$$

Equating the coefficients of $\sin\psi$ and $\cos\psi$ in the right-hand side of relation (2.20) to zero, we find the following expressions for the required functions $A_1(\tau, a)$ and $B_1(\tau, a)$:

$$\left.\begin{aligned} A_1(\tau, a) &= -\frac{a}{2m(\tau)\omega(\tau)} \frac{d[m(\tau)\omega(\tau)]}{d\tau} - \\ &\quad - \frac{1}{2\pi\omega(\tau) m(\tau)} \int_0^{2\pi} F_0(\tau, a, \psi) \sin\psi \, d\psi, \\ B_1(\tau, a) &= -\frac{1}{2\pi m(\tau)\omega(\tau) a} \int_0^{2\pi} F_0(\tau, a, \psi) \cos\psi \, d\psi. \end{aligned}\right\} \tag{2.23}$$

Having uniquely defined $A_1(\tau, a)$ and $B_1(\tau, a)$, and also derived an expression for the periodic function $u_1(\tau, a, \psi)$, we may find $A_2(\tau, a)$, $B_2(\tau, a)$ and $u_2(\tau, a, \psi)$ by solving equation (2.14), the right-hand side of which is now known, and keeping in mind the condition (2.8).

After some elementary calculations we obtain

$$u_2(\tau, a, \psi) = \frac{1}{2\pi c(\tau)} \sum_{n\neq\pm1} \frac{e^{in\psi}}{1-n^2} \int_0^{2\pi} F_1(\tau, a, \psi) e^{-in\psi} d\psi, \tag{2.24}$$

and also

$$\left.\begin{aligned}
A_2(\tau, a) &= -\frac{1}{2\omega(\tau)}\left[a\frac{\partial B_1}{\partial a}A_1 + 2A_1B_1 + \frac{a}{m(\tau)}\frac{d[m(\tau)B_1]}{d\tau}\right] - \\
&\quad - \frac{1}{2\pi m(\tau)\omega(\tau)}\int_0^{2\pi} F_1(\tau, a, \psi)\sin\psi\, d\psi, \\
B_2(\tau, a) &= \frac{1}{2\omega(\tau)a}\left[\frac{\partial A_1}{\partial a}A_1 - aB_1^2 + \frac{1}{m(\tau)}\frac{d[m(\tau)A_1]}{d\tau}\right] - \\
&\quad - \frac{1}{2\pi m(\tau)\omega(\tau)a}\int_0^{2\pi} F_1(\tau, a, \psi)\cos\psi\, d\psi.
\end{aligned}\right\} \tag{2.25}$$

Thus, we obtain a procedure for a successive unique derivation of the functions (2.7).

The method described allows us to determine

$$u_n(\tau, a, \psi), \quad A_n(\tau, a), \quad B_n(\tau, a) \quad (n = 1, 2, 3, \ldots)$$

up to any desired n and hence to construct approximate solutions which satisfy equation (2.1) up to quantities of any desired order of smallness with respect to ε.

Consider the first approximation

$$x = a\cos\psi + \varepsilon u_1(\tau; a, \psi), \tag{2.26}$$

where a and ψ are determined by the following system of equations

$$\frac{da}{dt} = \varepsilon A_1(\tau, a), \quad \frac{d\psi}{dt} = \omega(\tau) + \varepsilon B_1(\tau, a). \tag{2.27}$$

Proceeding from equations (2.27), one may write down

$$\left.\begin{aligned}
\Delta(a) &= a(t) - a(0) \sim \varepsilon t\widetilde{A}_1, \\
\Delta(\psi - \omega t) &= [\psi(t) - \omega t] - \psi(0) \sim \varepsilon t\widetilde{B}_1,
\end{aligned}\right\} \tag{2.28}$$

where $\widetilde{A}_1$ and $\widetilde{B}_1$ are some mean values of $A_1(\tau, a)$, $B_1(\tau, a)$ in the interval $[0, t]$. From the expressions (2.28) one can see that the time t, during which a and $\psi - \omega t$ may assume finite increments, should be of the order of $\frac{1}{\varepsilon}$.

On the other hand, the equations of the first approximation (2.27) are obtained after neglecting terms of the order of ε^2 in (2.6), and such an error in the values of the first derivatives $\frac{da}{dt}$ and $\frac{d\psi}{dt}$ leads during the time t to an error of the order of $\varepsilon^2 t$ in the values of the functions a and ψ themselves. Hence we see that in the time interval during which a and $\psi - \omega t$ depart considerably from their initial values (in the interval of the order of $\frac{1}{\varepsilon}$), the errors in the values of the vibration amplitudes and phases become of the order of ε, and it is therefore meaningless to retain in this interval the term $\varepsilon u_1(\tau, a, \psi)$ in (2.26), being of the first order of smallness, since the error of the formula (2.26), as well as the error of the simplified formula $x = a\cos\psi$ are also of the first order.

Keeping that in mind, we accept in what follows the following expression for the first approximation

$$x = a\cos\psi, \tag{2.29}$$

where a and ψ are defined by equations of the first approximation

$$\left.\begin{aligned}\frac{da}{dt} &= -\frac{\varepsilon a}{2m(\tau)\,\omega(\tau)}\,\frac{d\,[m(\tau)\,\omega(\tau)]}{d\tau} - \\ &\quad -\frac{\varepsilon}{2\pi m(\tau)\,\omega(\tau)}\int_0^{2\pi} F_0(\tau,\ a,\ \psi)\sin\psi\,d\psi, \\ \frac{d\psi}{dt} &= \omega(\tau) - \frac{\varepsilon}{2\pi m(\tau)\,\omega(\tau)\,a}\int_0^{2\pi} F_0(\tau,\ a,\ \psi)\cos\psi\,d\psi.\end{aligned}\right\} \qquad (2.30)$$

We note that the term $\varepsilon u_1(\tau,\ a,\ \psi)$ usually introduces higher harmonics of the angle ψ in the solution (2.26). The approximate solution defined by formula (2.26), where a and ψ are derived from equations (2.30), will be called, as is generally accepted, "the improved first approximation"; it will later be found to be convenient in a few cases for the construction of higher approximations.

Completely analogous arguments lead to the following expression for the second approximation

$$x = a\cos\psi + \varepsilon u_1(\tau,\ a,\ \psi), \qquad (2.31)$$

where the functions of time a and ψ satisfy the equations

$$\left.\begin{aligned}\frac{da}{dt} &= \varepsilon A_1(\tau,\ a) + \varepsilon^2 A_2(\tau,\ a), \\ \frac{d\psi}{dt} &= \omega(\tau) + \varepsilon B_1(\tau,\ a) + \varepsilon^2 B_2(\tau,\ a),\end{aligned}\right\} \qquad (2.32)$$

where $A_1(\tau,\ a)$ and $B_1(\tau,\ a)$ are determined by formulas (2.23), $A_2(\tau,\ a)$ $B_2(\tau,\ a)$— by formulas (2.25), $u_1(\tau,\ a,\ \psi)$— by formula (2.22). a and τ are in all these formulas considered as constant parameters inthe integration over ψ.

It should be mentioned that in order to simplify the calculations, it is sufficient to put the first approximation of a and ψ, defined by equations (2.30), in $\varepsilon u_1(\tau,\ a,\ \psi)$ and the required accuracy is guaranteed by the factor in the second term of (2.31).

Comparing our results for the first and second approximations with the generally known results /9/ in nonlinear mechanics, one easily verifies that the general scheme of constructing asymptotic solutions for vibrating systems, described by equation (2.1), is exactly the same as for nonlinear vibrating systems with constant parameters. The equations of the first approximation (2.30), which we derived, differ from the equations (1.32) of /9/ (p. 45) by the "slowing" time τ and the additional term

$$-\frac{\varepsilon a}{2m(\tau)\,\omega(\tau)}\,\frac{d\,[m(\tau)\,\omega(\tau)]}{d\tau}. \qquad (2.33)$$

Thus, a slow variation of the mass and the coefficient of elasticity, besides violating the harmonicity of oscillations, introduces in the first approximation additional "friction forces" with a sign depending on the variation of the parameters of the specific vibrating system.

When deriving asymptotic approximate solutions we reduce the integration of one second order equation (2.1) to the integration of two first order equations (2.30) or (2.32); these cannot be integrated by elementary functions in most cases and should, therefore, be solved by numerical methods. Applying numerical methods, one could integrate equation (2.1) directly. This is however a complicated problem which requires an extremely long time and is in many cases very difficult to complete in view of a possible accumulation of a large systematic error. Numerical integration of equations of the first (or the second) approximation does not involve any difficulties, since the variables of these equations are the amplitude and phase, and not the oscillating function x. To give a full demonstration of the procedure, it is sufficient to compute a small number of points located on a relatively "smooth" curve; this considerably simplifies numerical integration, while to integrate equation (2.1) one would have to derive directly a "sinusoid" and not the envelope.

§ 2. Equations of the first and second approximations and methods for their construction

It is known that the first approximation is sufficient in most important practical cases. Because of this, we shall discuss in greater detail the analysis and methods of derivation of the first-approximation equations (2.30).

Let us first present another method for deriving the equations of the first approximation, based on the previously mentioned method of averaging which is extensively applied in nonlinear mechanics.

To this end we note that putting $\varepsilon = 0$ in equation (2.1) and considering τ as constant, we obtain equation (2.2), which admits the solution

$$\left.\begin{aligned} x &= a\cos\psi, \\ \frac{dx}{dt} &= -a\omega(\tau)\sin\psi, \end{aligned}\right\} \tag{2.34}$$

where $\psi = \omega(\tau)t + \theta$, a and θ are arbitrary constants.

We consider the formulas (2.34) as a certain change of variables, where a and θ are the new unknown functions of time; performing the change of variables in equation (2.1), we already take $\tau = \varepsilon t$ in the differentiation. This leads after a few operations to the following system of equations, equivalent to equation (2.1),

$$\left.\begin{aligned} \frac{da}{dt} &= -\frac{\varepsilon}{m(\tau)\,\omega(\tau)}\left\{a\,\frac{d\,[m(\tau)\,\omega(\tau)]}{d\tau}\sin^2\psi + \right. \\ &\qquad \left. + F(\tau, a\cos\psi, -a\omega\sin\psi)\sin\psi\right\}, \\ \frac{d\theta}{dt} &= -\frac{\varepsilon}{m(\tau)\,\omega(\tau)}\left\{\frac{d\,[m(\tau)\,\omega(\tau)]}{d\tau}\sin\psi\cos\psi + \right. \\ &\qquad \left. + \frac{1}{a}F(\tau, a\cos\psi, -a\omega\sin\psi)\cos\psi\right\}. \end{aligned}\right\} \tag{2.35}$$

Equations (2.35) represent a system of equations in standard form. Applying to them the method of averaging (a and τ are, as usual, kept constant during integration) and using the notation (2.15), we obtain the system of equations

$$\left.\begin{aligned} \frac{da}{dt} &= -\frac{\varepsilon a}{2m(\tau)\,\omega(\tau)}\,\frac{d\,[m(\tau)\,\omega(\tau)]}{d\tau} - \frac{\varepsilon}{2\pi m(\tau)\,\omega(\tau)}\int_0^{2\pi} F_0(\tau, a, \psi)\sin\psi\,d\psi, \\ \frac{d\theta}{dt} &= -\frac{\varepsilon}{2\pi m(\tau)\,\omega(\tau)\,a}\int_0^{2\pi} F_0(\tau, a, \psi)\cos\psi\,d\psi, \end{aligned}\right\} \tag{2.36}$$

which becomes after the substitution $\frac{d\theta}{dt} = \frac{d\psi}{dt} - \omega(\tau)$ identical with the previously derived equations of the first approximation (2.30).

The second approximation is obtained by performing a change of variables in the system of equations (2.35), equivalent to equation (2.1), according to the formulas

$$\left.\begin{aligned} a &= \bar{a} + \varepsilon \sum_{n\neq 0} \frac{1}{n} [F_{n1}^{(1)}(\bar{a}, \tau)\sin n\bar{\psi} - F_{n2}^{(1)}(\bar{a}, \tau)\cos n\bar{\psi}], \\ \theta &= \bar{\theta} + \varepsilon \sum_{n\neq 0} \frac{1}{n} [F_{n1}^{(2)}(\bar{a}, \tau)\sin n\bar{\psi} - F_{n2}^{(2)}(\bar{a}, \tau)\cos n\bar{\psi}], \end{aligned}\right\} \tag{2.37}$$

where $\bar{\psi} = \omega t + \bar{\theta}$. The sums on the right-hand sides represent small oscillating terms (appearing in a and θ by virtue of the small terms) periodic in ψ, which appear in the right-hand sides of equations (2.35):

$$\left.\begin{aligned} &\varepsilon \sum_{n\neq 0} [F_{n1}^{(1)}(a, \tau)\cos n\psi + F_{n2}^{(1)}(a, \tau)\sin n\psi], \\ &\varepsilon \sum_{n\neq 0} [F_{n1}^{(2)}(a, \tau)\cos n\psi + F_{n2}^{(2)}(a, \tau)\sin n\psi], \end{aligned}\right\} \tag{2.38}$$

where $F_{nk}^{(i)}(a, \tau)$ $(i, k = 1, 2; n = 1, 2, 3, \ldots)$ are Fourier coefficients of the periodic functions in the right-hand sides of equations (2.35). These terms were neglected in passing from the system of exact equations (2.35) to the equations of the first approximation (2.36).

The expressions (2.37) correspond to the improved first approximation. In fact, inserting (2.37) in the right-hand side of (2.34), we obtain (2.31) up to and including quantities of the first order.

To derive the equations of the second approximation, defining a and θ up to quantities of the first order inclusive, it is necessary to substitute in equations (2.35) the values of a and θ given by (2.37), and to average over the explicitly contained time, i. e., over ψ. In substituting the values of a and θ in equations (2.35) we evidently take into account that $\tau = \varepsilon t$; $\bar{a}$ and $\bar{\theta}$ are the required functions of time. In averaging the integration, we consider $\bar{a}$, $\bar{\theta}$, and τ as constant parameters. Performing the averaging and retaining terms of the second order, we obtain the equations of the second approximation which coincide with equations (2.32).

Let us still consider one other form of notation and derivation of the equations of the first approximation (2.30), convenient in some cases.

We recall that the first approximation (2.29) is, according to the conditions (2.8), the principal harmonic of the approximate solution (2.5), and the amplitude a is the full amplitude of the principal harmonic.

We introduce functions of the amplitude a and slow time τ, defined by

$$\left.\begin{aligned} \lambda_e(a, \tau) &= \frac{\varepsilon}{\omega(\tau)} \frac{d[m(\tau)\omega(\tau)]}{d\tau} + \frac{\varepsilon}{\pi a \omega(\tau)} \int_0^{2\pi} F_0(\tau, a, \psi)\sin\psi\, d\psi, \\ c_e(a, \tau) &= c(\tau) - \frac{\varepsilon}{\pi a} \int_0^{2\pi} F_0(\tau, a, \psi)\cos\psi\, d\psi. \end{aligned}\right\} \tag{2.39}$$

We have in first approximation

$$x = a\cos\psi, \tag{2.40}$$

where a and ψ are defined by equations of the first approximation, which may be written in the form

$$\left.\begin{aligned} \frac{da}{dt} &= -\frac{\lambda_e(a, \tau)}{2m(\tau)} a, \\ \frac{d\psi}{dt} &= \omega_e(a, \tau). \end{aligned}\right\} \tag{2.41}$$

Here $\omega_e^2(a, \tau) = \frac{c_e(a, \tau)}{m(\tau)}$.

Substituting the first approximation (2.40) in equation (2.1), one easily obtains

$$m(\tau)\frac{d^2x}{dt^2}+\frac{\varepsilon}{\omega(\tau)}\frac{d[m(\tau)\omega(\tau)]}{d\tau}\frac{dx}{dt}+\frac{\varepsilon}{a\omega(\tau)}\int_0^{2\pi}F_0(\tau, a, \psi)\sin\psi\, d\psi\cdot\frac{dx}{dt}+$$

$$+c(\tau)x-\frac{\varepsilon}{\pi a}\int_0^{2\pi}F_0(\tau, a, \psi)\cos\psi\, d\psi\cdot x=O(\varepsilon^2)$$

or, using the notations (2.39),

$$m(\tau)\frac{d^2x}{dt^2}+\lambda_e(a, \tau)\frac{dx}{dt}+c_e(a, \tau)x=O(\varepsilon^2), \qquad (2.42)$$

where $O(\varepsilon^2)$ is a quantity of the order of ε^2.

Hence we find that the first approximation under consideration (2.40) satisfies a linear equation of the form

$$m(\tau)\frac{d^2x}{dt^2}+\lambda_e(a, \tau)\frac{dx}{dt}+c_e(a, \tau)x=0. \qquad (2.43)$$

with an accuracy of the order of ε^2. Consequently, the vibrations of the given nonlinear vibrating system with slowly varying parameters, described by the differential equation (2.1), are in the first approximation equivalent (up to terms of the order of ε^2, i. e., with an accuracy of magnitudes which are neglected in constructing the equations of the first approximation (2.30)) to the vibrations of a linear vibrating system with a damping coefficient $\lambda_e(a, \tau)$ and an elasticity coefficient $c_e(a, \tau)$.

Hence, we shall call $\lambda_e(a, \tau)$ the equivalent damping coefficient, $c_e(a, \tau)$, the equivalent elasticity coefficient, and the linear vibrating system itself, described by the differential equation (2.43), will be called the equivalent system; this is usual in nonlinear mechanics.

We write equation (2.1) in the form

$$m(\tau)\frac{d^2x}{dt^2}+c(\tau)x=\varepsilon\left\{F\left(\tau, x, \frac{dx}{dt}\right)-\frac{dm(\tau)}{d\tau}\frac{dx}{dt}\right\} \qquad (2.44)$$

Comparing equations (2.43) and (2.44), we see that the former is obtained from the latter by replacing the right-hand expression by the linear term with coefficients depending on the parameter a and the slowly varying time τ:

$$-\left[\lambda_e(a, \tau)\frac{dx}{dt}+c_1(a, \tau)x\right], \qquad (2.45)$$

where

$$c_1(a, \tau)=c_e(a, \tau)-c(\tau).$$

We note that the expression

$$\delta_e(a, \tau)=\frac{\lambda_e(a, \tau)}{2m(\tau)} \qquad (2.46)$$

represents the damping decrement of the equivalent linear system, and

$$\omega_e(a, \tau)=\sqrt{\frac{c_e(a, \tau)}{m(\tau)}} \qquad (2.47)$$

is the normal frequency of vibrations of this system (with an accuracy of terms of the first order inclusive).

Consequently, the equations of the first approximation (2.30) may be constructed as follows.

We consider instead of the vibrating system described by equation (2.1) its equivalent linearized system, obtained by replacing the right-hand side in equation (2.44) by the expression (2.45), where $\lambda_e(a, \tau)$ and $c_e(a, \tau)$ are defined by formulas (2.39). The damping decrement and the normal frequency of vibrations are derived in the usual way from formulas (2.46) and (2.47) for the new system with mass $m(\tau)$, damping coefficient $\lambda_e(a, \tau)$, and elasticity coefficient $c_e(a, \tau)$. The derived expressions are then substituted in the generally known formulas for linear vibrating systems

$$\frac{da}{dt} = -\delta a, \quad \frac{d\psi}{dt} = \omega, \tag{2.48}$$

according to which the damping decrement δ is the logarithmic derivative of the amplitude with negative sign, and the frequency ω is the angular velocity of rotation of the full phase of vibration.

Substituting the values of $\delta_e(a, \tau)$ and $\omega_e(a, \tau)$ from (2.46) and (2.47) in equations (2.48), we obtain equations which coincide with the previously deduced equations of the first approximation (2.41).

This method, called the method of equivalent linearization, may easily be given a physical interpretation and a rigorous foundation in application to a nonlinear differential equation with slowly varying parameters. We shall however not dwell upon this problem, remarking that all the arguments and operations may be performed in analogy with /9/, keeping in mind the additional considerations arising in connection with the presence of slowly varying parameters in the vibrating system.

We also note that it is possible to consider completely analogously, instead of a vibrating system described by the differential equation (2.43), equivalent to equation (2.1) up to terms of second order, an equivalent system from which we can construct the equations of the second approximation (2.32), etc.

Besides the presented method of equivalent linearization, the so-called method of harmonic balance is known to be very convenient for constructing the equations of the first, second, etc., approximations. Without giving a rigorous justification of this method, we formally quote the rules for deriving certain quantities, required in constructing approximate solutions. Thus, to derive the equations of the first (2.30) and the second (2.32) approximations we may directly use, instead of formulas (2.23) and (2.25), the following equations, called the equations of harmonic balance,

$$\left.\begin{aligned}
&\int_0^{2\pi} \left\{ \frac{d}{dt}\left[m(\tau) \frac{dx}{dt} \right] + \right. \\
&\qquad \left. + c(\tau) x - \varepsilon F\left(\tau, x, \frac{dx}{dt}\right)\right\}_{x = a\cos\psi + \dots} \cos\psi \, d\psi = 0, \\
&\int_0^{2\pi} \left\{ \frac{d}{dt}\left[m(\tau) \frac{dx}{dt} \right] + \right. \\
&\qquad \left. + c(\tau) x - \varepsilon F\left(\tau, x, \frac{dx}{dt}\right)\right\}_{x = a\cos\psi + \dots} \sin\psi \, d\psi = 0.
\end{aligned}\right\} \tag{2.49}$$

For obtaining the required expressions we insert into the integrands the values derived from (2.5) of x, $\frac{dx}{dt}$, $\frac{d^2x}{dt^2}$, up to an accuracy of the first order, taking into account that a and ψ are functions of time satisfying equations (2.27). Integrating, we obtain an expression for $A_1(\tau, a)$ and $B_1(\tau, a)$ coinciding with (2.23). Taking into consideration also quantities proportional to ε^2 in the substitutions for x, $\frac{dx}{dt}$, $\frac{d^2x}{dt^2}$, and considering a and ψ as functions of time satisfying equations (2.32), we obtain for $A_2(\tau, a)$ and $B_2(\tau, a)$ expressions which coincide with (2.25).

We note that τ and a are considered as constant parameters in calculating the integrals (2.49) and the integration is only over ψ. Substituting the values of x, $\frac{dx}{dt}$, $\frac{d^2x}{dt^2}$ in the integrands, we take $\tau = \varepsilon t$.

§ 3. Particular cases of equation (2.1)

We shall now examine some particular cases of equation (2.1) for which the equations of the first approximation assume quite simple forms.

The first particular case considered is a vibrating system without friction, described by the differential equation

$$\frac{d}{dt}\left\{m(\tau)\frac{dx}{dt}\right\} + c(\tau)x = \varepsilon F(\tau, x), \tag{2.50}$$

where $\tau = \varepsilon t$ and ε is a small positive parameter.

In this case the equations of the first approximation are

$$\left.\begin{aligned} \frac{da}{dt} &= -\frac{\varepsilon a}{2m(\tau)\omega(\tau)}\frac{d[m(\tau)\omega(\tau)]}{d\tau}, \\ \frac{d\psi}{dt} &= \omega(\tau) - \frac{\varepsilon}{2\pi m(\tau)\omega(\tau)a}\int_0^{2\pi} F(\tau, a\cos\psi)\cos\psi\, d\psi. \end{aligned}\right\} \tag{2.51}$$

This system of equations may be integrated completely. Indeed, we obtain from the first equation

$$a(\tau) = \frac{a_0}{[m(\tau)\omega(\tau)]^{1/2}}, \tag{2.52}$$

where a_0 is the initial value of the amplitude at $t=0$. Substituting this value of the amplitude in the second of equations (2.51), we obtain

$$\psi = \int_0^t \omega_e[a(\tau), \tau]\, dt, \tag{2.53}$$

where

$$\omega_e[a(\tau), \tau] = \omega(\tau) - \frac{\varepsilon}{2\pi[m(\tau)\omega(\tau)]^{1/2}a_0}\int_0^{2\pi} F\left(\tau, \frac{a_0}{[m(\tau)\omega(\tau)]^{1/2}}\cos\psi\right)\cos\psi\, d\psi.$$

Thus, the vibrations described by equation (2.50) will in the first approximation be "sinusoidal" with a slowly varying amplitude inversely proportional to $\sqrt{m(\tau)\omega(\tau)}$, and a phase varying according to formula (2.53).

As a second particular case we examine a vibrating system with a slowly varying mass $m(\tau)$ under the action of a linear elastic force $c(\tau)x$ with a slowly varying elasticity coefficient and a small nonlinear friction depending only on the velocity and the slowing time. In this case we obtain the equation

$$\frac{d}{dt}\left\{m(\tau)\frac{dx}{dt}\right\} + c(\tau)x = \varepsilon F\left(\tau, \frac{dx}{dt}\right), \tag{2.54}$$

for which the equations of the first approximation have the form

$$\left.\begin{aligned}\frac{da}{dt}&=-\frac{\varepsilon a}{2m(\tau)\,\omega(\tau)}\frac{d\,[m(\tau)\,\omega(\tau)]}{d\tau}-\\&\qquad-\frac{\varepsilon}{2\pi m(\tau)\,\omega(\tau)}\int_0^{2\pi}F(\tau,\,-a\omega\sin\psi)\sin\psi\,d\psi,\\\frac{d\psi}{dt}&=\omega(\tau).\end{aligned}\right\}\qquad(2.55)$$

The law of variation of the full phase of vibration is immediately found from the second equation

$$\psi=\int_0^t\omega(\tau)\,dt.\qquad(2.56)$$

The frequency of vibrations, described by equation (2.54), obviously does not depend in the first approximation on the amplitude, but only on the character of the slow variation of the mass and rigidity of the system. If the mass and rigidity were constant, we would obtain, as known, vibrations called quasi-isochronous, for which the frequency is in the first approximation constant and independent of the amplitude, as is the case in most non-linear vibrating systems.

We examine one other particular case. Let the right-hand side of equation (2.1) be a linear function of x and $\frac{dx}{dt}$ with slowly varying coefficients. We are then led to the following linear differential equation with slowly varying parameters

$$\frac{d}{dt}\left\{m(\tau)\frac{dx}{dt}\right\}+c(\tau)\,x=\varepsilon\left[c_1(\tau)\,x+\lambda(\tau)\frac{dx}{dt}\right].\qquad(2.57)$$

Linear differential equations with slowly varying parameters are studied in the works of S. F. Freshchenko /161/ and we shall therefore not discuss in detail equations of the type (2.57), which are also very important.

The equations of the first approximation for equation (2.57) are easily found to be

$$\left.\begin{aligned}\frac{da}{dt}&=-\varepsilon a\left[\frac{1}{2m(\tau)\,\omega(\tau)}\frac{d\,[m(\tau)\,\omega(\tau)]}{d\tau}-\frac{\lambda(\tau)}{2m(\tau)}\right],\\\frac{d\psi}{dt}&=\omega(\tau),\end{aligned}\right\}\qquad(2.58)$$

where

$$\omega(\tau)=\sqrt{\frac{c(\tau)-\varepsilon c_1(\tau)}{m(\tau)}}.$$

The integration of the system (2.58) is elementary, as it is linear with respect to a with separable variables.

We note that the equations of the second and higher approximations for equation (2.57) may also be completely integrated.

§ 4. Vibrations of a pendulum with variable length

As an example we shall consider the vibrations of a mathematical pendulum of a constant mass (additional difficulties do not arise in the case of variable mass) when there is a small damping proportional to the first power of the velocity, and when the length varies slowly /93/. Numerous practical problems reduce to this case. Denoting the angle of deviation of the pendulum from the vertical by x, the acceleration due to gravity by g, the mass of the pendulum by m, the slowly varying length by $l = l(\tau)$, and the friction coefficient by $2n$, we obtain the differential equation

$$\frac{d}{dt}\left[ml^2(\tau)\frac{dx}{dt}\right] + 2n\frac{d}{dt}[l(\tau)x] + mgl(\tau)\sin x = 0. \tag{2.59}$$

For small vibrations, we may replace $\sin x$ by the first two terms of its expansion in a power series. Equation (2.59) may then be written in the form

$$\frac{d}{dt}\left[ml^2(\tau)\frac{dx}{dt}\right] + mgl(\tau)x = \varepsilon F\left(\tau, x, \frac{dx}{dt}\right), \tag{2.60}$$

where

$$\varepsilon F\left(\tau, x, \frac{dx}{dt}\right) = \frac{mgl(\tau)}{6}x^3 - 2nl(\tau)\frac{dx}{dt} - 2\varepsilon n\frac{dl(\tau)}{d\tau}x.$$

Equation (2.60) contains, essentially, two small parameters: the small parameter ε, characterizing the slow change in the length of the pendulum and the presence of a small friction (the friction coefficient is small by assumption and we may therefore set $n = \varepsilon n_1$), and a small amplitude of vibrations. An estimation of the error is in this case somewhat complicated but, as is evident from a detailed examination of vibrations of a pendulum of constant length (cf. /9/, p. 54), may be obtained without considerable difficulties.

In order to apply the previously discussed scheme to a construction of the equations of the first and second approximations, and to treat appropriately terms of small order in the corresponding approximations, we once more transform equation (2.60).

Supposing that the amplitude of vibrations is small, we define

$$x = \sqrt{\varepsilon}\, x_1,$$

and write equation (2.60) in the form

$$\frac{d}{dt}\left[ml^2(\tau)\frac{dx_1}{dt}\right] + mgl(\tau)x_1 = \varepsilon F_1\left(\tau, x_1, \frac{dx_1}{dt}, \varepsilon\right), \tag{2.61}$$

where

$$F_1\left(\tau, x_1, \frac{dx_1}{dt}, \varepsilon\right) = \frac{mgl(\tau)}{6}x_1^3 - 2n_1 l(\tau)\frac{dx_1}{dt} - 2\varepsilon n_1\frac{dl(\tau)}{d\tau}x_1.$$

Applying formulas (2.29) and (2.30) to equation (2.61), we find in the first approximation

$$x_1 = a\cos\psi, \tag{2.62}$$

where a and ψ should be determined from the system of equations of the first approximation

$$\left.\begin{aligned} \frac{da}{dt} &= -\varepsilon\left[\frac{n_1}{ml(\tau)} + \frac{3l'(\tau)}{4l(\tau)}\right]a, \\ \frac{d\psi}{dt} &= \omega(\tau)\left(1 - \frac{\varepsilon a^2}{16}\right) \end{aligned}\right\} \tag{2.63}$$

(here $\omega(\tau) = \sqrt{\frac{g}{l(\tau)}}$).

Integrating the first of equations (2.63) with the initial values $t = 0$, $a = a_0$, we obtain for a the expression

$$a = a_0 e^{-\frac{\varepsilon n_1}{m}\int\limits_0^t \frac{dt}{l(\tau)}} \left[\frac{l(0)}{l(\tau)}\right]^{\frac{3}{4}}. \tag{2.64}$$

Substituting this value of a into the second equation of the system (2.63), we obtain

$$\psi = \int\limits_0^t \omega(\tau)\left[1 - \frac{\varepsilon a_0^2 e^{-\frac{2n_1\varepsilon}{m}\int\limits_0^t \frac{dt}{l(\tau)}} \left[\frac{l(0)}{l(\tau)}\right]^{\frac{3}{2}}}{16}\right] dt. \tag{2.65}$$

Formulas (2.64) and (2.65) enable us to plot the curve exhibiting the dependence of the amplitude and phase on time when the length of the pendulum varies slowly.

If we put $l = \text{const}$ in these formulas, we obtain

$$\left.\begin{aligned} a &= a_0 e^{-\frac{\lambda}{2}t}, \\ \psi &= \omega\left[t + \frac{\varepsilon a_0^2 (e^{-\lambda t} - 1)}{16\lambda}\right] + \psi_0, \end{aligned}\right\} \tag{2.66}$$

where $\lambda = \frac{2\varepsilon n_1}{ml}$, and ψ_0 is the initial value of the phase. Formulas (2.66) are identical with the well-known formulas for the vibrations of a pendulum with constant length.

Let us now assume that the length of the pendulum varies according to the linear law $l(\tau) = l_0 + l_1\tau$; l_0 is the value of the length at $t = 0$, εl_1, the rate of change of the length of the pendulum. In this case we have the following expressions for the amplitude and phase

$$a = a_0 \left(\frac{l_0}{l_0 + l_1\tau}\right)^{\frac{3}{4} + \frac{n}{ml_1}}, \tag{2.67}$$

$$\psi = \int\limits_0^t \sqrt{\frac{g}{l_0 + l_1\tau}} \left[1 - \frac{\varepsilon a_0^2}{16}\left(\frac{l_0}{l_0 + l_1\tau}\right)^{\frac{3}{2} + \frac{2n_1}{ml_1}}\right] dt \tag{2.68}$$

(we note that it is necessary here to take $\tau = \varepsilon t$ during the integration).

According to formula (2.67), when the length of the pendulum varies slowly the amplitude of vibrations will not change according to an exponential law as in the case of usual linear friction, but in inverse proportion to a certain power function of time. Here it is obvious that when $n_1 < 0$, $l_1 > 0$ and $\left|\frac{n_1}{ml_1}\right| < \frac{3}{4}$, and when $n_1 > 0$ and $l_1 > 0$, then the vibrations damp down. Thus, a slow increase in the length of the pendulum promotes, as could be expected, the damping of the vibrations. If $l_1 < 0$, $n_1 > 0$ and $\left|\frac{n_1}{ml_1}\right| < \frac{3}{4}$, the amplitude increases, and if $\left|\frac{n_1}{ml_1}\right| > \frac{3}{4}$ it decreases. When $l_1 < 0$ and $n_1 < 0$, the amplitude increases. In the absence of damping ($n_1 = 0$) the amplitude of vibrations increases with a reduction in length and decreases with an increase in length.

A similar analysis may be performed for the frequency of vibrations. Thus, for example, when there is no damping, the instantaneous frequency decreases (and hence the period increases) with an increase in the length of the pendulum, and increases (the period decreases) with a decrease in this length.

We now calculate the second approximation for the given example (vibrations of a pendulum with a slowly varying length).

According to formulas (2.31), (2.32) and (2.22) we obtain, after some calculations,

$$x_1 = a\cos\psi - \frac{\varepsilon a^3}{192}\cos 3\psi, \tag{2.69}$$

where a and ψ should be determined from the system of equations of the second approximation

$$\left.\begin{aligned} \frac{da}{dt} &= -\varepsilon\left[\frac{3l'(\tau)}{4l(\tau)} + \frac{n_1}{ml(\tau)}\right]\left(a + \frac{\varepsilon a^3}{16}\right), \\ \frac{d\psi}{dt} &= \omega(\tau)\left(1 - \frac{\varepsilon a^2}{16}\right) + \frac{\varepsilon^2}{2\omega(\tau)}\left\{\frac{n_1^2}{m^2 l^2(\tau)} + \frac{l'(\tau)n_1}{ml^2(\tau)} + \right. \\ &\left. + \frac{5l''(\tau)}{4l(\tau)} + \frac{5\omega^2(\tau)a^4}{3\cdot 2^3} - \frac{3l'^2(\tau)}{16l^2(\tau)}\right\}, \end{aligned}\right\} \tag{2.70}$$

which may be integrated completely. Thus, from the first equation of the system (2.70), we obtain the following relation between a and t:

$$\frac{a}{\sqrt{16+\varepsilon a^2}} = \frac{a_0}{\sqrt{16+\varepsilon a_0^2}}\, e^{-\frac{\varepsilon n_1}{m}\int\limits_0^t \frac{dt}{l(\tau)}} \left[\frac{l(0)}{l(\tau)}\right]^{\frac{3}{4}}, \tag{2.71}$$

after which we can also integrate the second equation of the system (2.70).

§ 5. Nonlinear differential equation with slowly varying parameters, close to an exactly integrable one

We have considered above nonlinear differential equations describing nonstationary vibrational processes in nonlinear systems with slowly varying parameters, close to linear vibrating systems. A general method of constructing approximate solutions for such systems was developed in § 1 of the present chapter. It is based on asymptotic methods of nonlinear mechanics and enables us to construct without essential difficulties approximate solutions of any order.

For many important practical problems in the theory of vibrations one must consider nonlinear differential equations with slowly varying parameters which essentially differ from linear differential equations. There do not exist any general effective methods of constructing approximate solutions for such equations in the general case, not even in the case of a vibrating system with one degree of freedom.

However, many cases involve nonlinear differential equations with slowly varying parameters which depend on a small parameter ε in such a way that they reduce for a zero value of the latter to differential equations which, although nonlinear, possess certain properties from which special methods of constructing approximate solutions for the initial equation with $\varepsilon \neq 0$ and $\tau = \varepsilon t$ can be worked out. Such equations have been examined by the author /94/ and by V. M. Volosov /30/.

Let us first consider a nonlinear differential equation with slowly varying parameters which differs essentially from a linear one, but contains a small parameter ε in such a way that it may be integrated exactly for a zero value of the latter and for $\tau = \text{const}$. Let this equation, describing a vibrating system with one degree of freedom, have the form*

$$\frac{d^2x}{dt^2}+f(\tau, x)=\varepsilon F\left(\tau, x, \frac{dx}{dt}, \varepsilon\right), \tag{2.72}$$

where, as before, ε is a small positive parameter; $\tau=\varepsilon t$ is the "slowing time"; and $F\left(\tau, x, \frac{dx}{dt}, \varepsilon\right)$ is an analytic function of ε, which may, for small values of ε, be expanded in a power series

$$F\left(\tau, x, \frac{dx}{dt}, \varepsilon\right)=\sum_{n=0}^{\infty}\varepsilon^n F_n\left(\tau, x, \frac{dx}{dt}\right), \tag{2.73}$$

where the coefficients possess the desired number of bounded derivatives for all finite values of their arguments τ, x, $\frac{dx}{dt}$.

We assume, according to previous arguments, that for the equation

$$\frac{d^2x}{dt^2}+f(\tau, x)=0 \tag{2.74}$$

(we shall call it henceforth the "unperturbed" equation, considering τ as a constant parameter) a periodic solution is known for any τ within the interval $0\leqslant\tau\leqslant L$

$$x=z(\tau, \psi, a), \tag{2.75}$$

where

$$\psi=\omega(\tau, a)t+\varphi,$$

where a and φ are two arbitrary constants, whose physical meaning is quite obvious: parameter a characterizes the form and amplitude of vibrations, parameter ψ, the initial phase of vibrations, parameter φ, the initial phase of vibrations.

Before working out a method of constructing approximate solutions for equation (2.72) in the case under consideration, we transform it to new variables, utilizing the known solution (2.75).

We introduce in equation (2.72) new variables a and ψ according to the formulas

$$\left.\begin{aligned} x&=z(\tau, \psi, a),\\ \frac{dx}{dt}&=\omega(\tau, a)z'_{\psi}(\tau, \psi, a),\end{aligned}\right\} \tag{2.76}$$

where $\tau=\varepsilon t$.

Differentiating formula (2.76), substituting the result into (2.72), and taking into account the identity

$$\omega^2(\tau, a)z''_{\psi^2}+f(\tau, z)=0, \tag{2.77}$$

we obtain the following system

$$\left.\begin{aligned} z'_a\frac{da}{dt}+z'_{\psi}\frac{d\psi}{dt}&=\omega(\tau, a)z'_{\psi}-\varepsilon z'_{\tau},\\ (\omega z'_{\psi})'_a\frac{da}{dt}+\omega z''_{\psi^2}\frac{d\psi}{dt}&=-\omega^2(\tau, a)z''_{\psi^2}+\varepsilon\{F(\tau, z, \omega z'_{\psi}, \varepsilon)-(\omega z'_{\psi})'_{\tau}\}.\end{aligned}\right\} \tag{2.78}$$

Solving the system of equations (2.78) for $\frac{da}{dt}$ and $\frac{d\psi}{dt}$, we obtain the following system of differential equations, equivalent to equation (2.72):

$$\frac{da}{dt}=\frac{\varepsilon\{-F(\tau, z, \omega z'_{\psi}, \varepsilon)z'_{\psi}+(\omega z'_{\psi})'_{\tau}z'_{\psi}-\omega z''_{\psi^2}z'_{\tau}\}}{\omega z'_a z''_{\psi^2}-z'_{\psi}(\omega z'_{\psi})'_a}, \tag{2.79}$$

* It is assumed here, without loss of generality, that the mass equals unity, since an equation for $m=m(\tau)$ may always be reduced to the form (2.72), as is done, for example, on p. 52.

$$\frac{d\psi}{dt}=\omega(\tau, a)+\frac{\varepsilon\{F(\tau, z, \omega z'_\psi, \varepsilon)z'_a-(\omega z'_\psi)'_\tau z'_a+(\omega z'_\psi)'_a z'_\tau\}}{\omega z'_a z''_{\psi^2}-z'_\psi(\omega z'_\psi)'_a}. \tag{2.79}$$

It is easily shown that the denominator in the right-hand sides of equations (2.79) does not depend on ψ. In fact, differentiating identity (2.77) with respect to ψ and a, we obtain

$$\left.\begin{aligned} \omega^2 z'''_{\psi^3}+f'_z(\tau, z)z'_\psi=0,\\ 2\omega\omega'_a z''_{\psi^2}+\omega^2 z'''_{\psi^2 a}+f'_z(\tau, z)z'_a=0. \end{aligned}\right\} \tag{2.80}$$

Multiplying the first identity in (2.80) by z'_a, the second by z'_ψ, and subtracting the second from the first, we obtain

$$\omega^2 z'''_{\psi^3}z'_a-\omega^2 z'''_{\psi^2 a}z'_\psi-2\omega\omega'_a z''_{\psi^2}z'_\psi=0 \tag{2.81}$$

or

$$\frac{d}{d\psi}[\omega z'_a z''_{\psi^2}-z'_\psi(\omega z'_\psi)'_a]=0. \tag{2.82}$$

Integrating this expression from 0 to 2π, we finally find

$$\omega z'_a z''_{\psi^2}-z'_\psi(\omega z'_\psi)'_a=[\omega z'_a z''_{\psi^2}-z'_\psi(\omega z'_\psi)'_a]_{\psi=0}=D_0(\tau, a), \tag{2.83}$$

where $D_0(\tau, a)$ does not depend on ψ.

We define

$$\left.\begin{aligned} \frac{1}{D_0}\{-F(\tau, z, \omega z'_\psi, \varepsilon)z'_\psi+(\omega z'_\psi)'_\tau z'_\psi-\omega z''_{\psi^2}z'_\tau\}=\Phi_1(\tau, \psi, a, \varepsilon),\\ \frac{1}{D_0}\{F(\tau, z, \omega z'_\psi, \varepsilon)z'_a-(\omega z'_\psi)'_\tau z'_a+(\omega z'_\psi)'_a z'_\tau\}=\Phi_2(\tau, \psi, a, \varepsilon). \end{aligned}\right\} \tag{2.84}$$

The system of equations (2.79) may now be written in the form

$$\left.\begin{aligned} \frac{da}{dt}=\varepsilon\Phi_1(\tau, \psi, a, \varepsilon),\\ \frac{d\psi}{dt}=\omega(\tau, a)+\varepsilon\Phi_2(\tau, \psi, a, \varepsilon). \end{aligned}\right\} \tag{2.85}$$

The following expansions are valid, on account of (2.73) and the periodicity of the function (2.75) in ψ,

$$\Phi_j(\tau, \psi, a, \varepsilon)=\sum_{m=0}^{\infty}\varepsilon^m\Phi_j^{(m)}(\tau, \psi, a) \qquad (j=1, 2), \tag{2.86}$$

$$\Phi_j^{(m)}(\tau, \psi, a)=\sum_{n=-\infty}^{\infty}\Phi_{j, n}^{(m)}(\tau, a)e^{in\psi} \qquad (j=1, 2;\ m=0, 1, 2, \ldots), \tag{2.87}$$

where

$$\Phi_{j, n}^{(m)}(\tau, a)=\frac{1}{2\pi}\int_0^{2\pi}\Phi_j^{(m)}(\tau, \psi, a)e^{-in\psi}d\psi \tag{2.88}$$

$$(j=1, 2;\ m=0, 1, 2, \ldots;\ \infty<n<-\infty).$$

We turn now to a presentation of the method of constructing approximate solutions for the system of equations (2.85), based on a particular change of variables /70/ which leads, after discarding small terms of higher order, to differential equations that are simpler than (2.85). This method is by its contents equivalent to the method of averaging; this will become evident from the following calculations.

Let us first consider the following functions

$$\left.\begin{aligned} u_1(\tau, \psi, a)&=\sum_{n\neq 0}\frac{\Phi_{1, n}^{(0)}(\tau, a)}{in\omega(\tau, a)}e^{in\psi}+u_{10}(\tau, a),\\ v_1(\tau, \psi, a)&=\sum_{n\neq 0}\left\{\frac{\Phi_{2, n}^{(0)}(\tau, a)}{in\omega(\tau, a)}-\frac{\omega'_a\Phi_{1, n}^{(0)}(\tau, a)}{n^2\omega^2(\tau, a)}\right\}e^{in\psi}+v_{10}(\tau, a). \end{aligned}\right\} \tag{2.89}$$

The functions $u_{10}(\tau, a)$ and $v_{10}(\tau, a)$ are for the time being unknown and will be determined later from the form of the solution (2.75) and some additional conditions imposed on the quantity a.

It is easily verified that the functions (2.89) satisfy the relations

$$\left.\begin{aligned} &\omega(\tau, a)\frac{\partial u_1}{\partial\psi}=\Phi_1^{(0)}(\tau, \psi, a)-\Phi_{1,0}^{(0)}(\tau, a),\\ &\omega(\tau, a)\frac{\partial v_1}{\partial\psi}=\Phi_2^{(0)}(\tau, \psi, a)-\Phi_{2,0}^{(0)}(\tau, a)+\\ &\qquad+\omega_a'(\tau, a)[u_1(\tau, \psi, a)-u_{10}(\tau, a)].\end{aligned}\right\} \tag{2.90}$$

Let us now perform the following change of variables in the system (2.85)

$$\left.\begin{aligned} a&=a_1+\varepsilon u_1(\tau, \psi_1, a_1),\\ \psi&=\psi_1+\varepsilon v_1(\tau, \psi_1, a_1).\end{aligned}\right\} \tag{2.91}$$

Substituting (2.91) into equations (2.85), we obtain

$$\left.\begin{aligned} &\frac{da_1}{dt}\left(1+\varepsilon\frac{\partial u_1}{\partial a_1}\right)+\varepsilon\frac{d\psi_1}{dt}\frac{\partial u_1}{\partial\psi_1}=\varepsilon\Phi_1^{(0)}(\tau, \psi_1, a_1)+\varepsilon^2\Big[\Phi_{1\psi}^{(0)\prime}(\tau, \psi_1, a_1)v_1+\\ &\qquad+\Phi_{1a}^{(0)\prime}(\tau, \psi_1, a_1)u_1+\Phi_1^{(1)\prime}(\tau, \psi_1, a_1)-\frac{\partial u_1}{\partial\tau}\Big]+\varepsilon^3\ldots,\\ &\varepsilon\frac{da_1}{dt}\frac{\partial v_1}{\partial a_1}+\frac{d\psi_1}{dt}\left(1+\varepsilon\frac{\partial v_1}{\partial\psi_1}\right)=\omega(\tau, a_1)+\varepsilon[\omega_a'(\tau, a_1)u_1+\\ &\qquad+\Phi_2^{(0)}(\tau, \psi_1, a_1)]+\varepsilon^2\Big[\Phi_{2\psi}^{(0)\prime}(\tau, \psi_1, a_1)v_1+\Phi_{2a}^{(0)\prime}(\tau, \psi_1, a_1)u_1+\\ &\qquad+\Phi_2^{(1)}(\tau, \psi_1, a_1)+\frac{1}{2}\omega_{a^2}''(\tau, a_1)u_1^2-\frac{\partial v_1}{\partial\tau}\Big]+\varepsilon^3\ldots\end{aligned}\right\} \tag{2.92}$$

Solving this system for $\frac{da_1}{dt}$ and $\frac{d\psi_1}{dt}$, and taking into account identities (2.90), we find

$$\left.\begin{aligned} &\frac{da_1}{dt}=\varepsilon\Phi_{10}^{(0)}(\tau, a_1)+\varepsilon^2R_1(\tau, \psi_1, a_1, \varepsilon),\\ &\frac{d\psi_1}{dt}=\omega(\tau, a_1)+\varepsilon[\omega_a'(\tau, a_1)u_{10}(\tau, a_1)+\\ &\qquad+\Phi_{20}^{(0)}(\tau, a_1)]+\varepsilon^2R_2(\tau, \psi_1, a_1, \varepsilon),\end{aligned}\right\} \tag{2.93}$$

where

$$\left.\begin{aligned} &R_1(\tau, \psi_1, a_1, 0)=R_1^{(0)}(\tau, \psi_1, a_1)=\Phi_{1\psi}^{(0)\prime}(\tau, \psi_1, a_1)v_1+\\ &+\Phi_{1a}^{(0)\prime}(\tau, \psi_1, a_1)u_1+\Phi_1^{(1)}(\tau, \psi_1, a_1)-\frac{\partial u_1}{\partial\tau}-\Phi_1^{(0)}(\tau, \psi_1, a_1)\frac{\partial u_1}{\partial a_1}-\\ &-[\omega_a'(\tau, a_1)u_1+\Phi_2^{(0)}(\tau, \psi_1, a_1)]\frac{\partial u_1}{\partial\psi_1}+\omega(\tau, a_1)\frac{\partial u_1}{\partial\psi_1}\left(\frac{\partial u_1}{\partial a_1}+\frac{\partial v_1}{\partial\psi_1}\right),\\ &R_2(\tau, \psi_1, a_1, 0)=R_2^{(0)}(\tau, \psi_1, a_1)=\Phi_{2\psi}^{(0)\prime}(\tau, \psi_1, a_1)v_1+\\ &\quad+\Phi_{2a}^{(0)\prime}(\tau, \psi_1, a_1)u_1+\Phi_2^{(1)}(\tau, \psi_1, a_1)-\frac{\partial v_1}{\partial\tau}-\frac{1}{2}\omega_{a^2}''(\tau, a_1)u_1^2-\\ &\qquad-[\omega_a'(\tau, a_1)u_1+\Phi_2^{(0)}(\tau, \psi_1, a_1)]\frac{\partial v_1}{\partial\psi_1}-\\ &\qquad-\omega(\tau, a_1)\frac{\partial u_1}{\partial a_1}\left(\frac{\partial u_1}{\partial a_1}-\frac{\partial u_1}{\partial\psi_1}\right)-\Phi_1^{(0)}(\tau, \psi_1, a_1)\frac{\partial v_1}{\partial a_1}.\end{aligned}\right\} \tag{2.94}$$

Here the functions $R_1(\tau, \psi_1, a_1, \varepsilon)$ and $R_2(\tau, \psi_1, a_1, \varepsilon)$ possess the same properties as the functions (2.84).

Neglecting terms of the second order in equations (2.93), we obtain equations with right-hand sides not depending on ψ, and hence their integration is much simpler than the integration of the exact equations (2.93).

Continuing this process, we make another change of variables in equations (2.93), according to the formulas

$$\left.\begin{aligned} a_1&=a_2+\varepsilon^2u_2(\tau, \psi_2, a_2),\\ \psi_1&=\psi_2+\varepsilon^2v_2(\tau, \psi_2, a_2),\end{aligned}\right\} \tag{2.95}$$

in which

$$
\left.\begin{aligned}
u_2(\tau, \psi_2, a_2) &= \sum_{n\neq 0} \frac{R_{1,n}^{(0)}(\tau, a_2)}{in\omega(\tau, a_2)} e^{in\psi_2} + u_{20}(\tau, a_2),\\
v_2(\tau, \psi_2, a_2) &= \sum_{n\neq 0} \left\{\frac{R_{2,n}^{(0)}(\tau, a_2)}{in\omega(\tau, a_2)} - \frac{\omega_a'(\tau, a_2) R_{1,n}^{(0)}(\tau, a_2)}{n^2\omega^2(\tau, a_2)}\right\} e^{in\psi_2} + v_{20}(\tau, a_2),
\end{aligned}\right\} \qquad (2.96)
$$

where

$$
R_{j,n}^{(0)}(\tau, a_2) = \frac{1}{2\pi}\int_0^{2\pi} R_j(\tau, \psi_2, a_2, 0)\, e^{-in\psi_2}\, d\psi_2 \qquad (j = 1,\ 2), \qquad (2.97)
$$

and $u_{20}(\tau, a_2)$ and $v_{20}(\tau, a_2)$ are unknown functions which will be determined below.

Substituting (2.95) in equations (2.93), and taking into account the identities

$$
\left.\begin{aligned}
\omega(\tau, a_2)\frac{\partial u_2}{\partial\psi_2} &= R_1^{(0)}(\tau, \psi_2, a_2) - R_{10}^{(0)}(\tau, a_2),\\
\omega(\tau, a_2)\frac{\partial v_2}{\partial\psi_2} &= R_2^{(0)}(\tau, \psi_2, a_2) - R_{20}^{(0)}(\tau, a_2) + \\
&+ \omega_a'(\tau, a_2)\left[u_2(\tau, \psi_2, a_2) - u_{20}(\tau, a_2)\right],
\end{aligned}\right\} \qquad (2.98)
$$

we obtain the following system of equations for the variables a_2 and ψ_2:

$$
\left.\begin{aligned}
\frac{da_2}{dt} &= \varepsilon\Phi_{10}^{(0)}(\tau, a_2) + \varepsilon^2 R_{10}^{(0)}(\tau, a_2) + \varepsilon^3 S_1(\tau, \psi_2, a_2, \varepsilon),\\
\frac{d\psi_2}{dt} &= \omega(\tau, a_2) + \varepsilon\left[\omega_a'(\tau, a_2)\, u_{10}(\tau, a_2) + \Phi_{20}^{(0)}(\tau, a_2)\right] + \\
&+ \varepsilon^2\left[\omega_a'(\tau, a_2)\, u_{20}(\tau, a_2) + R_{20}^{(0)}(\tau, a_2)\right] + \varepsilon^3 S_2(\tau, \psi_2, a_2, \varepsilon),
\end{aligned}\right\} \qquad (2.99)
$$

where the functions $S_1(\tau, \psi_2, a_2, \varepsilon)$ and $S_2(\tau, \psi_2, a_2, \varepsilon)$ possess the same properties as the functions (2.84) or (2.94).

Continuing this method of successive transformations, ψ appears on the right-hand sides of the transformed system of equations only in terms proportional to the $(m+1)$-th power of ε (after m successive transformations of the type (2.91)).

We note that instead of the just considered successive change of variables according to formulas (2.91), (2.95), etc., which reduce the system (2.85) to the form (2.93) or (2.99), one could apply the u-method, developed in the first section of this chapter, and substitute directly in equation (2.72)

$$
x = z(\tau, \psi, a) + \varepsilon u_1(\tau, \psi, a) + \varepsilon^2 u_2(\tau, \psi, a) + \varepsilon^3 \ldots, \qquad (2.100)
$$

where a and ψ should be determined from the system of equations

$$
\left.\begin{aligned}
\frac{da}{dt} &= \varepsilon A_1(\tau, a) + \varepsilon^2 A_2(\tau, a) + \varepsilon^3 \ldots,\\
\frac{d\psi}{dt} &= \omega(\tau, a) + \varepsilon B_1(\tau, a) + \varepsilon^2 B_2(\tau, a) + \varepsilon^3 \ldots
\end{aligned}\right\} \qquad (2.101)
$$

Inserting (2.100) in equation (2.72), and equating coefficients of equal powers of ε, we obtain, as before, a series of equations for $u_1(\tau, \psi, a)$, $u_2(\tau, \psi, a)$, ... Imposing conditions of uniqueness (of the type of the conditions (2.8)) on the functions $u_1(\tau, \psi, a)$, $u_2(\tau, \psi, a)$, ..., we obtain relations which allow us to define $A_1(\tau, a)$, $A_2(\tau, a)$, ..., $B_1(\tau, a)$, $B_2(\tau, a)$, ... However, the u-method is not always effective for equations of the type (2.72) (especially in constructing the second and higher approximations) because the equations defining $u_1(\tau, \psi, a)$, $u_2(\tau, \psi, a)$, ... are in the case of constructing approximate solutions for equation (2.72), generally speaking, equations with periodic coefficients. However, the method based on successive substitutions of variables, in spite of being somewhat cumbersome, always leads to the desired results.

Before analyzing the equations obtained from the systems (2.93) and (2.99) after discarding terms of the second and third order respectively, we discuss the problem of determining the functions $u_{10}(\tau, a), v_{10}(\tau, a), u_{20}(\tau, a)$, $v_{20}(\tau, a)$ entering in formulas (2.89) and (2.96), which in view of their arbitrariness still do not define uniquely the functions $u_1(\tau, \psi, a)$, $v_1(\tau, \psi, a)$ and $u_2(\tau, \psi, a)$, $v_2(\tau, \psi, a)$. Certain additional conditions which allow the functions $u_{10}(\tau, a), v_{10}(\tau, a), u_{20}(\tau, a), v_{20}(\tau, a)$ to be defined are necessary for the uniqueness of the expressions (2.89) and (2.96). These additional conditions may be various, and their choice depends on the form of the solution (2.75) of the unperturbed equation and on what is meant by the quantity a.

Let us consider in detail an example of such additional conditions and a method of determining actual expressions for the functions $u_{10}(\tau, a)$, $v_{10}(\tau, a)$.

We assume that for the "unperturbed" equation (2.74), any value of the solution

$$z(t) = z[\tau, \omega(\tau, a)t + \varphi, a]$$

($\tau = \text{const}$), between a minimum value $z_{\min}$ and a maximum value $z_{\max}$, is attained twice during a half-period of vibrations — at the time t and at the time $T - t$, where $T = \frac{2\pi}{\omega(\tau, a)}$.

In that case we have the relation

$$z(t) = z(T - t),$$

expressing the symmetry of vibrations about the half-period /60b/.

The Fourier expansion of $z(t)$ would contain in this case only cosine harmonics and no sine harmonics.

If, in addition to that, the minimum and maximum deviations are equal by modulus, i. e.,

$$z_{\min} = -z_{\max},$$

then besides the symmetry about the half-period, the vibrations would also be symmetrical with respect to the quarter-period, and the Fourier expansion of $z(t)$ would contain only odd cosine harmonics.

Let us also assume that a is the full amplitude of the principal harmonic. Then the Fourier expansion of a solution of the "unperturbed" equation has the form

$$z(\tau, \psi, a) = a\cos\psi + \sum_{(n=\pm 3, \pm 5, \ldots)} z_n(\tau, a)e^{in\psi}, \tag{2.102}$$

where

$$z_n(\tau, a) = \frac{1}{2\pi}\int_0^{2\pi} z(\tau, \psi, a)e^{-in\psi}\,d\psi.$$

After these introductory remarks and assumptions about the form of solution of the "unperturbed" equation, we impose some additional conditions which lead to a unique choice of the functions $u_{10}(\tau, a)$, $v_{10}(\tau, a)$, $u_{20}(\tau, a)$, $v_{20}(\tau, a)$, etc.

We require that the new variable a_1, appearing after the transformation of equations (2.85) according to formulas (2.91), is also the full amplitude of the principal harmonic of vibrations up to and including terms of the first order. This condition allows us to determine uniquely the unknown functions $u_{10}(\tau, a)$, $v_{10}(\tau, a)$. Indeed, substituting the values of a and ψ from (2.91) into the expression (2.102) for $z(\tau, \psi, a)$, we obtain

$$z(\tau, \psi, a) = (a_1 + \varepsilon u_1)\cos(\psi_1 + \varepsilon v_1) + \sum_{(n=\pm 3, \pm 5, \dots)} z_n(\tau, a_1 + \varepsilon u_1)\, e^{in(\psi_1 + \varepsilon v_1)} =$$
$$= a_1 \cos\psi_1 + \sum_{(n=\pm 3, \pm 5, \dots)} z_n(\tau, a_1)\, e^{in\psi_1} + \varepsilon\, \{u_1 \cos\psi_1 - a_1 v_1 \sin\psi_1 +$$
$$+ \sum_{(n=\pm 3, \pm 5, \dots)} [z'_{na}(\tau, a_1)\, u_1 + z_n(\tau, a_1)\, inv_1]\, e^{in\psi_1}\} + \varepsilon^2 \dots \quad (2.103)$$

The second term on the right-hand side of (2.103) does not contain the first harmonic. Therefore, if a_1 is to represent the full amplitude of the first harmonic of the angle ψ_1 up to terms of the first order inclusive, it is sufficient to require that the first harmonic be absent in the expression

$$u_1 \cos\psi_1 - av_1 \sin\psi_1 + \sum_{(n=\pm 3, \pm 5, \dots)} [z'_{na}(\tau, a_1)\, u_1 + z_n(\tau, a_1)\, inv_1]\, e^{in\psi_1}.$$

This condition is equivalent to the validity of the equalities

$$\left.\begin{aligned}
&\int_0^{2\pi} \Big\{u_1 \cos\psi_1 - a_1 v_1 \sin\psi_1 + \\
&+ \sum_{(n=\pm 3, \pm 5, \dots)} [z'_{na}(\tau, a_1)\, u_1 + z_n(\tau, a_1)\, inv_1]\, e^{in\psi_1}\Big\} \cos\psi_1\, d\psi_1 = 0, \\
&\int_0^{2\pi} \Big\{u_1 \cos\psi_1 - a_1 v_1 \sin\psi_1 + \\
&+ \sum_{(n=\pm 3, \pm 5, \dots)} [z'_{na}(\tau, a_1)\, u_1 + z_n(\tau, a_1)\, inv_1]\, e^{in\psi_1}\Big\} \sin\psi_1\, d\psi_1 = 0.
\end{aligned}\right\} \quad (2.104)$$

Substituting the values of $u_1(\tau, \psi_1, a_1)$ and $v_1(\tau, \psi_1, a_1)$ given by (2.89) into (2.104) and performing elementary manipulations, we find

$$\left.\begin{aligned}
u_{10}(\tau, a_1) = &-\frac{1}{\omega(\tau, a_1)} \Big\{\frac{1}{4i} [\Phi^{(0)}_{1, 2}(\tau, a_1) - \Phi^{(0)}_{1, -2}(\tau, a_1)] - \\
&- \frac{a_1}{4} [\Phi^{(0)}_{2, 2}(\tau, a_1) - \Phi^{(0)}_{2, -2}(\tau, a_1)] + \\
&+ \frac{a_1 \omega'_a(\tau, a_1)}{8i\omega(\tau, a_1)} [\Phi^{(0)}_{1, -2}(\tau, a_1) - \Phi^{(0)}_{1, 2}(\tau, a_1)] + \\
&+ 2 \sum_{\substack{n=-\infty \\ (n \neq \pm 1)}}^{\infty} \Big[\frac{1}{in} \Phi^{(0)}_{1, n}(\tau, a_1)(z'_{1-n, a}(\tau, a_1) + z'_{-1-n, a}(\tau, a_1)) + \\
&+ \frac{1}{n} \Phi^{(0)}_{2, n}(\tau, a_1)((1-n)\, z_{1-n}(\tau, a_1) - (1+n)\, z_{-1-n}(\tau, a_1)) - \\
&- \frac{i\omega'_a(\tau, a_1)\, \Phi^{(0)}_{1, n}(\tau, a_1)}{n^2 \omega(\tau, a_1)} ((1-n)\, z_{1-n}(\tau, a_1) - (1+n)\, z_{-1-n}(\tau, a_1))\Big]\Big\}, \\
v_{10}(\tau, a_1) = &\frac{1}{a_1 \omega(\tau, a_1)} \Big\{\frac{1}{4} [\Phi^{(0)}_{1, 2}(\tau, a_1) + \Phi^{(0)}_{1, -2}(\tau, a_1)] - \\
&- \frac{a_1 i}{4} [\Phi^{(0)}_{2, 2}(\tau, a_1) - \Phi^{(0)}_{2, -2}(\tau, a_1)] - \\
&- \frac{a_1 \omega'_a(\tau, a_1)}{8\omega(\tau, a_1)} [\Phi^{(0)}_{1, 2}(\tau, a_1) + \Phi^{(0)}_{1, -2}(\tau, a_1)] + \\
&+ 2 \sum_{\substack{n=-\infty \\ (n \neq \pm 1)}}^{\infty} \frac{1}{n} \Big[\Phi^{(0)}_{1, n}(\tau, a_1)(z'_{1-n, a}(\tau, a_1) - z'_{-1-n, a}(\tau, a_1)) + \\
&+ i\Phi^{(0)}_{2, n}(\tau, a_1)((1-n)\, z_{1-n}(\tau, a_1) + (1+n) z_{-1-n}(\tau, a_1)) - \\
&- \frac{\omega'_a(\tau, a_1)}{\omega(\tau, a_1)} \Phi^{(0)}_{1, n}(\tau, a_1)((1+n)\, z_{-1-n}(\tau, a_1) + (1-n)\, z_{1-n}(\tau, a_1))\Big]\Big\}.
\end{aligned}\right\} \quad (2.105)$$

The uniqueness of the transformation formulas in the second approximation (2.95) may be achieved by a similar additional condition requiring that a_2 also represent the full amplitude of the first harmonic of the angle ψ_2, but up to and including terms of the second order.

We note that $z(\tau, \psi, a)$ could be assumed to be expandable in a Fourier series containing all harmonics, or only odd sine harmonics, as will be shown by an actual example. No additional difficulties arise in all these cases, only the expressions of $u_{i0}(\tau, a)$, $v_{i0}(\tau, a)$ $(i=1, 2, \ldots)$ are changed.

Let us now consider the integration of the systems of equations (2.93) and (2.99), and the error introduced by neglecting part of the terms on the right-hand sides of these systems of equations.

Leaving out terms of second order on the right-hand side of equations (2.93), we obtain the system of equations

$$\left.\begin{aligned} \frac{da_1}{dt} &= \varepsilon\Phi_{10}^{(0)}(\tau, a_1), \\ \frac{d\psi_1}{dt} &= \omega(\tau, a_1) + \varepsilon[\omega_a'(\tau, a_1)u_{10}(\tau, a_1) + \Phi_{2}^{(0)}(\tau, a_1)]. \end{aligned}\right\} \qquad (2.106)$$

This system, like the previously deduced equations of the first (or second) approximation, are generally not integrable in a closed form and should therefore be integrated numerically. Integrating the first of equations (2.106) over the interval $0 \leqslant t \leqslant \frac{L}{\varepsilon}$ (our system may be considered only in a finite interval since $\tau \in [0, L]$), we obtain the values of a_1 with an error of the order of ε. Substituting these values of a_1 into the second of equations (2.106) and integrating it over the interval $0 \leqslant t \leqslant \frac{L}{\varepsilon}$, we obtain a finite error for the phase of vibration ψ, since the right-hand side of this equation contains the term $\omega(\tau, a_1)$ which is not a small quantity. Substituting the values of $a = a_1(t)$ and $\psi = \psi_1(t)$, obtained in this way, into (2.75), we obtain the expression

$$x_1 = z(\varepsilon t, \psi_1(t), a_1(t)),$$

which, generally speaking, deviates from the true value of x by a quantity of finite order.

It is however sufficient in most important practical cases to derive the quantity a_1 and the frequency of vibrations $\omega(\tau, a_1)$, the phase of vibration being of no interest here (the phase plays an important role in the presence of external periodic forces in the case of resonance, which will be discussed in detail in the following chapter). Numerical integration of the system (2.106) is therefore quite sufficient in solving many practical problems.

If it is desired to calculate in addition to the quantity a_1 and the frequency of vibrations $\omega(\tau, a_1)$ also the values of x up to terms of the order of ε inclusive, then it is necessary to consider in the given case the system of equations

$$\left.\begin{aligned} \frac{da_1}{dt} &= \varepsilon\Phi_{10}^{(0)}(\tau, a_1) + \varepsilon^2 R_{10}^{(0)}(\tau, a_1), \\ \frac{d\psi_1}{dt} &= \omega(\tau, a_1) + \varepsilon[\omega_a'(\tau, a_1)u_{10}(\tau, a_1) + \Phi_{20}^{(0)}(\tau, a_1)], \end{aligned}\right\} \qquad (2.107)$$

obtained from the system (2.99) by neglecting in the latter terms of third order in the first equation and of second order in the second.

The same arguments apply when it is desired to calculate x up to terms of the order of ε^2 inclusive.

§ 6. Example of an equation close to an exactly integrable one

Let us first consider the application of the previously derived formulas to the solution in first approximation of the extensively studied equation (2.1), which we now write in the form

$$\frac{d^2x}{dt^2}+\omega^2(\tau)x=\varepsilon\left\{\frac{1}{m(\tau)}F\left(\tau,\ x,\ \frac{dx}{dt}\right)-\frac{1}{m(\tau)}\frac{dm(\tau)}{d\tau}\frac{dx}{dt}\right\}. \tag{2.108}$$

The unperturbed equation

$$\frac{d^2x}{dt^2}+\omega^2(\tau)x=0,\qquad \tau=\text{const} \tag{2.109}$$

has in our case the pericdic solution

$$x=a\cos\psi\qquad (\psi=\omega(\tau)t+\varphi), \tag{2.110}$$

depending on the two arbitrary constants a and φ, and the parameter τ, while the frequency $\omega(\tau)$ depends only on τ and not on the amplitude a.

The change of variables (2.76) is in the given case

$$\left.\begin{aligned} x&=a\cos\psi,\\ \frac{dx}{dt}&=-a\omega(\tau)\sin\psi, \end{aligned}\right\} \tag{2.111}$$

and after transforming equation (2.110) to the new variables a and ψ, we obtain instead of the system (2.79) the following system

$$\left.\begin{aligned} \frac{da}{dt}&=-\frac{\varepsilon}{m(\tau)\omega(\tau)}\left\{\frac{d[m(\tau)\omega(\tau)]}{d\tau}a\sin^2\psi+\right.\\ &\qquad\left.+F(\tau,\ a\cos\psi,\ -a\omega\sin\psi)\sin\psi\right\},\\ \frac{d\psi}{dt}&=\omega(\tau)-\frac{\varepsilon}{m(\tau)\omega(\tau)}\left\{\frac{d[m(\tau)\omega(\tau)]}{d\tau}\sin\psi\cos\psi+\right.\\ &\qquad\left.+\frac{1}{a}F(\tau,\ a\cos\psi,\ -a\omega\sin\psi)\cos\psi\right\}, \end{aligned}\right\} \tag{2.112}$$

This system coincides, after introducing the new variable $\theta=\psi-\omega(\tau)t$, with the previously discussed equations (2.30) in standard form, which were solved approximately by means of the method of averaging.

The functions (2.89) assume for the equation considered the form

$$\left.\begin{aligned} u_1(\tau,\ a,\ \psi)&=-\frac{\varepsilon}{2m(\tau)\omega(\tau)}\left\{-\frac{1}{2}\frac{d[m(\tau)\omega(\tau)]}{d\tau}a\sin 2\psi+\right.\\ &\quad\left.+\frac{1}{\pi}\sum_{n\neq 0}\frac{e^{in\psi}}{in}\int_0^{2\pi}F(\tau,a\cos\psi,-a\omega\sin\psi)\sin\psi\, e^{-in\psi}d\psi\right\},\\ v_1(\tau,\ a,\ \psi)&=-\frac{\varepsilon}{2m(\tau)\omega(\tau)}\left\{\frac{1}{2}\frac{d[m(\tau)\omega(\tau)]}{d\tau}\cos 2\psi+\right.\\ &\quad\left.+\frac{1}{a\pi}\sum_{n\neq 0}\frac{e^{in\psi}}{in}\int_0^{2\pi}F(\tau,\ a\cos\psi,\ -a\omega\sin\psi)\cos\psi\, e^{-in\psi}d\psi\right\} \end{aligned}\right\} \tag{2.113}$$

$(u_{10}(\tau,\ a)=v_{10}(\tau,\ a)=0)$. Transforming equations (2.112) to the new variables a_1 and ψ_1, defined by

$$\left.\begin{aligned} a&=a_1+\varepsilon u_1(\tau,\ a_1,\ \psi_1),\\ \psi&=\psi_1+\varepsilon v_1(\tau,\ a_1,\ \psi_1), \end{aligned}\right\} \tag{2.114}$$

(where $u_1(\tau,\ a_1,\ \psi_1)$ and $v_1(\tau,\ a_1,\ \psi_1)$ are determined by the expressions (2.113)), and neglecting terms of second and higher order in ε, we obtain equations of the first approximation

$$\left.\begin{aligned}\frac{da_1}{dt} &= -\frac{\varepsilon a_1}{2m(\tau)\,\omega(\tau)}\frac{d[m(\tau)\,\omega(\tau)]}{d\tau} - \\ &\quad -\frac{\varepsilon}{2\pi m(\tau)\,\omega(\tau)}\int_0^{2\pi} F_0(\tau, a_1, \psi)\sin\psi\, d\psi, \\ \frac{d\psi_1}{dt} &= \omega(\tau) - \frac{\varepsilon}{2\pi m(\tau)\,\omega(\tau)\,a_1}\int_0^{2\pi} F_0(\tau, a_1, \psi)\cos\psi\, d\psi\end{aligned}\right\} \qquad (2.115)$$

$(F_0(\tau, a_1, \psi) = F(\tau, a, \cos\psi, -a, \omega\sin\psi))$, which coincide with the previously derived scheme (2.30).

As a second example we consider torsional vibrations of a shaft bearing at its ends time variable masses (Figure 2). The moment of inertia of the shaft is assumed to be small compared with the moments of inertia of the rotating masses. We may then neglect in this case the mass of the shaft in the equations of motion.

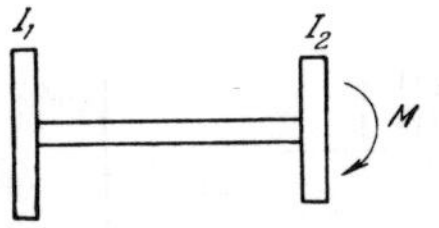

FIGURE 2

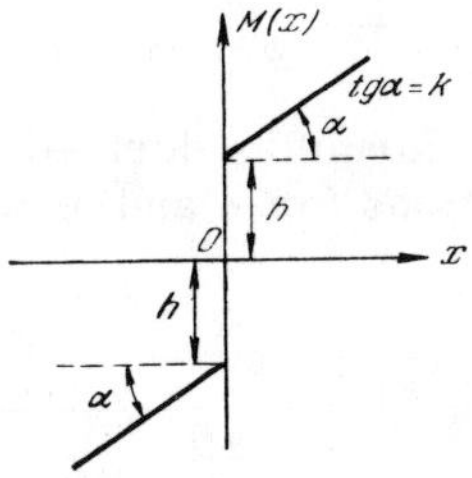

FIGURE 3

We denote by $I_1(\tau)$ and $I_2(\tau)$ the moments of inertia of the rotating masses, and by φ_1 and φ_2, the angles of rotation of these masses. Let $M(x)$, where $x = \varphi_1 - \varphi_2$, be the torsional moment of the elastic coupling (e.g., an elastic muff) depending on the torsional angle. We assume for definiteness that the function $M(x)$ has the form of the broken line shown in Figure 3, i.e., that

$$\left.\begin{aligned}M(x) &= h + kx, \quad \text{if} \quad x > 0, \\ M(x) &= -h + kx, \quad \text{if} \quad x < 0,\end{aligned}\right\} \qquad (2.116)$$

where h and k are constants.

We can now write the equations

$$\left.\begin{aligned}I_1(\tau)\frac{d^2\varphi_1}{dt^2} + M(\varphi_1 - \varphi_2) &= 0, \\ I_2(\tau)\frac{d^2\varphi_2}{dt^2} - M(\varphi_1 - \varphi_2) &= 0,\end{aligned}\right\} \qquad (2.117)$$

which lead immediately to the equation of torsional vibrations

$$I_1(\tau)\,I_2(\tau)\frac{d^2x}{dt^2} + [I_1(\tau) + I_2(\tau)]\,M(x) = 0. \qquad (2.118)$$

Let us assume that the considered vibrating system is also under the action of some frictional force proportional to the torsional velocity with a proportionality coefficient $2\lambda(\tau)$ slowly varying with time. Then, with the notation

$$\nu(\tau) = \frac{I_1(\tau) + I_2(\tau)}{I_1(\tau)\,I_2(\tau)}, \quad \nu(\tau)\,M(x) = f(\tau, x),$$

we obtain a differential equation of the type (2.72)

$$\frac{d^2x}{dt^2}+2\lambda(\tau)\frac{dx}{dt}+f(\tau, x)=0. \tag{2.119}$$

Here we have

$$\varepsilon F\left(\tau, x, \frac{dx}{dt}\right)=-2\lambda(\tau)\frac{dx}{dt}.$$

In our case the equation of "unperturbed" motion

$$\left.\begin{aligned}&\frac{d^2x}{dt^2}+\nu(\tau)(h+kx)=0, \quad \text{if } x>0,\\ &\frac{d^2x}{dt^2}+\nu(\tau)(-h+kx)=0, \quad \text{if } x<0\ (\tau=\text{const}),\end{aligned}\right\} \tag{2.120}$$

is easily seen to be satisfied by the periodic solution

$$x=z(\tau, \psi, a)=a\sin\psi+\frac{4h}{\pi}\sum_{n=3,\,5,\,7,\dots}\frac{\sin n\psi}{n\left[\frac{\omega^2(\tau, a)}{\nu(\tau)}n^2-k\right]}, \tag{2.121}$$

where

$$\psi=\omega(\tau, a)t+\varphi, \quad \omega^2(\tau, a)=\nu(\tau)k\left[1+\frac{4h}{\pi ka}\right]. \tag{2.122}$$

Using the formulas derived in the preceding section, we obtain the following equations for a and ψ with an accuracy up to and including the first order:

$$\left.\begin{aligned}&\frac{da_1}{dt}=-\frac{\varepsilon a_1\left[1+\frac{4h}{\pi ka_1}\right]\left[\nu(\tau)-\frac{\nu'(\tau)}{2\nu(\tau)}\right]}{2\left\{1+\frac{4h}{\pi ka_1}\sum\limits_{(n=3,\,5,\,7,\dots)}\frac{1}{\left[n^2\left(1+\frac{4h}{\pi ka_1}\right)-1\right]}\right\}},\\ &\frac{d\psi_1}{dt}=\nu^{\frac{1}{2}}(\tau)k^{\frac{1}{2}}\left(1+\frac{4h}{\pi ka_1}\right)^{\frac{1}{2}}-\frac{2\varepsilon hu_{10}(\tau, a_1)\left[\nu(\tau)+\frac{\nu'(\tau)}{2\nu(\tau)}\right]}{\pi k^{\frac{3}{2}}\nu^{\frac{1}{2}}(\tau)\left[1+\frac{4h}{\pi ka_1}\right]^{\frac{1}{2}}},\end{aligned}\right\} \tag{2.123}$$

where $u_{10}(\tau, a_1)$ should be determined from the condition requiring that a_1 is the full amplitude of the first harmonic of the angle ψ_1*. This necessarily demands that the expression

$$u_1\sin\psi_1+a_1v_1\cos\psi_1+\sum_{(n=\pm 3,\,\pm 5,\dots)}[\theta_n(\tau, a_1)inv_1+\theta_n'(\tau, a_1)u_1]e^{in\psi_1} \tag{2.124}$$

does not contain the first harmonic of the angle ψ_1. We have used here the following definitions

$$\theta_n(\tau, a_1)=\frac{2h}{\pi kin\left[n^2\left(1+\frac{4h}{\pi ka_1}\right)-1\right]}, \tag{2.125}$$

$$\left.\begin{aligned}&u_1(\tau, a_1, \psi_1)=\sum_{n\neq 0}\frac{\Phi_{1,n}^{(0)}(\tau, a_1)e^{in\psi_1}}{in}+u_{10}(\tau, a_1),\\ &v_1(\tau, a_1, \psi_1)=\sum_{n\neq 0}\left\{\frac{\Phi_{2,n}^{(0)}(\tau, a_1)}{in}-\frac{\omega_a'(\tau, a_1)\Phi_{1,n}^{(0)}(\tau, a_1)}{n^2}\right\}e^{in\psi_1}+\\ &\qquad\qquad+v_{10}(\tau, a_1),\end{aligned}\right\} \tag{2.126}$$

* The second of equations (2.123), as shown before, does not define any approximation. It is written here merely for the discussion of one of the methods of determining $u_{10}(\tau, a_1)$; we need this for the following approximations.

$$\left.\begin{aligned}\Phi_{1,n}^{(0)}(\tau, a_1) &= \frac{\lambda(\tau)}{\pi D_0(\tau, a_1)} \int_0^{2\pi} [z'_\psi(\tau, \psi_1, a_1)]^2 e^{-in\psi_1} d\psi_1, \\ \Phi_{2,n}^{(0)}(\tau, a_1) &= \frac{-\lambda(\tau)\,\omega(\tau, a_1)}{\pi D_0(\tau, a_1)} \int_0^{2\pi} z'_\psi(\tau, \psi_1, a_1) z'_a(\tau, \psi_1, a_1) e^{-in\psi_1} d\psi_1.\end{aligned}\right\} \quad (2.127)$$

After a series of calculations we find the following expression for $u_{10}(\tau, a_1)$:

$$\begin{aligned} u_{10}(\tau, a_1) = \frac{1}{4}\Big[\frac{1}{i}\Big(1 + \frac{a_1\omega'_a(\tau, a_1)}{2}\Big)(\Phi_{1,2}^{(0)}(\tau, a_1) - \Phi_{1,-2}^{(0)}(\tau, a_1)) - \\ - a_1(\Phi_{2,2}^{(0)}(\tau, a_1) - \Phi_{2,-2}^{(0)}(\tau, a_1))\Big] - \\ - \sum_{\binom{n+m\pm 1=0}{n\neq\pm 1}} \Big[\theta_n(\tau, a_1)\Big(\frac{\Phi_{2,m}^{(0)}(\tau, a_1)}{im} n - \frac{\Phi_{1,m}^{(0)}(\tau, a_1)}{m^2} n\omega_a(\tau, a_1)\Big) - \\ \frac{\theta'_{na}(\tau, a_1)\,\Phi_{1,m}^{(0)}(\tau, a_1)}{m}\Big]. \end{aligned} \quad (2.128)$$

An expression for $v_{10}(\tau, a_1)$ may be derived analogously.

The equations derived for a_1 and ψ_1 (2.123) may be integrated relatively easily.

For example, assuming for simplicity that the moments of inertia I_1 and I_2 do not depend on τ and, consequently, $\nu = \text{const}$, $\lambda = \text{const}$, we obtain the following relation between a_1 and t:

$$\frac{a_1\left(1 + \frac{4h}{\pi k a_1}\right)^2}{a_0\left(1 + \frac{4h}{\pi k a_0}\right)^2} \prod_{(n=3,5,7,\ldots)} \frac{\left(1 + \frac{4h}{\pi k a_0}\right) - \frac{1}{n^2}}{\left(1 + \frac{4h}{\pi k a_1}\right) - \frac{1}{n^2}} = e^{-\lambda t}, \quad (2.129)$$

where the initial value at $t = 0$ is taken as $a_1 = a_0$.

When $h = 0$, i. e., when equation (2.119) degenerates to a linear one, formula (2.129) gives, as expected, the known value of the amplitude

$$a_1 = a_0 e^{-\lambda t}. \quad (2.130)$$

If the moments of inertia, varying with time, satisfy the condition

$$I_1(\tau) = C I_2(\tau),$$

then the first equation of the system (2.123) implies

$$A(a_1) = \left[\frac{I_1(\tau)}{I_1(0)}\right]^{\frac{\varepsilon}{4}} \cdot e^{-\frac{\varepsilon}{2}\int_0^t \frac{1+C}{I_1(\tau)} dt}, \quad (2.131)$$

where $A(a_1)$ represents the left-hand side of (2.129)

We note that formula (2.129) is not applicable in practical calculations as it contains an infinite multiplication. We shall show in § 8 of the present chapter how one obtains a formula for the amplitude of the first harmonic $a_1(\tau)$, convenient for actual calculations.

§ 7. Investigation of the equation of first approximation for the amplitude, and methods for its construction

It was assumed in § 5 that the equation of "unperturbed" motion (2.74) admits the periodic solution (2.75) containing two arbitrary constants a and ψ, where the parameter a determines the form and amplitude of vibration, and the parameter ψ, its phase.

Knowing this periodic solution, we transformed equation (2.72) to the new variables a and ψ. To obtain simplified systems ((2.106) or (2.107)) we performed on the system of equations (2.85) (obtained for the new variables a and ψ) a substitution of variables, making the essential assumption that a represents the full amplitude of the first harmonic of the angle ψ in the expansion of the solution (2.75) in the series (2.102). We supposed in addition to this that the vibrations were symmetrical.

It will however be shown below, by means of results due to Volosov /27/, that many of the above restrictions may be abandoned, and equations of the type (2.106) and (2.107) be constructed not for the amplitude of the principal harmonic, but for the maximum and minimum deviation of the oscillating quantity. We do this by proceeding directly from the functions $f(\tau, x)$ and $F\left(\tau, x, \frac{dx}{dt}\right)$, characterizing equation (2.72).

Let us first examine, as in /114/, the particular case when the quantity a represents the maximum deviation $a = x_{\max}$, and the vibrations defined by the "unperturbed" equation are symmetrical, i. e., $x_{\max} = -x_{\min}$. It is easily shown that the equations of the first approximation (2.106) in this case be brought to a form in which the right-hand side is completely independent of the solution of the "unperturbed" equation (2.74) and is directly expressed by the functions $f(\tau, x)$, $F\left(\tau, x, \frac{dx}{dt}\right)$, characterizing equation (2.72).

Actually, the first equation of the system (2.106) may, with the help of (2.84), be written in the form

$$\frac{da}{dt} = \frac{1}{2\pi D_0(\tau, a)} \int_0^{2\pi} \{-F(\tau, z, \omega z'_\psi, 0) z'_\psi + (\omega z'_\psi)'_\tau z'_\psi - \omega z''_{\psi^2} z'_\tau\} \, d\psi, \tag{2.132}$$

where

$$D_0(\tau, a) = [\omega^2 z'_a z''_{\psi^2} - z'_\psi (\omega^2 z'_\psi)'_a]_{\psi=0}.$$

Consider now the equation of "unperturbed" motion (2.74)

$$\frac{d^2x}{dt^2} + f(\tau, x) = 0,$$

where τ is a fixed parameter.

We denote by $V(\tau, x)$ the potential energy of the "unperturbed" system

$$V(\tau, x) = \int^x f(\tau, x) \, dx. \tag{2.133}$$

The energy equation may now be written in the form

$$\frac{1}{2}\left(\frac{dx}{dt}\right)^2 + V(\tau, x) = V(\tau, x_{\max}) = E, \tag{2.134}$$

where E is the complete energy, being an arbitrary integration constant. We have from (2.134)

$$\left(\frac{dx}{dt}\right)^2 = 2[V(\tau, x_{\max}) - V(\tau, x)], \tag{2.135}$$

from which we find

$$\frac{dx}{dt} \equiv \omega z'_\psi = \sqrt{2[V(\tau, x_{\max}) - V(\tau, z)]}. \tag{2.136}$$

According to (2.129) we may also write

$$\omega^2 \frac{1}{2}\left(\frac{dz}{d\psi}\right)^2 + V(\tau, z) = V(\tau, x_{max}), \tag{2.137}$$

from which we obtain after differentiating with respect to τ

$$\omega'_\tau \omega (z'_\psi)^2 + \omega^2 z''_{\psi\tau} z'_\psi + V'_z(\tau, z) z'_\tau + V'_\tau(\tau, z) = V'_\tau(\tau, x_{max}).$$

Taking into account that

$$V'_z(\tau, z) = f(\tau, z) = -\omega^2 z''_{\psi^2},$$

we finally obtain

$$(\omega z'_\psi)'_\tau z'_\psi - \omega z''_{\psi^2} z'_\tau = \frac{V'_\tau(\tau, x_{max}) - V'_\tau(\tau, z)}{\omega}. \tag{2.138}$$

The integral on the right-hand side of equation (2.132) may, by virtue of (2.136) and (2.138), be written in the form

$$\int_0^{2\pi} \{-F(\tau, z, \omega z'_\psi, 0) z'_\psi + (\omega z'_\psi)'_\tau z'_\psi - \omega z''_{\psi^2} z'_\tau\} d\psi =$$

$$= \int_{-x_{max}}^{x_{max}} \Big\{F\left(\tau, z, \sqrt{2[V(\tau, x_{max}) - V(\tau, z)]}, 0\right) -$$

$$- F\left(\tau, z, -\sqrt{2[V(\tau, x_{max}) - V(\tau, z)]}, 0\right) + 2\frac{V'_\tau(\tau, x_{max}) - V'_\tau(\tau, z)}{\sqrt{2[V(\tau, x_{max}) - V(\tau, z)]}}\Big\} dz. \tag{2.139}$$

We shall now modify the expression of $D_0(\tau, a)$. We assume that x attains its maximum value when $\psi = 0$. This assumption does not affect generality, since if x would reach its maximum value at a certain $\psi = \bar{\psi}$, we would then calculate $D_0(\tau, a)$ for the value $\psi = \bar{\psi}$.

Thus, let

$$z(\tau, 0, a) = x_{max} = a. \tag{2.140}$$

Then, bearing in mind (2.74), (2.139), and (2.136), we may write

$$z''_{\psi^2}|_{\psi=0} = -\frac{1}{\omega^2} V'_z(\tau, x_{max}), \quad z'_\psi|_{\psi=0} = 0, \quad z'_a|_{\psi=0} = 1$$

and consequently,

$$D_0(\tau, x_{max}) = -V'_z(\tau, x_{max}). \tag{2.141}$$

Substituting (2.141) and (2.139) in the right-hand side of equation (2.132), we obtain the transformed equation in the form

$$\frac{da}{dt} = \frac{1}{2\pi V'_z(\tau, a)} \int_{-a}^{a} \Big\{F\left(\tau, z, -\sqrt{2[V(\tau, a) - V(\tau, z)]}, 0\right) -$$

$$- F\left(\tau, z, \sqrt{2[V(\tau, a) - V(\tau, z)]}, 0\right) - 2\frac{V'_\tau(\tau, a) - V'_\tau(\tau, z)}{\sqrt{2[V(\tau, a) - V(\tau, z)]}}\Big\} dz, \tag{2.142}$$

where a is the maximum value of x.

The right-hand side of equation (2.142) evidently contains only the known functions $f(\tau, x)$ and $F\left(\tau, x, \frac{dx}{dt}\right)$, and its construction does not require, as in the previous case, the knowledge of the periodic solution of the "unperturbed" equation.

Performing analogous, but slightly more complicated calculations, we may transform the second of equations (2.106), defining the phase of vibration, and also the equations of higher approximations in the same way. We do not, however, consider it here and rather pass on to the more general case and examine it by means of a method suggested by Volosov /26-28/.

§ 8. Construction of envelopes for the amplitudes of vibrations described by equations close to linear ones

We present in this section, without detailed calculations, which may be found in /26, 27/, some results of Volosov's method which allows one to construct the envelopes for amplitudes of vibrations described by a differential equation of the type

$$\frac{d}{dt}\left[m(\tau)\frac{dx}{dt}\right]+Q(\tau, x)=\varepsilon f\left(\tau, x, \frac{dx}{dt}\right). \tag{2.143}$$

We proceed directly from the expressions of the functions $m(\tau)$, $f\left(\tau, x, \frac{dx}{dt}\right)$, $Q(\tau, x)$, entering in this equation, without requiring, as in § 5 of the present chapter, an explicit expression of the periodic solution of the unperturbed equation

$$m(\tau)\frac{d^2x}{dt^2}+Q(\tau, x)=0. \tag{2.144}$$

The essence of Volosov's method consists in finding the so-called amplitude curves, $F_1(\tau)>0$ and $F_2(\tau)<0$, which pass through the maxima and minima of the oscillating solution.

In equation (2.143), as usually, ε is a small positive parameter; $\tau=\varepsilon t$, the "slowing" time; the functions $m(\tau)$, $Q(\tau, x)$, and $f\left(\tau, x, \frac{dx}{dt}\right)$ possess several properties (cf., for example, /28/), ensuring that equation (2.144) admits for $\tau=\text{const}$ a purely periodic solution on the interval $0\leqslant\tau\leqslant L$. Equation (2.143) has under these conditions ($\operatorname{sign} Q=\operatorname{sign} x$, $m(\tau)>0$, etc.) an oscillating solution with slowly varying amplitude and period.

The equations (2.143) and (2.144) coincide with the equations (2.72) and (2.74) in § 5 of the present chapter, if

$$\left.\begin{aligned} f(\tau, x)&=\frac{1}{m(\tau)}Q(\tau, x),\\ F\left(\tau, x, \frac{dx}{dt}, \varepsilon\right)&=f\left(\tau, x, \frac{dx}{dt}\right)-\frac{dm(\tau)}{d\tau}\frac{dx}{dt}. \end{aligned}\right\} \tag{2.145}$$

It is shown by means of a series of rigorous arguments, given in /26/ and /27/, that the amplitude curves $F_1(\tau)$ and $F_2(\tau)$ are connected by the relation

$$\int\limits_{F_1(\tau)}^{F_2(\tau)} Q(\tau, x)\,dx=0 \tag{2.146}$$

and are determined up to terms of the order of ε, by the differential equation

$$\frac{d}{dt}I[\tau, F_1(\tau), F_2(\tau)]=\varepsilon A[\tau, F_1(\tau), F_2(\tau)], \tag{2.147}$$

where

$$I(\tau, F_1, F_2) = 2\sqrt{m(\tau)}\left\{\int_0^{F_1}\left[2\int_x^{F_1} Q(\tau, z)\,dz\right]^{\frac{1}{2}} dx - \int_0^{F_2}\left[2\int_x^{F_2} Q(\tau, z)\,dz\right]^{\frac{1}{2}} dx\right\}, \qquad (2.148)$$

$$A(\tau, F_1, F_2) = -\sum_{i,\,k=1}^{2} (-1)^{k+i}\int_0^{F_k} f\left[\tau, x, (-1)^i m(\tau)^{-\frac{1}{2}}\left(2\int_x^{F_k} Q(\tau, z)\,dz\right)^{\frac{1}{2}}\right] dx. \qquad (2.149)$$

If the initial values of the solution of equation (2.143) are given as $x(0) = x_0$, $\dot{x}(0) = \dot{x}_0$, then the initial values $F_1(0)$ and $F_2(0)$ are determined by the formulas

$$\left.\begin{aligned} &2\int_{x_0}^{F_k} Q(0, x)\,dx = m(0)\,\dot{x}_0^2,\\ &\operatorname{sign} F_k(0) = \operatorname{sign}(-1)^{k+1} \qquad (k = 1, 2). \end{aligned}\right\} \qquad (2.150)$$

The period of vibrations T (the time between two consecutive maxima or minima) is determined with an accuracy of the order of ε by the formula

$$T_{\mathrm{I}}(\tau) = 2\sqrt{m(\tau)}\sum_{k=1}^{2}(-1)^{k+1}\int_0^{F_k(\tau)}\left[2\int_x^{F_k(\tau)} Q(\tau, z)\,dz\right]^{-\frac{1}{2}} dx, \qquad (2.151)$$

which is valid in the interval $0 \leqslant \tau \leqslant L$.

Equation (2.147), describing the amplitude curves, and also formula (2.151) may be expressed by the known functions $m(\tau)$, $Q(\tau, x)$ and $f\left(\tau, x, \frac{dx}{dt}\right)$, and may be constructed for a more complicated vibrating system than that considered in § 7 of the present chapter.

The method developed in the above-quoted articles allows one to construct equations for the amplitude curves describing the extrema of the solution of equation (2.143) up to and including terms of the order of ε, but the formulas obtained in this case are quite cumbersome. We write only the improved formula for the period of vibrations which involves an error of the order of ε^2 /28/

$$\begin{aligned} T_{\mathrm{II}}(\tau) = {} & 2\sqrt{m(\tau)}\sum_{k=1}^{2}(-1)^{k+1}\int_0^{G_k(\varepsilon\theta,\,\varepsilon)}\left[2\int_x^{G_k(\varepsilon\theta,\,\varepsilon)} Q(\varepsilon\theta, z)\,dz\right]^{\frac{1}{2}} - \\ & - \varepsilon\sqrt{m(\tau)}\sum_{i,\,k=1}^{2}(-1)^{k+i}\int_0^{F_k(\tau)}\left[2\int_x^{F_k(\tau)} Q(\tau, z)\,dz\right]^{-\frac{3}{2}} dx\, \times \\ & \times\left\{\int_x^{F_k(\tau)} f\left[\tau, z, (-1)^i m(\tau)^{-\frac{1}{2}}\left(2\int_z^{F_k(\tau)} Q(\tau, y)\,dy\right)^{\frac{1}{2}}\right] dz\right\}, \end{aligned} \qquad (2.152)$$

where $\theta = \frac{1}{2}T(\tau)$, $G_k(\tau, \varepsilon) \equiv F_k(\tau) + \varepsilon H_k(\tau)$ $(k = 1, 2)$ are the amplitude curves, accurate up to and including terms of the order of ε.

Let us examine some particular cases.

Let $Q(\tau, x)$ be an odd function of x. Then (2.146) implies that $F_1(\tau) = -F_2(\tau) = F(\tau)$. Equation (2.147) becomes in this case simpler, and $F(\tau)$ instead of being determined by (2.147), is determined by the differential equation

$$\begin{aligned} &\frac{d}{dt}\left\{\int_0^{F(\tau)}\left[2\int_x^{F(\tau)} Q(\tau, z)\,m(\tau)\,dz\right]^{\frac{1}{2}} dx\right\} = \\ &= \frac{\varepsilon}{4}\int_{-F(\tau)}^{F(\tau)}\sum_{i=1}^{2}(-1)^i f\left[\tau, x, (-1)^i m(\tau)^{-\frac{1}{2}}\left(2\int_x^{F(\tau)} Q(\tau, z)\,dz\right)^{\frac{1}{2}}\right] dx. \end{aligned} \qquad (2.153)$$

We obtain instead of (2.151) for the period

$$T_{\mathrm{I}}(\tau)=2\sqrt{m(\tau)}\int_{-F(\tau)}^{F(\tau)}\left[2\int_{x}^{F(\tau)}Q(\tau, z)\,dz\right]^{-\frac{1}{2}}dx. \tag{2.154}$$

Let the function $Q(\tau, x)$ depend linearly on x, i. e., $Q(\tau, x)=c(\tau)x$. Then equation (2.143) becomes an equation of the type (2.1)

$$\frac{d}{dt}\left[m(\tau)\frac{dx}{dt}\right]+c(\tau)x=\varepsilon f\left(\tau, x, \frac{dx}{dt}\right), \tag{2.155}$$

which was examined in detail in § 1 of the present chapter.

It is clear, on the grounds of the results obtained above, that the solution of equation (2.155) is in the first approximation

$$x=F(\tau)\cos\psi, \tag{2.156}$$

where $F(\tau)$ and ψ should, according to (2.153) and (2.152), be determined from the system of equations

$$\left.\begin{aligned}\frac{dF}{dt}&=-\frac{\varepsilon F}{2m(\tau)\omega(\tau)}\frac{d[m(\tau)\omega(\tau)]}{d\tau}-\\&\quad-\frac{\varepsilon}{2\pi m(\tau)\omega(\tau)}\int_{0}^{2\pi}f[\tau, F(\tau)\cos\psi, -F(\tau)\omega(\tau)\sin\psi]\sin\psi\,d\psi,\\\frac{d\psi}{dt}&=\omega(\tau)-\frac{\varepsilon}{2\pi m(\tau)\omega(\tau)F(\tau)}\times\\&\quad\times\int_{0}^{2\pi}f[\tau, F(\tau)\cos\psi, -F(\tau)\omega(\tau)\sin\psi]\cos\psi\,d\psi\end{aligned}\right\} \tag{2.157}$$

$\left(\text{here } \omega^2(\tau)=\frac{c(\tau)}{m(\tau)}\right)$. This system is identical with the previously derived system of equations of the first approximation (2.30).

In equation (2.155) let $\varepsilon f\left(\tau, x, \frac{dx}{dt}\right)=-\varepsilon\left[c_1(\tau)x+\lambda(\tau)\frac{dx}{dt}\right]$; then we obtain the linear differential equation

$$\frac{d}{dt}\left[m(\tau)\frac{dx}{dt}\right]+c(\tau)x=-\varepsilon\left[c_1(\tau)x+\lambda(\tau)\frac{dx}{dt}\right], \tag{2.158}$$

for which, according to (2.153), we find

$$\frac{dF}{dt}=-\varepsilon\left[\frac{1}{2m(\tau)\omega(\tau)}\frac{d[m(\tau)\omega(\tau)]}{d\tau}+\frac{\lambda(\tau)}{2m(\tau)}\right]F, \tag{2.159}$$

where $\omega^2(\tau)=\frac{c(\tau)+\varepsilon c_1(\tau)}{m(\tau)}$. This last equation coincides with the previously deduced equation (2.58).

Integrating (2.159), we obtain

$$F(\tau)=F_0[m(\tau)\omega(\tau)]^{-\frac{1}{2}}e^{-\frac{1}{2}\int_0^{\tau}\frac{\lambda(\tau)}{m(\tau)}d\tau}. \tag{2.160}$$

Of interest in solving many problems are the cases in which the right-hand side of equation (2.147) vanishes. This happens when $f\left(\tau, x, \frac{dx}{dt}\right)\equiv 0$, and also, for example, when $f\left(\tau, x, \frac{dx}{dt}\right)\neq 0$ but is even in $\frac{dx}{dt}$, or when $Q(\tau, x)$ and $f\left(\tau, x, \frac{dx}{dt}\right)$ are odd in x. Integrating equation (2.147) in these cases, we immediately obtain the relation

$$I(\tau, F_1, F_2)=\text{const}, \tag{2.161}$$

which expresses the adiabatic invariance of the action integral in the absence of friction and a slow variation of the system parameters.

Relations (2.161) together with (2.146) allow us to derive the expressions for the amplitude curves $F_1(\tau)$ and $F_2(\tau)$.

Let us consider one other particular case when a small friction, proportional to the first power of velocity, is present in the system, and equation (2.143) may, consequently, be written in the form

$$\frac{d}{dt}\left[m(\tau)\frac{dx}{dt}\right]+Q(\tau,\ x)+\varepsilon\cdot 2\lambda(\tau)\frac{dx}{dt}=0. \tag{2.162}$$

An equation for the amplitude is, as shown in /28/, easily obtained for equation (2.162). It has the form

$$I^*=2m^*(\tau)^{\frac{1}{2}}\int\limits_{F_2(\tau)}^{F_1(\tau)}dx\left\{2\int\limits_{x}^{F_1(\tau)}Q(\tau,\ y)\,dy\right\}^{\frac{1}{2}}=\mathrm{const}\qquad(\tau=\varepsilon t), \tag{2.163}$$

where

$$m^*(\tau)=m(\tau)\,e^{4\int\limits_0^{\tau}\frac{\lambda(\tau)}{m(\tau)}d\tau} \tag{2.164}$$

is the "reduced mass".

In this case the action integral, unlike the foregoing examples, is not conserved, but the quantity I^*, the "reduced action", is, as implied by (2.163), conserved.

We shall now apply the relation (2.163) to an actual example. As such an example we take equation (2.119). Since in this equation the function $Q(\tau,\ x)=f(\tau,\ x)$ is, according to (2.116), odd, we have $F_2(\tau)=-F_1(\tau)$ and consequently we obtain from (2.163), bearing in mind (2.116), the amplitude integral in the form

$$I^*=4e^{2\int\limits_0^{\tau}\lambda(\tau)d\tau}\nu^{\frac{1}{2}}(\tau)\int\limits_0^{F_1(\tau)}dx\left\{2\int\limits_x^{F_1(\tau)}[h+ky]\,dy\right\}^{\frac{1}{2}}=$$

$$=2e^{2\int\limits_0^{\tau}\lambda(\tau)d\tau}\nu^{\frac{1}{2}}(\tau)\,k^{-\frac{3}{2}}(h+kF_1(\tau))^2\left\{\arccos\frac{h}{h+kF_1(\tau)}-\right.$$

$$\left.-h\,(h+kF_1(\tau))^{-2}\sqrt{k^2F_1^2(\tau)+2khF_1(\tau)}\right\}. \tag{2.165}$$

We find from here

$$\left[(h+kF_1(\tau))^2k^{-\frac{3}{2}}\arccos\frac{h}{h+kF_1(\tau)}-\right.$$

$$\left.-k^{-1}h\sqrt{kF_1^2(\tau)+2hF_1(\tau)}\,\right]\nu^{\frac{1}{2}}(\tau)\,e^{2\int\limits_0^{\tau}\lambda(\tau)d\tau}=\mathrm{const}\qquad(\tau=\varepsilon t). \tag{2.166}$$

This relation contains only elementary functions and $F_1(\tau)$ may therefore be easily computed with any desired accuracy.

We show now that using formula (2.166) which characterizes the maximum deviation amplitude $F_1(\tau)$, one can derive a formula for the amplitude of the first harmonic $a_1(\tau)$ (cf. remark at the end of § 6).

The frequency and amplitude of the first harmonic for the unperturbed equation (2.120), corresponding to equation (2.119), are according to (2.122) related by

$$\omega^2(\tau,\ a_1)=\nu(\tau)\,k\left(1+\frac{4h}{\pi k a_1}\right). \tag{2.167}$$

By (2.154) we obtain for the period

$$T = 4k^{-\frac{1}{2}} \arccos \frac{h}{h + kF_1(\tau)}. \tag{2.168}$$

Since the frequency and the period are connected by the relation $T = \frac{2\pi}{\omega}$, we find from (2.167) and (2.168) the formula

$$\frac{h}{h + kF_1(\tau)} = \cos \frac{\pi}{2\sqrt{\nu(\tau)\left(1 + \frac{4h}{\pi k a_1}\right)}}, \tag{2.169}$$

connecting $F_1(\tau)$ and $a_1(\tau)$.

Substituting now the value of $F_1(\tau)$ given by (2.169) in (2.166), we obtain the following formula for $a_1(\tau)$:

$$\left(h^2 \Phi^{-2} k^{-\frac{3}{2}} \frac{\pi}{2\sqrt{\nu(\tau)\left(1 + \frac{4h}{\pi k a_1}\right)}} - k^{-\frac{3}{2}} h^2 \sqrt{\Phi^{-1} - 1}\right) \nu^{\frac{1}{2}}(\tau)\, e^{2\int\limits_0^{\tau} \lambda(\tau) d\tau} = \text{const}$$

$$(\tau = \varepsilon t), \tag{2.170}$$

where $\Phi = \Phi(\tau, a_1) \equiv \cos \frac{\pi}{2\sqrt{\nu(\tau)\left(1 + \frac{h}{\pi k a_1}\right)}}$.

Formula (2.170), exactly as (2.166), contains only elementary functions and allows us to calculate $a_1(\tau)$ with any desired accuracy.

In conclusion, the method expounded in this section was, as shown in /31/, developed successfully by Volosov for systems of a more general form

$$\left.\begin{aligned} \frac{d}{dt}\left[m(y)\frac{dx}{dt}\right] + Q(y, x) &= \varepsilon f\left(y, x, \frac{dx}{dt}\right), \\ \frac{dy}{dt} &= \varepsilon Y\left(y, x, \frac{dx}{dt}\right). \end{aligned}\right\} \tag{2.171}$$

Chapter III

THE ACTION OF "PERIODIC" FORCES ON NONLINEAR VIBRATING SYSTEMS WITH SLOWLY VARYING PARAMETERS

§ 1. General method for constructing asymptotic solutions

We now proceed to study the action of external "periodic" forces on nonlinear vibrating systems with slowly varying parameters, assuming that the "frequencies" and amplitudes of the external forces also vary slowly with time (slowly in the sense explained before).

We shall examine the nonlinear differential equation with slowly varying coefficients given by us at the beginning of the first chapter

$$\frac{d}{dt}\left[m(\tau)\frac{dx}{dt}\right]+c(\tau)x=\varepsilon F\left(\tau,\ \theta,\ x,\ \frac{dx}{dt}\right), \tag{3.1}$$

where, as everywhere, ε is a small positive parameter; $\tau=\varepsilon t$ is the "slowing" time; $F\left(\tau,\ \theta,\ x,\ \frac{dx}{dt}\right)$ is a function periodic in θ with period 2π which may be represented as

$$F\left(\tau,\ \theta,\ x,\ \frac{dx}{dt}\right)=\sum_{n=-N}^{N} e^{in\theta}F_n\left(\tau,\ x,\ \frac{dx}{dt}\right). \tag{3.2}$$

Here the coefficients $F_n\left(\tau,\ x,\ \frac{dx}{dt}\right)$ in the finite sum are certain polynomials in x and $\frac{dx}{dt}$. The coefficients in these polynomials depend on τ. We assume in addition to this that $\frac{d\theta}{dt}=\nu(\tau)$, i. e., the instantaneous frequency of the external periodic force also varies slowly with time. In order that asymptotic methods apply to the construction of approximate solutions of equation (3.1), we assume, as in the preceding chapter, that the coefficients $m(\tau)$ and $c(\tau)$ in equation (3.1), and also $F_n\left(\tau,\ x,\ \frac{dx}{dt}\right)$ and $\nu(\tau)$ have the desired number of derivatives with respect to τ for any finite value of τ, and that $m(\tau)$ and $c(\tau)$ are strictly positive on the interval $0\leqslant\tau\leqslant L$.

As known (cf., for example, /9/), resonance phenomena in nonlinear vibrating systems under the action of external forces may ensue upon the fulfillment of the condition

$$\nu\approx\frac{p}{q}\omega, \tag{3.3}$$

where ν is the frequency of the external periodic force; ω, a normal frequency of the system; p and q, mutually prime integers (usually small). The following classification of the various cases of resonance is generally accepted: 1) $p=1,\ q=1$ i. e., $\nu\approx\omega$; this case is called "main" or ordinary resonance; 2) $q = 1$, i. e., $\nu \approx p\omega$ or $\omega \approx \frac{\nu}{p}$; this case is called an overtone resonance of the normal frequency or a parametric resonance. Resonance of this type is, as known, also possible in linear systems with periodic coefficients; 3) $p = 1$, i. e., $\omega \approx q\nu$; this case is called an overtone resonance of the external frequency.

Basically, we shall keep to this terminology also when studying phenomena arising in nonlinear vibrating systems with slowly varying parameters which are subject to external periodic forces. We already know that in constructing approximate asymptotic solutions for nonlinear systems with constant parameters it is possible to discriminate the resonance and nonresonance case, and construct a solution appropriate to the study of the vibrations either only in a narrow resonance region or only away from resonance. Such a discrimination cannot be made in our case, in view of a possible dependence of the "normal" frequency ω and the instantaneous frequency of the external periodic force ν on the slowing time τ, since the vibrating system may, for example, pass in the course of vibrations from a resonant to a nonresonant state and conversely, because of the variation of these parameters (ω and ν).

Because of this, and bearing in mind the previously made assumptions, we shall elaborate a method of constructing approximate solutions for equation (3.1) in the most general form, suitable for the study of the resonance zone and also of transitions to the latter from the nonresonance zone for the case of any demultiplicative resonance.

The correct choice of the structure of an asymptotic solution of equation (3.1) is made, as usually in nonlinear mechanics, on the grounds of physical arguments similar to those presented at the beginning of the second chapter. Equation (3.1) degenerates in the absence of perturbation ($\varepsilon = 0$) and in the case of constant τ to the unperturbed equation (2.2) which is solved by a sinusoid of constant amplitude and phase determined by the initial values.

In the presence of perturbation there may appear in the solution overtones, harmonics of combination frequencies, the normal frequency will depend on the amplitude, various resonances may occur, etc. The presence of a slowly varying time τ (slow variation of the mass of the system, of the elasticity coefficient, of the frequency of the external periodic force, and of other parameters) gives rise to several additional phenomena not observed in nonlinear systems with constant parameters. Thus, for example, the generally accepted concept of a normal frequency already loses its sense here, as shown in the preceding chapter, since the "normal frequency" $\omega(\tau) = \sqrt{\frac{c(\tau)}{m(\tau)}}$ is in this case also slowly varying with time; the dependence of the instantaneous frequency of the external force $\nu(\tau)$ on time effects the magnitude of the amplitude, etc.

Taking into account all these physical arguments, keeping in mind the structure of asymptotic solutions for equation (2.1), and also the structure of solutions generally applied in nonlinear mechanics /9/, one finds it natural to look for an approximate solution of equation (3.1) in the form of the asymptotic series

$$x = a\cos\left(\frac{p}{q}\theta + \psi\right) + \varepsilon u_1\left(\tau, a, \theta, \frac{p}{q}\theta + \psi\right) + \varepsilon^2 u_2\left(\tau, a, \theta \ \frac{p}{q}\theta + \psi\right) + \ldots, \qquad (3.4)$$

where $u_1\left(\tau, a, \theta, \frac{p}{q}\theta+\psi\right)$, $u_2\left(\tau, a, \theta, \frac{p}{q}\theta+\psi\right)$, ..., are periodic functions of the angles θ, $\frac{p}{q}\theta+\psi$ with period 2π; p and q, as mentioned before, are some small mutually prime integers chosen according to the particular resonance we intend to examine; a and ψ are functions of time defined by the system of differential equations

$$\left.\begin{aligned} \frac{da}{dt} &= \varepsilon A_1(\tau, a, \psi)+\varepsilon^2 A_2(\tau, a, \psi)+\ldots, \\ \frac{d\psi}{dt} &= \omega(\tau)-\frac{p}{q}\nu(\tau)+\varepsilon B_1(\tau, a, \psi)+\varepsilon^2 B_2(\tau, a, \psi)+\ldots, \end{aligned}\right\} \tag{3.5}$$

where $\omega(\tau)=\sqrt{\frac{c(\tau)}{m(\tau)}}$ is the "normal" frequency of the system; $\frac{d\theta}{dt}=\nu(\tau)$, the instantaneous frequency of the external periodic perturbation; $\tau=\varepsilon t$; and the difference $\omega(\tau)-\frac{p}{q}\nu(\tau)$ characterizes the difference between the phases of the normal vibration and the external perturbation. This difference may vary in course of vibration.

To construct an approximate solution (in the first, second, etc., approximation) it is necessary, as in the previous chapter, to know the functions

$$\left.\begin{aligned} &u_1\left(\tau, a, \theta, \frac{p}{q}\theta+\psi\right), \quad u_2\left(\tau, a, \theta, \frac{p}{q}\theta+\psi\right), \ldots, \\ &A_1(\tau, a, \psi), \quad B_1(\tau, a, \psi), \quad A_2(\tau, a, \psi), \quad B_2(\tau, a, \psi), \ldots \end{aligned}\right\} \tag{3.6}$$

Their definition is generally speaking not unique (as indicated in Chapter II, § 1). A unique definition of these functions is attained , as in Chapter II, by an additional condition: that the periodic functions $u_1\left(\tau, a, \theta, \frac{p}{q}\theta+\psi\right)$, $u_2\left(\tau, a, \theta, \frac{p}{q}\theta+\psi\right)$, ..., do not contain the first harmonic of the angular argument $\frac{p}{q}\theta+\psi$. This condition may be written in the form

$$\left.\begin{aligned} &\int_0^{2\pi} u_1(\tau, a, \theta, p\varphi+\psi)\cos(p\varphi+\psi)\,d(p\varphi+\psi)=0, \\ &\int_0^{2\pi} u_1(\tau, a, \theta, p\varphi+\psi)\sin(p\varphi+\psi)\,d(p\varphi+\psi)=0, \\ &\int_0^{2\pi} u_2(\tau, a, \theta, p\varphi+\psi)\cos(p\varphi+\psi)\,d(p\varphi+\psi)=0, \\ &\int_0^{2\pi} u_2(\tau, a, \theta, p\varphi+\psi)\sin(p\varphi+\psi)\,d(p\varphi+\psi)=0, \\ &\ldots\ldots\ldots\ldots\ldots\ldots\ldots\ldots \end{aligned}\right\} \tag{3.7}$$

where $\varphi=\frac{1}{q}\theta$.

Condition (3.7) is from the physical point of view equivalent to choosing the quantity a as the full amplitude of the first principal vibration harmonic, which is the same as the absence of resonance terms (i. e., terms with denominators which may vanish) in the functions $u_1(\tau, a, \theta, p\varphi+\psi)$, $u_2(\tau, a, \theta, p\varphi+\psi)$,... for any τ within the interval $0 \leqslant \tau \leqslant L$.

We now proceed to solve the problem stated — to determine the functions (3.6), taking into consideration the condition (3.7).

Differentiating the right-hand sides of (3.5), we obtain the following expressions for $\frac{d^2a}{dt^2}$ and $\frac{d^2\psi}{dt^2}$:

$$\left.\begin{aligned}\frac{d^2a}{dt^2}&=\varepsilon\left[\omega(\tau)-\frac{p}{q}\nu(\tau)\right]\frac{\partial A_1}{\partial\psi}+\\&+\varepsilon^2\left\{\frac{\partial A_1}{\partial a}A_1+\frac{\partial A_1}{\partial\psi}B_1+\frac{\partial A_1}{\partial\tau}+\left[\omega(\tau)-\frac{p}{q}\nu(\tau)\right]\frac{\partial A_2}{\partial\psi}\right\}+\varepsilon^3\ldots,\\\frac{d^2\psi}{dt^2}&=\varepsilon\left\{\frac{d\omega}{d\tau}-\frac{p}{q}\frac{d\nu}{d\tau}+\left[\omega(\tau)-\frac{p}{q}\nu(\tau)\right]\frac{\partial B_1}{\partial\psi}\right\}+\\&+\varepsilon^2\left\{\frac{\partial B_1}{\partial a}A_1+\frac{\partial B_1}{\partial\psi}B_1+\frac{\partial B_1}{\partial\tau}+\left[\omega(\tau)-\frac{p}{q}\nu(\tau)\right]\frac{\partial B_2}{\partial\psi}\right\}+\varepsilon^3\ldots\end{aligned}\right\}\quad(3.8)$$

Differentiating the right-hand side of (3.4), and taking into account equations (3.5) and the expressions of $\frac{d^2a}{dt^2}$ and $\frac{d^2\psi}{dt^2}$, we obtain the following expressions for $\frac{dx}{dt}$ and $\frac{d^2x}{dt^2}$ (the result is expanded in powers of ε)

$$\begin{aligned}\frac{dx}{dt}&=-a\omega(\tau)\sin(p\varphi+\psi)+\\&+\varepsilon\left\{A_1\cos(p\varphi+\psi)-B_1a\sin(p\varphi+\psi)+\frac{\partial u_1}{\partial\theta}\nu(\tau)+\frac{\partial u_1}{\partial(p\varphi+\psi)}\omega(\tau)\right\}+\\&+\varepsilon^2\left\{A_2\cos(p\varphi+\psi)-B_2a\sin(p\varphi+\psi)+\frac{\partial u_1}{\partial\tau}+\frac{\partial u_1}{\partial a}A_1+\right.\\&\qquad\left.+\frac{\partial u_1}{\partial(p\varphi+\psi)}B_1+\frac{\partial u_2}{\partial\theta}\nu(\tau)+\frac{\partial u_2}{\partial(p\varphi+\psi)}\omega(\tau)\right\}+\varepsilon^3\ldots,\end{aligned}\quad(3.9)$$

$$\begin{aligned}\frac{d^2x}{dt^2}&=-a\omega^2(\tau)\cos(p\varphi+\psi)+\\&+\varepsilon\left\{\left[\left[\omega(\tau)-\frac{p}{q}\nu(\tau)\right]\frac{\partial A_1}{\partial\psi}-2a\omega(\tau)B_1\right]\cos(p\varphi+\psi)-\right.\\&-\left[\left[\omega(\tau)-\frac{p}{q}\nu(\tau)\right]a\frac{\partial B_1}{\partial\psi}+2\omega(\tau)A_1\right]\sin(p\varphi+\psi)-\\&-\frac{d\omega(\tau)}{d\tau}a\sin(p\varphi+\psi)+\frac{\partial^2u_1}{\partial\theta^2}\nu^2(\tau)+2\frac{\partial^2u_1}{\partial\theta\,\partial(p\varphi+\psi)}\nu(\tau)\omega(\tau)+\\&\left.+\frac{\partial^2u_1}{\partial(p\varphi+\psi)^2}\omega^2(\tau)\right\}+\varepsilon^2\left\{\left[\left[\omega(\tau)-\frac{p}{q}\nu(\tau)\right]\frac{\partial A_2}{\partial\psi}-2a\omega(\tau)B_2\right]\cos(p\varphi+\psi)-\right.\\&-\left[\left[\omega(\tau)-\frac{p}{q}\nu(\tau)\right]a\frac{\partial B_2}{\partial\psi}+2\omega(\tau)A_2\right]\sin(p\varphi+\psi)+\\&+\left[\frac{\partial A_1}{\partial a}A_1+\frac{\partial A_1}{\partial\psi}B_1+\frac{\partial A_1}{\partial\tau}-aB_1^2\right]\cos(p\varphi+\psi)-\\&-\left[2A_1B_1+a\frac{\partial B_1}{\partial a}A_1+a\frac{\partial B_1}{\partial\psi}B_1+a\frac{\partial B_1}{\partial\tau}\right]\sin(p\varphi+\psi)+\\&+2\frac{\partial^2u_1}{\partial\tau\,\partial(p\varphi+\psi)}\omega(\tau)+2\frac{\partial^2u_1}{\partial\tau\partial\theta}\nu(\tau)+2\frac{\partial^2u_1}{\partial a\,\partial\theta}\nu(\tau)A_1+2\frac{\partial^2u_1}{\partial a\,\partial(p\varphi+\psi)}\omega(\tau)A_1+\\&+2\frac{\partial^2u_1}{\partial\theta\,\partial(p\varphi+\psi)}\nu(\tau)B_1+2\frac{\partial^2u_1}{\partial(p\varphi+\psi)^2}\omega(\tau)B_1+\\&+\frac{\partial u_1}{\partial(p\varphi+\psi)}\left[\omega(\tau)-\frac{p}{q}\nu(\tau)\right]\frac{\partial B_1}{\partial\psi}+\frac{\partial u_1}{\partial a}\left[\omega(\tau)-\frac{p}{q}\nu(\tau)\right]\frac{\partial A_1}{\partial\psi}+\\&+\frac{\partial u_1}{\partial\theta}\frac{d\nu(\tau)}{d\tau}+\frac{\partial u_1}{\partial(p\varphi+\psi)}\frac{d\omega(\tau)}{d\tau}+\frac{\partial^2u_2}{\partial\theta^2}\nu^2(\tau)+\\&\left.+2\frac{\partial^2u_2}{\partial\theta\,\partial(p\varphi+\psi)}\nu(\tau)\omega(\tau)+\frac{\partial^2u_2}{\partial(p\varphi+\psi)^2}\omega^2(\tau)\right\}+\varepsilon^3\ldots\end{aligned}\quad(3.10)$$

Putting (3.4), (3.9), and (3.10) in the left-hand side of equation (3.1), we obtain

$$\begin{aligned}\frac{d}{dt}\left[m(\tau)\frac{dx}{dt}\right]+c(\tau)x&=\varepsilon\left\{m(\tau)\left[\left[\omega(\tau)-\frac{p}{q}\nu(\tau)\right]\frac{\partial A_1}{\partial\psi}-2a\omega(\tau)B_1\right]\cos(p\varphi+\psi)-\right.\\&-m(\tau)\left[\left[\omega(\tau)-\frac{p}{q}\nu(\tau)\right]a\frac{\partial B_1}{\partial\psi}+2\omega(\tau)A_1\right]\sin(p\varphi+\psi)-\\&-\frac{d[m(\tau)\omega(\tau)]}{d\tau}a\sin(p\varphi+\psi)+m(\tau)\left[\frac{\partial^2u_1}{\partial\theta^2}\nu^2(\tau)+\right.\\&\left.\left.+2\frac{\partial^2u_1}{\partial\theta\,\partial(p\varphi+\psi)}\nu(\tau)\omega(\tau)+\frac{\partial^2u_1}{\partial(p\varphi+\psi)^2}\omega^2(\tau)+\omega^2(\tau)u_1\right]\right\}+\varepsilon^2\ldots\end{aligned}\quad(3.11)$$

Substituting the value of x from (3.4) and of $\frac{dx}{dt}$ from (3.9) in the right-hand side of equation (3.1), and expanding in a Taylor series, we obtain

$$\varepsilon F\left(\tau,\ \theta,\ x,\ \frac{dx}{dt}\right)=\varepsilon F(\tau,\ \theta,\ a\cos(p\varphi+\psi),\ \ -a\omega\sin(p\varphi+\psi))+ \\ +\varepsilon^2\left\{F'_x u_1+F'_{x'}\left[A_1\cos(p\varphi+\psi)-B_1 a\sin(p\varphi+\psi)+\frac{\partial u_1}{\partial\theta}\nu(\tau)+\right.\right. \\ \left.\left.+\frac{\partial u_1}{\partial(p\varphi+\psi)}\omega(\tau)\right]\right\}+\varepsilon^3\ldots \quad (3.12)$$

In order that the series (3.4) may formally satisfy equation (3.1), it is necessary that the coefficients of equal powers of ε in (3.11) and (3.12) be equal. Equating them, we obtain the equations

$$m(\tau)\left[\frac{\partial^2 u_1}{\partial\theta^2}\nu^2(\tau)+2\frac{\partial^2 u_1}{\partial\theta\,\partial(p\varphi+\psi)}\nu(\tau)\,\omega(\tau)+\frac{\partial^2 u_1}{\partial(p\varphi+\psi)^2}\omega^2(\tau)+\omega^2(\tau)\,u_1\right]= \\ =F_0(\tau,\ a,\ \theta,\ p\varphi+\psi)-m(\tau)\left[\left[\omega(\tau)-\frac{p}{q}\nu(\tau)\right]\frac{\partial A_1}{\partial\psi}-2a\omega(\tau)B_1\right]\cos(p\varphi+\psi)+ \\ +m(\tau)\left[\left[\omega(\tau)-\frac{p}{q}\nu(\tau)\right]a\frac{\partial B_1}{\partial\psi}+2\omega(\tau)A_1+\frac{1}{m(\tau)}\frac{d[m(\tau)\,\omega(\tau)]}{d\tau}a\right]\sin(p\varphi+\psi), \quad (3.13)$$

$$m(\tau)\left[\frac{\partial^2 u_2}{\partial\theta^2}\nu^2(\tau)+2\frac{\partial^2 u_2}{\partial\theta\,\partial(p\varphi+\psi)}\nu(\tau)\,\omega(\tau)+\frac{\partial^2 u_2}{\partial(p\varphi+\psi)^2}\omega^2(\tau)+\omega^2(\tau)\,u_2\right]= \\ =F_1(\tau,\ a,\ \theta,\ p\varphi+\psi)-m(\tau)\left[\left[\omega(\tau)-\frac{p}{q}\nu(\tau)\right]\frac{\partial A_2}{\partial\psi}-2a\omega(\tau)B_2+\right. \\ \left.+\frac{\partial A_1}{\partial a}A_1+\frac{\partial A_1}{\partial\psi}B_1+\frac{\partial A_1}{\partial\tau}-aB_1^2+\frac{dm(\tau)}{d\tau}\frac{A_1}{m(\tau)}\right]\cos(p\varphi+\psi)+ \\ +m(\tau)\left[\left[\omega(\tau)-\frac{p}{q}\nu(\tau)\right]a\frac{\partial B_2}{\partial\psi}+2\omega(\tau)A_2+2A_1B_1+a\frac{\partial B_1}{\partial a}A_1+\right. \\ \left.+a\frac{\partial B_1}{\partial\psi}B_1+a\frac{\partial B_1}{\partial\tau}+\frac{dm(\tau)}{d\tau}\frac{a}{m(\tau)}B_1\right]\sin(p\varphi+\psi), \quad (3.14)$$

where

$$F_0(\tau,\ a,\ \theta,\ p\varphi+\psi)=F[\tau,\ \theta,\ a\cos(p\varphi+\psi),\ \ -a\omega\sin(p\varphi+\psi)], \quad (3.15)$$

$$F_1(\tau,\ a,\ \theta,\ p\varphi+\psi)=F'_x u_1+F'_{x'}\left[A_1\cos(p\varphi+\psi)-B_1 a\sin(p\varphi+\psi)+\right. \\ \left.+\frac{\partial u_1}{\partial\theta}\nu(\tau)+\frac{\partial u_1}{\partial(p\varphi+\psi)}\omega(\tau)\right]-m(\tau)\left[2\frac{\partial^2 u_1}{\partial\tau\,\partial(p\varphi+\psi)}\omega(\tau)+2\frac{\partial^2 u_1}{\partial\tau\,\partial\theta}\nu(\tau)+\right. \\ +2\frac{\partial^2 u_1}{\partial a\,\partial\theta}\nu(\tau)A_1+2\frac{\partial^2 u_1}{\partial a\,\partial(p\varphi+\psi)}\omega(\tau)A_1+2\frac{\partial^2 u_1}{\partial\theta\,\partial(p\varphi+\psi)}\nu(\tau)B_1+ \\ +2\frac{\partial^2 u_1}{\partial(p\varphi+\psi)^2}\omega(\tau)B_1+\frac{\partial u_1}{\partial(p\varphi+\psi)}\left[\omega(\tau)-\frac{p}{q}\nu(\tau)\right]\frac{\partial B_1}{\partial\psi}+ \\ +\frac{\partial u_1}{\partial a}\left[\omega(\tau)-\frac{p}{q}\nu(\tau)\right]\frac{\partial A_1}{\partial\psi}+\frac{\partial u_1}{\partial\theta}\frac{d\nu(\tau)}{d\tau}+\frac{\partial u_1}{\partial(p\varphi+\psi)}\frac{d\omega(\tau)}{d\tau}+ \\ \left.+\frac{1}{m(\tau)}\frac{\partial u_1}{\partial\theta}\nu(\tau)\frac{dm(\tau)}{d\tau}+\frac{\partial u_1}{\partial(p\varphi+\psi)}\frac{\omega(\tau)}{m(\tau)}\frac{dm(\tau)}{d\tau}\right]. \quad (3.16)$$

It is easily verified that the function $F_k(\tau,\ a,\ \theta,\ p\varphi+\psi)$ is periodic in θ and $p\varphi+\psi$ with period 2π, and that it also depends on τ and a. The explicit expression for this function may be obtained as soon as the functions

$$u_j(\tau,\ a,\ \theta,\ p\varphi+\psi),\quad A_j(\tau,\ a,\ \psi),\quad B_j(\tau,\ a,\ \psi)\qquad (j=1,\ 2,\ \ldots,\ k-1),$$

are found.

To determine $A_1(\tau,\ a,\ \psi)$, $B_1(\tau,\ a,\ \psi)$, and $u_1(\tau,\ a,\ \theta,\ p\varphi+\psi)$ from equation (3.13), we represent the function (3.15) as a trigonometric double sum

$$F_0(\tau,\ a,\ \theta,\ p\varphi+\psi)=\sum_n\sum_m F_{0,\,n,\,m}(\tau,\ a)\,e^{i\{n\theta+m(p\varphi+\psi)\}}, \quad (3.17)$$

where

$$F_{0,\,n,\,m}(\tau,\ a)=\frac{1}{4\pi^2}\int_0^{2\pi}\int_0^{2\pi}F_0(\tau,\ a,\ \theta,\ p\varphi+\psi)\,e^{-i\{n\theta+m(p\varphi+\psi)\}}\,d\theta\,d(p\varphi+\psi). \quad (3.18)$$

The required function $u_1(\tau, a, \theta, p\varphi+\psi)$, periodic in θ and $p\varphi+\psi$ with period 2π for both angular variables, may also be looked for in the form of a finite Fourier sum

$$u_1(\tau, a, \theta, p\varphi+\psi)=\sum_n\sum_m g_{n,m}(\tau, a)\, e^{i\{n\theta+m(p\varphi+\psi)\}}, \tag{3.19}$$

in which the coefficients $g_{n,m}(\tau, a)$ are to be determined.

Substituting (3.17) and (3.19) in the right-hand side of equation (3.13), we obtain

$$\begin{aligned} m(\tau)\sum_n\sum_m\{\omega^2(\tau)-[m\omega(\tau)+n\nu(\tau)]^2\}\,g_{n,m}(\tau, a)\,e^{i\{n\theta+m(p\varphi+\psi)\}} &= \\ =\sum_n\sum_m e^{i\{n\theta+m(p\varphi+\psi)\}}\frac{1}{4\pi^2}\int_0^{2\pi}\int_0^{2\pi}F_0(\tau, a, \theta, p\varphi+\psi)\,e^{-i\{n\theta+m(p\varphi+\psi)\}}\,d\theta\, d(p\varphi+\psi)&- \\ -m(\tau)\left\{\left[\omega(\tau)-\frac{p}{q}\nu(\tau)\right]\frac{\partial A_1}{\partial\psi}-2a\omega(\tau)B_1\right\}\cos(p\varphi+\psi)&+ \\ +m(\tau)\left\{\left[\omega(\tau)-\frac{p}{q}\nu(\tau)\right]a\frac{\partial B_1}{\partial\psi}+2\omega(\tau)A_1+\frac{1}{m(\tau)}\frac{d[m(\tau)\omega(\tau)]}{d\tau}a\right\}\sin(p\varphi+\psi).& \end{aligned} \tag{3.20}$$

We note that $u_1(\tau, a, \theta, p\varphi+\psi)$ should, according to the condition (3.7), not contain the first harmonic of the angle $p\varphi+\psi$. Hence, the sum (3.19) should contain only terms with indexes n and m satisfying the inequality

$$n\theta+m(p\varphi+\psi)\neq\pm(p\varphi+\psi)+p_1\psi, \tag{3.21}$$

where p_1 is some integer.

In other words, for condition (3.7) to be satisfied it is necessary that $g_{n,m}(\tau, a)=0$ for all pairs of indexes n, m satisfying the relation

$$nq+p(m\pm1)=0. \tag{3.22}$$

Equating the coefficients of the same harmonics in (3.20), we obtain

$$g_{n,m}(\tau, a)=\frac{\dfrac{1}{4\pi^2 m(\tau)}\displaystyle\int_0^{2\pi}\int_0^{2\pi}F_0(\tau, a, \theta, p\varphi+\psi)\,e^{-i\{n\theta+m(p\varphi+\psi)\}}\,d\theta\, d(p\varphi+\psi)}{\omega^2(\tau)-[m\omega(\tau)+n\nu(\tau)]^2} \tag{3.23}$$

for all pairs of indexes, n, m not satisfying (3.22).

Substituting the expressions (3.23) in the right-hand side of (3.19), we finally obtain

$$\begin{aligned} u_1(\tau, a, \theta, p\varphi+\psi)&=\frac{1}{4\pi^2 m(\tau)}\sum_{\substack{n\ \ m\\ [nq+p(m\pm1)\neq0]}}\sum\frac{e^{i\{n\theta+m(p\varphi+\psi)\}}}{\omega^2(\tau)-[m\omega(\tau)+n\nu(\tau)]^2}\times \\ &\times\int_0^{2\pi}\int_0^{2\pi}F_0(\tau, a, \theta, p\varphi+\psi)\,e^{-i\{n\theta+m(p\varphi+\psi)\}}\,d\theta\, d(p\varphi+\psi). \end{aligned} \tag{3.24}$$

Having determined the quantities (3.23), we obtain at the same time a relation for determining $A_1(\tau, a, \psi)$ and $B_1(\tau, a, \psi)$:

$$\begin{aligned} m(\tau)\left\{\left[\omega(\tau)-\frac{p}{q}\nu(\tau)\right]a\frac{\partial B_1}{\partial\psi}+2\omega(\tau)A_1+\frac{1}{m(\tau)}\frac{d[m(\tau)\omega(\tau)]}{d\tau}a\right\}\sin(p\varphi+\psi)&- \\ -m(\tau)\left\{\left[\omega(\tau)-\frac{p}{q}\nu(\tau)\right]\frac{\partial A_1}{\partial\psi}-2a\omega(\tau)B_1\right\}\cos(p\varphi+\psi)&+ \\ +\sum_{\substack{n\ \ m\\ [nq+p(m\pm1)=0]}}\sum\frac{1}{4\pi^2}e^{i\{n\theta+m(p\varphi+\psi)\}}&\times \\ \times\int_0^{2\pi}\int_0^{2\pi}F_0(\tau, a, \theta, p\varphi+\psi)\,e^{-i\{n\theta+m(p\varphi+\psi)\}}\,d\theta\, d(p\varphi+\psi)&=0. \end{aligned} \tag{3.25}$$

The summation runs over all integral values of n and m satisfying condition (3.22). Consequently, complex exponents of the form

$$e^{i\left\{n\theta+m\left(\frac{p}{q}\theta+\psi\right)\right\}}=e^{i\left\{\left(n+m\frac{p}{q}\right)\theta+m\psi\right\}}=e^{i\left\{\mp\frac{p}{q}\theta+m\psi\right\}}=$$
$$=e^{i\left\{\mp\left(\frac{p}{q}\theta+\psi\right)+(m\pm 1)\psi\right\}}=\{\cos(p\varphi+\psi)\mp i\sin(p\varphi+\psi)\}\,e^{i(m\pm 1)\psi}, \qquad (3.26)$$

appear in the double sum in (3.25).

It should be mentioned that $m \pm 1$ is, on account of the condition (3.22), divisible by q, so that we may write the factor $m \pm 1$ in the exponent $e^{i(m\pm 1)\psi}$ as $q\sigma$ $(-\infty < \sigma < \infty)$.

After these remarks we equate the coefficients of $\cos(p\varphi+\psi)$ and $\sin(p\varphi+\psi)$ in (3.25) to zero, and obtain the following system of equations

$$\left.\begin{aligned}&\left[\omega(\tau)-\frac{p}{q}\nu(\tau)\right]\frac{\partial A_1}{\partial\psi}-2a\omega(\tau)B_1=\\&\qquad=\frac{1}{2\pi^2 m(\tau)}\sum_{\sigma}e^{i\sigma q\psi}\int_0^{2\pi}\int_0^{2\pi}F_0(\tau,a,\theta,\vartheta)\,e^{-i\sigma q\psi_1}\cos\vartheta\,d\theta\,d\vartheta,\\&\left[\omega(\tau)-\frac{p}{q}\nu(\tau)\right]a\frac{\partial B_1}{\partial\psi}+2\omega(\tau)A_1=-\frac{1}{m(\tau)}\frac{d[m(\tau)\omega(\tau)]}{d\tau}a-\\&\qquad-\frac{1}{2\pi^2 m(\tau)}\sum_{\sigma}e^{i\sigma q\psi}\int_0^{2\pi}\int_0^{2\pi}F_0(\tau,a,\theta,\vartheta)\,e^{-i\sigma q\psi_1}\sin\vartheta\,d\theta\,d\vartheta\end{aligned}\right\} \qquad (3.27)$$

$(\psi_1 = \psi - p\varphi)$.

The summation in formulas (3.27) runs over all values of the index σ, positive and negative, for which the integrals under the summation sign do not vanish. These integrals do not vanish for those values of σ for which the total exponent in the corresponding exponential (obtained after expanding the integrand in a Fourier series) equals zero. Thus, assuming in our case that the right-hand side of equation (3.1) is a polynomial in x, $\frac{dx}{dt}$, $\cos\theta$, and $\sin\theta$, the index σ takes on a finite number of integral values.

We have thus derived equations (3.27) for the determination of $A_1(\tau, a, \psi)$ and $B_1(\tau, a, \psi)$. If $A_1(\tau, a, \psi)$ and $B_1(\tau, a, \psi)$ satisfy these equations, then the expression (3.24) for $u_1(\tau, a, \theta, p\varphi+\psi)$, defined as a particular solution of the inhomogeneous equation (3.13) corresponding to a doubly periodic right part, does not contain terms with vanishing denominators. Consequently, for the function $u_1(\tau, a, \theta, p\varphi+\psi)$ to be finite, it is sufficient to choose for $A_1(\tau, a, \psi)$ and $B_1(\tau, a, \psi)$ a particular solution of the system (3.27) that is periodic in ψ. Such a solution may be easily found. After a series of elementary calculations we obtain

$$\left.\begin{aligned}&A_1(\tau,a,\psi)=-\frac{a}{2m(\tau)\omega(\tau)}\frac{d[m(\tau)\omega(\tau)]}{d\tau}+\frac{1}{2\pi^2 m(\tau)}\times\\&\times\sum_{\sigma}e^{i\sigma q\psi}\left\{\frac{[q\omega(\tau)-p\nu(\tau)]\sigma i\int_0^{2\pi}\int_0^{2\pi}F_0(\tau,a,\theta,\vartheta)\,e^{-i\sigma q\psi_1}\cos\vartheta\,d\theta\,d\vartheta}{4\omega^2(\tau)-[q\omega(\tau)-p\nu(\tau)]^2\sigma^2}-\right.\\&\qquad\left.-\frac{2\omega(\tau)\int_0^{2\pi}\int_0^{2\pi}F_0(\tau,a,\theta,\vartheta)\,e^{-i\sigma q\psi_1}\sin\vartheta\,d\theta\,d\vartheta}{4\omega^2(\tau)-[q\omega(\tau)-p\nu(\tau)]^2\sigma^2}\right\},\end{aligned}\right\} \qquad (3.28a)$$

$$B_1(\tau, a, \psi) = -\frac{1}{2\pi^2 m(\tau) a} \times$$

$$\times \sum_{\sigma} e^{i\sigma q\psi} \left\{ \frac{[q\omega(\tau) - p\nu(\tau)]\sigma i \cdot \int_0^{2\pi}\int_0^{2\pi} F_0(\tau, a, \theta, \vartheta) e^{-i\sigma q\psi_1} \sin\vartheta \, d\theta \, d\vartheta}{4\omega^2(\tau) - [q\omega(\tau) - p\nu(\tau)]^2\sigma^2} + \right.$$

$$\left. + \frac{2\omega(\tau) \int_0^{2\pi}\int_0^{2\pi} F_0(\tau, a, \theta, \vartheta) e^{-i\sigma q\psi_1} \cos\vartheta \, d\theta \, d\vartheta}{4\omega^2(\tau) - [q\omega(\tau) - p\nu(\tau)]^2\sigma^2} \right\}. \tag{3.28b}$$

After we have obtained expressions for $u_1(\tau, a, \theta, p\varphi + \psi)$, $A_1(\tau, a, \psi)$, and $B_1(\tau, a, \psi)$ we may derive all the quantities necessary for the solution of equation (3.1) in the second approximation. We do this by means of equation (3.14), the right-hand side of which is already known, and the condition requiring that $u_2(\tau, a, \theta, p\varphi + \psi)$ does not contain the first harmonic of the angle $p\varphi + \psi$. Thus, after similar arguments and calculations, we obtain the following system of equations for $A_2(\tau, a, \psi)$ and $B_2(\tau, a, \psi)$:

$$\left\{\begin{aligned}
&\left[\omega(\tau) - \frac{p}{q}\nu(\tau)\right]\frac{\partial A_2}{\partial\psi} - 2\omega(\tau) a B_2 = \\
&\quad = -\left[\frac{\partial A_1}{\partial a} A_1 + \frac{\partial A_1}{\partial\psi} B_1 + \frac{\partial A_1}{\partial\tau} - a B_1^2 + \frac{dm(\tau)}{d\tau}\frac{A_1}{m(\tau)}\right] + \\
&\quad + \frac{1}{2\pi^2 m(\tau)} \sum_{\sigma} e^{i\sigma q\psi} \int_0^{2\pi}\int_0^{2\pi} F_1(\tau, a, \theta, \vartheta) e^{-i\sigma q\psi_1} \cos\vartheta \, d\theta \, d\vartheta, \\
&\left[\omega(\tau) - \frac{p}{q}\nu(\tau)\right] a \frac{\partial B_2}{\partial\psi} + 2\omega(\tau) A_2 = \\
&\quad = -\left[\frac{\partial B_1}{\partial a} a A_1 + \frac{\partial B_1}{\partial\psi} a B_1 + \frac{\partial B_1}{\partial\tau} a + 2A_1 B_1 + \frac{dm(\tau)}{d\tau}\frac{a B_1}{m(\tau)}\right] - \\
&\quad - \frac{1}{2\pi^2 m(\tau)} \sum_{\sigma} e^{i\sigma q\psi} \int_0^{2\pi}\int_0^{2\pi} F_1(\tau, a, \theta, \vartheta) e^{-i\sigma q\psi_1} \sin\vartheta \, d\theta \, d\vartheta,
\end{aligned}\right. \tag{3.29}$$

where $\psi_1 = \psi - p\varphi$, and $F_1(\tau, a, \theta, \vartheta)$ is a function periodic in θ and ϑ with period 2π, whose explicit form will be known as soon as we find $u_1(\tau, a, \theta, p\varphi + \psi)$, $A_1(\tau, a, \psi)$ and $B_1(\tau, a, \psi)$. The system (3.29) is of the same form as (3.27), and consequently no difficulties arise in deriving explicit expressions for $A_2(\tau, a, \psi)$ and $B_2(\tau, a, \psi)$.

Returning to the above method, we present a scheme for constructing asymptotic solutions of equation (3.1) in the first and second approximation.

Bearing in mind the arguments of the first section of the preceding chapter, we take for the first approximation the function

$$x = a\cos(p\varphi + \psi), \tag{3.30}$$

where a and ψ should be determined from the equations of the first approximation

$$\left.\begin{aligned}
\frac{da}{dt} &= \varepsilon A_1(\tau, a, \psi), \\
\frac{d\psi}{dt} &= \omega(\tau) - \frac{p}{q}\nu(\tau) + \varepsilon B_1(\tau, a, \psi),
\end{aligned}\right\} \tag{3.31}$$

where $\omega(\tau) = \sqrt{\frac{c(\tau)}{m(\tau)}}$; $\frac{d\theta}{dt} = \nu(\tau)$; and $A_1(\tau, a, \psi)$ and $B_1(\tau, a, \psi)$ are determined from formulas (3.28) or directly as particular solutions of the system of equations (3.27), periodic in ψ with period 2π.

If the difference $\omega(\tau) - \frac{p}{q}\nu(\tau)$ in equations (3.31) is not small (the nonresonance case), then the equations may be averaged over ψ as a system with rotating phase (cf. Chapter II, § 5). If the difference $\omega(\tau)-\frac{p}{q}\nu(\tau)$ is of the order of ε (the resonance zone), then the system may be considered as equations of slow motion.

For the second approximation we take

$$x=a\cos(p\varphi+\psi)+\varepsilon u_1(\tau,\ a,\ \theta,\ p\varphi+\psi), \tag{3.32}$$

where a and ψ should be determined from the equations of the second approximation

$$\left.\begin{aligned}\frac{da}{dt}&=\varepsilon A_1(\tau,\ a,\ \psi)+\varepsilon^2 A_2(\tau,\ a,\ \psi),\\ \frac{d\psi}{dt}&=\omega(\tau)-\frac{p}{q}\nu(\tau)+\varepsilon B_1(\tau,\ a,\ \psi)+\varepsilon^2 B_2(\tau,\ a,\ \psi).\end{aligned}\right\} \tag{3.33}$$

Here $A_1(\tau, a, \psi)$, $B_1(\tau, a, \psi)$, $A_2(\tau, a, \psi)$, $B_2(\tau, a, \psi)$ are particular (periodic in ψ with period 2π) solutions of equations (3.27) and (3.29), and $u_1(\tau, a, \theta, p\varphi+\psi)$ is determined from (3.24).

We recall that (cf. Chapter I, § 9) not the second, but the first approximation for a and ψ defined by the system (3.31) can be substituted in $\varepsilon u_1(\tau, a, \theta, p\varphi+\psi)$ the second term of (3.32).

As in the first section of Chapter II, the equations obtained here for a and ψ can in the general case not be integrated in a closed form, and one therefore has to apply numerical methods of integration or restrict oneself to qualitative investigation. Numerical integration of equations of the type (3.33) is much simpler than the numerical integration of equation (3.1); this has already been mentioned.

We now proceed to discuss in detail the first approximation.

Unlike equations (2.22), the variables in these equations of the first approximation are not separable, and we have a system of two mutually connected equations for the two unknowns a and ψ. In the equations obtained in the study of resonance phenomena in nonlinear vibrating systems with constant parameters, the phase plane method applies when investigating the equations of the first approximation. Here this is not the case since the right-hand sides in equations (3.27) depend not only on a and ψ but also on the slowing time τ. We assume that the right-hand side of equation (3.1) is a polynomial in $x, \frac{dx}{dt}$, $\cos\theta$ and $\sin\theta$. Then the expression of $u_1(\tau, a, \theta, p\varphi+\psi)$ entering in the improved first approximation or in the second approximation (all this refers also to $u_2(\tau, a, \theta, p\varphi+\psi)$, $u_3(\tau, a, \theta, p\varphi+\psi), \ldots$), consists of a finite sum, and the denominators of the separate terms are always different from zero for our choice of $A_1(\tau, a, \psi)$ and $B_1(\tau, a, \psi)$. This is because the mutually prime numbers p and q are completely determined depending on which resonance we intend to examine, and the number of combinations of various p and q is limited by the polynomial nature of the right-hand side of equation (3.1).

Regarding the right-hand side of equation (3.1), we make, for the sake of greater clarity, a more general assumption: the function $F\left(\tau, \theta, x, \frac{dx}{dt}\right)$ may, for any τ within the interval $0\leqslant\tau\leqslant L$, be represented by a uniformly convergent series

$$F\left(\tau,\ \theta,\ x,\ \frac{dx}{dt}\right)=\sum_{n=-\infty}^{\infty} e^{in\theta}F_n\left(\tau\ \ x,\ \frac{dx}{dt}\right), \tag{3.34}$$

where $F_n\left(\tau, x, \frac{dx}{dt}\right)$ are some regular functions of x and $\frac{dx}{dt}$, differentiable to any order with respect to τ for any finite values of τ. In this case, infinite double series of the form

$$\sum_{\substack{n=-\infty \\ [nq+p(m\pm 1)\neq 0]}}^{\infty} \sum_{m=-\infty}^{\infty} \frac{e^{i\{n\theta+m(p\varphi+\psi)\}}}{\omega^2(\tau)-[m\omega(\tau)+n\nu(\tau)]^2} \times$$

$$\times \int_0^{2\pi}\int_0^{2\pi} F_0(\tau, a, \theta, p\varphi+\psi)\, e^{-i\{n\theta+m(p\varphi+\psi)\}}\, d\theta\, d(p\varphi+\psi), \qquad (3.35)$$

appear in the expressions for $u_1(\tau, a, \theta, p\varphi+\psi)$, $u_2(\tau, a, \theta, p\varphi+\psi), \ldots$

These series, generally speaking, diverge because of the presence of denominators of the form $\omega^2(\tau) - [m\omega(\tau) - n\nu(\tau)]^2$. The points of divergence of such series constitute a dense set of the ν-axis (i. e., those values of ν for which the series (3.35) diverges). It may be shown that the series (3.35) will absolutely converge under certain additional conditions, for example, if it is assumed that $F_0(\tau, a, \theta, p\varphi+\psi)$ has a sufficient number of continuous partial derivatives with respect to the angular variables θ and $p\varphi+\psi$.

It is of practical convenience, however, not to allow infinite sums of harmonic terms in $u_1(\tau, a, \theta, p\varphi+\psi)$, $u_2(\tau, a, \theta, p\varphi+\psi), \ldots$, to appear, referring the remainder of the converging series of $F\left(\tau, \theta, x, \frac{dx}{dt}\right)$ (3.34) to higher orders of ε. It is also convenient to proceed, not from equation (3.1) with its right-hand side representable in the form (3.34), but from an equation of the type

$$\frac{d}{dt}\left[m(\tau)\frac{dx}{dt}\right]+c(\tau)x=\varepsilon F^{(0)}\left(\tau, \theta, x, \frac{dx}{dt}\right)+\varepsilon^2 F^{(1)}\left(\tau, \theta, x, \frac{dx}{dt}\right)+\ldots, \qquad (3.36)$$

where $F^{(0)}\left(\tau, \theta, x, \frac{dx}{dt}\right)$, $F^{(1)}\left(\tau, \theta, x, \frac{dx}{dt}\right)$, ... are finite sums of the form (3.2).

As mentioned before, no additional difficulties arise when extending the method for constructing approximate solutions to equation (3.36).

We shall not discuss in detail the theoretical aspect of this operation nor the estimation of the error resulting from cutting off the series (3.34). We only remark that this series may obviously be formally represented by the expression in the right-hand side of equation (3.36). In fact, since the series (3.34) converges uniformly, one may always find such n_1, n_2, $n_3, \ldots$, for which the following inequalities are satisfied

$$\left|\sum_{n=n_s}^{n_{s+1}} e^{in\theta} F_n\left(\tau, x, \frac{dx}{dt}\right)\right| < \varepsilon^s C^{(s)} \qquad (s=1, 2, 3, \ldots),$$

where $C^{(s)}$ $(s=1, 2, 3, \ldots)$ are constants satisfying, in the domain of the variables $\tau, \theta, x, \frac{dx}{dt}$, the inequalities

$$\max\left|F^{(s)}\left(\tau, \theta, x, \frac{dx}{dt}\right)\right| \leqslant C^{(s)}, \quad (s=1, 2, 3, \ldots).$$

Consequently, it is possible to choose finite sums $F^{(s)}\left(\tau, \theta, x, \frac{dx}{dt}\right)$ $(s=0, 1, 2, 3, \ldots)$ of the type (3.2) and introduce a small parameter ε, so that the series (3.34) is represented in the form of the right-hand side of (3.36).

We illustrate this by a very simple example. Consider the equation

$$\frac{d^2x}{dt^2}+x=\frac{x^3}{3!}-\frac{x^5}{5!}+\frac{x^7}{7!}-\ldots$$

Substituting $x=\sqrt{\varepsilon}x_1$ in this equation, we may write it in the form

$$\frac{d^2x_1}{dt^2}+x_1=\varepsilon\frac{x_1^3}{3!}-\varepsilon^2\frac{x^5}{5!}+\varepsilon^3\frac{x^7}{7!}-\ldots$$

We shall now consider one other method for deriving the first, or directly the second approximation for equation (3.1), based on the method of harmonic balance. This method is well known in nonlinear mechanics, and was applied in the preceding chapter when constructing approximate solutions for equation (2.1). By the formal method presented below we can derive solutions, especially in the first approximation; this generally answers practical demands.

In order to obtain the first approximation, we examine by the method of harmonic balance (the formalism of this method follows quite naturally from the above method of deducing approximate solutions) the principal harmonic of the solution

$$x=a\cos(p\varphi+\psi), \tag{3.37}$$

representing (for a and ψ satisfying the equations of the first approximation) the solution of equation (3.1) in the first approximation.

When (3.37) is substituted in equation (3.1) and account is taken of the system of equations (3.31), where it is assumed that $\tau=\varepsilon t$, the principal harmonics on both sides of equation (3.1) must be equal.

In order to arrive at the second approximation we must naturally take into account terms with ε^2 when determining the principal harmonic in the left-hand side of equation (3.1), and terms with $\varepsilon u_1(\tau, a, \theta, p\varphi+\psi)$ in the expression for $F\left(\tau, \theta, x, \frac{dx}{dt}\right)$, i.e., we must substitute the expression

$$x=a\cos(p\varphi+\psi)+\varepsilon u_1(\tau, a, \theta, p\varphi+\psi) \tag{3.38}$$

where $\tau=\varepsilon t$, and a and ψ satisfy equations (3.33).

Thus in the second approximation we immediately obtain for the principal harmonic of the left-hand side of equation (3.1):

$$\begin{aligned}
&\textit{principal harmonic of } \left\{\frac{d}{dt}\left[m(\tau)\frac{dx}{dt}\right]+c(\tau)x\right\}=\\
&=\varepsilon\left\{m(\tau)\left[\left[\omega(\tau)-\frac{p}{q}\nu(\tau)\right]\frac{\partial A_1}{\partial\psi}-2a\omega(\tau)B_1\right]\cos(p\varphi+\psi)-\right.\\
&-m(\tau)\left[\left[\omega(\tau)-\frac{p}{q}\nu(\tau)\right]a\frac{\partial B_1}{\partial\psi}+2\omega(\tau)A_1-\right.\\
&\left.\left.-\frac{d[m(\tau)\omega(\tau)]}{d\tau}\frac{a}{m(\tau)}\right]\sin(p\varphi+\psi)\right\}+\varepsilon^2\left\{m(\tau)\left[\left[\omega(\tau)-\frac{p}{q}\nu(\tau)\right]\frac{\partial A_2}{\partial\psi}-\right.\right.\\
&\left.-2a\omega(\tau)B_2+\frac{\partial A_1}{\partial a}A_1+\frac{\partial A_1}{\partial\psi}B_1+\frac{\partial A_1}{\partial\tau}-aB_1^2+\frac{dm(\tau)}{d\tau}\frac{A_1}{m(\tau)}\right]\cos(p\varphi+\psi)-\\
&-m(\tau)\left[\left[\omega(\tau)-\frac{p}{q}\nu(\tau)\right]a\frac{\partial B_2}{\partial\psi}+2\omega(\tau)A_2+\frac{\partial B_1}{\partial a}aA_1+\right.\\
&\left.\left.+\frac{\partial B_1}{\partial\psi}aB_1+\frac{\partial B_1}{\partial\tau}+2A_1B_1+\frac{dm(\tau)}{d\tau}\frac{aB_1}{m(\tau)}\right]\sin(p\varphi+\psi)\right\}.
\end{aligned} \tag{3.39}$$

Putting (3.38) in the right-hand side of equation (3.1) and taking account of (3.33), we obtain the following expression for the principal harmonic with an accuracy of the second order:

$$\textit{principal harmonic of } \left\{\varepsilon F\left(\tau,\ \theta,\ x,\ \frac{dx}{dt}\right)\right\} =$$

$$= \varepsilon\left\{\cos(p\varphi+\psi)\frac{1}{2\pi^2}\sum_{\sigma} e^{i\sigma q\psi}\int_0^{2\pi}\int_0^{2\pi} F_0(\tau,\ a,\ \theta,\ \vartheta)\, e^{-i\sigma q\psi_1}\cos\vartheta\, d\theta\, d\vartheta +\right.$$

$$\left. + \sin(p\varphi+\psi)\frac{1}{2\pi^2}\sum_{\sigma} e^{i\sigma q\psi}\int_0^{2\pi}\int_0^{2\pi} F_0(\tau,\ a,\ \theta,\ \vartheta)\, e^{-i\sigma q\psi_1}\sin\vartheta\, d\theta\, d\vartheta\right\} +$$

$$+ \varepsilon^2\left\{\cos(p\varphi+\psi)\frac{1}{2\pi^2}\sum_{\sigma} e^{i\sigma q\psi}\int_0^{2\pi}\int_0^{2\pi} F_1(\tau,\ a,\ \theta,\ \vartheta)\, e^{-i\sigma q\psi_1}\cos\vartheta\, d\theta\, d\vartheta +\right.$$

$$\left. + \sin(p\varphi+\psi)\frac{1}{2\pi^2}\sum_{\sigma} e^{i\sigma q\psi}\int_0^{2\pi}\int_0^{2\pi} F_1(\tau,\ a,\ \theta,\ \vartheta)\, e^{-i\sigma q\psi_1}\sin\vartheta\, d\theta\, d\vartheta\right\}, \qquad (3.40)$$

where $\psi_1 = \psi - p\varphi$; $\vartheta = p\varphi + \psi$; and $F_0(\tau, a, \theta, \vartheta)$ and $F_1(\tau, a, \theta, \vartheta)$ are defined by (3.15) and (3.16).

Equating the coefficients of the same harmonics in the expressions (3.39) and (3.40), we obtain in the first approximation equations (3.27), and in the second approximation equations (3.29); this may be verified easily.

The expression for $\varepsilon u_1(\tau, a, \theta, p\varphi + \psi)$ is determined as a forced vibration, excited in x by the action of higher harmonics of the external force $\varepsilon F\left(\tau, \theta, x, \frac{dx}{dt}\right)$ in the mode of harmonic vibrations ($x = a\cos(p\varphi + \psi)$, $\frac{dx}{dt} = -a\omega(\tau)\sin(p\varphi + \psi)$); it coincides with the previously derived expression (3.24).

The formalism of the method of harmonic balance may be written in a compact form by means of the equations of harmonic balance of the form

$$\left.\begin{aligned} &\int_0^{2\pi q}\left\{\frac{d}{dt}\left[m(\tau)\frac{dx}{dt}\right] + c(\tau)x - \right.\\ &\qquad\left. - \varepsilon F\left(\tau, \theta, x, \frac{dx}{dt}\right)\right\}_{x=a\cos(p\varphi+\psi)+\varepsilon u_1\ldots}\cos(p\varphi+\psi)\,d\theta = 0,\\ &\int_0^{2\pi q}\left\{\frac{d}{dt}\left[m(\tau)\frac{dx}{dt}\right] + c(\tau)x - \right.\\ &\qquad\left. - \varepsilon F\left(\tau, \theta, x, \frac{dx}{dt}\right)\right\}_{x=a\cos(p\varphi+\psi)+\varepsilon u_1\ldots}\sin(p\varphi+\psi)\,d\theta = 0. \end{aligned}\right\} \qquad (3.41)$$

Substituting the values of x, $\frac{dx}{dt}$, $\frac{d^2x}{dt^2}$ found from (3.30) with an accuracy of the first order in the integrands, bearing in mind that a and ψ are functions of time and that $\tau = \varepsilon t$, and integrating (the integration is over θ while a, ψ and τ are kept constant), we obtain equations (3.27) determining the functions $A_1(\tau, a, \psi)$ and $B_1(\tau, a, \psi)$.

Taking account of terms proportional to ε^2 in substituting the values of x, $\frac{dx}{dt}$, $\frac{d^2x}{dt^2}$ found from (3.32) (a and ψ must be considered as functions of time satisfying equations (3.33)), we obtain equations (3.29) determining $A_2(\tau, a, \psi)$ and $B_2(\tau, a, \psi)$.

We now proceed to examine some particular cases of equation (3.1) and illustrate at the same time the effectiveness of this formal method for deriving approximate solutions.

§ 2. Particular cases of equation (3.1)

As a first example of a particular case of equation (3.1) we investigate the differential equations of a nonlinear vibrator under the action of a "sinusoidal" force with slowly varying amplitude and instantaneous frequency. In this case we have the following differential equation

$$m\frac{d^2x}{dt^2}+cx=\varepsilon F\left(x, \frac{dx}{dt}\right)+\varepsilon E(\tau)\sin\theta, \tag{3.42}$$

where $\frac{d\theta}{dt}=\nu(\tau)$; $\tau=\varepsilon t$; and m and c are constants.

Vibrating systems described by an equation of this type are frequently met in various problems of physics and engineering.

It is easily verified that in the first approximation only fundamental resonance may appear in systems described by an equation of this type. Consequently, we shall now construct the equations of the first approximation for the case of fundamental resonance $p = 1$, $q = 1$. These equations will be derived by the method of harmonic balance just developed.

The solution of equation (3.42) for the case of fundamental resonance is in the first approximation looked for in the form

$$x=a\cos(\theta+\psi), \tag{3.43}$$

where a and ψ should be determined from the equations of the first approximation

$$\left.\begin{aligned}\frac{da}{dt}&=\varepsilon A_1(\tau, a, \psi),\\ \frac{d\psi}{dt}&=\omega-\nu(\tau)+\varepsilon B_1(\tau, a, \psi),\end{aligned}\right\} \tag{3.44}$$

in which $\omega=\sqrt{\frac{c}{m}}$.

For determining $A_1(\tau, a, \psi)$ and $B_1(\tau, a, \psi)$, we find (up to terms of the order of ε)

$$\begin{aligned}\frac{dx}{dt}&=\frac{da}{dt}\cos(\theta+\psi)-a\sin(\theta+\psi)\frac{d(\theta+\psi)}{dt}=\\ &=\varepsilon A_1\cos(\theta+\psi)-a\omega\sin(\theta+\psi)-\varepsilon aB_1\sin(\theta+\psi),\end{aligned} \tag{3.45}$$

$$\begin{aligned}\frac{d^2x}{dt^2}&=\varepsilon\left\{[\omega-\nu(\tau)]\frac{\partial A_1}{\partial\psi}-2a\omega B_1\right\}\cos(\theta+\psi)-a\omega^2\cos(\theta+\psi)-\\ &-\varepsilon\left\{[\omega-\nu(\tau)]\,a\frac{\partial B_1}{\partial\psi}+2\omega A_1\right\}\sin(\theta+\psi).\end{aligned} \tag{3.46}$$

Substituting (3.43) and (3.46) in the left-hand side of equation (3.42), we obtain

$$\begin{aligned}&\textit{principal harmonic of}\left\{m\frac{d^2x}{dt^2}+cx\right\}_{x=a\cos(\theta+\psi)}=\\ &=\varepsilon m\left\{[\omega-\nu(\tau)]\frac{\partial A_1}{\partial\psi}-2a\omega B_1\right\}\cos(\theta+\psi)-\\ &-\varepsilon m\left\{[\omega-\nu(\tau)]\,a\frac{\partial B_1}{\partial\psi}+2\omega A_1\right\}\sin(\theta+\psi).\end{aligned} \tag{3.47}$$

The principal harmonic of the angle $(\theta+\psi)$ for the right-hand side of equation (3.42) may, with the help of (3.43) and (3.45), be represented with an accuracy of the order of ε, in the form

$$\textit{principal harmonic of } \left\{\varepsilon F\left(x, \frac{dx}{dt}\right)+\varepsilon E(\tau)\sin\theta\right\}_{x=a\cos(\theta+\psi)}=$$

$$=\frac{\varepsilon\cos(\theta+\psi)}{\pi}\int_0^{2\pi} F[a\cos(\theta+\psi), -a\omega\sin(\theta+\psi)]\cos(\theta+\psi)\,d(\theta+\psi)+$$

$$+\frac{\varepsilon\sin(\theta+\psi)}{\pi}\int_0^{2\pi} F[a\cos(\theta+\psi), -a\omega\sin(\theta+\psi)]\sin(\theta+\psi)\,d(\theta+\psi)+$$

$$+\varepsilon E(\tau)[\cos\psi\cdot\sin(\theta+\psi)-\sin\psi\cos(\theta+\psi)]. \qquad (3.48)$$

Equating coefficients of the same powers of ε and of the same harmonics (correspondingly for sines and cosines) in the right-hand sides of the expressions (3.47) and (3.48), we find the following system of equations for $A_1(\tau, a, \psi)$ and $B_1(\tau, a, \psi)$:

$$\left.\begin{aligned} m\left\{[\omega-\nu(\tau)]\frac{\partial A_1}{\partial\psi}-2a\omega B_1\right\}&=\frac{1}{\pi}\int_0^{2\pi} F_0(a, \vartheta)\cos\vartheta\,d\vartheta-E(\tau)\sin\psi,\\ m\left\{[\omega-\nu(\tau)]\,a\frac{\partial B_1}{\partial\psi}+2\omega A_1\right\}&=-\frac{1}{\pi}\int_0^{2\pi} F_0(a, \vartheta)\sin\vartheta\,d\vartheta-E(\tau)\cos\psi,\end{aligned}\right\} \qquad (3.49)$$

from which we easily obtain with the desired accuracy

$$\left.\begin{aligned} A_1(\tau, a, \psi)&=-\frac{1}{2\pi m\omega}\int_0^{2\pi} F_0(a, \vartheta)\sin\vartheta\,d\vartheta-\frac{E(\tau)}{m[\omega+\nu(\tau)]}\cos\psi,\\ B_1(\tau, a, \psi)&=-\frac{1}{2\pi m\omega a}\int_0^{2\pi} F_0(a, \vartheta)\cos\vartheta\,d\vartheta+\frac{E(\tau)}{ma[\omega+\nu(\tau)]}\sin\psi,\end{aligned}\right\} \qquad (3.50)$$

where $F_0(a, \vartheta)=F(a\cos\vartheta, -a\omega\sin\vartheta)$.

The equations of the first approximation now become

$$\left.\begin{aligned} \frac{da}{dt}&=-\frac{\varepsilon}{2\pi m\omega}\int_0^{2\pi} F_0(a, \vartheta)\sin\vartheta\,d\vartheta-\frac{\varepsilon E(\tau)}{m[\omega-\nu(\tau)]}\cos\psi,\\ \frac{d\psi}{dt}&=\omega-\nu(\tau)-\frac{\varepsilon}{2\pi m\omega a}\int_0^{2\pi} F_0(a, \vartheta)\cos\vartheta\,d\vartheta+\frac{\varepsilon E(\tau)}{ma[\omega+\nu(\tau)]}\sin\psi.\end{aligned}\right\} \qquad (3.51)$$

We consider the equivalent damping decrement $\delta_e(a)$ and the equivalent frequency $\omega_e(a)$ for a nonlinear vibrating system described by the equation

$$m\frac{d^2x}{dt^2}+cx=\varepsilon F\left(x, \frac{dx}{dt}\right). \qquad (3.52)$$

The equivalent damping decrement and frequency are, as known, defined by the formulas

$$\left.\begin{aligned} \delta_e(a)&=\frac{\varepsilon}{2\pi m\omega a}\int_0^{2\pi} F_0(a, \vartheta)\sin\vartheta\,d\vartheta,\\ \omega_e^2(a)&=\frac{c}{m}-\frac{\varepsilon}{\pi m a}\int_0^{2\pi} F_0(a, \vartheta)\cos\vartheta\,d\vartheta.\end{aligned}\right\} \qquad (3.53)$$

The equations of the first approximation may then be written in the form

$$\left.\begin{aligned} \frac{da}{dt}&=-\delta_e(a)\,a-\frac{\varepsilon E(\tau)}{m[\omega+\nu(\tau)]}\cos\psi,\\ \frac{d\psi}{dt}&=\omega_e(a)-\nu(\tau)+\frac{\varepsilon E(\tau)}{ma[\omega+\nu(\tau)]}\sin\psi.\end{aligned}\right\} \qquad (3.54)$$

$\varepsilon u_1(\tau, a, \theta, p\varphi+\psi)$ may be determined either from (3.24) or directly as forced vibrations excited in the system under investigation by higher harmonics of the external force $\varepsilon F\left(x, \frac{dx}{dt}\right)$ in the mode of sinusoidal vibrations, i. e., by the sum

$$\varepsilon \sum_{n\neq 1} \{F_n^{(1)}(a)\cos n(\theta+\psi)+F_n^{(2)}(a)\sin n(\theta+\psi)\}, \tag{3.55}$$

where

$$\left.\begin{aligned} F_n^{(1)}(a) &= \frac{1}{\pi}\int_0^{2\pi} F_0(a, \vartheta)\cos n\vartheta\, d\vartheta, \\ F_n^{(2)}(a) &= \frac{1}{\pi}\int_0^{2\pi} F_0(a, \vartheta)\sin n\vartheta\, d\vartheta. \end{aligned}\right\} \tag{3.56}$$

The explicit expression of $\varepsilon u_1(\tau, a, \theta, p\varphi+\psi)$ will be

$$\varepsilon u_1(\tau, a, \theta, p\varphi+\psi) = \frac{1}{\pi\omega^2}\sum_{n\neq 1}\frac{1}{1-n^2}\left[\cos n(\theta+\psi)\int_0^{2\pi} F_0(a, \vartheta)\cos n\vartheta\, d\vartheta + \right.$$

$$\left. + \sin n(\theta+\psi)\int_0^{2\pi} F_0(a, \vartheta)\sin n\vartheta\, d\vartheta\right]. \tag{3.57}$$

For the second approximation it is necessary to find $A_2(\tau, a, \psi)$ and $B_2(\tau, a, \psi)$; this is again conveniently done by the method of harmonic balance. However, we shall not consider this problem here.

As a second particular case we consider the Mathieu equation, extensively applied in engineering, in the case when the modulation frequency varies slowly with time. We have the equation

$$\frac{d^2x}{dt^2}+\omega^2(1-h\cos\theta)x=0, \tag{3.58}$$

where $\frac{d\theta}{dt}=\nu(\tau)$; $\tau=\varepsilon t$; and ω and h are constants. It is also assumed that $h \ll 1$. Then, putting $h=\varepsilon h_1$, we may represent equation (3.58) in the form

$$\frac{d^2x}{dt^2}+\omega^2 x=\varepsilon\omega^2 h_1 x\cos\theta. \tag{3.59}$$

It is known that for equation (3.59) a demultiplicative resonance may be considered even in the first approximation. We therefore construct an asymptotic solution in the first approximation for the case $p = 1, q = 2$, applying, as before, the method of harmonic balance.

Thus, we are looking for a solution of (3.59) in the form

$$x=a\cos\left(\frac{1}{2}\theta+\psi\right), \tag{3.60}$$

where a and ψ should be determined from the system

$$\left.\begin{aligned} \frac{da}{dt} &= \varepsilon A_1(\tau, a, \psi), \\ \frac{d\psi}{dt} &= \omega-\frac{1}{2}\nu(\tau)+\varepsilon B_1(\tau, a, \psi). \end{aligned}\right\} \tag{3.61}$$

Let us determine $A_1(\tau, a, \psi)$ and $B_1(\tau, a, \psi)$. To do this we substitute (3.60) in the left-hand side of equation (3.59) and obtain up to terms of the first order:

$$\textit{principal harmonic of } \left\{\frac{d^2x}{dt^2}+\omega^2 x\right\}_{x=a\cos\left(\frac{1}{2}\theta+\psi\right)} = \varepsilon\left\{\left[\omega-\frac{1}{2}\nu(\tau)\right]\frac{\partial A_1}{\partial\psi}-\right.$$

$$\left.-2a\omega B_1\right\}\cos\left(\frac{1}{2}\theta+\psi\right)-\varepsilon\left\{\left[\omega-\frac{1}{2}\nu(\tau)\right]a\frac{\partial B_1}{\partial\psi}+2\omega A_1\right\}\sin\left(\frac{1}{2}\theta+\psi\right). \quad (3.62)$$

The principal harmonic of the right-hand side of equation (3.59) may be represented in the form

$$\textit{principal harmonic of } \{\varepsilon\omega^2 h_1 x\cos\theta\}_{x=a\cos\left(\frac{1}{2}\theta+\psi\right)} =$$

$$= a\frac{\varepsilon\omega^2 h_1}{2}\cos 2\psi\cos\left(\frac{1}{2}\theta+\psi\right)+a\frac{\varepsilon\omega^2 h_1}{2}\sin 2\psi\sin\left(\frac{1}{2}\theta+\psi\right). \quad (3.63)$$

Equating the coefficients of the same harmonics in the right-hand sides of (3.62) and (3.63), we obtain the system of equations

$$\left.\begin{aligned}\left[\omega-\frac{1}{2}\nu(\tau)\right]\frac{\partial A_1}{\partial\psi}-2a\omega B_1&=\frac{a\omega^2 h_1}{2}\cos 2\psi,\\ \left[\omega-\frac{1}{2}\nu(\tau)\right]a\frac{\partial B_1}{\partial\psi}+2\omega A_1&=-\frac{a\omega^2 h_1}{2}\sin 2\psi,\end{aligned}\right\} \quad (3.64)$$

from which

$$\left.\begin{aligned}A_1(\tau,\ a,\ \psi)&=-\frac{ah_1\omega^2}{2\nu(\tau)}\sin 2\psi,\\ B_1(\tau,\ a,\ \psi)&=-\frac{h_1\omega^2}{2\nu(\tau)}\cos 2\psi.\end{aligned}\right\} \quad (3.65)$$

Consequently the equations of the first approximation for a and ψ are in the case under consideration

$$\left.\begin{aligned}\frac{da}{dt}&=-\frac{ah\omega^2}{2\nu(\tau)}\sin 2\psi,\\ \frac{d\psi}{dt}&=\omega-\frac{\nu(\tau)}{2}-\frac{h\omega^2}{2\nu(\tau)}\cos 2\psi.\end{aligned}\right\} \quad (3.66)$$

§ 3. Stationary modes and their stability in nonlinear vibrating systems

In the study of a vibrating system described by a differential equation with slowly varying coefficients it is frequently desired to compare the observed nonstationary vibrational mode with a stationary mode which is possible in the same system for constant values of its parameters. We shall therefore discuss briefly in the present section some statements and rules of nonlinear mechanics for determining the stationary amplitudes and phases of vibration of a nonlinear vibrating system with one degree of freedom, and the question of stability of the stationary mode.

In (3.1) let τ be a constant parameter: $\tau=\text{const}$. The equation then degenerates to a nonlinear differential equation of the form

$$m\frac{d^2x}{dt^2}+cx=\varepsilon F\left(\nu t,\ x,\ \frac{dx}{dt}\right), \quad (3.67)$$

where m and c are positive constants, and $F\left(\nu t,\ x,\ \frac{dx}{dt}\right)$ is a periodic function of t with period $\frac{2\pi}{\nu}$.

Equation (3.67) is thoroughly studied in /9/, and its asymptotic solution in the first approximation has the form

$$x = a\cos\left(\frac{p}{q}\nu t + \psi\right), \tag{3.68}$$

where a and ψ should be determined from the system of equations of the first approximation

$$\left.\begin{aligned} \frac{da}{dt} &= \varepsilon A_1(a, \psi), \\ \frac{d\psi}{dt} &= \omega - \frac{p}{q}\nu + \varepsilon B_1(a, \psi), \end{aligned}\right\} \tag{3.69}$$

The expressions of $A_1(a, \psi)$ and $B_1(a, \psi)$ may be obtained from formulas (3.28), by putting τ = const.

The following states of a vibrating system are referred to in /9/ as stationary modes: the first type of a stationary mode is the state of the system for which a and ψ are constants determined by the equations

$$A_1(a, \psi) = 0, \quad \omega - \frac{p}{q}\nu + \varepsilon B_1(a, \psi) = 0; \tag{3.70}$$

the second type is the state of the system for which a and ψ are periodic solutions of the system (3.69).

We now proceed to study the first type of stationary vibrations, when a and ψ are constants determined by equations (3.70). For this type we obtain vibrations with a frequency exactly equal to $\frac{p}{q}\nu$. Such a mode of stationary vibrations is called synchronous since the frequency of vibrations is in simple rational proportion to the frequency of excitation.

We denote a constant solution of the system (3.70) by a_0 and ψ_0, and derive the stability conditions for this stationary synchronous mode. To do this it is necessary to set up the corresponding variational equations.

Inserting $a = a_0 + \delta a$, $\psi = \psi_0 + \delta\psi$ in the equations of the first approximation (3.69), we obtain the following system of variational equations

$$\left.\begin{aligned} \frac{d\delta a}{dt} &= \varepsilon A'_{1a}(a_0, \psi_0)\,\delta a + \varepsilon A'_{1\psi}(a_0, \psi_0)\,\delta\psi_0, \\ \frac{d\delta\psi}{dt} &= \varepsilon B'_{1a}(a_0, \psi_0)\,\delta a + \varepsilon B'_{1\psi}(a_0, \psi_0)\,\delta\psi_0. \end{aligned}\right\} \tag{3.71}$$

The characteristic equation for this system is

$$\begin{aligned} \lambda^2 - \varepsilon\,[A'_{1a}(a_0, \psi_0) + B'_{1\psi}(a_0, \psi_0)]\,\lambda + \\ + \varepsilon^2\,[A'_{1a}(a_0, \psi_0)\,B'_{1\psi}(a_0, \psi_0) - A'_{1\psi}(a_0, \psi_0)\,B'_{1a}(a_0, \psi_0)] = 0. \end{aligned} \tag{3.72}$$

From equation (3.72) we obtain the following stability conditions for the synchronous stationary vibrations

$$\left.\begin{aligned} & A'_{1a}(a_0, \psi_0) + B'_{1\psi}(a_0, \psi_0) < 0, \\ & A'_{1a}(a_0, \psi_0)\,B'_{1\psi}(a_0, \psi_0) - A'_{1\psi}(a_0, \psi_0)\,B'_{1a}(a_0, \psi_0) > 0. \end{aligned}\right\} \tag{3.73}$$

Applying these results we now derive a few formulas required when examining stationary modes and their stability in vibrating systems described by the equation

$$m\frac{d^2x}{dt^2} + cx = \varepsilon F\left(x, \frac{dx}{dt}\right) + \varepsilon E\sin\nu t, \tag{3.74}$$

where m, c, E, and ν are constants. This equation is obtained by putting $\tau = \text{const}$ in (3.42).

Equating the right-hand sides of the equations of the first approximation for the equation (3.74) (i. e., for equations (3.54) with $\tau = \text{const}$) to zero, we obtain the system

$$\left.\begin{aligned} \delta_e(a)\,a + \frac{\varepsilon E}{m(\omega+\nu)}\cos\psi = 0, \\ \omega_e(a) - \nu + \frac{\varepsilon E}{am(\omega+\nu)}\sin\psi = 0, \end{aligned}\right\} \tag{3.75}$$

determining the stationary values of the amplitude and phase. This system may be written with an accuracy of the second order of ε in the form

$$\left.\begin{aligned} 2m\nu a\delta_e(a) = -\varepsilon E\cos\psi, \\ ma\,[\omega_e^2(a) - \nu^2] = -\varepsilon E\sin\psi. \end{aligned}\right\} \tag{3.76}$$

Eliminating the phase ψ from these relations, we obtain the relation

$$m^2a^2\{[\omega_e^2(a) - \nu^2]^2 + 4\delta_e^2(a)\,\nu^2\} = \varepsilon^2E^2, \tag{3.77}$$

from which one may deduce the stationary values of the amplitude depending on the frequency (i. e., construct the so-called resonance curve) or on the parameter τ, on which the mass of the system, the elasticity coefficients, the amplitude of external force, and other parameters may depend in some fashion.

For the stationary values of the phases of vibrations we obtain the relation

$$\psi = \operatorname{arc\,tg}\frac{\omega_e(a) - \nu}{\delta_e(a)}. \tag{3.78}$$

Since the relations (3.77) and (3.78) were obtained from equation (3.75), the difference, $\omega_e(a) - \nu$, between the normal frequency of vibrations and the frequency of the external force is a quantity of the first order. Consequently, these relations may be used only when examining the resonance zone and a sufficiently close approach to it, for which the difference $\omega_e(a) - \nu$ is of the order of ε; this in practice is the case of interest.

We shall now derive stability conditions for the stationary values of amplitudes and phases of vibration. For investigating the resonance zone and transitions to such a zone, we write the equations of the first approximation (3.54) (where $\tau = \text{const}$) with an accuracy of the second order in the form

$$\left.\begin{aligned} 2\nu\frac{da}{dt} = -2\nu a\delta_e(a) - \frac{\varepsilon E}{m}\cos\psi, \\ 2\nu a\frac{d\psi}{dt} = [\omega_e^2(a) - \nu^2]\,a + \frac{\varepsilon E}{m}\sin\psi. \end{aligned}\right\} \tag{3.79}$$

The variational equations of the type (3.71) may then be written as

$$\left.\begin{aligned} 2\nu\frac{d\delta a}{dt} = -2\nu\frac{\partial[\delta_e(a)a]}{\partial a}\bigg|_{a=a_0}\delta a + \frac{\varepsilon E}{m}\sin\psi_0\,\delta\psi, \\ 2\nu a_0\frac{d\delta\psi}{dt} = \frac{\partial\{[\omega_e^2(a) - \nu^2]\,a\}}{\partial a}\bigg|_{a=a_0}\delta a + \frac{\varepsilon E}{m}\cos\psi_0\,\delta\psi. \end{aligned}\right\} \tag{3.80}$$

The stability conditions for stationary values of the amplitude and phase of vibration a_0 and ψ_0 (conditions (3.73)) are the inequalities

$$\frac{\varepsilon E}{m}\cos\psi_0 - 2\nu a_0\frac{\partial[\delta_e(a)\,a]}{\partial a}\bigg|_{a=a_0} < 0, \tag{3.81}$$

$$2\nu\frac{\partial[\delta_e(a)\,a]}{\partial a}\bigg|_{a=a_0}\cos\psi_0 + \frac{\partial\{[\omega_e^2(a) - \nu^2]\,a\}}{\partial a}\bigg|_{a=a_0}\sin\psi_0 < 0. \tag{3.82}$$

Using the first equation of the system (3.76), we may write condition (3.81) in the form

$$-2\nu a_0 \frac{\partial [\delta_e(a)\, a]}{\partial a}\bigg|_{a=a_0} - 2\nu a_0 \delta_e(a_0) < 0 \qquad (3.83)$$

or

$$-2\nu \frac{\partial [\delta_e(a)\, a^2]}{\partial a}\bigg|_{a=a_0} < 0.$$

Using (2.41) we have

$$2\nu a_0^2 \delta_e(a_0) = \frac{a_0^2 \lambda_e(a_0)}{m} \nu. \qquad (3.84)$$

Thus the inequality (3.81) is always satisfied for a system under a usual law of friction, since $a^2\lambda_e(a_0)$ increases together with the amplitude.

We shall now examine the inequality (3.82). Let us consider the dependence of the stationary solutions a_0 and ψ_0 on the external force frequency ν.

Differentiating equations (3.76) with respect to ν we obtain

$$\left.\begin{aligned} 2\nu \frac{\partial [\delta_e(a)\, a]}{\partial a} \frac{da}{d\nu} - \frac{\varepsilon E}{m} \sin\psi \frac{d\psi}{d\nu} &= -2a\delta_e(a), \\ \frac{\partial [\omega_e^2(a) - \nu^2]\, a}{\partial a} \frac{da}{d\nu} + \frac{\varepsilon E}{m} \cos\psi \frac{d\psi}{d\nu} &= 2\nu a, \end{aligned}\right\} \qquad (3.85)$$

from which we find

$$\left\{2\nu \frac{\partial [\delta_e(a)\, a]}{\partial a} \cos\psi + \frac{\partial \{[\omega_e^2(a) - \nu^2]\, a\}}{\partial a} \sin\psi\right\} \frac{da}{d\nu} = -2a\delta_e(a)\cos\psi + 2\nu a \sin\psi. \qquad (3.86)$$

The right-hand side of (3.86) may, by virtue of the equations (3.76), be written in the form

$$-2a\delta_e(a)\cos\psi + 2\nu a \sin\psi = -2\nu a^2 \{[\omega_e^2(a) - \nu^2] - 2\delta_e^2(a)\} \frac{m}{\varepsilon E}. \qquad (3.87)$$

Comparing (3.82), (3.86), and (3.87), we obtain the stability conditions for stationary values of the amplitudes and phases of vibration a_0 and ψ_0 in the form

$$\left.\begin{aligned} \frac{da}{d\nu} > 0, &\quad \text{if} \quad \omega_e^2(a) > \nu^2 + 2\delta_e^2(a), \\ \frac{da}{d\nu} < 0, &\quad \text{if} \quad \omega_e^2(a) < \nu^2 + 2\delta_e^2(a) \end{aligned}\right\} \qquad (3.88)$$

or, up to terms of the first order,

$$\left.\begin{aligned} \frac{da}{d\nu} > 0, &\quad \text{if} \quad \omega_e(a) > \nu, \\ \frac{da}{d\nu} < 0, &\quad \text{if} \quad \omega_e(a) < \nu. \end{aligned}\right\} \qquad (3.89)$$

These relations are very convenient when representing graphically the dependence of the amplitude on frequency. We construct together with the resonance curve (representing the relation between amplitude and frequency) also the so-called skeleton curve, which is described by the equation

$$\omega_e(a) = \nu \qquad (3.90)$$

and determines the form of all the resonance curves. In this case, according to conditions (3.89), those sections on the branch of the resonance curve (defined by equation (3.77)) lying to the left of the curve (3.90) for which a increases with ν will be stable (i. e., correspond to stable amplitude values);

those sections on the branch lying to the right of (3.90) for which a decreases with increasing ν will be stable.

We shall frequently have to apply this rule when discussing actual examples for determining the stable and unstable values of stationary vibration amplitudes.

For determining the stable and unstable values of the phases of vibrations, we may derive criteria similar to the inequalities (3.89). To this end we eliminate $\frac{da}{d\nu}$ from the system (3.85), and obtain

$$\frac{\varepsilon E}{m}\left\{2\nu\frac{\partial[\delta_e(a)a]}{\partial a}\cos\psi+\frac{\partial\{[\omega_e^2(a)-\nu^2]a\}}{\partial a}\sin\psi\right\}\frac{d\psi}{d\nu}=$$
$$=4\nu^2 a\frac{\partial[\delta_e(a)a]}{\partial a}+2a\delta_e(a)\frac{\partial\{[\omega_e^2(a)-\nu^2]a\}}{\partial a}. \quad (3.91)$$

Since the second term on the right-hand side is a quantity of the second order, those branches on the curve $\psi=\psi(\nu)$, defined by (3.78) (where we must put $a=a(\nu)$), will correspond to stable values of the phase for which

$$\frac{d\psi}{d\nu}<0, \quad (3.92)$$

and to unstable — those for which

$$\frac{d\psi}{d\nu}>0. \quad (3.93)$$

§ 4. Linear second order equations with slowly varying coefficients

We shall now examine the vibrations of a system with one degree of freedom described by a linear differential equation with slowly varying coefficients of the form

$$\frac{d}{dt}\left[m(\tau)\frac{dx}{dt}\right]+c(\tau)x=\varepsilon\lambda(\tau)\frac{dx}{dt}+\varepsilon E(\tau)\sin\theta, \quad (3.94)$$

where $m(\tau)$, $c(\tau)$, $\lambda(\tau)$, $E(\tau)$, $\frac{d\theta}{dt}=\nu(\tau)$ satisfy the previously indicated conditions. If m, c, and λ are constant (i. e., only the amplitude and frequency of the external force are changing), then this equation is integrable in a closed form although the quadrature is not realized by means of elementary functions. The works of S. F. Feshchenko, N. G. Gorchakov, A. M. Kats and others are devoted to the construction of asymptotic and approximate solutions for equations of this type.

As equation (3.94) is frequently encountered in solving many vibration problems of engineering, it is advisable to derive its solution and compare it with exact solutions available in some particular cases.

Using the results of the preceding section, we construct the solution of equation (3.94) in the first approximation (only the fundamental resonance may occur in linear systems and higher harmonics do not appear in the solution).

Comparing (3.94) with (3.42), and taking account of the formulas (3.43) and (3.51), we obtain

$$x=a\cos(\theta+\psi), \quad (3.95)$$

where a and ψ should be determined from the system of equations

$$\left.\begin{aligned}\frac{da}{dt} &= -\frac{\varepsilon}{2m\omega}\left[\frac{d(m\omega)}{d\tau} - \lambda(\tau)\right]a - \frac{\varepsilon E(\tau)}{m(\omega+\nu)}\cos\psi,\\ \frac{d\psi}{dt} &= \omega - \nu + \frac{\varepsilon E(\tau)}{m(\omega+\nu)\,a}\sin\psi.\end{aligned}\right\} \tag{3.96}$$

This system may be reduced to quadratures. In fact, changing over to the new variables

$$u = a\cos\psi,\quad v = a\sin\psi, \tag{3.97}$$

we obtain in terms of u and v the system of equations

$$\left.\begin{aligned}\frac{du}{dt} &= -\frac{\varepsilon}{2m\omega}\left[\frac{d(m\omega)}{d\tau} - \lambda(\tau)\right]u - (\omega-\nu)v - \frac{\varepsilon E(\tau)}{m(\omega+\nu)},\\ \frac{dv}{dt} &= -\frac{\varepsilon}{2m\omega}\left[\frac{d(m\omega)}{d\tau} - \lambda(\tau)\right]v + (\omega-\nu)u.\end{aligned}\right\} \tag{3.98}$$

Introducing now the complex function

$$z = u + iv, \tag{3.99}$$

instead of the system (3.98) we obtain the inhomogeneous first order linear equation

$$\frac{dz}{dt} + \left\{\frac{\varepsilon}{2m\omega}\left[\frac{d(m\omega)}{d\tau} - \lambda(\tau)\right] - i(\omega-\nu)\right\}z = -\frac{\varepsilon E(\tau)}{m(\omega+\nu)}. \tag{3.100}$$

Integrating this equation, we find

$$z = e^{-\varepsilon T}\left[C - \varepsilon\int_{t_0}^{t}\frac{E(\tau)}{m(\omega+\nu)}e^{\varepsilon T}\,dt\right] \tag{3.101}$$

where

$$T = \int_{t_0}^{t}\left\{\frac{1}{2m\omega}\left[\frac{d(m\omega)}{d\tau} - \lambda(\tau)\right] - i(\omega-\nu)\right\}dt.$$

We take $x = a$, $\frac{dx}{dt} = 0$ for $t = t_0$. Then according to (3.95) and (3.99), $z = 0$ for $t = t_0$; and hence the arbitrary constant C on the right-hand side of (3.101) equals a. Therefore

$$z = e^{-\varepsilon T}\left[a - \varepsilon\int_{t_0}^{t}\frac{E(\tau)}{m(\omega+\nu)}e^{\varepsilon T}\,dt\right]. \tag{3.102}$$

Separating the real and imaginary parts, we find u and v:

$$\begin{aligned}u &= e^{-\int_{t_0}^{t}\frac{1}{2m\omega}\left[\frac{d(m\omega)}{d\tau} - \lambda(\tau)\right]d\tau}\times\\ &\times\left\{a\cos\varphi - \int_{t_0}^{t}\frac{E(\tau)}{m(\omega+\nu)}e^{\int_{t_0}^{\tau}\frac{1}{2m\omega}\left[\frac{d(m\omega)}{d\tau} - \lambda(\tau)\right]d\tau}\cos\left[\varphi - \int_{t_0}^{\tau}(\omega-\nu)\,d\tau\right]d\tau\right\},\end{aligned} \tag{3.103}$$

$$\begin{aligned}v &= e^{-\int_{t_0}^{t}\frac{1}{2m\omega}\left[\frac{d(m\omega)}{d\tau} - \lambda(\tau)\right]d\tau}\times\\ &\times\left\{a\sin\varphi - \int_{t_0}^{t}\frac{E(\tau)}{m(\omega+\nu)}e^{\int_{t_0}^{\tau}\frac{1}{2m\omega}\left[\frac{d(m\omega)}{d\tau} - \lambda(\tau)\right]d\tau}\sin\left[\varphi - \int_{t_0}^{\tau}(\omega-\nu)\,d\tau\right]d\tau\right\},\end{aligned}$$

where

$$\varphi = \int_{t_0}^{t}(\omega-\nu)\,d\tau.$$

Substituting the values of u and v in the right-hand side of (3.95), we obtain the solution of equation (3.94) expressed in quadratures

$$x = u\cos\theta - v\sin\theta. \tag{3.104}$$

The quadratures on the right-hand sides of the formulas (3.103) are not resolvable in the general case, and they should therefore be calculated either by numerical methods or with the help of asymptotic formulas.

Let us now consider the simple case when m, c, and E in equation (3.94) are constant, $\lambda = 0$, and the frequency of the external force $\frac{d\theta}{dt} = \nu(\tau)$ varies according to the law

$$\nu(\tau) = \nu_0(1 - \varepsilon t). \tag{3.105}$$

Then, if $\frac{c}{m} = \nu_0^2$, $\frac{E}{m} = h$, and the vibrating system under investigation passes at $t = 0$ through resonance, and the equation of motion assumes the form

$$\frac{d^2x}{dt^2} + \nu_0^2 x = h\sin\left(\nu_0 t - \frac{\varepsilon\nu_0 t^2}{2}\right). \tag{3.106}$$

This equation is according to preceding formulas solved by

$$x = a\cos(\theta + \psi)$$

or

$$x = u\cos\theta - v\sin\theta, \tag{3.107}$$

where

$$u + iv = z,$$

and z is determined from the equation

$$\frac{dz}{dt} - i[\nu_0 - \nu(\tau)]z = -\frac{h}{\nu_0 + \nu(\tau)}. \tag{3.108}$$

Integrating this equation, we find the following expressions for u and v (up to and including terms of the first order)

$$\left.\begin{aligned} u &= -\frac{h}{2\nu_0}\int_{t_0}^{t}\cos\frac{\varepsilon\nu_0}{2}(t^2 - t_1^2)\,dt_1, \\ v &= -\frac{h}{2\nu_0}\int_{t_0}^{t}\sin\frac{\varepsilon\nu_0}{2}(t^2 - t_1^2)\,dt_1; \end{aligned}\right\} \tag{3.109}$$

where the terms $a\cos\varphi$ and $a\sin\varphi$ have been left out since they damp down in the presence of friction.

Substituting these values of u and v in the right-hand side of (3.107), we obtain

$$x = -\frac{h}{2\nu_0}\left\{\cos\theta\int_{t_0}^{t}\cos\frac{\varepsilon\nu_0}{2}(t^2 - t_1^2)\,dt_1 - \sin\theta\int_{t_0}^{t}\sin\frac{\varepsilon\nu_0}{2}(t^2 - t_1^2)\,dt_1\right\} \tag{3.110}$$

or, taking into account that $\theta = \nu_0\left(t - \frac{\varepsilon t^2}{2}\right)$,

$$x = -\frac{h}{2\nu_0}\int_{t_0}^{t}\cos\left(\nu_0 t - \frac{\varepsilon\nu_0 t_1^2}{2}\right)dt_1. \tag{3.111}$$

The integral on the right-hand side of (3.111) cannot be solved by means of elementary functions, but may be reduced to the Fresnel integrals*, which have been tabulated. Actually, putting $-\frac{\varepsilon\nu_0 t_1^2}{2}=\frac{\pi}{2}\varrho^2$, we obtain

$$x=-\frac{h}{2\nu_0}\sqrt{-\frac{\pi}{\varepsilon\nu_0}}\int\limits_{t_0\sqrt{-\frac{\varepsilon\nu_0}{\pi}}}^{t\sqrt{-\frac{\varepsilon\nu_0}{\pi}}}\cos\left(\frac{\pi}{2}\varrho^2+\nu_0 t\right)d\varrho, \tag{3.112}$$

and consequently, in order to determine x it is sufficient to find from tables the values of the integrals

$$\int\limits_{x_0}^{x}\cos\frac{\pi\varrho^2}{2}\,d\varrho \text{ and } \int\limits_{x_0}^{x}\sin\frac{\pi\varrho^2}{2}\,d\varrho. \tag{3.113}$$

We shall now analyze the solution (3.111).

The system of equations of the first approximation for the amplitude of phase of vibration has in the case of equation (3.106), according to our theory, the form

$$\left.\begin{aligned}\frac{da}{dt}&=-\frac{h}{\nu_0+\nu(\tau)}\cos\psi,\\ \frac{d\psi}{dt}&=\nu_0-\nu(\tau)+\frac{h}{a[\nu_0+\nu(\tau)]}\sin\psi.\end{aligned}\right\} \tag{3.114}$$

This system may be derived by means of the previously expounded method of averaging. To do this we first perform in equation (3.106) a change of variables according to the formulas

$$\left.\begin{aligned}x&=a\cos(\theta+\psi),\\ \frac{dx}{dt}&=-a\nu(\tau)\sin(\theta+\psi).\end{aligned}\right\} \tag{3.115}$$

Considering a and ψ as the new variables, we find for them a system of two first order equations in standard form

$$\left.\begin{aligned}\frac{da}{dt}&=\frac{a}{\nu}(\omega^2-\nu^2)\cos(\theta+\psi)\sin(\theta+\psi)-\\ &\quad-\frac{h}{\nu}[\sin^2\theta\cos\psi+\sin\theta\cos\theta\sin\psi],\\ \frac{d\psi}{dt}&=\frac{1}{\nu}(\omega^2-\nu^2)\cos^2(\theta+\psi)-\\ &\quad-\frac{h}{\nu}[\sin\theta\cos\theta\cos\psi-\sin^2\theta\sin\psi],\end{aligned}\right\} \tag{3.116}$$

which is equivalent to equation (3.106). In order to deduce from this system the equations of the first approximation (3.114) (with an accuracy of the second order), the right-hand sides of equations (3.116) must be averaged over the explicitly involved time. Some of the terms fall out in averaging: the first and third terms on the right-hand side of the first of equations (3.116), the second term on the right-hand side of the second equation, and $\sin^2\theta$ and $\cos^2(\theta+\psi)$ are replaced by their average values in a period of vibrations, i. e., by 1/2.

* The results of calculations do not agree in this case with experimental data because of neglecting dissipative forces. The introduction of friction leads to integrals of the type

$$\int\limits_{u_1(t)}^{\infty} e^{-u^2}\,du,$$

which may be calculated by approximate or asymptotic formulas /79/.

Let us examine what is lost in an exact solution of equation (3.106) or, which is the same, in equations (3.116) upon passing to the solution of the simpler approximate equations (3.114). It will be shown below that such a replacement is associated with neglecting in the solution those terms which for small h and a slow passage through resonance (i. e., for a slow variation of the external force frequency, as stipulated before) are insignificant and damp down rapidly in the presence of friction.

The exact solution of equation (3.106) for the same initial conditions $\left(t = t_0,\, x = a,\, \frac{dx}{dt} = 0\right)$ is known to be of the form

$$x = \frac{h}{\nu_0} \int_{t_0}^{t} \sin \nu_0 (t - t_1) \sin \nu_0 \left(t_1 - \frac{\varepsilon t_1^2}{2}\right) dt_1 \tag{3.117}$$

or

$$x = \frac{h}{2\nu_0} \int_{t_0}^{t} \cos\left(2\nu_0 t_1 - \frac{\varepsilon \nu_0 t_1^2}{2} - \nu_0 t\right) dt_1 - \frac{h}{2\nu_0} \int_{t_0}^{t} \cos\left(\nu_0 t - \frac{\varepsilon \nu_0 t_1^2}{2}\right) dt_1. \tag{3.118}$$

Thus our approximate solution (3.111) coincides with the second term of the exact solution (3.118).

To facilitate the estimation of the integrals appearing in (3.118) we recall that for the same initial conditions (we put for simplicity $t_0 = 0$) the equation

$$\frac{d^2x}{dt^2} + \nu_0^2 x = h \cos \nu_0 t \tag{3.119}$$

is solved by

$$x = \frac{h}{\nu_0} \int_{0}^{t} \sin \nu_0 (t - t_1) \sin \nu_0 t_1 \, dt_1 = \frac{h}{2\nu_0} \left\{ \int_{0}^{t} \cos (2\nu_0 t_1 - \nu_0 t)\, dt_1 - \int_{0}^{t} \cos \nu_0 t \, dt_1 \right\} \tag{3.120}$$

or

$$x = \frac{h}{2\nu_0^2} \sin \nu_0 t - \frac{ht}{2\nu_0} \cos \nu_0 t. \tag{3.121}$$

Here the term $\frac{h}{2\nu_0^2} \sin \nu_0 t$ represents, as known, forced vibrations arising on account of the perturbing force and having the frequency of free vibrations. The term $\frac{ht}{2\nu_0} \cos \nu_0 t$ represents purely forced vibrations. The time in front of the trigonometric function appears there because friction was neglected in this term; this is in fact not possible. The first term in (3.121) damps down rapidly in the presence of even a small friction and there remains only the second term corresponding to a stationary mode.

The first integral in (3.118) may for small h and a slow variation of the external force frequency be approximately considered as a vibration caused by the perturbing force and having the frequency of a free vibration. The second integral may be considered as a purely forced vibration. For $\varepsilon \to 0$ and $t_0 = 0$ we actually have

$$\frac{h}{2\nu_0} \int_{0}^{t} \cos\left(2\nu_0 t_1 - \frac{\varepsilon \nu_0 t_1^2}{2} - \nu_0 t\right) dt_1 \to \frac{h}{2\nu_0^2} \sin \nu_0 t, \tag{3.122}$$

$$-\frac{h}{2\nu_0} \int_{0}^{t} \cos\left(\nu_0 t - \frac{\varepsilon \nu_0 t_1^2}{2}\right) dt_1 \to \frac{ht}{2\nu_0} \cos \nu_0 t. \tag{3.123}$$

Thus the integral (3.111) representing our approximate solution is the main term in the exact solution on which the intensive increase of the vibration amplitude during a passage through resonance (especially for a slow passage) predominantly depends. The integral (3.122) is for a sufficiently slow passage bounded from above by the constant $\frac{h}{2\nu_0^2}$ and attenuates in the exact solution (3.118) due to the presence of friction which cannot be ignored.

§ 5. Numerical integration of the systems of equations of the first, second, etc. approximations

It was already mentioned before that the system of equations of the first approximation (3.31) (and also of higher approximations) can in the general case not be integrated in a closed form and have to be integrated numerically in the time interval (t_0, t_1) under consideration; this gives the dependence of a and ψ on time. The functions $a = a(t)$, $\psi = \psi(t)$ may then be plotted graphically for the sake of clarity.

In order to facilitate the application of the methods presented above to the solution of actual problems and their reduction to numerical results, we give below one of the methods of numerical integration of systems of differential equations.

The numerical integration of the system (3.31) or (3.33) is conveniently accomplished by means of A. N. Krylov's method. Denoting the right-hand sides of equations (3.31) or (3.33) correspondingly by $f_1(\tau, a, \psi)$ and $f_2(\tau, a, \psi)$, we obtain the system

$$\left.\begin{aligned} \frac{da}{dt} &= f_1(\tau, a, \psi), \\ \frac{d\psi}{dt} &= f_2(\tau, a, \psi). \end{aligned}\right\} \tag{3.124}$$

Let the initial conditions be given as $a(t_0) = a_0$, $\psi(t_0) = \psi_0$. We find by means of a Taylor series expansion of the functions $a = a(t)$ and $\psi = \psi(t)$ in the neighborhood of $t = t_0$ three other pairs of values $a_1, \psi_1, a_2, \psi_2, a_3, \psi_3$ for three moments of time taken at equal intervals.

We now simultaneously construct four tables, the first two of which contain consecutive values of a and ψ and their first differences (Tables 1 and 2), and the other two (Tables 3 and 4) the consecutive values of the quantities

$$\xi_i = f_1(\tau_i, a_i, \psi_i)\Delta t, \quad \eta_i = f_2(\tau_i, a_i, \psi_i)\Delta t \qquad (i = 0, 1, 2, 3) \tag{3.125}$$

and their first, second, third, etc. differences (in practice it is fully sufficient to confine oneself to the third order).

The differences are for convenience written in the intermediate lines between the two numbers for which they are calculated according to the formulas

$$\left.\begin{gathered} \Delta\xi_i = \xi_{i+1} - \xi_i, \quad \Delta^2\xi_i = \Delta\xi_{i+1} - \Delta\xi_i, \quad \Delta^3\xi_i = \Delta^2\xi_{i+1} - \Delta^2\xi_i \\ (i = 0, 1, 2, 3, \ldots) \end{gathered}\right\} \tag{3.126}$$

and analogously

$$\left.\begin{gathered} \Delta\eta_i = \eta_{i+1} - \eta_i, \quad \Delta^2\eta_i = \Delta\eta_{i+1} - \Delta\eta_i, \quad \Delta^3\eta_i = \Delta^2\eta_{i+1} - \Delta^2\eta_i \\ (i = 0, 1, 2, 3, \ldots). \end{gathered}\right\} \tag{3.127}$$

Having constructed these tables we proceed to the numerical integration of the system (3.125), which consists in adding at the bottom of each table one oblique row at a time. To do this we derive Δa_n and $\Delta\psi_n$ from the formulas

$$\left.\begin{aligned}\Delta a_n &= \xi_n + \frac{1}{2}\Delta\xi_{n-1} + \frac{5}{12}\Delta^2\xi_{n-2} + \frac{3}{8}\Delta^3\xi_{n-3},\\ \Delta\psi_n &= \eta_n + \frac{1}{2}\Delta\eta_{n-1} + \frac{5}{12}\Delta^2\eta_{n-2} + \frac{3}{8}\Delta^3\eta_{n-3}\end{aligned}\right\} \tag{3.128}$$

and insert them correspondingly in the first and second tables. Adding these values to a_n and ψ_n, we find

$$a_{n+1} = a_n + \Delta a_n, \quad \psi_{n+1} = \psi_n + \Delta\psi_n. \tag{3.129}$$

It is now possible to derive ξ_{n+1} and η_{n+1} by the formulas (3.125), and, using formulas (3.126), (3.127), to add one oblique row in Tables 3 and 4.

We can now again continue Tables 1 and 2, etc.

TABLE 1

n	a	Δa
0	a_0	
		Δa_0
1	a_1	
		Δa_1
2	a_2	
		Δa_2
3	a_3	

TABLE 2

n	ψ	$\Delta\psi$
0	ψ_0	
		$\Delta\psi_0$
1	ψ_1	
		$\Delta\psi_1$
2	ψ_2	
		$\Delta\psi_2$
3	ψ_3	

TABLE 3

n	ξ	$\Delta\xi$	$\Delta^2\xi$	$\Delta^3\xi$
0	ξ_0			
		$\Delta\xi_0$		
1	ξ_1		$\Delta^2\xi_0$	
		$\Delta\xi_1$		$\Delta^3\xi_0$
2	ξ_2		$\Delta^2\xi_1$	
		$\Delta\xi_2$		
3	ξ_3			

TABLE 4

n	η	$\Delta\eta$	$\Delta^2\eta$	$\Delta^3\eta$
0	η_0			
		$\Delta\eta_0$		
1	η_1		$\Delta^2\eta_0$	
		$\Delta\eta_1$		$\Delta^3\eta_0$
2	η_2		$\Delta^2\eta_1$	
		$\Delta\eta_2$		
3	η_3			

The technique of numerical integration of systems of equations of the type (3.125) will be illustrated below by a numerical example. We shall now consider certain remarks which considerably reduce the scale of calculations.

The system of equations (3.125) is usually to be integrated not on the entire time interval, but only on such an interval on which a and ψ may differ considerably from the values obtained for them by some other less precise method of approximation. Thus, for example, in order to obtain the complete behavior of a vibrating system during a transition through a resonance (this problem will be considered in detail below) it is sufficient to integrate the system of equations (3.125) starting from that time for which the frequency of the external force is sufficiently close to a normal frequency of the system but is still outside the resonance zone.

Practice shows that the curves of a transition through resonance only differ slightly from stationary resonance curves even for a quite large variation of the external force frequency, if the stationary resonance curve is for these frequency values close to a horizontal line. The curves of transition through resonance may, therefore, be replaced outside the resonance zone by the corresponding sections of the stationary resonance curve. The numerical integration of the system of equations (3.125) should be started by taking as the initial values of a and ψ, values corresponding to a stationary mode close to the resonance zone, but at the same time not in a zone of rapidly increasing amplitude.

§ 6. Forced vibrations of a nonlinear vibrator during a transition through resonance

We shall now illustrate the theory developed in the preceding sections by the simplest example of a transition through resonance of a nonlinear vibrator, noticing at the same time that the example considered below reveals quite new phenomena arising in a transition through resonance, characteristic of nonlinear vibrating systems. These phenomena do not occur in linear vibrating systems and could not have been observed until the present method was elaborated.

The theory of nonstationary vibrating processes occurring in the presence of a perturbation with variable frequency and amplitude (connected mainly with phenomena arising in the transition through resonance) is, as known, a relatively new branch of the theory of vibrations. Following the first works in this field (T. Poschl /138/ and F. Lewis /78/) which appeared more than thirty years ago, the study of problems of nonstationary vibrations was confined to linear vibrating systems with one degree of freedom; it was also assumed that the perturbation frequency varies linearly. This is the nature of the works of A. M. Kats /60/, N. G. Gorchakov /35/, N. O. Bykova /20/, G. Hok /54/, N. Barber and F. Ursell /4/, Hasselgruber and Schwinges /53/, A. A. Smelkov /152/, A. P. Filippov /166/, and others. Their schematization is due to the considerable mathematical complexity of the problems of nonstationary vibrating processes and the complete absence of an appropriate mathematical apparatus, convenient for solving problems even in a few particular cases that are of practical importance. In considering the problems of a transition through resonance one usually aimed at a schematization of the vibrating system in question; this simplified the solution of the problem but distorted the real picture of the nonstationary vibrational process. Thus, for example, the neglect of even a small amount of friction in a transition through resonance greatly facilitates the calculations (since it is then possible to express the solutions by means of tabulated functions, like the Fresnel integrals for example), but it changes the picture of the vibrational process considerably.

In view of the mathematical complexity of the problem, use was sometimes made of experimental investigations of nonstationary vibrations, which gave a more correct description of the features of the vibrating process. Nevertheless, it is impossible to detect and explain by means of an experiment all the phenomena occurring in various nonlinear vibrating systems in nonstationary modes, still less to be able to predict new phenomena. Among the experimental works devoted to the problem of transition through resonance we mention the fundamental studies of K. T. Shatalov /172/, which will be discussed below in greater detail.

New interesting phenomena occurring in a nonstationary process (transition through resonance) in nonlinear vibrating systems with a single as well as with many degrees of freedom, close to linear ones, were for the first time revealed (and some phenomena observed experimentally described) by means of the presented asymptotic method /100, 101, 103, 104, 108/. In the present section we quote the characteristic results obtained for a nonlinear vibrating system with one degree of freedom, and further give a few examples characterizing nonstationary processes in nonlinear vibrating systems.

Consider the transition through resonance in a nonlinear vibrator with a rigid characteristic of a nonlinear restoring force, a small friction proportional to the first power of the velocity, and subjected to the action of an external "sinusoidal" force of constant amplitude and variable frequency. Let this system be described by the following differential equation

$$m\frac{d^2x}{dt^2}+2n\frac{dx}{dt}+cx+dx^3=E\sin\theta, \tag{3.130}$$

where x is a coordinate defining the position of the system; t, the time; m, the mass of the system; n, the coefficient of resistivity; $F(x)=cx+dx^3$, the nonlinear restoring force; E, is the amplitude of the perturbing force, $\theta(t)$, a certain function of time; $\frac{d\theta}{dt}=\nu(\tau)$, the instantaneous frequency of the perturbing force, depending on the "slowing" time.

We assume that the coefficients m, n, c, d, E are positive constants.

It is convenient for further calculations and also for the exposition of terms proportional to a small parameter, to define a dimensionless coordinate x_1 and time t_1 by the relations

$$x_1=\sqrt{\frac{d}{c}}\,x,\qquad t_1=\sqrt{\frac{c}{m}}\,t. \tag{3.131}$$

Equation (3.130) assumes now the form (the index of the variables x and t is omitted)

$$\frac{d^2x}{dt^2}+2\delta\frac{dx}{dt}+x+x^3=E_1\sin\theta, \tag{3.132}$$

where

$$\delta=\frac{n}{\sqrt{mc}},\quad E_1=\frac{E}{c}\sqrt{\frac{d}{c}}.$$

We assume that the friction in the system is small $(\delta\ll 1)$, and also that the amplitude of the external force and the deviations of the system from equilibrium are small $(E_1\ll 1, x^2\ll 1)$. Equation (3.132) then belongs to the previously considered type (3.42), and it is possible to construct its approximate solution by means of previously derived formulas. Comparing equation (3.132) with (3.42), we have

$$\varepsilon f\left(\tau,\ x,\ \frac{dx}{dt}\right)+\varepsilon E(\tau)\sin\theta=-x^3-2\delta\frac{dx}{dt}+E_1\sin\theta. \tag{3.133}$$

Using formulas (3.43) and (3.51), we obtain the solution of equation (3.132) in the first approximation in the form

$$x=a\cos(\theta+\psi), \tag{3.134}$$

where a and ψ should be determined from the system of differential equations of the first approximation

$$
\left.
\begin{aligned}
\frac{da}{dt} &= -\delta a - \frac{E_1}{1+\nu(\tau)}\cos\psi,\\
\frac{d\psi}{dt} &= 1-\nu(\tau)+\frac{3a^2}{8}+\frac{E_1}{a(1+\nu(\tau))}\sin\psi\\
&\left(\nu(\tau)=\frac{d\theta}{dt},\ \tau=\varepsilon t\right).
\end{aligned}
\right\}
\tag{3.135}
$$

Applying formulas (3.32) and equations (3.33) according to the general method, we obtain in the second approximation, after a series of elementary operations,

$$
x = a\cos(\theta+\psi)+\frac{a^3}{32}\cos 3(\theta+\psi),
\tag{3.136}
$$

where a and ψ should be determined from the system of equations of the second approximation

$$
\left.
\begin{aligned}
\frac{da}{dt} &= -\delta a+\frac{3a^2\delta}{8}-E_1\left[\frac{1}{1+\nu}-\frac{3a^2(7-\nu)}{8(3-\nu)(1+\nu)^2}\right]\cos\psi+\\
&\quad +E_1\left[\frac{1}{(1+\nu)^3}\frac{d\nu}{dt}-\frac{\delta}{(1+\nu)^2}\right]\sin\psi,\\
\frac{d\psi}{dt} &= 1-\nu+\frac{3a^2}{8}-\frac{\delta^2}{2}-\frac{15a^4}{256}+\\
&\quad +\frac{E_1}{a}\left[\frac{1}{1+\nu}-\frac{3a^2(5-3\nu)}{8(3-\nu)(1+\nu)^2}\right]\sin\psi+\frac{E_1}{a}\left[\frac{1}{(1+\nu)^3}\frac{d\nu}{dt}-\frac{\delta}{(1+\nu)^2}\right]\cos\psi.
\end{aligned}
\right\}
\tag{3.137}
$$

Before examining the nonstationary process using equations (3.135) or (3.137) (in the given case — the behavior of the amplitude and phase of vibration during the transition of the external frequency through resonance values), we consider the stationary mode of a vibrating system described by the nonlinear equation (3.132). For this it is necessary, as shown before, to equate the right-hand sides of equations (3.135) or (3.137) to zero and eliminate ψ. One must also put $\frac{d\nu}{dt}=0$ since the frequency of the external force is considered as constant in a stationary mode.

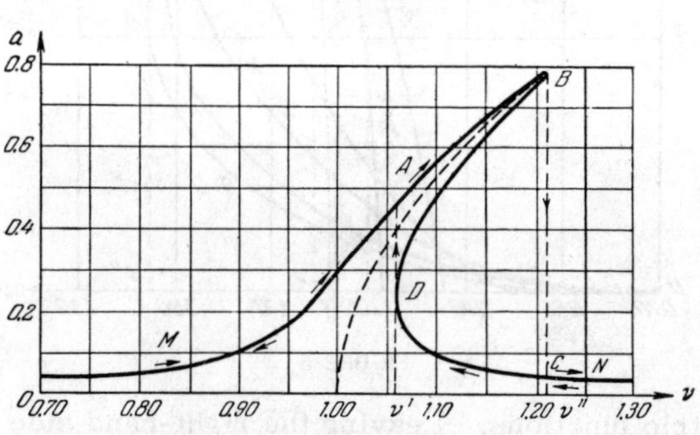

FIGURE 4

After the elimination we obtain in the first approximation the following relation between the amplitude and frequency of the external force

$$
a^2\left\{\left[\left(1+\frac{3a^2}{8}\right)^2-\nu^2\right]^2+4\delta^2\right\}=E_1^2
\tag{3.138}
$$

or

$$
\nu^2=\left(1+\frac{3a^2}{8}\right)^2\pm\sqrt{\frac{E_1^2}{a^2}-4\delta^2}.
\tag{3.139}
$$

Let us now plot the resonance curve defined by (3.139) (Figure 4, where we take $\delta = 0.01$, $E_1 = 0.02$), and also the skeleton curve defined by the relation

$$\omega_e(a) = 1 + \frac{3a^2}{8} \tag{3.140}$$

(the dotted line in Figure 4).

We denote the intersection point of the skeleton curve and the curve of forced amplitudes by B and call it the resonance point (its phase difference is $\frac{\pi}{2}$). It is evident from the figure that to each amplitude value there correspond two frequencies: ν_1 and ν_2, of which $\nu_1 > \omega_e(a)$ and $\nu_2 < \omega_e(a)$. The higher the amplitude of the external sinusoidal force E, the greater the difference between ν_1, ν_2, and ω_e, and this difference decreases with increasing a. To each frequency ν within the interval (ν^I, ν^{II}) there correspond three amplitude values, consequently, three different stationary modes of vibrations exist for one and the same frequency of the external force.

Having determined the stationary amplitudes, we may derive the phase of vibration by means of the formula

$$\operatorname{tg}\psi = \frac{1 + \frac{3a^2}{8} - \nu}{2\delta}. \tag{3.141}$$

Before plotting the relation between ψ and ν it is more convenient to express the perturbing sinusoidal force and also the solution by means of the

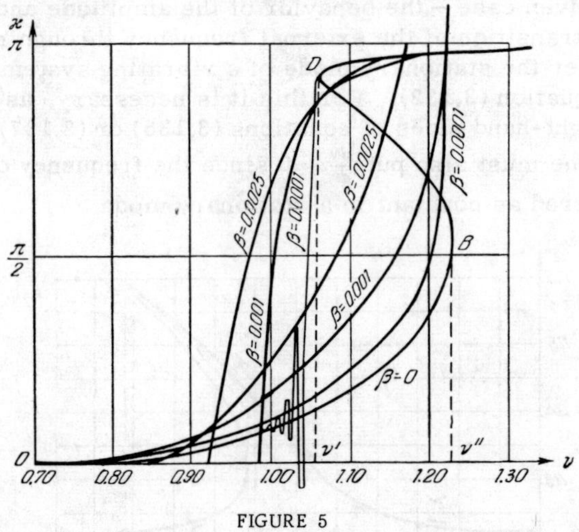

FIGURE 5

same trigonometric functions. Leaving the right-hand side of the original equation $E_1 \sin\theta$ unchanged, we represent the solution $x = a\cos(\theta + \psi)$ as $x = a\sin(\theta - \varkappa)$, where $\varkappa = \frac{3\pi}{2} - \psi$. We then plot (Figure 5) the dependence between the phase of vibration $\varkappa$ and the external force frequency ν. To each frequency within the interval (ν^I, ν^{II}) there again correspond three values of $\varkappa$, as well as three values of a, i. e., each of the stationary modes occurs with its own phase of vibration.

In equation (3.130) we put $d = 0$, $m = 1$, $c = 1$, $n = \delta$ and compare, for the sake of clarity, the obtained linear equation

$$\frac{d^2x}{dt^2}+2\delta\frac{dx}{dt}+x=E\sin\theta \tag{3.142}$$

with the one under investigation. The solution of this equation corresponding to a stationary mode is

$$x=a\sin(\theta-\varkappa), \tag{3.143}$$

and the graph describing the dependence of the amplitude on frequency (the resonance curve) has the form shown in Figure 6. In Figure 7 we plot the relation $\varkappa=\varkappa(\nu)$, where the friction coefficient is in both cases taken equal to 0.01. The skeleton curve becomes the straight line $\omega=1$ in the case of a linear system.

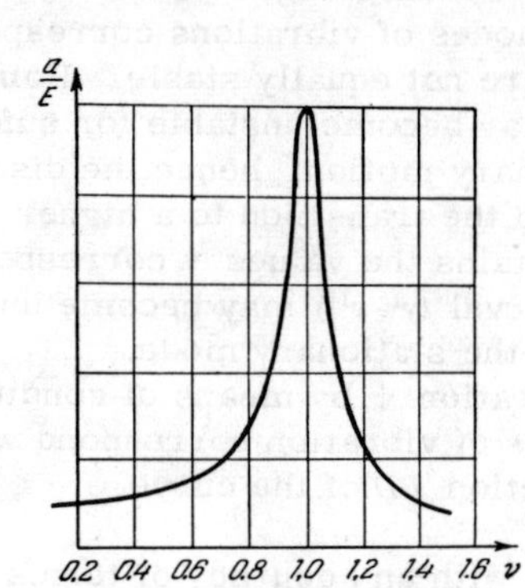

FIGURE 6

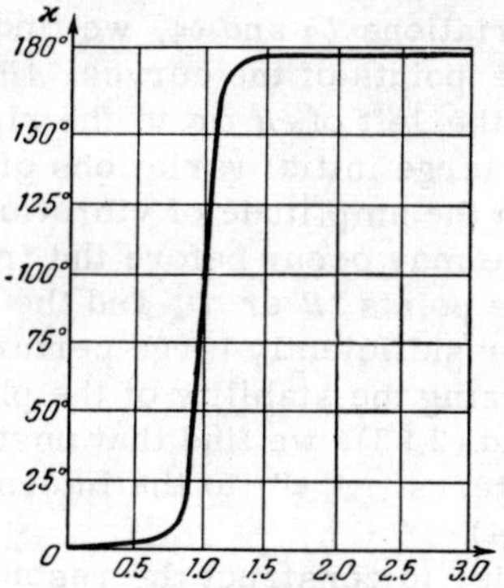

FIGURE 7

The maximum amplitude of a stationary mode in linear as well as nonlinear vibrating systems is in the presence of friction not given by the point *B*, i.e., the resonance point. The presence of resistance results in a certain (very small) shift of the maximum, which may be analyzed if a more accurate formula is applied in place of (3.139) for the construction of the resonance curve. However, we shall not concern ourselves with this question; it is discussed in the literature and is not essential for what follows.

Let us now discuss the stability of the considered stationary mode.

It is not difficult to locate the zones of stable and unstable amplitudes according to the conditions (3.89) worked out before. The frequency intervals, in which only one amplitude value corresponds to each point, are zones of stable vibrations; within the interval (ν^{I}, ν^{II}), in which there correspond (in the given example) three amplitude values to each frequency ν, the intermediate amplitude is unstable. At the points *B* and *D* (Figure 4) we get a transition from the stable to the unstable state and conversely. Thus the sections *MAB* and *DCN* of the resonance curve correspond to stable amplitudes, and the section *DB* corresponds to unstable amplitudes.

The unstable amplitude can practically not become stationary and hence there may occur within the interval (ν^{I}, ν^{II}) vibrations with either a large or a small amplitude, depending on the initial conditions of motion.

Physically we observe the following phenomena. When the frequency of the external force increases slowly (from small values) the amplitude of the forced vibrations increases, taking on large values along the branch *MAB* of the resonance curve. At the point *B* there occurs a discontinuity in the amplitude: the amplitude value jumps to the point *C* and when the frequency increases further it varies along the curve *CN*.

If the experiment is performed in the opposite direction, then as the frequency decreases slowly (from sufficiently large values) the amplitude of the forced vibrations changes along the curve *NCD* (the system performs vibrations with a small amplitude). Reaching the point *D* the value of the amplitude passes over to the point *A*, i. e., the amplitude increases instantaneously; as the frequency continues to decrease the amplitude varies along the branch *AM* of the resonance curve. The result is the picture displayed in Figure 4.

In deriving the stability conditions (3.89) the variations of the stationary motion were assumed to be infinitely small. Considering the stability for finite variations δa and $\delta\psi$, we find that the modes of vibrations corresponding to the points of the curves *AB* and *CD* are not equally stable. Points lying to the left of *B* or to the right of *D* may become unstable for sufficiently large initial variations of the stationary motion; hence the discontinuity in the amplitude of vibrations and also the transition to a higher amplitude may occur before the frequency attains the values ν corresponding to the points *B* or *D*, and the entire interval (ν^{I}, ν^{II}) may become unstable for sufficiently large perturbations of the stationary mode.

Analyzing the stability of the phase of vibration ψ by means of conditions (3.92) and (3.93), we find that unstable phases of vibration correspond within the interval (ν^{I}, ν^{II}) to the intermediate section *BD* of the curve $\psi = \psi(\nu)$ (Figure 4).

In order to construct the resonance curve with an accuracy of terms of the second order, it is necessary to apply the equations of the second approximation (3.137). Eliminating ψ from these equations, we obtain (up to and including terms of the second order) the following relation between the amplitude of vibration and the frequency of the external force:

$$\begin{aligned}(1-\nu)^2\{a^2\varphi_2^2(a) - E_1^2[\varphi_3^2(a)\varphi_6^2(a) + 2\varphi_2(a)\varphi_4(a)\varphi_6^2(a)]\} +\\ +(1-\nu)[2a^2\varphi_2^2(a)\varphi_5(a) - 2\varphi_2(a)\varphi_3(a)\varphi_6^2(a)E_1^2] +\\ +\varphi_1^2(a)\varphi_6^2(a) + \varphi_5^2(a)\varphi_2^2(a)a^2 - \varphi_2^2(a)\varphi_6^2(a)E_1^2 = 0,\end{aligned} \tag{3.144}$$

where

$$\left.\begin{aligned}&\varphi_1(a) = -\delta a + \frac{3a^2\delta}{8}, \quad \varphi_2(a) = \frac{1}{2} - \frac{9a^2}{32}, \quad \varphi_3(a) = \frac{1}{8} - \frac{21}{128}a^2,\\ &\varphi_4(a) = \frac{1}{8} - \frac{21}{128}a^2, \quad \varphi_5(a) = \frac{3a^2}{8} - \frac{\delta^2}{2} - \frac{15a^4}{256},\\ &\varphi_6(a) = \frac{1}{2} - \frac{3a^2}{32}, \quad \varphi_7(a) = \frac{1}{8} - \frac{15a^2}{128}.\end{aligned}\right\} \tag{3.145}$$

The second approximation, as indicated above, introduces only insignificant quantitative changes without modifying the qualitative picture.

We shall examine the behavior of the given system during various modes of transition through resonance. It is assumed for simplicity that the instantaneous frequency of the external force is a linear function of time, i.e.,

$$\nu(\tau) = \nu_0 + \beta t \tag{3.146}$$

(we could have assumed that $\nu(\tau)$ varies according to any law, it only being necessary that the time derivative of $\nu(\tau)$ is proportional to the small parameter ε).

To construct curves characterizing the amplitude variation for various modes of transition through resonance, the value of $\nu(\tau)$ must be substituted into the system of equations (3.137) which must then be integrated numerically according to the method described in § 5.

The velocity of transition through resonance depends on the value of β — the larger the absolute value of β, the faster will the system pass through resonance. It is clearer to characterize the velocity of transition through resonance by the tangent of the straight line Oa (Figure 8), and also by the

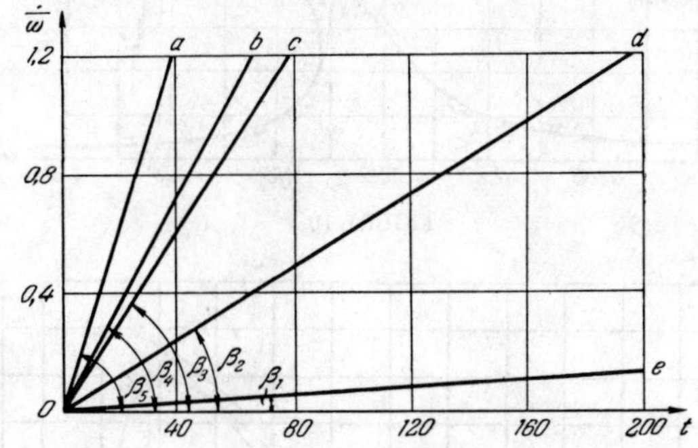

FIGURE 8

number of cycles* performed by the system either during the transition through the resonance zone, or from the beginning of motion to the moment at which the maximum amplitude is attained. Characterizing the velocity of

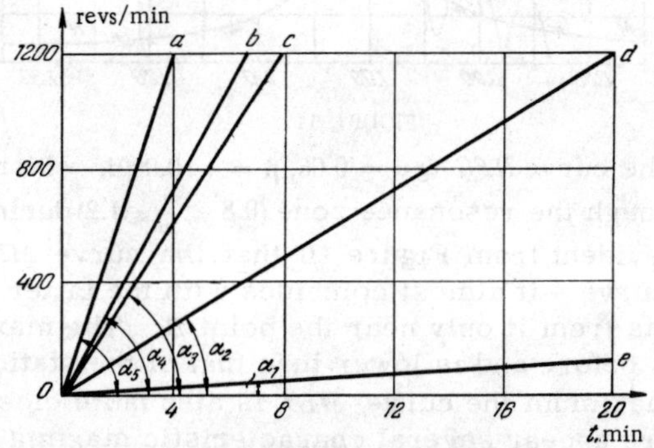

FIGURE 9

transition through resonance, we shall in the sequel give in parallel the tangent of the line Oa and also the number of cycles during which the system passes through the resonance zone.

We note that formula (3.146) may be written in the form

$$n \text{ revs/min} = n_0 \text{ revs/min} + t \cdot \operatorname{tg} \alpha, \tag{3.147}$$

where t is the time measured in minutes; $\operatorname{tg} \alpha = \frac{\beta \cdot 60^2 \omega^2}{2\pi}$.

In Figures 8 and 9 we show graphs describing the dependence of $\frac{\nu}{\omega}$ on t, and also of n on t for various α and β, according to the formulas (3.146) and (3.147) with $\nu_0 = 0$ and $n_0 = 0$.

* By a cycle we mean here the quantity $T = \frac{2\pi}{\omega}$, where $\omega = 1 \text{ sec}^{-1}$

Figure 10 shows the curves *MLT* and *NFG* which characterize a very slow transition through resonance, namely, for the curve *MLT* $\operatorname{tg}\alpha = 0.06$,

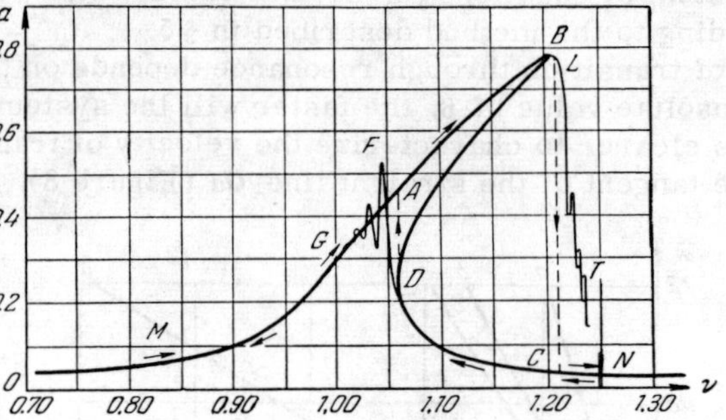

FIGURE 10

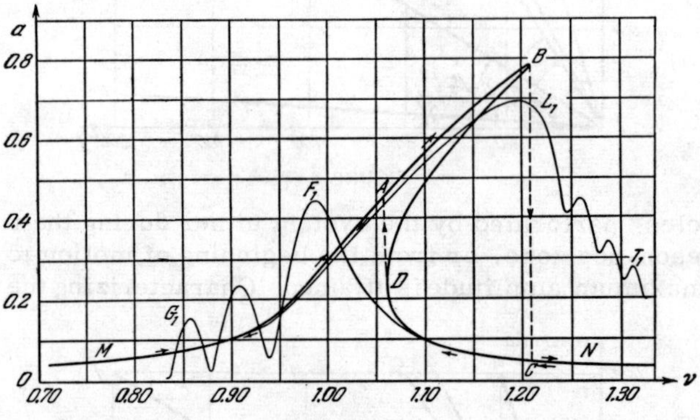

FIGURE 11

$\beta = 0.0001$ and for the curve *NFG* $\operatorname{tg}\alpha = 0.06$, $\beta = -0.0001$, which corresponds to a transition through the resonance zone $(0.8 \leqslant \frac{\nu}{\omega} \leqslant 1.2)$ during approximately 640 cycles. It is evident from Figure 10 that the curve *MLT* is very close to the stationary curve — it almost coincides with the latter along the section *MB* and departs from it only near the point *B*. The maximum of the curve *MLT* occurs before and is lower than that of the stationary curve. Beyond the first maximum the curve *MLT* is also quite close to the stationary mode, but there appear several characteristic maxima of smaller magnitude. When the frequency decreases $(\beta < 0)$ we obtain the curve *NFG* which is also very close to the stationary curve.

Figure 11 shows the curves ML_1T_1 and NF_1G_1 which characterize a faster transition through the resonance zone. Here $\beta = \pm\, 0.001$, $\operatorname{tg}\alpha = 0.6$, which corresponds to a transition through the resonance zone during approximately 64 cycles. The curves in this case, as is obvious from Figure 11, differ considerably from the stationary curve.

For the curves ML_2T_2 and NF_2G_2 (Figure 12) $\beta = \pm\, 0.0025$, $\operatorname{tg}\alpha = 1.5$; the transition through the resonance zone lasts over 26 cycles.

These amplitude variation curves for a transition through resonance (Figures 10, 11, 12, and also Figure 13 showing curves for various velocities of transition through resonance) resemble the similar curves in the case of a linear system, constructed by a method of F. M. Lewis (Figure 14, taken from the book by S. V. Serensen "Dinamicheskaya prochnost' v mashinostroenii" (Dynamical Strength in Machine Engineering) /150/). When the variation velocity of the frequency of the external force increases, then, exactly as in the case of a linear system, the maximum amplitude decreases.

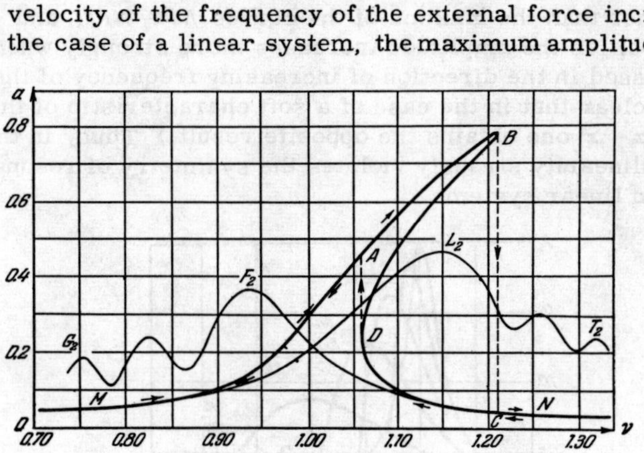

FIGURE 12

The faster the transition through resonance (in presence of a small friction $\delta \ll 1$), the more pronounced are the few maxima of smaller magnitude following the first maximum. Thus here also the vibrations proceed like damped beats.

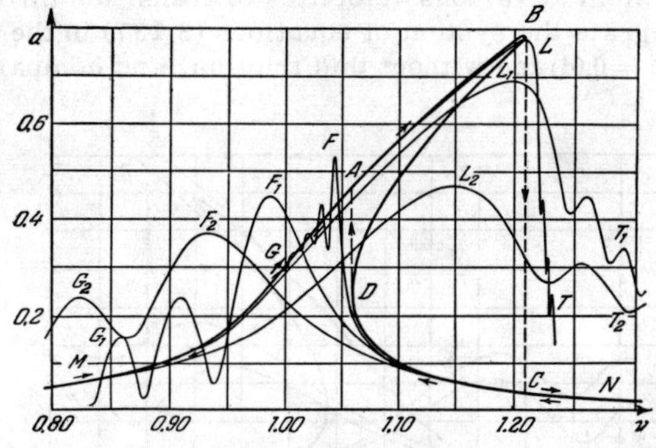

FIGURE 13

In addition, a few modifications occur on account of the nonlinearity (it is clearly demonstrated in our example — the maximum correction of the frequency is about 20 %). The maximum amplitude in linear systems is attained, as indicated before, not at the moment when the frequency of the external force coincides with a normal frequency of the system but later; besides, an increase in the variation velocity of the frequency implies an increase in the shift of the maximum. In our case, when the frequency of the external force passes through the resonance zone while increasing ($\beta > 0$) the maxima of the amplitude occur the earlier, the faster the frequency increases. In a transition in the opposite direction ($\beta < 0$) the maximum is shifted as in the linear case, but not as intensively.

However, for various velocities of transition through the resonance zone the maximum amplitudes are located with respect to the skeleton curve $\omega = \omega(a)$, which is the symmetry axis of the stationary resonance curve, exactly as in the linear case. The maxima of the curves MLT, ML_1T_1 and ML_2T_2 lie higher than the maxima of the curves NFG, NF_1G_1 and NF_2G_2, i.e., for equal velocities the amplitude increases more strongly when the resonance is passed in the direction of increasing frequency of the external force. (It is clear that in the case of a soft characteristic of the elastic force $F(x) = x - x^3$ one obtains the opposite result.) Thus, in the given example the nonlinearity strongly violates the symmetry of resonance curves encountered in linear systems.

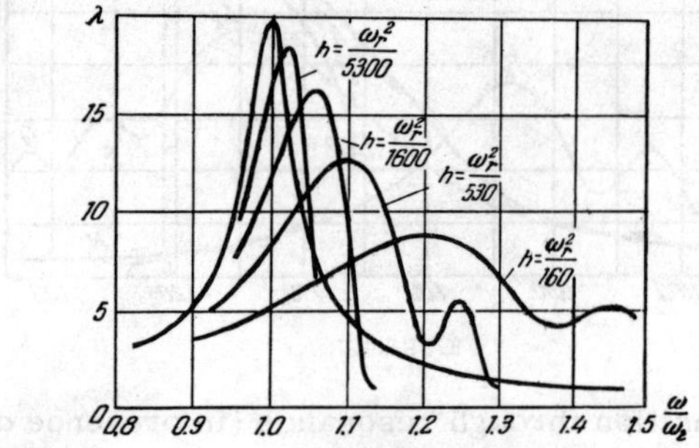

FIGURE 14

We examine, in passing, the affect of friction on the amplitude for the same vibrating system at various velocities of transition through resonance. To do this we integrate the system of equations (3.137) in the presence of a small friction ($\delta = 0.01$) and without this friction, and compare the results.

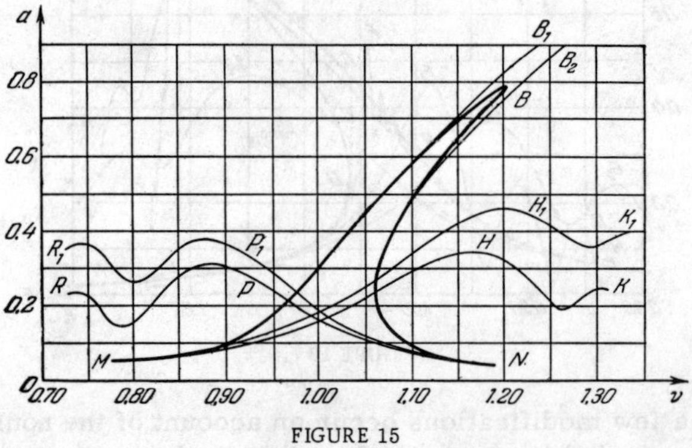

FIGURE 15

Figure 15 shows the resonance curves MHK and NPR for a transition through the resonance zone during 13 cycles ($\beta = \pm 0.005$, $\operatorname{tg} \alpha = 3$) with a small friction, and the curves MH_1K_1 and NP_1R_1 characterizing a transition through resonance (at the same velocity, the friction being neglected ($\delta = 0$)). Similar curves corresponding to a transition through the resonance zone during 18 cycles ($\beta = \pm 0.0035$, $\operatorname{tg} \alpha = 2.1$) are plotted in Figure 16. Both graphs also contain the stationary resonance curve with friction not taken into account (the curve MB_1B_2N).

Comparing Figure 15 with Figure 16, we conclude that when friction is absent, the amplitude of vibration increases sharply for a small decrease in the velocity of transition through the resonance zone in the direction of increasing frequency of the external force (in the case of the rigid characteristic of the elastic force $F(x) = x + x^3$), taken by us).

In the case of a transition through resonance in the opposite direction the vibration amplitude increases less sharply — here the nonlinearity plays the role of "passive" friction restricting the amplitude.

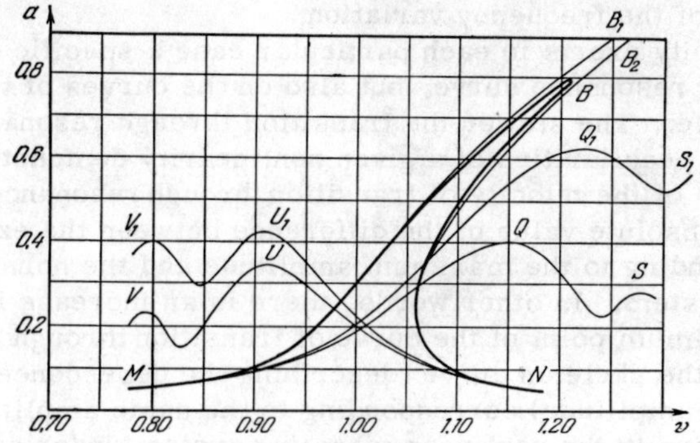

FIGURE 16

Thus in a slow transition through resonance even a small friction strongly reduces the maximum of the resonance curve.

The investigation of the obtained curves describing the transition through resonance for various characteristics of a nonlinear restoring force (and also examples to be considered below) leads to the conclusion that the linearity has in each actual case a specific affect not only on the stationary resonance curve (which may assume most diverse forms), but also on the curves of transition through resonance, making them (especially in the case of a slow transition) completely different from those obtained in linear systems.

We mention the following characteristic features encountered in nonlinear systems during the transition through resonance:

1) When the variation velocity of the frequency of the external force increases then the maximum amplitude decreases, exactly as in the case of transition through resonance in linear systems. The sharpness of the first maximum of the resonance curve is less pronounced than in the stationary resonance curve the faster the transition through resonance.

2) After the first maximum is reached beats of the amplitude appear. The faster the transition through resonance (in presence of a small friction), the more clearly are the several maxima of smaller magnitude following the first maximum distinguished. The swing of the beats, and also their periods decrease with time (these phenomena are also observed in linear systems).

3) The resonance curves are not symmetric as in linear systems. The curves obtained when the resonance zone is passed in the direction of increasing frequency of the external force differ sharply (especially in the case of a slow passage) from the curves obtained in the opposite transition.

4) If a system, by virtue of nonlinearity, possesses in the stationary resonance mode discontinuity points of the amplitude, then sharp changes in the amplitude occur during a transition through resonance at the corresponding frequency values (especially in the case of a slow passage).

5) The maximum amplitude occurs in linear systems, as mentioned before, not at the moment when the frequency of the perturbing force coincides with a normal frequency of the system, but later. This shifting of the maximum increases with an increase in the variation velocity of the frequency. Such a regularity is not observed in nonlinear systems. Here the location of maxima depends on the nature of the nonlinearity, and also on the direction of the frequency variation.

The nonlinearity exerts in each particular case a specific effect not only on the stationary resonance curve, but also on the curves of the transition through resonance. The slower the transition through resonance, the stronger are the peculiarities of a given nonlinearity demonstrated.

6) An increase of the velocity of transition through resonance implies the increase of the absolute value of the difference between the external frequency corresponding to the maximum amplitude and the normal frequency of a nonlinear system. In other words, there is an increase in the distance between the maximum point of the curve of transition through resonance and the point on the skeleton curve (describing the dependence of the normal frequency on the amplitude) corresponding to the same amplitude value.

We now examine in the nonlinear vibrating system under consideration more complicated nonstationary modes, directly associated with the transition through resonance. Integrating numerically the system of equations (3.137) for the values

$$\left.\begin{aligned} \nu(\tau) &= \nu_0 + \beta t \quad \text{for } 0 \leqslant t \leqslant t_1, \\ \nu(\tau) &= \nu(\tau_1) + \gamma t \quad \text{for } t_1 \leqslant t, \end{aligned}\right\} \tag{3.148}$$

we obtain curves describing the time variation of the amplitude (Figure 17). These curves are plotted using the values $\beta = 0.0035$, $\gamma = -\beta$, $\nu(\tau_1) = 1.13$ for the curve MPC, and $\nu(\tau_1) = 1.18$ for the curve MP_1C_1.

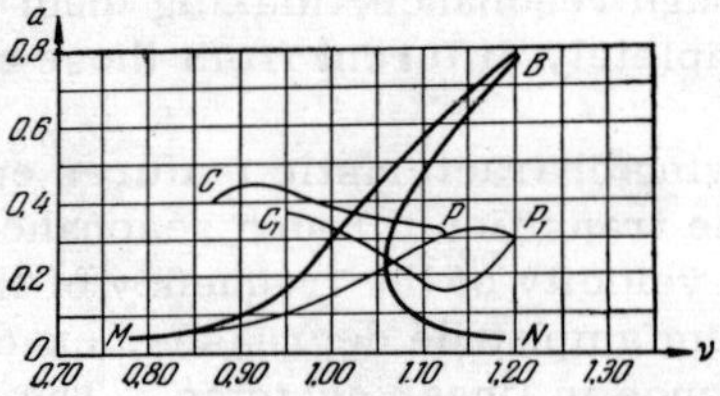

FIGURE 17

It is easily seen that the nature of the curve essentially depends not only on the variation velocity of the frequency of the external force, but also on the choice of the value of $\nu(\tau_1)$, at which the direction of the frequency variation is reversed.

We shall now solve the following problem. Let the frequency of the external perturbing force $\nu(\tau)$ vary in the resonance zone with a certain constant velocity β. Then the vibration amplitude takes on values corresponding to the resonance curve for a transition through resonance. We assume that at some moment $t = t_1$ we stop the variation of the frequency of the external force, so that $\nu^{\mathrm{I}} \leqslant \nu(\tau_1) \leqslant \nu^{\mathrm{II}}$, where ν^{I} is the frequency corresponding to the point D on the resonance curve (Figure 14), ν^{II} is the frequency corresponding to the point B. The question is: does the amplitude established in the system correspond to the lower branch of the resonance curve ND or to the upper branch MB. This problem may in the general case be treated as the question of stability of these branches for arbitrary perturbations of the amplitude and phase of vibration (and not small, as we have done before).

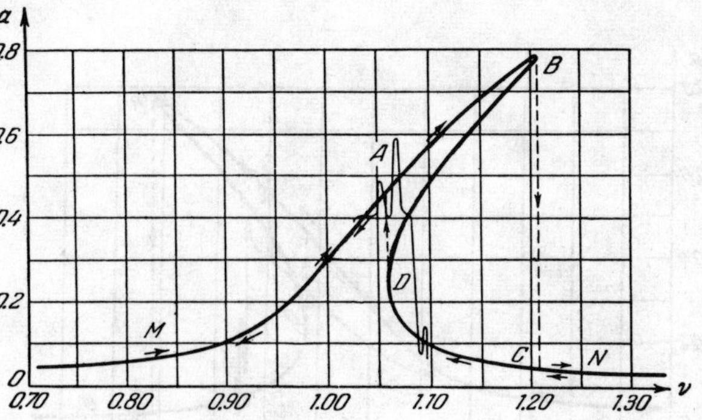

FIGURE 18

Considering the stability of stationary vibrations in the resonant domain, it is impossible to decide on the grounds of the generally known criteria what amplitude will be established if the initial values correspond to the unstable branch of the resonance curve.

The nature of unstability of the branch *DB* may be explained by the following reasoning. Noting that the resonance curves for the transition through resonance are in the case of a very slow transition very close to the stationary mode, we may integrate the system of equations (3.137) for a very slow variation of the frequency of the external force, taking the corresponding values of the amplitude and phase as the initial values (the values of the amplitude, phase, and frequency corresponding to the section *DB* of the resonance curve will in the sequel be denoted by a_{DB}, ψ_{DB}, and ν_{DB}). The results of such an integration are shown in Figure 18, which clearly demonstrates that when the initial values correspond to the branch *DB*, then the system may perform vibrations with a low, as well as with a high amplitude, both cases being equally probable. If the initial values are a_{DB}, ψ_{DB}, then when $\nu > \nu_{DB}$, a stationary mode with a small amplitude will be established in the system, and when $\nu < \nu_{DB}$, a large amplitude.

We apply these arguments to the problem stated above. Let the frequency of the external force vary according to the first formula in (3.148). Integrating the system (3.137) numerically, we obtain the curve of transition through resonance. Let the variation of the frequency be stopped at the moment $t = t_1$. To determine the further variation of the amplitude we integrate the system (3.137) for the initial values $a = a(t_1)$, $\psi = \psi(t_1)$, $\nu = \nu(\tau_1)$ assuming that ν varies according to the second formula of (3.148) with a very small γ (in the case considered here $\gamma = 0.0001$). To be safe it is desirable to perform the integration for positive and negative values of γ. The results of computations are shown in Figures 19 and 20 (where $\beta = 0.025$ and $\beta = 0.005$, respectively). Analyzing these graphs, we conclude that whether the one or the other amplitude of vibrations will be established in the transition of the system to a stationary state depends essentially on the variation velocity of the frequency of the external force, also on the value

$\nu(\tau_1)$ at which the system goes over to a stationary state. We may always choose for any given value of $\nu(\tau_1)$ such β that the amplitude of the stationary mode will correspond to a definite stable branch of the stationary resonance curve.

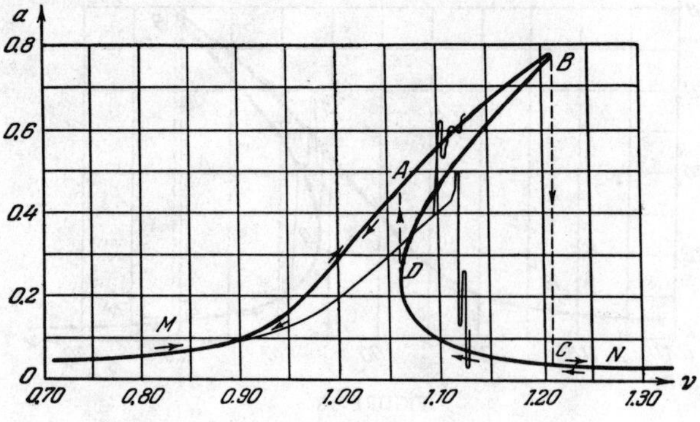

FIGURE 19

Concluding the present section, devoted to an illustration of the basic phenomena observed in a nonlinear vibrating system during the transition through resonance, we note that the specific properties indicated here by no means cover all the interesting phenomena which may be observed in

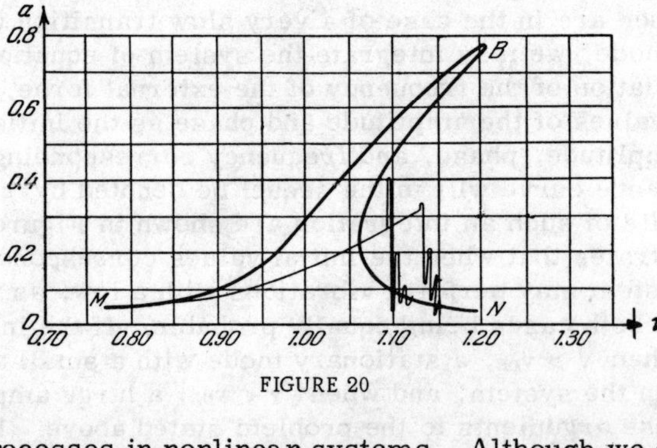

FIGURE 20

nonstationary processes in nonlinear systems. Although we do not intend to give a complete analysis of nonstationary vibrations in nonlinear systems in the present book, we shall, as far as possible, consider this problem in the following sections while illustrating the developed method.

§ 7. Comparison of theoretical and experimental resonance curves for transition through resonance

We shall now apply our method to the construction of resonance curves for a transition through resonance of an actual mechanical system, for which the experimental vibrogramms have been obtained. We can thus compare the results of theoretical investigation with experimental data, and illustrate visually the effectiveness of the method developed.

As such a mechanical system we take the experimental device constructed by K. T. Shatalov in the year 1933 (cf., "Dizelestroenie", No. 8. 1935); the system allows various studies of torsional vibrations to be performed. The device contained a mechanism of a strictly harmonic excitation, and also a specially designed mechanism ensuring the linear variation of the velocity of rotation. Furthermore, the system as a whole could be regarded as a system with one degree of freedom. With the help of this device Shatalov conducted experiments determining the amplitudes of forced torsional vibrations during the transition through resonance (particularly for samples of untempered steel 1035 of 1.2 cm diameter and 24 cm length) /172/. The curves of transition through resonance obtained experimentally by Shatalov may easily be derived theoretically by means of the above method.

Denoting the deviation angles from a uniform rotation by φ_1 and φ_2, we may write the differential equation of torsional vibrations of the considered system, under several restrictions, in the form

$$I_1 I_2 \ddot{\varphi} + c(I_1 + I_2)\varphi = \varepsilon \overleftrightarrow{f(\varphi)} - M I_2 \sin\left(\omega_0 t \pm \frac{\beta t^2}{2}\right), \tag{3.149}$$

where $\varphi = \varphi_1 - \varphi_2$; β is the acceleration (or deceleration) coefficient during the transition through resonance; and

$$\varepsilon \overleftrightarrow{f(\varphi)} = \mp (I_1 + I_2) \frac{4 r_0^{n-1} \nu_1}{n(n+3)\, l^{n-1}} [(a \mp \varphi)^n - 2^{n-1} a^n] \tag{3.150}$$

is a linear function taking into account the damping in the sample material. The values of this function are distinct for an ascending or descending motion, which is marked correspondingly by two oppositely directed arrows*. The parameters n and ν_1 entering into the right-hand side of (3.150) are usually determined experimentally. For the steel brand under investigation it is accurate enough if we take $n = 2$, $\nu_1 = 10$.

Solving equation (3.149), we confine ourselves only to the first approximation. According to (3.43) and (3.51) we have

$$\varphi = a \cos(\theta + \psi), \tag{3.151}$$

where, as usual, a and ψ are to be determined from the system of equations of the first approximation

$$\left.\begin{aligned} \frac{da}{dt} &= -\frac{c(I_1+I_2)\, 2^{n+2} r_0^{n-1} \nu_1 a^n}{I_1 I_2 \pi \omega n (n+3)\, l^{n-1}} + \frac{M}{I_1 I_2 (\omega + \nu)} \cos\psi, \\ \frac{d\psi}{dt} &= \omega - \nu + \frac{4c(I_1+I_2)\, r_0^{n-1} \nu_1 a^{n-1}}{\pi I_1 I_2 \omega n (n+3)\, l^{n-1}} \int_0^{\pi} (1-\cos\varphi)^n \cos\varphi \, d\varphi - \frac{M}{I_1 I_2 a (\omega+\nu)} \sin\psi, \end{aligned}\right\} \tag{3.152}$$

where $\omega = \sqrt{\frac{(I_1+I_2)c}{I_1 I_2}}$ is the frequency of the normal torsional vibrations of the sample; $\nu = \frac{d\theta}{dt} = \omega_0 \pm \beta t$ is the frequency of the external moment.

The above-mentioned experimental device was characterized by the

* The expression of the nonlinear function $\varepsilon \overleftrightarrow{f(\varphi)}$ taking into account the damping for torsional vibrations was taken from the works of G. S. Pisarenko /132/.

following numerical data:

$$I_1 = 179.6\,\text{kg}\cdot\text{cm}\cdot\text{sec}^2,\quad c = 7000\,\text{kg}\cdot\text{cm},\quad I_2 = 5.3\,\text{kg}\cdot\text{cm}\cdot\text{sec}^2,$$

the external moment of the centrifugal exciter

$$\frac{M}{\omega^2} = M_0 = 0.088\,\text{kg}\cdot\text{cm}\cdot\text{sec}^2,\quad l = 24\,\text{cm},\quad 2r_0 = 1.2\,\text{cm}.$$

Inserting these values into the right-hand sides of (3.152), we obtain

$$\left.\begin{aligned}\frac{da}{dt} &= -4.698\,a^2 + \frac{0.00049\nu^2}{36.87+\nu}\cos\psi,\\ \frac{d\psi}{dt} &= 36.87 - \nu - 3.688a - \frac{0.00049\nu^2}{a\,(36.87+\nu)}\sin\psi,\end{aligned}\right\} \tag{3.153}$$

where $\nu = \omega_0 \pm \beta t$.

To obtain the curves of transition through resonance the system (3.152) must be integrated numerically, assigning to β the same values as in the experiment. We give the detailed procedure of the numerical integration of the system (3.152) for the value $\beta = -2.3$.

1. We derive, as was shown above, the initial values satisfying the stationary mode, so that the stationary resonance curve would be close to a horizontal line in the neighborhood of the initial frequency value of the external moment. To do this we determine a_0 and ψ_0 from the equations

$$\left.\begin{aligned}-4.698\,a^2 + \frac{0.00049\omega_0^2}{36.87+\omega_0}\cos\psi &= 0,\\ 36.87 - \omega_0 - 3.688a - \frac{0.00049\omega_0^2}{a\,(36.87+\omega_0)}\sin\psi &= 0\end{aligned}\right\} \tag{3.154}$$

(The terms $4.698\,a^2$ and $3.688\,a$ may be taken as quantities of higher order of smallness, and neglected for sufficiently small a since the friction due to hysteresis does not affect considerably the nature of vibrations for small amplitudes.) Taking for ν the value $\omega_0 = 40.31\,\text{rad/sec}^2$ at $t = 0$, we find $a_0 = 0.00300$, $\psi_0 = 4.712$.

2. Differentiating the right-hand sides of the expressions for $\frac{da}{dt}$ and $\frac{d\psi}{dt}$, and putting $t = 0$, $a_0 = 0{,}00300$, $\psi_0 = 4.712$, $\omega_0 = 40.31$, we obtain

$$\frac{d^2a}{dt^2} = 0,\quad \frac{d^2\psi}{dt^2} = 2.010,\quad \frac{d^3a}{dt^3} = 0.02074,\quad \frac{d^3\psi}{dt^3} = 0.0024.$$

It is then possible to expand $a = a(t)$ and $\psi = \psi(t)$ in a Taylor expansion in the vicinity of the point $t = 0$:

$$\left.\begin{aligned}a &= 0.00300 + 0.00346\,(\Delta t)^3 + \cdots,\\ \psi &= 4.712 + 1.005\,(\Delta t)^2 + 0.0006\,(\Delta t)^3 + \cdots\end{aligned}\right\} \tag{3.155}$$

Taking $\Delta t = 0.1;\ 0.2;\ 0.3$, we obtain for a and ψ

$$\left.\begin{aligned}a_1 &= 0.00300, & \psi_1 &= 4.722,\\ a_2 &= 0.00303, & \psi_2 &= 4.752,\\ a_3 &= 0.00309, & \psi_3 &= 4.802.\end{aligned}\right\} \tag{3.156}$$

3. With the notation

$$\left.\begin{aligned}f_1(a,\ \psi,\ \nu) &= -4.698\,a^2 + \frac{0.00049\nu^2}{36.87+\nu}\cos\psi,\\ f_2(a,\ \psi,\ \nu) &= 36.87 - \nu - 3.688\,a - \frac{0.00049\nu^2}{a\,(36.87+\nu)}\sin\psi,\end{aligned}\right\} \tag{3.157}$$

we obtain

$$\xi_i = f_1(a_i,\ \psi_i,\ \nu_i)\,\Delta t,\quad \eta_i = f_2(a_i,\ \psi_i,\ \nu_i)\,\Delta t \qquad (i = 0,\ 1,\ 2,\ 3), \tag{3.158}$$

and also (according to (2.80)) their first, second, third, etc. differences (it is sufficient in practice to confine oneself to the second or third order).

Next, as was shown before, we construct simultaneously three tables, one of which contains the consecutive values of t (or $\nu(t)$), a, ψ, and their first differences $\Delta a_i = a_{i+1} - a_i$, $\Delta\psi_i = \psi_{i+1} - \psi_i$, and the second and third contain correspondingly the values of ξ_i and η_i, and their differences (cf., Tables 5, 6 and 7).

TABLE 5

n	$\frac{\nu}{\omega}$	a	Δa	ψ	$\Delta\psi$
0	1.093	0.00300		4.712	
1	1.087	0.00300	0.00000	4.722	0.010
2	1.081	0.00303	0.00003	4.752	0.030
3	1.075	0.00309	0.00006	4.802	0.050
4	1.068	0.00319	0.00010	4.847	0.045
5	1.062	0.00335	0.00016	4.901	0.054
6	1.056	0.00355	0.00020	4.956	0.055
7	1.050	0.00381	0.00026	5.009	0.053
8	1.043	0.00411	0.00030	5.065	0.056
9	1.037	0.00446	0.00035	5.119	0.054
10	1.031	0.00486	0.00040	5.175	0.056
11	1.018	0.00577	0.00091	5.287	0.112
12	1.006	0.00684	0.00107	5.310	0.123
13	0.993	0.00804	0.00120	5.454	0.144
14	0.981	0.00935	0.00131	5.639	0.185
15	0.969	0.01074	0.00139	5.856	0.217
16	0.956	0.01219	0.00145	6.118	0.262
17	0.944	0.01363	0.00144	6.438	0.320
18	0.932	0.01490	0.00127	6.824	0.386
19	0.919	0.01568	0.00078	7.288	0.464
20	0.906	0.01571	0.00003	7.842	0.554
21	0.894	0.01474	−0.00097	8.498	0.656
22	0.882	0.01300	−0.00174	9.285	0.787
23	0.870	0.01167	−0.00133	10.212	0.927
24	0.858	0.01177	0.00010	11.293	1.081
25	0.845	0.01355	0.00178	12.538	1.245

TABLE 6

n	ξ	$\Delta\xi$	$\Delta^2\xi$	$\Delta^3\xi$	Increment
0	0				
1	0.00001	0.00001			
2	0.00004	0.00003	0.00002		
3	0.00008	0.00004	0.00001	−0.00001	
4	0.00013	0.00005	0.00001	0.00000	
5	0.00018	0.00005	0.00000	−0.00001	
6	0.00023	0.00005	0.00000	0.00000	
7	0.00028	0.00005	0.00000	0.00000	$\Delta t=0.1$
8	0.00033	0.00005	0.00000	0.00000	
9	0.00038	0.00005	0.00000	0.00000	
10	0.00042	0.00004	−0.00001	−0.00001	
4	0.00026				
6	0.00046	0.00020	0.00000		
8	0.00066	0.00020	−0.00002	−0.00002	
10	0.00084	0.00018	−0.00002	0.00000	
11	0.00100	0.00016	−0.00002	0.00000	
12	0.00114	0.00014	−0.00002	0.00000	
13	0.00126	0.00012	−0.00003	−0.00001	
14	0.00135	0.00009	−0.00002	0.00001	$\Delta t=0.2$
15	0.00142	0.00007	−0.00004	−0.00002	
16	0.00145	0.00003	−0.00010	−0.00006	
17	0.00138	−0.00007	−0.00023	−0.00015	
18	0.00108	−0.00030	−0.00030	−0.00007	
19	0.00048	−0.00060	−0.00028	0.00002	
20	−0.00040	−0.00088	−0.00004	0.00024	
21	−0.00132	−0.00092	0.00059	0.00063	
22	−0.00165	0.00033	0.00106	0.00047	
23	−0.00092	0.00073	0.00087	−0.00019	
24	−0.00068	0.00160			

TABLE 7

η	η	Δη	$\Delta^2\eta$	$\Delta^3\eta$	Increment
1	0				
2	0.018	0.018			
3	0.030	0.012	−0.006		
4	0.040	0.010	−0.002	0.004	
5	0.049	0.009	−0.001	0.001	
6	0.055	0.006	−0.003	−0.002	
7	0.056	0.001	−0.005	−0.002	
8	0.056	0.000	−0.001	0.004	$\Delta t = 0.1$
9	0.055	−0.001	−0.001	0.000	
10	0.055	0.000	0.001	0.002	
4	0.098				
6	0.112	0.014			
8	0.110	−0.002	−0.016		
10	0.110	0.000	0.002	0.018	
11	0.115	0.005	0.005	0.003	
12	0.130	0.015	0.010	0.005	
13	0.157	0.027	0.012	0.002	
14	0.194	0.037	0.010	−0.002	
15	0.238	0.044	0.007	−0.003	$\Delta t = 0.2$
16	0.290	0.052	0.008	0.001	
17	0.351	0.061	0.009	0.001	
18	0.423	0.072	0.011	0.002	
19	0.507	0.084	0.012	0.001	
20	0.603	0.096	0.012	0.000	
21	0.722	0.119	0.013	0.001	
22	0.855	0.133	0.014	0.001	
23	1.002	0.147	0.014	0.000	
24	1.161	0.159	0.012	−0.002	

4. To continue these tables we proceed as indicated above. Namely, we derive Δa_n and $\Delta\psi_n$ from Tables 6 and 7 according to the formulas (3.128) and insert their values, and also the values of

$$a_{n+1} = a_n + \Delta a_n, \quad \psi_{n+1} = \psi_n + \Delta\psi_n \tag{3.159}$$

in the corresponding columns of Table 5.

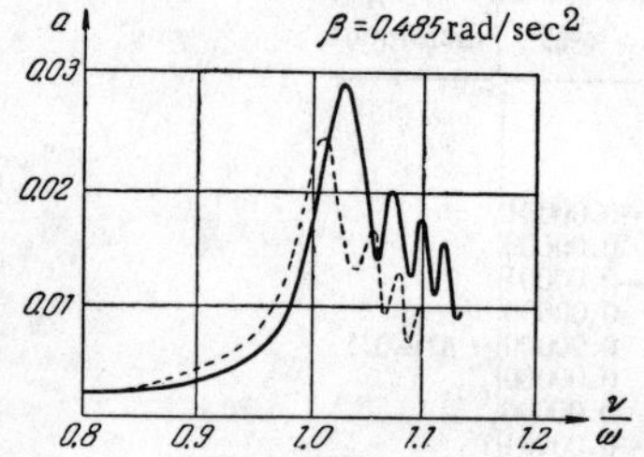

FIGURE 21

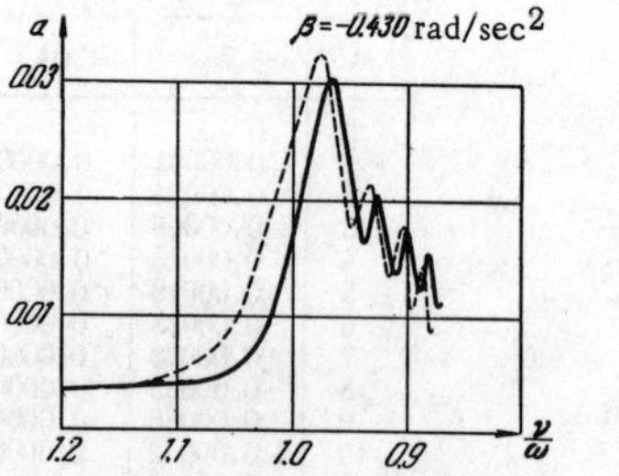

FIGURE 22

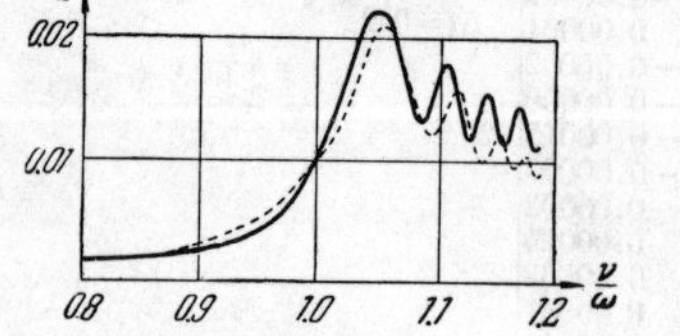

FIGURE 23

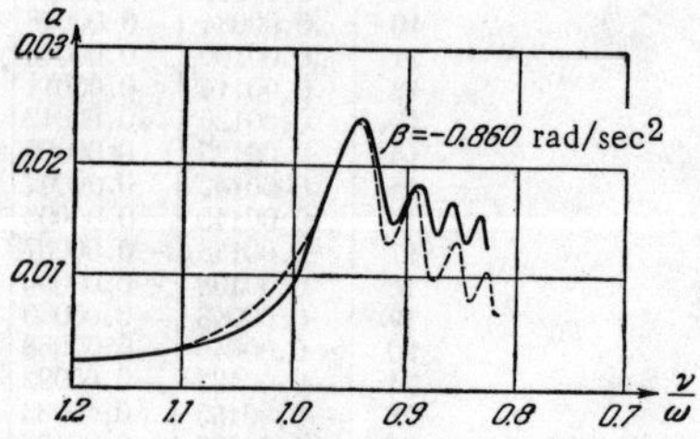

FIGURE 24

Next, using the formulas (3.129), we may add one oblique row in Tables 6 and 7, and then calculate similarly a_{n+2} and ψ_{n+2}. Continuing this procedure, we obtain a series of values of a and ψ (Table 5) characterizing the variation in amplitude and phase as a function of time.

According to the tables given, only 28 points were required for the calculation of the curve of transition through resonance for the acceleration $\beta = -2.30\ \text{rad/sec}^2$. Their number increases in the case of a slower transition through resonance, but not considerably (for example, for the acceleration $\beta = 0.485\ \text{rad/sec}^2$, i. e., almost five times smaller, it is necessary to calculate only 50–60 points).

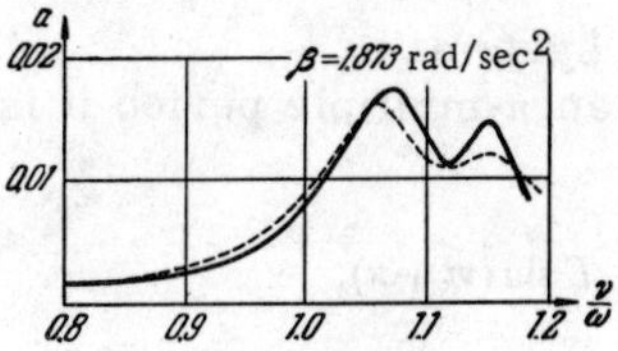

FIGURE 25

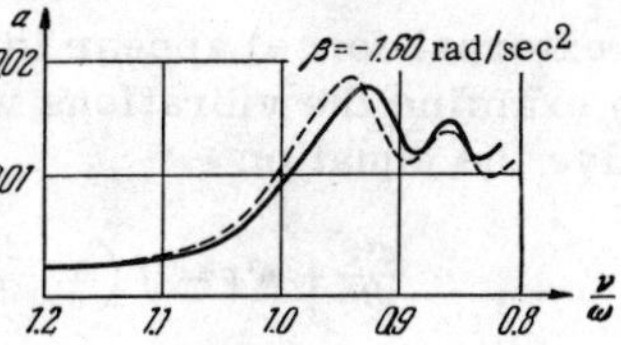

FIGURE 26

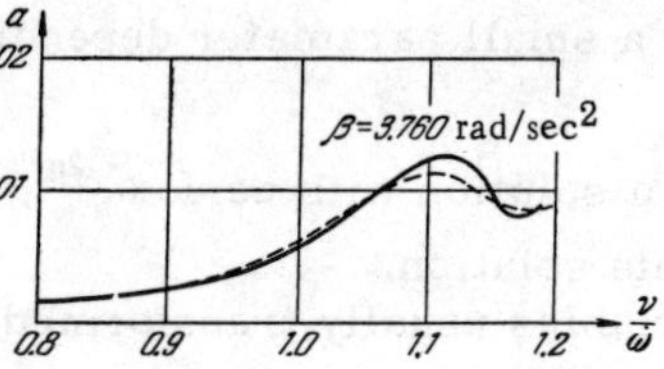

FIGURE 27

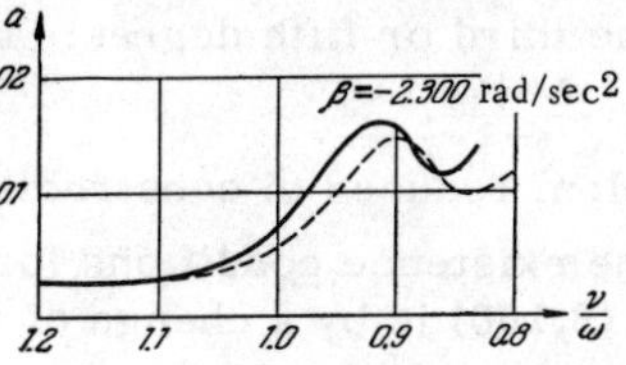

FIGURE 28

Integrating the system (3.153) numerically for various values of β, we obtain a series of resonance curves. Figures 21–28 show such curves calculated by our method for various velocities of transition through resonance when the frequency of the external force increases as well as decreases*, and also the experimental resonance curves (dotted lines) obtained by Shatalov for the same modes.

The graphs plotted display a good qualitative agreement between theoretical and experimental results. Some quantitative disagreement should be attributed to an insufficient account of friction and choice of initial values. The values of n and ν_1 chosen by us (from experimental data), and possibly even the relation (3.150) itself, do not account exactly for all the friction losses during the vibrations.

§ 8. Examples of transition through demultiplicative and parametric resonances

We shall discuss the application of our method to the investigation of a nonlinear vibrating system with slowly varying parameters, in which a more complicated resonance may possibly occur.

As the first example we study the behavior of the amplitude during a

* We plot only the dependence of the amplitude of the forced vibrations on time.

resonance of the n-th class depending upon the state of change in the detuning of an oscillating circuit with feed-back through a tube. The phenomenon of an n-th class resonance observed in nonlinear vibrating systems consists, as known /88/, in the fact that periodic vibrations with a period which is a multiple of the period of the exciting force may be enhanced.

Physically the phenomenon of an n-th class resonance consists in the following: the vibrations excited in a definite state in a potentially-self-vibrating system are very small until the normal period approaches a multiple of the period of the external periodic force; when the normal period comes sufficiently close to the n-multiple of the period of the external force, then intensive vibrations with a frequency exactly equal to $\frac{\nu}{n}$ (ν is the frequency of the external force) appear in the system.

In order to examine the vibrations with an n-multiple period it is necessary to solve the equation

$$\frac{d^2x}{dt^2}+\omega^2 x=\varepsilon f\left(x,\ \frac{dx}{dt}\right)+E\sin(\nu t+\varkappa), \tag{3.160}$$

where $\omega=\frac{\nu}{n}$; $f\left(x,\ \frac{dx}{dt}\right)$ is in the case of a tube generator a polynomial in x and $\frac{dx}{dt}$ of the third or fifth degree; ε is a small parameter depending upon the detuning $\frac{\nu^2-n^2\omega^2}{n^2\omega^2}$.

The problem reduces to constructing a solution with period $\frac{2\pi}{\omega}$, and clarifying the existence conditions for this solution.

Equation (3.160) is by a change of variables usually transformed into the equation

$$\frac{d^2x}{dt^2}+x=\varepsilon f\left(x,\ \frac{dx}{dt}\right)+E\sin nt, \tag{3.161}$$

for which one finds a periodic solution with period 2π in the form

$$x=a\cos t-b\sin t+\frac{E}{1-n^2}\sin nt, \tag{3.162}$$

and then determines the dependence (resonance curve) between its amplitude $X=\sqrt{a^2+b^2}$ and the detuning $\xi=\frac{\nu^2-n^2\omega^2}{n^2\omega^2}$ in a stationary mode.

To investigate a nonstationary mode (dependence of the amplitude of resonance (of the n-th class) vibrations upon various states of change of the detuning for small values of the latter) one has to study, instead of (3.160), the equation

$$\frac{d^2x}{dt^2}+x=\varepsilon(\tau)f\left(x,\ \frac{dx}{dt},\ \xi(\tau)\right)+E\sin nt, \tag{3.163}$$

where, unlike equation (3.161), the small parameter ε depends on τ for a slow variation in the detuning $\xi=\xi(\tau)$; it is however clear that in the first approximation one may always assume $\varepsilon=\text{const}$. Performing the change of variables

$$x=z+\frac{E}{1-n^2}\sin nt, \tag{3.164}$$

in equation (3.163) we obtain the equation

$$\frac{d^2z}{dt^2}+z=\varepsilon(\tau)f\left(z+\frac{E}{1-n^2}\sin nt,\ \frac{dz}{dt}+\frac{nE}{1-n^2}\cos nt,\ \xi(\tau)\right) \tag{3.165}$$

or

$$\frac{d^2z}{dt^2}+z=\varepsilon(\tau)f_1\left(\tau,\ z,\ \frac{dz}{dt},\ \theta\right),\quad \frac{d\theta}{dt}=n, \tag{3.166}$$

which may be solved approximately without difficulty, assuming herewith $p=1$, $q=n$ in (3.30) and (3.31), the formulas for the first approximation.

We consider an actual example taken from the previously mentioned work of L. I. Mandel'shtam and N. D. Papaleksi. We examine a linear oscillating circuit with feed-back through an electronic tube. Let the characteristic of the tube P-7 be of the form

$$i_a=0.95+3.35x+2.25x^2-1.5x^3, \tag{3.167}$$

where $V_0=12$ v, $I_0=142$ ma. After a few transformations* we obtain the following expression for the function $f\left(x,\ \frac{dx}{dt},\ \xi(\tau)\right)$ in the right-hand side of equation (3.163)

$$f\left(x,\ \frac{dx}{dt},\ \xi(\tau)\right)=(k+2x+\gamma x^2)\frac{dx}{dt}+\frac{\xi(\tau)}{0{,}016}x, \tag{3.168}$$

where

$$\varepsilon(\tau)=\frac{0.016}{1+\xi(\tau)},\ k=k_0+2\vartheta\frac{\xi(\tau)}{\beta},\ \beta=0.016,\ \vartheta=0.013,\ \gamma=-2,\ k_0=-0.05.$$

We derive an approximate solution for the nonlinear equation

$$\frac{d^2z}{dt^2}+z=\frac{0.016}{1+\xi(\tau)}\left\{k(\tau)+2\left(z+\frac{E}{1-n^2}\sin nt\right)+\gamma\left(z+\frac{E}{1-n^2}\sin nt\right)^2\right\}\times$$
$$\times\left(\frac{dz}{dt}+\frac{nE}{1-n^2}\cos nt\right)+\frac{\xi(\tau)}{1+\xi(\tau)}\left(z+\frac{E}{1-n^2}\sin nt\right) \tag{3.169}$$

for the case $n=2$. Using (3.30) and (3.31) and assuming $p=1$, $q=2$, we obtain after a few calculations

$$z=a\cos(t+\psi), \tag{3.170}$$

where a and ψ are to be determined from the system of equations of the first approximation

$$\left.\begin{aligned}\frac{da}{dt}&=\varepsilon(\tau)\left[\frac{1}{2}a\left(k(\tau)+\frac{\gamma a^2}{4}\right)+\frac{\gamma E^2a}{36}+\frac{aE}{6}\sin 2\psi\right],\\ \frac{d\psi}{dt}&=\varepsilon(\tau)\left[-\frac{\xi(\tau)}{2\beta}+\frac{E}{6}\cos 2\psi\right].\end{aligned}\right\} \tag{3.171}$$

Equating the right-hand sides of this system to zero, we obtain the relations

$$k+\frac{\gamma a^2}{4}+\frac{\gamma E^2}{18}+\frac{E}{3}\sin 2\psi=0,\quad -\frac{\xi}{\beta}+\frac{E}{3}\cos 2\psi=0, \tag{3.172}$$

defining the stationary values of the vibration amplitude and phase. Eliminating from these the phase ψ, we obtain the well-known relation

$$a^2=-\frac{2E^2}{9}-\frac{4}{\gamma}\left[k\pm\sqrt{\frac{E^2}{9}-\frac{\xi^2}{\beta^2}}\right], \tag{3.173}$$

which may be utilized to construct resonance curves characterizing the dependence of a on ξ.

For the phase ψ we obtain the formula

$$\operatorname{tg}2\psi=-\frac{k+\frac{\gamma a^2}{4}+\frac{\gamma E^2}{18}}{\frac{\xi}{\beta}}. \tag{3.174}$$

* Cf. the work /88/ mentioned above.

To construct the graphs showing the variation in the vibration amplitude for a second class resonance and various states of changes in the detuning $\xi=\frac{\nu^2-4\omega^2}{4\omega^2}$ the system of equations of the first approximation (3.171) must be integrated numerically, putting $\xi=\xi(\tau)$, $k=k(\tau)$.

We assume that the detuning varies due to the changes in the normal frequency ω of the oscillating system, it being assumed for simplicity that ξ varies linearly according to the formula

$$\xi=\xi_0+\alpha t. \tag{3.175}$$

Putting this expression for ξ into the equations of the first approximation and integrating, we obtain curves (Figure 29) characterizing the dependence of a on ξ, for three different values of α (0.000003; 0.000001; 0.0000003). Values of a and ψ corresponding to some stationary mode

$$t=0,\ \xi(0)=-0.006522,\ a=0.0174814,\ \psi=1.2620,$$

are taken as the initial values.

The same figure contains the stationary resonance curve 1 constructed according to (3.173).

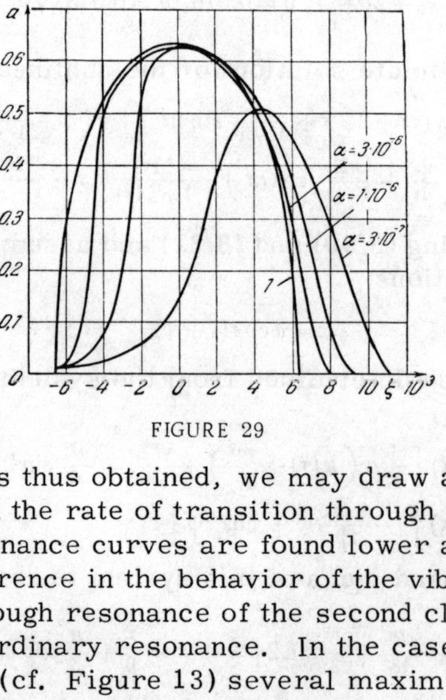

FIGURE 29

Analyzing the curves thus obtained, we may draw a number of conclusions. As usual, when the rate of transition through resonance increases, the maxima of the resonance curves are found lower and shifted. There appears a radical difference in the behavior of the vibration amplitude during a transition through resonance of the second class in contrast to a transition through an ordinary resonance. In the case of a transition through an ordinary resonance (cf. Figure 13) several maxima of smaller magnitude appeared on the amplitude curve after the first maximum, and thus the vibrations were found to take up the character of damping beats. In a transition through a second class resonance there are no beats — the amplitude curve decreases continuously after attaining a maximum value, resembling by its form a curve corresponding to a stationary mode, although a little bit shifted to the right.

As a second example we study the transition through a parameteric resonance.

Let us consider the mechanical system — a rod of length l with hinged ends (Figure 30) under the action of a "periodic" longitudinal force $E\cos\theta$, whose instantaneous frequency $\frac{d\theta}{dt}=\nu(\tau)$ is slowly varying with time. We assume that this frequency, while varying, passes through the double value of the first (lowest) normal frequency of the rod.

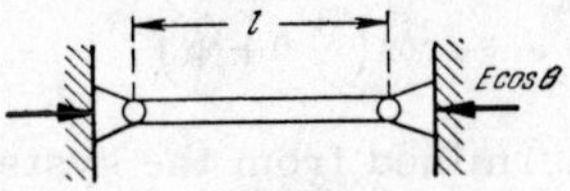

FIGURE 30

The differential equation for the transversal vibrations of the rod may be written in the form

$$EI\frac{\partial^4 y}{\partial z^4}+\frac{\gamma A}{g}\frac{\partial^2 y}{\partial t^2}+E_0\cos\theta\,\frac{\partial^2 y}{\partial z^2}=0, \tag{3.176}$$

where A is the area of the transverse cross section; EI, the rigidity; γ, the density of the material of the rod; g, the acceleration due to gravity.

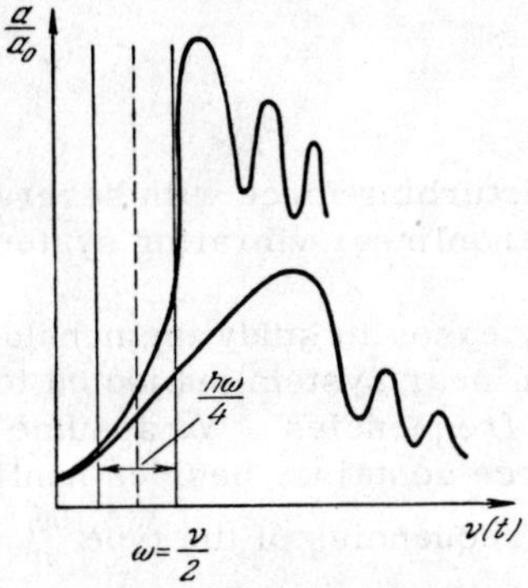

FIGURE 31

The boundary conditions for equations (3.176), under our assumptions, are

$$\left.\begin{aligned} y\,|_{z=0}=0, &\quad \left.\frac{\partial^2 y}{\partial z^2}\right|_{z=0}=0,\\ y\,|_{z=l}=0, &\quad \left.\frac{\partial^2 y}{\partial z^2}\right|_{z=l}=0. \end{aligned}\right\} \tag{3.177}$$

Assuming that vibrations at the fundamental frequency dominate in the system, while higher harmonics are either rapidly attenuated or not excited at all, we substitute

$$y = x\sin\pi\frac{z}{l} \tag{3.178}$$

and may thereby reduce the study of the first tone of motion described by equation (3.176) to the investigation of the following second order differential equation with slowly varying coefficients

$$\frac{d^2x}{dt^2}+\omega^2(1-h\cos\theta)\,x=0, \tag{3.179}$$

where

$$h=\frac{E_0 l^2}{EI\pi^2},\quad \omega^2=\frac{EIg\pi^4}{\gamma A l^4}.$$

For investigating the stationary oscillatory process we construct the first approximation corresponding to the fundamental parameteric resonance

$p=1,\ q=2$. Using formulas (3.60) and (3.66) and assuming $\nu(\tau)=\nu_0+\alpha t$, we obtain after elementary calculations

$$x=a\cos\left(\frac{1}{2}\theta+\psi\right), \tag{3.180}$$

where a and ψ are to be determined from the system of equations

$$\left.\begin{aligned}\frac{da}{dt}&=-\frac{ah\omega^2}{2(\nu_0+\alpha t)}\sin 2\psi,\\ \frac{d\psi}{dt}&=\omega-\frac{\nu_0+\alpha t}{2}-\frac{h\omega^2}{2(\nu_0+\alpha t)}\cos 2\psi.\end{aligned}\right\} \tag{3.181}$$

Taking numerical values of $h,\ \omega,\ \nu_0,\ \alpha$ and integrating the system (3.181) numerically, we obtain curves of transition through parametric resonance, shown in Figure 31.

We shall return below to consider a few interesting phenomena appearing during nonstationary vibrations of a system in the state of parametric resonance.

§ 9. Action of perturbing force with several harmonics on a nonlinear vibrating system

It is of interest in many cases to study the nonstationary vibrational phenomena occurring in a nonlinear system subjected to the action of a perturbing force with several frequencies. We assume for the sake of simplicity that the perturbing force contains, besides nonlinear terms, also terms with instantaneous frequencies of the type $\frac{d\theta_i}{dt}=\nu_i(\tau)\ (i=1,\ 2)$.

In this case one has to deal with a differential equation of the form

$$\frac{d}{dt}\left\{m(\tau)\frac{dx}{dt}\right\}+c(\tau)x=\varepsilon F\left(\tau,\ x,\ \frac{dx}{dt},\ \theta_1,\ \theta_2\right)^*, \tag{3.182}$$

where $F\left(\tau,\ x,\ \frac{dx}{dt},\ \theta_1,\ \theta_2\right)$ is a periodic function of θ_1 and θ_2 with period 2π; $\tau=\varepsilon t$; $\frac{d\theta_1}{dt}=\nu_1(\tau)$; $\frac{d\theta_2}{dt}=\nu_2(\tau)$; and, in addition, all the conditions imposed on equation (3.1) (p. 63) are satisfied.

To avoid long calculations we examine the resonance zone and close approaches to it, and assume that $\nu_1(\tau)$ and $\nu_2(\tau)$ are within the resonance zone or close to it for any τ in the interval $0\leqslant\tau\leqslant L$, i. e., that the following equalities hold

$$\left.\begin{aligned}\nu_1(\tau)&=\omega(\tau)+\varepsilon\Delta_1(\tau),\\ \nu_2(\tau)&=\omega(\tau)+\varepsilon\Delta_2(\tau).\end{aligned}\right\} \tag{3.183}$$

From these we obtain

$$\nu_2(\tau)=\nu_1(\tau)+\varepsilon\sigma(\tau), \tag{3.184}$$

where

$$\sigma(\tau)=\Delta_2(\tau)-\Delta_1(\tau),$$

* The investigation of an equation of the type (3.182) and an analysis of the mutual influence of several harmonics during a transition through resonance are given in the work of Rubanik /146/. The stationary mode in vibrating systems described by equation (3.182) was studied theoretically and experimentally by A. I. Chekmarev /169/.

and consequently,

$$\theta_2 = \theta_1 + \varepsilon\bar{\theta} \quad \left(\frac{d\bar{\theta}}{dt} = \sigma(\tau)\right) \tag{3.185}$$

in the interval $0 \leqslant \tau \leqslant L$.

We confine ourselves, without loss of generality, to the examination of the fundamental resonance. For constructing an approximate solution of equation (3.182) it is possible to make use directly of the results of § 1 of the present chapter.

Actually, substituting the value of θ_2 given by (3.185) into the right-hand side of equation (3.182), we obtain

$$F\left(\tau,\ x,\ \frac{dx}{dt},\ \theta_1,\ \theta_2\right) = F\left(\tau,\ x,\ \frac{dx}{dt},\ \theta_1,\ \theta_1 + \varepsilon\bar{\theta}\right) = F_1\left(\tau,\ x,\ \frac{dx}{dt},\ \theta_1\right)$$

and consequently equation (3.182) assumes the form

$$\frac{d}{dt}\left\{m(\tau)\frac{dx}{dt}\right\} + c(\tau)\,x = \varepsilon F_1\left(\tau,\ x,\ \frac{dx}{dt},\ \theta_1\right), \tag{3.186}$$

which is of the type (3.1) examined in § 1.

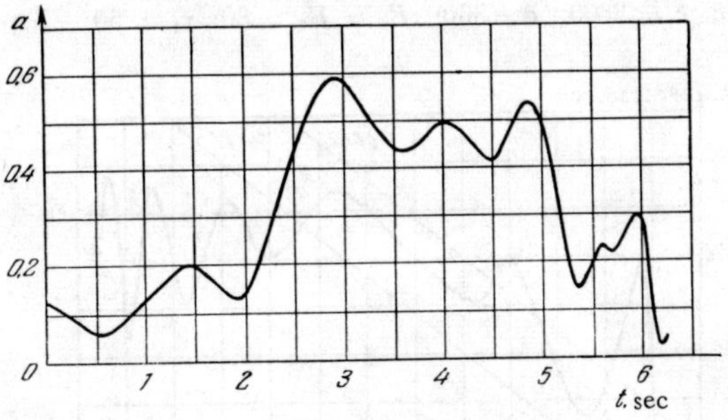

FIGURE 32

As an example illustrating the nonstationary mode in a nonlinear vibrating system subjected to the action of a perturbing force containing several harmonic components, we consider a system with a single degree of freedom described by the differential equation

$$m\frac{d^2x}{dt^2} + 2n\frac{dx}{dt} + cx + dx^3 = E_1\cos\theta_1 + E_2\cos\theta_2, \tag{3.187}$$

where $m,\ n,\ c,\ d,\ E_1,\ E_2$ are constants; $\frac{d\theta_1}{dt} = \nu_1(\tau)$; $\frac{d\theta_2}{dt} = \nu_2(\tau)$; and $\nu_2(\tau) = \nu_1(\tau) + \varepsilon\sigma(\tau)$.

Performing a change of variables

$$\sqrt{\frac{c}{m}}\,t = t_1,\quad \sqrt{\frac{d}{c}}\,x = x_1,$$

we obtain the equation

$$\frac{d^2x_1}{dt_1^2} + 2\delta\frac{dx_1}{dt_1} + x_1 + x_1^3 = h_1\cos\theta_1 + h_2\cos\theta_2, \tag{3.188}$$

where $\delta = \frac{n}{\sqrt{mc}}$, $h_1 = \frac{E_1}{c}\sqrt{\frac{d}{c}}$, $h_2 = \frac{E_2}{c}\sqrt{\frac{d}{c}}$.

Making use of the formulas of § 1, we find after a few calculations a solution in the first approximation for equation (3.188) in the form

$$x = a\cos(\theta_1 + \psi), \tag{3.189}$$

where a and ψ are to be determined from the system of equations of the first approximation

$$\left.\begin{aligned} \frac{da}{dt} &= -\frac{\delta a}{2} - \frac{h_1}{2}\sin\psi - \frac{h_2}{2}\sin(\psi - \varepsilon\theta), \\ \frac{d\psi}{dt} &= -\varepsilon\Delta_1(\tau) + \frac{3}{8}a^2 - \frac{h_1}{2a}\cos\psi - \frac{h_2}{2a}\cos(\psi - \varepsilon\theta). \end{aligned}\right\} \tag{3.190}$$

Let us now assume that the frequencies of the external perturbing force, while varying with time, cross successively the value of the normal frequency of the system.

For constructing graphs characterizing the nonstationary process, the system of equations (3.190) must be integrated numerically. This results in a resonance curve characterizing the successive transitions of the first and second frequency of the external perturbing forces (Figure 32). Figure 33 contains a graph showing the relation between $\frac{d\theta}{dt}$ and t. The following numerical values of the system parameters were taken in the integration $m = 1$; $n = 0.5$; $c = 3600$; $d = 360$; $E_1 = E_2 = 80$; $\nu_1 = 50 + 4t$; $\nu_2 = 45 + 4t$.

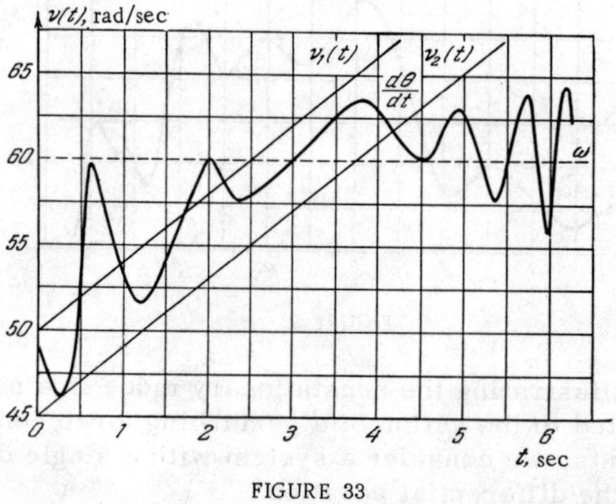

FIGURE 33

It is convenient in some cases to apply in the analysis of stationary modes, and also in the study of nonstationary modes in resonance zones a different method for constructing approximate solutions /146/. Assuming that the right-hand side of equation (3.182) may be written in the form

$$F\left(\tau, x, \frac{dx}{dt}, \theta_1, \theta_2\right) = -2\lambda\frac{dx}{dt} + \varepsilon f\left(\tau, x, \frac{dx}{dt}\right) + E_1\cos\theta_1 + E_2\cos\theta_2, \tag{3.191}$$

one can look for an approximate solution of equation (3.182) in the form of an asymptotic series

$$x = a_1\cos(\theta_1 + \psi_1) + a_2\cos(\theta_2 + \psi_2) + \varepsilon u_1(\tau, a_1, a_2, \theta_1, \theta_2, \psi_1, \psi_2) + \ldots, \tag{3.192}$$

where a_k and ψ_k $(k = 1, 2)$ are functions of time determined from the system of equations

$$\left.\begin{aligned} \frac{da_k}{dt} &= \varepsilon A_1^{(k)}(\tau, a_1, a_2, \psi_1, \psi_2) + \ldots, \\ \frac{d\psi_k}{dt} &= \varepsilon B_1^{(k)}(\tau, a_1, a_2, \psi_1, \psi_2) + \ldots \\ &(k = 1, 2), \end{aligned}\right\} \tag{3.193}$$

while in the zero approximation a_k and ψ_k $(k=1, 2)$ are determined from the system

$$\left.\begin{array}{c} 2\lambda a_k\nu_k + E_k \sin\psi_k = 0, \\ (1-\nu_k^2)\, a_k - E_k \cos\psi_k = 0 \\ (k=1,\ 2). \end{array}\right\} \tag{3.194}$$

Thus, in order to derive approximate solutions the values of $u_1(\tau, a_1, a_2, \theta_1, \theta_2, \psi_1, \psi_2), \ldots, A_1^{(k)}(\tau, a_1, a_2, \psi_1, \psi_2), \ldots, B_1^{(k)}(\tau, a_1, a_2, \psi_1, \psi_2), \ldots$ must be found. We shall derive these quantities in the general case, and construct only the solution in the first approximation for the nonlinear differential equation (3.187).

In the first approximation we have

$$x = a_1 \cos(\theta_1 + \psi_1) + a_2 \cos(\theta_2 + \psi_2), \tag{3.195}$$

where a_1, a_2, ψ_1, ψ_2 are to be determined from the system of equations

$$\left.\begin{array}{l} \frac{da_i}{dt} = -\frac{\delta a_i}{2} - \frac{E_i}{2\nu_i}\sin\psi_i - \frac{d\nu_i}{dt}\, a_i, \\ \frac{d\psi_i}{dt} = 1 - \nu_i + \frac{3}{8}(a_i^2 + 2a_{i\pm 1}^2) - \frac{E_i}{2\nu_i a_i}\cos\psi_i \\ \qquad (i=1,\ 2). \end{array}\right\} \tag{3.196}$$

Equating the right-hand sides of the system (3.196) to zero, we obtain the relations for determining the stationary values of the vibration amplitude and phase

$$\left.\begin{array}{l} \delta a_i \nu_i + E_i \sin\psi_i = 0, \\ (1-\nu_i^2)\, a_i + \frac{3}{4}\nu_i (a_i^2 + 2a_{i\pm 1}^2)\, a_i - E_i \cos\psi_i = 0 \\ \qquad (i=1,\ 2). \end{array}\right\} \tag{3.197}$$

Figures 34–37 contain the stationary resonance curves determined by these relations. The first two figures characterize the stationary amplitude and phase values in the case of a rigid characteristic of the nonlinear restoring force $(d>0)$, and the second two — in the case of a soft characteristic $(d<0)$.

Setting up the variational equations, it is possible to determine the stable stationary amplitude and phase values. The corresponding sections on the given graphs are marked by a continuous thick line. The sections corresponding to unstable values are marked by a continuous thin line. The dotted curves correspond to the separate action of the harmonic components of the external perturbing force.

The resonance curves thus obtained allow us to derive a number of conclusions concerning the character of vibrations in nonlinear systems under the combined action of two harmonic perturbing forces. As is known, one of the basic conclusions is (cf. /169/) the impossibility of simultaneous excitation of high vibration amplitudes of several harmonics in nonlinear vibrating systems. If high resonance amplitudes of a definite harmonic are built up, then the amplitudes of other harmonics are small compared to their values for a separate operation — their growth is choked by the building up of resonant vibrations of the first harmonic.

Further, in the case of combined action of two harmonic perturbing forces on a nonlinear vibrating system, their mutual influence depends strongly on the nonlinear character of the system. Thus, for example, the development of a large resonance amplitude of a higher frequency harmonic in the case of a rigid characteristic of the nonlinear elastic force causes a decrease in the amplitude of a lower frequency harmonic, and the

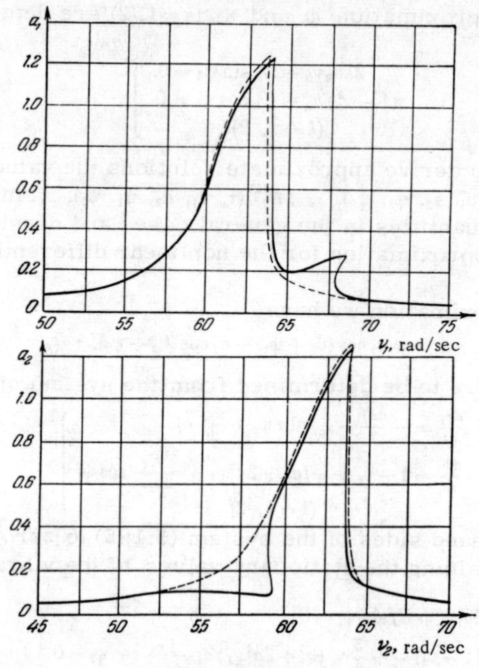

FIGURE 34

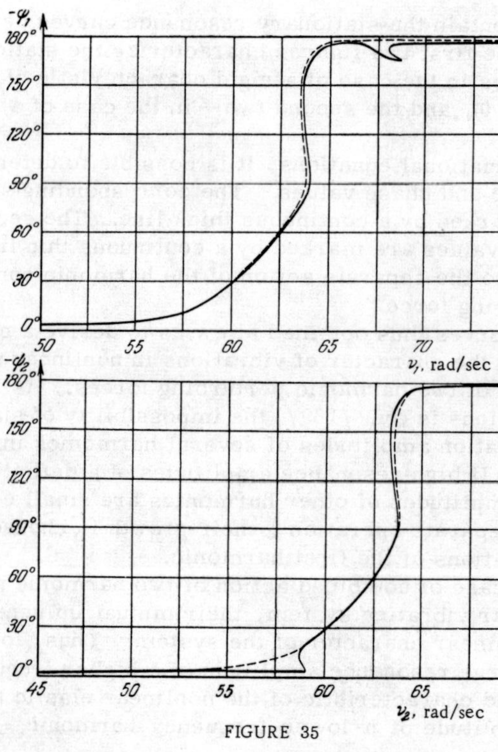

FIGURE 35

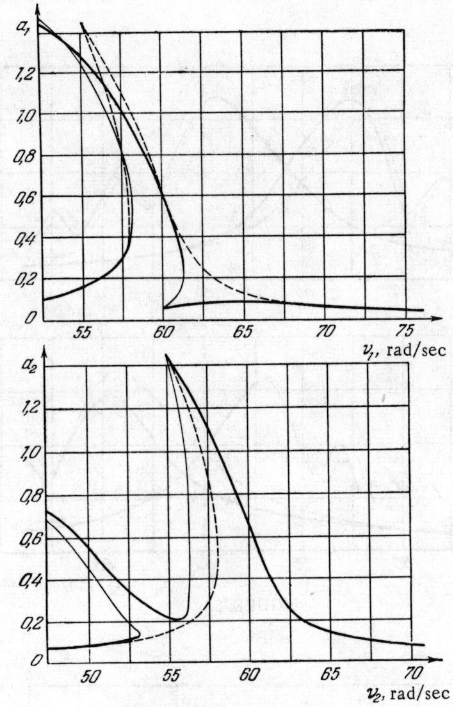

FIGURE 36

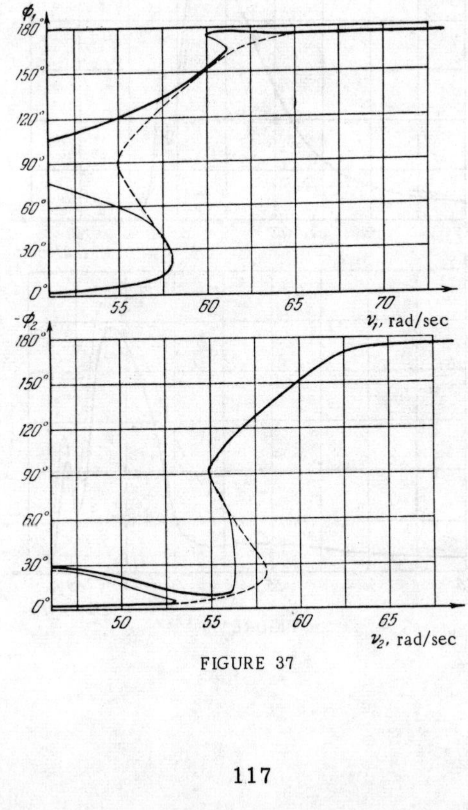

FIGURE 37

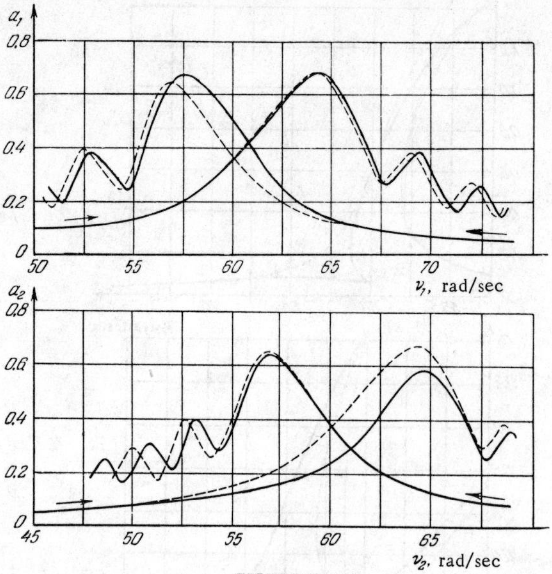

FIGURE 38

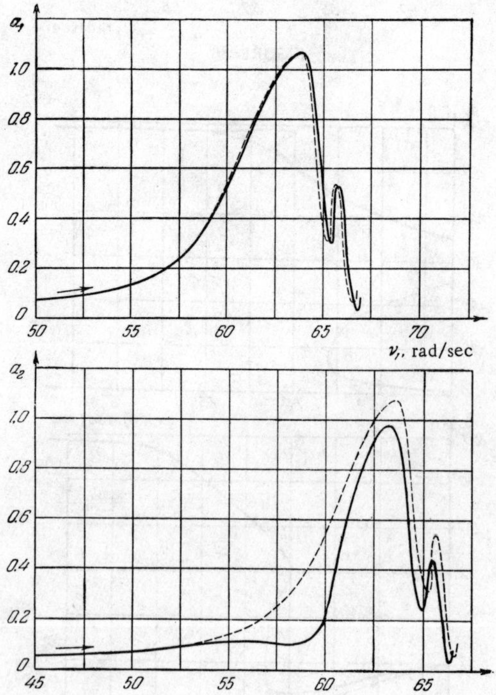

FIGURE 39

development of a large resonance amplitude of a lower frequency harmonic causes not a decrease, but an increase in the amplitude of a higher frequency harmonic. On the other hand, in the case of a soft characteristic of the nonlinear restoring force, a development of a resonance amplitude of a higher frequency harmonic favors an increase in the amplitude of a lower frequency harmonic and the development of a large resonance amplitude of a lower frequency harmonic now favors a decrease in the amplitude of a lower frequency harmonic.

An essential conclusion is that the development of large amplitudes of each of the harmonics during their combined action almost coincides with the development of the amplitudes of the resonating harmonics in the case of their separate action. Besides, the influence of the harmonics far from resonance is insignificant; but if the regions of development of large resonance amplitudes of two or more harmonics overlap, then the mutual influence of these harmonics is considerable.

All these conclusions are also verified by experiment /169/.

We shall now discuss a successive transition of the harmonics of external force through a resonance (as the most interesting nonstationary mode). Integrating the system of equations (3.196) for the same numerical values (cf. p. 114), we obtain curves of transition through resonance shown in Figures 38, 39 for the case of a rigid characteristic of the nonlinear restoring force, and in Figures 40, 41 for the case of a soft characteristic. The dotted lines represent the resonance curves corresponding to the separate action of the harmonics (we note that the envelope of the total combined vibration corresponding to Figure 38 was obtained before in Figure 32).

The curves in Figures 39, 41 characterize a transition through resonance at a four times slower rate than in the preceding case.

Analyzing the curves thus obtained, it is possible to derive a number of conclusions.

Exactly as in the stationary case, the harmonics far from resonance do not affect each other significantly. During the successive transition of the two harmonics through resonance, the harmonic which is first to cross the resonance does not experience the influence of the other; on the other hand, the second harmonic experiences a considerable influence from the side of the first, which gives rise, depending upon the nonlinearity character of the system and the direction of transition through resonance, to an increase or decrease in its amplitude.

Thus, for example, during a successive transition of a system with a rigid characteristic of the nonlinear restoring force through resonance towards the direction of increasing frequencies of the perturbing forces, the higher harmonic (which, naturally, is the first to pass through resonance) does not acquire a significant deviation, while the amplitude of the lower harmonic is considerably diminished. On the other hand, during a successive transition through resonance towards the direction of decreasing frequencies of the perturbing forces, the lower harmonic (which in this case crosses the resonance first) does not experience a considerable influence, while the amplitude of the higher harmonic increases considerably.

However, in a system with a soft characteristic of the nonlinear restoring forces, the amplitude of the lower harmonic increases during a transition through resonance in the direction of increasing frequency, while the amplitude of the higher harmonic decreases significantly during a transition in the direction of decreasing frequency of the perturbing force.

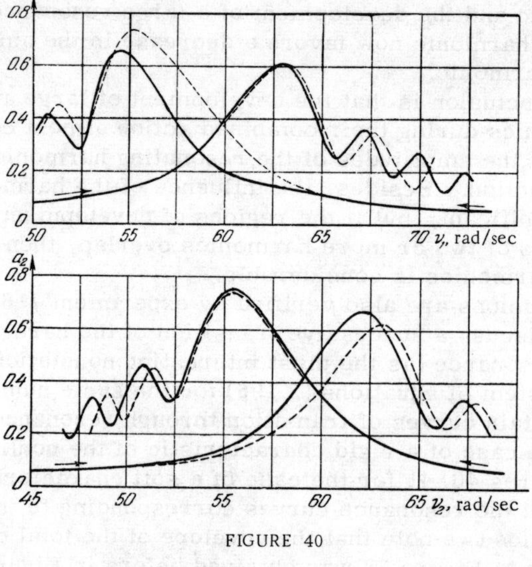

FIGURE 40

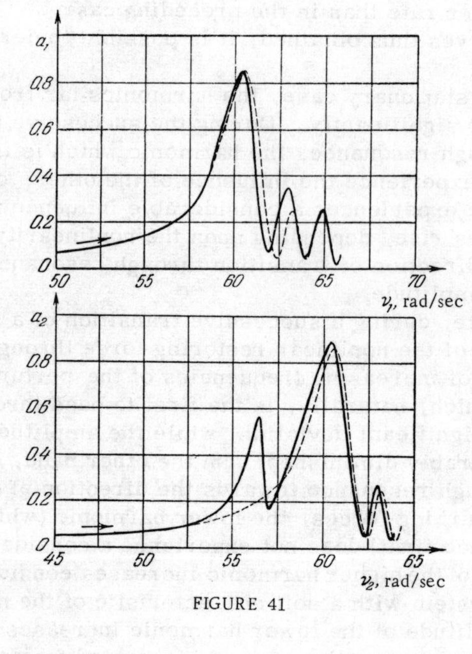

FIGURE 41

The mutual influence of harmonics during a nonstationary mode in nonlinear vibrating systems depends essentially on the nonlinearity character of the system, and also on the direction and velocity of change in the frequencies of the perturbing forces. An increase of the latter implies a weaker mutual influence between the harmonics, which is physically clear; the occurring resonant vibrations corresponding to one of the harmonics do not have enough time during a fast transition through resonance to exert a significant influence on the nature of development of the other harmonics.

These conclusions allow us in many cases to confine ourselves, while calculating the vibrations of nonlinear systems (in stationary as well as in nonstationary modes), to a computation of the amplitudes of the separate harmonics of the external force, and then to take into account their mutual influence. This is very important, since the solution of the problem in the case of a polyharmonic excitation is, even when the nonlinearity characteristic is given analytically, extremely complicated already in the first approximation. The solution for a separate action of each of the harmonics is, on the other hand, easily obtained in the first approximation for any given nonlinearity characteristic.

§ 10. The effect of external periodic forces on strongly nonlinear vibrating systems

The study of resonance phenomena in essentially nonlinear vibrating systems is known to be a difficult and, despite its importance, a poorly explored problem.

In the present section we present only a few concepts concerning the construction of asymptotic solutions for more complicated cases of nonlinear vibrating systems then those considered before. We confine ourselves to formal operations, not claiming in several places to an exhaustive completeness of presentation, but supposing that they may become useful in a further more profound study of resonance phenomena in essentially nonlinear systems.

Thus, let a vibrating system be described by a differential equation of the type

$$\frac{d^2x}{dt^2}+f(\tau, x)=\varepsilon F\left(\tau, \theta, x, \frac{dx}{dt}, \varepsilon\right), \tag{3.198}$$

where, as usual, ε is a small positive parameter; $\tau=\varepsilon t$, the "slowing" time; $\frac{d\theta}{dt}=\nu(\tau)$; and the function $F\left(\tau, \theta, x, \frac{dx}{dt}, \varepsilon\right)$ is periodic in θ with period 2π and analytic in ε:

$$F\left(\tau, \theta, x, \frac{dx}{dt}, \varepsilon\right)=\sum_{n=0}^{\infty} \varepsilon^n F_n\left(\tau, \theta, x, \frac{dx}{dt}\right). \tag{3.199}$$

Furthermore, the functions $f(\tau, x)$, $\nu(\tau)$, $F\left(\tau, \theta, x, \frac{dx}{dt}, \varepsilon\right)$ are unlimitedly differentiable with respect to τ for any τ within the interval $0\leqslant\tau\leqslant L$. We also assume that $F_n\left(\tau, \theta, x, \frac{dx}{dt}\right)$ $(n=0, 1, 2, \ldots)$ may be represented in the form of finite Fourier sums

$$F_n\left(\tau, \theta, x, \frac{dx}{dt}\right) = \sum_{\sigma=-N}^{N} e^{i\sigma\theta} F_n^{(\sigma)}\left(\tau, x, \frac{dx}{dt}\right) \quad (n=0, 1, 2, \dots), \tag{3.200}$$

where the coefficients are certain polynomials of x, $\frac{dx}{dt}$.

We shall consider a simpler case for investigation when, for the "unperturbed" equation

$$\frac{d^2x}{dt^2} + f(\tau, x) = 0 \qquad (\tau = \text{const}) \tag{3.201}$$

a periodic solution

$$\left.\begin{aligned} x &= z(\tau, \psi, a), \\ z(\tau, \psi+2\pi, a) &\equiv z(\tau, \psi, a), \end{aligned}\right\} \tag{3.202}$$

is known, in which $\psi = \omega(\tau, a)t + \varphi$; a and φ are arbitrary constants; τ is some constant parameter.

During the investigation of vibrations described by equation (3.198) we shall be interested in the following two cases: the "resonance" case when the frequency of the external force $\nu(\tau)$ and the normal frequency of the system $\omega(\tau, a)$ satisfy for some values of the parameters τ and a the condition

$$\nu(\tau) \approx \frac{p}{q}\,\omega(\tau, a), \tag{3.203}$$

where p and q are mutually prime integers, and the "nonresonance" case when condition (3.203) is not satisfied for any values of τ and a.

The frequency $\omega(\tau, a)$ in (3.203) depends not only on τ, the domain of variation of which is known beforehand, but also on the quantity $a(\tau)$— an unknown function of time, the variation domain of which is unknown as long as equation (3.198) is not solved. The theoretical solution of the problem of whether a given case is the "resonance" or the "nonresonance" case is very complicated. It is therefore necessary to assume one of the two cases on the grounds of physical considerations, construct the equations of the first approximation for a and ψ, and, solving them, to verify whether condition (3.203) is satisfied or not.

We begin the study of equation (3.198) with the nonresonance case as the simpler of the two.

We assume that condition (3.203) is not satisfied for any values of τ and a in their domain of variation. We transform equation (3.198), as in the foregoing chapter, to the new variables a and ψ by the substitution

$$\left.\begin{aligned} x &= z(\tau, \psi, a), \\ \frac{dx}{dt} &= \omega(\tau, a)\, z'_\psi(\tau, \psi, a), \end{aligned}\right\} \tag{3.204}$$

where $\tau = \varepsilon t$.

After a few operations we obtain

$$\left.\begin{aligned} \frac{da}{dt} &= \varepsilon\Phi_1(\tau, \theta, \psi, a, \varepsilon), \\ \frac{d\psi}{dt} &= \omega(\tau, a) + \varepsilon\Phi_2(\tau, \theta, \psi, a, \varepsilon), \end{aligned}\right\} \tag{3.205}$$

where

$$\Phi_1(\tau, \theta, \psi, a, \varepsilon) = \\ = \frac{1}{D_0(\tau, a)}\{-F(\tau, \theta, z, \omega z'_\psi, \varepsilon)\, z'_\psi + (\omega z'_\psi)'_\tau z'_\psi - \omega z''_{\psi^2} z'_\tau\}, \tag{3.206}$$

$$\left.\begin{aligned}\Phi_2(\tau, \theta, \psi, a, \varepsilon) &= \\ &= \frac{1}{D_0(\tau, a)}\{F(\tau, \theta, z, \omega z'_\psi, \varepsilon) z'_a - (\omega z'_\psi)'_\tau z'_a + (\omega z'_\psi)'_a z'_\tau\}, \\ D_0(\tau, a) &= [\omega z'_a z''_{\psi^2} - z'_\psi(\omega z'_\psi)'_a]_{\psi=0}.\end{aligned}\right\} \quad (3.206)$$

The following expansions are obviously valid:

$$\Phi_j(\tau, \theta, \psi, a, \varepsilon) = \sum_{k=0}^{\infty} \varepsilon^k \Phi_j^{(k)}(\tau, \theta, \psi, a) \qquad (j=1, 2), \quad (3.207)$$

and also

$$\Phi_j^{(k)}(\tau, \theta, \psi, a) = \sum_n \sum_m \Phi_{j,n,m}^{(k)}(\tau, a) e^{i(n\theta+m\psi)} \qquad (j=1, 2; k=0, 1, 2, \ldots), \quad (3.208)$$

where

$$\Phi_{j,n,m}^{(k)}(\tau, a) = \frac{1}{4\pi^2}\int_0^{2\pi}\int_0^{2\pi} \Phi_j^{(k)}(\tau, \theta, \psi, a) e^{-i(n\theta+m\psi)} d\theta\, d\psi. \quad (3.209)$$

To make calculations easier we assume in what follows that the right-hand sides of (3.208) are finite Fourier sums for any value of the index k. This may always be achieved in practice, by truncating the infinite convergent Fourier series in the angular variable ψ (resulting from the expansion of the right-hand side of (3.198) after the substitution of the periodic solution (3.202), and of some terms appearing on the right-hand sides of (3.206)) so that the remainder would be of the order of ε^2, and referring this remainder to terms of the second order with respect to the parameter ε.

Let us now write down the expressions

$$\left.\begin{aligned}u_1(\tau, \theta, \psi, a) &= \sum_{\substack{n \\ (n^2+m^2\neq 0)}}\sum_m \frac{e^{i(n\theta+m\psi)}}{i[n\nu(\tau)+m\omega(\tau, a)]}\Phi_{1,n,m}^{(0)}(\tau, a) + u_{10}(\tau, a), \\ v_1(\tau, \theta, \psi, a) &= \sum_{\substack{n \\ (n^2+m^2\neq 0)}}\sum_m \left\{\frac{\Phi_{2,n,m}^{(0)}(\tau, a)}{i[n\nu(\tau)+m\omega(\tau, a)]} - \frac{\Phi_{1,n,m}^{(0)}(\tau, a)\,\omega'_a(\tau, a)}{[n\nu(\tau)+m\omega(\tau, a)]^2}\right\}\times \\ &\qquad \times e^{i(n\theta+m\psi)} + v_{10}(\tau, a),\end{aligned}\right\} \quad (3.210)$$

which obviously satisfy the indentities

$$\left.\begin{aligned}\nu(\tau)\frac{\partial u_1}{\partial\theta} + \omega(\tau, a)\frac{\partial u_1}{\partial\psi} &= \Phi_1^{(0)}(\tau, \theta, \psi, a) - \Phi_{1,0,0}^{(0)}(\tau, a), \\ \nu(\tau)\frac{\partial v_1}{\partial\theta} + \omega(\tau, a)\frac{\partial v_1}{\partial\psi} &= \Phi_2^{(0)}(\tau, \theta, \psi, a) - \Phi_{2,0,0}^{(0)}(\tau, a) + \\ &\quad + \omega'_a(\tau, a)[u_1(\tau, \theta, \psi, a) - u_{10}(\tau, a)].\end{aligned}\right\} \quad (3.211)$$

We then perform the following change of variables in equations (3.205)

$$\left.\begin{aligned}a &= a_1 + \varepsilon u_1(\tau, \theta, \psi_1, a_1), \\ \psi &= \psi_1 + \varepsilon v_1(\tau, \theta, \psi_1, a_1).\end{aligned}\right\} \quad (3.212)$$

Then, after a few calculations we obtain instead of the system (3.205) the system

$$\left.\begin{aligned}\frac{da_1}{dt} &= \varepsilon\Phi_{1,0,0}^{(0)}(\tau, a_1) + \varepsilon^2 R_1(\tau, \theta, \psi_1, a_1, \varepsilon), \\ \frac{d\psi_1}{dt} &= \omega(\tau, a_1) + \varepsilon\{\Phi_{2,0,0}^{(0)}(\tau, a_1) + \omega'_a(\tau, a_1)u_{10}(\tau, a_1)\} + \\ &\quad + \varepsilon^2 R_2(\tau, \theta, \psi_1, a_1, \varepsilon),\end{aligned}\right\} \quad (3.213)$$

where we have used the notation

$$
\left.
\begin{aligned}
R_1(\tau, \theta, \psi_1, a_1, 0) = {} & \Phi_{1\psi}^{(0)\prime}(\tau, \theta, \psi_1, a_1) v_1 + \Phi_{1a}^{(0)\prime}(\tau, \theta, \psi_1, a_1) u_1 + \\
& + \Phi_1^{(1)}(\tau, \theta, \psi_1, a_1) - \omega_a'(\tau, a_1) u_1 \frac{\partial u_1}{\partial \psi_1} - \\
& - \Phi_2^{(0)}(\tau, \theta, \psi_1, a_1) \frac{\partial u_1}{\partial \psi_1} + \nu(\tau) \frac{\partial v_1}{\partial \theta} \frac{\partial u_1}{\partial \psi_1} - \\
& - \Phi_1^{(0)}(\tau, \theta, \psi_1, a_1) \frac{\partial u_1}{\partial a_1} + \nu(\tau) \frac{\partial u_1}{\partial \theta} \frac{\partial u_1}{\partial a_1} + \\
& + \omega(\tau, a_1) \frac{\partial u_1}{\partial \psi_1} \left(\frac{\partial u_1}{\partial a_1} + \frac{\partial v_1}{\partial \psi_1} \right) - \frac{\partial u_1}{\partial \tau}, \\
R_2(\tau, \theta, \psi_1, a_1, 0) = {} & \Phi_{2\psi}^{(0)\prime}(\tau, \theta, \psi_1, a_1) v_1 + \Phi_{2a}^{(0)\prime}(\tau, \theta, \psi_1, a_1) u_1 + \\
& + \Phi_2^{(1)}(\tau, \theta, \psi_1, a_1) - \omega_a'(\tau, a_1) u_1 \frac{\partial v_1}{\partial \psi_1} - \\
& - \Phi_1^{(0)}(\tau, \theta, \psi_1, a_1) \frac{\partial v_1}{\partial \psi_1} + \nu(\tau) \frac{\partial v_1}{\partial \theta} \frac{\partial v_1}{\partial \psi_1} - \\
& - \Phi_1^{(0)}(\tau, \theta, \psi_1, a_1) \frac{\partial v_1}{\partial a_1} + \nu(\tau) \frac{\partial u_1}{\partial \theta} \frac{\partial v_1}{\partial a_1} + \\
& + \frac{1}{2} \omega_{a^2}''(\tau, a_1) u_1^2 - \omega(\tau, a_1) \frac{\partial u_1}{\partial a_1} \left(\frac{\partial u_1}{\partial a_1} + \frac{\partial v_1}{\partial \psi_1} \right) - \frac{\partial v_1}{\partial \tau}.
\end{aligned}
\right\} \tag{3.214}
$$

The functions $R_1(\tau, \theta, \psi_1, a_1, \varepsilon)$ and $R_2(\tau, \theta, \psi_1, a_1, \varepsilon)$ possess the same properties as the right-hand sides of equations (3.205).

Discarding in the right-hand sides of equations (3.213) terms of the order of ε^2, we obtain the system of equations

$$
\left.
\begin{aligned}
\frac{da_1}{dt} &= \varepsilon \Phi_{1,0,0}^{(0)}(\tau, a_1), \\
\frac{d\psi_1}{dt} &= \omega(\tau, a_1) + \varepsilon \{ \Phi_{2,0,0}^{(0)}(\tau, a) + \omega_a'(\tau, a_1) u_{10}(\tau, a_1) \},
\end{aligned}
\right\} \tag{3.215}
$$

where the right-hand sides do not depend on θ and ψ.

The obtained system of equations (3.215) has the same structure as the system (2.106), and may be integrated according to the scheme given in § 5 of the present chapter.

However, the integration of the system (3.215) over the interval $[0, T]$ gives the values of a_1 and $\omega(\tau, a_1)$ with an error of the order of ε, and the value of the required solution of equation (3.198)

$$x_1 = z(\tau, \psi_1, a_1)$$

will differ from the value of the correct solution by quantities of finite order.

In order to improve the accuracy it is necessary, exactly as in the second chapter, to perform another change of variables in equations (3.213) defined by the formulas

$$
\left.
\begin{aligned}
a_1 &= a_2 + \varepsilon^2 u_2(\tau, \theta, \psi_2, a_2), \\
\psi_1 &= \psi_2 + \varepsilon^2 v_2(\tau, \theta, \psi_2, a_2),
\end{aligned}
\right\} \tag{3.216}
$$

where

$$
\left.
\begin{aligned}
u_2(\tau, \theta, \psi_2, a_2) = {} & \sum_{\substack{n, m \\ (n^2+m^2 \neq 0)}} \frac{R_{1,n,m}^{(0)}(\tau, a_2)}{i[n\nu(\tau) + m\omega(\tau, a_2)]} e^{i[n\theta + m\psi_2]} + u_{20}(\tau, a_2), \\
v_2(\tau, \theta, \psi_2, a_2) = {} & \sum_{\substack{n, m \\ (n^2+m^2 \neq 0)}} \left\{ \frac{R_{2,n,m}^{(0)}(\tau, a_2)}{i[n\nu(\tau) + m\omega(\tau, a_2)]} - \right. \\
& \left. - \frac{\omega_a(\tau, a_2) R_{1,n,m}^{(0)}(\tau, a_2)}{[n\nu(\tau) + m\omega(\tau, a_2)]^2} \right\} e^{i[n\theta + m\psi_2]} + v_{20}(\tau, a_2),
\end{aligned}
\right\} \tag{3.217}
$$

$$
R_{i,n,m}^{(0)}(\tau, a_2) = \frac{1}{4\pi^2} \int_0^{2\pi} \int_0^{2\pi} R_i(\tau, \theta, \psi, a_2, 0) e^{-i[n\theta + m\psi]} d\theta \, d\psi \quad (i = 1, 2). \tag{3.218}
$$

Then, neglecting terms of higher order (than ε^3 in the first equation, and ε^2 in the second), we obtain the system of equations

$$\left.\begin{aligned}\frac{da_2}{dt} &= \varepsilon\Phi^{(0)}_{1,0,0}(\tau, a_2) + \varepsilon^2 R^{(0)}_{1,0,0}(\tau, a_2),\\ \frac{d\psi_2}{dt} &= \omega(\tau, a_2) + \varepsilon\{\Phi^{(0)}_{2,0,0}(\tau, a_2) + \omega'_a(\tau, a_2)\, u_{10}(\tau, a_2)\},\end{aligned}\right\} \tag{3.219}$$

from which the quantity a_2 can be calculated with an error of the order of ε^2, the phase of vibrations ψ_2 with an error of the order of ε, and with this the value x with an error of the order of ε.

The functions $u_{10}(\tau, a)$ and $v_{10}(\tau, a)$ appearing on the right-hand sides of (3.210) are necessarily chosen so that the substitution formulas (3.212) be unique. Their choice depends essentially on the particular problem under consideration.

In order to avoid cumbersome calculations and simplify the resultant formulas, we assume that the solution of the "unperturbed" equation (3.204) may be represented as a finite sum (we have not checked the validity of this assumption on actual examples)

$$z(\tau, \psi, a) = a\cos\psi + \sum_n z_n(\tau, a)\, e^{in\psi} \qquad [n = \pm 3, \pm 5, \ldots, \pm(2m+1)]. \tag{3.220}$$

We also require that after the change of variables according to formulas (3.212) a_1 also be the full amplitude of the first harmonic of the angular variable ψ_1. After a few operations we then obtain expressions for $u_{10}(\tau, a)$, $v_{10}(\tau, a)$ which resemble by their structure the expressions (2.105) derived in the preceding chapter.

We now turn to the "resonance" case as the more interesting of the two. The condition (3.203) is assumed to be satisfied for certain values of τ and a, and for simplicity we consider the case when $p = q = 1$, i. e., the case when a fundamental resonance may occur in the system for certain values of τ and a from a definite interval.

As in the foregoing case, we introduce into equations (3.198) the new variables a and ϑ defined by

$$\left.\begin{aligned}x &= z(\tau, \theta+\vartheta, a),\\ \frac{dx}{dt} &= \omega(\tau, a)\, z'_\psi(\tau, \theta+\vartheta, a) \qquad (\theta+\vartheta = \psi),\end{aligned}\right\} \tag{3.221}$$

which are obtained by replacing ψ in formulas (3.204) by the sum $\theta+\vartheta$.

After the substitution of (3.221) in equations (3.198), we obtain for the new variables a and ϑ the system of equations

$$\left.\begin{aligned}\frac{da}{dt} &= \varepsilon\Phi_1(\tau, \theta, \theta+\vartheta, a, \varepsilon),\\ \frac{d\vartheta}{dt} &= \omega(\tau, a) - \nu(\tau) + \varepsilon\Phi_1(\tau, \theta, \theta+\vartheta, a, \varepsilon),\end{aligned}\right\} \tag{3.222}$$

where we have used the notation

$$\left.\begin{aligned}\Phi_1(\tau, \theta, \theta+\vartheta, a, \varepsilon) &= \frac{1}{D_0(\tau, a)}\Big\{-F(\tau, \theta, z, \omega z'_\psi, \varepsilon)\, z'_\psi - \\ &\qquad -(\omega z'_\psi)'_\tau z'_\psi - \omega z''_{\psi 2} z'_\tau\Big\},\\ \Phi_2(\tau, \theta, \theta+\vartheta, a, \varepsilon) &= \frac{1}{D_0(\tau, a)}\Big\{F(\tau, \theta, z, \omega z'_\psi, \varepsilon)\, z'_a - \\ &\qquad -(\omega z'_\psi)'_\tau z'_a + (\omega z'_\psi)'_a z'_\tau\Big\}\end{aligned}\right\} \tag{3.223}$$

and assumed that

$$\Phi_j(\tau,\ \theta,\ \theta+\vartheta,\ a,\ \varepsilon)=\sum_{k=0}^{\infty}\varepsilon^k\Phi_j^{(k)}(\tau,\ \theta,\ \theta+\vartheta,\ a)\qquad (j=1,\ 2), \tag{3.224}$$

and also

$$\Phi_j^{(k)}(\tau,\ \theta,\ \theta+\vartheta,\ a)=\sum_{n,m}\Phi^{(k)}{}_{j,n,m}(\tau,\ a)\,e^{i[n\theta+m(\theta+\vartheta)]}, \tag{3.225}$$

where

$$\Phi_{j,n,m}^{(k)}(\tau,\ a)=\frac{1}{4\pi^2}\int_0^{2\pi}\int_0^{2\pi}\Phi_j^{(k)}(\tau,\ \theta,\ \theta+\vartheta,\ a)\,e^{-i[n\theta+m(\theta+\vartheta)]}\,d\theta\,d(\theta+\vartheta)$$
$$(j=1,\ 2;\quad k=0,\ 1,\ 2,\ \ldots). \tag{3.226}$$

We now consider the functions

$$\left.\begin{aligned}
u_1(\tau,\ \theta,\ \theta+\vartheta,\ a)&=\sum_{\substack{n,m\\ \left(\substack{n^2+m^2\neq 0\\ n+m\neq 0}\right)}}\frac{\Phi_{1,n,m}^{(0)}(\tau,\ a)}{i\,[n\nu(\tau)+m\omega(\tau,\ a)]}\,e^{i[n\theta+m(\theta+\vartheta)]}+\\
&\qquad+u_{10}(\tau,\ a,\ \vartheta),\\
v_1(\tau,\ \theta,\ \theta+\vartheta,\ a)&=\sum_{\substack{n,m\\ \left(\substack{n^2+m^2\neq 0\\ n+m\neq 0}\right)}}\left\{\frac{\Phi_{2,n,m}^{(0)}(\tau,\ a)}{i\,[n\nu(\tau)+m\omega(\tau,\ a)]}-\right.\\
&\left.\quad-\frac{\omega_a'(\tau,\ a)\,\Phi_{1,n,m}^{(0)}(\tau,\ a)}{[n\nu(\tau)+m\omega(\tau,\ a)]^2}\right\}e^{i[n\theta+m(\theta+\vartheta)]}+v_{10}(\tau,\ a,\ \vartheta),
\end{aligned}\right\} \tag{3.227}$$

which satisfy the identities

$$\left.\begin{aligned}
\nu(\tau)\frac{\partial u_1}{\partial\theta}+\omega(\tau,\ a)\frac{\partial u_1}{\partial(\theta+\vartheta)}&=\Phi_1^{(0)}(\tau,\ \theta,\ \theta+\vartheta,\ a)-\\
&\quad-\sum_{\substack{n,\ m\\ (n+m=0)}}\Phi_{1,n,m}^{(0)}(\tau,\ a)\,e^{im\vartheta},\\
\nu(\tau)\frac{\partial v_1}{\partial\theta}+\omega(\tau,\ a)\frac{\partial v_1}{\partial(\theta+\vartheta)}&=\Phi_2^{(0)}(\tau,\ \theta,\ \theta+\vartheta,\ a)-\\
&\quad-\sum_{\substack{n,\ m\\ (n+m=0)}}\Phi_{2,n,m}(\tau,\ a)\,e^{im\vartheta}+\omega_a'(\tau,\ a)\,[u_1-u_0].
\end{aligned}\right\} \tag{3.228}$$

We then perform on the system (3.222) the change of variables

$$\left.\begin{aligned}
a&=a_1+\varepsilon u_1(\tau,\ \theta,\ \theta+\vartheta_1,\ a_1),\\
\vartheta&=\vartheta_1+\varepsilon v_1(\tau,\ \theta,\ \theta+\vartheta_1,\ a_1).
\end{aligned}\right\} \tag{3.229}$$

Substituting (3.229) in (3.222) and keeping in mind identities (3.228), we obtain

$$\left.\begin{aligned}
\frac{da_1}{dt}&=\varepsilon\sum_{\substack{n,\ m\\ (n+m=0)}}\Phi_{1,\ n,\ m}^{(0)}(\tau,\ a_1)\,e^{im\vartheta_1}+\varepsilon^2R_1(\tau,\ \theta,\ \theta+\vartheta_1,\ a_1,\ \varepsilon),\\
\frac{d\vartheta_1}{dt}&=\omega(\tau,\ a_1)-\nu(\tau)+\varepsilon\left\{\sum_{\substack{n,\ m\\ (n+m=0)}}\Phi_{2,\ n,\ m}^{(0)}(\tau,\ a_1)\,e^{im\vartheta_1}+\right.\\
&\qquad\left.+\omega_a'(\tau,\ a_1)\ u_{10}(\tau,\ a_1,\ \vartheta_1)\right\}+\varepsilon^2R_2(\tau,\ \theta,\ \theta+\vartheta_1,\ a_1,\ \varepsilon),
\end{aligned}\right\} \tag{3.230}$$

where

$$\begin{aligned}
R_1(\tau,\ \theta,\ \theta+\vartheta_1,\ a_1,\ 0)&=\Phi_1^{(1)}(\tau,\ \theta,\ \theta+\vartheta_1,\ a_1)+\\
&+\Phi_{1,\ \theta+\vartheta}^{(0)\prime}(\tau,\ \theta,\ \theta+\vartheta_1,\ a_1)\,v_1+\Phi_{1,a}^{(0)\prime}(\tau,\ \theta,\ \theta+\vartheta_1,\ a_1)\,u_1-\frac{\partial u_1}{\partial\tau}+\\
&+\omega_a'(\tau,\ a_1)\,u_1\frac{\partial u_1}{\partial(\theta+\vartheta_1)}-\Phi_2^{(0)}(\tau,\ \theta,\ \ \theta+\vartheta_1,\ a_1)\,\frac{\partial u_1}{\partial(\theta+\vartheta_1)}+\\
&+\nu(\tau)\frac{\partial v_1}{\partial\theta}\,\frac{\partial u_1}{\partial(\theta+\vartheta_1)}-\Phi_1^{(0)}(\tau,\ \theta,\ \theta+\vartheta_1,\ a_1)\,\frac{\partial u_1}{\partial a_1}+\nu(\tau)\frac{\partial u_1}{\partial\theta}\frac{\partial u_1}{\partial a_1}+\\
&\qquad+\omega(\tau,\ a_1)\frac{\partial u_1}{\partial(\theta+\vartheta_1)}\left(\frac{\partial u_1}{\partial a_1}+\frac{\partial v_1}{\partial(\theta+\vartheta_1)}\right),
\end{aligned} \tag{3.231a}$$

$$R_2(\tau, \theta, \theta+\vartheta_1, a_1, 0)=\frac{1}{2}\omega''_{a^2}(\tau, a_1)u_1^2+$$
$$+\Phi^{(0)}_{2,\theta+\vartheta}(\tau, \theta, \theta+\vartheta_1, a_1)v_1+\Phi^{(0)}_{2a}(\tau, \theta, \theta+\vartheta_1, a_1)u_1+$$
$$+\Phi^{(1)}_2(\tau, \theta, \theta+\vartheta_1, a_1)-\frac{\partial v_1}{\partial \tau}-$$
$$-\Phi^{(0)}_1(\tau, \theta, \theta+\vartheta_1, a_1)\frac{\partial v_1}{\partial a_1}+$$
$$+\nu(\tau)\frac{\partial v_1}{\partial a_1}\left(\frac{\partial u_1}{\partial \theta}+\frac{\partial u_1}{\partial(\theta+\vartheta_1)}\right)-$$
$$-\omega'_a(\tau, a_1)u_1\frac{\partial v_1}{\partial(\theta+\vartheta_1)}-$$
$$-\Phi^{(0)}_2(\tau, \theta, \theta+\vartheta_1, a_1)\frac{\partial v_1}{\partial(\theta+\vartheta_1)}+$$
$$+\nu(\tau)\frac{\partial v_1}{\partial \theta}\frac{\partial v_1}{\partial(\theta+\vartheta_1)}+\left[\nu(\tau)\frac{\partial v_1}{\partial(\theta+\vartheta_1)}-\right.$$
$$\left.-\omega(\tau, a)\frac{\partial u_1}{\partial a_1}\right]\left(\frac{\partial u_1}{\partial a_1}+\frac{\partial v_1}{\partial(\theta+\vartheta_1)}\right)+\nu(\tau)\left(\frac{\partial u_1}{\partial a_1}\right)^2 \qquad (3.231b)$$

From the expressions (3.231a), (3.231b) it is possible to obtain functions $u_2(\tau, \theta, \theta+\vartheta, a)$ and $v_2(\tau, \theta, \theta+\vartheta, a)$ of the type (3.227), and then perform in equations (3.230) a change of variables

$$\left.\begin{aligned} a_1 &= a_2+\varepsilon^2 u_2(\tau, \theta, \theta+\vartheta_2, a_2),\\ \vartheta_1 &= \vartheta_2+\varepsilon^2 v_2(\tau, \theta, \theta+\vartheta_2, a_2).\end{aligned}\right\} \qquad (3.232)$$

We finally obtain the system of equations

$$\left.\begin{aligned} \frac{da_2}{dt} &= \varepsilon\sum_{\substack{n,m\\(n+m=0)}}\Phi^{(0)}_{1,n,m}(\tau, a_2)e^{im\vartheta_2}+\varepsilon^2\sum_{\substack{n,m\\(n+m=0)}}R^{(0)}_{1,n,m}(\tau, a_2)e^{im\vartheta_2}+\\ &\qquad +\varepsilon^3 S_1(\tau, \theta, \theta+\vartheta_2, a_2, \varepsilon),\\ \frac{d\vartheta_2}{dt} &= \omega(\tau, a_2)-\nu(\tau)+\varepsilon\Big\{\sum_{\substack{n,m\\(n+m=0)}}\Phi^{(0)}_{2,n,m}(\tau, a_2)e^{im\vartheta_2}+\\ &\qquad +\omega'_a(\tau, a_2)u_{10}(\tau, a_2, \vartheta_2)\Big\}+\varepsilon^2\Big\{\sum_{\substack{n,m\\(n+m=0)}}R^{(0)}_{2,n,m}(\tau, a_2)e^{im\vartheta_2}+\\ &\qquad +\omega'_a(\tau, a_2)u_{20}(\tau, a_2, \vartheta_2)\Big\}+\varepsilon^3 S_2(\tau, \theta, \theta+\vartheta_2, a_2, \varepsilon),\end{aligned}\right\} \qquad (3.233)$$

where S_1 and S_2 are functions which possess the same properties as the functions R_1 and R_2 (3.231) but are more complicated.

To ensure a unique definition of the substitution formulas (3.229), (3.232), etc., we also impose an additional condition which allows us to derive the functions $u_{10}(\tau, a, \vartheta)$, $v_{10}(\tau, a, \vartheta)$, $u_{20}(\tau, a, \vartheta)$, $v_{20}(\tau, a, \vartheta)$ etc.

Requiring that after the change of variables according to formulas (3.229) the parameter a_1 remains the full amplitude of the first harmonic of the angle $\theta+\vartheta_1$ in the expansion for $x=z(\tau, \theta+\vartheta_1, a_1)$*, we obtain after a few manipulations the following expression for $u_{10}(\tau, a_1, \vartheta_1)$:

$$u_{10}(\tau, a_1, \vartheta_1)=\sum_{\substack{n\\(n\neq 0,\ 2)}}\left\{\frac{1}{2\,[n\nu(\tau)+(2-n)\,\omega(\tau, a_1)]}\times\right.$$
$$\times\left[\left(\frac{a_1\,\omega'_a(\tau, a_1)}{[n\nu(\tau)+(2-n)\,\omega(\tau, a_1)]}-1\right)\frac{1}{i}\Phi^{(0)}_{1,n,2-n}(\tau, a_1)+\right.$$

* It is again assumed for simplicity that the solution of the "unperturbed" equation may be expanded in a finite Fourier sum of cosines of odd harmonics.

$$+ a_1\Phi^{(0)}_{2,\,n,\,2-n}(\tau,\ a_1)\Big]\Big\}e^{-in\vartheta_1} - \sum_{\substack{n\\(n\neq 0,\,-2)}} \Big\{\frac{1}{2\,[n\nu(\tau)-(2+n)\,\omega(\tau,\ a_1)]}\times$$

$$\times\Big[\Big(\frac{a_1\omega'_a(\tau,\ a_1)}{[n\nu(\tau)-(2+n)\,\omega(\tau,\ a_1)]}+1\Big)\frac{1}{i}\Phi^{(0)}_{1,\,n,\,-2-n}(\tau,\ a_1)+$$

$$+ a_1\Phi^{(0)}_{2,\,n,\,-2-n}(\tau,\ a_1)\Big]\Big\}e^{-in\vartheta_1} + \sum_{\substack{n,\,m\\(n+m=2,\,\pm 4,\,\pm 6,\dots)}} \Big\{\frac{(1+n+m)}{[n\nu(\tau)+m\omega(\tau,\ a_1)]}\times$$

$$\times z_{-1-n-m}(\tau,\ a_1)\Big[\Phi^{(0)}_{2,\,n,\,m}(\tau,\ a_1)-\frac{i\omega'_a(\tau,\ a_1)}{[n\nu(\tau)+m\omega(\tau,\ a_1)]}\Phi^{(0)}_{1,\,n,\,m}(\tau,\ a_1)\Big]-$$

$$-\frac{z'_{-1-n-m}(\tau,\ a_1)}{i\,[n\nu(\tau)+m\omega(\tau,\ a_1)]}\Phi^{(0)}_{1,\,n,\,m}(\tau,\ a_1)\Big\}e^{-in\vartheta_1}-$$

$$-\sum_{\substack{n,\,m\\(n+m=-2,\,\pm 4,\,\pm 6,\dots)}} \Big\{\frac{(1-n-m)}{[n\nu(\tau)+m\omega(\tau,\ a_1)]}z_{1-n-m}(\tau,\ a_1)\Big[\Phi^{(0)}_{2,\,n,\,m}(\tau,\ a_1)-$$

$$-\frac{i\omega'_a(\tau,\ a_1)}{[n\nu(\tau)+m\omega(\tau,\ a_1)]}\Phi^{(0)}_{1,\,n,\,m}(\tau,\ a_1)\Big]-$$

$$-\frac{z'_{1-n-m}(\tau,\ a_1)}{i\,[n\nu(\tau)+m\omega(\tau,\ a_1)]}\Phi^{(0)}_{1,\,n,\,m}(\tau,\ a_1)\Big\}e^{-in\vartheta}, \qquad (3.234)$$

and a similar expression for $v_{10}(\tau, a)$, which is not given here as it is very cumbersome.

We shall now discuss the problem of integration of the system (3.233).

In the previously considered nonresonance case we obtain, after neglecting in the right-hand sides of equations (3.213) terms of the second order, the system (3.215) which is of the same type as the system (2.106) (since in the nonresonance case the phase of vibration does not affect the amplitude). Consequently, the previous arguments concerning the obtained degree of accuracy are valid.

In the resonance case the phase of vibration ϑ, as is known, affects strongly the amplitude of vibration. This is in our case evident directly from equations (3.233). Discarding in these equations the terms $\varepsilon^3 S_1(\tau,\theta,\theta+\vartheta_2,a_2,\varepsilon)$ and $\varepsilon^3 S_2(\tau,\theta,\theta+\vartheta_2,a_2,\varepsilon)$, we obtain a system containing the phase ϑ on the right-hand sides. Hence, the error involved in an approximated determination of the phase affects the accuracy of derivation of the parameter a, defining the form and amplitude of vibrations.

Furthermore, examining the resonance case, the vibration frequency depends for systems of the type (3.198) on the vibration amplitude (in our case on the parameter a) and, consequently, a change in a may affect the frequency $\omega(\tau, a)$ considerably. In view of this it is impossible in the resonance case to make assumptions about the smallness of the detuning

$$\Delta(\tau,\ a)=\omega(\tau,\ a)-\nu(\tau).$$

From the above arguments we conclude that in order to obtain the values of a and ϑ by an integration over the interval $\left[0, \frac{L}{\varepsilon}\right]$, with an accuracy of terms of the first order, the system of equations

$$\left.\begin{aligned}\frac{da_1}{dt}&=\varepsilon\sum_{\substack{n,\,m\\(n+m=0)}}\Phi^{(0)}_{1,\,n,\,m}(\tau,\ a_1)\,e^{im\vartheta_1}+\varepsilon^2\sum_{\substack{n,\,m\\(n+m=0)}}R_{1,n,m}(\tau,\ a_1)\,e^{im\vartheta_1},\\ \frac{d\vartheta_1}{dt}&=\omega(\tau,\ a_1)-\nu(\tau)+\varepsilon\Big\{\sum_{\substack{n,\,m\\(n+m=0)}}\Phi^{(0)}_{2,\,n,\,m}(\tau,\ a_1)\,e^{im\vartheta_1}+\omega'_a(\tau,\ a_1)\,u_{10}(\tau,\ a_1,\ \vartheta_1)\Big\},\end{aligned}\right\}\qquad(3.235)$$

must be integrated numerically.

Calculating a_1 and ϑ_1 from this system, substituting them in (3.202) (obviously keeping in mind that $\psi = \theta + \vartheta_1$), we obtain a solution of the equation (3.198)

$$x = z(\tau,\ \theta + \vartheta_1,\ a_1),$$

which differs from the exact solution by terms of the order of ε.

In conclusion we note that the quantity a could, in the considered case of external perturbing forces explicitly depending on time, also be taken to represent in the nonresonance and in the resonance cases the maximum value of x instead of the amplitude of the fundamental harmonic of vibration. It is in that case possible to transform the right-hand sides of the equations determining a and ϑ or a and ψ so that they will not contain a solution of the "unperturbed" motion, and will be expressed directly by known functions entering in equation (3.198).

§ 11. Investigation of differential equations close to equations with "periodic" coefficients

The solution of a number of important problems associated with the study of nonstationary vibrational processes involves the consideration of nonlinear differential equations with slowly varying parameters and coefficients, which are periodic functions of θ, where $\frac{d\theta}{dt} = \nu(\tau)$.

Whenever it is possible to find a small parameter in the equation, such that the equation degenerates for its zero value and $\tau = \text{const}$ to a linear differential equation with constant coefficients, we have the case of equation (3.1) which was examined in detail in this chapter. However, in many problems a small parameter may be introduced only in such a way that the nonlinear differential equation under investigation (describing a stationary vibrational process) degenerates for a zero value of the parameter and $\tau = \text{const}$ to a linear differential equation with periodic coefficients.

It is possible to derive approximate solutions for such equations, as was shown in /64, 65/, by a scheme based on the asymptotic methods of nonlinear mechanics.

In this section we present a few results obtained by the author together with O. B. Lykova /84/, concerning a nonlinear differential equation with slowly varying coefficients of the type

$$\frac{d^2x}{dt^2} + p(\tau,\ \theta)x = \varepsilon F\left(\tau,\ \theta,\ x,\ \frac{dx}{dt}\right), \tag{3.236}$$

where the functions $p(\tau, \theta)$ and $F\left(\tau, \theta, x, \frac{dx}{dt}\right)$ are periodic in θ with period 2π, $\frac{d\theta}{dt} = \nu(\tau)$; $p(\tau, \theta)$, $\nu(\tau)$; and $F\left(\tau, \theta, x, \frac{dx}{dt}\right)$ are unlimitedly differentiable with respect to τ for any finite τ; $\tau = \varepsilon t$. Furthermore, $F\left(\tau, \theta, x, \frac{dx}{dt}\right)$ satisfies the conditions formulated in the first section of the present chapter. We note that equation (1.20) mentioned before may be reduced to (3.236). To obtain an approximate solution it is possible, as was done in /65/, to reduce equation (3.236) on the grounds of well-known results of Flouquet-Lyapunov to an equation with slowly varying coefficients, and then apply, for example, our u-method.

It is however sometimes convenient to transform the original differential equation with "periodic" coefficients to the standard form directly by means of special substitutions of variables, and apply the method of averaging. This is especially convenient when only the first approximation is being constructed /84/.

Before discussing equation (3.236) we present some well-known propositions of the theory of equations with periodic coefficients.

Consider the second order differential equation

$$\frac{d^2x}{dt^2}+p(t)\,x=0, \tag{3.237}$$

where $p(t)$ is a periodic function of t with period 2π.

We assume that $x_1(t)$ and $x_2(t)$, two linearly independent particular solutions of equation (3.237), satisfying the initial conditions

$$x_1(0)=1,\quad \dot{x}_1(0)=0,$$
$$x_2(0)=0,\quad \dot{x}_2(0)=1,$$

are known.

Then, as is known*, the nature of the general solution of equation (3.237) is completely determined by the roots of the equation

$$\sigma^2-[x_1(2\pi)+\dot{x}_2(2\pi)]\,\sigma+1=0. \tag{3.238}$$

In particular, if the condition

$$A=|x_1(2\pi)+\dot{x}_2(2\pi)|<2, \tag{3.239}$$

is satisfied, then equation (3.238) will have complex conjugate roots

$$\sigma_1=e^{i\alpha},\quad \sigma_2=e^{-i\alpha},$$

and the general solution of equation (3.237) may be represented in the form

$$x(t)=C_1e^{i\alpha\frac{t}{2\pi}}u_1(t)+C_2e^{-i\alpha\frac{t}{2\pi}}u_2(t), \tag{3.240}$$

where $u_1(t)$ and $u_2(t)$ are periodic functions of t with period 2π determined by

$$\left.\begin{aligned} u_1(t)&=\left\{x_1(t)+\frac{\sigma_1-x_1(2\pi)}{x_2(2\pi)}\,x_2(t)\right\}e^{-i\alpha\frac{t}{2\pi}},\\ u_2(t)&=\left\{x_1(t)+\frac{\sigma_2-x_1(2\pi)}{x_2(2\pi)}\,x_2(t)\right\}e^{i\alpha\frac{t}{2\pi}},\end{aligned}\right\} \tag{3.241}$$

and C_1, C_2 are arbitrary constants.

We now proceed to solve the problem stated above.

Assuming $\varepsilon=0$ and $\tau=\text{const}$ in equation (3.236), we obtain the equation

$$\frac{d^2x}{dt^2}+p[\tau,\ \nu(\tau)\,t]\,x=0, \tag{3.242}$$

which we call, as usual, the unperturbed equation. Here $p[\tau, \nu(\tau)t]$ is a periodic function of t with period $\frac{2\pi}{\nu(\tau)}$.

We assume that $x_1(\tau, t)$ and $x_2(\tau, t)$, two independent solutions of equation (3.242) depending on τ as a parameter and satisfying the initial conditions

$$x_1(\tau,\ 0)=1,\quad \dot{x}_1(\tau,\ 0)=0, \tag{3.243}$$

* Cf., for example, Loitsyanskii, A. G. and A. I. Lur'e, Teoreticheskaya mekhanika (Theoretical Mechanics). — GTTI, Part III, p. 613. 1934.

$$x_2(\tau,\ 0)=0,\quad \dot{x}_2(\tau,\ 0)=1, \tag{3.243}$$

are known.

In the considered case equation (3.238) is of the form

$$\sigma^2-\left[x_1\left(\tau,\ \frac{2\pi}{\nu(\tau)}\right)+\dot{x}_2\left(\tau,\ \frac{2\pi}{\nu(\tau)}\right)\right]\sigma+1=0. \tag{3.244}$$

In what follows we consider the case when a condition similar to (3.239),

$$A(\tau)=\left|x_1\left(\tau,\ \frac{2\pi}{\nu(\tau)}\right)+\dot{x}_2\left(\tau,\ \frac{2\pi}{\nu(\tau)}\right)\right|<2, \tag{3.245}$$

is satisfied for any constant τ in the interval $0\leqslant\tau\leqslant L$.

The roots of equation (3.244) are in this case

$$\sigma_1=\sigma_1(\tau)=e^{i\alpha(\tau)},\quad \sigma_2=\sigma_2(\tau)=e^{-i\alpha(\tau)},$$

and the general solution of equation (3.242) has the form

$$x(t)=C_1e^{i\alpha(\tau)\frac{\nu(\tau)}{2\pi}t}u_1(\tau,\ t)+C_2e^{-i\alpha(\tau)\frac{\nu(\tau)}{2\pi}t}u_2(\tau,\ t), \tag{3.246}$$

where $u_1(\tau,\ t)$ and $u_2(\tau,\ t)$ are periodic functions of t with period $\frac{2\pi}{\nu(\tau)}$, determined by the expressions

$$\left.\begin{aligned} u_1(\tau,\ t)&=\left[x_1(\tau,\ t)+\frac{\sigma_1(\tau)-x_1\left(\tau,\ \frac{2\pi}{\nu(\tau)}\right)}{x_2\left(\tau,\ \frac{2\pi}{\nu(\tau)}\right)}x_2(\tau,\ t)\right]e^{-i\alpha(\tau)\frac{\nu(\tau)}{2\pi}t},\\ u_2(\tau,\ t)&=\left[x_1(\tau,\ t)+\frac{\sigma_2(\tau)-x_1\left(\tau,\ \frac{2\pi}{\nu(\tau)}\right)}{x_2\left(\tau,\ \frac{2\pi}{\nu(\tau)}\right)}x_2(\tau,\ t)\right]e^{i\alpha(\tau)\frac{\nu(\tau)}{2\pi}t}. \end{aligned}\right\} \tag{3.247}$$

The expression (3.246) satisfies equation (3.242) for $\tau=\text{const}$. If one puts $\tau=\varepsilon t$ in equation (3.242) and also in the expression (3.246), then (3.246) satisfies approximately equation (3.242) in the interval $0\leqslant t\leqslant\frac{L}{\varepsilon}$ up to terms of the order of ε.

After these preliminary considerations concerning the solution of the unperturbed equation (3.242) we turn to the derivation of an approximate solution for equation (3.236).

We assume that condition (3.245) is satisfied in the time interval $0\leqslant t\leqslant\frac{L}{\varepsilon}$ under consideration. This corresponds to the problem of constructing an approximate solution of equation (3.236) in the domain of stable solutions of the unperturbed equation (3.242). We use the notation $\alpha(\tau)\frac{\nu(\tau)}{2\pi}=\alpha_1(\tau)$.

We shall consider the resonance case when $\alpha_1(\tau)$ may in the time interval $0\leqslant\tau\leqslant L$ come close to, or even coincide with the quantity $\frac{p}{q}\nu(\tau)$, where p and q are mutually prime integers.

Under these conditions we transform equation (3.236) to the standard form. To do this we introduce in equation (3.236) the new variables a and ψ by means of the formulas

$$x=ae^{i\left(\frac{p}{q}\theta+\psi\right)}u_1(\tau,\ t)+ae^{-i\left(\frac{p}{q}\theta+\psi\right)}u_2(\tau,\ t), \tag{3.248}$$

$$\frac{da}{dt} = ae^{i\left(\frac{p}{q}\theta+\psi\right)}[i\alpha_1(\tau)u_1(\tau,\ t)+\dot{u}_1(\tau,\ t)] -$$
$$- ae^{-i\left(\frac{p}{q}\theta+\psi\right)}[i\alpha_1(\tau)u_2(\tau,\ t)-\dot{u}_2(\tau,\ t)], \qquad (3.249)$$

where $\dot{u}_i(\tau,\ t)$ $(i = 1, 2)$ denotes the derivative with respect to the explicitly appearing time. These substitution formulas follow quite naturally from (3.246).

We differentiate the right-hand side of (3.248), keeping in mind that $a = a(t)$, $\psi = \psi(t)$, $\tau = \varepsilon t$, and equate the result to the right-hand side of (3.249) We also differentiate (3.249), and substitute the result, and also the value of x given by (3.248) into equation (3.236). Finally, taking into account that (3.246) is a solution of the unperturbed equation (3.242), we obtain the system of equations

$$\left.\begin{aligned}
&\frac{da}{dt}\{e^{i(p\varphi+\psi)}u_1(\tau,\ t)+e^{-i(p\varphi+\psi)}u_2(\tau,\ t)\}+\frac{d\psi}{dt}ia\{e^{i(p\varphi+\psi)}u_1(\tau,\ t)-\\
&-e^{-i(p\varphi+\psi)}u_2(\tau,\ t)\}=i\left[\alpha_1(\tau)-\frac{p}{q}\nu(\tau)\right]a\,[e^{i(p\varphi+\psi)}u_1(\tau,t)-\\
&-e^{-i(p\varphi+\psi)}u_2(\tau,\ t)]+\varepsilon a\,[e^{i(p\varphi+\psi)}u'_{1\tau}(\tau,\ t)+e^{-i(p\varphi+\psi)}u'_{2\tau}(\tau,\ t)],\\
&\frac{da}{dt}\{e^{i(p\varphi+\psi)}[i\alpha_1(\tau)u_1(\tau,\ t)+\dot{u}_1(\tau,\ t)]-e^{-i(p\varphi+\psi)}[i\alpha_1(\tau)u_2(\tau,\ t)-\\
&-\dot{u}_2(\tau,\ t)]\}+\frac{d\psi}{dt}ia\{e^{i(p\varphi+\psi)}[i\alpha_1(\tau)u_1(\tau,\ t)+\dot{u}_1(\tau,\ t)]-\\
&-e^{-i(p\varphi+\psi)}[i\alpha_1(\tau)u_2(\tau,\ t)-\dot{u}_2(\tau,\ t)]\}=ia\left[\alpha_1(\tau)-\frac{p}{q}\nu(\tau)\right]\times\\
&\times\{e^{i(p\varphi+\psi)}[i\alpha_1(\tau)u_1(\tau,\ t)+\dot{u}_1(\tau,\ t)]+e^{-i(p\varphi+\psi)}[i\alpha_1(\tau)u_2(\tau,\ t)-\\
&-\dot{u}_2(\tau,\ t)]\}+\varepsilon\{F(\tau,\ a,\ \theta,\ \psi)-ae^{i(p\varphi+\psi)}[i\alpha'_{1\tau}(\tau)u_1(\tau,\ t)+\\
&+i\alpha_1(\tau)u'_{1\tau}(\tau,\ t)+\dot{u}'_{1\tau}(\tau,\ t)]+ae^{-i(p\varphi+\psi)}[i\alpha'_{1\tau}(\tau)u_2(\tau,\ t)+\\
&\qquad\qquad+i\alpha_1(\tau)u'_{2\tau}(\tau,\ t)-\dot{u}_{2\tau}(\tau,\ t)]\},
\end{aligned}\right\} \qquad (3.250)$$

where $\varphi=\frac{1}{q}\theta$, and

$$F(\tau,\ a,\ \theta,\ \psi)=F\{\tau,\ \theta,\ ae^{i(p\varphi+\psi)}u_1(\tau,\ \theta)+ae^{-i(p\varphi+\psi)}u_2(\tau,\ t),$$
$$ae^{i(p\varphi+\psi)}[i\alpha_1(\tau)u_1(\tau,\ t)+\dot{u}_1(\tau,\ t)]-$$
$$-ae^{-i(p\varphi+\psi)}[i\alpha_1(\tau)u_2(\tau,\ t)-u_2(\tau,\ t)]\}. \qquad (3.251)$$

Before solving the system (3.250) for $\frac{da}{dt}$ and $\frac{d\psi}{dt}$ we make a few remarks concerning the determinant of this system

$$D=2ai\,[u_1(\tau,\ t)\dot{u}_2(\tau,\ t)-\dot{u}_1(\tau,\ t)u_2(\tau,\ t)+i2\alpha_1(\tau)u_1(\tau,\ t)u_2(\tau,\ t)]. \qquad (3.252)$$

Since the expressions

$$v_1(\tau,\ t)=e^{i\alpha_1(\tau)t}u_1(\tau,\ t),\quad v_2(\tau,\ t)=e^{-i\alpha_2(\tau)t}u_2(\tau,\ t) \qquad (3.253)$$

are particular solutions of the unperturbed equation (3.242), for any τ from the interval $0\leqslant\tau\leqslant L$ we have the following identities

$$\left.\begin{aligned}
&\ddot{v}_1(\tau,\ t)+p\,[\tau,\ \nu(\tau)\,t]\,v_1(\tau,\ t)=0,\\
&\ddot{v}_2(\tau,\ t)+p\,[\tau,\ \nu(\tau)\,t]\,v_2(\tau,\ t)=0.
\end{aligned}\right\} \qquad (3.254)$$

Multiplying them by $v_2(\tau,\ t)$ and $v_1(\tau,\ t)$ respectively, and subtracting the second from the first, we obtain

$$\ddot{v}_1(\tau,\ t)\,v_2(\tau,\ t)-\ddot{v}_2(\tau,\ t)\,v_1(\tau,\ t)=0$$

or

$$\frac{d}{dt}[\dot{v}_1(\tau, t)\, v_2(\tau, t) - v_1(\tau, t)\, \dot{v}_2(\tau, t)] = 0. \tag{3.255}$$

We find from (3.255), because of (3.254),

$$\frac{d}{dt}[u_1(\tau, t)\, \dot{u}_2(\tau, t) + 2i\alpha_1(\tau)\, u_1(\tau, t)\, u_2(\tau, t) - \dot{u}_1(\tau, t)\, u_2(\tau, t)] = 0. \tag{3.256}$$

Comparing (3.252) with (3.256), we see that the determinant of the system (3.250) does not depend on t and is a slowly varying nonvanishing quantity. This is because $v_1(\tau, t)$ and $v_2(\tau, t)$ are linearly independent particular solutions, and the expression within the square brackets in (3.256) is the Wronskian. Comparing (3.252) with (3.256), we obtain

$$\begin{aligned} &2ai\,[u_1(\tau, t)\, \dot{u}_2(\tau, t) - \dot{u}_1(\tau, t)\, u_2(\tau, t) + 2i\alpha_1(\tau)\, u_1(\tau, t)\, u_2(\tau, t)] = \\ &= 2ai\,[u_1(\tau, 0)\, \dot{u}_2(\tau, 0) - \dot{u}_1(\tau, 0)\, u_2(\tau, 0) + 2i\alpha_1(\tau)\, u_1(\tau, 0)\, u_2(\tau, 0)] = D(\tau, a). \end{aligned} \tag{3.257}$$

Next, solving the system (3.250) for $\frac{da}{dt}$ and $\frac{d\psi}{dt}$, we find

$$\left.\begin{aligned} &\frac{da}{dt} = \varepsilon F_1(\tau, \theta, a, \psi), \\ &\frac{d\psi}{dt} = \alpha_1(\tau) - \frac{p}{q}\nu(\tau) + \varepsilon F_2(\tau, \theta, a, \psi), \end{aligned}\right\} \tag{3.258}$$

where we have used the notation

$$\left.\begin{aligned} F_1(\tau, \theta, a, \psi) = {} & \frac{ia}{D(\tau,a)} F(\tau, \theta, a, \psi)\, [u_1(\tau, t)\, e^{i(p\varphi+\psi)} - \\ & - u_2(\tau, t)\, e^{-i(p\varphi+\psi)}] - \frac{ia}{D(\tau,a)} \{e^{2i(p\varphi+\psi)}\, [\dot{u}_1(\tau, t)\, u'_{1\tau}(\tau, t) - \\ & - u_1(\tau, t)\, \dot{u}'_{1\tau}(\tau, t) - iu_1^2(\tau, t)\, \alpha'_{1\tau}(\tau)] - e^{-2i(p\varphi+\psi)} \times \\ & \times [u_2(\tau, t)\, \dot{u}'_{2\tau}(\tau, t) - u_2(\tau, t)\, \dot{u}'_{2\tau}(\tau, t) + iu_2^2(\tau, t)\, \alpha'_{1\tau}(\tau)] + \\ & + [i2\alpha_1(\tau)\, u_1(\tau, t)\, u'_{2\tau}(\tau, t) + i2\alpha_1(\tau)\, u_2(\tau, t)\, u'_2(\tau, t) + \\ & + i2\alpha'_{1\tau}(\tau)\, u_1(\tau, t)\, u_2(\tau, t) + \dot{u}_1(\tau, t)\, u'_{2\tau}(\tau, t) - \\ & - \dot{u}_2(\tau, t)\, u'_{2\tau}(\tau, t) - u_1(\tau, t)\, \dot{u}'_{2\tau}(\tau, t)]\}, \\ F_2(\tau, \theta, a, \psi) = {} & \frac{1}{D(\tau, a)} F(\tau, \theta, a, \psi)\, [u_1(\tau, t)\, e^{i(p\varphi+\psi)} + \\ & + u_2(\tau, t)\, e^{-i(p\varphi+\psi)}] + \frac{1}{D(\tau, a)} \{e^{2i(p\varphi+\psi)}\, [\dot{u}_1(\tau, t)\, u'_{1\tau}(\tau, t) - \\ & - \dot{u}'_{1\tau}(\tau, t)\, u_1(\tau, t) - iu_1^2(\tau, t)\, \alpha'_{1\tau}(\tau)] - \\ & - e^{-2i(p\varphi+\psi)}\, [\dot{u}_2(\tau, t)\, u'_{2\tau}(\tau, t) - \dot{u}'_{2\tau}(\tau, t)\, u_2(\tau, t) + \\ & + iu_2^2(\tau, t)\, \alpha'_{1\tau}(\tau)] - [i2\alpha_1(\tau)\, u_1(\tau, t)\, u'_{2\tau}(\tau, t) + \\ & + i2\alpha_1(\tau)\, u_2(\tau, t)\, u'_{1\tau}(\tau, t) + \dot{u}_1(\tau, t)\, u'_{2\tau}(\tau, t) - \\ & - u_1(\tau, t)\, \dot{u}'_{2\tau}(\tau, t) + u_2(\tau, t)\, \dot{u}'_{1\tau}(\tau, t)]\}, \end{aligned}\right\} \tag{3.259}$$

where $F(\tau, \theta, a, \psi)$ and $D(\tau, a)$ are determined by (3.251) and (3.257).

Averaging the right-hand sides of the system (3.258) over t, while τ, a, ψ are being considered as constant parameters, and $\theta = \int^t \nu(\tau)\, dt$, we obtain in the resonance case when the difference $\alpha_1(\tau) - \frac{p}{q}\nu(\tau)$ is a quantity of the order of ε, the following system of equations of the first approximation:

$$\left.\begin{aligned} &\frac{da_1}{dt} = \varepsilon \underset{t}{M}\{F_1(t, \tau, \theta, a_1, \psi_1)\}, \\ &\frac{d\psi_1}{dt} = \alpha_1(\tau) - \frac{p}{q}\nu(\tau) + \varepsilon \underset{t}{M}\{F_2(t, \tau, \theta, a_1, \psi_1)\}. \end{aligned}\right\} \tag{3.260}$$

Deriving $a_1 = a_1(t)$ and $\psi_1 = \psi_1(t)$ from this system and putting them in the right-hand side of the expression (3.248) for x, we obtain a solution of equation (3.236) in the first approximation

$$x_1(t) = a_1(t)\, e^{i[p\varphi + \psi_1(t)]} u_1(\tau, t) + a_1(t)\, e^{-i[p\varphi + \psi_1(t)]} u_2(\tau, t). \tag{3.261}$$

If we wish to investigate not only the resonance region, but also transitions to it from the nonresonance region, then the difference $\alpha_1(\tau) - \frac{p}{q}\nu(\tau)$ may also be a not small quantity. The equations of the first approximation are in that case conveniently constructed by means of the method of successive changes of variables, which was developed in the preceding section of the present chapter, i. e., by performing in the system (3.260) the substitution

$$a = a_1 + \varepsilon u_1(\tau, \theta, \psi_1, a_1),$$
$$\psi = \psi_1 + \varepsilon v_1(\tau, \theta, \psi_1, a_1),$$

where the functions $u_1(\tau, \theta, \psi_1, a_1)$ and $v_1(\tau, \theta, \psi_1, a_1)$ are determined by formulas of the type (3.210).

In certain cases it is possible to construct asymptotic solutions for a nonlinear equation with periodic coefficients and slowly varying parameters with the help of the method expounded in /64/, which is based on known results (cf., for example, /89/) concerning the reducibility of equations with periodic coefficients, and a subsequent application of the u-method developed in /9, 104/.

In the general case a single second order equation of the type (3.236) may, as known, be replaced by a system of two nonlinear first order equations of the form

$$\left.\begin{aligned} \frac{dx_1}{dt} &= p_{11}(\tau, \theta)\, x_1 + p_{12}(\tau, \theta)\, x_2 + \varepsilon F_1(\tau, \theta, x_1, x_2), \\ \frac{dx_2}{dt} &= p_{21}(\tau, \theta)\, x_1 + p_{22}(\tau, \theta)\, x_2 + \varepsilon F_2(\tau, \theta, x_1, x_2), \end{aligned}\right\} \tag{3.262}$$

where $p_{ij}(\tau, \theta)$ $(i, j = 1, 2)$ are periodic functions of θ with period 2π; $\tau = \varepsilon t$; $\frac{d\theta}{dt} = \nu(\tau)$; and the functions $p_{ij}(\tau, \theta)$, $F_i(\tau, \theta, x_1, x_2)$ $(i, j = 1, 2)$ satisfy the conditions imposed on the functions $p(\tau, \theta)$ and $F\left(\tau, \theta, x, \frac{dx}{dt}\right)$ (cf. p. 129).

We transform the system of equations (3.262) to new variables. To do this we first consider the system of unperturbed equations corresponding to the system (3.262)

$$\left.\begin{aligned} \frac{dx_1}{dt} &= p_{11}(\tau, \theta)\, x_1 + p_{12}(\tau, \theta)\, x_2, \\ \frac{dx_2}{dt} &= p_{21}(\tau, \theta)\, x_1 + p_{22}(\tau, \theta)\, x_2, \end{aligned}\right\} \tag{3.263}$$

where τ is regarded as a constant parameter.

Together with (3.263) we consider the conjugate system

$$\left.\begin{aligned} \frac{dy_1}{dt} + p_{11}(\tau, \theta)\, y_1 + p_{21}(\tau, \theta)\, y_2 &= 0, \\ \frac{dy_2}{dt} + p_{12}(\tau, \theta)\, y_1 + p_{22}(\tau, \theta)\, y_2 &= 0, \end{aligned}\right\} \tag{3.264}$$

where it is also assumed that τ is a constant parameter.

We assume that the following solution is known for the system (3.264)

$$\left.\begin{aligned} y_1^{(1)} &= e^{\beta_1}\psi_1^{(1)}(\tau, \theta), \quad y_2^{(1)} = e^{\beta_1}\psi_2^{(1)}(\tau, \theta), \\ y_1^{(2)} &= e^{\beta_2}\psi_1^{(2)}(\tau, \theta), \quad y_2^{(2)} = e^{\beta_2}\psi_2^{(2)}(\tau, \theta), \end{aligned}\right\} \tag{3.265}$$

in which $\frac{d\beta_i}{dt}=\alpha_i(\tau)\,(i=1, 2)$ are the characteristic indexes, $\psi_i^{(\sigma)}(\tau, \theta)$ $(i, \sigma=1, 2)$ are periodic functions of θ with period 2π which depend on the parameter τ.

Multiplying the first equation of the system (3.263) by $y_1^{(1)}$, the second by $y_2^{(1)}$, the first equation of the system (3.264) by x_1, the second by x_2, and adding all four equations, we obtain

$$\frac{d}{dt}[x_1y_1^{(1)}+x_2y_2^{(1)}]=0 \tag{3.266}$$

and similarly

$$\frac{d}{dt}[x_1y_1^{(2)}+x_2y_2^{(2)}]=0. \tag{3.267}$$

We restrict our investigation to the domain in which the solutions (3.265) are stable, i. e., $\alpha_1(\tau)=-\alpha_2(\tau)=i\alpha(\tau)$.

We now substitute into the system (3.262) the new variables z_1 and z_2 defined by

$$\left.\begin{aligned} z_1&=\psi_1^{(1)}(\tau, \theta)\,x_1+\psi_2^{(1)}(\tau, \theta)\,x_2,\\ z_2&=\psi_1^{(2)}(\tau, \theta)\,x_1+\psi_2^{(2)}(\tau, \theta)\,x_2, \end{aligned}\right\} \tag{3.268}$$

where it has already been assumed that $\tau=\varepsilon t$.

After a few operations taking the expressions (3.266) and (3.267) into account, we obtain in place of the system of equations (3.262) the following system of differential equations for the new variables z_1 and z_2

$$\left.\begin{aligned} \frac{dz_1}{dt}+i\alpha(\tau)\,z_1&=\varepsilon\Phi_1(\tau, \theta, z_1, z_2, \varepsilon),\\ \frac{dz_2}{dt}-i\alpha(\tau)\,z_2&=\varepsilon\Phi_2(\tau, \theta, z_1, z_2, \varepsilon), \end{aligned}\right\} \tag{3.269}$$

where we have used the notation

$$\begin{aligned} \Phi_i(\tau, \theta, z_1, z_2, \varepsilon)=F_1\psi_1^{(i)}+F_2\psi_2^{(i)}-\frac{1}{\Delta}\Big\{\Big[\frac{\partial}{\partial\tau}\psi_1^{(i)}\psi_2^{(2)}-\frac{\partial}{\partial\tau}\psi_2^{(i)}\psi_1^{(2)}\Big]z_1+\\ +\Big[\frac{\partial}{\partial\tau}\psi_2^{(i)}\psi_1^{(1)}-\frac{\partial}{\partial\tau}\psi_1^{(i)}\psi_2^{(1)}\Big]z_2\Big\}\quad(i=1, 2), \end{aligned} \tag{3.270}$$

$$\left.\begin{aligned} F_i&=F_i\Big[\tau, \theta, \frac{1}{\Delta}(z_1\psi_2^{(2)}-z_2\psi_2^{(1)}), \frac{1}{\Delta}(z_2\psi_1^{(1)}-z_1\psi_1^{(2)})\Big]\qquad(i=1, 2),\\ \Delta&=\psi_1^{(1)}\psi_2^{(2)}-\psi_1^{(2)}\psi_2^{(1)},\ \psi_j^{(i)}=\psi_j^{(i)}(\tau, \theta)\qquad(i, j=1, 2). \end{aligned}\right\} \tag{3.271}$$

Since $z_1=z_2^*$ and $\Phi_1(\tau, \theta, z_1, z_2, \varepsilon)=\Phi_2^*(\tau, \theta, z_1, z_2, \varepsilon)$, where * designates a complex conjugate quantity, it is sufficient to consider only the first equation of (3.269) in what follows. The asymptotic solution may be deduced by applying the u-method presented in the first section of the present chapter directly, or by performing another change of variables defined by the formula

$$u=z_1e^{i\int\alpha(\tau)\,dt}. \tag{3.272}$$

We then obtain instead of equation (3.269) the following equation in the standard form

$$\frac{du}{dt}=\varepsilon\Phi_1\Big(\tau, \theta, ue^{-i\int\alpha(\tau)\,dt}, ue^{i\int\alpha(\tau)\,dt}, \varepsilon\Big)e^{i\int\alpha(\tau)\,d\tau}, \tag{3.273}$$

which may be solved approximately by means of the previously developed method of averaging.

We note that in order to avoid small denominators in the expressions for $u_1, u_2, \ldots,$ we require that the functions on the right-hand sides of the corresponding equations be expandable in finite Fourier sums when constructing the second and higher approximations. In the present section we have required that the right-hand sides of equations (3.236) and (3.262) should be represented as finite Fourier sums in the angular variables θ with coefficients which are polynomials in x, $\frac{dx}{dt}$ or x_1, x_2. However, the right-hand sides of equations (3.258) and (3.269) may as the result of the substitutions of variables (3.248), (3.249) and (3.268), (3.272) lose this property, and be expandable in converging Fourier series with convergent coefficients. In such case it is necessary, as was shown in § 1 of the present chapter, to eliminate the appearance of infinite sums of harmonic terms in u_1, u_2, . . . and refer the remainder of the series on the right-hand sides of equations (3.258), (3.269) to higher orders of ε.

§ 12. Examples of nonstationary vibrations in systems described by equations close to an equation with "periodic" coefficients

As the simplest example illustrating the theory of the preceding section, we consider a vibrating system described by a differential equation of the type

$$\frac{d^2x}{dt^2}+2n(\tau)\frac{dx}{dt}+p(t)\,x=0, \qquad (3.274)$$

where $n(\tau)$ is a small slowly varying friction coefficient; $\tau=\varepsilon t$; $p(t)$ is a periodic function of t with period 2π which varies according to the law

$$\left.\begin{aligned} &p(t)=k^2, \quad 2n\pi\leqslant t\leqslant(2n+1)\pi,\\ &p(t)=0, \quad (2n+1)\pi\leqslant t\leqslant(2n+2)\pi\\ &\qquad (n=0,\,1,\,2,\,3,\,\ldots), \end{aligned}\right\} \qquad (3.275)$$

where k is a constant.

The unperturbed equation corresponding to the perturbed one (3.274) has the form

$$\frac{d^2x}{dt^2}+p(t)\,x=0. \qquad (3.276)$$

We derive particular solutions for equation (3.276) which satisfy the initial conditions (3.299). We have

$$\left.\begin{aligned} &x_1(t)=\cos kt, && 0\leqslant t\leqslant\pi,\\ &x_1(t)=\cos k\pi+(\pi-t)\,k\sin k\pi, && \pi\leqslant t\leqslant 2\pi, \end{aligned}\right\} \qquad (3.277)$$

$$\left.\begin{aligned} &x_2(t)=\frac{\sin kt}{k}, && 0\leqslant t\leqslant\pi,\\ &x_2(t)=\frac{\sin k\pi}{k}+(t-\pi)\cos k\pi, && \pi\leqslant t\leqslant 2\pi. \end{aligned}\right\} \qquad (3.278)$$

Equation (3.238) which determines the nature of the general solution of equation (3.276) is in our case

$$\sigma^2-(2\cos k\pi-k\pi\sin k\pi)\,\sigma+1=0, \qquad (3.279)$$

and the condition for complex roots (condition (3.239)) assumes for this equation the form

$$|\,2\cos k\pi-k\pi\sin k\pi\,|<2. \qquad (3.280)$$

In particular, condition (3.280) will be satisfied for $k=\frac{1}{2}$. Equation

(3.279) is in this case written as

$$\sigma^2+\frac{\pi}{2}\sigma+1=0, \tag{3.281}$$

whence we find

$$\sigma_1=e^{2.475i}, \quad \sigma_2=e^{-2.475i}. \tag{3.282}$$

The solutions (3.277) and (3.278) have in the case $k=\frac{1}{2}$ the form

$$\left.\begin{aligned} &x_1(t)=\cos\frac{t}{2}, && x_2(t)=2\sin\frac{t}{2}, && 0\leqslant t\leqslant\pi,\\ &x_1(t)=\frac{1}{2}(\pi-t), && x_2(t)=2, && \pi\leqslant t\leqslant 2\pi. \end{aligned}\right\} \tag{3.283}$$

Let us now derive the general solution of equation (3.276). According to (3.240) we obtain, on the grounds of (3.282),

$$x(t)=C_1e^{2.475i\frac{t}{2\pi}}u_1(t)+C_2e^{-2.475i\frac{t}{2\pi}}u_2(t), \tag{3.284}$$

where C_1 and C_2 are arbitrary constants; $u_1(t)$ and $u_2(t)$, periodic functions of t with period 2π, defined according to (3.241) by

$$\left.\begin{aligned} u_1(t)&=\left\{x_1(t)+\left(\frac{\pi}{8}+i\cdot 0.309\right)x_2(t)\right\}e^{-i\frac{2.475t}{2\pi}},\\ u_2(t)&=\left\{x_1(t)+\left(\frac{\pi}{8}-i\cdot 0.309\right)x_2(t)\right\}e^{i\frac{2.475t}{2\pi}}. \end{aligned}\right\} \tag{3.285}$$

Substituting (3.285) into (3.284), we obtain

$$\begin{aligned} x(t)=C_1&\left\{x_1(t)+\left(\frac{\pi}{8}+i\cdot 0.309\right)x_2(t)\right\}+\\ &+C_2\left\{x_1(t)+\left(\frac{\pi}{8}-i\cdot 0.309\right)x_2(t)\right\}, \end{aligned} \tag{3.286}$$

where $x_1(t)$, $x_2(t)$ are given by (3.283).

When t is replaced by $2\pi n$ in (3.286), the expressions within the curled brackets evidently become respectively multiplied by the factors

$$e^{2.475in} \text{ and } e^{-2.475in}.$$

Thus the general solution giving the value of $x(t)$ at any time may be written in the form

$$\begin{aligned} x(t+2\pi n)=C_1e^{2.475in}&\left\{x_1(t)+\left(\frac{\pi}{8}+i\cdot 0.309\right)x_2(t)\right\}+\\ &+C_2e^{-2.475in}\left\{x_1(t)+\left(\frac{\pi}{8}-i\cdot 0.309\right)x_2(t)\right\}, \end{aligned} \tag{3.287}$$

where $x_1(t)$ and $x_2(t)$ are given by (3.283).

Let there be given the initial conditions

$$x(0)=x_0, \quad \dot{x}(0)=0. \tag{3.288}$$

Then the constants C_1 and C_2 take the values

$$C_1\simeq\frac{x_0}{2}(1+1.271i), \quad C_2\simeq\frac{x_0}{2}(1-1.271i),$$

and the solution of the unperturbed equation (3.276), satisfying the initial conditions (3.288), has for t varying from 0 to 2π the form

$$x(t)=x_0x_1(t),$$

and for later time the form

$$x(t+2\pi n)=x_0\{x_1(t)[\cos 2.475n-1.271\sin 2.475n]-\\ -x_2(t)\cdot 0.808\sin 2.475n\}, \qquad (3.289)$$

where $x_1(t)$, $x_2(t)$ are given by (3.283), $n=1, 2, 3, \ldots$

It is not difficult to plot with the help of (3.289) the time-dependence $x(t)$ for the unperturbed motion (cf. Figure 42, where the ordinate measures the ratio x/x_0).

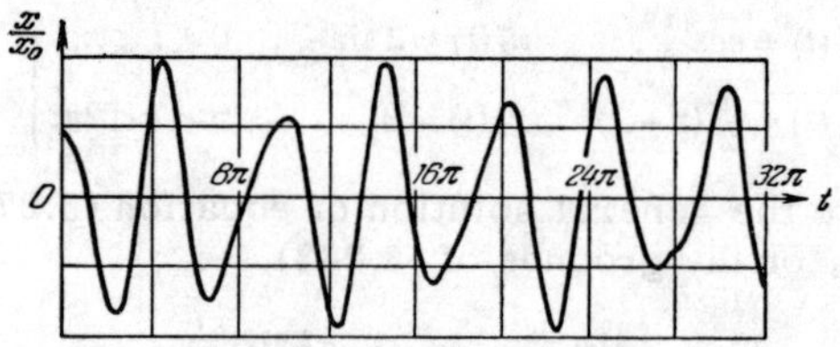

FIGURE 42

Let us now examine the influence of a small friction, which in our case is a slowly varying function of time (the friction coefficient $n=n(\tau)$), on the vibrating system under consideration.

Equation (3.274) may be solved approximately (in the first approximation) by a direct application of the formulas of the preceding section. However, because the perturbing function is linear and $p(t)$ is periodic with period 2π, it is not difficult to transform equation (3.274) so as to bring it directly to the standard form.

To do this we introduce new variables η_1 and η_2 defined by

$$\left.\begin{aligned} x&=\eta_1 e^{\frac{i\alpha}{2\pi}t}u_1(t)+\eta_2 e^{-\frac{i\alpha}{2\pi}t}u_2(t),\\ \frac{dx}{dt}&=\eta_1\cdot\frac{i\alpha}{2\pi}e^{\frac{i\alpha}{2\pi}t}u_1(t)+\eta_1 e^{\frac{i\alpha}{2\pi}t}\dot{u}_1(t)-\\ &\quad-\eta_2\cdot\frac{i\alpha}{2\pi}e^{-\frac{i\alpha}{2\pi}t}u_2(t)+\eta_2 e^{-\frac{i\alpha}{2\pi}t}\dot{u}_2(t),\end{aligned}\right\} \qquad (3.290)$$

where $\alpha=2.475$; and $u_1(t)$ and $u_2(t)$ are determined by (3.285).

Substituting (3.290) into equation (3.274) and taking into account that

$$e^{\frac{i\alpha}{2\pi}t}u_1(t) \quad\text{and}\quad e^{-\frac{i\alpha}{2\pi}t}u_2(t) \qquad (3.291)$$

are particular solutions of the unperturbed equation (3.276), which implies the identities

$$\left.\begin{aligned} -\frac{\alpha^2}{4\pi^2}u_1(t)+\frac{i\alpha}{\pi}\dot{u}_1(t)+\ddot{u}_1(t)+p(t)u_1(t)&=0,\\ -\frac{\alpha^2}{4\pi^2}u_2(t)-\frac{i\alpha}{\pi}\dot{u}_2(t)+\ddot{u}_2(t)+p(t)u_2(t)&=0,\end{aligned}\right\} \qquad (3.292)$$

we obtain after a series of operations the following system of equations in the standard form, equivalent to equation (3.274),

$$\left.\begin{aligned} \frac{d\eta_1}{dt}&=\frac{2n(\tau)}{D(0)}u_2(t)\left\{\frac{i\alpha}{2\pi}\eta_1u_1(t)+\eta_1\dot{u}_1(t)-\right.\\ &\qquad\left.-\frac{i\alpha}{2\pi}\eta_2u_2(t)+\eta_2\dot{u}_2(t)\right\},\\ \frac{d\eta_2}{dt}&=-\frac{2n(\tau)}{D(0)}u_1(t)\left\{\frac{i\alpha}{2\pi}\eta_1u_1(t)+\eta_1\dot{u}_1(t)-\right.\\ &\qquad\left.-\frac{i\alpha}{2\pi}\eta_2u_2(t)+\eta_2\dot{u}_2(t)\right\},\end{aligned}\right\} \qquad (3.293)$$

where $\alpha = 2.475$, $u_1(t)$ and $u_2(t)$ are determined by (3.285), and $D(0)$ has the form

$$D(0) = 2i\left[u_1(0)\dot{u}_2(0) - \dot{u}_1(0)u_2(0) + \frac{i\alpha}{\pi}u_1(0)u_2(0)\right]. \tag{3.294}$$

Averaging the right-hand sides of equations (3.293), we obtain the following system of equations of the first approximation

$$\left.\begin{aligned} \frac{d\overline{\eta_1}}{dt} &= \overline{\eta_1}[n(\tau)(a+ib)] + \overline{\eta_2}[n(\tau)(c+id)], \\ \frac{d\overline{\eta_2}}{dt} &= \overline{\eta_2}[n(\tau)(c-id)] + \overline{\eta_2}[n(\tau)(a-ib)], \end{aligned}\right\} \tag{3.295}$$

where

$$a = -1, \quad b = 0.1929, \quad c = -0.3125, \quad d = 0.04537.$$

We find from the system of equations (3.295) with an accuracy of terms of the first order (integrating the system (3.295) we take $\tau = \text{const}$)

$$\left.\begin{aligned} \overline{\eta_1} &= C_1(1+iN_1)e^{\lambda_1(\tau)t} + C_2(1+iN_2)e^{\lambda_2(\tau)t}, \\ \overline{\eta_2} &= C_1(1-iN_1)e^{\lambda_1(\tau)t} + C_2(1-iN_2)e^{\lambda_2(\tau)t}, \end{aligned}\right\} \tag{3.296}$$

where $\lambda_1(\tau)$, $\lambda_2(\tau)$ are roots of the characteristic equation of the system (3.295), given by the expressions

$$\lambda_{1,2}(\tau) = n(\tau)[a \pm \sqrt{c^2+d^2-b^2}], \tag{3.297}$$

and the constants N_1 and N_2 are determined by the formulas

$$N_1 = \frac{c-\sqrt{c^2+d^2-b^2}}{b-d}, \quad N_2 = \frac{c+\sqrt{c^2+d^2-b^2}}{b-d},$$

C_1 and C_2 being arbitrary constants.

The general solution of equation (3.274) (in the first approximation) for any time t may according to (3.290) and (3.287) be written in the form

$$\begin{aligned} x(t+2\pi n) = 2C_1\Big\{e^{\lambda_1(\tau)(t+2\pi n)}\Big[x_1(t)(\cos 2.475n - N_1\sin 2.475n) + \\ + x_2(t)\Big(\Big(\frac{\pi}{8} - N_1\cdot 0.309\Big)\cos 2.475n - \Big(0.309 + N_1\frac{\pi}{8}\Big)\sin 2.475n\Big)\Big]\Big\} + \\ + 2C_2\Big\{e^{\lambda_2(\tau)(t+2\pi n)}\Big[x_1(t)(\cos 2.475n - N_2\sin 2.475n) + \\ + x_2(t)\Big(\Big(\frac{\pi}{8} - N_2\cdot 0.309\Big)\cos 2.475n - \Big(0.309 + N_2\frac{\pi}{8}\Big)\sin 2.475n\Big)\Big]\Big\} \\ (n = 0, 1, 2, 3, \ldots). \end{aligned} \tag{3.298}$$

Taking into account the initial values

$$x(0) = x_0, \quad \dot{x}(0) = 0,$$

and the initial conditions for the particular solutions

$$\left.\begin{aligned} x_1(0) &= 1, \quad \dot{x}_1(0) = 0, \\ x_2(0) &= 0, \quad \dot{x}_2(0) = 1, \end{aligned}\right\} \tag{3.299}$$

we determine the arbitrary constants C_1 and C_2.

The expression (3.298) may now finally be expressed in the form

$$\begin{aligned} x(t+2\pi n) = -0.4577x_0\{e^{\lambda_1(\tau)(t+2\pi n)}[x_1(t)(\cos 2.475n + 3.8129\sin 2.475n) + \\ + x_2(t)(1.5707\cos 2.475n + 1.1876\sin 2.475n)]\} + \\ + 1.4577x_0\{e^{\lambda_2(\tau)(t+2\pi n)}[x_1(t)(\cos 2.475n + 0.4239\sin 2.475n) + \\ + x_2(t)(0.5236\cos 2.475n - 0.1425\sin 2.475n)]\}, \end{aligned} \tag{3.300}$$

where $x_1(t)$ and $x_2(t)$ are given by (3.283), and $\lambda_1(\tau)$ and $\lambda_2(\tau)$ are, according to (3.297),

$$\lambda_1(\tau) = -0.7500\, n(\tau), \quad \lambda_2(\tau) = -1.2500\, n(\tau).$$

It is not difficult to plot with the help of (3.300) a graph describing the vibrational process for a slow increase or decrease of the friction coefficient. Figure 43 shows such a graph for the case $n(\tau) = 0.03 + 0.003t$; we

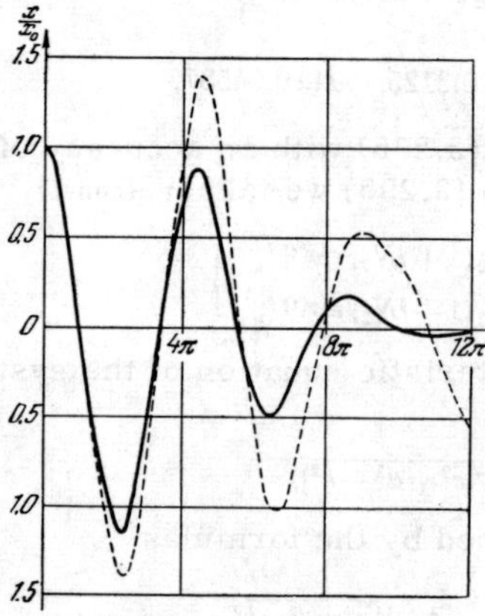

FIGURE 43

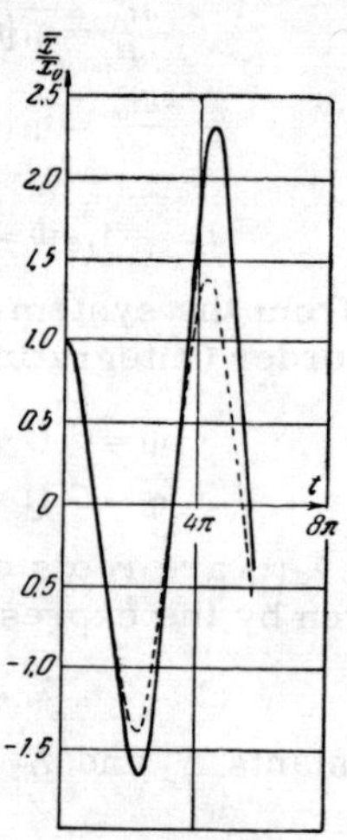

FIGURE 44

have plotted for the sake of clearness also the curve which characterizes the motion when friction is not taken into account (dotted curve). Figure 44 shows a graph for $n(\tau) = 0.03 - 0.003t$ and the dotted curve describes the case of constant friction $n = 0.03$.

§ 13. Construction of asymptotic solutions for a nonlinear differential equation with slowly varying parameters of the type (1.16)

We shall now consider the construction of asymptotic solutions in the case when the slow variation of parameters of a nonlinear vibrating system depends not only on time, but also on the nature of the motion of the system (on the coordinates and velocities). In this case, as mentioned already before, we are led to the system of equations (1.16)

$$\left.\begin{aligned} \frac{d}{dt}\left[m(y)\frac{dx}{dt}\right] + c(y)\,x &= \varepsilon F\left(y,\ \theta,\ x,\ \frac{dx}{dt}\right), \\ \frac{dy}{dt} &= \varepsilon f\left(y,\ \theta,\ x,\ \frac{dx}{dt}\right), \end{aligned}\right\} \qquad (3.301)$$

where ε is a small positive parameter; $\frac{d\theta}{dt} = \nu(y) > 0$; $m(y)$ and $c(y)$ are positive for any y derived from the system (3.301); the functions $F\left(y,\ \theta,\ x, \frac{dx}{dt}\right)$ and $f\left(y,\ \theta,\ x,\ \frac{dx}{dt}\right)$ are periodic in θ with period 2π and satisfy the same conditions as the right-hand side of equation (3.1).

Generally speaking, the system (3.301) could have been discussed in one of the following chapters that deal with an approximate solution of systems of nonlinear differential equations with slowly varying parameters. However, we consider it suitable to examine this system here, where nonlinear vibrating systems with a single degree of freedom are studied.

In the following chapters we shall consider systems of differential equations of the more general type (1.12) in a formulation suggested by V. M. Volosov. Here we give the formalism of constructing approximate solutions for the system (3.301) and examine a few particular cases.

Let us first of all recall that if the right-hand side of the second of equations (3.301), characterizing the slowly varying quantity y, does not depend on x and $\frac{dx}{dt}$, then its integration yields explicitly the solution $y=y(t, \varepsilon)$ which is a slowly varying function of time, and the problem is reduced to the problem which was considered in detail above. In particular, when $f\left(y, \theta, x; \frac{dx}{dt}\right) \equiv 1$ (and, consequently, $y=\tau$), then even the form of notation of the first equation of the system (3.301) is identical to that of (3.1).

We shall apply the u-method for constructing asymptotic approximate formulas.

The system (3.301) degenerates for $\varepsilon=0$ to

$$m(y)\frac{d^2x}{dt^2}+c(y)x=0, \quad y=y_0=\text{const}. \tag{3.302}$$

Proceeding from physical arguments which have already been applied at the beginning of the present section, we are looking for an approximate solution of the first equation of the system (3.301) in the form of an asymptotic series (we consider for simplicity the case of fundamental resonance)

$$x=a\cos(\theta+\psi)+\varepsilon u_1(y, a, \theta, \theta+\psi)+\varepsilon^2 u_2(y, a, \theta, \theta+\psi)+\varepsilon^3\ldots, \tag{3.303}$$

where $u_1(y, a, \theta, \theta+\psi)$, $u_2(y, a, \theta, \theta+\psi)$, ... are periodic functions of the angular variables $\theta, \theta+\psi$ with period 2π; y, a, ψ are certain functions of time which are to be determined.

Taking into account that the second equation of the system (3.301) is in the standard form, its right-hand side being a periodic function of θ with period 2π, it is natural to look for a solution of this equation in the form of the series

$$y=y_0+\varepsilon v_1(y_0, a, \theta, \theta+\psi)+\varepsilon^2 v_2(y_0, a, \theta, \theta+\psi)+\varepsilon^3\ldots, \tag{3.304}$$

where $v_1(y_0, a, \theta, \theta+\psi)$, $v_2(y_0, a, \theta, \theta+\psi)$, ... are periodic functions of the angular variables θ, $\theta+\psi$ with period 2π, and y_0 is a smoothly varying quantity.

We consider directly the region of the fundamental resonance; we then have

$$\omega(y)-\nu(y)\approx\varepsilon\Delta(y).$$

In this case one easily finds the improved first approximation for the system (3.301) in the form

$$\left.\begin{aligned} x&=a\cos(\theta+\psi)+\varepsilon u_1(y_0, a, \theta \ \theta+\psi),\\ y&=y_0+\varepsilon v_1(y_0, a, \theta, \theta+\psi). \end{aligned}\right\} \tag{3.305}$$

Here y_0, a, ψ are smoothly varying functions of time, determined by the system of differential equations of the first approximation

$$\left.\begin{aligned}\frac{da}{dt}&=\varepsilon A_1(y_0,\ a,\ \psi),\\ \frac{d\psi}{dt}&=\omega(y_0)-\nu(y_0)+\varepsilon B_1(y_0,\ a,\ \psi),\\ \frac{dy_0}{dt}&=\varepsilon D_1(y_0,\ a,\ \psi),\end{aligned}\right\}\tag{3.306}$$

where $\omega(y_0)=\sqrt{\frac{c(y_0)}{m(y_0)}}$. When the difference $\omega(y)-\nu(y)$ is not small, and also when it is desired to construct higher approximations, it is more convenient to derive equations of the type (3.306) by means of the method of successive substitutions of variables, presented in § 10 of the present chapter.

We differentiate (3.305), keeping in mind equations (3.306), and obtain up to and including terms of the first order

$$\frac{dx}{dt}=-a\,\omega(y_0)\sin(\theta+\psi)+\varepsilon\left\{A_1\cos(\theta+\psi)-aB_1\sin(\theta+\psi)+\frac{\partial u_1}{\partial\theta}\nu(y_0)+\right.$$
$$\left.+\frac{\partial u_1}{\partial(\theta+\psi)}\omega(y_0)\right\},\tag{3.307}$$

$$\frac{d^2x}{dt^2}=-a\,\omega^2(y_0)\cos(\theta+\psi)+\varepsilon\left\{-2A_1\,\omega(y_0)\sin(\theta+\psi)-\right.$$
$$-2aB_1\,\omega(y_0)\cos(\theta+\psi)-a\frac{d\,\omega(y_0)}{dy}D_1\sin(\theta+\psi)+\frac{\partial^2u_1}{\partial\theta^2}\nu^2(y_0)+$$
$$\left.+2\frac{\partial^2u_1}{\partial\theta\,\partial(\theta+\psi)}\nu(y_0)\,\omega(y_0)+\frac{\partial^2u_1}{\partial(\theta+\psi)^2}\omega^2(y_0)\right\},\tag{3.308}$$

and also

$$\frac{dy}{dt}=\varepsilon\left\{D_1+\frac{\partial v_1}{\partial\theta}\nu(y_0)+\frac{\partial v_1}{\partial(\theta+\psi)}\omega(y_0)\right\}.\tag{3.309}$$

Substituting (3.309) into the second equation of the system (3.301), expanding the right-hand side of this equation in a power series of ε (taking account of (3.305) and (3.307)), and discarding terms of a higher order than the first with respect to ε, we obtain

$$D_1+\frac{\partial v_1}{\partial\theta}\nu(y_0)+\frac{\partial v_1}{\partial(\theta+\psi)}\omega(y_0)=f[y_0,\ \theta,\ a\cos(\theta+\psi),\ -a\,\omega\sin(\theta+\psi)].\tag{3.310}$$

Hence, taking into account the periodicity of $v_1(y_0,\ a,\ \theta,\ \theta+\psi)$ in the angular variables θ and $\theta+\psi$ with period 2π, we find uniquely

$$v_1(y_0,\ a,\ \theta,\ \theta+\psi)=\sum_{n+m\neq0}\frac{f_{n,m}(y_0,\ a)\,e^{i[n\theta+m(\theta+\psi)]}}{i[n\,\nu(y_0)+m\,\omega(y_0)]},\tag{3.311}$$

where

$$f_{n,m}(y_0,\ a)=\frac{1}{4\pi^2}\int_0^{2\pi}\int_0^{2\pi}f_0(y_0,\ a,\ \theta,\ \theta+\psi)\,e^{-i[n\theta+m(\theta+\psi)]}\,d\theta\,d(\theta+\psi),\tag{3.312}$$

$$f_0(y_0,\ a,\ \theta,\ \theta+\psi)=f(y_0,\ \theta,\ a\cos(\theta+\psi),\ -a\,\omega\sin(\theta+\psi)),\tag{3.313}$$

and also

$$D_1(y_0,\ a,\ \psi)=\sum_{n+m=0}f_{n,m}(y_0,\ a)\,e^{im\psi},\tag{3.314}$$

where

$$f_{n,m}(y_0,\ a)=\frac{1}{4\pi^2}\int_0^{2\pi}\int_0^{2\pi}f_0(y_0,\ a,\ \theta,\ \theta+\psi)\,e^{-im\psi}\,d\theta\,d(\theta+\psi).\tag{3.315}$$

We turn now to derive the functions $u_1(y_0,\ a,\ \theta,\ \theta+\psi)$, $A_1(y_0,\ a,\ \psi)$, and $B_1(y_0,\ a,\ \psi)$.

Inserting the values of x from (3.305), $\frac{dx}{dt}$ from (3.307), and $\frac{d^2x}{dt^2}$ from (3.308) into the left-hand side of the first equation of the system (3.301), we obtain with an accuracy up to and including terms of the first order

$$\frac{d}{dt}\left[m(y)\frac{dx}{dt}\right]+c(y)x=+\varepsilon\Big\{m(y_0)\Big[-2\omega(y_0)A_1\sin(\theta+\psi)-$$
$$-2a\omega(y_0)B_1\cos(\theta+\psi)-\frac{1}{m(y_0)}\frac{d[m(y_0)\omega(y_0)]}{dy}aD_1\sin(\theta+\psi)\Big]+$$
$$+\frac{dc(y_0)}{dy}v_1(y_0,\ a,\ \theta,\ \theta+\psi)\,a\cos(\theta+\psi)+$$
$$+\frac{dm(y_0)}{dy}v_1(y_0,\ a,\ \theta,\ \theta+\psi)\,a\omega^2(y_0)\cos(\theta+\psi)+m(y_0)\Big[\frac{\partial^2 u_1}{\partial\theta^2}\nu(y_0)+$$
$$+2\frac{\partial^2 u_1}{\partial\theta\,\partial(\theta+\psi)}\nu(y_0)\,\omega(y_0)+\frac{\partial^2 u_1}{\partial(\theta+\psi)^2}\omega^2(y_0)\Big]\Big\}. \tag{3.316}$$

Inserting the values of x and y from (3.305), and also $\frac{dx}{dt}$ from (3.307) into the right-hand side of the first equation of the system (3.301), and retaining only terms of the first order, we obtain

$$\varepsilon F\left(y,\ \theta,\ x,\ \frac{dx}{dt}\right)=\varepsilon F(y_0,\ \theta,\ a\cos(\theta+\psi),\ -a\omega\sin(\theta+\psi))=$$
$$=\varepsilon F_0(y_0,\ a,\ \theta,\ \theta+\psi). \tag{3.317}$$

Equating the right-hand sides of the expressions (3.316) and (3.317), we obtain the following relation for the determination of $u_1(y_0,\ a,\ \theta,\ \theta+\psi)$, $A_1(y_0,\ a,\ \psi)$, and $B_1(y_0,\ a,\ \psi)$:

$$m(y_0)\Big[\frac{\partial^2 u_1}{\partial\theta^2}\nu(y_0)+2\frac{\partial^2 u_1}{\partial\theta\,\partial(\theta+\psi)}\nu(y_0)\,\omega(y_0)+\frac{\partial^2 u_1}{\partial(\theta+\psi)^2}\omega^2(y_0)+\omega^2(y_0)\,u_1\Big]=$$
$$=F_1(y_0,\ a,\ \theta,\ \theta+\psi)+m(y_0)\Big[2\omega(y_0)A_1\sin(\theta+\psi)+$$
$$+2\omega(y_0)\,a\,B_1\cos(\theta+\psi)+\frac{1}{m(y_0)}\frac{d[m(y_0)\omega(y_0)]}{dy}aD_1\sin(\theta+\psi)\Big], \tag{3.318}$$

where

$$F_1(y_0,\ a,\ \theta,\ \theta+\psi)=$$
$$=F_0(y_0,\ a,\ \theta,\ \theta+\psi)-\frac{c(y_0)}{m(y_0)}\frac{d[m(y_0)c(y_0)]}{dy}v_1(y_0,\ a,\ \theta,\ \theta+\psi)\,a\cos(\theta+\psi), \tag{3.319}$$

and $v_1(y_0,\ a,\ \theta,\ \theta+\psi)$ and $D_1(y_0,\ a,\ \psi)$ are known functions determined by the expressions (3.311) and (3.314).

Following the procedure in § 1 of the present chapter (p. 68), we obtain, after a series of operations, an expression for $u_1(y_0,\ a,\ \theta,\ \theta+\psi)$:

$$u_1(y_0,\ a,\ \theta,\ \theta+\psi)=\frac{1}{m(y_0)}\sum_{n}\sum_{\substack{m\\(n+m\neq\pm1)}}\frac{F_{1,n,m}(y_0,\ a)\,e^{i[n\theta+m(\theta+\psi)]}}{\omega^2(y_0)-[n\nu(y_0)+m\omega(y_0)]^2}, \tag{3.320}$$

where

$$F_{1,n,m}(y_0,\ a)=\frac{1}{4\pi^2}\int_0^{2\pi}\int_0^{2\pi}F_1(y_0,\ a,\ \theta,\ \theta+\psi)\,e^{-i[n\theta+m(\theta+\psi)]}\,d\theta\,d(\theta+\psi), \tag{3.321}$$

and also expressions for $A_1(y_0,\ a,\ \psi)$ and $B_1(y_0,\ a,\ \psi)$:

$$\left.\begin{aligned}
&A_1(y_0,\ a,\ \psi)=\\
&\quad=-\frac{1}{2m(y_0)\,\omega(y_0)}\frac{d[m(y_0)\,\omega(y_0)]}{dy}a\sum_{n+m=0}f_{n,m}(y_0,\ a)\,e^{im\psi}-\\
&\qquad-\frac{1}{4\pi^2m(y_0)\,\omega(y_0)}\sum_{\sigma}e^{i\sigma\psi}\int_0^{2\pi}\int_0^{2\pi}F_1(y_0,\ a,\ \theta,\ \vartheta)\,e^{-i\sigma\psi_1}\sin\vartheta\,d\theta\,d\vartheta,\\
&B_1(y_0,a,\psi)=-\frac{1}{4\pi^2m(y_0)\,\omega(y_0)a}\sum_{\sigma}e^{i\sigma\psi}\int_0^{2\pi}\int_0^{2\pi}F_1(y_0,\ a,\ \theta,\ \vartheta)\,e^{-i\sigma\psi_1}\cos\vartheta\,d\theta\,d\vartheta,
\end{aligned}\right\} \tag{3.322}$$

where $\psi_1=\vartheta-\theta$.

Before undertaking an analysis of the first approximation, we consider another method of deriving the equations of the first approximation (3.306), based on the method of averaging; this is convenient in many cases.

Let us introduce in the system of equations (3.301) instead of x and $\frac{dx}{dt}$ the new variables a and ψ defined by the relations (for the sake of simplicity we consider again the fundamental resonance $p=1,\ q=1$)

$$\left.\begin{aligned} x &= a\cos(\theta+\psi), \\ \frac{dx}{dt} &= -a\omega(y)\sin(\theta+\psi), \end{aligned}\right\} \tag{3.323}$$

where $\omega(y)=\sqrt{\frac{c(y)}{m(y)}}$.

We then obtain, after a few calculations, instead of the system (3.301) the following equivalent system of equations

$$\left.\begin{aligned} \frac{da}{dt} &= -\frac{\varepsilon}{m(y)\,\omega(y)}\Big\{a\frac{d\,[m(y)\,\omega(y)]}{dy} f\,[y,\theta,a\cos(\theta+\psi),-a\omega\sin(\theta+\psi)]\sin^2(\theta+\psi)+ \\ &\qquad + F\,[y,\ \theta,\ a\cos(\theta+\psi),\ -a\omega\sin(\theta+\psi)]\sin(\theta+\psi)\Big\}, \\ \frac{d\psi}{dt} &= \omega(y)-\nu(y)- \\ &\quad -\frac{\varepsilon}{m(y)\,\omega(y)}\Big\{\frac{d\,[m(y)\,\omega(y)]}{dy} f\,[y,\ \theta,\ a\cos(\theta+\psi),\ -a\omega\sin(\theta+\psi)]\times \\ &\qquad \times\sin(\theta+\psi)\cos(\theta+\psi)+ \\ &\qquad +\frac{1}{a}F\,[y,\ \theta,\ a\cos(\theta+\psi),\ -a\omega\sin(\theta+\psi)]\cos(\theta+\psi)\Big\}, \\ \frac{dy}{dt} &= \varepsilon f\,[y,\ \theta,\ a\cos(\theta+\psi),\ -a\omega\sin(\theta+\psi)]. \end{aligned}\right\} \tag{3.324}$$

If, as above, we confine our investigation of the vibrational process to the resonance zone and, consequently, take $\omega(y)-\nu(y)=\varepsilon\Delta(y)$, then the system of equations (3.324) is a system of equations in the standard form. Applying the principle of averaging (during the averaging we consider $y=y_0$, and a and ψ as constants), we obtain the system of equations of the first approximation

$$\left.\begin{aligned} \frac{da}{dt} &= -\frac{\varepsilon}{m(y_0)\,\omega(y_0)}\Big\{\frac{d\,[m(y_0)\,\omega(y_0)]}{dy}\underset{\theta}{M}\{f\,[y_0,\ \theta,\ a\cos(\theta+\psi),\ -a\omega\sin(\theta+\psi)]\times \\ &\qquad \times\sin^2(\theta+\psi)\}+\underset{\theta}{M}\{F\,[y_0,\ \theta,\ a\cos(\theta+\psi),\ -a\omega\sin(\theta+\psi)]\sin(\theta+\psi)\}\Big\}, \\ \frac{d\psi}{dt} &= \varepsilon\Delta(y_0)-\frac{\varepsilon}{m(y_0)\,\omega(y_0)}\Big\{\frac{d\,[m(y_0)\,\omega(y_0)]}{dy}\times \\ &\qquad \times\underset{\theta}{M}\{f\,[y_0,\ \theta,\ a\cos(\theta+\psi),\ -a\omega\sin(\theta+\psi)]\sin(\theta+\psi)\cos(\theta+\psi)\}+ \\ &\qquad +\frac{1}{a}\underset{\theta}{M}\{F\,[y_0,\ \theta,\ a\cos(\theta+\psi),\ -a\omega\sin(\theta+\psi)]\cos(\theta+\psi)\}\Big\}, \\ \frac{dy_0}{dt} &= \varepsilon\underset{\theta}{M}\{f\,[y_0,\ \theta,\ a\cos(\theta+\psi),\ -a\omega\sin(\theta+\psi)]\}, \end{aligned}\right\} \tag{3.325}$$

where the symbol $\underset{\theta}{M}$ denotes the averaging operation over θ.

It is easily verified that taking the average of the right-hand sides of the system (3.325), one obtains a system of equations identical to the system (3.306).

When the difference $\omega(y)-\nu(y)$ is not small, then the approximate solutions of the system (3.324) may again be derived successfully with the help of the method of successive substitutions of variables, developed in § 10 of the present chapter.

We pass now to an analysis of the system of equations of the first approximation.

According to the theory presented in the present section, the system of equations (3.301) is in the general case replaced in the first approximation by a system of three coupled first order equations. However, its integration (even in the general case when it is necessary to apply numerical integration) is considerably simpler than the integration of the original system (3.301), since here one deals with smoothly varying functions y_0, a, ψ.

We shall examine several particular cases when the integration of the system of equations of the first approximation (3.306) is greatly simplified and may even be reduced to quadratures.

Let the right-hand side of the last equation of the system (3.306) be independent of the choice of the trajectory (3.323) (solution of the unperturbed system (3.302)), along which we perform the averaging.

Then the last equation of the system (3.306) assumes the form

$$\frac{dy_0}{dt} = \varepsilon \bar{f}(y_0), \tag{3.326}$$

where

$$\bar{f}(y_0) = \underset{\theta}{M}\{f[y_0, \theta, a\cos(\theta+\psi), -a\omega\sin(\theta+\omega)]\}.$$

We derive from equation (3.326) $y_0 = y_0(\tau)$, after which the system of equations of the first approximation coincides in its structure with the equations of the first approximation, which were obtained for the equation (3.1) and studied in detail in § 1 of the present chapter.

Let us still consider the case when the right-hand sides of the system (3.301) do not depend on θ, i. e., the case when the vibrating system with slowly varying parameters is described by the system of equations

$$\left.\begin{aligned} \frac{d}{dt}\left[m(y)\frac{dx}{dt}\right] + c(y)x &= \varepsilon F\left(y, x, \frac{dx}{dt}\right), \\ \frac{dy}{dt} &= \varepsilon f\left(y, x, \frac{dx}{dt}\right). \end{aligned}\right\} \tag{3.327}$$

Then the equations of the first approximation assume the form

$$\left.\begin{aligned} \frac{da}{dt} &= -\frac{\varepsilon}{m(y_0)\omega(y_0)}\left\{\frac{d[m(y_0)\omega(y_0)]}{dy} a \underset{\psi}{M}\{f(y_0, a\cos\psi, -a\omega\sin\psi)\sin^2\psi\} + \right. \\ &\qquad \left. + \underset{\psi}{M}\{F(y_0, a\cos\psi, -a\omega\sin\psi)\sin\psi\}\right\}, \\ \frac{d\psi}{dt} &= \omega(y_0) - \frac{\varepsilon}{m(y_0)\omega(y_0)}\left\{\frac{d[m(y_0)\omega(y_0)]}{dy}\underset{\psi}{M}\{f(y_0, a\cos\psi, -a\omega\sin\psi)\times\right. \\ &\qquad \left. \times\sin\psi\cos\psi\} + \frac{1}{a}\underset{\psi}{M}\{F(y_0, a\cos\psi, -a\omega\sin\psi)\cos\psi\}\right\}, \\ \frac{dy_0}{dt} &= \varepsilon \underset{\psi}{M}\{f(y, a\cos\psi, -a\omega\sin\psi)\}. \end{aligned}\right\} \tag{3.328}$$

It is easily verified that the right-hand sides of the system of equations of the first approximation do not in this case depend on ψ and consequently the first and third equations may be treated by means of the phase plane method (in the plane (a, y_0)).

We do not consider other particular cases, nor stationary modes and problems of their stability, remarking that the analysis of these cases and of the stationary modes may be performed without essential difficulties, as was done before for the systems (2.1) and (3.1).

§ 14. Examples of vibrating systems with slowly varying parameters described by equations of the type (1.16)

As the first simplest example illustrating the method of the preceding section, we consider the particular case when the second equation of the system (3.301) may be integrated independently of the first equation /103, 104/.

Let the vibrations of a nonlinear system with a single degree of freedom, subjected to an external sinusoidal force, be described by the system of differential equations

$$\left.\begin{aligned} &\frac{d^2x}{dt^2}+2\delta\frac{dx}{dt}+x+x^3=\varepsilon E\sin(\nu_0 t+y),\\ &\frac{dy}{dt}=\varepsilon\nu_1\cos\alpha t,\end{aligned}\right\} \tag{3.329}$$

where ν_0, ν_1, δ, and E are constants; ε is a small positive parameter.

Integrating the second equation of the system (3.329), we obviously obtain $y(t)=-\frac{\varepsilon\nu_1}{\alpha}\sin\alpha t+\text{const}$, and the law of slow variation of the phase in the right-hand side of the first equation of the system (3.329) is thus known; this brings us to the problem studied in detail in the present chapter.

Since the investigation of a vibrational process described by the system of equations (3.329) is important for the study of nonstationary resonance phenomena in nonlinear systems, we construct, in passing, its solution in the first approximation, and analyze the approximation.

The solution of the system (3.329) for the case of a fundamental resonance is in the first approximation

$$x=a\cos(\theta+\psi), \tag{3.330}$$

where a and ψ are to be determined from the system of equations of the first approximation

$$\left.\begin{aligned} &\frac{da}{dt}=-\delta a+\frac{\varepsilon E}{1+\nu(t)}\cos\psi,\\ &\frac{d\psi}{dt}=1-\nu(t)+\frac{3a^2}{8}+\frac{\varepsilon E}{a\,[1+\nu(t)]}\sin\psi,\end{aligned}\right\} \tag{3.331}$$

where $\nu(t)=\nu_0+\frac{dy}{dt}$.

Integrating this system numerically for various values of the parameters ν_0, $\varepsilon\nu_1$, and α, characterizing the variation of the frequency of the external force, we obtain a series of graphs representing the dependence of the amplitude of vibration on the frequency or time; these supplement our previous results concerning nonstationary vibrational processes in the resonance zone.

The curves in Figures 45 and 46 represent the dependence of the amplitude on the frequency for $\nu_0=1.0$, $\varepsilon\nu_1=0.09841$, $\delta=0.02$, $E=0.02$; $\alpha=0.04$ for Figure 45, and $\alpha=0.08$ for Figure 46. As the initial values we have taken $\nu(0)=\nu_0$, and the corresponding stationary values of a and ψ. These figures (as well as those that follow) contain for comparison the stationary resonance curves.

It is possible to draw several conclusions from an analysis of the amplitude curves in Figures 45 and 46. For example, in the case of a small vibration frequency of the external frequency, the nonlinearity exerts an essential influence on the nature of the amplitude variation, which is, in particular, demonstrated by an "attraction" of the vibration amplitude values to the stationary values (especially in regions where $\cos \alpha t$ passes through zero).

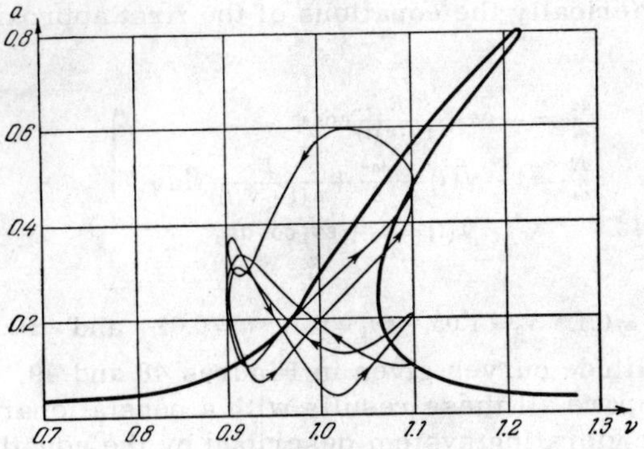

FIGURE 45

In Figure 47 we show for clarity the dependence of the amplitude on time, in other words, the "unfolded" amplitude curves of Figures 45 and 46.

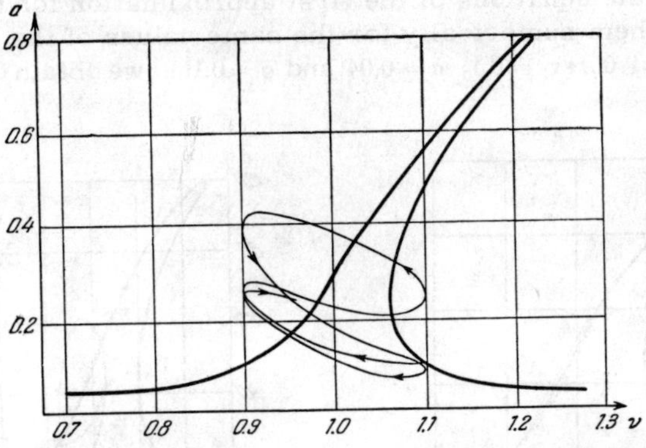

FIGURE 46

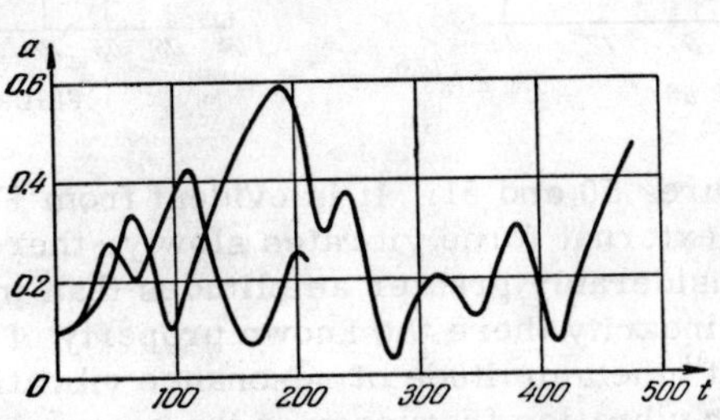

FIGURE 47

Let us also consider the case of a nonlinear vibrating system with a soft characteristic of the nonlinear restoring force, the vibrations of which are described by the system of equations

$$\left.\begin{aligned}\frac{d^2x}{dt^2}+2\delta\frac{dx}{dt}+x-x^3=E\sin(\nu_0 t+y),\\ \frac{dy}{dt}=\varepsilon\nu_1\cos\alpha t.\end{aligned}\right\}\tag{3.332}$$

Integrating numerically the equations of the first approximation for this case

$$\left.\begin{aligned}\frac{da}{dt}&=-\delta a+\frac{E}{1+\nu(t)}\cos\psi,\\ \frac{d\psi}{dt}&=1-\nu(\tau)-\frac{3a^2}{8}+\frac{E}{a(1+\nu(t))}\sin\psi,\\ \nu(t)&=\nu_0+\varepsilon\nu_1\cos\alpha t\end{aligned}\right\}\tag{3.333}$$

for the values

$$E=1,\quad \delta=0.1,\quad \nu_0=1.05,\quad \varepsilon\nu_1=0.1,\quad \alpha=0.04\quad\text{and}\quad \alpha=0.16,$$

we obtain the amplitude curves given in Figures 48 and 49.

Let us now compare all these results with a nonstationary process occurring in a linear vibrating system described by the equations

$$\left.\begin{aligned}\frac{d^2x}{dt^2}+2\delta\frac{dx}{dt}+x=E\sin(\nu_0 t+y),\\ \frac{dy}{dt}=\varepsilon\nu_1\cos\alpha t.\end{aligned}\right\}\tag{3.334}$$

Constructing the equations of the first approximation for this system, and integrating them numerically for the same values of the parameters ($E=1.\ \delta=0.1,\ \nu_0=1.05,\ \varepsilon\nu_1=0.1,\ \alpha=0.04$ and $\alpha=0.16$), we obtain the amplitude

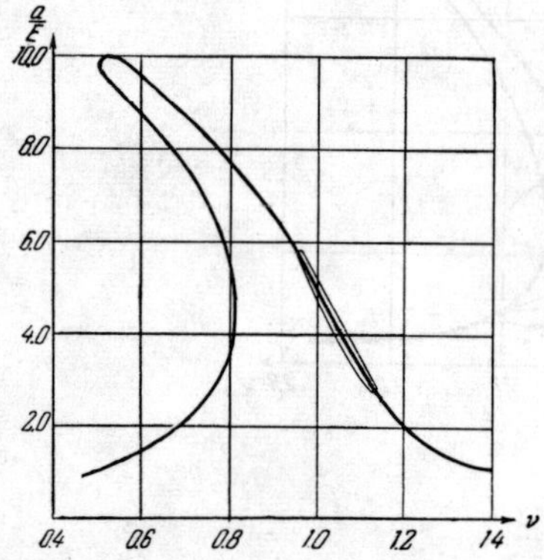

FIGURE 48

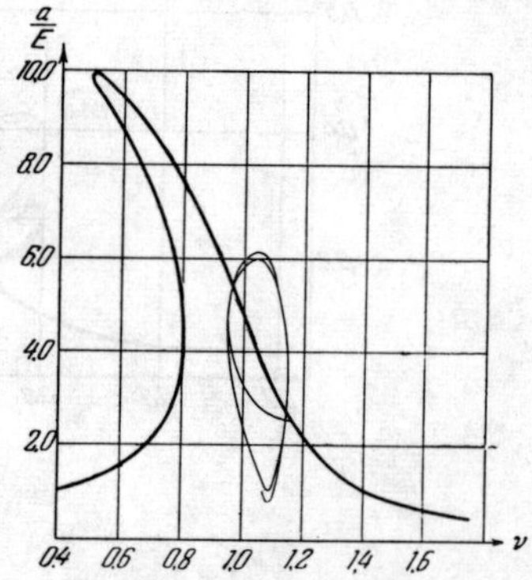

FIGURE 49

curves shown in Figures 50 and 51. It is evident from Figure 50 that when the frequency of the external force vibrates slowly, there appear in a linear vibrating system considerably greater amplitudes than in the same mode of a system with a nonlinearity (here the known property of a nonlinearity consisting in "cutting off" the amplitude of resonance vibrations is displayed again). For a larger vibration frequency of the external frequency the vibration amplitude will perform a vibrational motion about a certain average value.

In Figures 52 and 53 we compare the curves describing the dependence of the amplitude on the time in the linear and the nonlinear (the lower curve) system.

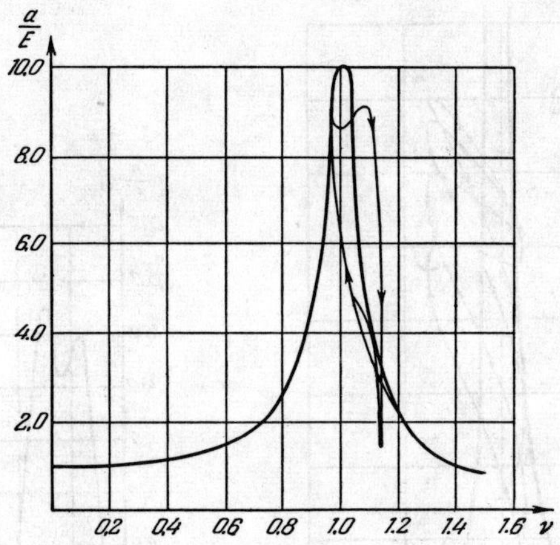

FIGURE 50

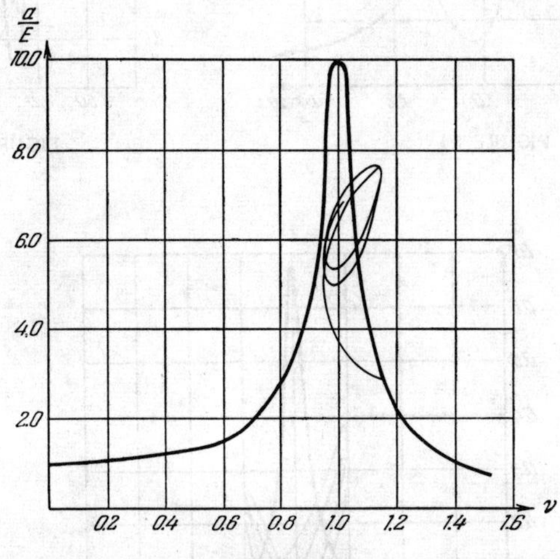

FIGURE 51

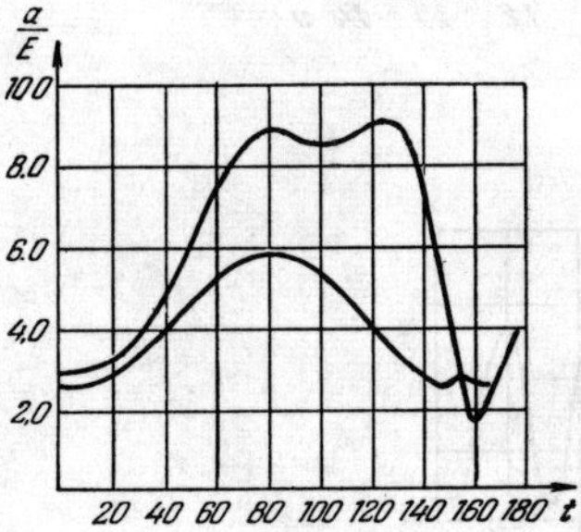

FIGURE 52

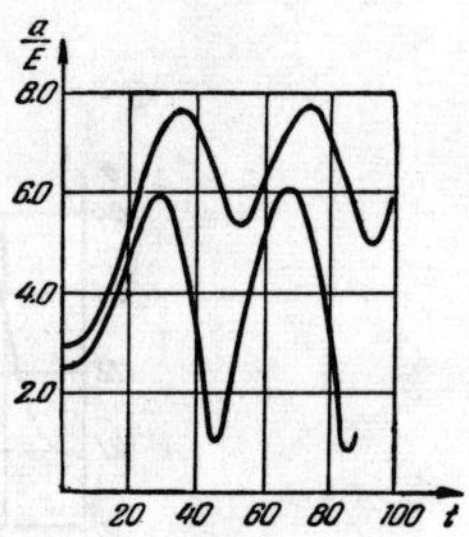

FIGURE 53

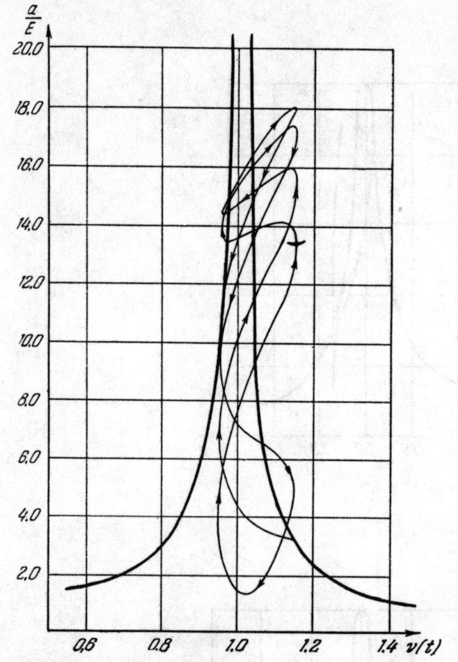

FIGURE 54

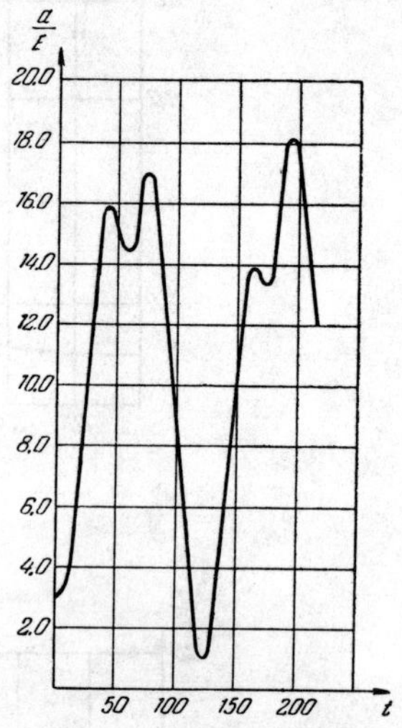

FIGURE 55

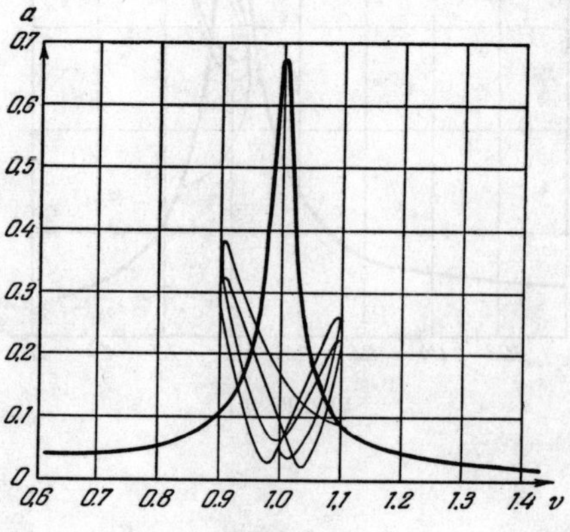

FIGURE 56

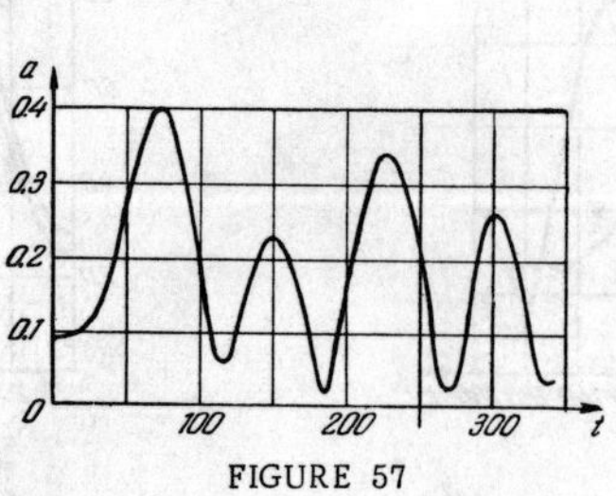

FIGURE 57

If, finally, there is no friction (i. e., $\delta = 0$) in the linear system under consideration, subjected to the action of an external force with a vibrating frequency, then, as is shown in Figures 54 and 55, a very strong "tightening" of the vibration amplitude is possible.

We give another example of the action of an external perturbing "sinusoidal" force of modulated frequency on a vibrating system with nonlinear friction, which is for simplicity assumed to be proportional to the square of the velocity. The equations of motion may in this case be written in the form

$$\left.\begin{aligned} &\frac{d^2x}{dt^2} + \delta\left(\frac{dx}{dt}\right)^2 + x = E\sin(\nu_0 t + y), \quad \text{if} \quad \frac{dx}{dt} > 0, \\ \text{and} \quad &\frac{d^2x}{dt^2} - \delta\left(\frac{dx}{dt}\right)^2 + x = E\sin(\nu_0 t + y), \quad \text{if} \quad \frac{dx}{dt} < 0, \\ &\frac{dy}{dt} = \varepsilon\nu_1\cos\alpha t, \end{aligned}\right\} \qquad (3.335)$$

and consequently the equations of the first approximation are

$$\left.\begin{aligned} \frac{da}{dt} &= -\frac{4\delta a^2}{3\pi} - \frac{E}{1+\nu(\tau)}\cos\psi, \\ \frac{d\psi}{dt} &= 1 - \nu(\tau) + \frac{E}{a(1+\nu(\tau))}\sin\psi, \\ \nu(\tau) &= \nu_0 + \varepsilon\nu_1\cos\alpha t. \end{aligned}\right\} \qquad (3.336)$$

We note that since the friction affects the frequency of vibrations only in the second approximation, the second equation is identical with the equation for the instantaneous phase in the case of a linear system.

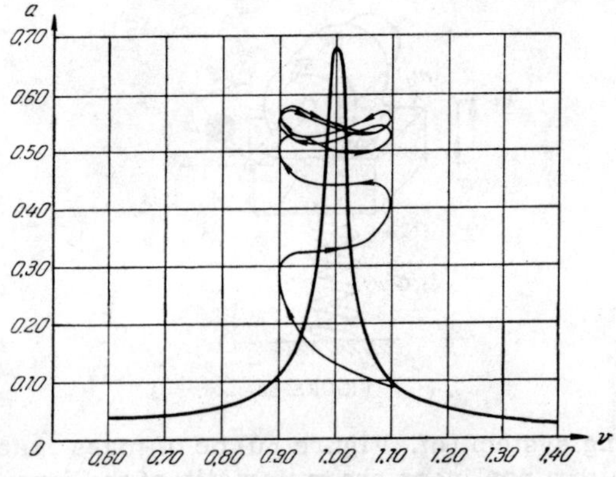

FIGURE 58

Assuming in the system (3.336) $\delta = 0.05$, $E = 0.02$, $\nu_0 = 1$, $\varepsilon\nu_1 = 0.1$, $\alpha = 0.04$; 0.08; 0.16, and integrating numerically, we obtain graphs characterizing the vibrating amplitude which are given in Figures 56, 58, and 59. (Figure 57 shows the amplitude of Figure 56 as a function of time). Figures 56, 58 and 59 also contain the resonance curves corresponding to the stationary mode. It is obvious that vibrations of a constant (Figure 59) or almost constant (Figure 58) amplitude may be established in the system when the parameters ν_0, $\varepsilon\nu_1$, α are chosen appropriately.

As a second characteristic example leading to the study of a system of differential equations of the type (3.301), we consider the problem of interaction between a nonlinear vibrator and an energy source during forced vibrations. This problem was fully examined by V. O. Kononenko /66, 67/.

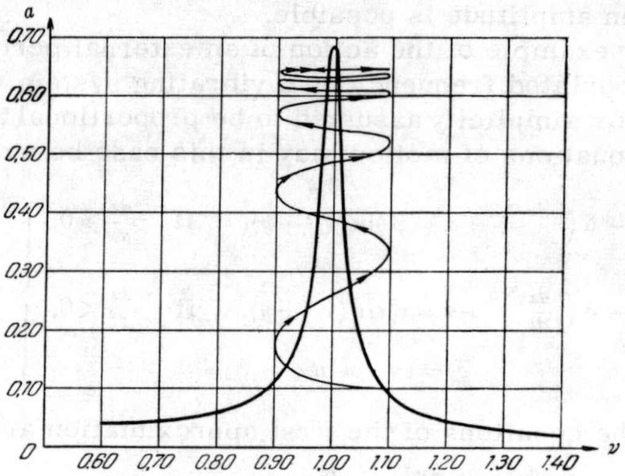

FIGURE 59

As an actual example we consider a vibrating system with one degree of freedom which performs vibrations excited by the inertial forces of an unbalanced mass m rotating at the distance r /67/.

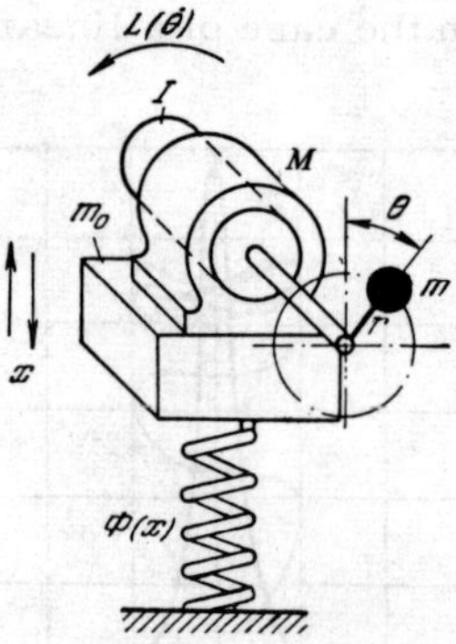

FIGURE 60

Let the vibrating system (cf. Figure 60) be of mass m_0 and have an elastic coupling with a nonlinear characteristic of the form

$$\Phi(x) = cx + f(x),$$

where $f(x)$ is a small nonlinear function of x.

The energy source in the system under consideration is the motor M. The mass is attached to the motor's rotor. The moment of inertia of the rotor is denoted by I. The rotating moment produced by the motor is denoted by $L(\theta)$, the moment of the forces of resistance to rotation by $H(\dot{\theta})$, and the force of resistance to vibration by $\beta\dot{x}$.

We assume that the quantities $\frac{m}{m_0}$, $\frac{m}{I}$, and $\frac{1}{I}[L(\dot{\theta}) - H(\dot{\theta})]$ are small; then the study of the vibrational process reduces to the following system of differential equations

$$\left.\begin{aligned}\frac{d^2x}{dt^2}+\omega^2x&=\varepsilon\left[-h\frac{dx}{dt}-f_1(x)+q_2\dot{\theta}^2\cos\theta+q_2\ddot{\theta}\sin\theta\right],\\ \frac{d^2\theta}{dt^2}&=\varepsilon\left[M_1(\dot{\theta})+q_3\frac{d^2x}{dt^2}\sin\theta+q_4\sin\theta\right],\end{aligned}\right\}\tag{3.337}$$

where

$$\omega^2=\frac{c}{m_0},\quad \frac{m}{m_0}r=\varepsilon q_2,\quad \frac{\beta}{m_0}=\varepsilon h,\quad \frac{m}{I}r=\varepsilon q_3,$$

$$\frac{m}{I}gr=\varepsilon q_4,\quad \frac{1}{I}[L(\dot{\theta})-H(\dot{\theta})]=\varepsilon M_1(\dot{\theta}),\quad \frac{1}{m_0}f(x)=\varepsilon f_1(x).$$

Let us introduce into the system (3.337) the new variables a, ψ, and ν defined by the formulas

$$\left.\begin{aligned}x&=a\cos(\theta+\psi),\\ \frac{dx}{dt}&=-a\omega\sin(\theta+\psi),\\ \frac{d\theta}{dt}&=\nu.\end{aligned}\right\}\tag{3.338}$$

Equations (3.337) are thereby transformed into the following system of differential equations in the standard form

$$\left.\begin{aligned}\frac{da}{dt}&=-\frac{\varepsilon}{\omega}\{ah\omega\sin(\theta+\psi)-f_1[a\cos(\theta+\psi)]+q_2\nu^2\cos\theta\}\sin(\theta+\psi)+\varepsilon^2\ldots,\\ \frac{d\psi}{dt}&=\omega-\nu-\frac{\varepsilon}{a\omega}\{ah\omega\sin(\theta+\psi)-f_1[a\cos(\theta+\psi)]+q_2\nu^2\cos\theta\}\cos(\theta+\psi)+\\ &\qquad+\varepsilon^2\ldots,\\ \frac{d\nu}{dt}&=\varepsilon[M_1(\nu)+q_3\ddot{x}\sin\theta+q_4\sin\theta].\end{aligned}\right\}\tag{3.339}$$

It is not difficult to see that the obtained system of equations (3.339) is of the type (3.324), and hence by means of the method of averaging it may be replaced in the case of fundamental resonance ($\omega-\nu=\varepsilon\alpha$) in the first approximation by the following averaged system

$$\left.\begin{aligned}\frac{da}{dt}&=-\frac{\varepsilon}{2}\left(ha+\frac{q_2\nu^2}{\omega}\sin\psi\right),\\ \frac{d\psi}{dt}&=\omega-\nu+\frac{\varepsilon}{2\pi a\omega}\int_0^{2\pi}f_1(a\cos(\theta+\psi))\cos(\theta+\psi)\,d(\theta+\psi)-\frac{\varepsilon q_2\nu^2}{2\omega a}\cos\psi,\\ \frac{d\nu}{dt}&=\varepsilon\left[M_1(\nu)+\frac{1}{2}q_3\omega\nu a\sin\psi\right].\end{aligned}\right\}\tag{3.340}$$

The system of equations (3.340) represents, as indicated in the preceding section, a system of three coupled equations. The investigation and also the numerical integration of this system are, however, simpler than those of the original system of equations (3.337).

The system of equations (3.340) enables us to study without difficulties the stationary mode of a nonlinear vibrator coupled to an energy source.

Thus, equating the right-hand sides of the system (3.340) to zero, we obtain the following system of equations for the stationary values of a, ψ, and ν:

$$\left.\begin{aligned}h\omega a+q_2\nu^2\sin\psi&=0,\\ (\omega-\nu)+F(a)-\frac{\varepsilon q_2\nu^2}{2\omega a}\cos\psi&=0,\\ M(\nu)+\frac{1}{2}q_3\omega\nu a\sin\psi&=0,\end{aligned}\right\}\tag{3.341}$$

where

$$F(a)=\frac{\varepsilon}{2\pi a\omega}\int_0^{2\pi}f_1(a\cos(\theta+\psi))\cos(\theta+\psi)\,d(\theta+\psi),$$

from which we derive an equation for the stationary frequency ν

$$M(\nu)-\frac{\beta}{2\nu}\omega^2 a^2=0, \qquad (3.342)$$

and also an expression for the vibration amplitude

$$a=\frac{m}{m_0}\frac{r\nu^2}{\sqrt{4\omega^2(\omega_e-\nu)^2+\frac{\beta^2\omega^2}{m_0^2}}}, \qquad (3.343)$$

where

$$\omega_e=\omega+F(a).$$

Analyzing equation (3.342) and formula (3.343), one may find a number of interesting features characteristic of nonlinear vibrating systems coupled to an energy source. For example, it is possible to expose the role of the motor characteristic and its influence on the stability of stationary vibrations, etc. We do not consider this in detail, sending the interested reader to the specific literature on this problem (cf. /67/).

Let us examine another example — the vibrations of a mathematical pendulum of slowly varying length but unlike the case discussed in Chapter II, § 4, where it was assumed that the pendulum length varies slowly only due to the action of external forces, we now assume that the rate of change of the pendulum length $\frac{dy}{dt}$ depends also upon the state of the vibrating system. We assume that the pendulum thread is being stretched "plastically" under the action of the weight and the centrifugal force, and its deformation velocity is small and proportional to the force imposed, i. e.,

$$\frac{dy}{dt}=\varepsilon\lambda P,$$

where P is the tension in the thread due to the pendulum weight and the centrifugal force. The friction force is neglected.

The problem of vibrations of such a pendulum with an account of the centrifugal force was stated and solved by V. M. Volosov /34/.

The vibrations of the pendulum are under these assumptions described by the system of equations

$$\left.\begin{aligned}&\frac{d}{dt}\left[y^2\frac{dx}{dt}\right]+gy\sin x=0,\\&\frac{dy}{dt}=\varepsilon\lambda\left[mg\cos x+my\left(\frac{dx}{dt}\right)^2\right],\end{aligned}\right\} \qquad (3.344)$$

where m is the mass of the vibrating point; y, the slowly varying length; x, the angle of deviation from the vertical; g, the acceleration due to gravity; $\varepsilon\lambda$, the coefficient of the plastic deformation of the pendulum thread.

The system of equations (3.344) is, for small angles of deviation, of the type (3.301).

For the sake of simplicity, we shall take into account in the system (3.344) terms up to and including x^3. Then, discarding terms of the order x^4 and higher, we obtain instead of the system of equations (3.344) the system

$$\left.\begin{aligned}&\frac{d}{dt}\left[y^2\frac{dx}{dt}\right]+gyx=\frac{gyx^3}{6},\\&\frac{dy}{dt}=\varepsilon\lambda\left[mg-\frac{mgx^2}{2}+my\left(\frac{dx}{dt}\right)^2\right].\end{aligned}\right\} \qquad (3.345)$$

Let us derive the improved first approximation for the system of equations (3.345). Making use of the formulas (3.311), (3.320), and (3.325), we obtain after a series of operations

$$\left.\begin{aligned} x_I &= a\cos\psi - \frac{a^3}{192}(1+3\varepsilon\lambda m g^2 y)\cos 3\psi - \frac{\varepsilon\lambda m g y a^2\omega}{2}\sin 2\psi,\\ y_I &= y_0 - \frac{3}{4}\varepsilon\lambda m g a^2\cos 2\psi, \end{aligned}\right\} \tag{3.346}$$

where y_0, a, and ψ are to be determined from the system of equations of the first approximation

$$\left.\begin{aligned} \frac{da}{dt} &= -\frac{3\varepsilon\lambda m g a}{4y_0}\left(1+\frac{5}{8}a^2\right),\\ \frac{d\psi}{dt} &= \sqrt{\frac{g}{y_0}}\left\{\left(1-\frac{a^2}{16}\right)-\frac{\varepsilon\lambda m g}{2y_0}\left(1+\frac{a^2}{4}\right)\right\},\\ \frac{dy_0}{dt} &= \varepsilon\lambda m g\left(1+\frac{a^2}{4}\right). \end{aligned}\right\} \tag{3.347}$$

The first and third equations of this system yield the following relation between the pendulum vibration amplitude and its length in the first approximation

$$a^2\left(1+\frac{5}{8}a^2\right)^{-\frac{3}{5}} y_0^{\frac{3}{2}} = \text{const.} \tag{3.348}$$

Chapter IV

MONOFREQUENCY VIBRATIONS IN NONLINEAR SYSTEMS WITH MANY DEGREES OF FREEDOM AND SLOWLY VARYING PARAMETERS

§ 1. Construction of asymptotic solutions corresponding to a monofrequency mode in a vibrating system

We proceed now to elaborate a method of constructing asymptotic approximations for vibrating systems with many degrees of freedom and slowly varying parameters. These systems are constantly met in actual problems of physics and engineering.

As we know, even in the case where the vibrations in a nonlinear system with many degrees of freedom are described by differential equations close to linear ones, the application of the usual methods of nonlinear mechanics first requires the solution of a set of linear differential equations with a number of unknowns proportional to the number of degrees of freedom; this creates considerable difficulties in the practical application of these methods. Also, in many cases in nonlinear vibrating systems with many degrees of freedom, the presence of some factors (friction, external perturbing periodic force, etc.) may lead to a rapid elimination of higher harmonics and to the establishment of the basic tone of vibrations.

Hence, during the investigation of a vibrating system with many degrees of freedom it is expedient to study the monofrequency state, for which all the coordinates of the given system vibrate with one and the same frequency. As we shall see in the sequel, we can construct asymptotic solutions corresponding to the monofrequency state in the same manner as if we were to deal with a vibrating system with a single degree of freedom.

We now develop a method of constructing asymptotic solutions corresponding to the monofrequency state in a nonlinear vibrating system with N degrees of freedom and slowly varying parameters; this was first presented in /95, 98, 104/. The asymptotic approximate solutions will be derived directly from the Lagrange equations of the second kind. As indicated before, this considerably facilitates the application of the results to the solution of many practical problems.

Thus, we examine a vibrating system with N degrees of freedom, the kinetic and potential energies of which may be written in the form

$$T(\dot{q})=\frac{1}{2}\sum_{i,j=1}^{N} a_{ij}(\tau)\dot{q}_i\dot{q}_j, \quad V(q)=\frac{1}{2}\sum_{i,j=1}^{N} b_{ij}(\tau)q_iq_j, \tag{4.1}$$

where $q_1, q_2, \ldots, q_N$ are generalized coordinates; $\tau=\varepsilon t$, the "slowing" time; ε is a small positive parameter; $a_{rs}(\tau)=a_{sr}(\tau)$, $b_{rs}(\tau)=b_{sr}(\tau)$ $(s, r=1, 2, \ldots, N)$ are some functions of the "slowing" time τ, having derivatives of any order for any finite τ.

We also assume that in the finite interval $0 \leqslant t \leqslant T$, where $T = \frac{L}{\varepsilon}$ (and L may, as before, be made as large as desired for any small values of ε), the quadratic forms of T and V are positive definite.

Let the vibrating system under investigation be subjected to the action of a small perturbation given by the generalized forces

$$\varepsilon Q_j(\tau, \theta, q_1, \ldots, q_N, \dot{q}_1, \ldots, \dot{q}_N, \varepsilon) = \varepsilon Q_j^{(1)}(\tau, \theta, q_1, \ldots, q_N, \dot{q}_1, \ldots, \dot{q}_N) + \\ + \varepsilon^2 Q_j^{(2)}(\tau, \theta, q_1, \ldots, q_N, \dot{q}_1, \ldots, \dot{q}_N) + \varepsilon^3 \ldots \qquad (j = 1, 2, \ldots, N), \tag{4.2}$$

which are periodic in θ with period 2π. It is assumed that the functions on the right-hand side of (4.2) can be expanded in finite Fourier sums

$$Q_j^{(k)}(\tau, \theta, q_1, \ldots, q_N, \dot{q}_1, \ldots, \dot{q}_N) = \\ = \sum_{n=-M}^{M} e^{in\theta} Q_{jn}^{(k)}(\tau, q_1, \ldots, q_N, \dot{q}_1, \ldots, \dot{q}_N) \qquad (j = 1, 2, \ldots, N;\ k = 1, 2, \ldots),$$

where the coefficients are certain polynomials in q_s and $\dot{q}_s$ $(s = 1, 2, \ldots, N)$. We also assume that $\frac{d\theta}{dt} = \nu(\tau)$ and that the functions $\nu(\tau)$, $Q_j(\tau, \theta, q_1, \ldots, q_N, \dot{q}_1, \ldots, \dot{q}_N, \varepsilon)$ $(j = 1, 2, \ldots, N)$ are differentiable with respect to τ to any desired order in the interval $0 \leqslant \tau \leqslant L$.

Then, according to the well-known principles of mechanics, we are led to the system of N nonlinear second order differential equations with slowly varying coefficients

$$\frac{d}{dt}\left\{\sum_{i=1}^{N} a_{ij}(\tau)\dot{q}_i\right\} + \sum_{i=1}^{N} b_{ij}(\tau) q_i = \\ = \varepsilon Q_j(\tau, \theta, q_1, \ldots, q_N, \dot{q}_1, \ldots, \dot{q}_N, \varepsilon) \qquad (j = 1, 2, \ldots, N). \tag{4.3}$$

We shall study simultaneously with (4.3) the system of linear second order differential equations with constant coefficients

$$\sum_{i=1}^{N} a_{ij}(\tau)\ddot{q}_i + \sum_{i=1}^{N} b_{ij}(\tau) q_i = 0 \qquad (j = 1, 2, \ldots, N). \tag{4.4}$$

This system is obtained by putting $\varepsilon = 0$ in (4.3) and regarding τ as a constant parameter.

The system (4.4), which plays in the sequel the role of an auxiliary system for constructing the solutions of equations (4.3), will be called the system of differential equations of the unperturbed motion, or simply — the unperturbed system.

It is possible to construct, with the help of the usual methods, solutions of the equations (4.4) corresponding to normal vibrations

$$q_i^{(k)} = \varphi_i^{(k)} a \cos(\omega_k t + \alpha_k) \qquad (i, k = 1, 2, \ldots, N), \tag{4.5}$$

where ω_k $(k = 1, 2, \ldots, N)$ are the normal frequencies determined by the equation

$$D\,\| - a_{ij}\omega^2 + b_{ij} \| = 0, \tag{4.6}$$

$\varphi_i^{(k)}$ $(i, k = 1, 2, \ldots, N)$ are normal functions which are nontrivial solutions of the system of the homogeneous algebraic equations

$$\sum_{i=1}^{N} \{- a_{ij}\omega_k^2 + b_{ij}\}\, \varphi_i^{(k)} = 0 \qquad (j, k = 1, 2, \ldots, N) \tag{4.7}$$

and possess the orthogonality properties

$$\left.\begin{aligned}&\sum_{i,\,j=1}^{N} a_{ij}\varphi_i^{(k)}\varphi_j^{(l)}=0,\\&\sum_{i,\,j=1}^{N} b_{ij}\varphi_i^{(k)}\varphi_j^{(l)}=0\qquad(k\neq l),\end{aligned}\right\}\tag{4.8}$$

and a and α_k are arbitrary real constants.

The quantities ω_k and $\varphi_i^{(k)}$ also depend on τ as a parameter. If we now set $\tau=\varepsilon t$ in (4.4) and (4.5), then the functions (4.5) will only approximately (with an accuracy of the order of ε) satisfy equations (4.4) representing vibrations with a slowly varying frequency and form.

We now construct for the system (4.3) asymptotic solutions corresponding to monofrequency vibrations, close (for sufficiently small ε) to one of the normal unperturbed vibrations (4.5) (to be definite — to the first). We assume that the following conditions are satisfied for all values of the parameter τ:

1) undamped harmonic vibrations with a frequency $\omega_1(\tau)$, depending only on two arbitrary constants, are possible in the unperturbed system;

2) the single solution of the system (4.4), corresponding to equilibrium, is the trivial solution $q_1=q_2=\ldots=q_N=0$;

3) neither the frequency $\omega_1(\tau)$, nor any of its overtones $2\omega_1(\tau),\ \ldots,\ k\omega_1(\tau)$ is equal to any normal frequency $\omega_2(\tau),\ \omega_3(\tau),\ \ldots,\ \omega_N(\tau)$ of the unperturbed system (there is no internal resonance).

Under these assumptions we search for the solution of the perturbed equations (4.3) in the form of the asymptotic series*

$$\begin{aligned}q_i=\varphi_i^{(1)}(\tau)\,a\cos(p\varphi+\psi)+\varepsilon u_i^{(1)}(\tau,\ a,\ \theta,\ p\varphi+\psi)+\\+\varepsilon^2u_i^{(2)}(\tau,\ a,\ \theta,\ p\varphi+\psi)+\ldots\qquad(i=1,\ 2,\ \ldots,\ N),\end{aligned}\tag{4.9}$$

where $\tau=\varepsilon t$; the functions $u_i^{(l)}(\tau,\ a,\ \theta,\ p\varphi+\psi)$ $\begin{pmatrix}i=1,\ 2,\ \ldots,\ N\\ l=1,\ 2,\ \ldots\end{pmatrix}$ are periodic in θ and $p\varphi+\psi$ with period 2π; and the quantities a and ψ are functions of time to be determined from the system of differential equations

$$\left.\begin{aligned}\frac{da}{dt}&=\varepsilon A_1(\tau,\ a,\ \psi)+\varepsilon^2A_2(\tau,\ a,\ \psi)+\ldots,\\ \frac{d\psi}{dt}&=\omega_1(\tau)-\frac{p}{q}\nu(\tau)+\varepsilon B_1(\tau,\ a,\ \psi)+\varepsilon^2B_2(\tau,\ a,\ \psi)+\ldots,\end{aligned}\right\}\tag{4.10}$$

where $\varphi=\frac{1}{q}\theta$; p and q are mutually prime small integers whose choice depends on the resonance which is examined; $\omega_1(\tau)$ is a root of the equation (4.6); $\varphi_j^{(1)}(\tau)$ $(j=1,\ 2,\ \ldots,\ N)$ are nontrivial solutions of the algebraic equations (4.7).

In order to solve our problem it is also necessary, as in the cases considered before, to derive expressions for the functions

$$\left.\begin{aligned}&u_i^{(1)}(\tau,\ a,\ \theta,\ p\varphi+\psi),\quad u_i^{(2)}(\tau,\ a,\ \theta,\ p\varphi+\psi),\ \ldots\qquad(i=1,\ 2,\ \ldots,\ N),\\&A_1(\tau,\ a,\ \psi),\quad A_2(\tau,\ a,\ \psi),\ \ldots,\ B_1(\tau,\ a,\ \psi),\quad B_2(\tau,\ a,\ \psi),\ \ldots,\end{aligned}\right\}\tag{4.11}$$

such that the asymptotic series (4.9) become, after the substitution of a and ψ by the functions of time derived from equations (4.10), solutions of equations (4.3).

Thus, the integration of the system (4.3) is in our case reduced to a simpler integration of equations (4.10).

* The upper index of q_i will in future be omitted, but one should keep in mind that we are examining vibrations close to the first normal vibration.

The construction of the asymptotic series (4.9) creates, as mentioned before, no essential difficulties, but in view of the rapidly increasing complexity of the formulas, in practice it is only possible to derive the first few terms. Confining ourselves to the first m terms in (4.9), we obtain approximations of the m-th order

$$q_i^{(m)} = \varphi_i^{(1)}(\tau)\, a\cos(p\varphi+\psi) + \varepsilon u_i^{(1)}(\tau,\ a,\ \theta,\ p\varphi+\psi) + \ldots \\ \ldots + \varepsilon^m u_i^{(m)}(\tau,\ a,\ \theta,\ p\varphi+\psi) \qquad (i=1,\ 2,\ \ldots,\ N), \tag{4.12}$$

where a and ψ are to be determined from the system of equations of the m-th approximation

$$\left.\begin{aligned} \frac{da}{dt} &= \varepsilon A_1(\tau,\ a,\ \psi) + \ldots + \varepsilon^m A_m(\tau,\ a,\ \psi), \\ \frac{d\psi}{dt} &= \omega_1(\tau) - \frac{p}{q}\nu(\tau) + \varepsilon B_1(\tau,\ a,\ \psi) + \ldots + \varepsilon^m B_m(\tau,\ a,\ \psi). \end{aligned}\right\} \tag{4.13}$$

The series (4.9) and (4.10) should here also be considered as formal expansions, required for the construction of the asymptotic approximations (4.12), which tend for a fixed m to a particular solution of the unperturbed system (4.4) as $\varepsilon \to 0$.

On the grounds of the arguments of the second chapter it is meaningless to retain in the right-hand sides of (4.12) terms of the m-th order

$$\varepsilon^m u_i^{(m)}(\tau,\ a,\ \theta,\ p\varphi+\psi) \quad (i=1,\ 2,\ \ldots,\ N). \tag{4.14}$$

Hence, as usual, the m-th approximation is taken in what follows to be the expression

$$q_i^{(m-1)} = \varphi_i^{(1)}(\tau)\, a\cos(p\varphi+\psi) + \varepsilon u_1^{(1)}(\tau,\ a,\ \theta,\ p\varphi+\psi) + \ldots \\ \ldots + \varepsilon^{m-1} u_i^{(m-1)}(\tau,\ a,\ \theta,\ p\varphi+\psi) \qquad (i=1,\ 2,\ \ldots,\ N), \tag{4.15}$$

where a and ψ are some functions of time which are to be determined from the system of equations of the m-th approximation (4.13). The series (4.12) are the so-called "improved" m-th approximations.

Since the integration of equations (4.13) introduces only two arbitrary constants, it is possible to obtain with the help of (4.15) an approximate expression only for some two-parameteric family of particular solutions and not for the general solution of equations (4.3), which should depend on $2N$ arbitrary constants. The method of constructing the general solution will be given below.

The principle of superposition does not hold in nonlinear systems and it is therefore impossible to derive the general solution from various particular solutions. However, in a series of important cases, the family of particular solutions (4.15) possesses a property of strong stability, i.e., any solution of the system (4.3) tends in course of time to the derived family of particular solutions (4.15) (this question is considered in detail in the last chapter); the investigation of the latter is then of physical interest.

We turn to the solution of our problem — the derivation of the functions (4.11).

Differentiating (4.10), we have

$$\frac{d^2a}{dt^2} = \varepsilon\left\{\left[\omega_1(\tau) - \frac{p}{q}\nu(\tau)\right]\frac{\partial A_1}{\partial\psi}\right\} + \varepsilon^2\left\{\frac{\partial A_1}{\partial a}A_1 + \frac{\partial A_1}{\partial\psi}B_1 + \right. \\ \left. + \frac{\partial A_1}{\partial\tau} + \left[\omega_1(\tau) - \frac{p}{q}\nu(\tau)\right]\frac{\partial A_2}{\partial\psi}\right\} + \varepsilon^3\{\ldots\} + \ldots, \tag{4.16}$$

$$\frac{d^2\psi}{dt^2} = \varepsilon\left\{\frac{d\omega_1(\tau)}{d\tau} - \frac{p}{q}\frac{d\nu(\tau)}{d\tau} + \left[\omega_1(\tau) - \frac{p}{q}\nu(\tau)\right]\frac{\partial B_1}{\partial\psi}\right\} + \\ + \varepsilon^2\left\{\frac{\partial B_1}{\partial a}A_1 + \frac{\partial B_1}{\partial\psi}B_1 + \frac{\partial B_1}{\partial\tau} + \left[\omega_1(\tau) - \frac{p}{q}\nu(\tau)\right]\frac{\partial B_2}{\partial\psi}\right\} + \varepsilon^3\{\ldots\} + \ldots \tag{4.17}$$

Differentiating (4.9) and taking into account equations (4.10), (4.16), and (4.17), we find expressions for $\frac{dq_j}{dt}$ and $\frac{d^2q_j}{dt^2}$ $(j=1, 2, \ldots, N)$. Substituting these expressions and also the expressions (4.11) for q_j into the left-hand side of equation (4.3), we obtain

$$\sum_{i=1}^{N}\left\{a_{ij}(\tau)\frac{d^2q_i^{(1)}}{dt^2}+\frac{da_{ij}(\tau)}{dt}\frac{dq_i^{(1)}}{dt}+b_{ij}(\tau)\,q_i^{(1)}\right\}=$$
$$=\sum_{i=1}^{N}\left[-a_{ij}(\tau)\,\omega_1^{(2)}(\tau)+b_{ij}(\tau)\right]\varphi_i^{(1)}(\tau)\,a\cos(p\varphi+\psi)+$$
$$+\sum_{i=1}^{N}\Big\{\varepsilon\Big[a_{ij}(\tau)\Big]\varphi_i^{(1)}(\tau)\Big[\omega_1(\tau)-\frac{p}{q}\nu(\tau)\Big]\frac{\partial A_1}{\partial\psi}-2\varphi_i^{(1)}(\tau)\,a\omega_1(\tau)B_1\Big]\cos(p\varphi+\psi)-$$
$$-a_{ij}(\tau)\Big[\varphi_i^{(1)}(\tau)\,a\Big[\omega_1(\tau)-\frac{p}{q}\nu(\tau)\Big]\frac{\partial B_1}{\partial\psi}+2\varphi_i^{(1)}(\tau)\,\omega_1(\tau)A_1\Big]\sin(p\varphi+\psi)-$$
$$-\Big[2\omega_1(\tau)\frac{d\varphi_i^{(1)}(\tau)}{d\tau}a_{ij}(\tau)+\frac{d\omega_1(\tau)}{d\tau}\varphi_i^{(1)}(\tau)\,a_{ij}(\tau)+\omega_1(\tau)\,\varphi_i^{(1)}(\tau)\frac{da_{ij}(\tau)}{d\tau}\Big]\times$$
$$\times a\sin(p\varphi+\psi)+a_{ij}(\tau)\Big[\nu^2(\tau)\frac{\partial^2u_i^{(1)}}{\partial\theta^2}+2\nu(\tau)\,\omega_1(\tau)\frac{\partial^2u_i^{(1)}}{\partial\theta\,\partial(p\varphi+\psi)}+$$
$$+\omega_1^2(\tau)\frac{\partial^2u_i^{(1)}}{\partial(p\varphi+\psi)^2}\Big]+b_{ij}(\tau)\,u_i^{(1)}\Big]+\varepsilon^2\Big[a_{ij}(\tau)\Big[\varphi_i^{(1)}(\tau)\Big[\omega_1(\tau)-\frac{p}{q}\nu(\tau)\;\frac{\partial A_2}{\partial\psi}-$$
$$-2\varphi_i^{(1)}(\tau)\,a\omega_1(\tau)B_2\Big]\cos(p\varphi+\psi)-a_{ij}(\tau)\Big[\varphi_i^{(1)}(\tau)\Big[\omega_1(\tau)-\frac{p}{q}\nu(\tau)\Big]\frac{\partial B_2}{\partial\psi}a+$$
$$+2\varphi_i^{(1)}(\tau)\,\omega_1(\tau)A_2\Big]\sin(p\varphi+\psi)+a_{ij}(\tau)\Big[\varphi_i^{(1)}(\tau)\frac{\partial A_1}{\partial a}A_1+\varphi_i^{(1)}(\tau)\frac{\partial A_1}{\partial\psi}B_1+$$
$$+\varphi_i^{(1)}(\tau)\frac{\partial A_1}{\partial\tau}-\varphi_i^{(1)}(\tau)\,aB_1^2\Big]\cos(p\varphi+\psi)-a_{ij}(\tau)\Big[2\varphi_i^{(1)}(\tau)A_1B_1+$$
$$+\varphi_i^{(1)}(\tau)\,a\frac{\partial B_1}{\partial a}A_1+\varphi_i^{(1)}(\tau)\,a\frac{\partial B_1}{\partial\psi}B_1+\varphi_i^{(1)}(\tau)\frac{\partial B_1}{\partial\tau}\Big]\sin(p\varphi+\psi)+$$
$$+\Big[a_{ij}(\tau)\frac{d^2\varphi_i^{(1)}(\tau)}{d\tau^2}a+2\frac{d\varphi_i^{(1)}(\tau)}{d\tau}A_1a_{ij}(\tau)+\frac{d\varphi_i^{(1)}(\tau)}{d\tau}\frac{da_{ij}(\tau)}{d\tau}a+$$
$$+\varphi_i^{(1)}(\tau)\frac{da_{ij}(\tau)}{d\tau}A_1\Big]\cos(p\varphi+\psi)-\Big[2\frac{d\varphi_i^{(1)}(\tau)}{d\tau}a_{ij}(\tau)B_1+$$
$$+\varphi_i^{(1)}(\tau)\frac{da_{ij}(\tau)}{d\tau}B_1\Big]a\sin(p\varphi+\psi)+\ldots\Big]+\varepsilon^3\ldots\Big\}\tag{4.18}$$
$$(j=1,\ 2,\ \ldots,\ N).$$

Substituting for q_j and $\frac{dq_j}{dt}$ $(j=1, 2, \ldots, N)$ in the right-hand sides of equations (4.3) and expanding in a Taylor series, we find

$$\varepsilon Q_j=\varepsilon Q_j^{(1)}(\tau,\,\theta,\,q_{10},\,\ldots,\,q_{N0},\,\dot q_{10},\,\ldots,\,\dot q_{N0})+$$
$$+\varepsilon^2\Big\{Q_j^{(2)}(\tau,\,\theta,\,q_{10},\,\ldots,\,q_{N0},\,\dot q_{10},\,\ldots,\,\dot q_{N0})+$$
$$+\sum_{i=1}^{N}\Big[\frac{\partial Q_j^{(1)}}{\partial q_i}u_i^{(1)}+\frac{\partial Q_j^{(1)}}{\partial\dot q_i}\Big(\frac{d\varphi_i^{(1)}(\tau)}{d\tau}a\cos(p\varphi+\psi)+$$
$$+\varphi_i^{(1)}(\tau)\cos(p\varphi+\psi)A_1-\varphi_i^{(1)}(\tau)\,a\sin(p\varphi+\psi)B_1+$$
$$+\frac{\partial u_i^{(1)}}{\partial\theta}\nu(\tau)+\frac{\partial u_i^{(1)}}{\partial(p\varphi+\psi)}\omega_1(\tau)\Big)\Big]\Big\}+\varepsilon^3\ldots,\tag{4.19}$$
$$(j=1,\ 2,\ \ldots,\ N),$$

where we have used the notation

$$q_{i0}=\varphi_i^{(1)}(\tau)\,a\cos(p\varphi+\psi),\tag{4.20}$$
$$\dot q_{i0}=-\varphi_i^{(1)}(\tau)\,\omega_1(\tau)\,a\sin(p\varphi+\psi)\qquad(i=1,2,\ldots,N).\tag{4.21}$$

In order that the asymptotic series (4.11) satisfy equations (4.3) up to terms of the order of ε^m, the coefficients of equal powers of ε in the right-hand sides of (4.18) and (4.19) must be equated, up to and including terms of the $(m-1)$-th order. This results in the equations

$$
\begin{aligned}
&\sum_{i=1}^{N}\Big\{a_{ij}(\tau)\Big[\omega_1^2(\tau)\frac{\partial^2 u_i^{(1)}}{\partial(p\varphi+\psi)^2}+2\nu(\tau)\,\omega_1(\tau)\frac{\partial^2 u_i^{(1)}}{\partial\theta\,\partial(p\varphi+\psi)}+\nu^2(\tau)\frac{\partial^2 u_i^{(1)}}{\partial\theta^2}\Big]+\\
&\quad+b_{ij}(\tau)\,u_i^{(1)}\Big\}=Q_{j0}^{(1)}(\tau,\theta,a,p\varphi+\psi)-\sum_{i=1}^{N}\Big\{a_{ij}(\tau)\Big[\varphi_i^{(1)}(\tau)\Big[\omega_1(\tau)-\\
&\quad-\frac{p}{q}\nu(\tau)\Big]\frac{\partial A_1}{\partial\psi}-2\varphi_i^{(1)}(\tau)\,\omega_1(\tau)\,aB_1\Big]\cos(p\varphi+\psi)-\\
&\quad-a_{ij}(\tau)\Big[\varphi_i^{(1)}(\tau)\,a\Big[\omega_1(\tau)-\frac{p}{q}\nu(\tau)\Big]\frac{\partial B_1}{\partial\psi}+\\
&\quad+2\varphi_i^{(1)}(\tau)\,\omega_1(\tau)\,A_1\Big]\sin(p\varphi+\psi)-\\
&\quad-\Big[2\omega_1(\tau)\frac{d\varphi_i^{(1)}(\tau)}{d\tau}a_{ij}(\tau)+\frac{d\omega_1(\tau)}{d\tau}\varphi_i^{(1)}(\tau)\,a_{ij}(\tau)+\\
&\quad+\omega_1(\tau)\,\varphi_i^{(1)}(\tau)\frac{da_{ij}(\tau)}{d\tau}\Big]a\sin(p\varphi+\psi)\Big\}=G_{j0}^{(1)}(\tau,\theta,a,p\varphi+\psi),
\end{aligned}
\tag{4.22}
$$

$$
\begin{aligned}
&\sum_{i=1}^{N}\Big\{a_{ij}(\tau)\Big[\omega_1^2(\tau)\frac{\partial^2 u_i^{(2)}}{\partial(p\varphi+\psi)^2}+2\nu(\tau)\,\omega_1(\tau)\frac{\partial^2 u^{(2)}}{\partial\theta\,\partial(p\varphi+\psi)}+\\
&+\nu^2(\tau)\frac{\partial^2 u_i^{(2)}}{\partial\theta^2}\Big]+b_{ij}(\tau)\,u_i^{(2)}\Big\}=\Phi_{j0}^{(2)}(\tau,\theta,a,p\varphi+\psi)-\\
&-\sum_{i=1}^{N}\Big\{a_{ij}(\tau)\Big[\varphi_i^{(1)}(\tau)\Big[\omega_1(\tau)-\frac{p}{q}\nu(\tau)\Big]\frac{\partial A_2}{\partial\psi}-2\varphi_i^{(1)}(\tau)\,a\omega_1(\tau)\,B_2\Big]\cos(p\varphi+\psi)-\\
&-a_{ij}(\tau)\Big[\varphi_i^{(1)}(\tau)\Big[\omega_1(\tau)-\frac{p}{q}\nu(\tau)\Big]\frac{\partial B_2}{\partial\psi}a+2\varphi_i^{(1)}(\tau)\,\omega_1(\tau)\,A_2\Big]\sin(p\varphi+\psi)+\\
&+a_{ij}(\tau)\Big[\varphi_i^{(1)}(\tau)\frac{\partial A_1}{\partial a}A_1+\varphi_i^{(1)}(\tau)\frac{\partial A_1}{\partial\psi}B_1+\varphi_i^{(1)}(\tau)\frac{\partial A_1}{\partial\tau}-\\
&-\varphi_i^{(1)}(\tau)\,aB_1^2\Big]\cos(p\varphi+\psi)-a_{ij}(\tau)\Big[2\varphi_i^{(1)}(\tau)\,A_1B_1+\varphi_i^{(1)}(\tau)\,a\frac{\partial B_1}{\partial a}A_1+\\
&+\varphi_i^{(1)}(\tau)\,a\frac{\partial B_1}{\partial\psi}B_1+\varphi_i^{(1)}(\tau)\frac{\partial B_1}{\partial\tau}\Big]\sin(p\varphi+\psi)+\Big[a_{ij}(\tau)\frac{d^2\varphi_i^{(1)}(\tau)}{d\tau^2}a+\\
&+\frac{d\varphi_i^{(1)}(\tau)}{d\tau}\frac{da_{ij}(\tau)}{d\tau}a+\varphi_i^{(1)}(\tau)\frac{da_{ij}(\tau)}{d\tau}A_1\Big]\cos(p\varphi+\psi)-\\
&-\Big[2\frac{d\varphi_i^{(1)}(\tau)}{d\tau}a_{ij}(\tau)\,B_1+\varphi_i^{(1)}(\tau)\frac{da_{ij}(\tau)}{d\tau}B_1\Big]a\sin(p\varphi+\psi)\Big\}=\\
&\qquad=G_{j0}^{(2)}(\tau,\theta,a,p\varphi+\psi),
\end{aligned}
\tag{4.23}
$$

. . . .

$$(j=1,2,\ldots,N),$$

where we have used the notation

$$
\begin{aligned}
&Q_{j0}^{(1)}(\tau,\theta,a,p\varphi+\psi)=Q_j^{(1)}(\tau,\theta,q_{10},\ldots,q_{N0},\dot q_{10},\ldots,\dot q_{N0}),
\end{aligned}
\tag{4.24}
$$

$$
\begin{aligned}
&\Phi_{j0}^{(2)}(\tau,\theta,a,p\varphi+\psi)=Q_j^{(2)}(\tau,\theta,q_{10},\ldots,q_{N0},\dot q_{10},\ldots,\dot q_{N0})+\\
&+\sum_{i=1}^{N}\Big[\frac{\partial Q_j^{(1)}}{\partial q_i}u_i^{(1)}+\frac{\partial Q_j^{(1)}}{\partial\dot q_i}\Big(\frac{d\varphi_i^{(1)}(\tau)}{d\tau}a\cos(p\varphi+\psi)+\varphi_i^{(1)}(\tau)\cos(p\varphi+\psi)\,A_1-\\
&-\varphi_i^{(1)}(\tau)\,a\sin(p\varphi+\psi)\,B_1\Big)+\frac{\partial u_i^{(1)}}{\partial\theta}\nu(\tau)+\frac{\partial u_i^{(1)}}{\partial(p\varphi+\psi)}\omega_1(\tau)\Big]-\\
&-\sum_{i=1}^{N}\Big\{a_{ij}(\tau)\Big[2\frac{\partial^2 u_i^{(1)}}{\partial\tau\,\partial(p\varphi+\psi)}\omega_1(\tau)+2\frac{\partial^2 u_i^{(1)}}{\partial\tau\,\partial\theta}\nu(\tau)+2\frac{\partial^2 u_i^{(1)}}{\partial a\,\partial\theta}\nu(\tau)\,A_1+
\end{aligned}
$$

$$+2\frac{\partial^2 u_i^{(1)}}{\partial\theta\,\partial(p\varphi+\psi)}\nu(\tau)B_1+2\frac{\partial^2 u_i^{(1)}}{\partial(p\varphi+\psi)^2}\omega_1(\tau)B_1+$$

$$+2\frac{\partial^2 u_i^{(1)}}{\partial a\,\partial(p\varphi+\psi)}\omega_1(\tau)A_1+\frac{\partial u_i^{(1)}}{\partial(p\varphi+\psi)}\left[\omega_1(\tau)-\frac{p}{q}\nu(\tau)\right]\frac{\partial B_1}{\partial\psi}+$$

$$+\frac{\partial u_i^{(1)}}{\partial a}\left[\omega_1(\tau)-\frac{p}{q}\nu(\tau)\right]\frac{\partial A_1}{\partial\psi}+\frac{\partial u_i^{(1)}}{\partial\theta}\frac{d\nu(\tau)}{d\tau}+\frac{\partial u_i^{(1)}}{\partial(p\varphi+\psi)}\frac{d\omega_1(\tau)}{d\tau}\Big]+$$

$$+\frac{da_{ij}(\tau)}{d\tau}\frac{\partial u_i^{(1)}}{\partial\theta}\nu(\tau)+\frac{da_{ij}(\tau)}{d\tau}\frac{\partial u_i^{(1)}}{\partial(p\varphi+\psi)}\omega_1(\tau)\Big\}\quad(j=1,2,\ldots,N). \tag{4.25}$$

The right-hand sides of equations (4.22), which are periodic functions of θ and $p\varphi+\psi$ with period 2π, are conveniently represented in the form of the Fourier series

$$G_{i0}^{(1)}(\tau,a,\theta,p\varphi+\psi)=\sum_{n,m}g_{nm}^{(i1)}(\tau,a)\,e^{i\{n\theta+m(p\varphi+\psi)\}} \tag{4.26}$$

$$(i=1,2,\ldots,N),$$

where

$$g_{nm}^{(i1)}(\tau,a)=\frac{1}{4\pi^2}\int_0^{2\pi}\int_0^{2\pi}G_{i0}^{(1)}(\tau,a,\theta,p\varphi+\psi)\,e^{-i\{n\theta+m(p\varphi+\psi)\}}\,d\theta\,d(p\varphi+\psi)$$

$$(i=1,2,\ldots,N).$$

Naturally, the unknown functions $u_j^{(1)}(\tau,a,\theta,p\varphi+\psi)$ $(j=1,2,\ldots,N)$ are also sought for in the form of the sums

$$u_j^{(1)}(\tau,a,\theta,p\varphi+\psi)=\sum_{n,m}K_{nm}^{(j1)}(\tau,a)\,e^{i\{n\theta+m(p\varphi+\psi)\}} \tag{4.27}$$

$$(j=1,2,\ldots,N),$$

where the coefficients $K_{nm}^{(j1)}(\tau,a)$ $(j=1,2,\ldots,N;\ -\infty<n,\ m<\infty)$ are still to be determined. Differentiating (4.27), we obtain

$$\frac{\partial^2 u_j^{(1)}}{\partial(p\varphi+\psi)^2}=-\sum_{n,m}K_{nm}^{(j1)}(\tau,a)\,m^2e^{i\{n\theta+m(p\varphi+\psi)\}}, \tag{4.28}$$

$$\frac{\partial^2 u_j^{(1)}}{\partial\theta\,\partial(p\varphi+\psi)}=-\sum_{n,m}K_{nm}^{(j1)}(\tau,a)\,nme^{i\{n\theta+m(p\varphi+\psi)\}}, \tag{4.29}$$

$$\frac{\partial^2 u_j^{(1)}}{\partial\theta^2}=-\sum_{n,m}K_{nm}^{(j1)}(\tau,a)\,n^2e^{i\{n\theta+m(p\varphi+\psi)\}}. \tag{4.30}$$

Substituting in equations (4.22) the expressions for

$$u_j^{(1)}(\tau,a,\theta,p\varphi+\psi),\ \frac{\partial^2 u_j^{(1)}}{\partial(p\varphi+\psi)^2},\ \frac{\partial^2 u_j^{(1)}}{\partial\theta\,\partial(p\varphi+\psi)},\ \frac{\partial^2 u_j^{(1)}}{\partial\theta^2}\ \text{and}\ G_{j0}^{(1)}(\tau,a,\theta,p\varphi+\psi)$$

$$(j=1,2,\ldots,N)$$

and equating the coefficients of equal harmonics, we obtain for the coefficients $K_{nm}^{(j1)}(\tau,a)$ the following system of algebraic equations:

$$\sum_{j=1}^{N}\{a_{ij}(\tau)[-\omega_1^2(\tau)m^2-2\nu(\tau)\omega_1(\tau)mn-\nu^2(\tau)n^2]+$$
$$+b_{ij}(\tau)\}K_{nm}^{(j1)}(\tau,a)=g_{nm}^{(i1)}(\tau,a) \tag{4.31}$$

$$(i=1,2,\ldots,N),$$

which will be solved with the help of normal coordinates. We are looking for an expression for $K_{nm}^{(i1)}(\tau,a)$ in the form of the sum

$$K_{nm}^{(i1)}(\tau,a)=\sum_{k=1}^{N}c_k\varphi_i^{(k)}(\tau)\qquad(i=1,2,\ldots,N), \tag{4.32}$$

where $\varphi_i^{(k)}(\tau)$ are normal functions; and c_k are unknown coefficients which are to be determined. Inserting (4.32) into the system of equations (4.31) and taking into account that $\varphi_j^{(k)}(\tau)$ $(j=1,2,\ldots,N)$ satisfy the system of homogeneous algebraic equations

$$\sum_{j=1}^{N}[b_{ij}(\tau)-a_{ij}(\tau)\omega_k^2(\tau)]\,\varphi_j^{(k)}(\tau)=0\qquad(i=1,2,\ldots,N), \tag{4.33}$$

we obtain

$$\sum_{k=1}^{N}\sum_{j=1}^{N} a_{ij}(\tau)\{\omega_k^2(\tau)-[\omega_1(\tau)m+\nu(\tau)n]^2\}c_k\varphi_i^{(k)}(\tau)=g_{nm}^{(i1)}(\tau,a) \tag{4.34}$$

$$(i=1,\ 2,\ \ldots,\ N).$$

Multiplying these equations by $\varphi_1^{(k)}(\tau), \varphi_2^{(k)}(\tau), \ldots, \varphi_N^{(k)}(\tau)$ respectively, and summing the result over i, we have

$$\sum_{k=1}^{N} c_k\{\omega_k^2(\tau)-[\omega_1(\tau)m+\nu(\tau)n]^2\}\sum_{i=1}^{N} a_{ij}(\tau)\varphi_i^{(k)}(\tau)\varphi_j^{(k_1)}(\tau)=$$

$$=\sum_{i=1}^{N} g_{nm}^{(i1)}(\tau,a)\varphi_i^{(k_1)}(\tau). \tag{4.35}$$

Keeping in mind the orthogonality of the normal functions (relations (4.8)) and using the notation

$$\sum_{i,\,j=1}^{N} a_{ij}(\tau)\varphi_i^{(k)}(\tau)\varphi_j^{(k)}(\tau)=m_k(\tau)\qquad(k=1,\ 2,\ \ldots,\ N), \tag{4.36}$$

we obtain

$$c_k=\frac{\sum_{i=1}^{N} g_{nm}^{(i1)}(\tau,a)\varphi^{(k)}(\tau)}{m_k(\tau)\{\omega_k^2(\tau)-[\omega_1(\tau)m+\nu(\tau)n]^2\}}\qquad(k=1,\ 2,\ \ldots,\ N), \tag{4.37}$$

thus for $u_j^{(1)}(\tau, a, \theta, p\varphi+\psi)$ $(j=1, 2, \ldots, N)$ we obtain the expressions

$$u_j^{(1)}(\tau,a,\theta,p\varphi+\psi)=\sum_{n,m}\sum_{k=1}^{N}\varphi_j^{(k)}(\tau)\frac{\sum_{i=1}^{N} g_{nm}^{(i1)}(\tau,a)\varphi_i^{(k)}(\tau)e^{i\{n\theta+m(p\varphi+\psi)\}}}{m_k(\tau)\{\omega_k^2(\tau)-[\omega_1(\tau)m+\nu(\tau)n]^2\}} \tag{4.38}$$

$$(j=1,\ 2,\ \ldots,\ N).$$

The periodic functions $u_j^{(1)}(\tau, a, \theta, p\varphi+\psi)$ $(j=1, 2, \ldots, N)$ should be finite (since the coordinates of the mechanical system under investigation cannot take on unlimitedly large values). For this the right-hand sides of the formulas (4.38) should not contain terms with denominators that may vanish for some values of τ $(0\leqslant\tau\leqslant L)$; this may occur for values of n and m satisfying the relations

$$m\omega_1(\tau)+n\nu(\tau)=\omega_1(\tau),\quad m\omega_1(\tau)+n\nu(\tau)=-\omega_1(\tau) \tag{4.39}$$

or

$$mq+p(m-1)=0,\quad nq+p(m+1)=0 \tag{4.40}$$

(because $\omega_1(\tau)=\frac{p}{q}\nu(\tau)$ is possible for some values of τ $(0\leqslant\tau\leqslant L)$).

Hence the conditions of finiteness for $u_j^{(1)}(\tau, a, \theta, p\varphi+\psi)$ $(j=1, 2, \ldots, N)$ take the form

$$\sum_{\substack{i=1\\ [nq+p(m\pm1)=0]}}^{N} g_{nm}^{(i1)}(\tau,a)\varphi_i^{(1)}(\tau)=0 \tag{4.41}$$

identically in a and τ within the considered domain of their variation. It should be emphasized that at the same time when the domain of τ $(0\leqslant\tau\leqslant L)$ is known a priori, the domain of the amplitude $a(\tau)$, determined by the equations of the asymptotic approximations, is a priori not known. Consequently the possible domain of variation of the function $a(\tau)$ must be estimated before solving the equations, and the validity of the condition (4.41) in this $a(\tau)$ domain must be tested.

Taking into account (4.27) and (4.38), we obtain, after a series of transformations, the following expression for $u_i^{(1)}(\tau, a, \theta, p\varphi+\psi)$:

$$u_i^{(1)}(\tau, a, \theta, p\varphi+\psi)=$$

$$=\frac{1}{4\pi^2}\sum_{\substack{n,m\\ [nq+p(m\pm 1)\neq 0]\\ \text{для } k=1}}\sum_{k=1}^{N}\varphi_i^{(k)}(\tau)\frac{\int\limits_0^{2\pi}\int\limits_0^{2\pi}\sum\limits_{j=1}^{N}Q_{j0}^{(1)}\varphi_j^{(k)}(\tau)e^{-i\{n\theta+m(p\varphi+\psi)\}}\,d\theta\,d(p\varphi+\psi)}{m_k(\tau)\{\omega_k^2(\tau)-[\omega_1(\tau)m+\nu(\tau)n]^2\}}\times$$

$$\times e^{i\{n\theta+m(p\varphi+\psi)\}}-2\omega_1(\tau)a\sum_{k=2}^{N}\frac{\sum\limits_{j=1}^{N}\frac{d}{d\tau}\left[\frac{\partial T(\dot{q})}{\partial\dot{q}_j}\right]_{\dot{q}_i=\varphi_i^{(1)}(\tau)}\cdot\varphi_j^{(k)}(\tau)}{m_k(\tau)[\omega_k^2(\tau)-\omega_1^2(\tau)]}\sin(p\varphi+\psi), \qquad (4.42)$$

where

$$m_k(\tau)=\sum_{i,j=1}^{N}a_{ij}(\tau)\varphi_i^{(k)}(\tau)\varphi_j^{(k)}(\tau)=2T[\varphi^{(k)}(\tau)],$$

$$Q_{j0}^{(1)}(\tau, a, \theta, p\varphi+\psi)=Q_j^{(1)}[\tau, \theta, \varphi_1^{(1)}(\tau)a\cos(p\varphi+\psi), \ldots$$
$$\ldots, -\varphi_1^{(1)}(\tau)a\omega_1(\tau)\sin(p\varphi+\psi), \ldots] \qquad (4.43)$$
$$(i, k, j=1, 2, \ldots, N).$$

We now formulate a rule for constructing expressions for $u_i(\tau, a, \theta, p\varphi+\psi)$ $(i=1, 2, \ldots, N)$ proceeding directly from the expressions for the perturbing forces and for the kinetic and potential energies. The sum $\sum\limits_{j=1}^{N}Q_{j0}^{(1)}(\tau, a, \theta, p\varphi+\psi)\varphi_j^{(k)}(\tau)$ represents a generalized force acting upon the k-th normal coordinate. The expression $\sum\limits_{j=1}^{N}\frac{d}{d\tau}\left[\frac{\partial T(\dot{q})}{\partial\dot{q}_j}\right]_{\dot{q}_i=\varphi_i^{(1)}(\tau)}\varphi_j^{(k)}(\tau)$ may also be interpreted as a generalized force acting upon the k-th normal coordinate. The presence of this force is explained by the appearance (as a result of the dependence of the inertial coefficients a_{ij} on the "slowing" time τ) of the "force" $\frac{d}{d\tau}\left[\frac{\partial T(\dot{q})}{\partial\dot{q}_j}\right]_{\dot{q}_i=\varphi_i^{(1)}(\tau)}$. Thus, in deriving an expression for $u_i^{(1)}(\tau, a, \theta, p\varphi+\psi)$ the zeroeth approximations for q_i, $\dot{q}_i$ $(i=1, 2, \ldots, N)$ must be put in the functions $Q_j^{(1)}(\tau, \theta, q_1, \ldots, q_N, \dot{q}_1, \ldots, \dot{q}_N)$ and then the (n, m)-th term in the Fourier sum of the generalized force acting upon the k-th normal coordinate must be calculated. The derivative with respect to the velocity of the kinetic energy must also be calculated, $\dot{q}_i$ must be replaced by $\varphi_i^{(1)}(\tau)$, and the expressions thus obtained should be substituted into formulas (4.42). The quantity $m_k(\tau)$ represents the double form of the kinetic energy in which the velocities $\dot{q}_i$ are replaced by the "normal" functions $\varphi_i^{(k)}(\tau)$.

We now proceed to derive expressions for the functions $A_1(\tau, a, \psi)$, $B_1(\tau, a, \psi)$ which should be determined so that $u_i^{(1)}(\tau, a, \theta, p\varphi+\psi)$ be finite, i. e., that they do not contain denominators which can vanish. In other words, $A_1(\tau, a, \psi)$ and $B_1(\tau, a, \psi)$ should satisfy (4.41).

Introducing the notation $n=-\sigma p$, $m=\sigma q\pm 1$, it is possible to represent (4.41) in the form

$$\sum_{j=1}^{N}g_{-\sigma p,\,\sigma q\pm 1}^{(j1)}(\tau, a)\varphi_j^{(1)}(\tau)e^{\pm i(p\varphi+\psi)+i\sigma q\psi}=0 \qquad (\sigma=-\infty, \ldots, \infty). \qquad (4.44)$$

Substituting the values of $g_{-\sigma p,\,\sigma q\pm 1}^{(j1)}$ in the σ-th equality and equating the coefficients of equal harmonics, we obtain the system of equations

$$\left.\begin{aligned}
&\left[\omega_1(\tau)-\frac{p}{q}\nu(\tau)\right]\frac{\partial A_1}{\partial\psi}-2a\omega_1(\tau)B_1=\\
&\quad=\frac{1}{2\pi^2 m_1(\tau)}\sum_\sigma e^{i\sigma q\psi}\int_0^{2\pi}\int_0^{2\pi}\sum_{j=1}^{N}Q_{j0}^{(1)}(\tau,a,\theta,p\varphi+\psi)\varphi_j^{(1)}(\tau)\times\\
&\qquad\times e^{-i\sigma q\psi}\cos(p\varphi+\psi)\,d\theta\,d(p\varphi+\psi),\\
&\left[\omega_1(\tau)-\frac{p}{q}\nu(\tau)\right]a\frac{\partial B_1}{\partial\psi}+2\omega_1(\tau)A_1=-\frac{a}{m_1(\tau)}\frac{d[m_1(\tau)\omega_1(\tau)]}{d\tau}-\\
&\quad-\frac{1}{2\pi^2 m_1(\tau)}\sum_\sigma e^{i\sigma q\psi}\int_0^{2\pi}\int_0^{2\pi}\sum_{j=1}^{N}Q_{j0}^{(1)}(\tau,a,\theta,p\varphi+\psi)\varphi_j^{(1)}(\tau)\times\\
&\qquad\times e^{-i\sigma q\psi}\sin(p\varphi+\psi)\,d\theta\,d(p\varphi+\psi).
\end{aligned}\right\}\qquad(4.45)$$

As shown before, it is sufficient to take for $A_1(\tau, a, \psi)$ and $B_1(\tau, a, \psi)$ (in order that the conditions (4.41) be satisfied), some particular solution of the system (4.45), periodic in ψ. Such a solution may be found without difficulty. Exactly as in the third chapter, we obtain after a few transformations

$$\left.\begin{aligned}
&A_1(\tau,a,\psi)=-\frac{a}{2\omega_1(\tau)m_1(\tau)}\frac{d[\omega_1(\tau)m_1(\tau)]}{d\tau}+\frac{1}{2\pi^2 m_1(\tau)}\sum_\sigma e^{i\sigma q\psi}\times\\
&\times\left[\frac{[q\omega_1(\tau)-p\nu(\tau)]\sigma i\int_0^{2\pi}\int_0^{2\pi}\sum_{j=1}^{N}Q_{j0}^{(1)}(\tau,a,\theta,p\varphi+\psi)\varphi_j^{(1)}(\tau)e^{-i\sigma q\psi}\cos(p\varphi+\psi)d\theta\,d(p\varphi+\psi)}{4\omega_1^2(\tau)-[q\omega_1(\tau)-p\nu(\tau)]^2\sigma^2}\right.\\
&\left.-\frac{2\omega_1(\tau)\int_0^{2\pi}\int_0^{2\pi}\sum_{j=1}^{N}Q_{j0}^{(1)}(\tau,a,\theta,p\varphi+\psi)\varphi_j^{(1)}(\tau)e^{-i\sigma q\psi}\sin(p\varphi+\psi)\,d\theta\,d(p\varphi+\psi)}{4\omega_1^2(\tau)-[q\omega_1(\tau)-p\nu(\tau)]^2\sigma^2}\right],\\
&B_1(\tau,a,\psi)=-\frac{1}{2\pi^2 m_1(\tau)}\sum_\sigma e^{i\sigma q\psi}\times\\
&\times\left[\frac{[q\omega_1(\tau)-p\nu(\tau)]\sigma i\int_0^{2\pi}\int_0^{2\pi}\sum_{j=1}^{N}Q_{j0}^{(1)}(\tau,a,\theta,p\varphi+\psi)\varphi_j^{(1)}(\tau)e^{-i\sigma q\psi}\sin(p\varphi+\psi)d\theta\,d(p\varphi+\psi)}{4\omega_1^2(\tau)-[q\omega_1(\tau)-p\nu(\tau)]^2\sigma^2}+\right.\\
&\left.+\frac{2\omega_1(\tau)\int_0^{2\pi}\int_0^{2\pi}\sum_{j=1}^{N}Q_{j0}^{(1)}(\tau,a,\theta,p\varphi+\psi)\varphi_j^{(1)}(\tau)e^{-i\sigma q\psi}\cos(p\varphi+\psi)\,d\theta\,d(p\varphi+\psi)}{4\omega_1^2(\tau)-[q\omega_1(\tau)-p\nu(\tau)]^2\sigma^2}\right].
\end{aligned}\right\}\qquad(4.46)$$

Having derived the expressions for $u_j^{(1)}(\tau, a, \theta, p\varphi+\psi)$, $A_1(\tau, a, \psi)$ and $B_1(\tau, a, \psi)$ we may apply equations (4.23), the right-hand sides of which are already known. Keeping in mind the condition that $u_j^{(2)}(\tau, a, \theta, p\varphi+\psi)$ $(j = 1, 2, \ldots, N)$ do not contain terms whose denominators can vanish, i. e., the condition

$$\sum_{\substack{j=1\\ [nq+p(m\pm1)=0]}}^{N} g_{n,m}^{(j2)}(\tau, a)\,\varphi_j^{(1)}(\tau)=0, \qquad (4.47)$$

where $g_{n,m}^{(j2)}(\tau, a)$ are Fourier coefficients of the functions (4.25), we can also derive all the functions required for the construction of the solutions of the system of equations (4.3) in the second approximation. Thus, after a series of calculations we obtain the following system of equations for determining $A_2(\tau, a, \psi)$ and $B_2(\tau, a, \psi)$

$$\left.\begin{aligned}
&\left[\omega_1(\tau) - \frac{p}{q}\nu(\tau)\right]\frac{\partial A_2}{\partial\psi} - 2a\omega_1(\tau)B_2 = \\
&= -\left\{\frac{\partial A_1}{\partial a}A_1 + \frac{\partial A_1}{\partial\psi}B_1 + \frac{\partial A_1}{\partial\tau} - aB_1^2 + \frac{dm_1(\tau)}{d\tau}\frac{A_1}{m(\tau)} - \frac{\gamma(\tau)a}{m_1(\tau)}\right\} + \\
&+ \frac{1}{2\pi^2 m_1(\tau)}\sum_\sigma e^{i\sigma q\psi}\int_0^{2\pi}\int_0^{2\pi}\sum_{j=1}^N \Phi_{j0}^{(2)}(\tau, a, \theta, p\varphi+\psi)\varphi_j^{(1)}(\tau)e^{-i\sigma q\psi}\times \\
&\qquad\qquad \times\cos(p\varphi+\psi)\,d\theta\,d(p\varphi+\psi), \\
&\left[\omega_1(\tau) - \frac{p}{q}\nu(\tau)\right]a\frac{\partial B_2}{\partial\psi} + 2\omega_1(\tau)A_2 = \\
&= -\left\{a\frac{\partial B_1}{\partial a}A_1 + a\frac{\partial B_1}{\partial\psi}B_1 + a\frac{\partial B_1}{\partial\tau} + 2A_1B_1 + \frac{dm_1(\tau)}{d\tau}a\frac{B_1}{m_1(\tau)}\right\} - \\
&- \frac{1}{2\pi^2 m_1(\tau)}\sum_\sigma e^{i\sigma q\psi}\int_0^{2\pi}\int_0^{2\pi}\sum_{j=1}^N \Phi_{j0}^{(2)}(\tau, a, \theta, p\varphi+\psi)\varphi_j^{(1)}(\tau)e^{-i\sigma q\psi}\times \\
&\qquad\qquad \times\sin(p\varphi+\psi)\,d\theta\,d(p\varphi+\psi),
\end{aligned}\right\} \tag{4.48}$$

which has the same form as the system (4.45); hence the derivation of explicit expressions for $A_2(\tau, a, \psi)$ and $B_2(\tau, a, \psi)$ also presents no difficulties.

Thus, we have derived the functions $u_j^{(1)}(\tau, a, \theta, p\varphi+\psi)$ $(j = 1, 2, \ldots, N)$, $A_1(\tau, a, \psi)$, $A_2(\tau, a, \psi)$, $B_1(\tau, a, \psi)$, $B_2(\tau, a, \psi)$ and we can construct the solution of equations (4.3) in the first and second approximation.

Resuming, we present a scheme for constructing the first and second approximations for the solutions of the system of equations (4.3) corresponding to the monofrequency mode. We first single out the unperturbed system (4.4) and test whether nondamping harmonic normal vibrations with the frequency $\omega_1(\tau)$, depending on two arbitrary constants, are possible in this system or not. We also test the absence of nontrivial static solutions and internal resonance. We then derive the normal frequencies $\omega_k(\tau)$ $(k = 1, 2, \ldots, N)$ and the eigenfunctions $\varphi_j^{(1)}(\tau)$ $(j = 1, 2, \ldots, N)$. Following this we take as the first approximation the expression

$$q_j = \varphi_j^{(1)}(\tau)a\cos(p\varphi+\psi) \qquad (j = 1, 2, \ldots, N), \tag{4.49}$$

in which a and ψ are certain functions which should be determined from the equations of the first approximation

$$\left.\begin{aligned}
\frac{da}{dt} &= \varepsilon A_1(\tau, a, \psi), \\
\frac{d\psi}{dt} &= \omega_1(\tau) - \frac{p}{q}\nu(\tau) + \varepsilon B_1(\tau, a, \psi),
\end{aligned}\right\} \tag{4.50}$$

where $A_1(\tau, a, \psi)$ and $B_1(\tau, a, \psi)$ are particular solutions of the system (4.45), periodic in ψ, which are determined by formulas (4.46), and p and q are some mutually prime (generally speaking, small) integers depending on the resonance under investigation.

As the second approximation we take the expression

$$q_j = \varphi_j^{(1)}(\tau)a\cos(p\varphi+\psi) + \varepsilon u_j^{(1)}(\tau, a, \theta, p\varphi+\psi) \tag{4.51}$$
$$(j = 1, 2, \ldots, N),$$

where a and ψ are determined by the equations of the second approximation

$$\left.\begin{aligned}
\frac{da}{dt} &= \varepsilon A_1(\tau, a, \psi) + \varepsilon^2 A_2(\tau, a, \psi), \\
\frac{d\psi}{dt} &= \omega_1(\tau) - \frac{p}{q}\nu(\tau) + \varepsilon B_1(\tau, a, \psi) + \varepsilon^2 B_2(\tau, a, \psi),
\end{aligned}\right\} \tag{4.52}$$

where $A_1(\tau, a, \psi)$, $B_1(\tau, a, \psi)$, $A_2(\tau, a, \psi)$, and $B_2(\tau, a, \psi)$ are obtained from the systems (4.45) and (4.48), and $u_j^{(1)}(\tau, a, \theta, p\varphi + \psi)$, from the formula (4.42).

Thus we have reduced the construction of an approximate solution for the system (4.3), corresponding to the monofrequency mode, to the integration of equations (4.50) or (4.52) which, as pointed out before, cannot in general be integrated in a closed form and should be integrated by numerical methods.

In the preceding chapters we have indicated the advantages of numerical integration of the system of equations of the first (or the second) approximation determining a and ψ over the direct numerical integration of the equations of motion determining the oscillating quantity x. In the given case there is a manifold increase in this advantage because during the numerical integration we do not deal with a system of N second order equations, but only two first order equations.

§ 2. Asymptotic expansions for particular cases of the system (4.3). Stationary modes and their stability

We shall now examine a few particular cases of the system of nonlinear differential equations with slowly varying parameters (4.3), for which the equations of the first approximation assume a particularly simple form. Sufficiently simple formal ways of their construction may be indicated.

As a first particular case, we consider "free" monofrequency vibrations of a system with many degrees of freedom and slowly varying parameters, i. e., when the right-hand sides of equations (4.3) do not depend on θ. In this case we have to deal with the following system of N second order equations

$$\frac{d}{dt}\left\{\sum_{i=1}^{N} a_{ij}(\tau)\dot{q}_i\right\} + \sum_{i=1}^{N} b_{ij}(\tau) q_i = \varepsilon Q_j(\tau, q_1, \ldots, q_N, \dot{q}_1, \ldots, \dot{q}_N, \varepsilon) \tag{4.53}$$

$$(j = 1, 2, \ldots, N),$$

where

$$\varepsilon Q_j(\tau, q_1, \ldots, q_N, \dot{q}_1, \ldots, \dot{q}_N, \varepsilon) = \varepsilon Q_j^{(1)}(\tau, q_1, \ldots, q_N, \dot{q}_1, \ldots, \dot{q}_N) + \varepsilon^2 Q_j^{(2)}(\tau, q_1, \ldots, q_N, \dot{q}_1, \ldots, \dot{q}_N) + \varepsilon^3 \ldots \quad (j = 1, 2, \ldots, N) \tag{4.54}$$

and, as in the preceding section, q_j are generalized coordinates; $\tau = \varepsilon t$ (ε is a small positive parameter); $a_{ij}(\tau) = a_{ji}(\tau)$, $b_{ij}(\tau) = b_{ji}(\tau)$ $(i, j = 1, 2, \ldots, N)$ for any τ; and the functions $a_{ij}(\tau)$, $b_{ij}(\tau)$, $Q_j^{(k)}(\tau, q_1, \ldots, q_N, \dot{q}_1, \ldots, \dot{q}_N)$ are differentiable to the desired order with respect to τ, q_i, $\dot{q}_i$ $(i = 1, 2, \ldots, N)$.

Then, assuming the validity of the conditions formulated in the first section of the present chapter (p. 158) for the corresponding unperturbed system (4.4), we obtain for the approximate two-parameteric family of particular solutions, close to the first normal mode (in the first approximation), the expression

$$q_j = \varphi_j^{(1)}(\tau)\, a \cos\psi \qquad (j = 1, 2, \ldots, N), \tag{4.55}$$

where a and ψ are to be determined from the system of equations of the first approximation

$$\left.\begin{aligned}\frac{da}{dt}&=-\frac{\varepsilon a}{2m_1(\tau)\,\omega_1(\tau)}\frac{d[m_1(\tau)\,\omega_1(\tau)]}{d\tau}-\\&-\frac{\varepsilon}{2\pi m_1(\tau)\,\omega_1(\tau)}\int_0^{2\pi}\sum_{j=1}^{N}Q_{j0}^{(1)}(\tau,\ a,\ \psi)\,\varphi_j^{(1)}(\tau)\sin\psi\,d\psi,\\\frac{d\psi}{dt}&=\omega_1(\tau)-\frac{\varepsilon}{2\pi m_1(\tau)\,\omega_1(\tau)\,a}\int_0^{2\pi}\sum_{j=1}^{N}Q_{j0}^{(1)}(\tau,\ a,\ \psi)\,\varphi_j^{(1)}(\tau)\cos\psi\,d\psi.\end{aligned}\right\}\tag{4.56}$$

Here

$$\begin{aligned}&Q_{j0}^{(1)}(\tau,\ a,\ \psi)=Q_j^{(1)}[\tau,\ \varphi_1^{(1)}(\tau)\,a\cos\psi,\ \ldots\\&\ldots,\ \varphi_N^{(1)}(\tau)\,a\cos\psi,\ -\varphi_1^{(1)}(\tau)\,a\omega_1(\tau)\sin\psi,\ \ldots\ -\varphi_N^{(1)}(\tau)\,a\omega_1(\tau)\sin\psi].\end{aligned}\tag{4.57}$$

Equations (4.56) are obtained directly from the equations (4.50) and (4.45), in which θ is assumed to be absent and, consequently, A_1 and B_1 are regarded as independent of ψ (since the amplitude and frequency of vibration do not depend on the phase when there are no external periodic forces), and dependent only on a and τ.

We introduce in analogy with Chapter II, § 1 (formulas (2.34)) the notation

$$\left.\begin{aligned}\lambda_e^{(1)}(a,\tau)&=\frac{\varepsilon}{\omega_1(\tau)}\frac{d[m_1(\tau)\,\omega_1(\tau)]}{d\tau}+\\&+\frac{\varepsilon}{\pi a\omega_1(\tau)}\int_0^{2\pi}\sum_{j=1}^{N}Q_{j0}^{(1)}(\tau,\ a,\ \psi)\,\varphi_j^{(1)}(\tau)\sin\psi\,d\psi,\\{}[\omega_e^{(1)}(a,\tau)]^2&=\omega_1^2(\tau)-\frac{\varepsilon}{\pi a m_1(\tau)\,\omega_1(\tau)}\int_0^{2\pi}\sum_{j=1}^{N}Q_{j0}^{(1)}(\tau,a,\psi)\,\varphi_j^{(1)}(\tau)\cos\psi\,d\psi.\end{aligned}\right\}\tag{4.58}$$

The equations of the second approximation may consequently be written in the form

$$\left.\begin{aligned}\frac{da}{dt}&=-\delta_e^{(1)}(a,\ \tau)\,a,\\\frac{d\psi}{dt}&=\omega_e^{(1)}(a,\ \tau),\end{aligned}\right\}\tag{4.59}$$

where

$$\delta_e^{(1)}(a,\ \tau)=\frac{\lambda_e^{(1)}(a,\ \tau)}{2m_1(\tau)}.$$

Thus in the case of a monofrequency mode, the equations of the first approximation for the system of N differential equations (4.53) will be the same as for a system with a single degree of freedom described by the equation

$$\begin{aligned}&m_1(\tau)\left[\frac{d^2x}{dt^2}+\omega_1^2(\tau)\,x\right]=\\&=\varepsilon\sum_{j=1}^{N}Q_j^{(1)}[\tau,\ \varphi_1^{(1)}(\tau)\,x,\ \ldots,\ \varphi_N^{(1)}(\tau)\,x,\ \varphi_1^{(1)}(\tau)\,\dot{x},\ \ldots,\ \varphi_N^{(1)}(\tau)\,\dot{x}]\,\varphi_j^{(1)}(\tau),\end{aligned}\tag{4.60}$$

i. e., for a vibrating system with a single degree of freedom.

The mass of this vibrating system

$$m_1(\tau)=\sum_{i,\,j=1}^{N}a_{ij}(\tau)\,\varphi_i^{(1)}(\tau)\,\varphi_j^{(1)}(\tau)=2T\,[\varphi^{(1)}(\tau)]\tag{4.61}$$

represents the double form of the kinetic energy of the system, where the generalized velocities $\dot{q}_i$ $(i=1,\ 2,\ \ldots,\ N)$ are replaced by the "normal" functions $\varphi_i^{(1)}(\tau)$ corresponding to the first frequency $\omega_1(\tau)$. The rigidity of this system is determined by the expression

$$m_1(\tau)\,\omega_1^2(\tau)=\sum_{i,j}^{N} c_{ij}(\tau)\,\varphi_i^{(1)}(\tau)\,\varphi_j^{(1)}(\tau). \tag{4.62}$$

The system is also subjected to the action of a small external perturbation determined by the force

$$\varepsilon\sum_{j=1}^{N} Q_j^{(1)}\varphi_j^{(1)}(\tau), \tag{4.63}$$

representing, from the point of view of the general theory of small oscillations, a generalized force acting on the first normal coordinate.

It is therefore evident that the equations of the first and also the second approximation may be derived with the help of the formulas of Chapter II, § 1, or of the formulas of harmonic balance, which in our case have the form

$$\left.\begin{aligned}&\int_0^{2\pi}\Big\{m_1(\tau)\Big[\frac{d^2x}{dt^2}+\omega_1^2(\tau)x\Big]-\varepsilon\sum_{j=1}^{N}Q_j\,[\tau,\ \varphi_1^{(1)}(\tau)x,\ \ldots\\&\qquad\ldots,\ \varphi_N^{(1)}(\tau)x,\ \varphi_1^{(1)}(\tau)\dot{x},\ \ldots,\ \varphi_N^{(1)}(\tau)\dot{x},\ \varepsilon]\,\varphi_j^{(1)}(\tau)\Big\}\cos\psi\,d\psi=0,\\&\int_0^{2\pi}\Big\{m_1(\tau)\Big[\frac{d^2x}{dt^2}+\omega_1^2(\tau)x\Big]-\varepsilon\sum_{j=1}^{N}Q_j\,[\tau,\ \varphi_1^{(1)}(\tau)x,\ \ldots\\&\qquad\ldots,\ \varphi_N^{(1)}(\tau)x,\ \varphi_1^{(1)}(\tau)\dot{x},\ \ldots,\ \varphi_N^{(1)}(\tau)\dot{x},\ \varepsilon]\,\varphi_j^{(1)}(\tau)\Big\}\sin\psi\,d\psi=0,\end{aligned}\right\} \tag{4.64}$$

where it is necessary before integrating to put $x = a\cos\psi$ in order to derive the equations of the first approximation, while for deriving the equations of the second approximation it is necessary to replace the product $\varphi_j^{(1)}(\tau)\,x$ under the summation sign by $\varphi_j^{(1)}a\cos\psi+\varepsilon u_j^{(1)}(\tau,\ a,\ \psi)$ $(j = 1,\ 2,\ \ldots,\ N)$, where $u_j^{(1)}(\tau,\ a,\ \psi)$ may be determined from formula (4.38), in which it is taken into account that the right-hand sides of equations (4.53) do not depend on θ.

Let us consider one other particular example of the system (4.3), frequently encountered in the computation of vibrating systems.

Let the kinetic and potential energies of a vibrating system with N degrees of freedom, for the sake of simplicity, be given by the positive definite quadratic forms

$$T(\dot{q})=\frac{1}{2}\sum_{i,j=1}^{N}a_{ij}\dot{q}_i\dot{q}_j,\quad V=\frac{1}{2}\sum_{i,j=1}^{N}b_{ij}q_iq_j, \tag{4.65}$$

where $q_1,\ q_2,\ \ldots,\ q_N$ are generalized coordinates, $a_{ij}=a_{ji}$, $b_{ij}=b_{ji}$ are constants.

Let the system be subjected to the action of a small perturbation determined by generalized forces of the form

$$\begin{aligned}\varepsilon Q_j(\theta,\ q_1,\ \ldots,\ q_N,\ \dot{q}_1,\ \ldots,\ \dot{q}_N,\ \varepsilon)=\varepsilon Q_j(q_1,\ \ldots,\ q_N,\ \dot{q}_1,\ \ldots,\ \dot{q}_N,\ \varepsilon)+\\+\varepsilon E_j(\tau)\cos\theta\qquad(j=1,\ 2,\ \ldots,\ N),\end{aligned} \tag{4.66}$$

where $\frac{d\theta}{dt}=\nu(\tau)$; $\tau=\varepsilon t$; and the functions $\nu(\tau)$, $E_j(\tau)$ $(j=1,\ 2,\ \ldots,\ N)$ are unlimitedly differentiable with respect to τ. In addition, we assume that the following expansion exists:

$$\varepsilon Q_j(q_1,\ \ldots,\ q_N,\ \dot{q}_1,\ \ldots,\ \dot{q}_N,\ \varepsilon)=\sum_{n=1}^{\infty}\varepsilon^n Q_j^{(n)}(q_1,\ \ldots,\ q_N,\ \dot{q}_1,\ \ldots,\ \dot{q}_N), \tag{4.67}$$

where the coefficients $Q_j^{(n)}(q_1,\ \ldots,\ q_N,\ \dot{q}_1,\ \ldots,\ \dot{q}_N)$ $(j=1,\ 2,\ \ldots,\ N;\ n=1,\ 2,\ \ldots)$ are polynomials in q_s, $\dot{q}_s$ $(s=1,\ 2,\ \ldots,\ N)$.

The investigation of the vibrational process reduces to the study of a system of nonlinear differential equations of the form

$$\sum_{i=1}^{N} a_{ij}\ddot{q}_i + \sum_{i=1}^{N} b_{ij}q_i = \varepsilon Q_j(q_1, \ldots, q_N, \dot{q}_1, \ldots, \dot{q}_N, \varepsilon) + \varepsilon E_j(\tau)\sin\theta \qquad (4.68)$$

$$(j = 1, 2, \ldots, N).$$

Assuming the validity of the conditions stated in § 1 of the present chapter (p. 158), we examine the case of the fundamental resonance ($p = 1$, $q = 1$), and confine ourselves for simplicity to the first approximation only.

According to the general method presented in the preceding section, the particular solution of the system (4.68) corresponding to the monofrequency mode which is close to the first normal mode is in the first approximation

$$q_i = \varphi_i^{(1)} a\cos(\theta + \psi) \qquad (i = 1, 2, \ldots, N), \qquad (4.69)$$

where the functions of time a and ψ are to be determined from the system of equations of the first approximation

$$\left.\begin{aligned} \frac{da}{dt} &= -\frac{\varepsilon}{2\pi m_1\omega_1}\int_0^{2\pi}\sum_{j=1}^{N} Q_{j0}^{(1)}(a, \psi)\varphi_j^{(1)}\sin\psi\, d\psi - \frac{\varepsilon\sum_{j=1}^{N} E_j(\tau)\varphi_j^{(1)}}{m_1[\omega_1 + \nu(\tau)]}\cos\psi, \\ \frac{d\psi}{dt} &= \omega_1 - \nu(\tau) - \frac{\varepsilon}{2\pi a m_1\omega_1}\int_0^{2\pi}\sum_{j=1}^{N} Q_{j0}^{(1)}(a, \psi)\varphi_j^{(1)}\cos\psi\, d\psi + \\ &\quad + \frac{\varepsilon\sum_{j=1}^{N} E_j(\tau)\varphi_j^{(1)}}{a m_1[\omega_1 + \nu(\tau)]}\sin\psi. \end{aligned}\right\} \qquad (4.70)$$

In this system ω_1 is a normal frequency of the unperturbed system (the system (4.68) with $\varepsilon = 0$), $\varphi_j^{(1)}$ $(j = 1, 2, \ldots, N)$ are nontrivial solutions of the system of algebraic homogeneous equations

$$\sum_{i=1}^{N}\{b_{ij} - a_{ij}\omega_1^2\}\varphi_i^{(1)} = 0 \qquad (j = 1, 2, \ldots, N), \qquad (4.71)$$

and

$$m_1 = \sum_{i,\, j=1}^{N} a_{ij}\varphi_i^{(1)}\varphi_j^{(1)},$$

$$Q_{j0}^{(1)}(a, \psi) = Q_j^{(1)}[\varphi_1^{(1)} a\cos\psi, \ldots, \varphi_N^{(1)} a\cos\psi, -\varphi_1^{(1)} a\omega_1\sin\psi, \ldots, -\varphi_N^{(1)} a\omega_1\sin\psi]$$

$$(j = 1, 2, \ldots, N).$$

The system (4.70) may with the help of the notations (4.58) (taking there $m_1 = \text{const}$, $\omega_1 = \text{const}$) be written in the form

$$\left.\begin{aligned} \frac{da}{dt} &= -\delta_e^{(1)}(a)\, a - \frac{E^{(1)}(\tau)}{m_1[\omega_1 + \nu(\tau)]}\cos\psi, \\ \frac{d\psi}{dt} &= \omega_e^{(1)}(a) - \nu(\tau) + \frac{E^{(1)}(\tau)}{m_1 a[\omega_1 + \nu(\tau)]}\sin\psi, \end{aligned}\right\} \qquad (4.72)$$

where

$$E^{(1)}(\tau) = \varepsilon\sum_{j=1}^{N} E_j(\tau)\varphi_j^{(1)}. \qquad (4.73)$$

As known, the parameters $\lambda_e^{(1)}(a) = 2m_1\delta_e^{(1)}(a)$ and $\omega_e^{(1)}(a)$ respectively represent the equivalent attenuation coefficient and the full equivalent normal frequency of the vibrating system described by the equation

$$m_1\left[\frac{d^2x}{dt^2} + \omega_1^2 x\right] = \varepsilon\sum_{j=1}^{N} Q_j\varphi_j^{(1)}. \qquad (4.74)$$

Thus, the equations of the first approximation for a vibrating system with N degrees of freedom described by equations (4.68) will, in the case of a monofrequency mode close to one of the normal modes of vibrations of the unperturbed system (the system (4.68) with $\varepsilon = 0$), be the same as for a system of a single degree of freedom with the mass m_1 and a normal frequency ω_1 under the action of a perturbing force $\varepsilon \sum_{j=1}^{N} Q_j \varphi_j^{(1)}$ (generalized force acting on the first normal coordinate) and a perturbing sinusoidal force with the amplitude $E^{(1)}(\tau)$.

While illustrating the method just presented by an actual example, we shall sometimes find it advisable to construct resonance curves characterizing the stationary mode.

With regard to this we shall now derive the formulas characterizing the stationary values of the amplitude and phase corresponding to the stationary monofrequency mode (with constant amplitude and frequency) in a vibrating system described by the system of equations (4.68) with $\tau = \text{const}$.

Equating the right-hand sides of the equations of the first approximation (4.72) to zero, we obtain for the determination of the stationary values of the amplitude a and phase ψ the system

$$\left.\begin{aligned} \delta_e^{(1)}(a)\, a + \frac{E^{(1)}}{m_1(\omega_1+\nu)} \cos\psi = 0, \\ \omega_e^{(1)}(a) - \nu + \frac{E^{(1)}}{m_1 a(\omega_1+\nu)} \sin\psi = 0. \end{aligned}\right\} \tag{4.75}$$

Eliminating ψ from these equations, we find, with an accuracy of terms of the second order, the relation between the amplitude a and the frequency of the external force ν:

$$m_1^2 a^2 \{[\omega_e^{(1)2}(a) - \nu^2]^2 + 4\delta_e^{(1)2}(a)\,\nu^2\} = E^{(1)2}. \tag{4.76}$$

Equations (4.75) have the structure of the equations of the classical linear theory, which determine the amplitude and phase of forced vibrations in a system of mass m_1, normal frequency $\omega_e^{(1)}(a)$, damping decrement $\delta_e^{(1)}(a)$, subjected to the action of an external sinusoidal force with amplitude $E^{(1)}(\nu)$. Hence, the relations (4.75) and (4.76) may be deduced from a rule formulated for a nonlinear system with one degree of freedom subjected to the action of an external sinusoidal force. This rule consists in our case in the following.

Let the vibrations of a certain system with N degrees of freedom be described by a system of nonlinear differential equations of the type (4.68), and let the frequency of the external force ν be close to the fundamental normal frequency of the unperturbed system ω_1. It is required to derive the values of the amplitude and phase for a stationary vibration. To do this we consider a vibrating system with one degree of freedom, of mass m_1, and normal frequency ω_1, which is under the action of the force $\sum_{j=1}^{N} \varepsilon Q_{j0}^{(1)} \varphi_j^{(1)}$ (generalized force acting on the first normal coordinate). Linearizing this system, we determine the equivalent damping decrement $\delta_e^{(1)}(a)$ and the equivalent frequency of normal vibrations $\omega_e^{(1)}(a)$ as functions of the amplitude, and substitute the expressions obtained into the classical relations of the linear theory of forced vibrations (4.75) and (4.76). In order to determine the stationary values of the phase of vibration from the relations (4.75) we find the formula

$$\psi = \operatorname{arctg} \frac{\omega_e^{(1)2}(a) - \nu^2}{2\nu\delta_e^{(1)}(a)}. \tag{4.77}$$

The rules for locating the stable and unstable sections of the resonance curve are analogous to those derived in Chapter III, § 3 (p. 82).

Thus, the stability conditions (up to terms of the first order) are the inequalities

$$\left.\begin{aligned} \frac{da}{dt} > 0, \quad \text{if} \quad \omega_e^{(1)}(a) > \nu, \\ \frac{da}{dt} < 0, \quad \text{if} \quad \omega_e^{(1)}(a) < \nu \end{aligned}\right\} \tag{4.78}$$

In conclusion we note that if the right-hand sides of equations (4.68) do not depend on q_i or $\dot{q}_i$ ($i = 1, 2, \ldots, N$), then the equations of the first approximation are easily deduced by an analogy with a system of one degree of freedom in which $m(\tau)$, $\omega(\tau)$, and $F_0(\tau, a, \psi)$ are replaced by m_1, ω_1, and $\sum_{j=1}^{N} Q_{j0}^{(1)} \varphi_j^{(1)}$.

We consider one other particular case of the system of differential equations (4.3), when the latter degenerates to a system of linear differential equations with slowly varying coefficients of the form

$$\frac{d}{dt}\left\{\sum_{i=1}^{N} a_{ij}(\tau)\dot{q}_i\right\} + \sum_{i=1}^{N} b_{ij}(\tau) q_i = \varepsilon \sum_{i=1}^{N} \lambda_{ij}(\tau)\dot{q}_i + \varepsilon Q_j(\tau, \theta) \quad (j = 1, 2, \ldots, N), \tag{4.79}$$

where it is assumed for definiteness that

$$\varepsilon Q_j(\tau, \theta) = \varepsilon E_j(\tau) \sin\theta \quad (j = 1, 2, \ldots, N), \tag{4.80}$$

where $\frac{d\theta}{dt} = \nu(\tau)$. We assume that the conditions stated in § 1 of the present chapter (p. 158) are valid. The works of S. F. Feshchenko /161, 162/ deal with the construction of approximate solutions for linear systems of differential equations with slowly varying parameters. However, for the conformity of presentation we derive approximate solutions of the system (4.79) with the help of the method given in the present chapter*.

After a series of elementary operations we obtain in the first approximation the following expression for the approximate solution of the system of equations (4.79) corresponding to the monofrequency mode

$$q_i = \varphi_i^{(1)}(\tau) a \cos(\theta + \psi) \quad (i = 1, 2, \ldots, N), \tag{4.81}$$

where a and ψ should be determined from the system of equations

$$\left.\begin{aligned} \frac{da}{dt} &= \frac{\varepsilon\left(n_1(\tau)\omega_1(\tau) - \frac{d[m_1(\tau)\omega_1(\tau)]}{d\tau}\right)}{2m_1(\tau)\omega_1(\tau)} a - \frac{\varepsilon \sum_{j=1}^{N} E_j(\tau)\varphi_j^{(1)}(\tau)}{m_1(\tau)[\omega_1(\tau) + \nu(\tau)]}\cos\psi, \\ \frac{d\psi}{dt} &= \omega_1(\tau) - \nu + \frac{\varepsilon \sum_{j=1}^{N} E_j(\tau)\varphi_j^{(1)}(\tau)}{a m_1(\tau)[\omega_1(\tau) + \nu(\tau)]}\sin\psi, \end{aligned}\right\} \tag{4.82}$$

in which $\omega_1(\tau)$ is a root of the secular equation (4.6); $\varphi_j^{(1)}(\tau)$ are the eigenfunctions determined by the system of equations (4.7); $m_1(\tau)$ is found from formula (4.43); and

$$n_1(\tau) = \sum_{i,j=1}^{N} \lambda_{ij}(\tau)\varphi_i^{(1)}(\tau)\varphi_j^{(1)}(\tau). \tag{4.83}$$

We obtain these equations by putting $p = 1$, $q = 1$ in (4.45) (only the fundamental resonance may occur in linear systems with nonperiodic coefficients).

The obtained system of equations of the first approximation is, as was

* B. I. Moseenkov /120/ gives a detailed investigation of this case.

already shown in Chapter III (p. 83), reducible to quadratures. Indeed, defining

$$\gamma_1(\tau) = \frac{\varepsilon \sum_{j=1}^{N} E_j(\tau)\varphi_j^{(1)}(\tau)}{m_1(\tau)[\omega_1(\tau)+\nu(\tau)]}, \qquad \delta_1(\tau) = \varepsilon \frac{n_1(\tau)\omega_1(\tau) - \frac{d[m_1(\tau)\omega_1(\tau)]}{d\tau}}{2m_1(\tau)\omega_1(\tau)}, \tag{4.84}$$

we obtain, instead of the system (4.82)

$$\left.\begin{aligned} \frac{da}{dt} &= \delta_1(\tau)a - \gamma_1(\tau)\cos\psi, \\ \frac{d\psi}{dt} &= \omega_1(\tau) - \nu(\tau) + \frac{\gamma_1(\tau)}{a}\sin\psi. \end{aligned}\right\} \tag{4.85}$$

Changing over to the new variables

$$u = a\cos\psi, \quad v = a\sin\psi \tag{4.86}$$

and introducing the complex function

$$z = u + iv, \tag{4.87}$$

we obtain instead of these equations the following linear inhomogeneous first order differential equation

$$\frac{dz}{dt} - [\delta_1(\tau) - i(\omega_1(\tau) - \nu(\tau))]z = -\gamma_1(\tau). \tag{4.88}$$

After the integration of this equation*, we pass to the original quantities a and ψ given by the formulas

$$a^2 = u^2 + v^2, \quad \psi = \operatorname{arctg}\frac{v}{u}. \tag{4.89}$$

We now give the solution in the second approximation. Applying relation (4.42), we find

$$\begin{aligned} q_i = {} & \varphi_i^{(1)}(\tau)a\cos(\theta+\psi) + 2\varepsilon\sum_{k=2}^{N}\varphi_i^{(k)}(\tau)\frac{\sum_{j=1}^{N}E_j(\tau)\varphi_j^{(k)}(\tau)\sin\theta}{m_k(\tau)[\omega_k^2(\tau)-\nu^2(\tau)]} - \\ & -\varepsilon\sum_{k=2}^{N}\frac{\varphi_i^{(k)}(\tau)a\omega_1(\tau)}{m_k(\tau)[\omega_k^2(\tau)-\nu^2(\tau)]}\left\{\sum_{s,\,j=1}^{N}\left[\lambda_{sj}(\tau)\varphi_s^{(1)}(\tau)\varphi_j^{(k)}(\tau) + \right.\right. \\ & \left.\left. + a_{sj}(\tau)\left[\varphi_j^{(k)}(\tau)\frac{d\varphi_s^{(1)}(\tau)}{d\tau} - \varphi_s^{(1)}(\tau)\frac{d\varphi_j^{(k)}(\tau)}{d\tau}\right]\right]\right\}\sin(\theta+\psi) \quad (i=1,2,\ldots,N), \end{aligned} \tag{4.90}$$

where a and ψ are to be determined from the system of equations of the second approximation

$$\left.\begin{aligned} \frac{da}{dt} &= \delta_1(\tau)a - \gamma_1(\tau)\cos\psi - r_1(\tau)\sin\psi, \\ \frac{d\psi}{dt} &= \omega_1(\tau) - \nu(\tau) + \frac{\gamma_1(\tau)}{a}\sin\psi + s_1(\tau) - \frac{r_1(\tau)}{a}\cos\psi. \end{aligned}\right\} \tag{4.91}$$

Here

$$\begin{aligned} r_1(\tau) = {} & \delta_1(\tau)\gamma_1(\tau) + \frac{d\gamma_1(\tau)}{dt} + \frac{dm_1(\tau)}{dt}\cdot\frac{\gamma_1(\tau)}{m_1(\tau)} + \\ & + \frac{1}{m_1(\tau)}\sum_{i,\,j=1}^{N}\varphi_j^{(1)}(\tau)\left\{2\nu(\tau)\left[\varepsilon\lambda_{ij}(\tau)\alpha_i(\tau) - a_{ij}(\tau)\frac{d\alpha_i(\tau)}{dt}\right] - \right. \\ & \left. - a_{ij}(\tau)\beta_i(\tau)\gamma_1(\tau)[\omega_1(\tau)+\nu(\tau)] - 2\varepsilon\frac{d[\nu(\tau)a_{ij}(\tau)\alpha_i(\tau)]}{dt}\right\}, \end{aligned} \tag{4.92}$$

* It was pointed out that the integration of (4.88) leads to quadratures which cannot be performed analytically, and it is therefore simpler to perform a direct numerical integration on the system (4.85).

$$s_1(\tau)=\delta_1^2(\tau)+\frac{d\delta_1(\tau)}{dt}+\frac{dm_1(\tau)}{dt}\,\frac{\delta_1(\tau)}{m_1(\tau)}+\frac{\varepsilon\gamma_1(\tau)}{m_1(\tau)}-$$
$$-\frac{1}{m_1(\tau)}\sum_{i,\,j=1}^{N}\varphi_j^{(1)}(\tau)\left\{\omega_1(\tau)\left[a_{ij}(\tau)\frac{d\beta_i(\tau)}{dt}-\varepsilon\lambda_{ij}(\tau)\beta_i(\tau)+\right.\right.$$
$$\left.\left.+2a_{ij}(\tau)\beta_i(\tau)\delta_1(\tau)\right]+\frac{d\left[\omega_1(\tau)\,a_{ij}(\tau)\,\beta_i(\tau)\right]}{d\tau}\right\}, \qquad (4.92)$$

$$\left.\begin{aligned}
&\alpha_i(\tau)=\varepsilon\sum_{k=2}^{N}\varphi_i^{(k)}(\tau)\frac{\sum_{j=1}^{N}E_j\varphi_j^{(k)}(\tau)}{m_k(\tau)\left[\omega_k^2(\tau)-\nu^2(\tau)\right]},\\
&\beta_i(\tau)=\varepsilon\sum_{k=2}^{N}\varphi_i^{(k)}(\tau)\frac{\omega_1(\tau)\left(\sum_{j,\,s=1}^{N}\lambda_j(\tau)\varphi_s^{(1)}(\tau)\,\varphi_j^{(k)}(\tau)+k_1(\tau)\right)}{m_k(\tau)\left[\omega_k^2(\tau)-\nu^2(\tau)\right]},\\
&k_1(\tau)=\sum_{s,\,j=1}^{N}a_{sj}(\tau)\left[\varphi_j^{(k)}(\tau)\frac{d\varphi_s^{(1)}(\tau)}{d\tau}-\varphi_s^{(1)}(\tau)\frac{d\varphi_j^{(k)}(\tau)}{d\tau}\right]\\
&\qquad (i=1,\,2,\,\ldots,\,N).
\end{aligned}\right\} \qquad (4.93)$$

The system (4.91), as was shown in /120/, may also be reduced to quadratures. but this will not be considered here.

§ 3. Torsional vibrations of a crankshaft in a nonstationary mode

We shall now illustrate the application of our results to the study of forced vibrations in an actual mechanical system with many degrees of freedom and slowly varying parameters. Let there be given the previously

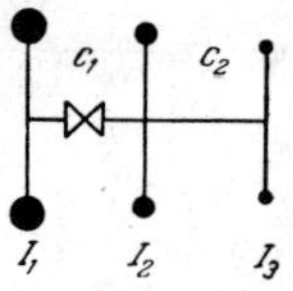

FIGURE 61

mentioned system of a crankshaft (see Figure 61), with a nonlinear coupling between the first and second mass. We assume, for the sake of simplicity, that a periodic torsional moment of the form

$$M=E\sin\theta, \qquad (4.94)$$

acts on the intermediate mass; here $E=\text{const}$, $\frac{d\theta}{dt}=\nu$ is the frequency of the moment proportional to the number of revolutions of the motor, while the moments acting on the masses located at the ends of the given shaft are equal to zero.

Let I_1, I_2, I_3 be the moments of inertia of the masses of the motor and φ_1, φ_2, φ_3 the angles of deviation from uniform rotation. The rigidity of the section of the shaft between the first and second mass depends on the characteristic of the nonlinear coupling. The rigidity of the section of the shaft between the second and the third mass will be denoted by c_2. The elastic moment depending on the difference between the angles of rotation of adjacent masses is, for the first section,

$$F(\varphi_2-\varphi_1)=c_1'(\varphi_2-\varphi_1)+\varepsilon f(\varphi_2-\varphi_1), \tag{4.95}$$

where the function $\varepsilon f(\varphi_2-\varphi_1)$ is determined by the given characteristic of the nonlinear coupling, and c_1' is some constant. The elastic moment for the second section is

$$c_2(\varphi_3-\varphi_2). \tag{4.96}$$

We also assume the presence of internal friction in the second section of the shaft. This friction is assumed to be proportional to the velocity (with the proportionality coefficient α).

The equations of torsional vibrations of the system under consideration are then

$$\left.\begin{aligned} &I_1\ddot{\varphi}_1-F(\varphi_2-\varphi_1)=0,\\ &I_2\ddot{\varphi}_2+F(\varphi_2-\varphi_1)-c_2(\varphi_3-\varphi_2)=M+\alpha(\dot{\varphi}_3-\dot{\varphi}_2),\\ &I_3\ddot{\varphi}_3+c_2(\varphi_3-\varphi_2)=-\alpha(\dot{\varphi}_3-\dot{\varphi}_2). \end{aligned}\right\} \tag{4.97}$$

(The study of the same differential equations is involved, for example, in the problem of vibration damping.)

Introducing the notation

$$\varphi_2-\varphi_1=x,\quad \varphi_3-\varphi_2=y,$$

one may reduce equations (4.97) to the following system of equations (the third degree of freedom, that of rotation, is eliminated)

$$\left.\begin{aligned} &I_1I_2\ddot{x}+c_1'(I_1+I_2)x-c_2I_1y=-(I_1+I_2)\varepsilon f(x)+\alpha I_1\dot{y}+EI_1\sin\theta,\\ &I_2I_3\ddot{y}-c_1'I_3x+c_2(I_2+I_3)y=I_3\varepsilon f(x)-\alpha(I_2+I_3)\dot{y}-I_3E\sin\theta. \end{aligned}\right\} \tag{4.98}$$

In order to derive approximate solutions for this system, we assume that the degree of nonlinearity and the friction coefficient are small (i. e., that in the free state the vibrations of the system will be close to linear vibrations), and that the amplitude of the external moment is small compared with the other parameters of the system.

Hence, according to (4.69) and (4.70), the particular solution corresponding to monofrequency vibrations close to one of the normal modes of unperturbed vibrations, is in the first approximation

$$\left.\begin{aligned} x&=\varphi_1^{(1)}a\cos(\theta+\psi),\\ y&=\varphi_2^{(1)}a\cos(\theta+\psi), \end{aligned}\right\} \tag{4.99}$$

where a and ψ are to be determined from the system of equations of the first approximation

$$\left.\begin{aligned} \frac{da}{dt}&=-\frac{\alpha}{m_1}\left[(I_2+I_3)\varphi_2^{(1)2}-I_1\varphi_1^{(1)}\varphi_2^{(1)}\right]a-\frac{E\left(I_1\varphi_1^{(1)}-I_3\varphi_2^{(1)}\right)}{m_1(\omega_1+\nu)}\cos\psi,\\ \frac{d\psi}{dt}&=\omega_1-\nu-\frac{\left[I_3\varphi_2^{(1)}-(I_1+I_2)\varphi_1^{(1)}\right]}{m_1\omega_1a}\frac{1}{2\pi}\int_0^{2\pi}\varepsilon f(\varphi_1^{(1)}a\cos\beta)\cos\beta\,d\beta-\\ &\qquad-\frac{E\left(I_1\varphi_1^{(1)}-I_3\varphi_2^{(1)}\right)}{m_1a(\omega_1+\nu)}\sin\psi. \end{aligned}\right\} \tag{4.100}$$

Here ω_1 is a root of the secular equation of the "unperturbed" system

$$\begin{vmatrix} c_1'(I_1+I_2)-\omega^2I_1I_2 & -c_2I_1\\ -c_1'I_3 & c_2(I_2+I_3)-\omega^2I_2I_3 \end{vmatrix}=0, \tag{4.101}$$

$\varphi_1^{(1)}$ and $\varphi_2^{(1)}$ are basic functions constituting a nontrivial solution of the system of homogeneous algebraic equations

$$\left.\begin{aligned}[c_1'(I_1+I_2)-\omega_1^2 I_1 I_2]\,\varphi_1^{(1)}-c_2 I_1\varphi_2^{(1)}=0,\\ -c_1' I_3\varphi_1^{(1)}+[c_2(I_2+I_3)-\omega_1^2 I_2 I_3]\,\varphi_2^{(1)}=0,\end{aligned}\right\} \tag{4.102}$$

and

$$m_1=I_1 I_2\varphi_1^{(1)^2}+I_2 I_3\varphi_2^{(1)^2}.$$

We shall now calculate the integral on the right-hand side of the second equation of the system (4.100) for the following actual example. The

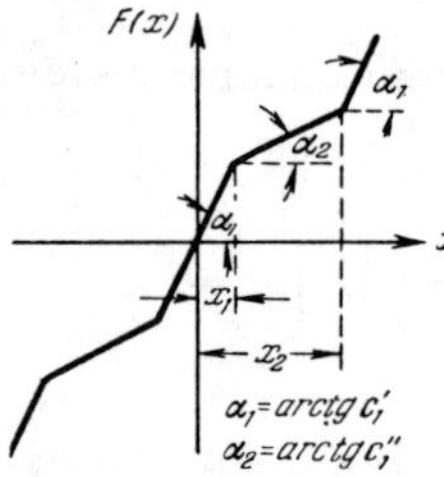

FIGURE 62

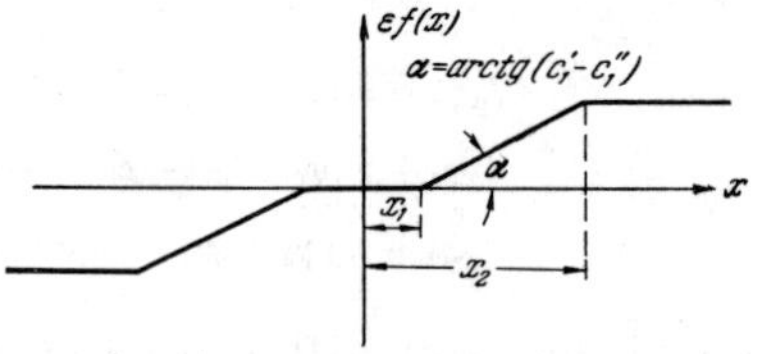

FIGURE 63

characteristic of the nonlinear coupling in the section of the shaft between the first and second mass is assumed to be of the form

$$F(x)=\begin{cases} c_1'x, & -x_1\leqslant x\leqslant x_1,\\ c_1'x-(c_1'-c_1'')\,x+(c_1'-c_1'')\,x_1, & x_1\leqslant x\leqslant x_2,\\ c_1'x-(c_1'-c_1'')\,x-(c_1'-c_1'')\,x_1, & -x_2\leqslant x\leqslant -x_1,\\ c_1'x-(c_1'-c_1'')\,(x_2-x_1), & x_2\leqslant x\leqslant \infty,\\ c_1'x+(c_1'-c_1'')\,(x_2-x_1), & -\infty\leqslant x\leqslant -x_2.\end{cases} \tag{4.103}$$

The function $F(x)$ is represented graphically in Figure 62.

Denoting by β_1 and β_2 the smallest roots of the equations $\bar{x}_1=a\cos\beta$, $\bar{x}_2=a\cos\beta$ $(x_1=\bar{x}_1\varphi_1^{(1)},\ x_2=\bar{x}_2\varphi_1^{(1)})$, we have

$$\varepsilon f\,(a\cos\beta)=\begin{cases} 0, & \beta_1\leqslant\beta\leqslant\pi-\beta_1,\\ -(c_1'-c_1'')\,a\cos\beta+(c_1'-c_1'')\,a\cos\beta_1, & \beta_2\leqslant\beta\leqslant\beta_1,\\ -(c_1'-c_1'')\,a\cos\beta-(c_1'-c_1'')\,a\cos\beta_1, & \pi-\beta_1\leqslant\beta\leqslant\pi-\beta_2,\\ -(c_1'-c_1'')\,(a\cos\beta_2-a\cos\beta_1), & 0\leqslant\beta\leqslant\beta_2,\\ (c_1'-c_1'')\,(a\cos\beta_2-a\cos\beta_1), & \pi-\beta_2\leqslant\beta\leqslant\pi.\end{cases} \tag{4.104}$$

The function $y=\varepsilon f(x)$ is represented graphically in Figure 63.

Dividing the integration interval into five sections and substituting the values of β_1 and β_2, we obtain

$$\frac{1}{\pi}\int_0^{2\pi}\varepsilon f\,(\varphi_1^{(1)}a\cos\beta)\cos\beta\,d\beta=\frac{2\,(c_1'-c_1'')}{\pi}\,a\varphi_1^{(1)}\left\{\arccos\frac{\bar{x}_2}{a}-\right.$$
$$\left.-\arccos\frac{\bar{x}_1}{a}+\frac{\bar{x}_1}{a}\sqrt{1-\left(\frac{\bar{x}_1}{a}\right)^2}-\frac{\bar{x}_2}{a}\sqrt{1-\left(\frac{\bar{x}_2}{a}\right)^2}\right\}. \tag{4.105}$$

Taking $x_1=\frac{x_2}{4}$, we reduce the system of equations of the first approximation to the form

$$\frac{d\left(\frac{a}{\bar{x}_2}\right)}{dt}=-\frac{a\,[(I_2+I_3)\,\varphi_2^{(1)^2}-I_1\varphi_1^{(1)}\varphi_2^{(1)}]\,\frac{a}{\bar{x}_2}}{m_1}-\frac{\frac{E}{\bar{x}_2}\,(I_1\varphi_1^{(1)}-I_3\varphi_2^{(1)})}{m_1(\omega_1+\nu)}\sin\psi, \tag{4.106}$$

$$\frac{d\psi}{dt} = \omega_1 - \nu - F\left(\frac{a}{\bar{x}_2}\right) - \frac{\frac{E}{\bar{x}_2}(I_1\varphi_1^{(1)} - I_3\varphi_2^{(1)})}{m_1 \frac{a}{\bar{x}_2}(\omega_1 + \nu)} \cos\psi, \tag{4.106}$$

where

$$F\left(\frac{a}{\bar{x}_2}\right) = \frac{I_3\varphi_1^{(1)}\varphi_2^{(1)} - (I_1 + I_2)\varphi_1^{(1)2}}{\pi m_1 \omega_1} \left\{ \arccos\frac{\bar{x}_2}{a} - \arccos\frac{\bar{x}_2}{4a} + \right.$$
$$\left. + \frac{\bar{x}_2}{4a}\sqrt{1 - \left(\frac{\bar{x}_2}{4a}\right)^2} - \frac{\bar{x}_2}{a}\sqrt{1 - \left(\frac{\bar{x}_2}{a}\right)^2} \right\} (c_1' - c_1''). \tag{4.107}$$

We restrict our examination to the behavior of the system near the resonance at the eigenfrequency $\omega_1 \left(0.8 \leqslant \frac{\nu}{\omega_1} \leqslant 1.0\right)$.

Expanding the last terms on the right-hand sides of equations (4.106) in powers of $(\omega_1 - \nu)$, it is possible to replace these equations (still preserving accuracy), by the system

$$\left.\begin{aligned} \frac{d\left(\frac{a}{\bar{x}_2}\right)}{dt} &= -\frac{\alpha\left[(I_2 + I_3)\varphi_2^{(1)2} - I_1\varphi_1^{(1)}\varphi_2^{(1)}\right]}{m_1} \frac{a}{\bar{x}_2} - \\ &- \frac{\frac{E}{\bar{x}_2}(I_1\varphi_1^{(1)} - I_3\varphi_2^{(1)})}{m_1\omega_1}\left[\frac{1}{2} + \frac{1}{4\omega_1}(\omega_1 - \nu) + \frac{1}{8\omega_1^2}(\omega_1 - \nu)^2\right]\sin\psi, \\ \frac{d\psi}{dt} &= (\omega_1 - \nu) - F_1\left(\frac{a}{\bar{x}_2}\right) - \\ &- \frac{\frac{E}{\bar{x}_2}(I_1\varphi_1^{(1)} - I_3\varphi_2^{(1)})}{m_1\left(\frac{a}{\bar{x}_2}\right)\omega_1}\left[\frac{1}{2} + \frac{1}{4\omega_1}(\omega_1 - \nu) + \frac{1}{8\omega_1^2}(\omega_1 - \nu)^2\right]\cos\psi. \end{aligned}\right\} \tag{4.108}$$

Before investigating the behavior of the system during a transition through resonance, we consider the resonance in the stationary mode.

Using (4.76), we derive, after a number of operations, the following relation between the vibration amplitude and the frequency of the external force, with the specified degree of accuracy (while constructing the resonance curve for the second coordinate y it is necessary to multiply the obtained values of $\frac{\varphi_1^{(1)}a}{\bar{x}_2}$ by the ratio $\frac{\varphi_2^{(1)}}{\varphi_1^{(1)}}$):

$$(\omega_1 - \nu)^2\left[\left(\frac{a}{\bar{x}_2}\right)^2 - \frac{3E_1^2}{16\omega_1^2}\right] - (\omega_1 - \nu)\left[2F_1\left(\frac{a}{\bar{x}_2}\right) - \frac{E_1^2}{4\omega_1}\right] +$$
$$+ F_1^2\left(\frac{2}{\bar{x}_2}\right) + \delta^2\left(\frac{a}{\bar{x}_2}\right)^2 - \frac{E_1^2}{4} = 0, \tag{4.109}$$

where

$$\delta = \frac{\alpha\left[(I_1 + I_2)\varphi_1^{(1)2} - I_1\varphi_1^{(1)}\varphi_2^{(1)}\right]}{m_1}, \quad E_1 = -\frac{E(I_1\varphi_1^{(1)} - I_3\varphi_2^{(1)})}{\bar{x}_2 m_1 \omega_1}.$$

We assign the following numerical values to the parameters of the system

$$\left.\begin{aligned} &I_1 = 120 \text{ kg}\cdot\text{cm}\cdot\text{sec}^2, \quad I_2 = 9 \text{ kg}\cdot\text{cm}\cdot\text{sec}^2, \quad I_3 = 3 \text{ kg}\cdot\text{cm}\cdot\text{sec}^2, \\ &c_1' = 3.7\cdot 10^6 \text{ kg}\cdot\text{cm}, \quad c_1'' = 2\cdot 10^6 \text{ kg}\cdot\text{cm}, \quad c_2 = 0.13\cdot 10^6 \text{ kg}\cdot\text{cm}, \\ &\alpha = 85, \quad \frac{E}{\bar{x}_2} = 3.5\cdot 10^0 \text{ kg}\cdot\text{cm}. \end{aligned}\right\} \tag{4.110}$$

Using (4.109) we can now construct the resonance curve corresponding to the monofrequency mode with the eigenfrequency $\omega_1 = 675.9 \text{ sec}^{-1}$ (Figure 64). Replacing $\omega_1, \varphi_1^{(1)}, \varphi_2^{(1)}$ and m_1 in (4.109) by $\omega_2, \varphi_1^{(2)}, \varphi_2^{(2)}$, and m_2, one can construct the resonance curve corresponding to the monofrequency mode with the eigenfrequency $\omega_2 = 207.1\ \sec^{-1}$ (Figure 65). Figures 64 and 65 also contain the skeleton curves (the curves *abcC*) characterizing the dependence of the eigenfrequency of vibration on the amplitude. (These curves determine the form of the family of curves of forced vibrations for different amplitudes of external force.)

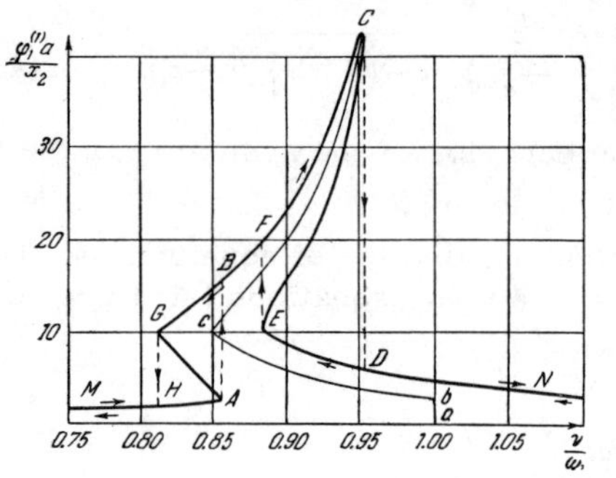

FIGURE 64

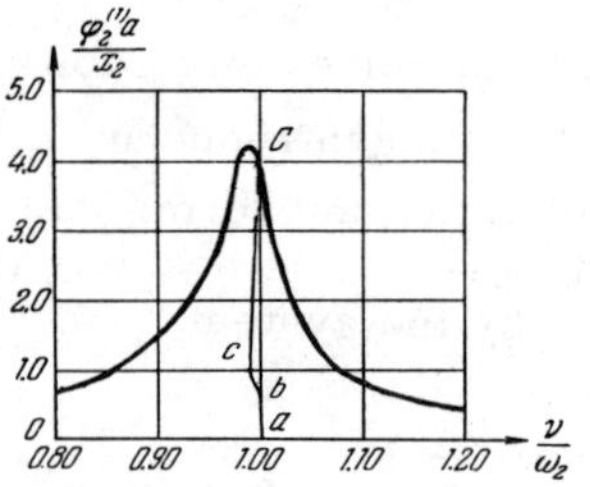

FIGURE 65

Let us consider the resonance curve shown in Figure 64, assuming that monofrequency vibrations at the frequency ω_1 may occur in the system under consideration. According to the stability criteria (4.78), the branches *MA*, *GC*, and *EN* of the resonance curve correspond to stable, and the branches *AG* and *CE*— to unstable vibration amplitude. The points *A*, *G*, *C*, and *E* are points of discontinuity and jump, characteristic of the hysteresis phenomena; these appear because of the nonlinear coupling.

In the given example (see Figure 64), when the frequency of the external force (in a stationary mode) varies adiabatically from small values, the amplitude of the forced vibrations increases first along the curve *MA* and thereafter jumps from the point *A* to the point *B*, varying further along the curve *BC*. At the point *C* the amplitude jumps to the point *D*, and when the frequency continues to increase it varies along the curve *DN*.

If the frequency of the impressed force now begins to decrease, then the amplitude of the forced vibrations varies along the curve *NDE*. At the point *E* the value of the amplitude jumps to the point *F*, and proceeds to change along the curve *FBG*. Upon reaching the point *G* the amplitude jumps to the point *H*, varying thereafter along the curve *HM*.

A similar picture is obtained for the relation between the amplitude and frequency, when the latter varies in the vicinity of ω_2 (Figure 65) (assuming, obviously, that the harmonics of the frequency ω_1 have either damped down or not been excited at all). The situation in that case differs only very slightly from that encountered in a linear system, since the nonlinear characteristic of the elastic coupling exerts an insignificant influence on the vibrations at the frequency ω_2. Hysteresis phenomena may in this case be detected only for sufficiently small amplitudes of the external perturbing force.

We shall now examine the behavior of the specified mechanical system during various modes of transition through resonance, confining ourselves to the case when the frequency of the external force, while varying with time, passes through values equal to the eigenfrequency ω_1. This case is of the greatest interest in view of the characteristic nonlinearity, whose influence demonstrates itself in the nature of the resonance curves in a stationary mode as well as during a transition through resonance.

We assume for simplicity that the instantaneous frequency of the external force depends linearly on time, i.e.,

$$\nu(\tau) = \nu_0 + \beta\tau. \tag{4.111}$$

Substituting this value of $\nu(\tau)$ into the equations of the first approximation (4.106), we obtain (for the numerical values of the parameters (4.110))

$$\left.\begin{aligned} \frac{d\left(\frac{a}{x_2}\right)}{dt} &= -5.2\left(\frac{a}{x_2}\right) - \frac{22.698}{675.9+\nu_0+\beta\tau}\sin\psi, \\ \frac{d\psi}{dt} &= 675.9 - \nu_0 - \beta\tau - F\left(\frac{a}{x_2}\right) - \frac{22.698}{\frac{a}{x_2}(675.9+\nu_0+\beta\tau)}\cos\psi. \end{aligned}\right\} \tag{4.112}$$

To obtain the resonance curves during a transition through resonance the system (4.112) must be integrated numerically according to the method described in Chapter III. To simplify the calculations, it is advisable to take as the initial values of $\frac{a}{x_2}$, ψ, and ν_0, values corresponding to a stationary mode near the resonance zone, namely, corresponding to the section of the resonance curve (Figure 64) located either on the left of point A or on the right of point D. We obtain the initial values as follows: putting $\nu = \nu_0$ (where ν_0 is a value of the frequency of the external force chosen by us slightly to the left of the projection of point A) in (4.109), we find that initially

$$\left(\frac{a}{x_2}\right)_0 = \frac{a}{x_2}\bigg|_{\nu=\nu_0}$$

We then derive ψ_0 from the formula

$$\sin\psi_0 = -\frac{5.2\,(675.9+\nu_0)\left(\frac{a}{x_2}\right)_0}{22.698} \tag{4.113}$$

or from the formula

$$\cos\psi_0 = \frac{\left\{675.9-\nu_0-F\left(\frac{a}{x_2}\right)_0\right\}\left(\frac{a}{x_2}\right)_0}{22.698},$$

which are obtained by equating the right-hand sides of the system (4.112) to zero at $t = 0$.

Let us now pass to the numerical integration. We obtain a number of curves describing the variation of the amplitude during different modes of transitions through resonance.

Figure 66 shows the curves *MALK* and *NERS* that characterize a very low transition through resonance, corresponding to a passage through the resonance zone $(0.8 \leqslant \frac{\nu}{\omega_1} \leqslant 1.0)$ during approximately 375 cycles. (The numerical integration was performed with the following values

$$\beta = 40 \text{ rad/sec}^2,\ \nu_0 = 578.3,\ \frac{\varphi_1^{(1)}a}{x_2} = 0.2501,\ \psi_0 = 0.0572$$

for the curve *MALK*;

$$\beta = -40 \text{ rad/sec}^2,\ \nu_0 = 596.5,\ \frac{\varphi_1^{(1)}a}{x_2} = 1.0000,\ \psi_0 = 3.3783$$

for the curve *NERS*.)

The curves ML_1K_1 and NR_1S_1 in Figure 67 describe a faster transition through resonance — $\beta = \pm 400$ rad/sec^2, which corresponds to a passage through the resonance zone during 37.5 cycles. Here we have taken the initial values:

$$\nu_0 = 551.1, \quad \frac{\varphi_1^{(1)}a}{x_2} = 0.2000, \quad \psi_0 = -0.0418 \quad \text{for the curve } ML_1K_1,$$

$$\nu_0 = 618.4, \quad \frac{\varphi_1^{(1)}a}{x_2} = 0.7500, \quad \psi_0 = 3.4075 \quad \text{for the curve } NR_1S_1.$$

Analyzing the curves thus obtained, we deduce, as should have been expected, that in the case of a very slow transition through resonance these curves are very close to the curves describing the stationary mode (Figure 66). Despite the complicated structure of the resonance curve for the

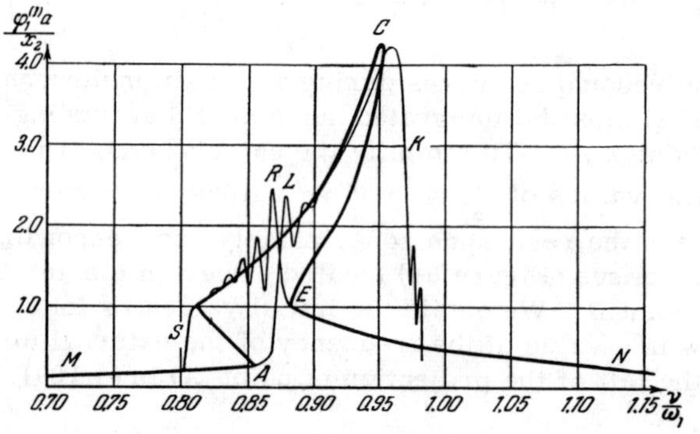

FIGURE 66

stationary mode, the resonance curves obtained (*MALK* and *NERS*) almost coincide with the curve *MAGBCEDN* (Figure 64) along all its stable sections.

In the case of a faster transition through resonance (Figure 67) we obtain

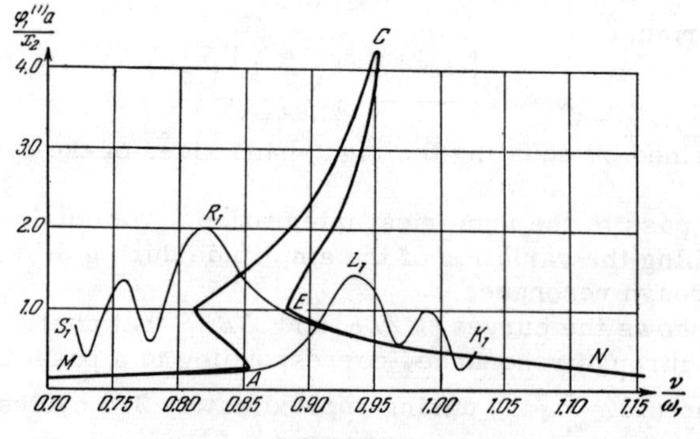

FIGURE 67

curves resembling the curves of transition through resonance in a linear system. Here the nonlinearity exerts a smaller influence on the character of the curves. (Figures 66 and 67 also contain, for the sake of clearness, the stationary resonance curves *MACEN*, and Figure 68 — the resonance curves corresponding to various velocities of transition through resonance.)

We shall not give a detailed analysis of the affect of the nonlinearity on the character of the resonance curves and other specific features characterizing the transition through resonance. We only remark that by analyzing curves of transition through resonance, similar to those given in Figures 66 and 67, we can avoid at the planning stage the appearance of vibrations of high amplitudes in the construction. It is easily seen from Figure 66 that during a slow passage through resonance the vibration amplitude may attain almost the same values as for a resonance in a stationary mode.

Figure 67 indicates the existence of such minimum (and even large) velocities of transition through resonance, for which the maximum amplitude is considerably smaller than in the case of resonance in a stationary mode. (Evidently, one should also take into account here the value of the amplitude of the external perturbing sinusoidal force, and also the nonlinearity character of the elastic coupling.)

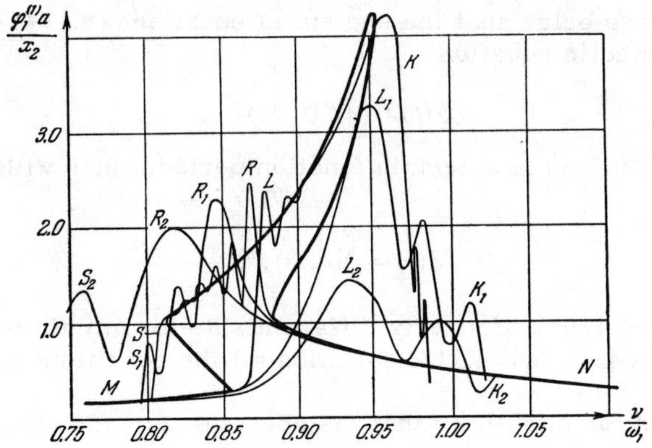

FIGURE 68

We may always distinguish such modes of transition through resonance (corresponding to a given construction of the elastic coupling with a nonlinear characteristic, and given magnitude of perturbing force), for which vibrations of large amplitude will not occur. Conversely, the parameters of elastic coupling may for definite modes of transition through resonance be adjusted so that vibrations of large amplitudes will not develop. In the last case the elastic coupling with nonlinear characteristic plays the role of a "passive damper".

§ 4. Construction of asymptotic solutions in the case of a vibrating system with a single nonlinear element

While studying vibrational processes in nonlinear systems with many degrees of freedom and slowly varying parameters, one frequently encounters systems involving a nonlinearity of such a nature that a nonlinear function depending on only one generalized coordinate appears in just one equation of the corresponding system of differential equations.

In that case we have to deal with the following system of differential equations

$$\frac{d}{dt}\left\{\sum_{i=1}^{N} a_{ij}(\tau)\dot{q}_i\right\}+\sum_{i=1}^{N} b_{ij}(\tau)q_i=\varepsilon Q_j(\tau,\theta)+\varepsilon f_k(\tau,\theta,q_k) \qquad (4.114)$$

$$(j=1, 2, \ldots, N),$$

where k is one of the integers $1, 2, \ldots, N$ depending on which equation contains the nonlinear function, and $\varepsilon f_k(\tau, \theta, q_k)$ is a certain nonlinear function satisfying the conditions imposed on the right-hand sides of equations (4.3). The system of equations (4.114) constitutes a particular case of the system (4.3), and hence may be solved approximately according to the general scheme developed in § 1 of the present chapter.

In order to construct approximate solutions for the system of equations (4.114), it is in some cases advisable to apply a symbolic method which leads to considerable simplifications in general derivations, as well as in actual numerical calculations /96/.

It will be shown below that the system of equations (4.114) may be reduced to the symbolic equation

$$Z(p)x=\varepsilon F(\tau,\theta,x), \qquad (4.115)$$

where $p=\frac{d}{dt}$; $F(\tau, \theta, x)$ is a certain function periodic in θ with period 2π; $\frac{d\theta}{dt}=\nu(\tau)$; and

$$Z(p)=\sum_{n=1}^{N} a_n(\tau)p^n. \qquad (4.116)$$

$F(\tau, \theta, x)$ and $a_n(\tau)$ are unlimitedly differentiable for any finite τ, if the original system of equations (4.114) satisfies all the conditions imposed on equations (4.3).

It is assumed for simplicity that the operator $Z(p)$ may be represented as the sum

$$Z(p)=Q(\tau,p)+\varepsilon R(\tau,p), \qquad (4.117)$$

where $R(\tau, i\omega)\neq 0$, and $Q(\tau, p)$ has an imaginary root, i. e.,

$$Q(\tau,p)=(p^2+\omega^2)Q_1(\tau,p), \qquad (4.118)$$

where $Q_1(\tau, i\omega)\neq 0$ for any τ.

The vibrating system in question performs, by virtue of the assumption made, only one form of normal vibrations at the frequency ω.

Putting $\frac{1}{Q_1(\tau, p)}=P(\tau, p)$, we can represent equation (4.115) in the form

$$(p^2+\omega^2)x=\varepsilon P(\tau,p)\{-R(\tau,p)x+F(\tau,\theta,x)\}. \qquad (4.119)$$

To solve this equation we define a function

$$y=y(\tau,a,\theta,p\varphi+\psi,\varepsilon), \qquad (4.120)$$

such that the expression

$$x(t)=\{y(\tau,a,\theta,p\varphi+\psi,\varepsilon)\}_{\substack{\tau=\varepsilon t\\ a=a(t)\\ \theta=\theta(t)\\ p\varphi+\psi=p\varphi(t)+\psi(t)}} \qquad (4.121)$$

would be an approximate solution of equation (4.119). Here, as before, $\varphi=\frac{1}{q}\theta$, and p and q are some mutually prime integers depending on the resonance under investigation.

One can write down the identity

$$p\{y(\tau, a, \theta, p\varphi+\psi, \varepsilon)\}=\{p_c y(\tau, a, \theta, p\varphi+\psi, \varepsilon)\}_{\substack{\tau=\varepsilon t \\ a=a(t) \\ \theta=\theta(t) \\ p\varphi+\psi=p\varphi(t)+\psi(t)}}, \tag{4.122}$$

where

$$p_c=p_a\frac{da}{dt}+p_\theta\frac{d\theta}{dt}+p_{p\varphi+\psi}\frac{d(p\varphi+\psi)}{dt}+p_\tau\frac{d\tau}{dt}. \tag{4.123}$$

Hence, if the function (4.120) satisfies the symbolic equation

$$(p_c^2+\omega^2)\,y=\varepsilon P(\tau, p_c)\{-R(\tau, p_c)\,y+F(\tau, \theta, y)\}, \tag{4.124}$$

the expression (4.121) (for $\tau=\varepsilon t$, $a=a(t)$, $\theta=\theta(t)$, $p\varphi+\psi=p\varphi(t)+\psi(t)$) is a solution of equation (4.119).

So instead of looking for the solution of equation (4.119) equivalent to a certain ordinary differential equation, we are looking for a solution of the equation (4.124) equivalent to a partial differential equation.

The vibrating system described by the equation

$$Q(\tau, p)\,x=0, \tag{4.125}$$

where τ is a certain parameter, will be called, as usual, the "unperturbed" system. It may sustain, according to (4.118), nondamping vibrations at the frequency $\omega(\tau)$.

We assume, as in § 1, that there is no internal resonance in our system (i. e., no nondamping vibrations at frequencies which for any τ are multiples of $\omega(\tau)$ may occur in the system), and that the unique static solution is the trivial solution $y=0$.

These conditions being satisfied, we look for an approximate solution of equation (4.124), according to the method given above, in the form of an asymptotic series

$$y=u_0(\tau, a, \theta, p\varphi+\psi)+\varepsilon u_1(\tau, a, \theta, p\varphi+\psi)+\varepsilon^2\ldots, \tag{4.126}$$

where a and ψ are functions of time determined from the system of differential equations

$$\left.\begin{aligned} \frac{da}{dt}&=\varepsilon A_1(\tau, a, \psi)+\varepsilon^2 A_2(\tau, a, \psi)+\ldots, \\ \frac{d\psi}{dt}&=\omega(\tau)-\frac{p}{q}\nu(\tau)+\varepsilon B_1(\tau, a, \psi)+\varepsilon^2 B_2(\tau, a, \psi)+\ldots, \end{aligned}\right\} \tag{4.127}$$

where $\frac{d\theta}{dt}=\nu(\tau)$, $\tau=\varepsilon t$, and $\omega(\tau)$— the "eigenfrequency" of the system — is found from the equation $Q(\tau, \omega)=0$.

We now proceed to determine the functions $u_0(\tau, a, \theta, p\varphi+\psi)$, $u_1(\tau, a, \theta, p\varphi+\psi)$, . . ., $A_1(\tau, a, \psi)$, $B_1(\tau, a, \psi)$, $A_2(\tau, a, \psi)$, $B_2(\tau, a, \psi)$, ... In order to determine these functions uniquely, some additional conditions must, as usual, be introduced. As in the case of a vibrating system with a single degree of freedom, they may be chosen to be the absence of the first harmonic of $p\varphi+\psi$ in the functions $u_1(\tau, a, \theta, p\varphi+\psi)$, $u_2(\tau, a, \theta, p\varphi+\psi)$, . . .

Constructing approximate solutions, we use the notation

$$p_\theta\,\nu(\tau)+p_{p\varphi+\psi}\,\omega(\tau)=p_t. \tag{4.128}$$

Further, substituting (4.126) into equation (4.124), and taking (4.127) into account, we obtain, after equating the coefficients of equal powers of ε, the equations

$$[p_t^2+\omega^2(\tau)]\,u_0=0, \tag{4.129}$$

$$[p_t^2+\omega^2(\tau)]\,u_1=-P(\tau,\,p_t)\,R(\tau,\,p_t)\,u_0+P(\tau,\,p_t)\,F(\tau,\,\theta,\,u_0)-$$
$$-p_a\left\{\left[\omega(\tau)-\frac{p}{q}\nu(\tau)\right]\frac{\partial A_1}{\partial\psi}+2p_tA_1\right\}u_0-p_{p\varphi+\psi}\left\{\left[\omega(\tau)-\right.\right.$$
$$\left.\left.-\frac{p}{q}\nu(\tau)\right]\frac{\partial B_1}{\partial\psi}+2p_tB_1\right\}u_0-p_{p\varphi+\psi}\frac{d\omega(\tau)}{d\tau}u_0-p_\theta\frac{d\nu(\tau)}{d\tau}u_0-2p_\tau p_tu_0=$$
$$=F_1(\tau,\,a,\,\theta,\,p\varphi+\psi), \tag{4.130}$$

$$[p_t^2+\omega^2(\tau)]\,u_2=-[P'R+PR']\,(p_aA_1+p_{p\varphi+\psi}B_1+p_\tau)\,u_0+$$
$$+P'\,(p_aA_1+p_{p\varphi+\psi}B_1+p_\tau)\,F(\tau,\,\theta,\,u_0)-P(\tau,\,p_t)\,R(\tau,\,p_t)\,u_1+$$
$$+P(\tau,\,p_t)\left[\frac{dF(\tau,\theta,y)}{dy}u_1\right]_{y=u_0}-p_a\left\{\left[\omega(\tau)-\frac{p}{q}\nu(\tau)\right]\frac{\partial A_2}{\partial\psi}+2p_tA_2\right\}u_0-$$
$$-p_{p\varphi+\psi}\left\{\left[\omega(\tau)-\frac{p}{q}\nu(\tau)\right]\frac{\partial B_2}{\partial\psi}+2p_tB_2\right\}u_0-p_a\left[\frac{\partial A_1}{\partial a}A_1+\frac{\partial A_1}{\partial\psi}B_1+\frac{\partial A_1}{\partial\tau}+\right.$$
$$\left.+2p_{p\varphi+\psi}A_1B_1+p_aA_1^2+2p_\tau A_1\right]u_0-p_{p\varphi+\psi}\left[\frac{\partial B_1}{\partial a}A_1+\frac{\partial B_1}{\partial\psi}B_1+\frac{\partial B_1}{\partial\tau}+\right.$$
$$\left.+2p_\tau B_1+p_{p\varphi+\psi}B_1^2\right]u_0-p_a\left\{\left[\omega(\tau)-\frac{p}{q}\nu(\tau)\right]\frac{\partial A_1}{\partial\psi}+2p_tA_1\right\}u_1-$$
$$-p_{p\varphi+\psi}\left\{\left[\omega(\tau)-\frac{p}{q}\nu(\tau)\right]\frac{\partial B_1}{\partial\psi}+2p_tB_1\right\}u_1-p_{p\varphi+\psi}\frac{d\omega(\tau)}{d\tau}u_1-$$
$$-p_\theta\frac{d\nu(\tau)}{d\tau}u_1-2p_tp_\tau u_1=F_2(\tau,\,a,\,\theta,\,p\varphi+\psi). \tag{4.131}$$

Taking into account the second equation of the system (4.127) for $\varepsilon=0$, we obtain from (4.129) the zeroeth approximation in the form

$$u_0(\tau,\,a,\,\theta,\,p\varphi+\psi)=a\cos(p\varphi+\psi). \tag{4.132}$$

Introducing the notation

$$H_k\{F(x)\}=H_{ks}\{F(x)\}+H_{kc}\{F(x)\}=$$
$$=\frac{1}{\pi}\int_0^{2\pi}F(x)\sin kx\,dx+\frac{1}{\pi}\int_0^{2\pi}F(x)\cos kx\,dx \tag{4.133}$$

and keeping in mind that $u_1(\tau,\,a,\,\theta,\,p\varphi+\psi)$ does not contain the first harmonic of $p\varphi+\psi$, it is possible to write, on the grounds of equation (4.130), the two equalities

$$\left.\begin{aligned}H_{1s}\{F_1(\tau,\,a,\,\theta,\,p\varphi+\psi)\}=0,\\ H_{1c}\{F_1(\tau,\,a,\,\theta,\,p\varphi+\psi)\}=0.\end{aligned}\right\} \tag{4.134}$$

From these equalities (τ, a, and ψ are regarded as constants during the integration) we obtain the following system of partial differential equations for determining $A_1(\tau,\,a,\,\psi)$ and $B_1(\tau,\,a,\,\psi)$:

$$\left.\begin{aligned}\left[\omega(\tau)-\frac{p}{q}\nu(\tau)\right]\frac{\partial A_1}{\partial\psi}-2\omega(\tau)\,aB_1=L_1(\tau,\,a,\,\psi),\\ \left[\omega(\tau)-\frac{p}{q}\nu(\tau)\right]a\frac{\partial B_1}{\partial\psi}+2\omega(\tau)\,A_1=-\frac{d\omega(\tau)}{d\tau}a-L_2(\tau,\,a,\,\psi),\end{aligned}\right\} \tag{4.135}$$

where

$$\left.\begin{aligned}L_1(\tau,\,a,\,\psi)=H_{1c}\{-\varepsilon P(\tau,\,p_t)\,a\cos(p\varphi+\psi)+\varepsilon P(\tau,\,p_t)\,F(\tau,\,\theta,\,u_0)\},\\ L_2(\tau,\,a,\,\psi)=H_{1s}\{-\varepsilon P(\tau,\,p_t)\,a\cos(p\varphi+\psi)+\varepsilon P(\tau,\,p_t)\,F(\tau,\,\theta,\,u_0)\}.\end{aligned}\right\} \tag{4.136}$$

From equation (4.130) we find

$$u_1(\tau,\,a,\,\theta,\,p\varphi+\psi)=\frac{1}{4\pi^2}\sum_{\substack{n,\,m\\(qn+pm\neq\pm1)}}e^{i\{n\theta+m(p\varphi+\psi)\}}\times$$
$$\times\frac{\int_0^{2\pi}\int_0^{2\pi}P(\tau,\,p_t)\,F(\tau,\,\theta,\,u_0)\,e^{-i\{n\theta+m(p\varphi+\psi)\}}\,d\theta\,d(p\varphi+\psi)}{\omega^2(\tau)-[\omega(\tau)\,m+\nu(\tau)\,n]^2}. \tag{4.137}$$

To obtain the second approximation, $A_2(\tau, a, \psi)$ and $B_2(\tau, a, \psi)$ must be determined. Since $u_2(\tau, a, \theta, p\varphi+\psi)$ does not contain the fundamental harmonic, we obtain, on the grounds of equation (4.131), the two equalities

$$\left.\begin{aligned} &H_{1s}\{F_2(\tau, a, \theta, p\varphi+\psi)\}=0, \\ &H_{1c}\{F_2(\tau, a, \theta, p\varphi+\psi)\}=0, \end{aligned}\right\} \tag{4.138}$$

from which we have the system of equations

$$\left.\begin{aligned} &\left[\omega(\tau)-\frac{p}{q}\nu(\tau)\right]\frac{\partial B_2}{\partial\psi}a+2\omega(\tau)A_2= \\ &=-\left[\frac{\partial B_1}{\partial a}aA_1+\frac{\partial B_1}{\partial\psi}aB_1+\frac{\partial B_1}{\partial\tau}+2A_1B_1\right]-M_2(\tau, a, \psi), \\ &\left[\omega(\tau)-\frac{p}{q}\nu(\tau)\right]\frac{\partial A_2}{\partial\psi}-2\omega(\tau)aB_2= \\ &=-\left[\frac{\partial A_1}{\partial a}A_1+\frac{\partial A_1}{\partial\psi}B_1+\frac{\partial A_1}{\partial\tau}-aB_1^2\right]+M_1(\tau, a, \psi), \end{aligned}\right\} \tag{4.139}$$

where $M_1(\tau, a, \psi)$ and $M_2(\tau, a, \psi)$ are defined by the relations

$$\left.\begin{aligned} M_1(\tau, a, \psi)=H_{1c}\Big\{&P(\tau, p_t)\left[p_aA_1+p_{p\varphi+\psi}B_1+p_\tau\right]F(\tau, \theta, u_0)- \\ &-P(\tau, p_t)\left|\frac{dF(\tau,\theta,y)}{dy}u_1\right|_{y=u_0}-\left[P'R+PR'\right](p_aA_1+p_{p\varphi+\psi}B_1+p_\tau)\,u_0\Big\}, \\ M_2(\tau, a, \psi)=H_{1s}\Big\{&P(\tau, p_t)\left[p_aA_1+p_{p\varphi+\psi}B_1+p_\tau\right]F(\tau, \theta, u_0)- \\ &-P(\tau, p_t)\left[\frac{dF(\tau,\theta,y)}{dy}u_1\right]_{y=u_0}-\left[P'R+PR'\right](p_aA_1+p_{p\varphi+\psi}B_1+p_\tau)\,u_0\Big\}. \end{aligned}\right\} \tag{4.140}$$

In order that the functions $u_1(\tau, a, \theta, p\varphi+\psi)$, $u_2(\tau, a, \theta, p\varphi+\psi)$ do not contain the first harmonic it is sufficient to take for the required functions $A_1(\tau, a, \psi)$, $A_2(\tau, a, \psi)$, ..., $B_1(\tau, a, \psi)$, $B_2(\tau, a, \psi)$, ... some particular solution of the systems (4.135) and (4.139) that is periodic in ψ (see Chapter III).

To construct the system of partial differential equations for $A_1(\tau, a, \psi)$, $B_1(\tau, a, \psi)$ (correspondingly $A_2(\tau, a, \psi)$, $B_2(\tau, a, \psi)$) the coefficient of the first harmonic in the Fourier expansion of the right-hand side of equation (4.124) must be found and substituted into equation (4.135) (or correspondingly in equation (4.139)), taking into account the required degree of accuracy.

Summing up the obtained results, we present schematically the construction of the first and second approximations for the symbolic equation (4.115).

We first separate out the "unperturbed" system (4.125) and check the possibility of harmonic vibrations at the frequency $\omega(\tau)$ (for any values of the parameter τ in the interval $0\leqslant\tau\leqslant L$), and also the nonexistence of nontrivial "static" solutions and of internal resonance.

We then take as the first approximation the expression

$$x=a\cos(p\varphi+\psi), \tag{4.141}$$

where a and ψ are determined by the equations of the first approximation

$$\left.\begin{aligned} &\frac{da}{dt}=\varepsilon A_1(\tau, a, \psi), \\ &\frac{d\psi}{dt}=\omega(\tau)-\frac{p}{q}\nu(\tau)+\varepsilon B_1(\tau, a, \psi) \end{aligned}\right\} \tag{4.142}$$

(the quantities $A_1(\tau, a, \psi)$ and $B_1(\tau, a, \psi)$ are derived according to the above-mentioned rule).

For the second approximation we take the expression

$$x=a\cos(p\varphi+\psi)+\varepsilon u_1(\tau, a, \theta, p\varphi+\psi). \tag{4.143}$$

Here a and ψ are determined by the equations of the second approximation

$$\left.\begin{aligned}\frac{da}{dt}&=\varepsilon A_1(\tau, a, \psi)+\varepsilon^2 A_2(\tau, a, \psi),\\ \frac{d\psi}{dt}&=\omega(\tau)-\frac{p}{q}\nu(\tau)+\varepsilon B_1(\tau, a, \psi)+\varepsilon^2 B_2(\tau, a, \psi),\end{aligned}\right\} \qquad (4.144)$$

in which $A_1(\tau, a, \psi)$, $B_1(\tau, a, \psi)$, $A_2(\tau, a, \psi)$ and $B_2(\tau, a, \psi)$ are defined by equations (4.135), (4.139), and $u_1(\tau, a, \theta, p\varphi+\psi)$ by the formula (4.137).

The method just given simplifies considerably the derivation of the equations of the first (or second) approximation and the construction of resonance curves for various modes. Other advantages of this method will be indicated below.

§ 5. Methods of constructing approximate solutions for the fundamental resonance and stationary modes

Let us consider a particular case of the symbolic equation (4.115). Let the function $F(\tau, \theta, x)$ on the right-hand side of equation (4.115) have the form

$$F(\tau, \theta, x)=f(x)+E_1(\tau)\cos\theta+E_2(\tau)\sin\theta, \qquad (4.145)$$

and let $Z(p)$ be an even function of p, i. e.,

$$Z(p)=[p^2+\omega^2(\tau)]\,Q(p^2). \qquad (4.146)$$

Then instead of equation (4.119) we have

$$[p^2+\omega^2(\tau)]\,x=P(p^2)\{\varepsilon f(x)+\varepsilon E_1(\tau)\cos\theta+\varepsilon E_2(\tau)\sin\theta\}. \qquad (4.147)$$

Vibrating systems described by such an equation play an important role in mechanical engineering. For example, in the problem of torsional vibrations during a transition through resonance, these vibrating systems occur in the crankshafts that contain a nonlinear elastic coupling in one of their sections connecting two masses, when the friction depending on the first power of the velocity is not taken into account. (The friction has to be introduced by some relation containing only a displacement.) In this case we obtain the symbolic equation (4.115), in which the right-hand side (if only one harmonic of the external torsional moment is taken into account) represents

$$\varepsilon f(x)+M(\tau)+\varepsilon E_1^{(1)}(\tau)\cos\theta+\varepsilon E_2^{(1)}(\tau)\sin\theta, \qquad (4.148)$$

where $M(\tau)$ is a "constant" moment depending, for a slow transition through resonance, on τ, and the equation

$$Z(p)\,x=\varepsilon f(x)+M(\tau)+\varepsilon E_1(\tau)\cos\theta+\varepsilon E_2(\tau)\sin\theta \qquad (4.149)$$

is after the substitution

$$x=y+z(\tau),$$

where $z(\tau)$ is defined by the condition

$$Z(p)\,z(\tau)=M(\tau), \qquad (4.150)$$

transformed to an equation of the type (4.147)

$$Z(p)\,y=\varepsilon f(y+z(\tau))+\varepsilon E_1(\tau)\cos\theta+\varepsilon E_2(\tau)\sin\theta. \qquad (4.151)$$

We solve equation (4.151) in the first and second approximations for the case of the fundamental resonance ($p=1,\ q=1$).

Inserting the value of $F(\tau, \theta, x)$ from (4.145) into the expressions (4.136) and deriving $A_1(\tau, a, \psi)$ and $B_1(\tau, a, \psi)$ from equations (4.135), we obtain the equations of the first approximation in the form

$$\left.\begin{aligned}
&\frac{da}{dt}=-\varepsilon\frac{P[-\nu^2(\tau)]}{\omega(\tau)+\nu(\tau)}[E_1(\tau)\sin\psi-E_2(\tau)\cos\psi],\\
&\frac{d\psi}{dt}=\omega(\tau)-\nu(\tau)-\frac{P[-\omega^2(\tau)]}{2\pi a\omega(\tau)}\int_0^{2\pi}\varepsilon f\,[a\cos(\theta+\psi)]\times\\
&\times\cos(\theta+\psi)\,d(\theta+\psi)-\frac{\varepsilon P[-\nu^2(\tau)]}{[\omega(\tau)+\nu(\tau)]\,a}[E_1(\tau)\cos\psi-E_2(\tau)\sin\psi].
\end{aligned}\right\}\qquad(4.152)$$

After a number of operations we obtain

$$x=a\cos(\theta+\psi)+\varepsilon u_1(\tau, a, \theta, \theta+\psi),$$

for the second approximation, where a and ψ are to be determined from the system of equations

$$\left.\begin{aligned}
&\frac{da}{dt}=-\left\{\frac{P[-\nu^2(\tau)]}{\omega(\tau)+\nu(\tau)}+\frac{[P(-\omega^2(\tau))-2\omega(\tau)P'(-\omega^2(\tau))]\,P[-\nu^2(\tau)]}{2\omega(\tau)[\omega(\tau)+\nu(\tau)]^2[3\omega(\tau)-\nu(\tau)]}\times\right.\\
&\left.\times\left[\frac{\omega(\tau)-\nu(\tau)}{a}F_1(a)+\omega(\tau)(F_2'(a)-F_0'(a))\right]\right\}\times\\
&\qquad\times[\varepsilon E_1(\tau)\sin\psi-\varepsilon E_2(\tau)\cos\psi],\\
&\frac{d\psi}{dt}=\omega(\tau)-\nu(\tau)-\frac{P[-\omega^2(\tau)]}{2a\omega(\tau)}F_1(a)-\\
&-\frac{P[-\omega^2(\tau)]}{4a\omega(\tau)}\sum_{\substack{n=0\\n\neq1}}^{\infty}\frac{P[-n^2\omega^2(\tau)]}{\omega^2(\tau)(1-n^2)}F_n(a)[F_{n-1}'(a)+F_{n+1}'(a)]-\\
&-\left\{\frac{P[-\nu^2(\tau)]}{\omega(\tau)+\nu(\tau)}+\frac{\{P[-\omega^2(\tau)]-2\omega(\tau)P'[-\omega^2(\tau)]\}\,P[-\nu^2(\tau)]}{2\omega(\tau)[\omega(\tau)+\nu(\tau)]^2[3\omega(\tau)-\nu(\tau)]}\times\right.\\
&\left.\times\left[\frac{2\omega(\tau)}{a}F_1(a)+\frac{\omega(\tau)-\nu(\tau)}{2}(F_2'(a)+F_0'(a))\right]\right\}\times\\
&\qquad\times[\varepsilon E_1(\tau)\cos\psi-\varepsilon E_2(\tau)\sin\psi],
\end{aligned}\right\}\qquad(4.153)$$

and $u_1(\tau, a, \theta, \theta+\psi)$— from the formula

$$u_1(\tau, a, \theta, \theta+\psi)=\sum_{\substack{n=0\\(n\neq1)}}^{\infty}\frac{F_n(a)\cos n(\theta+\varphi)}{Z[-n^2\omega^2(\tau)]}.\qquad(4.154)$$

We have introduced here the notation

$$\left.\begin{aligned}
&F_n(a)=\frac{1}{\pi}\int_0^{2\pi}\varepsilon f[a\cos(\theta+\psi)]\cos n(\theta+\psi)\,d(\theta+\psi),\\
&F_n'(a)=\frac{1}{\pi}\int_0^{2\pi}\varepsilon f'[a\cos(\theta+\psi)]\cos n(\theta+\psi)\,d(\theta+\psi)\\
&\qquad(n=0, 1, 2, \ldots).
\end{aligned}\right\}\qquad(4.155)$$

We shall now derive relations for the amplitude and phase in the case of stationary vibrations of a constant amplitude (for constant τ) and the total phase rotating at a constant angular velocity. To obtain these relations in the first approximation the right-hand sides of equations (4.152) must be equated to zero. After a number of operations we obtain, up to an accuracy of the first order, the relations

$$\left.\begin{aligned} E_1 \sin\psi - E_2 \cos\psi = 0, \\ Z(-\nu^2)\,a - F_1(a) - \varepsilon E_1 \cos\psi + \varepsilon E_2 \sin\psi = 0. \end{aligned}\right\} \quad (4.156)$$

Eliminating ψ, we obtain a relation between the amplitude and the frequency of the external force; this relation may be utilized to construct the stationary resonance curve.

To determine the stationary values of a and ψ in the second approximation the right-hand sides of equations (4.153) must be equated to zero.

After a series of transformations we obtain, with an accuracy of second order terms, the relations

$$\left.\begin{aligned} E_1 \sin\psi - E_2 \cos\psi = 0, \\ Z(-\nu^2)\,a - F_1(a) - \frac{1}{2}\sum_{\substack{n=0 \\ (n\neq 1)}}^{\infty} \frac{F_n(a)\,[F'_{n+1}(a) + F'_{n-1}(a)]}{Z(-n^2\nu^2)} - \varepsilon E_1 \cos\psi + \varepsilon E_2 \sin\psi = 0. \end{aligned}\right\} \quad (4.157)$$

Expressions (4.156) and (4.157) may be derived (avoiding a large number of operations) without recourse to the equations of the first and second approximation (i. e., to the equations (4.152) and (4.153)).

To do this we apply the following formal method. Consider the equation

$$Z(p^2)\,x = \varepsilon f(x) + \varepsilon E_1 \cos\theta + \varepsilon E_2 \sin\theta \quad (4.158)$$

($E_i = E_i(\tau)$ for $\tau = \text{const}$, $i = 1, 2$). We look for its approximate solutions for the case of principal resonance in the form

$$x = a\cos(\theta + \psi). \quad (4.159)$$

Substituting this value of x into (4.158), we obtain

$$Z(-\nu^2)\,a\cos(\theta+\psi) = \sum_{n=0}^{\infty} \varepsilon f_n(a)\cos(\theta+\psi) + \varepsilon E_1[\cos(\theta+\psi)\cos\psi + \sin(\theta+\psi)\sin\psi] + \varepsilon E_2[\sin(\theta+\psi)\cos\psi - \cos(\theta+\psi)\sin\psi]. \quad (4.160)$$

Equating the coefficients of the first harmonics of the angle $\theta+\psi$ in this equation we find the previously derived expressions (4.156) for determining a and ψ.

The stationary solution of (4.158) is sought for in the form

$$x = a\cos(\theta+\psi) + \varepsilon u_1(\tau, a, \theta, \theta+\psi), \quad (4.161)$$

where a, ψ, and $u_1(\tau, a, \theta, \theta+\psi)$ are still to be determined.

Substituting this value of x into (4.158), we obtain

$$\begin{aligned} Z(-\nu^2)\,a\cos(\theta+\psi) + Z(p^2)\,\varepsilon u_1 = \sum_{n=0}^{\infty} F_n(a)\cos n(\theta+\psi) + \\ + \sum_{n=0}^{\infty}\left\{\frac{\varepsilon^2}{\pi}\int_0^{2\pi} f'(a\cos(\theta+\psi))\,u_1 \cos n(\theta+\psi)\,d(\theta+\psi)\right\}\cos n(\theta+\psi) + \\ + \varepsilon E_1[\cos(\theta+\psi)\cos\psi + \sin(\theta+\psi)\sin\psi] + \\ + \varepsilon E_2[\sin(\theta+\psi)\cos\psi - \cos(\theta+\psi)\sin\psi]. \end{aligned} \quad (4.162)$$

Equating the coefficients of the first harmonics, we obtain the previously derived expressions (4.141) for determining the stationary values of a and ψ.

We then find $u_1(\tau, a, \theta, \theta+\psi)$ from the equation

$$Z(p^2)\,\varepsilon u_1 = \sum_{\substack{n=0\\(n\neq 1)}}^{\infty} F_n(a)\cos n(\theta+\psi). \tag{4.163}$$

In order to obtain the equations determining the stationary values of the amplitude and phase of vibration, we must substitute $x = a\cos(\theta+\psi)$ (or $x = a\cos(\theta+\psi) + \varepsilon u_1$) in the original equation (4.158), regarding a and ψ as constants, and equate to zero the coefficients of the first harmonic, neglecting terms of order higher than the first (or second). If $E_2 = 0$ in equation (4.158) (this occurs, for example, when one considers the torsional vibrations of the crankshaft in a four-stroke engine taking into account only the third harmonic of the external torsional moment), then the following, greatly simplified, expressions are obtained for determining the stationary values of a and ψ

$$\left.\begin{aligned} &\psi = 0,\ \pi,\\ &Z(-\nu^2)\,a - F_1(a) \pm E = 0 \end{aligned}\right\} \tag{4.164}$$

in the first approximation, and

$$\left.\begin{aligned} &\psi = 0,\ \pi,\\ &Z(-\nu^2)\,a - F_1(a) - \frac{1}{2}\sum_{\substack{n=0\\(n\neq 1)}}^{\infty} \frac{F_n(a)\,[F'_{n+1}(a) + F'_{n-1}(a)]}{Z(-n^2\nu^2)} \pm E = 0 \end{aligned}\right\} \tag{4.165}$$

in the second approximation.

The stationary resonance curves described by (4.164) and (4.165) can be plotted without difficulty.

§ 6. Torsional vibrations of the crankshaft of an aircraft engine in a nonstationary mode

As an example illustrating the effectiveness of the symbolic method given above we consider the nonlinear system consisting of the crankshaft of an in-line aircraft engine with a reductor and an elastic coupling connecting the propeller with the crankshaft.

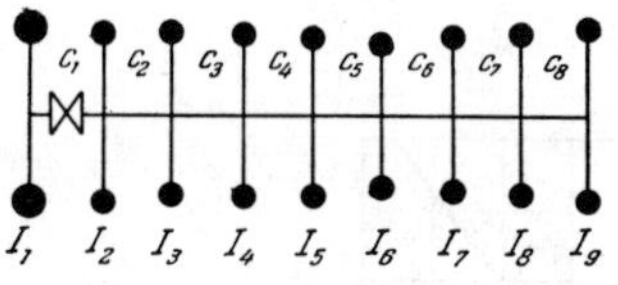

FIGURE 69

The purpose of the elastic coupling (we consider only nonlinear coupling) is to change suitably the properties of the system of the engine crankshaft with regard to torsional vibrations, and to avoid an excessive increase of stress in the system during transient motion (while the engine is started, during the transition from one mode of operation to another, etc.). The calculation of torsional vibrations in the system crankshaft-propeller with a nonlinear elastic coupling reduces to the calculation of torsional vibrations in a system with many degrees of freedom. The construction of resonance curves for such systems was studied in detail for the case of stationary modes by a number of authors (I. Sh. Neiman /125/, B. Natanson /124/, A. I. Lur'e and A. I. Chekmarev /80/, and others). Nonstationary processes were considered for the first time in /96, 97, 99/.

The construction of resonance curves for the stationary mode as well as for a transition through resonance may be carried out without difficulty in the considered example with the help of the method expounded in the preceding section.

Let us assume that the vibrating system under consideration is replaced by a reduced shaft with a rectilinear axis and concentrated masses (Figure 69) having the following moments of inertia

$$\left.\begin{aligned} I_9 &= 120 \text{ kg} \cdot \text{cm} \cdot \text{sec}^2, \quad I_8 = I_1 = 3 \text{ kg} \cdot \text{cm} \cdot \text{sec}^2, \\ I_2 &= I_3 = I_4 = I_5 = I_6 = I_7 = 1 \text{ kg} \cdot \text{cm} \cdot \text{sec}^2. \end{aligned}\right\} \tag{4.166}$$

Let the coefficients of rigidity have the values.

$$c_2 = c_3 = c_4 = c_5 = c_6 = c_7 = 15 \cdot 10^6 \text{ kg.cm}, \quad c_1 = 2 \cdot 10^6 \text{ kg} \cdot \text{cm}, \tag{4.167}$$

while the rigidity coefficients c_8 may assume two values:

$$c_8' = 3.7 \cdot 10^6 \text{ kg.cm}, \quad c_8'' = 2 \cdot 10^6 \text{ kg} \cdot \text{cm}, \tag{4.168}$$

which corresponds to an elastic coupling between the first and second mass, whose nonlinear characteristic is composed of rectilinear segments (coupling with preliminary compression of the springs). The calculation could be performed with the same success for any other characteristic of the elastic coupling given, for instance, by a quadratic or cubic parabola, etc.

We denote the angles of deviation from uniform rotation by

$$\varphi_1, \varphi_2, \varphi_3, \ldots, \varphi_9. \tag{4.169}$$

The elastic moment in each section of the shaft will be

$$c_i(\varphi_{i+1} - \varphi_i) \qquad (i = 2, 3, \ldots, 8),$$

but in the eighth section it depends on the characteristic of the elastic coupling, and is in our case

$$F(x) = c_8''x + \varepsilon f(x), \tag{4.170}$$

where $x = \varphi_9 - \varphi_8$, and $\varepsilon f(x)$ has the values

$$\varepsilon f(x) = \begin{cases} (c_8' - c_8'')\,x, & -x_0 \leqslant x \leqslant x_0, \\ (c_8' - c_8'')\,x_0, & x_0 \leqslant x \leqslant \infty, \\ -(c_8' - c_8'')\,x_0, & -\infty \leqslant x \leqslant -x_0. \end{cases} \tag{4.171}$$

The function $y = \varepsilon f(x)$ is plotted in Figure 70.

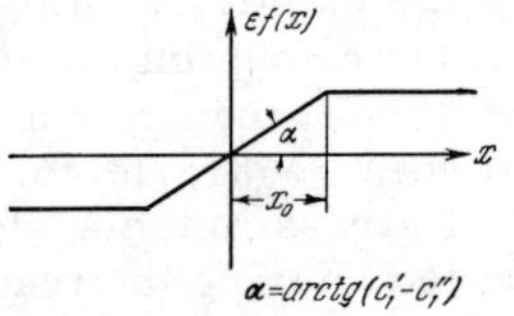

FIGURE 70

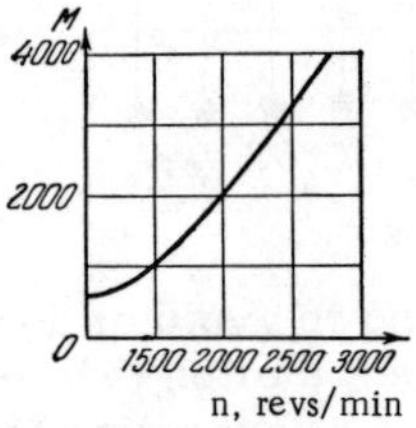

FIGURE 71

For simplicity we neglect moments which suppress the vibrations of the system, and consider in the exciting moment only the most "dangerous" harmonic.

Let the internal masses of the system be subjected to the action of periodic torsional moments (we also discard for simplicity constant moments)

$$M_{\text{per}} = M(\nu)\cos\theta, \tag{4.172}$$

where the frequency of the moment, $\frac{d\theta}{dt} = \nu$, is proportional to the number of revolutions of the engine (in a nonstationary state ν changes with time), and $M(\nu)$ is determined experimentally. (The dependence of M on ν in the present calculation is shown in Figure 71.)

The equations of torsional vibrations in the system of the reduced shaft (Figure 69) have the form

$$\left.\begin{aligned}
&I_1\frac{d^2\varphi_1}{dt^2} + c_1(\varphi_1-\varphi_2) = 0,\\
&I_2\frac{d^2\varphi_2}{dt^2} + c_1(\varphi_2-\varphi_1) + c_2(\varphi_2-\varphi_3) = M_{\text{per}},\\
&I_3\frac{d^2\varphi_3}{dt^2} + c_2(\varphi_3-\varphi_2) + c_3(\varphi_3-\varphi_4) = M_{\text{per}},\\
&I_4\frac{d^2\varphi_4}{dt^2} + c_3(\varphi_4-\varphi_3) + c_4(\varphi_4-\varphi_5) = M_{\text{per}},\\
&I_5\frac{d^2\varphi_5}{dt^2} + c_4(\varphi_5-\varphi_4) + c_5(\varphi_5-\varphi_6) = M_{\text{per}},\\
&I_6\frac{d^2\varphi_6}{dt^2} + c_5(\varphi_6-\varphi_5) + c_6(\varphi_6-\varphi_7) = M_{\text{per}},\\
&I_7\frac{d^2\varphi_7}{dt^2} + c_6(\varphi_7-\varphi_6) + c_7(\varphi_7-\varphi_8) = M_{\text{per}},\\
&I_8\frac{d^2\varphi_8}{dt^2} + c_7(\varphi_8-\varphi_7) + F(\varphi_8-\varphi_9) = M_{\text{per}},\\
&I_9\frac{d^2\varphi_9}{dt^2} + F(\varphi_9-\varphi_8) = 0.
\end{aligned}\right\} \tag{4.173}$$

This system may be reduced to the form

$$Z(p^2)x = F(x) + E(p)\cos\theta, \tag{4.174}$$

which was examined in detail in the preceding section. Indeed, putting $p = \frac{d}{dt}$ and taking into account (4.171), one may write the system (4.173) in the form

$$\left.\begin{aligned}
&-I_1p^2\varphi_1 + c_1(\varphi_1-\varphi_2) = 0,\\
&-I_2p^2\varphi_2 + c_1(\varphi_2-\varphi_1) + c_2(\varphi_2-\varphi_3) = M_{\text{per}},\\
&-I_3p^2\varphi_3 + c_2(\varphi_3-\varphi_2) + c_3(\varphi_3-\varphi_4) = M_{\text{per}},\\
&-I_4p^2\varphi_4 + c_3(\varphi_4-\varphi_3) + c_4(\varphi_4-\varphi_5) = M_{\text{per}},\\
&-I_5p^2\varphi_5 + c_4(\varphi_5-\varphi_4) + c_5(\varphi_5-\varphi_6) = M_{\text{per}},\\
&-I_6p^2\varphi_6 + c_5(\varphi_6-\varphi_5) + c_6(\varphi_6-\varphi_7) = M_{\text{per}},\\
&-I_7p^2\varphi_7 + c_6(\varphi_7-\varphi_6) + c_7(\varphi_7-\varphi_8) = M_{\text{per}},\\
&-I_8p^2\varphi_8 + c_7(\varphi_8-\varphi_7) + c_8''(\varphi_8-\varphi_9) = \varepsilon f(\varphi_9-\varphi_8),\\
&-I_9p^2\varphi_9 + c_8''(\varphi_9-\varphi_8) = -\varepsilon f(\varphi_9-\varphi_8).
\end{aligned}\right\} \tag{4.175}$$

To transform this system to the form (4.174) we apply the method of Tolle*. To do this we first consider the homogeneous linear system of equations

$$\left.\begin{aligned}
&-I_1p^2\lambda_1 + c_1(\lambda_1-\lambda_2) = 0,\\
&-I_2p^2\lambda_2 + c_1(\lambda_2-\lambda_1) + c_2(\lambda_2-\lambda_3) = 0,\\
&\quad\cdots\cdots\cdots\cdots\cdots\cdots\\
&-I_8p^2\lambda_8 + c_7(\lambda_8-\lambda_7) + c_8(\lambda_8-\lambda_9) = 0,\\
&-I_9p^2\lambda_9 + c_8(\lambda_9-\lambda_8) = 0.
\end{aligned}\right\} \tag{4.176}$$

* A detailed exposition of this method may be found in B. Natanson's book /124/.

Using the notation

$$E_i = c_i(\lambda_{i+1} - \lambda_i) \qquad (i = 1,\ 2,\ \ldots,\ 9), \tag{4.177}$$

we can write (4.176) in the form

$$E_i = -I_i p^2 \lambda_i + E_{i-2} \qquad (i = 1,\ 2,\ \ldots,\ 9), \tag{4.178}$$

where $E_0 = 0$, $E_9 = 0$, provided that p is an eigenfrequency. We then obtain the recurrence relation

$$\lambda_{i+1} = \lambda_i \left\{1 - \frac{I_i}{c_i} p^2 + \frac{c_{i-1}}{c_i}\right\} - \frac{c_{i-1}}{c_i} \lambda_{i-1} \qquad (i = 1,\ 2,\ \ldots,\ 9). \tag{4.179}$$

In order to solve the homogeneous equations (4.176) we must derive two sets of numbers associated with them (or, which is the same, with the equivalent equations (4.178)). To do this we assume that $\lambda_1 = 1$ and then calculate by means of the recurrence relations (4.179) the values of

$$\lambda_2,\ \lambda_3,\ \ldots,\ \lambda_9.$$

For an arbitrary p which is not an eigenfrequency of the system, $E_9 \neq 0$. We put $E_9 = R(p)$, where

$$R(p) = -I_9 p^2 \lambda_9 + c_8(\lambda_9 - \lambda_8). \tag{4.180}$$

We now define another sequence of values λ_i' $(i = 1,\ 2,\ \ldots,\ 9)$. Using the notation

$$E_i = c_i(\lambda_{i+1}' - \lambda_i'),$$

we find from (4.176)

$$E_i = I_{i+1} p^2 \lambda_{i+1}' + E_{i+1}, \tag{4.181}$$

whence we obtain for λ_i' the recurrence relation

$$\lambda_i' = \lambda_{i+1}' \left\{1 - \frac{I_{i+1}}{c_i} p^2 + \frac{c_{i+1}}{c_i}\right\} - \frac{c_{i+1}}{c_i} \lambda_{i+2}'. \tag{4.182}$$

Setting $\lambda_9' = 1$, we find successively

$$\lambda_8',\ \lambda_7',\ \ldots,\ \lambda_1'.$$

In the case $E_0 \neq 0$, we put $E_0 = R'(p)$, where

$$R'(p) = I_1 p^2 \lambda_1' + c_1(\lambda_2' - \lambda_1'). \tag{4.183}$$

It is not difficult to show that $R(p) = -R'(p)$.

Having defined in this manner the two series of values λ_i and λ_i' $(i = 1,\ 2,\ \ldots,\ 9)$ as functions of p and calculated $R = R(p)$, we can proceed to solve the inhomogeneous system (4.175). We first consider the particular case when the right-hand sides of all its equations, except the $(k+1)$-th, are zero. Denoting a solution of the system (4.175) by φ_{ik}, we have

$$\left.\begin{array}{l}
-I_1 p^2 \varphi_{1k} - c_1''(\varphi_{1k} - \varphi_{2k}) = 0, \\
-I_2 p^2 \varphi_{2k} + c_1''(\varphi_{1k} - \varphi_{2k}) + c_3(\varphi_{2k} - \varphi_{3k}) = 0, \\
\cdots\cdots\cdots\cdots\cdots\cdots\cdots\cdots\cdots\cdots \\
-I_i p^2 \varphi_{ik} + c_i(\varphi_{i-1,\,k} - \varphi_{ik}) + c_{i+1}(\varphi_{ik} - \varphi_{i+1,\,k}) = 0, \\
\cdots\cdots\cdots\cdots\cdots\cdots\cdots\cdots\cdots\cdots \\
-I_k p^2 \varphi_{kk} + c_k(\varphi_{k-1,\,k} - \varphi_{kk}) + c_{k+1}(\varphi_{kk} - \varphi_{k+1,\,k}) = M_k, \\
\cdots\cdots\cdots\cdots\cdots\cdots\cdots\cdots\cdots\cdots \\
-I_9 p^2 \varphi_{9k} + c_8(\varphi_{9k} - \varphi_{8k}) = 0.
\end{array}\right\} \tag{4.184}$$

Assuming first $i < k$, we multiply equations (4.184) by $\lambda_1',\ \lambda_2',\ \ldots,\ \lambda_9$ respectively, and add them to each other. We then put $\lambda_1',\ \lambda_2',\ \ldots,\ \lambda_9' = 1$ in equations (4.176), multiply them correspondingly by $\varphi_{1k},\ \varphi_{2k},\ \ldots,\ \varphi_{9k}$ and add them to each other; we obtain two sums with equal left-hand sides. Hence, the right-hand sides are equal, i. e.,

$$-\varphi_{1k}R'=M_k\lambda_k'. \tag{4.185}$$

We derive now any of the φ_{ik} $(i\neq 1)$. As the first $k-1$ equations of the system (4.184) are homogeneous, we obtain, taking $\varphi_{1k}=1$,

$$\varphi_{1k}=1,\ \varphi_{2k}=\lambda_2,\ \ldots,\ \varphi_{kk}=\lambda_k,$$

or

$$\varphi_{ik}=\varphi_{1k}\lambda_i \qquad (i\leqslant k).$$

According to (4.185) we have

$$\varphi_{ik}=M_k\frac{\lambda_k'\lambda_i}{R} \qquad (i\leqslant k). \tag{4.186}$$

If $i\geqslant k$, then by similar reasoning we obtain

$$\varphi_{ik}=M_k\frac{\lambda_k\lambda_i'}{R} \qquad (i\geqslant k). \tag{4.187}$$

Substituting the values of k from 1 to 9 into (4.186) and (4.187) we obtain 9 particular solutions, and their sum

$$\varphi_i=\sum_{k=1}^{9}\varphi_{ik} \qquad (i=1,\ 2,\ \ldots,\ 9) \tag{4.188}$$

is a solution of the inhomogeneous system (4.175). (φ_{ik} are derived from formula (4.186) or (4.187) depending on the value of i.

According to formula (4.188)

$$\varphi_9=\sum_{k=2}^{7}\frac{M_k\lambda_k}{R}+\frac{\varepsilon f(\varphi_9-\varphi_8)}{R}(\lambda_8-\lambda_9), \tag{4.189}$$

and also

$$\varphi_8=\sum_{k=2}^{7}\frac{M_k\lambda_k'\lambda_8}{R}+\frac{\varepsilon f(\varphi_9-\varphi_8)}{R}(\lambda_8'\lambda_8-\lambda_9'\lambda_8). \tag{4.190}$$

Subtracting (4.190) from (4.189) and using the notation $\varphi_9-\varphi_8=x$, we obtain the following symbolic equation

$$\frac{R}{\lambda_8-\lambda_9-\lambda_8(\lambda_8'-\lambda_9')}x=\varepsilon f(x)+\sum_{k=2}^{7}\frac{M_k(\lambda_k-\lambda_k'\lambda_8)}{\lambda_8-\lambda_9-\lambda_8(\lambda_8'-\lambda_9')}, \tag{4.191}$$

which belongs to the type considered in the preceding section.

We proceed to construct approximate solutions of equation (4.191) for the case of the fundamental resonance ($p=1$, $q=1$). Using the results of the preceding section, we have

$$x=a\cos(\theta+\psi), \tag{4.192}$$

in the first approximation, where a and ψ are to be determined from the system of equations of the first approximation

$$\left.\begin{aligned}\frac{da}{dt}&=-\frac{P(-\nu)}{\omega+\nu}E(\nu)\sin\psi,\\ \frac{d\psi}{dt}&=\omega-\nu-\frac{P(-\omega^2)}{2\omega}F_1\left(\frac{a}{x_0}\right)-\frac{P(-\nu^2)}{(\omega+\nu)a}E(\nu)\cos\psi,\end{aligned}\right\} \tag{4.193}$$

where ω is an eigenfrequency of the rigid system (the system (4.176) with $c_1=c_1'$),

$$F_1\left(\frac{a}{x_0}\right)=\frac{2}{\pi a}\int_0^{\pi}\varepsilon f[a\cos\vartheta]\cos\vartheta\,d\vartheta, \tag{4.194}$$

and $\varepsilon f(a\cos\vartheta)$ has, according to (4.155), the values

$$\varepsilon f(a\cos\vartheta)=\begin{cases}(c_1'-c_1'')\,a\cos\vartheta, & \vartheta_0\leqslant\vartheta\leqslant\pi-\vartheta_0 \quad (a>0,\ a>x_0),\\ (c_1'-c_1'')\,a\cos\vartheta_0, & 0\leqslant\vartheta\leqslant\vartheta_0,\\ -(c_1'-c_1'')\,a\cos\vartheta_0, & \pi-\vartheta_0\leqslant\vartheta\leqslant\pi.\end{cases} \tag{4.195}$$

Dividing the integration interval into these three sections, and substituting $\vartheta_0=\arccos\frac{x_0}{a}$, we obtain

$$F_1\left(\frac{a}{x_0}\right)=\frac{2}{\pi}(c_1'-c_1'')\left\{\frac{x_0}{a}\sqrt{1-\left(\frac{x_0}{a}\right)^2}+\arcsin\frac{x_0}{a}\right\}. \tag{4.196}$$

The dependence of $F_1\left(\frac{a}{x_0}\right)$ on $\frac{a}{x_0}$ is shown graphically in Figure 72.

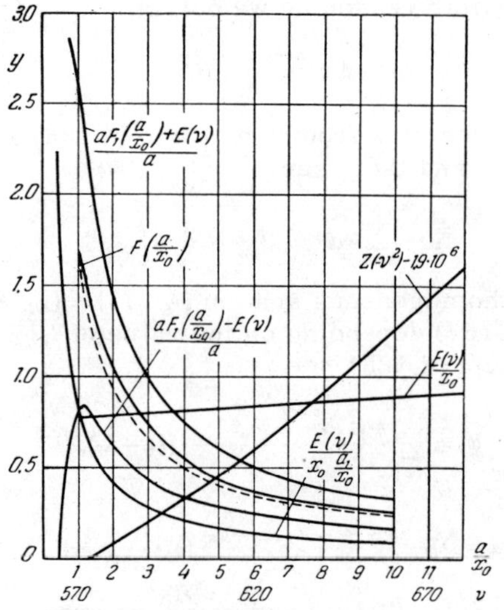

FIGURE 72

If we limit our examination to the resonance zone only $(0.8\leqslant\frac{\nu}{\omega_{\text{rig}}}\leqslant 1.1)$, then, without affecting the accuracy of calculation, we may consider the system

$$\left.\begin{aligned}\frac{d\left(\frac{a}{x_0}\right)}{dt}&=-\frac{\omega-\nu}{Z(-\nu^2)}\frac{E(\nu)}{x_0}\sin\psi,\\ \frac{d\psi}{dt}&=\omega-\nu-\frac{\omega-\nu}{Z(-\nu^2)}F_1\left(\frac{a}{x_0}\right)-\frac{\omega-\nu}{Z(-\nu^2)\frac{a}{x_0}}\frac{E(\nu)}{x_0}\cos\psi,\end{aligned}\right\} \tag{4.197}$$

instead of (4.193), where a is replaced for convenience by the ratio $\frac{a}{x_0}$. To construct the resonance curve for a stationary state we must equate the right-hand sides of equations (4.197) to zero. Eliminating ψ, we obtain the following relation between $\frac{a}{x_0}$ and ν:

$$Z(-\nu^2)-F_1\left(\frac{a}{x_0}\right)\mp\frac{E(\nu)}{x_0\frac{a}{x_0}}=0. \tag{4.198}$$

(This relation may be deduced directly, as was shown in the preceding section, by substituting $x = a\cos(\theta + \psi)$ into equation (4.191) and equating the coefficients of the same harmonics.)

We plot the dependence of $\frac{a}{x_0}$ on ν using the given numerical data.

We derive first of all the fundamental frequency of normal vibrations for c_1' and c_1'', i. e., the maximum and minimum frequency. We have $\omega_{\text{rig}} = 437\ \text{sec}^{-1}$, $\omega_{\text{soft}} = 364\ \text{sec}^{-1}$. We next calculate a number of values of $Z(-\nu^2)$ and $E(\nu)$ ($\omega_{\text{soft}} \leqslant \nu \leqslant \omega_{\text{rig}}$) from the formulas

$$Z(-\nu^2) = \frac{R}{\lambda_8 - \lambda_9 - \lambda_8(\lambda_8' - \lambda_9')}, \tag{4.199}$$

$$E(\nu)\cos\theta = \sum_{k=2}^{7} \frac{M_{\text{tran}}(\lambda_k - \lambda_k'\lambda_2)}{\lambda_8 - \lambda_9 - \lambda_8(\lambda_8' - \lambda_9')}, \tag{4.200}$$

in which the quantities $\lambda_1, \lambda_2, \ldots, \lambda_9, \lambda_1', \lambda_2', \ldots, \lambda_9'$ are calculated for a number of values of ν by the recurrence relations (4.179) and (4.182) (cf. Table 8), and

$$R = -I_9 p^2 \lambda_9 + c_8(\lambda_9 - \lambda_8). \tag{4.201}$$

In order to construct the resonance curve by (4.198) it is convenient to plot first several auxiliary curves (Figure 72)

$$\left.\begin{aligned} &y = Z(-\nu^2), \quad y = F_1\left(\frac{a}{x_0}\right), \quad y = \frac{E(\nu)}{x_0}, \\ &y = \frac{E(\nu)}{x_0 \frac{a}{x_0}}, \quad y = F_1\left(\frac{a}{x_0}\right) \pm \frac{E(\nu)}{x_0 \frac{a}{x_0}}, \end{aligned}\right\} \tag{4.202}$$

which are helpful when drawing the skeleton curve describing the forms of graphs of the forced amplitudes, and the resonance curves (Figure 73).

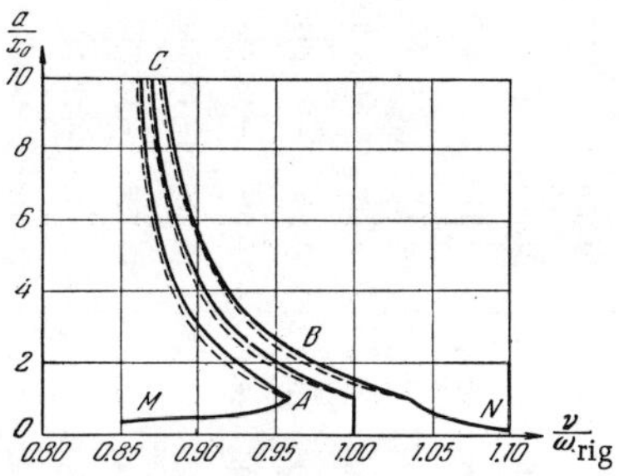

FIGURE 73

In order to construct the resonance curves in the second approximation, we calculate the sum

$$y = F_1\left(\frac{a}{x_0}\right) + \frac{1}{2}\sum_{\substack{n=0\\(n\neq 1)}}^{\infty} \frac{F_n\left(\frac{a}{x_0}\right)\left[F_{n+1}'\left(\frac{a}{x_0}\right) + F_{n-1}'\left(\frac{a}{x_0}\right)\right]}{Z(-n^2\nu^2)} \tag{4.203}$$

and plot it by a dotted line on the auxiliary graph (Figure 72). We then construct on the graph describing the dependence of the amplitude on the frequency (Figure 73) the skeleton curve and the resonance curves (dotted lines) corresponding to the second approximation.

When the frequency of the external moments (or of the revolutions of the shaft) increases in a stationary mode, then the amplitude changes first along the line *MA* (at the point *A* it discontinuously goes over to the point *B*), and then continues to vary along the right branch of the resonance curve *BN*.

TABLE 8

λ_k \ ν	364	374	384	394	404	414	424	434	444	454	464
λ_1	1	1	1	1	1	1	1	1	1	1	1
λ_2	0.801	0.710	0.779	0.767	0.755	0.743	0.730	0.717	0.704	0.691	0.677
λ_3	0.768	0.755	0.742	0.728	0.714	0.700	0.686	0.671	0.656	0.640	0.642
λ_4	0.727	0.712	0.697	0.682	0.666	0.649	0.633	0.616	0.598	0.581	0.563
λ_5	0.681	0.663	0.646	0.628	0.610	0.591	0.572	0.553	0.533	0.518	0.493
λ_6	0.628	0.608	0.588	0.568	0.547	0.526	0.505	0.483	0.461	0.439	0.416
λ_7	0.569	0.547	0.525	0.502	0.479	0.455	0.431	0.407	0.383	0.358	0.333
λ_8	0.506	0.481	0.456	0.431	0.405	0.379	0.353	0.326	0.300	0.273	0.245
λ_9	— 0.070	— 0.115	— 0.159	— 0.203	— 0.246	— 0.290	— 0.332	— 0.373	— 0.413	— 0.453	— 0.491
λ'_1	—13.801	—14.165	—14.501	—14.793	—15.046	—15.248	—15.405	—15.513	—15.567	—15.562	—15.497
λ'_2	—11.036	—11.560	—12.085	—12.606	—13.125	—13.638	—14.146	—14.648	—15.142	—15.626	—16.099
λ'_3	—10.570	—11.105	—11.644	—12.184	—12.726	—13.268	—13.808	—14.349	—14.886	—15.419	—15.978
λ'_4	—10.011	—10.546	—11.089	—11.636	—12.189	—12.745	—13.306	—13.869	—14.435	—15.001	—15.569
λ'_5	— 9.363	— 9.889	—10.424	—10.968	—11.519	—12.078	—12.643	—13.215	—13.754	—14.377	—14.966
λ'_6	— 8.632	— 9.140	— 9.658	—10.186	—10.724	—11.272	—11.829	—12.396	—12.971	—13.556	—14.148
λ'_7	— 7.826	— 8.305	— 8.796	— 9.298	— 9.812	—10.337	—10.873	—11.420	—11.979	—12.548	—13.127
λ'_8	— 6.950	— 7.393	— 7.848	— 8.314	— 8.793	— 9.284	— 9.787	—10.302	—10.828	—11.367	—11.918
λ'_9	1	1	1	1	1	1	1	1	1	1	1

If the frequency of the impressed torsional moments begins to decrease, then the amplitude varies first along the curve NBC, passing at the point C to the curve AM, and thereafter continues to change along the latter. The position of the point C is not defined, since the resonance curve is not closed (it was constructed without taking into account friction).

We shall now examine the behavior of the system during various modes of transition through resonance.

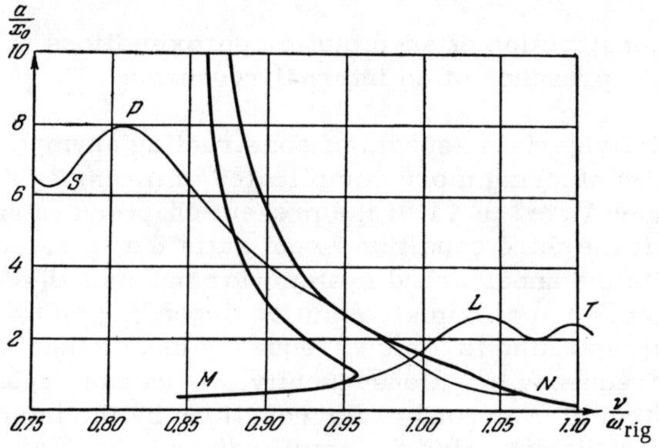

FIGURE 74

Assuming that the frequency of the external force changes linearly, and integrating numerically the system of equations of the first approximation for various rates of frequency variation, we obtain the curves shown in Figures 74 and 75, characterizing the variation of the amplitude during various modes of transition through resonance. (When integrating (4.197) numerically, it is not necessary to calculate the value of $Z(-\nu^2)$ for each ν, as we can use the graph given in Figure 72.)

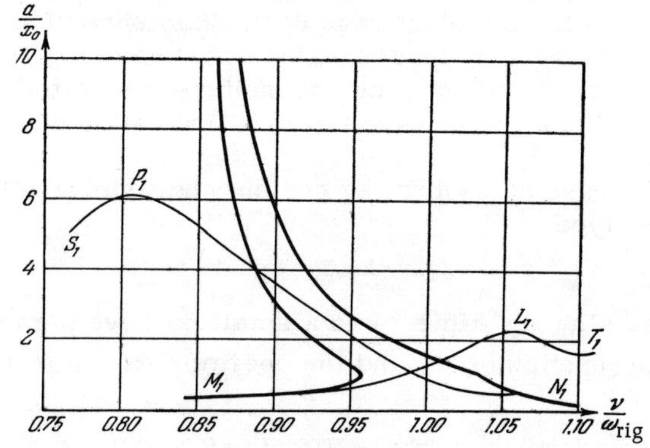

FIGURE 75

The curves MLT and NPS in Figure 74 characterize a quite slow transition through resonance — $\beta = \pm 200$ rad/sec^2, which corresponds to a transition through the resonance zone $(0.8 \leqslant \frac{\nu}{\omega_{\text{rig}}} \leqslant 1.1)$ during approximately 45 cycles.

The curves $M_1L_1T_1$ and $N_1P_1S_1$ characterize a faster passage through resonance — $\beta = \pm 300$ rad/sec^2, which corresponds to a transition through the resonance zones during 30 cycles. The amplitudes grow sharply when the resonance zone is crossed to the left (cf. curves NPS and $N_1P_1S_1$ in Figures 74 and 75) due to the neglect of friction, but when the resonance is swept to the right the nonlinearity plays the role of passive friction and prevents the growth of large amplitudes.

§ 7. Construction of asymptotic approximations in presence of an internal resonance

We discuss briefly certain aspects of constructing asymptotic approximate solutions when studying more complicated vibrating systems for which not all the conditions listed in § 1 of the present chapter are satisfied.

For example, if the third condition is not satisfied, i. e., if an internal resonance exists in the unperturbed system (for instance $\omega_1 = \omega_2$, then we are not able to construct an approximate solution depending on two arbitrary constants and corresponding to a monofrequency mode close to normal vibrations at the frequency ω_1. Consequently, we cannot replace the integration of the system of N second order equations by the integration of two first order equations determining the amplitude and phase of vibration.

The study of nonlinear vibrating systems with internal resonance gives rise to several additional difficulties. In particular, one has to apply asymptotic expansions, as was shown in /104, 109/ and in a number of other works (for instance /87,139/, etc.), not in powers of the small parameter ε, but in fractional powers of the parameter ε, depending on the multiplicity of the corresponding eigenfrequency.

The study of nonlinear systems with internal resonance is in some cases carried out conveniently with the help of the symbolic treatment of the nonlinear equation; this was worked out in the preceding sections.

We consider below the simplest case of the existence of an internal resonance in a vibrating system described by a nonlinear fourth order differential equation with slowly varying parameters, and attempt to illustrate by this simple case the difficulties that arise, indicating some ways of overcoming them.

We consider the case of a vibrating system characterized by a differential equation of the type

$$Z(p)\,x = \varepsilon F(\tau, x), \tag{4.204}$$

where $\tau = \varepsilon t$ is the "slowing" time; ε is a small positive parameter; $p = \frac{d}{dt}$; $F(\tau, x)$ is a certain function of x; and the operator $Z(p)$ may be represented as the sum

$$Z(p) = \sum_{n=1}^{4} a_n(\tau)\,p^n = Q(\tau, p) + \varepsilon R(\tau, p), \tag{4.205}$$

where $Q(\tau, p)$ has for any τ in the interval $0 \leqslant \tau \leqslant L$ a double root $i\omega(\tau)$, i.e.,

$$Q(\tau, p) = c\,[p^2 + \omega^2(\tau)]^2. \tag{4.206}$$

Using the notation

$$\frac{1}{c} F(\tau, x) - \frac{1}{c} R(\tau, p)\,x = F(x, x', x'', x''', \tau),$$

we may write in place of (4.204) the following equation

$$[p^2+\omega^2(\tau)]^2 x=\varepsilon F(x, x', x'', x''', \tau). \tag{4.207}$$

We now solve equation (4.207). We define a function $y=y(t, \tau, \varrho)$, where $\tau=\varrho t$, $\varrho=\sqrt{\varepsilon}$*, such that the expression $x(t)=\{y(\tau, t, \varrho)\}$ $(\tau=\varrho t)$ satisfies equation (4.207).

Thus instead of equation (4.207) we consider, as above, the equation

$$[p_c^2+\omega^2(\tau)]^2 y=\varrho^2 F(y, y', y'', y''', \tau), \tag{4.208}$$

where

$$p_c=p_t+\varrho p_\tau. \tag{4.209}$$

The solution of the symbolic equation (4.208) will be sought for in the form of the asymptotic series

$$y=u_0(\xi, \eta, t)+\varrho^2 u_1(\xi, \eta, t)+\varrho^3 \ldots, \tag{4.210}$$

where ξ and η are to be determined from the following system of second order differential equations

$$\left.\begin{aligned}\frac{d^2\xi}{dt^2}&=\varrho^2 A_1(\tau, \xi, \eta, \dot{\xi}, \dot{\eta})+\varrho^3 A_2(\tau, \xi, \eta, \dot{\xi}, \dot{\eta})+\ldots,\\ \frac{d^2\eta}{dt^2}&=\varrho^2 B_1(\tau, \xi, \eta, \dot{\xi}, \dot{\eta})+\varrho^3 B_2(\tau, \xi, \eta, \dot{\xi}, \dot{\eta})+\ldots\end{aligned}\right\} \tag{4.211}$$

or from the system

$$\left.\begin{aligned}&\frac{d\xi}{dt}=\varrho\xi_1, \quad \frac{d\eta}{dt}=\varrho\eta_1,\\ &\frac{d\xi_1}{dt}=\varrho A_1(\tau, \xi, \eta, \xi_1, \eta_1)+\varrho^2 A_2(\tau, \xi, \eta, \xi_1, \eta_1)+\ldots,\\ &\frac{d\eta_1}{dt}=\varrho B_1(\tau, \xi, \eta, \xi_1, \eta_1)+\varrho^2 B_2(\tau, \xi, \eta, \xi_1, \eta_1)+\ldots\end{aligned}\right\} \tag{4.212}$$

In order to solve our problem we have to find the functions $u_1(\tau, \xi, \eta, t)$, $u_2(\tau, \xi, \eta, \xi_1, \eta_1, t), \ldots$ and the quantities $A_1(\tau, \xi, \eta, \xi_1, \eta_1)$, $A_2(\tau, \xi, \eta, \xi_1, \eta_1), \ldots, B_1(\tau, \xi, \eta, \xi_1, \eta_1)$, $B_2(\tau, \xi, \eta, \xi_1, \eta_1), \ldots$

As additional conditions we require, as usual, that the functions $u_1(\tau, \xi, \eta, t)$, $u_2(\tau, \xi, \eta, \xi_1, \eta_1, t), \ldots$ do not contain the first harmonic of the argument ωt. We may write, according to (4.209),

$$[p_c^2+\omega^2(\tau)]^2=[p_t^2+\omega^2(\tau)]^2+2[p_t^2+\omega^2(\tau)](2\varrho p_t p_\tau+\varrho^2 p_\tau^2)+4\varrho^2 p_t^2 p_\tau^2+4\varrho^3 p_t p_\tau^3+\varrho^4 p_\tau^4,$$

from which, using (4.210), we obtain for the left-hand side of equation (4.208)

$$\begin{aligned}[p_c^2+\omega^2(\tau)]^2 y=&[p_t^2+\omega^2(\tau)]^2 u_0+4\varrho p_t p_\tau[p_t^2+\omega^2(\tau)]u_0+\\ &+\varrho^2\{4p_t^2 p_\tau^2 u_0^2+2[p_t^2+\omega^2(\tau)]p_\tau^2 u_0^2+[p_t^2+\omega^2(\tau)]^2 u_1\}+\\ &+\varrho^3\{4p_t p_\tau^3 u_0+4[p_t^2+\omega^2(\tau)]p_t p_\tau u_1+4p_t^2 p_\tau^2 u_0^3+\\ &\qquad+2[p_t^2+\omega^2(\tau)]p_\tau^2 u_0^3+[p_t^2+\omega^2(\tau)]^2 u_2\}+\ldots\end{aligned} \tag{4.213}$$

We may further write, using (4.210) and (4.212),

$$\begin{aligned}y'&=\frac{\partial u_0}{\partial t}+\varrho\left[\frac{\partial u_0}{\partial\xi}\xi_1+\frac{\partial u_0}{\partial\eta}\eta_1\right]+\varrho^2\frac{\partial u_1}{\partial t}+\varrho^3\ldots,\\ y''&=\frac{\partial^2 u_0}{\partial t^2}+2\varrho\left[\frac{\partial^2 u_0}{\partial\xi\,\partial t}\xi_1+\frac{\partial^2 u_0}{\partial\eta\,\partial t}\eta_1\right]+\varrho^2\left[\frac{\partial^2 u_1}{\partial t^2}+\frac{\partial u_0}{\partial\xi}A_1+\frac{\partial u_0}{\partial\eta}B_1\right]+\varrho^3\ldots,\\ y'''&=\frac{\partial^3 u_0}{\partial t^3}+3\varrho\left[\frac{\partial^3 u_0}{\partial\xi\,\partial t^2}\xi_1+\frac{\partial^3 u_0}{\partial\eta\,\partial t^2}\eta_1\right]+\varrho^2\ldots\end{aligned}$$

* The introduction of expansions in fractional powers of a small parameter proves to be convenient in many cases, and was often used by several authors; cf., for example, /87, 109, 139, 141/.

The right-hand side of (4.208) may now be given in the form

$$\varrho^2 F(y, y', y'', y''', \tau) =$$
$$= \varrho^2 F(u_0, u_0', u_0'', u_0''', \tau) + \varrho^3 \Big\{ F'_{y'}(u_0, u_0', u_0'', u_0''', \tau) \left[\frac{\partial u_0}{\partial \xi} \xi_1 + \frac{\partial u_0}{\partial \eta} \eta_1 \right] +$$
$$+ 2F'_{y''}(u_0, u_0', u_0'', u_0''', \tau) \left[\frac{\partial^2 u_0}{\partial \xi\, \partial t} \xi_1 + \frac{\partial^2 u_0}{\partial \eta\, \partial t} \eta_1 \right] +$$
$$+ 3F'_{y'''}(u_0, u_0', u_0'', u_0''', \tau) \left[\frac{\partial^3 u_0}{\partial \xi\, \partial t^2} \xi_1 + \frac{\partial^3 u_0}{\partial \eta\, \partial t^2} \eta_1 \right] \Big\} + \varrho^4 \ldots$$

Denoting the coefficients of ϱ^2, ϱ^3, ... in the right-hand side of the last equality correspondingly by $F_0(\tau, \xi, \eta, t)$, $F_1(\tau, \xi, \eta, \xi_1, \eta_1, t)$, ... we obtain

$$\varrho^2 F(y, y', y'', y''', \tau) = \varrho^2 F_0(\tau, \xi, \eta, t) + \varrho^3 F_1(\tau, \xi, \eta, \xi_1, \eta_1, t) + \varrho^4 \ldots \tag{4.214}$$

Equating the coefficients of the same powers of ϱ in the right-hand sides of (4.213) and (4.214), we obtain the following system of equations

$$\left.\begin{aligned}
&[p_c^2 + \omega^2(\tau)]^2 u_0 = 0,\\
&2[p_t^2 + \omega^2(\tau)]\, 2p_t p_\tau u_0 = 0,\\
&[p_t^2 + \omega^2(\tau)]^2 u_1 = -4p_t^2 p_\tau^2 u_0 - 2[p_t^2 + \omega^2(\tau)]\, p_\tau^2 u_0 + F_0(\tau, \xi, \eta, t),\\
&[p_t^2 + \omega^2(\tau)]^2 u_2 = -4p_t^2 p_\tau^2 u_0 - 2[p_t^2 + \omega^2(\tau)]\, p_\tau^2 u_0 -\\
&\quad - 4p_t p_\tau^3 u_0 - 4[p_t^2 + \omega^2(\tau)]\, p_t p_\tau u_1 + F_1(\tau, \xi, \eta, \xi_1, \eta_1, t),\\
&\ldots\ldots\ldots\ldots\ldots\ldots\ldots\ldots
\end{aligned}\right\} \tag{4.215}$$

From the first of these equations (regarding τ as a parameter) we obtain

$$u_0 = \xi \cos \omega t + \eta \sin \omega t. \tag{4.216}$$

The second equation, and also the second terms on the right-hand sides of the third and fourth equation vanish identically by virtue of the first equation.

The third equation gives us an expression for $u_1(\tau, \xi, \eta, t)$:

$$u_1(\tau, \xi, \eta, t) = F_{00} + \sum_{\substack{n \\ (n \neq \pm 1)}} \frac{1}{(n^2 - 1)^2 \omega^4} \{F_{0n} \cos n\omega t + P_{0n} \sin n\omega t\}, \tag{4.217}$$

where

$$F_{00} = \frac{1}{\pi} \int_0^{2\pi} F_0(\tau, \xi, \eta, t)\, dt, \quad F_{0n} = \frac{1}{\pi} \int_0^{2\pi} F_0(\tau, \xi, \eta, t) \cos n\omega t\, dt,$$

$$P_{0n} = \frac{1}{\pi} \int_0^{2\pi} F_0(\tau, \xi, \eta, t) \sin n\omega t\, dt,$$

and also expressions for $A_1(\tau, \xi, \eta)$ and $B_1(\tau, \xi, \eta)$:

$$\left.\begin{aligned}
A_1(\tau, \xi, \eta) &= -\frac{1}{4\pi\omega^2} \int_0^{2\pi} F_0(\tau, \xi, \eta, t) \cos \omega t\, dt,\\
B_1(\tau, \xi, \eta) &= -\frac{1}{4\pi\omega^2} \int_0^{2\pi} F_0(\tau, \xi, \eta, t) \sin \omega t\, dt.
\end{aligned}\right\} \tag{4.218}$$

In order to determine $A_2(\tau, \xi, \eta, \xi_1, \eta_1)$ and $B_2(\tau, \xi, \eta, \xi_1, \eta_1)$ we use the fourth equation and the condition that $u_2(\tau, \xi, \eta, \xi_1, \eta_1, t)$ does not contain the first harmonic. We have

$$4\omega^2 A_2 \cos \omega t + 4\omega^2 B_2 \sin \omega t + 4\omega \dddot{\xi} \sin \omega t - 4\omega \dddot{\eta} \cos \omega t =$$
$$= -\frac{\cos \omega t}{\pi} \int_0^{2\pi} F_1(\tau, \xi, \eta, \xi_1, \eta_1, t) \cos \omega t\, dt -$$
$$- \frac{\sin \omega t}{\pi} \int_0^{2\pi} F_1(\tau, \xi, \eta, \xi_1, \eta_1, t) \sin \omega t\, dt.$$

Taking into account that

$$\frac{d^3\xi}{d\tau^3}=\frac{\partial A_1}{\partial \xi}\xi_1+\frac{\partial A_1}{\partial \eta}\eta_1+\varrho\ldots,\quad \frac{d^3\eta}{d\tau^3}=\frac{\partial B_1}{\partial \xi}\xi_1+\frac{\partial B_1}{\partial \eta}\eta_1+\varrho\ldots,$$

we obtain the following expressions for $A_2(\tau, \xi, \eta, \xi_1, \eta_1)$ and $B_2(\tau, \xi, \eta, \xi_1, \eta_1)$:

$$\left.\begin{aligned} &A_2(\tau, \xi, \eta, \xi_1, \eta_1)=\\ &=\frac{1}{\omega}\left[\frac{\partial B_1}{\partial \xi}\xi_1+\frac{\partial B_1}{\partial \eta}\eta_1\right]-\frac{1}{4\pi\omega^2}\int_0^{2\pi} F_1(\tau, \xi, \eta, \xi_1, \eta_1, t)\cos\omega t\, dt,\\ &B_2(\tau, \xi, \eta, \xi_1, \eta_1)=\\ &=-\frac{1}{\omega}\left[\frac{\partial A_1}{\partial \xi}\xi_1+\frac{\partial A_1}{\partial \eta}\eta_1\right]-\frac{1}{4\pi\omega^2}\int_0^{2\pi} F_1(\tau, \xi, \eta, \xi_1, \eta_1, t)\sin\omega t\, dt. \end{aligned}\right\} \quad (4.219)$$

Thus the solution of the fourth order equation (4.207) has been reduced to the solution of the system of two second order equations (4.211). These equations cannot be integrated in a closed form, and even their qualitative study involves a few difficulties. However, these equations are in some particular cases integrable in quadratures. In addition, equations (4.211) facilitate the investigation of stationary modes (τ = const).

We consider as an example the vibrating system (Figure 76) consisting of a nonlinear conductor N included in the circuit LRC which is inductively coupled to the circuit $L_1R_1C_1$ (we take for simplicity τ = const).

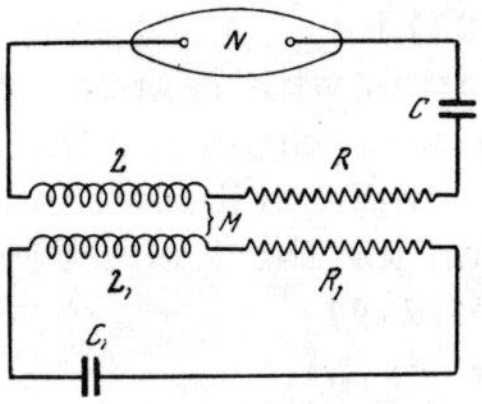

FIGURE 76

Let $L = L_1$, $R = R_1$, $C = C_1$; for the current in the first circuit we obtain the following differential equation:

$$\{(LCp^2+RCp+1)^2-C^2M^2p^4\}x=-\sqrt{\varepsilon}C\{LCp^3+RCp^2+p\}f(x), \quad (4.220)$$

where M is the coefficient of mutual induction, and $\sqrt{\varepsilon}\, f(x)$ is the nonlinear voltage-current characteristic of the conductor N.

We introduce the notation:

$\frac{1}{\sqrt{LC}}=\nu$, the resonant frequency of the circuit;

$\frac{M}{L}=q$, the coupling coefficient of the circuits;

$\frac{R}{2L}=\delta$, the damping decrement.

For the case of weak coupling and small damping, we may write

$$q=\varrho q_1, \quad \delta=\varrho\delta_1;$$

$\varrho=\sqrt{\varepsilon}$ is a certain small parameter.

Equation (4.220) assumes, using the above notation and after a few operations, the form

$$(p^2+\omega^2)x = -\varrho 4\delta_1 p(p^2+\omega^2)x - -\varrho^2\left\{p^2[4\delta_1^2+q_1^2\nu^2]x+\frac{p}{L}(p^2+\nu^2)f(x)+\frac{\delta_1 p^2}{L}f(x)\right\}+\varrho^3\ldots, \tag{4.221}$$

where

$$\omega^2 = \frac{\nu^2}{(1-q^2)^{\frac{1}{2}}} = \nu^2\left(1+\frac{p^2q_1^2}{2}+\ldots\right). \tag{4.222}$$

Equations (4.221) have the same structure as (4.207), and hence one can study nonstationary processes with the help of (4.218) and (4.219). We obtain

$$x = \xi\cos\omega t + \eta\sin\omega t, \tag{4.223}$$

in the first approximation, where ξ and η are determined from the following system of differential equations:

$$\left.\begin{aligned} \frac{d^2\xi}{dt^2} &= \varrho^2 S_1(\xi, \eta) - \frac{\varrho^2(4\delta_1^2+q_1^2\nu^2)}{4}\xi, \\ \frac{d^2\eta}{dt^2} &= \varrho^2 S_2(\xi, \eta) - \frac{\varrho^2(4\delta_1^2+q_1^2\nu^2)}{4}\eta, \end{aligned}\right\} \tag{4.224}$$

in which

$$\left.\begin{aligned} S_1(\xi, \eta) &= \frac{1}{4\pi\omega^2 L}\int_0^{2\pi}[p(p^2+\nu^2)+\delta_1 p^2]f(\xi\cos\omega t+\eta\sin\omega t)\cos\omega t\,dt, \\ S_2(\xi, \eta) &= \frac{1}{4\pi\omega^2 L}\int_0^{2\pi}[p(p^2+\nu^2)+\delta_1 p^2]f(\xi\cos\omega t+\eta\sin\omega t)\sin\omega t\,dt. \end{aligned}\right\} \tag{4.225}$$

Considering the system (4.224) for $f(x) = 0$, we note that the system of equations of the first approximation with respect to ϱ introduces into the required solution (4.223) only a correction for the frequency, and yields no correction characterizing the damping vibrations in the system.

Let us evaluate the correction for the frequency. When $f(x) = 0$ and $\delta = 0$, we have instead of the system (4.224)

$$\left.\begin{aligned} \frac{d^2\xi}{dt^2} + \frac{\varrho^2 q_1^2\nu^2}{4}\xi &= 0, \\ \frac{d^2\eta}{dt^2} + \frac{\varrho^2 q_1^2\nu^2}{4}\eta &= 0. \end{aligned}\right\} \tag{4.226}$$

Let $\xi = \xi_0$, $\xi' = \frac{q\nu}{2}\eta_0$, $\eta = \eta_0$, $\eta' = \frac{q\nu}{2}\xi_0$ at $t = 0$; x will then be of the form

$$x = \xi_0\cos\left(\omega - \frac{q\nu}{2}\right)t + \eta_0\sin\left(\omega + \frac{q\nu}{2}\right)t. \tag{4.227}$$

Thus the correction for the frequency introduced by equations (4.226) is, as should have been expected, equal to $\pm\frac{\varrho q_1\nu}{2}$, which fully agrees with the well-known rule that two circuits which are separately tuned to one and the same frequency ν being electrically connected, give not one but two resonance frequencies, one of which is higher and the other lower than ν.

We assume that the nonlinear characteristic is represented by a third degree polynomial

$$\varrho f(x) = S_0 x - \frac{4}{3}S_1 x^3. \tag{4.228}$$

Then the equations of the first approximation have the form

$$\left.\begin{aligned} \frac{d^2\xi}{dt^2} &= \frac{\varrho^2\delta_1}{4L}[-S_0+S_1(\xi^2+\eta^2)]\xi - \frac{\varrho^2(4\delta_1^2+q_1^2\nu^2)}{4}\xi, \\ \frac{d^2\eta}{dt^2} &= \frac{\varrho^2\delta_1}{4L}[-S_0+S_1(\xi^2+\eta^2)]\eta - \frac{\varrho^2(4\delta_1^2+q_1^2\nu^2)}{4}\eta. \end{aligned}\right\} \tag{4.229}$$

This system cannot be integrated in a closed form. However, very simple equations are obtained for a stationary mode

$$\left.\begin{aligned}\frac{\delta_1}{L}[-S_0+S_1(\xi^2+\eta^2)]\,\xi-(4\delta_1^2+q_1^2\nu^2)\,\xi=0,\\ \frac{\delta_1}{L}[-S_0+S_1(\xi^2+\eta^2)]\,\eta-(4\delta_1^2+q_1^2\nu^2)\,\eta=0,\end{aligned}\right\} \tag{4.230}$$

which yield one solution $\xi=0,\ \eta=0$ corresponding to the static state, and the other one determining the second power of the stationary amplitude

$$\xi^2+\eta^2=\frac{4\delta_1^2+q_1^2\nu^2}{\delta_1 S_1}+\frac{S_0}{S_1}. \tag{4.231}$$

The next approximation for a characteristic of the form (4.228) is

$$x=\xi\cos\omega t+\eta\sin\omega t+\varrho^2 u_1(\xi,\ \eta,\ t), \tag{4.232}$$

where

$$u_1(\xi,\ \eta,\ t)=\frac{\delta_1 S_1}{64\omega^2 L}[-(3\xi^2+\eta^2)\,\xi\cos 3\omega t+(\eta^2-3\xi^2)\,\eta\sin 3\omega t], \tag{4.233}$$

and ξ and η satisfy the system of equations

$$\left.\begin{aligned}\frac{d^2\xi}{dt^2}&=\varrho^2\left\{\frac{\delta_1}{4L}[-S_0+S_1(\xi^2+\eta^2)]\,\xi-\frac{1}{4}(4\delta_1^2+q_1^2\nu^2)\,\xi\right\}+\\ &+\varrho^3\left\{\frac{\delta_1}{4\omega L}\left[2S_1\xi\eta\frac{d\xi}{dt}+\right.\right.\\ &\left.\left.+[-S_0+S_1(\xi^2+\eta^2)-(4\delta_1^2+q_1^2\nu^2)+2S_1\eta^2]\frac{d\eta}{dt}\right]-\frac{\delta_1 q_1\nu^2}{2\omega}\eta\right\},\\ \frac{d^2\eta}{dt^2}&=\varrho^2\left\{\frac{\delta_1}{4L}[-S_0+S_1(\xi^2+\eta^2)]\eta-\frac{1}{4}(4\delta_1^2+q_1^2\nu^2)\,\eta\right\}-\\ &-\varrho^3\left\{\frac{\delta_1}{4\omega L}\left[2S_1\xi\eta\frac{d\eta}{dt}+\right.\right.\\ &\left.\left.+[-S_0+S_1(\xi^2+\eta^2)-(4\delta_1^2+q_1^2\nu^2)+2S_1\xi^2]\frac{d\xi}{dt}\right]+\frac{\delta_1 q_1^2}{2\omega}\xi\right\}.\end{aligned}\right\} \tag{4.234}$$

To determine the stationary values we must equate the first and second derivatives in the system (4.234) to zero— this leads to relatively simple equations.

The next approximation corresponding to the second approximation with respect to ε may be written without essential difficulties.

To investigate the stability of the stationary values of ξ and η (determined, for example, from the system (4.230)) we must set up the variational equations, proceeding from the system (4.229) or (4.234), and study the characteristic equation.

The characteristic equation in the given example should be constructed on the grounds of the second approximation, as the first approximation, in view of the existence of zero roots, does not resolve the problem of stability.

§ 8. The action of a perturbing force with several frequencies on a nonlinear vibrating system with many degrees of freedom

We consider a nonlinear vibrating system with N degrees of freedom, subjected to the action of an external perturbing force containing several harmonic components of different frequencies. Such vibrating systems are encountered in various fields of vibration engineering, e. g., in calculations of torsional vibrations of crankshafts, in investigations of the vibrations of complex systems of rotors in gas-turbine installations, etc.

The problem reduces to the consideration of a system of nonlinear differential equations of the type

$$\frac{d}{dt}\left\{\sum_{i=1}^{N} a_{ij}(\tau)\dot{q}_i\right\}+\sum_{i=1}^{N} b_{ij}(\tau)q_i=$$
$$=\varepsilon Q_j(\tau, \theta_1, \theta_2, \ldots, \theta_k, q_1, \ldots, q_N, \dot{q}_1, \ldots, \dot{q}_N, \varepsilon) \qquad (j=1, 2, \ldots, N), \tag{4.235}$$

where $\frac{d\theta_s}{dt}=\nu_s(\tau)$ $(s=1, 2, \ldots, k)$, and $Q_j(\tau, \theta_1, \ldots, \theta_k, q_1, \ldots, q_N, \dot{q}_1, \ldots, \dot{q}_N, \varepsilon)$ $(j=1, 2, \ldots, N)$ are periodic functions of $\theta_1, \theta_2, \ldots, \theta_k$ with period 2π. We assume that the functions $a_{ij}(\tau)$, $b_{ij}(\tau)$, $Q_j(\tau, \theta_1, \ldots, \theta_k, q_1, \ldots, q_N, \dot{q}_1, \ldots, \dot{q}_N, \varepsilon)$, $\nu_s(\tau)$ $(i, j=1, 2, \ldots, N;\ s=1, 2, \ldots, k)$ satisfy the conditions listed in § 1 of the present chapter, k is a not large integer (this is practically always satisfied), and the unperturbed system corresponding to equations (4.235) satisfies the three conditions given on p. 158.

The asymptotic expansions corresponding to the monofrequency and also to the polyfrequency modes may be obtained without difficulty with the help of the above method. We shall now consider the case of a nonlinear vibrating system with two degrees of freedom subjected to a perturbation containing three harmonic components of different frequencies. For this we examine the following system of differential equations, met in electroacoustics,

$$\left.\begin{aligned}
&m\ddot{q}_1+mh\ddot{q}_2+(c_{11}-c_1)q_1+c_{12}q_2=\\
&\qquad=-b_{11}\dot{q}_1-b_{12}\dot{q}_2+c_2q_1^3+\sum_{k=1}^{3} M_k\nu_k^2\cos\theta_k,\\
&mh\ddot{q}_1+m(h^2+\varrho^2)\ddot{q}_2+(c_{21}-c_1h)q_1+c_{22}q_2=\\
&\qquad=-b_{21}\dot{q}_1-b_{22}\dot{q}_2+hc_2q_1^3+h\sum_{k=1}^{3} M_k\nu_k^2\cos\theta_k\ ^*.
\end{aligned}\right\} \tag{4.236}$$

After a series of operations we obtain the solution in the first approximation in the form

$$\left.\begin{aligned}
q_1&=\varphi_1^{(1)}a_1\cos(\theta_1+\psi_1)+\varphi_1^{(2)}a_2\cos(\theta_2+\psi_2),\\
q_2&=\varphi_2^{(1)}a_1\cos(\theta_1+\psi_1)+\varphi_2^{(2)}a_2\cos(\theta_2+\psi_2),
\end{aligned}\right\} \tag{4.237}$$

where a_1, a_2, ψ_1, ψ_2 are to be determined from the system of equations

$$\left.\begin{aligned}
\frac{da_1}{dt}&=-\frac{\delta_1}{2}a_1-\frac{h_1}{2\omega_1}\sin\psi_1,\\
\frac{da_2}{dt}&=-\frac{\delta_2}{2}a_2-\frac{h_2}{2\omega_2}\sin\psi_2-\frac{h_3}{2\omega_2}\sin(\psi_2-\Delta\tau),\\
\frac{d\psi_1}{dt}&=\omega_1-\nu_1(\tau)-\frac{1}{2\omega_1}\left\{\frac{3}{4}\vartheta_1(\varphi_1^{(1)2}a_1^2+2\varphi_1^{(2)2}a_2^2)+\frac{h_1}{a_1}\cos\psi_1\right\},\\
\frac{d\psi_2}{dt}&=\omega_2-\nu_2(\tau)-\frac{1}{2\omega_2}\left\{\frac{3}{4}\vartheta_2(2\varphi_1^{(1)2}a_1^2+\varphi_1^{(2)2}a_2^2)+\right.\\
&\qquad\left.+\frac{h_2}{a_2}\cos\psi_2+\frac{h_3}{a_2}\cos(\psi_2-\Delta\tau)\right\}.
\end{aligned}\right\} \tag{4.238}$$

Here

$$\left.\begin{aligned}
&\varepsilon\Delta(\tau)=\nu_1(\tau)-\nu_2(\tau),\\
&\vartheta_i=c_2(\chi_1^{(i)}+h\chi_2^{(i)}) \qquad (i=1, 2),\\
&\delta_i=(b_{11}\varphi_1^{(i)}+b_{12}\varphi_2^{(i)})\chi_1^{(i)}+(b_{21}\varphi_1^{(i)}+b_{22}\varphi_2^{(i)})\chi_2^{(i)} \qquad (i=1, 2),\\
&h_1=M_1\nu_1^2(\chi_1^{(1)}+h\chi_2^{(1)}),
\end{aligned}\right\} \tag{4.239}$$

* A detailed investigation of this system is found in the article by V. P. Rubakin /147/.

$$\left.\begin{aligned} h_2 &= M_2\nu_2^2(\chi_1^{(2)} + h\chi_2^{(2)}), \\ h_3 &= M_3\nu_3^2(\chi_1^{(2)} + h\chi_2^{(2)}), \end{aligned}\right\} \tag{4.239}$$

and $\nu_k(\tau) = \frac{d\theta_k}{dt}$, where $\nu_2(\tau) = 6\nu_1(\tau)$, $\nu_3(\tau) = 7\nu_1(\tau)$, $m, h, c_{11}, c_1, c_{21} = c_{12}, b_{11}, b_{12} = b_{21}$, b_{22} are positive and satisfy the inequalities

$$c_{11} - c_1 > 0, \quad \begin{vmatrix} c_{11} - c_1 & c_{12} \\ c_{21} - c_1 h & c_{22} \end{vmatrix} > 0, \quad \begin{vmatrix} b_{11} & b_{12} \\ b_{21} & b_{22} \end{vmatrix} > 0. \tag{4.240}$$

Assuming that the right-hand sides of equations (4.236) are proportional to the small parameter ε, and introducing "quasi-normal" coordinates defined by the formulas

$$\left.\begin{aligned} q_1 &= \varphi_1^{(1)} x_1 + \varphi_1^{(2)} x_2, \\ q_2 &= \varphi_2^{(1)} x_1 + \varphi_2^{(2)} x_2, \end{aligned}\right\} \tag{4.241}$$

we obtain a system of the form

$$\left.\begin{aligned} \ddot{x}_1 + \omega_1^2 x_1 &= \varepsilon f_1(x_1, x_2, \dot{x}_1, \dot{x}_2) + \varepsilon \sum_{k=1}^{3} E_{1k}\cos\theta_k, \\ \ddot{x}_2 + \omega_2^2 x_2 &= \varepsilon f_2(x_1, x_2, \dot{x}_1, \dot{x}_2) + \varepsilon \sum_{k=1}^{3} E_{2k}\cos\theta_l. \end{aligned}\right\} \tag{4.242}$$

We assume that the following conditions are satisfied in the considered interval $0 \leqslant \tau \leqslant L$

$$\left.\begin{aligned} \nu_1(\tau) &\approx \omega_1, \\ \nu_2(\tau) &\approx \omega_2, \\ \nu_3(\tau) &\approx \omega_2, \end{aligned}\right\} \tag{4.243}$$

i. e., that the frequency $\nu_1(\tau)$ resonates with the first eigenfrequency, and the frequencies $\nu_2(\tau)$ and $\nu_3(\tau)$ — with the second. $\chi_1^{(i)}$, $\chi_2^{(i)}$ $(i = 1, 2)$ are the conjugate eigenfunctions.

It is possible to examine the stationary mode as well as various cases of transition through resonance with the help of the obtained equations of the first approximation /147/.

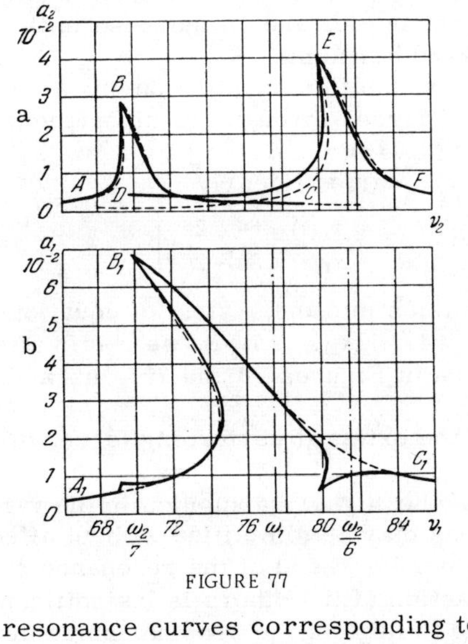

FIGURE 77

Figure 77 shows resonance curves corresponding to a stationary mode of vibrations characterized by the following values of parameters of the system (4.236):

$$\left.\begin{array}{c} m=1, \quad h=10, \quad \varrho=5, \quad c_{11}=5.1\cdot10^4, \\ c_{12}=c_{21}=4.5\cdot10^4, \quad c_{22}=75\cdot10^4, \quad c_1=3\cdot10^3, \\ c_2=12\cdot10^6, \quad b_{11}=4, \quad b_{12}=b_{21}=15, \quad b_{22}=150, \\ M_1=0.05, \quad M_2=0.0006, \quad M_3=0.0004. \end{array}\right\} \tag{4.244}$$

The continuous curve *ABC* (Figure 77a) corresponds to stationary amplitudes a_2 occurring because of the combined action of the first and third components of the external force. The line *DEF* corresponds to stationary amplitudes caused by the combined action of the first and second harmonic components. The dotted lines describe the stationary resonance curves corresponding to the separate action of these components.

The continuous line $A_1B_1C_1$ (Figure 77b) corresponds to the amplitudes a_1, caused by the combined action of all the three harmonic components of the perturbing force. The dotted curve corresponds to the action of only the first component of the external force.

Using these resonance curves we can draw a number of conclusions concerning the character of the vibrations caused by the combined action of several harmonic components of the perturbing force.

Thus, when large resonance amplitudes corresponding to the action of the 2-nd harmonic component build up, the vibration amplitudes caused by the 3-rd harmonic component become so small that they may be neglected. Conversely, when large resonance amplitudes corresponding to the action of the 3-rd harmonic component build up, the vibrations due to the 2-nd harmonic component may be neglected.

Moreover, the development of large resonance amplitudes corresponding to the action of the 2-nd harmonic causes a decrease in the vibration amplitudes caused by the 1-st harmonic component of the external force. The development of large resonance vibration amplitudes corresponding to the 1-st harmonic provokes a decrease in the vibration amplitudes caused by the action of the 3-rd harmonic, and an increase in the vibration amplitudes corresponding to the 2-nd harmonic.

We now examine the nonstationary state during a transition of the frequencies of the external forces through the eigenfrequencies of a system described by equations (4.236).

It is assumed that $\nu_1(\tau)$ varies linearly

$$\left.\begin{array}{l} \nu_1(\tau)=66+2t, \\ \nu_1(\tau)=85-2t. \end{array}\right\} \tag{4.245}$$

Substituting these values into the system of equations (4.238) and taking into account that $\nu_2(\tau)=6\nu_1(\tau)$, $\nu_3(\tau)=7\nu_1(\tau)$, we obtain after numerical integration the graphs given in Figures 78 and 79, characterizing the nonstationary vibrations.

The components of the external perturbing force will from now on be simply called harmonics.

By neglecting one of the higher harmonics in the various stages of integration, the calculations may be simplified without affecting the accuracy too much, since in the neighborhood of the resonance corresponding to one of the harmonics the action of the others is insignificant. Thus, for example, it is possible to neglect the 2-nd harmonic during the integration of the system (4.238) along the intervals *AB* (Figure 78a) and A_1B_1 (Figure 78b).

On the other hand, the 3-rd harmonic should be neglected in the intervals BC (Figure 79a) and B_1C_1 (Figure 79b).

Analyzing the graphs obtained (Figures 78 and 79), we arrive at the following conclusions characteristic of the vibrations of the system under investigation during a nonstationary process.

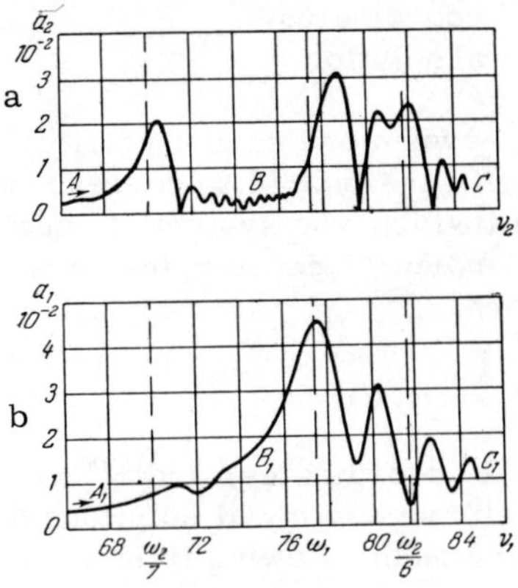

FIGURE 78

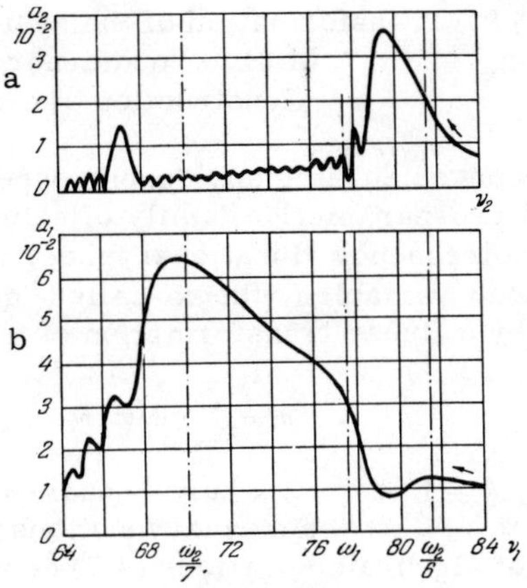

FIGURE 79

When the vibrating system crosses the resonance in the direction of increasing frequencies of the perturbing forces, there appear first the resonance vibrations caused by the action of the 3-rd harmonic. The development of their amplitudes provokes an increase in the amplitudes of vibrations excited by the 1-st harmonic. Then, as the vibrations caused by the action of the 3-rd harmonic become small and behave like beats, vibrations caused by the action of the first harmonic begin to develop. The vibrations excited by the 3-rd harmonic become even smaller, favoring the growth of vibrations due to the action of the 2-nd harmonic. As the vibrations excited by the 1-st harmonic decrease, there again occurs a resonance due to the 2-nd harmonic, and thereafter the vibrations show damping.

The double resonance due to the action of the 2-nd harmonic may be explained as follows: since the eigenfrequencies of a nonlinear system depend on the vibration amplitude, the growth of large resonance amplitudes, because of the action of the 1-st harmonic, causes the eigenfrequency to decrease (by virtue of the soft characteristic of the nonlinear elastic force), and hence the 2-nd harmonic enters into the resonance zone earlier. When the vibrations excited by the 1-st harmonic decrease, the eigenfrequencies increase and hence the 2-nd harmonic resonates once more.

If the resonance is passed in the direction of decreasing frequencies of the perturbing forces (Figure 79), the 2-nd harmonic is the first to resonate (Figure 79a) provoking the decrease of the vibration amplitudes of the 1-st harmonic (Figure 79b). Resonance vibrations due to the 1-st harmonic develop after the 2-nd harmonic has passed through resonance. These in turn suppress the growth of vibrations of the 3-rd harmonic, which begins to develop only when the vibration amplitudes of the 1-st harmonic decrease considerably. However, since the resonance zone of the 3-rd harmonic is essentially already crossed, its vibrations will not grow sufficiently and assume the nature of beats.

In the next chapter we consider nonstationary vibrations of coaxial rotors. We will need to examine the vibrations for double-frequency excitation of the k-th resonating eigenfrequency.

§ 9. Transformation of the system of differential equations (4.3) to "quasi-normal" coordinates. Construction of a general solution

It is convenient in many cases, especially when one cannot restrict oneself to a two-parametric family of solutions of the type (4.9) corresponding to a monofrequency vibrational mode, to transform the system of equations (4.3) to new variables, the so-called "quasi-normal" coordinates, which are defined by a linear transformation of the form

$$q_j = \sum_{k=1}^{N} \varphi_j^{(k)}(\tau) x_k \qquad (j = 1, 2, \ldots, N), \tag{4.246}$$

where x_k $(k = 1, 2, \ldots, N)$ are "quasi-normal" coordinates, and $\varphi_j^{(k)}(\tau)$ $(j, k = 1, 2, \ldots, N)$ are normal functions which are nontrivial solutions of the system of algebraic equations (4.7) depending on the slowing time $\tau = \varepsilon t$, and satisfying conditions (4.8).

Substituting (4.246) into equations (4.3), we obtain after a number of simple manipulations the following system of equations with respect to the new variables x_k $(k = 1, 2, \ldots, N)$:

$$\frac{d^2 x_k}{dt^2} + \omega_k^2(\tau) x_k = \\ = \frac{\varepsilon}{m_k(\tau)} X_k(\tau, \theta, x_1, \ldots, x_N, \dot{x}_1, \ldots, \dot{x}_N, \varepsilon) \qquad (k = 1, 2, \ldots, N), \tag{4.247}$$

where $\omega_k(\tau)$ $(k = 1, 2, \ldots, N)$ are "eigenfrequencies" determined from equation (4.6), and $m_k(\tau)$ $(k = 1, 2, \ldots, N)$ are determined according to (4.36).

The expressions for the kinetic and potential energies (4.1) in the new variables x_k $(k = 1, 2, \ldots, N)$ assume, to an accuracy of first order terms, the form

$$T(\dot{x}) = \frac{1}{2} \sum_{k=1}^{N} m_k(\tau) \dot{x}_k^2, \quad V(x) = \frac{1}{2} \sum_{k=1}^{N} m_k(\tau) \omega_k^2(\tau) x_k^2. \tag{4.248}$$

The right-hand sides of the system of equations (4.247) satisfy the same conditions as the functions (4.2) characterizing the generalized forces acting on the vibrating system under consideration.

The m-th approximation corresponding to a monofrequency vibrational mode in terms of the variables x_k $(k = 1, 2, \ldots, N)$ has, as is easily seen, the form

$$x_k^{(m-1)} = \frac{1}{m_k(\tau)} \sum_{i, j=1}^{N} a_{ij}(\tau) \varphi_i^{(k)}(\tau) [\varphi_j^{(1)}(\tau) a \cos(p\varphi + \psi) + \\ + \varepsilon u_j^{(1)}(\tau, a, \theta, p\varphi + \psi) + \ldots + \varepsilon^{m-1} u_j^{(m-1)}(\tau, a, \theta, p\varphi + \psi)] \\ (k = 1, 2, \ldots, N), \tag{4.249}$$

where a and ψ are to be determined from the system of equations of the m-th approximation (4.13).

To construct the general solution, it is advisable to reduce the system of equations (4.247) to the standard form. To do this we introduce the new variables z_k $(k = \pm 1, \pm 2, \ldots, \pm N)$, being generally speaking complex conjugate $(z_k = z_{-k}^*)$, by means of the formulas

$$\left.\begin{aligned} x_k &= z_k e^{i\int_0^t \omega_k(\tau)dt} + z_{-k} e^{-i\int_0^t \omega_k(\tau)dt}, \\ \dot{x}_k &= i\omega_k(\tau)\, z_k e^{i\int_0^t \omega_k(\tau)dt} - i\omega_k(\tau)\, z_{-k} e^{-i\int_0^t \omega_k(\tau)dt} \\ & (k = 1, 2, \ldots, N). \end{aligned}\right\} \qquad (4.250)$$

Substituting (4.250) into equations (4.247), assuming herewith for the simplicity of notation that

$$-\omega_{-k} = \omega_k, \quad X_{-k} = X_k \qquad (k = 1, 2, \ldots, N), \qquad (4.251)$$

we obtain, after a series of ordinary transformations, the following system in the standard form

$$\frac{dz_s}{dt} = \varepsilon Z_s(\tau, \theta, z_{-N}, \ldots, z_{-1}, z_1, \ldots, z_N, \varepsilon) \qquad (s = \pm 1, \pm 2, \ldots, \pm N), \quad (4.252)$$

instead of (4.247). The right-hand sides of (4.252) satisfy the same conditions as the right-hand sides of (4.247).

Approximate solutions of the system (4.252) may be derived without difficulty by means of the method of averaging given in the second and third chapters. We shall not discuss this problem here, since in the following chapter the question of constructing the general solution will be considered in detail for systems of equations of a more general type.

If the original system of equations (4.3) depends not on τ, but on k slowly varying parameters y_s $(s = 1, 2, \ldots, k)$, defined by equations of the type

$$\frac{dy_s}{dt} = \varepsilon f_s(y_1, \ldots, y_k, \theta, q_1, \ldots, q_N, \dot{q}_1, \ldots, \dot{q}_N, \varepsilon) \qquad (s = 1, 2, \ldots, k), \quad (4.253)$$

where $\frac{d\theta}{dt} = \nu(y_1, \ldots, y_k)$, and the functions $f_s(y_1, \ldots, y_s, \theta, q_1, \ldots, q_N, \dot{q}_1, \ldots, \dot{q}_N, \varepsilon)$ $(s = 1, 2, \ldots, k)$ are bounded, periodic in θ with period 2π, and differentiable to the desired order, then the problem of constructing approximate asymptotic solutions may be solved in exactly the same way as was done in Chapter III — either by introducing new variables which transfrom the whole system of equations to the standard form, or (for example in the nonresonant case) by applying the method of successive substitutions of variables given in Chapter III, § 10. This problem will be considered again later in greater detail.

Chapter V

NONLINEAR VIBRATING SYSTEMS WITH GYROSCOPIC TERMS

§ 1. Construction of asymptotic solutions describing monofrequency vibrations in nonlinear gyroscopic systems

Nonlinear differential equations with gyroscopic terms and slowly varying coefficients are known to play an important role in the study of nonstationary vibrational processes in various systems of gyroscopes, turbine rotors, centrifuges, etc. It is thus worthwhile discussing the equations involved quite thoroughly.

Let us first of all consider a vibrating system of N degrees of freedom, the Lagrangian of which may be represented in the form

$$\mathscr{L}=\frac{1}{2}\Big\{\sum_{i,j=1}^{N} a_{ij}(\tau)\dot{q}_i\dot{q}_j+2\sum_{i,j=1}^{N} g_{ij}(\tau)q_i\dot{q}_j-\sum_{i,j=1}^{N} b_{ij}(\tau)q_iq_j\Big\}, \tag{5.1}$$

where, as in the preceding chapter, $q_1, q_2, \ldots, q_N$ are generalized coordinates; $\tau=\varepsilon t$; t is the time; ε is a small positive parameter; and the functions $a_{ij}(\tau)=a_{ji}(\tau)$, $b_{ij}(\tau)=b_{ji}(\tau)$ and $g_{ij}(\tau)$ $(i, j=1, 2, \ldots, N)$ possess the desired number of derivates for all finite values of τ.

We assume that the complete energy of the considered vibrating system

$$H=\sum_{i=1}^{N}\dot{q}_i\frac{\partial\mathscr{L}}{\partial\dot{q}_i}-\mathscr{L} \tag{5.2}$$

is a positive definite quadratic form on any finite interval $0\leqslant\tau\leqslant L$, and that the system is subject to the action of a small perturbation determined by the generalized forces

$$\varepsilon Q_i(\tau, \theta, q_1, \ldots, q_N, \dot{q}_1, \ldots, \dot{q}_N, \varepsilon)=\varepsilon Q_i^{(1)}(\tau, \theta, q_1, \ldots, q_N, \dot{q}_1, \ldots, \dot{q}_N)+$$
$$+\varepsilon^2 Q_i^{(2)}(\tau, \theta, q_1, \ldots, q_N, \dot{q}_1, \ldots, \dot{q}_N)+\ldots \qquad (i=1, 2, \ldots, N), \tag{5.3}$$

periodic in θ with period 2π. We suppose that $\frac{d\theta}{dt}=\nu(\tau)\geqslant 0$, and that the functions $\nu(\tau)$, $Q_j(\tau, \theta, q_1, \ldots, q_N, \dot{q}_1, \ldots, \dot{q}_N, \varepsilon)$ $(j=1, 2, \ldots, N)$ satisfy the conditions listed in the first section of the foregoing chapter for the functions (4.2) (p. 158).

Inserting the values of $\mathscr{L}$ and $\varepsilon Q_i(\tau, \theta, q_1, \ldots, q_N, \dot{q}_1, \ldots, \dot{q}_N, \varepsilon)$ into the Lagrange equations, we obtain a system of nonlinear differential equations with slowly varying coefficients

$$\frac{d}{dt}\left\{\sum_{j=1}^{N} a_{ij}(\tau)\dot{q}_j\right\}+\sum_{j=1}^{N} c_{ij}(\tau)\dot{q}_j+\sum_{j=1}^{N} b_{ij}(\tau)q_j=$$

$$=\varepsilon\left[Q_i(\tau, \theta, q_1, \ldots, q_N, \dot{q}_1, \ldots, \dot{q}_N, \varepsilon)-\sum_{j=1}^{N}\frac{dg_{ij}}{d\tau}q_j\right] \quad (i=1, 2, \ldots, N), \quad (5.4)$$

in which $c_{ij}(\tau)=g_{ij}(\tau)-g_{ji}(\tau)$ and, consequently, $c_{ij}(\tau) = -c_{ji}(\tau)$ $(i, j=1, 2, \ldots, N)$.

Because of the structure of the Lagrangian (5.1), the system (5.4) contains, together with the slowly varying inertial and quasi-elastic coefficients $a_{ij}(\tau)$ and $b_{ij}(\tau)$, gyroscopic terms with the slowly varying coefficients $c_{ij}(\tau)$.

We consider simultaneously with this system the following system of linear second order differential equations whose constant coefficients depend on τ as a parameter,

$$\sum_{j=1}^{N} a_{ij}(\tau)\ddot{q}_j+\sum_{j=1}^{N} c_{ij}(\tau)\dot{q}_j+\sum_{j=1}^{N} b_{ij}(\tau)q_j=0 \quad (i=1, 2, \ldots, N). \quad (5.5)$$

Equations (5.5) are obtained, as shown before, by putting $\varepsilon = 0$ in (5.4) and regarding τ as a constant parameter.

Particular solutions of the auxiliary system of equations (5.5), each of which depends on two arbitrary constants, may be derived by means of usual methods in the form

$$q_j^{(k)}=a_k\varphi_j^{(k)}e^{i(\omega_k t+\varphi_k)}+a_k\overset{*}{\varphi}{}_j^{(k)}e^{-i(\omega_k t+\varphi_k)} \quad (j, k=1, 2, \ldots, N), \quad (5.6)$$

where ω_k $(k=1, 2, \ldots, N)$ are eigenfrequencies determined by the equation

$$D(\omega)=D\|-a_{ij}\omega^2+ic_{ij}\omega+b_{ij}\|=0, \quad (5.7)$$

$\varphi_j^{(k)}$ $(j, k=1, 2, \ldots, N)$ are basic functions being nontrivial solutions of the system of homogeneous algebraic equations

$$\sum_{j=1}^{N}\{-a_{ij}\omega_k^2+ic_{ij}\omega_k+b_{ij}\}\varphi_j^{(k)}=0 \quad (i, k=1, 2, \ldots, N), \quad (5.8)$$

and (because the quantities $q_j^{(k)}$ $(j, k=1, 2, \ldots, N)$ are real) the functions $\overset{*}{\varphi}{}_j^{(k)}$ $(j, k=1, 2, \ldots, N)$ are conjugate to $\varphi_j^{(k)}$ $(j, k=1, 2, \ldots, N)$ and satisfy the system of algebraic equations

$$\sum_{j=1}^{N}\{-a_{ij}\omega_k^2-ic_{ij}\omega_k+b_{ij}\}\overset{*}{\varphi}{}_j^{(k)}=0 \quad (i, k=1, 2, \ldots, N). \quad (5.9)$$

Since the coefficients a_{ij}, b_{ij}, c_{ij} in the system of equations (5.5) depend, by assumption, on the parameter τ, the quantities ω_k, $\varphi_j^{(k)}$, $\overset{*}{\varphi}{}_j^{(k)}$ $(j, k = 1, 2, \ldots, N)$ determined by equations (5.7), (5.8), and (5.9) also depend on τ as a parameter. Putting $\tau = \varepsilon t$ in equations (5.5) and in their solutions, the functions (5.6) will, as known, satisfy equations (5.5) only approximately (up to terms of the order of ε), representing vibrations with a slowly varying frequency and form.

We first construct, as in Chapter IV, § 1, asymptotic approximate formulas for a particular solution of the system (5.4) corresponding to monofrequency vibrations close (for sufficiently small ε) to a vibration at the frequency ω_1. Constructing such a solution, we assume that the conditions formulated in § 1 of the preceding chapter (p. 158) are satisfied for all values of the parameter τ $(0 \leqslant \tau \leqslant L)$.

With these assumptions we look for a solution of the perturbed equations (5.4) in the form of asymptotic series

$$q_j = u_j^{(0)}(\tau, a, p\varphi+\psi) + \varepsilon u_j^{(1)}(\tau, a, \theta, p\varphi+\psi) + \\ + \varepsilon^2 u_j^{(2)}(\tau, a, \theta, p\varphi+\psi) + \varepsilon^3 \ldots \qquad (j=1, 2, \ldots, N), \qquad (5.10)$$

in which $\tau=\varepsilon t$, $u_j^{(0)}(\tau, a, p\varphi+\psi) = a\varphi_j^{(1)}(\tau)e^{i(p\varphi+\psi)} + a\overset{*}{\varphi}{}_j^{(1)}(\tau)e^{-i(p\varphi+\psi)}$, $u_j^{(l)}(\tau, a, \theta, p\varphi+\psi)$ $(j=1, 2, \ldots, N; l=1, 2, \ldots)$ are functions periodic in θ and $p\varphi+\psi$ with period 2π, and the quantities a and ψ are determined by the equations

$$\left.\begin{aligned} \frac{da}{dt} &= \varepsilon A_1(\tau, a, \psi) + \varepsilon^2 A_2(\tau, a, \psi) + \ldots, \\ \frac{d\psi}{dt} &= \omega_1(\tau) - \frac{p}{q}\nu(\tau) + \varepsilon B_1(\tau, a, \psi) + \varepsilon^2 B_2(\tau, a, \psi) + \ldots, \end{aligned}\right\} \qquad (5.11)$$

where $\varphi = \frac{\theta}{q}$, p and q are mutually prime small integers whose choice depends on the resonance near which we intend to study the behavior of the given system, and $\varphi_j^{(1)}(\tau)$ and $\overset{*}{\varphi}{}_j^{(1)}(\tau)$ $(j=1, 2, \ldots, N)$ are nontrivial solutions of the algebraic equations (5.8) and (5.9).

To construct the m-th approximation we must derive such expressions for the functions

$$\left.\begin{aligned} &u_j^{(1)}(\tau, a, \theta, p\varphi+\psi), \ldots, u_j^{(m-1)}(\tau, a, \theta, p\varphi+\psi) \qquad (j=1, 2, \ldots, N), \\ &A_1(\tau, a, \psi), \ldots, A_m(\tau, a, \psi),\ B_1(\tau, a, \psi), \ldots, B_m(\tau, a, \psi), \end{aligned}\right\} \qquad (5.12)$$

so that the series

$$q_j^{(m)} = u_j^{(0)}(\tau, a, p\varphi+\psi) + \varepsilon u_j^{(1)}(\tau, a, \theta, p\varphi+\psi) + \ldots \\ \ldots + \varepsilon^{m-1} u_j^{(m-1)}(\tau, a, \theta, p\varphi+\psi) \qquad (j=1, 2, \ldots, N), \qquad (5.13)$$

in which a and ψ are functions of time determined by the system

$$\left.\begin{aligned} \frac{da}{dt} &= \varepsilon A_1(\tau, a, \psi) + \ldots + \varepsilon^m A_m(\tau, a, \psi), \\ \frac{d\psi}{dt} &= \omega_1(\tau) - \frac{p}{q}\nu(\tau) + \varepsilon B_1(\tau, a, \psi) + \ldots + \varepsilon^m B_m(\tau, a, \psi), \end{aligned}\right\} \qquad (5.14)$$

satisfy the initial system of equations (5.4) up to terms of the order of ε^m.

Differentiating the expressions (5.10), and taking into account equations (5.11), (4.16), and (4.17), we find expressions for $\frac{dq_j}{dt}$ and $\frac{d^2 q_j}{dt^2}$ $(j=1, 2, \ldots, N)$.

Substituting the values of q_j from (5.10) and the expressions obtained for $\frac{dq_j}{dt}$ into the right-hand sides of equations (5.4), and expanding these right-hand sides in Taylor series, we obtain

$$\varepsilon\left[Q_i(\tau, \theta, q_1, \ldots, q_N, \dot q_1, \ldots, \dot q_N, \varepsilon) - \frac{1}{2}\sum_{j=1}^{N}\frac{dg_{ij}}{d\tau}q_j\right] =$$

$$= \varepsilon\left[Q_i^{(1)}(\tau, \theta, q_{10}, \ldots, q_{N0}, \dot q_{10}, \ldots, \dot q_{N0}) - \frac{1}{2}\sum_{j=1}^{N}\frac{dg_{ij}}{d\tau}q_{j0}\right] +$$

$$+ \varepsilon^2\left\{\sum_{j=1}^{N}\left[\left(\frac{\partial Q_i^{(1)}}{\partial q_j}\right)_{\substack{q_i=q_{i0}\\ \dot q_i=\dot q_{i0}}} u_j^{(1)} + \left(\frac{\partial Q_i^{(1)}}{\partial \dot q_j}\right)_{\substack{q_i=q_{i0}\\ \dot q_i=\dot q_{i0}}}\left(A_1\frac{\partial u_j^{(0)}}{\partial a} + B_1\frac{\partial u_j^{(0)}}{\partial(p\varphi+\psi)} + \right.\right.\right.$$

$$\left.\left. + \frac{\partial u_j^{(0)}}{\partial \tau} + \frac{\partial u_j^{(1)}}{\partial \theta}\nu(\tau) + \frac{\partial u_j^{(1)}}{\partial(p\varphi+\psi)}\omega_1(\tau)\right)\right] +$$

$$\left. + Q_i^{(2)}(\tau, \theta, q_{10}, \ldots, q_{N0}, \dot q_{10}, \ldots, \dot q_{N0}) - \frac{1}{2}\sum_{j=1}^{N}\frac{dg_{ij}}{d\tau}u_i^{(1)}\right\} +$$

$$+ \varepsilon^3\{\ldots\} + \ldots \qquad (i=1, 2, \ldots, N), \qquad (5.15)$$

where we have used the notation

$$q_{i0}=u_i^{(0)}(\tau,\ a,\ p\varphi+\psi),\ \dot{q}_{i0}=\frac{\partial u_i^{(0)}(\tau,\ a,\ p\varphi+\psi)}{\partial(p\varphi+\psi)}\ \omega_1(\tau) \qquad (i=1,\ 2,\ \ldots,\ N). \tag{5.16}$$

We now substitute the values of q_j from (5.10) and also the expressions derived for $\frac{dq_j}{dt}$ and $\frac{d^2q_j}{dt^2}$ into the left-hand sides of (5.4), and replace the right-hand sides by their values according to the expressions (5.15). Equating the coefficients of equal powers of ε, we obtain the equations

$$\sum_{j=1}^{N}\left[a_{ij}(\tau)\frac{d^2u_j^{(l)}}{dt^2}+c_{ij}(\tau)\frac{du_j^{(l)}}{dt}+b_{ij}(\tau)\,u_j^{(l)}\right]=\bar{G}_{i0}^{(l)}(\tau,\ a,\ \theta,\ p\varphi+\psi) \tag{5.17}$$

$$(i=1,\ 2,\ \ldots,\ N;\ l=1,\ 2,\ 3,\ \ldots).$$

Here

$$\frac{du_j^{(l)}}{dt}=\frac{\partial u_j^{(l)}}{\partial\theta}\,\nu(\tau)+\frac{\partial u_j^{(l)}}{\partial(p\varphi+\psi)}\,\omega_1(\tau), \tag{5.18}$$

$$\frac{d^2u_j^{(l)}}{dt^2}=\frac{\partial^2u_j^{(l)}}{\partial\theta^2}\,\nu^2(\tau)+2\frac{\partial^2u_j^{(l)}}{\partial\theta\,\partial(p\varphi+\psi)}\,\nu(\tau)\,\omega_1(\tau)+\frac{\partial^2u_j^{(l)}}{\partial(p\varphi+\psi)^2}\,\omega_1^2(\tau), \tag{5.19}$$

$$\begin{aligned}
&\bar{G}_{i0}^{(1)}(\tau,\ a,\ \theta,\ p\varphi+\psi)=\\
&=Q_{i0}^{(1)}(\tau,\ a,\ \theta,\ p\varphi+\psi)-\sum_{j=1}^{N}\left\{a_{ij}(\tau)\left[\frac{\partial A_1}{\partial\psi}\left[\omega_1(\tau)-\frac{p}{q}\nu(\tau)\right]\frac{\partial u_j^{(0)}}{\partial a}+\right.\right.\\
&\quad+2\omega_1(\tau)\,B_1\frac{\partial^2u_j^{(0)}}{\partial(p\varphi+\psi)^2}+\left[\omega_1(\tau)-\frac{p}{q}\nu(\tau)\right]\frac{\partial B_1}{\partial\psi}\frac{\partial u_j^{(0)}}{\partial(p\varphi+\psi)}+\\
&\quad\left.+2\omega_1(\tau)\,A_1\frac{\partial^2u_j^{(0)}}{\partial a\,\partial(p\varphi+\psi)}+2\frac{\partial^2u_j^{(0)}}{\partial\tau\,\partial(p\varphi+\psi)}\,\omega_1(\tau)+\frac{\partial u_j^{(0)}}{\partial(p\varphi+\psi)}\frac{d\omega_1(\tau)}{d\tau}\right]+\\
&\quad\left.+\dot{c}_{ij}(\tau)\left[A_1\frac{\partial u_j^{(0)}}{\partial a}+B_1\frac{\partial u_j^{(0)}}{\partial(p\varphi+\psi)}+\frac{\partial u_j^{(0)}}{\partial\tau}\right]+\frac{1}{2}\frac{dg_{ij}}{d\tau}u^{(0)}+\frac{da_{ij}(\tau)}{d\tau}\frac{\partial u_j^{(0)}}{\partial(p\varphi+\psi)}\,\omega_1(\tau)\right\},
\end{aligned} \tag{5.20}$$

$$\begin{aligned}
&\bar{G}_{i0}^{(2)}(\tau,\ a,\ \theta,\ p\varphi+\psi)=\\
&=\bar{\Phi}_{i0}^{(2)}(\tau,\ a,\ \theta,\ p\varphi+\psi)-\sum_{j=1}^{N}\left\{a_{ij}(\tau)\left[\left[\omega_1(\tau)-\frac{p}{q}\nu(\tau)\right]\frac{\partial A_2}{\partial\psi}\frac{\partial u_j^{(0)}}{\partial a}+\right.\right.\\
&\quad+2\omega_1(\tau)B_2\frac{\partial^2u_j^{(0)}}{\partial(p\varphi+\psi)^2}+\left[\omega_1(\tau)-\frac{p}{q}\nu(\tau)\right]\frac{\partial B_2}{\partial\psi}\frac{\partial u_j^{(0)}}{\partial(p\varphi+\psi)}+\\
&\quad+2\omega_1(\tau)\,A_2\frac{\partial^2u_j^{(0)}}{\partial a\,\partial(p\varphi+\psi)}+A_1\frac{\partial A_1}{\partial a}\frac{\partial u_j^{(0)}}{\partial a}+\\
&\quad+\frac{\partial A_1}{\partial\tau}\frac{\partial u_j^{(0)}}{\partial a}+\frac{\partial A_1}{\partial\psi}B_1\frac{\partial u_j^{(0)}}{\partial a}+\frac{\partial^2u_j^{(0)}}{\partial a^2}A_1^2+2\frac{\partial^2u_j^{(0)}}{\partial a\,\partial\tau}A_1+\\
&\quad+2\frac{\partial^2u_j^{(0)}}{\partial a\,\partial(p\varphi+\psi)}A_1B_1+2\frac{\partial^2u_j^{(0)}}{\partial\tau\,\partial(p\varphi+\psi)}B_1+\frac{\partial^2u_j^{(0)}}{\partial(p\varphi+\psi)^2}B_1^2+\\
&\quad\left.+\frac{\partial^2u_j^{(0)}}{\partial\tau^2}+\frac{\partial B_1}{\partial a}A_1\frac{\partial u_j^{(0)}}{\partial(p\varphi+\psi)}+\frac{\partial B_1}{\partial\psi}B_1\frac{\partial u_j^{(0)}}{\partial(p\varphi+\psi)}+\frac{\partial B_1}{\partial\tau}\frac{\partial u_j^{(0)}}{\partial(p\varphi+\psi)}\right]+\\
&\quad\left.+c_{ij}(\tau)\left[A_2\frac{\partial u_j^{(0)}}{\partial a}+B_2\frac{\partial u_j^{(0)}}{\partial(p\varphi+\psi)}\right]+\frac{da_{ij}(\tau)}{d\tau}\left[A_1\frac{\partial u_j^{(0)}}{\partial a}+B_1\frac{\partial u_j^{(0)}}{\partial(p\varphi+\psi)}+\frac{\partial u_j^{(0)}}{\partial\tau}\right]\right\},
\end{aligned} \tag{5.21}$$

where

$$\begin{aligned}
&Q_{i0}^{(1)}(\tau,\ a,\ \theta,\ p\varphi+\psi)=Q_i^{(1)}(\tau,\ \theta,\ q_{10},\ \ldots,\ q_{N0},\ \dot{q}_{10},\ \ldots,\ \dot{q}_{N0}),\\
&\bar{\Phi}_{i0}^{(2)}(\tau,\ a,\ \theta,\ p\varphi+\psi)=Q_i^{(2)}(\tau,\ \theta,\ q_{10},\ \ldots,\ q_{N0},\ \dot{q}_{10},\ \ldots,\ \dot{q}_{N0})+\\
&\quad+\sum_{j=1}^{N}\left[\left(\frac{\partial Q_i^{(1)}}{\partial q_j}\right)_{\substack{q_i=q_{i0}\\ \dot{q}_i=\dot{q}_{i0}}}u_j^{(1)}+\left(\frac{\partial Q_i}{\partial\dot{q}_i}\right)_{\substack{q_i=q_{i0}\\ \dot{q}_i=\dot{q}_{i0}}}\left[A_1\frac{\partial u_j^{(0)}}{\partial a}+B_1\frac{\partial u_j^{(0)}}{\partial(p\varphi+\psi)}+\right.\right.\\
&\quad\left.\left.+\frac{\partial u_j^{(0)}}{\partial\tau}+\frac{\partial u_j^{(0)}}{\partial\theta}\,\nu(\tau)+\frac{\partial u_j^{(1)}}{\partial(p\varphi+\psi)}\,\omega_1(\tau)\right]-\frac{1}{2}\frac{dg_{ji}}{d\tau}u_i^{(1)}\right]-
\end{aligned} \tag{5.22}$$

$$-\sum_{j=1}^{N}\left\{a_{ij}(\tau)\left[2\frac{\partial^2 u_j^{(1)}}{\partial\theta\,\partial\tau}\nu(\tau)+2\frac{\partial^2 u_j^{(1)}}{\partial(p\varphi+\psi)\,\partial\tau}\omega_1(\tau)+\right.\right.$$
$$+2\frac{\partial^2 u_j^{(1)}}{\partial\theta\,\partial(p\varphi+\psi)}\nu(\tau)B_1+2\frac{\partial^2 u_j^{(1)}}{\partial(p\varphi+\psi)^2}\omega_1(\tau)B_1+2\frac{\partial^2 u_j^{(1)}}{\partial\theta\partial a}\nu(\tau)A_1+$$
$$+2\frac{\partial^2 u_j^{(1)}}{\partial a\,\partial(p\varphi+\psi)}\omega_1(\tau)A_1+\frac{\partial u_j^{(1)}}{\partial\theta}\frac{d\nu}{d\tau}+\frac{\partial u_j^{(1)}}{\partial(p\varphi+\psi)}\frac{d\omega(\tau)}{d\tau}+$$
$$\left.+\frac{\partial u_j^{(1)}}{\partial(p\varphi+\psi)}\frac{\partial B_1}{\partial\psi}\left[\omega_1(\tau)-\frac{p}{q}\nu(\tau)\right]+\frac{\partial u_j^{(1)}}{\partial a}\frac{\partial A_1}{\partial\psi}\left[\omega_1(\tau)-\frac{p}{q}\nu(\tau)\right]\right]+$$
$$+c_{ij}(\tau)\left[\frac{\partial u_j^{(1)}}{\partial\tau}+\frac{\partial u_j^{(1)}}{\partial(p\varphi+\psi)}B_1+\frac{\partial u_j^{(1)}}{\partial a}A_1\right]+$$
$$\left.+\frac{da_{ij}(\tau)}{d\tau}\left[\frac{\partial u_j^{(1)}}{\partial\theta}\nu(\tau)+\frac{\partial u_j^{(1)}}{\partial(p\varphi+\psi)}\omega_1(\tau)\right]\right\}. \qquad (5.22)$$

Before determining the functions $u_j^{(1)}(\tau,\ a,\ \theta,\ p\varphi+\psi)\ (j=1,\ 2,\ldots,\ N)$, we note that, because of the restrictions imposed on the vibrating system under consideration (cf. p. 158), the determinant $D(\omega)$ of the system of algebraic equations

$$\sum_{j=1}^{N}[-a_{ij}\omega^2+ic_{ij}\omega+b_{ij}]\,y_j=0 \qquad (i=1,\ 2,\ldots,\ N) \qquad (5.23)$$

(a_{ij}, b_{ij}, c_{ij} are constants depending on the parameter τ) has two simple purely imaginary roots: $\omega=i\omega_1$ and $\omega=-i\omega_1$, and that $\omega=0$, $\omega=\pm in\omega_1$, where n is any integer distinct from unity, are not roots of $D(\omega)$, i. e.,

$$D(in\omega_1)\neq 0 \qquad (-\infty<n<\infty),\quad n\neq\pm 1. \qquad (5.24)$$

Also, the adjoint system of (5.23) is, by virtue of the symmetry properties of the coefficients, the system

$$\sum_{j=1}^{N}[-a_{ij}\omega^2+ic_{ji}\omega+b_{ij}]\,y_j=0 \qquad (i=1,\ 2,\ldots,\ N), \qquad (5.25)$$

which coincides with the previously derived system of equations (5,9).

To investigate the forced vibrations in the unperturbed system (5.4) caused by the external harmonic forces

$$E_i^{(n)}e^{in\omega_1 t} \qquad (i=1,\ 2,\ldots,\ N), \qquad (5.26)$$

we must find a particular solution of the system of inhomogeneous equations

$$\sum_{j=1}^{N}[a_{ij}\ddot{q}_j+c_{ij}\dot{q}_j+b_{ij}q_j]=E_i^{(n)}e^{in\omega_1 t} \qquad (i=1,\ 2,\ldots,\ N), \qquad (5.27)$$

which in the case $n\neq\pm 1$ is determined by the formulas

$$q_j=e^{in\omega_1 t}\sum_{i=1}^{N}Z_{ij}(in\omega_1)E_i^{(n)} \qquad (j=1,\ 2,\ldots,\ N), \qquad (5.28)$$

where

$$Z_{ij}(\omega)=\frac{D_{ij}(\omega)}{D(\omega)}, \qquad (5.29)$$

and $D_{ij}(\omega)$ is the corresponding minor of the determinant $D(\omega)$.

If $n=\pm 1$, then secular terms may appear in the solutions of equations (5.27), and a necessary condition for the existence of periodic solutions is

$$\left.\begin{aligned}&\sum_{i=1}^{N}E_i^{(1)}\varphi_i^*=0 \quad \text{for } n=1,\\ &\sum_{i=1}^{N}E_i^{(-1)}\varphi_i=0 \quad \text{for } n=-1\end{aligned}\right\} \qquad (5.30)$$

(we write φ_i^* and φ_i directly, because in our case the solutions of the adjoint system (5.9) are complex conjugates of the solutions of the system of algebraic equations (5.23)).

When these conditions are satisfied, the forced vibrations of a system described by equations (5.27) are in the case $n=1$ determined by the formulas

$$q_j = e^{i\omega_1 t} \sum_{i=1}^{N} S_{ij}(i\omega_1) E_i^{(1)} + C\varphi_j^{(1)} e^{i\omega_1 t} \qquad (j=1, 2, \ldots, N), \tag{5.31}$$

where

$$S_{ij}(i\omega_1) = \left\{ \frac{\dfrac{\partial D_{ij}(\omega)}{\partial \omega}}{\dfrac{\partial D(\omega)}{\partial \omega}} \right\}_{\omega = i\omega_1}. \tag{5.32}$$

We obtain analogously for $n=-1$

$$q_j = e^{-i\omega_1 t} \sum_{i=1}^{N} S_{ij}(-i\omega_1) E_i^{(-1)} + C^* \varphi_j^{(1)} e^{-i\omega_1 t} \qquad (j=1, 2, \ldots, N), \tag{5.33}$$

where

$$S_{ij}(-i\omega_1) = \left\{ \frac{\dfrac{\partial D_{ij}(\omega)}{\partial \omega}}{\dfrac{\partial D(\omega)}{\partial \omega}} \right\}_{\omega = -i\omega_1}, \tag{5.34}$$

and

$$S_{ij}(-i\omega_1) = S_{ij}^*(i\omega_1). \tag{5.35}$$

Here C and C^* are arbitrary constants which for the sake of simplicity will be assumed in what follows to be equal to zero.

If the amplitudes of the external harmonic forces satisfy the condition

$$E_i^{(-1)} = E_i^{*(1)} \qquad (i=1, 2, \ldots, N), \tag{5.36}$$

then one of the equalities (5.30) follows from the other.

We shall now derive the functions $u_j^{(1)}(\tau, a, \theta, p\varphi+\psi)$ $(j=1, 2, \ldots, N)$ determined by the system of equations (5.18) (for $l=1$). As the right-hand sides of these equations are periodic in θ and $p\varphi+\psi$ with period 2π, we represent them in the form of Fourier sums

$$\bar{G}_{i0}^{(1)}(\tau, a, \theta, p\varphi+\psi) = \sum_{n, m} g_{nm}^{(i1)}(\tau, a)\, e^{i\{n\theta + m(p\varphi+\psi)\}}, \tag{5.37}$$

where

$$g_{nm}^{(i1)}(\tau, a) = \frac{1}{4\pi^2} \int_0^{2\pi}\int_0^{2\pi} \bar{G}_{i0}^{(1)}(\tau, a, \theta, p\varphi+\psi)\, e^{-i\{n\theta+m(p\varphi+\psi)\}}\, d\theta\, d(p\varphi+\psi) \tag{5.38}$$
$$(i=1, 2, \ldots, N).$$

The required functions are also sought for in the form of double sums

$$u_j^{(1)}(\tau, a, \theta, p\varphi+\psi) = \sum_{n, m} k_{nm}^{(j1)}(\tau, a)\, e^{i\{n\theta+m(p\varphi+\psi)\}} \qquad (j=1, 2, \ldots, N), \tag{5.39}$$

where the coefficients $k_{nm}^{(j1)}(\tau, a)$ are still to be determined.

Substituting the values of $u_j^{(1)}(\tau, a, \theta, p\varphi+\psi)$ and $\bar{G}_i^{(1)}(\tau, a, \theta, p\varphi+\psi)$ $(i, j=1, 2, \ldots, N)$ into equations (5.18) and equating the coefficients of equal harmonics, we obtain the following system of algebraic equations for the unknown $k_{nm}^{(j1)}(\tau, a)$

$$\sum_{j=1}^{N} \{-a_{ij}(\tau)\,[m\omega_1(\tau) + n\nu(\tau)]^2 + ic_{ij}(\tau)\,[m\omega_1(\tau) + n\nu(\tau)] + b_{ij}(\tau)\}\, k_{nm}^{(j1)}(\tau, a) =$$
$$= g_{nm}^{(i1)}(\tau, a) \qquad (i=1, 2, \ldots, N). \tag{5.40}$$

It should be kept in mind during the derivation of these coefficients that the functions $u_j^{(1)}(\tau, a, \theta, p\varphi+\psi)$ have to be finite. For this to hold, it is necessary that

$$\left.\begin{aligned}\sum_{\substack{i=1\\(nq+p(m-1)=0)}}^{N} g_{nm}^{(i1)}(\tau, a)\,\varphi_i^{*(1)}(\tau) = 0,\\ \sum_{\substack{i=1\\(nq+p(m+1)=0)}}^{N} g_{nm}^{(i1)}(\tau, a)\,\varphi_i^{(1)}(\tau) = 0,\end{aligned}\right\} \tag{5.41}$$

where one of the equalities is implied by the other since the reality condition

$$g_{nm}^{(i1)}(\tau, a) = g_{nm}^{*(i1)}(\tau, a) \qquad (nq + p(m-1) = 0), \tag{5.42}$$

should be satisfied.

The conditions we imposed on the vibrating system under investigation (p. 158) imply that the following inequalities

$$m\omega_1(\tau) + n\nu(\tau) \neq \pm\,\omega_k(\tau) \qquad (k = 2, 3, \ldots, N), \tag{5.43}$$

are valid for any n, m, at all τ, and hence periodic terms may appear in the expressions for $u_j^{(1)}(\tau, a, \theta, p\varphi+\psi)$ only when n and m satisfy the relation $nq + p(m \pm 1) = 0$.

Thus according to (5.40), (5.41), and (5.42), we obtain expressions for $k_{nm}^{(j1)}(\tau, a)$ (conditions (5.43) being satisfied)

$$\begin{aligned} k_{nm}^{(j1)}(\tau,a) = &\sum_{\substack{i=1\\(nq+p(m\pm1)\neq 0)}}^{N} Z_{ij}\,[mi\omega_1(\tau) + ni\nu(\tau)]\, g_{nm}^{(i1)}(\tau,a) + \\ &+ \sum_{\substack{i=1\\(nq+p(m-1)=0)}}^{N} S_{ij}\,[i\omega_1(\tau)]\, g_{nm}^{(i1)}(\tau,a) + \sum_{\substack{i=1\\(nq+p(m+1)=0)}}^{N} S_{ij}^{*}\,[i\omega_1(\tau)]\, g_{nm}^{(i1)}(\tau,a) \\ &\qquad\qquad (j = 1, 2, \ldots, N), \end{aligned} \tag{5.44}$$

where

$$Z_{ij}\,[mi\omega_1(\tau) + ni\nu(\tau)] = \frac{D_{ij}\,[mi\omega_1(\tau) + ni\nu(\tau)]}{D\,[mi\omega_1(\tau) + ni\nu(\tau)]}\,. \tag{5.45}$$

Substituting the values of $k_{nm}^{(j1)}(\tau, a)$ into (5.39), we find

$$\begin{aligned} u_j^{(1)}(\tau, a, \theta, p\varphi+\psi) = &\sum_{\substack{n,m\\(nq+p(m\pm1)\neq 0)}} \Big\{\sum_{i=1}^{N} Z_{ij}\,[mi\omega_1(\tau) + ni\nu(\tau)]\, g_{nm}^{(i1)}(\tau,a)\Big\} \times \\ &\times e^{i\{n\theta + m(p\varphi+\psi)\}} + e^{i(p\varphi+\psi)} \sum_{\sigma} \Big\{\sum_{i=1}^{N} S_{ij}\,[i\omega_1(\tau)]\, g_{-\sigma p,\,\sigma q+1}^{(i1)}(\tau,a)\Big\}\, e^{i\sigma q\psi} + \\ &+ e^{-i(p\varphi+\psi)} \sum_{\sigma} \Big\{\sum_{i=1}^{N} S_{ij}^{*}\,[i\omega_1(\tau)]\, g_{-\sigma p,\,\sigma q-1}^{(i1)}(\tau, a)\Big\}\, e^{i\sigma q\psi} \quad (j = 1, 2, \ldots, N). \end{aligned} \tag{5.46}$$

We now derive the functions $A_1(\tau, a, \psi)$ and $B_1(\tau, a, \psi)$ from conditions (5.41). (As one of the conditions follows from the other it is sufficient to use only one.)

Putting in the first of them the values of $g_{nm}^{(i1)}(\tau, a)$ (where n and m satisfy the relation $nq + p(m-1) = 0$) and keeping in mind the notation of (5.20), (5.37), and also that

$$u_j^{(0)}(\tau, a, p\varphi+\psi) = \varphi_j^{(1)}(\tau)\, a e^{i(p\varphi+\psi)} + \varphi_j^{*(1)}(\tau)\, a e^{-i(p\varphi+\psi)} \qquad (j = 1, 2, \ldots, N), \tag{5.47}$$

we obtain after several operations the relation

$$\begin{aligned} \sum_{i,j=1}^{N} \Big\{ a_{ij}(\tau)\,\varphi_j^{(1)}(\tau)\,\varphi_i^{*(1)}(\tau) \Big[\Big[\omega_1(\tau) - \frac{p}{q}\nu(\tau)\Big] \frac{\partial A_1}{\partial\psi} - 2a\omega_1(\tau) B_1 + \\ + i\Big[\omega_1(\tau) - \frac{p}{q}\nu(\tau)\Big] a \frac{\partial B_1}{\partial\psi} + 2\omega_1(\tau)\, iA_1 \Big] + ia\Big[\frac{d\omega_1(\tau)}{d\tau} a_{ij}(\tau)\,\varphi_j^{(1)}(\tau)\,\varphi_i^{*(1)}(\tau) + \end{aligned}$$

$$+\omega_1(\tau)\frac{da_{ij}(\tau)}{d\tau}\varphi_j^{(1)}(\tau)\varphi_i^{*(1)}+2a_{ij}(\tau)\omega_1(\tau)\frac{d\varphi_j^{(1)}(\tau)}{d\tau}\varphi_i^{*(1)}(\tau)\Big]+$$

$$+c_{ij}(\tau)\varphi_j^{(1)}(\tau)\varphi_i^{*(1)}(\tau)[A_1+iaB_1]+ac_{ij}(\tau)\frac{d\varphi_j^{(1)}(\tau)}{d\tau}\varphi_i^{*(1)}(\tau)+$$

$$+\frac{1}{2}a\frac{dg_{ij}(\tau)}{d\tau}\varphi_j^{(1)}(\tau)\varphi_i^{*(1)}(\tau)\Big\}=$$

$$=\sum_\sigma\frac{e^{i\sigma q\psi}}{4\pi^2}\int_0^{2\pi}\int_0^{2\pi}\sum_{i=1}^{N}Q_{i0}^{(1)}(\tau,\ a,\ \theta,\ p\varphi+\psi)\ \varphi_i^{*(1)}(\tau)\,e^{-i\sigma q\psi}e^{-i(p\varphi+\psi)}\,d\theta\,d(p\varphi+\psi),\quad(5.48)$$

which should be satisfied by $A_1(\tau,\ a,\ \psi)$ and $B_1(\tau,\ a,\ \psi)$. To simplify this expression we use the notation

$$\varphi_j^{(1)}(\tau)=\bar{\varphi}_j^{(1)}(\tau)+i\bar{\bar{\varphi}}_j^{(1)}(\tau),$$

then

$$\varphi_i^{*(1)}(\tau)=\bar{\varphi}_i^{(1)}(\tau)-i\bar{\bar{\varphi}}_i^{(1)}(\tau)\qquad(i,\ j=1,\ 2,\ \ldots,\ N),\qquad(5.49)$$

where $\bar{\varphi}_j^{(1)}(\tau)$, $\bar{\bar{\varphi}}_j^{(1)}(\tau)$ are real quantities. We also use the notation

$$\sum_{i,\ j=1}^{N}a_{ij}(\tau)\varphi_j^{(1)}(\tau)\varphi_i^{*(1)}(\tau)=\sum_{i,\ j=1}^{N}a_{ij}(\tau)\varphi_j^{*(1)}(\tau)\varphi_i^{(1)}(\tau)=m_1(\tau),\qquad(5.50)$$

$$\sum_{i,\ j=1}^{N}c_{ij}(\tau)\varphi_j^{*(1)}(\tau)\varphi_i^{(1)}(\tau)=-\sum_{i,\ j=1}^{N}c_{ij}(\tau)\varphi_i^{*(1)}(\tau)\varphi_j^{(1)}(\tau)=im_2(\tau),\qquad(5.51)$$

$$\frac{1}{2}\sum_{i,\ j=1}^{N}\frac{dg_{ij}(\tau)}{d\tau}\varphi_j^{(1)}(\tau)\varphi_i^{*(1)}(\tau)=\bar{m}_3(\tau)+i\bar{\bar{m}}_3(\tau),\qquad(5.52)$$

$$\sum_{i,\ j=1}^{N}c_{ij}(\tau)\frac{d\varphi_j^{(1)}(\tau)}{d\tau}\varphi_i^{*(1)}(\tau)=\bar{m}_4(\tau)+i\bar{\bar{m}}_4(\tau),\qquad(5.53)$$

noting that $m_1(\tau)$ and $m_2(\tau)$ are real.

Then

$$\sum_{i,\ j=1}^{N}\Big[\frac{d\omega_1(\tau)}{d\tau}a_{ij}(\tau)\varphi_j^{(1)}(\tau)\varphi^{*(1)}(\tau)+\omega_1(\tau)\frac{da_{ij}(\tau)}{d\tau}\varphi_j^{(1)}(\tau)\varphi_i^{*(1)}(\tau)+$$

$$+2a_{ij}(\tau)\omega_1(\tau)\frac{d\varphi_j^{(1)}(\tau)}{d\tau}\varphi_i^{*(1)}(\tau)\Big]=\frac{d}{d\tau}[\omega_1(\tau)m_1(\tau)].\qquad(5.54)$$

The expression (5.48) may now be written in the form

$$\Big\{m_1(\tau)\Big[\omega_1(\tau)-\frac{p}{q}\nu(\tau)\Big]\frac{\partial A_1}{\partial\psi}-[2\omega_1(\tau)m_1(\tau)+m_2(\tau)]aB_1+$$

$$+a(\bar{m}_3(\tau)+\bar{m}_4(\tau))\Big\}+i\Big\{m_1(\tau)\Big[\omega_1(\tau)-\frac{p}{q}\nu(\tau)\Big]a\frac{\partial B_1}{\partial\psi}+[2\omega_1(\tau)m_1(\tau)+$$

$$+m_2(\tau)\Big]A_1+a\frac{d}{d\tau}[\omega_1(\tau)m_1(\tau)]+a[\bar{\bar{m}}_3(\tau)+\bar{\bar{m}}_4(\tau)]\Big\}=$$

$$=\sum_\sigma\frac{e^{i\sigma q\psi}}{4\pi^2}\int_0^{2\pi}\int_0^{2\pi}\sum_{i=1}^{N}Q_{i0}(\tau,\ a,\ \theta,\ p\varphi+\psi)\varphi_i^{*(1)}(\tau)\,e^{-i\sigma q\psi}e^{-i(p\varphi+\psi)}\,d\theta\,d(p\varphi+\psi)\qquad(5.55)$$

or, separating the real and imaginary parts,

$$m_1(\tau)\Big[\omega_1(\tau)-\frac{p}{q}\nu(\tau)\Big]\frac{\partial A_1}{\partial\psi}-[2\omega_1(\tau)m_1(\tau)+m_2(\tau)]aB_1=$$

$$=-a[\bar{m}_3(\tau)+\bar{m}_4(\tau)]+\sum_\sigma\frac{e^{i\sigma q\psi}}{4\pi^2}\int_0^{2\pi}\int_0^{2\pi}\sum_{i=1}^{N}Q_{i0}(\tau,\ a,\ \theta,\ p\varphi+\psi)\times$$

$$\times e^{-i\sigma q\psi}[\bar{\varphi}_i^{(1)}(\tau)\cos(p\varphi+\psi)-\bar{\bar{\varphi}}_i^{(1)}(\tau)\sin(p\varphi+\psi)]\,d\theta\,d(p\varphi+\psi),\qquad(5.56)$$

$$m_1(\tau)\left[\omega_1(\tau)-\frac{p}{q}\nu(\tau)\right]a\frac{\partial B_1}{\partial\psi}+[2\omega_1(\tau)m_1(\tau)+m_2(\tau)]A_1=$$
$$=-a\frac{d}{d\tau}[\omega_1(\tau)m_1(\tau)]-a[\overset{=}{m}_3(\tau)+\overset{=}{m}_4(\tau)]-$$
$$-\sum_\sigma\frac{e^{i\sigma q\psi}}{4\pi^2}\int_0^{2\pi}\int_0^{2\pi}\sum_{i=1}^{N}Q_{i0}(\tau,\ a,\ \theta,\ p\varphi+\psi)\,e^{-i\sigma q\psi}\times$$
$$\times[\bar{\varphi}_i^{(1)}(\tau)\sin(p\varphi+\psi)+\overset{=}{\varphi}_i^{(1)}(\tau)\cos(p\varphi+\psi)]\,d\theta\,d(p\varphi+\psi). \tag{5.56}$$

The derived system of two equations determines the functions $A_1(\tau, a, \psi)$ and $B_1(\tau, a, \psi)$ completely. This is shown by demonstrating that the coefficients of $A_1(\tau, a, \psi)$ and $B_1(\tau, a, \psi)$ cannot vanish for any τ in the interval $0\leqslant\tau\leqslant L$.

Putting $\omega=i\omega_1(\tau)$ in the identities

$$D(\omega)=\sum_{j=1}^{N}\{b_{ij}-a_{ij}\omega^2+ic_{ij}\omega\}D_{ij}(\omega), \tag{5.57}$$

$$D(\omega)=\sum_{i=1}^{N}\{b_{ij}-a_{ij}\omega^2+ic_{ji}\omega\}D_{ji}(\omega) \tag{5.58}$$

and comparing them with equations (5.23) and (5.25), we find that the quantities $D_{ji}(\omega)$ satisfy equation (5.57) as well as equation (5.23). Consequently they should be proportional to $\varphi_j^{(1)}(\tau)$ for any i, and to $\overset{*}{\varphi}_i^{(1)}(\tau)$ for any j, i.e.,

$$D_{ji}(i\omega_1)=k_1\varphi_j^{(1)}(\tau), \tag{5.59}$$
$$D_{ji}(i\omega_1)=k_2\varphi_i^{*(1)}(\tau), \tag{5.60}$$

where k_1 and k_2 are the proportionality coefficients.

Differentiating identity (5.57) with respect to ω, and then setting $\omega=i\omega_1$, we obtain

$$D'(i\omega_1)=\sum_{j=1}^{N}\{-2a_{ij}\omega_1+ic_{ij}\}D_{ji}(i\omega_1)+\sum_{j=1}^{N}\{b_{ij}-a_{ij}\omega_1^2+ic_{ij}\omega_1\}D'_{ji}(i\omega_1) \tag{5.61}$$

or

$$1=\sum_{j=1}^{N}\{-2a_{ij}\omega_1+ic_{ij}\}\frac{k_1\varphi_j^{(1)}}{D'(i\omega_1)}+\sum_{j=1}^{N}\{b_{ij}-a_{ij}\omega_1^2+ic_{ij}\omega_1\}S_{ji}(i\omega_1).$$

The quantities $Y_j=S_{ji}(i\omega_1)$ satisfy for any i the system of equations

$$\sum_{j=1}^{N}\{b_{ij}-a_{ij}\omega_1^2+ic_{ij}\omega_1\}Y_j=1-\sum_{j=1}^{N}(-2a_{ij}\omega_1+ic_{ij})\frac{k_1\varphi_j^{(1)}(\tau)}{D'(i\omega_1)} \tag{5.62}$$
$$(i=1,\ 2,\ \ldots,\ N).$$

As $Y_j=S_{ji}(i\omega_1)$ are finite ($D'(i\omega_1)\neq 0$, since $i\omega_1(\tau)$ is a simple root of the equation $D(\omega)=0$), the right-hand sides of these equations should satisfy the condition (cf. (5.41))

$$\sum_{i=1}^{N}\left[1-\sum_{j=1}^{N}(-2a_{ij}\omega_1-ic_{ij})\frac{k_1\varphi_j^{(1)}(\tau)}{D'(i\omega_1)}\right]\varphi_i^{*(1)}(\tau)=0, \tag{5.63}$$

from which we find

$$\sum_{i,\,j=1}^{N}(-2a_{ij}\omega_1+ic_{ij})\,\varphi_j^{(1)}(\tau)\,\varphi_i^{*(1)}(\tau)=\frac{D'(i\omega_1)}{k_1}\neq 0. \tag{5.64}$$

We deduce analogously that

$$\sum_{i,\,j=1}^{N}(-2a_{ij}\omega_1-ic_{ij})\,\varphi_j^{*(1)}(\tau)\,\varphi_i^{(1)}(\tau)=\frac{D'(i\omega_1)}{k_2}\neq 0. \tag{5.65}$$

Equating the left-hand side of (5.65) with the coefficient of $A_1(\tau, a, \psi)$, we obtain

$$2\omega_1(\tau)\, m_1(\tau) + m_2(\tau) \neq 0$$

for any τ in the interval $0 \leqslant \tau \leqslant L$; this is what we wanted to prove. We shall not derive explicit expressions for $A_1(\tau, a, \psi)$ and $B_1(\tau, a, \psi)$.

We turn now to the second approximation. Having derived the expressions of $u_j^{(1)}(\tau, a, \theta, p\varphi+\psi)$, $A_1(\tau, a, \psi)$, and $B_1(\tau, a, \psi)$, the functions $G_{i0}^{(2)}(\tau, a, \theta, p\varphi+\psi)$ $(i=1, 2, \ldots, N)$ are already known, and it is easily verified that they are periodic in the variables θ and $p\varphi+\psi$ with period 2π. Substituting their values into the system of equations (5.18) for $l=2$, we derive $u_j^{(2)}(\tau, a, \theta, p\varphi+\psi)$ $(j=1, 2, \ldots, N)$, so that the finiteness conditions

$$\left.\begin{aligned} &\sum_{\substack{i=1 \\ (nq+p(m-1)=0)}}^{N} \bar{g}_{nm}^{(i2)}(\tau, a)\, \varphi_i^{*(1)}(\tau) = 0, \\ &\sum_{\substack{i=1 \\ (nq+p(m+1)=0)}}^{N} \bar{g}_{nm}^{(i2)}(\tau, a)\, \varphi_i^{(1)}(\tau) = 0, \end{aligned}\right\} \qquad (5.66)$$

are satisfied for these functions. In the above equation

$$\bar{g}_{nm}^{(i2)}(\tau, a) = \frac{1}{4\pi^2}\int_0^{2\pi}\int_0^{2\pi} \bar{G}_{i0}^{(2)}(\tau, a, \theta, p\varphi+\psi)\, e^{-i\{n\theta+m(p\varphi+\psi)\}}\, d\theta\, d(p\varphi+\psi). \qquad (5.67)$$

Using one of the conditions we obtain, after a number of operations, the following system of equations for $A_2(\tau, a, \psi)$ and $B_2(\tau, a, \psi)$:

$$\left.\begin{aligned} &m_1(\tau)\left[\omega_1(\tau) - \frac{p}{q}\nu(\tau)\right]\frac{\partial A_2}{\partial\psi} - [2\omega_1(\tau)\, m_1(\tau) + m_2(\tau)]\, aB_2 = \\ &\quad = -m_1(\tau)\left[\frac{\partial A_1}{\partial a}A_1 + \frac{\partial A_1}{\partial\psi}B_1 + \frac{\partial A_1}{\partial\tau} - aB_1^2\right] - \\ &\quad - \frac{dm_1(\tau)}{d\tau}A_1 - a\bar{m}_5(\tau) + \sum_{\sigma}\frac{e^{i\sigma q\psi}}{4\pi^2}\int_0^{2\pi}\int_0^{2\pi}\sum_{i=1}^{N}\Phi_{i0}^{(2)}(\tau, a, \theta, p\varphi+\psi)\, \times \\ &\quad \times e^{-i\sigma q\psi}\,[\bar{\varphi}_i^{(1)}\cos(p\varphi+\psi) - \bar{\bar{\varphi}}_i^{(1)}\sin(p\varphi+\psi)]\, d\theta\, d(p\varphi+\psi), \\ &m_1(\tau)\left[\omega_1(\tau) - \frac{p}{q}\nu(\tau)\right] a\frac{\partial B_2}{\partial\psi} + [2\omega_1(\tau)\, m_1(\tau) + m_2(\tau)]\, A_2 = \\ &\quad = -m_1(\tau)\left[a\frac{\partial B_1}{\partial a}A_1 + a\frac{\partial B_1}{\partial\psi}B_1 + a\frac{\partial B_1}{\partial\tau} + 2A_1B_1\right] - \\ &\quad - \frac{dm_1(\tau)}{d\tau}aB_1 + a\bar{\bar{m}}_5(\tau) - \sum_{\sigma}\frac{e^{i\sigma q\psi}}{4\pi^2}\int_0^{2\pi}\int_0^{2\pi}\sum_{i=1}^{N}\Phi_{i0}^{(2)}(\tau, a, \theta, p\varphi+\psi)\, \times \\ &\quad \times e^{-i\sigma q\psi}\,[\bar{\varphi}_i^{(1)}\sin(p\varphi+\psi) + \bar{\bar{\varphi}}_i^{(1)}\cos(p\varphi+\psi)]\, d\theta\, d(p\varphi+\psi), \end{aligned}\right\} \qquad (5.68)$$

where

$$\bar{m}_5(\tau) + i\bar{\bar{m}}_5(\tau) = \sum_{i,j=1}^{N}\frac{d}{d\tau}[a_{ij}(\tau)\,\varphi_j^{(1)}(\tau)]\,\varphi_i^{*(1)}(\tau). \qquad (5.69)$$

In order to satisfy conditions (5.66) it is sufficient to choose for the required functions $A_2(\tau, a, \psi)$ and $B_2(\tau, a, \psi)$ some particular solutions of the system (5.68) that are periodic in ψ. Their derivation, as in the preceding case, presents no difficulty. Indeed, after having determined $A_1(\tau, a, \psi)$, $B_1(\tau, a, \psi)$, and $u_j^{(1)}(\tau, a, \theta, p\varphi+\psi)$ $(j=1, 2, \ldots, N)$, the right-hand sides of equations (5.68) become known functions, periodic in ψ, and the coefficients of $A_2(\tau, a, \psi)$ and $B_2(\tau, a, \psi)$, as shown before, cannot vanish in the interval $0 \leqslant \tau \leqslant L$.

We have thus derived the functions $u_j^{(1)}(\tau, a, \theta, p\varphi+\psi)$ $(j=1, 2, \ldots N)$, $A_1(\tau, a, \psi)$, $A_2(\tau, a, \psi)$, $B_1(\tau, a, \psi)$ and $B_2(\tau, a, \psi)$, and, consequently, we can construct the approximate solution of equation (5.43) in the first and second approximation according to the general scheme given in § 1 of the foregoing chapter.

§ 2. Derivation of differential equations describing the nonstationary mode in a gyroscopic system of centrifugal type

In this section we derive a nonlinear differential equation of the type (5.4). We take as an actual example a gyroscopic system applicable in certain centrifugal devices, separators, etc. /104/.

Let the material axis of the gyroscope be held vertically by two bearings O and C (Figure 80), where the upper bearing C has at all radii the same rigidity depending nonlinearly on the displacement of the axis. (This dependence may be given graphically or analytically.)

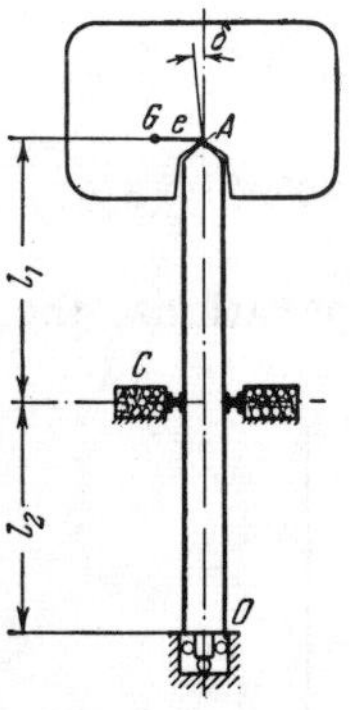

FIGURE 80

We assume that the rotor of the gyroscope is neither statically nor dynamically unbalanced, and that the mass of the rotor exceeds the mass of the axis considerably; this allows us to neglect the latter. We shall ignore possible torsional vibrations of the axis.

In a transient state, when the motion is started or stopped, or during a transition from one mode to another, many parameters of the centrifuge will vary with time and depend on the rotation frequency. Thus, for example, depending upon the setting, the mass of the rotor, the position of the center of gravity, the rotor's moments of inertia, and the amplitudes and frequencies of the external perturbing forces may vary together with the rotation frequency. The variation of some parameters (mass, moments of inertia of the rotor, etc.) depend, in an ideal formulation of the problem, not only on the rotation frequency of the entire system, but also on the vibrations occurring, i.e., on the whole complicated motion performed by the gyroscopic device under investigation. It should be assumed in that case, as indicated above, that the slowly varying parameters are determined in their turn by a system of equations. The problem reduces to considering the system (1.72). In practice, however, we can determine with sufficient accuracy the mass variation law, the moments of inertia and other parameters, by proceeding from the variation law of the rotation frequency.

We introduce the following notation: $m = m(t)$, mass of the rotor; $I_z = I_z(t)$, moment of inertia of the rotor with respect to the symmetry axis; $I = I(t)$, moment of inertia of the rotor with respect to a transversal axis perpendicular to the symmetry axis and passing through the center of gravity; $P = P(t)$, weight of the rotor; l, distance from the step bearing to the center of gravity of the rotor (we assume that the upper fixing point A lies on the symmetry axis of the rotor in the plane passing through the center of gravity perpendicular to symmetry axis); l_2, distance between the lower and upper bearings; l_1, distance from the upper bearing to the fixing point of the rotor ($l_1 + l_2 = l$); $AG = e$, displacement of the rotor's center of gravity with respect to the symmetry axis (the linear eccentricity of the rotor); and δ, angle between the symmetry axis of the rotor and the axis of the shaft on which it is fixed (the angular eccentricity of the rotor).

Before deriving the equations of motion, we quote some well-known results relating to the kinematics of rotation of a rigid body about a fixed point. The fixed point O of the body is taken as the origin of coordinates of three fixed orthogonal axes ξ, η, ζ, and of three orthogonal axes x, y, z associated with the body. The line ON is the intersection of the plane $O\xi\eta$ with the plane Oxy. This line (the line of nodes) is perpendicular to the axis $O\zeta$. The position of the body is completely defined by the position of the axes x, y, z associated with it; the position of these axes is determined by the Euler angles $(\zeta, z) = \vartheta$, $(\xi, N) = \psi$, $(N, x) = \varphi$, these angles being positive for a left-handed coordinate system if they are measured as shown in Figure 81.

The so-called Résal angles α, β, φ are frequently used instead of the Euler angles. These are defined as follows (Figure 82). In the plane $O\zeta z$ we draw the line OK perpendicular to the axis Oz. The axes NKz form an

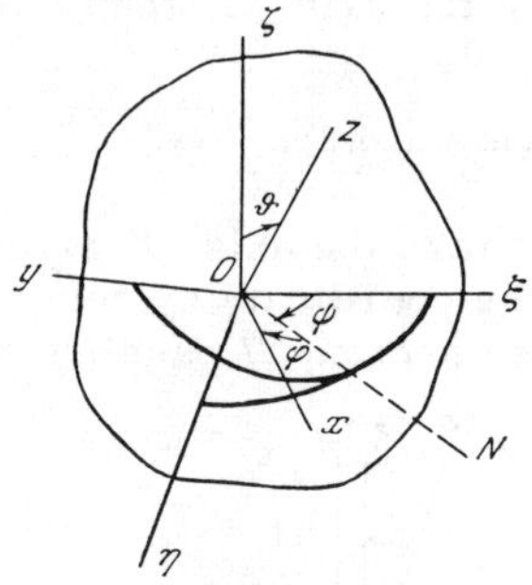

FIGURE 81

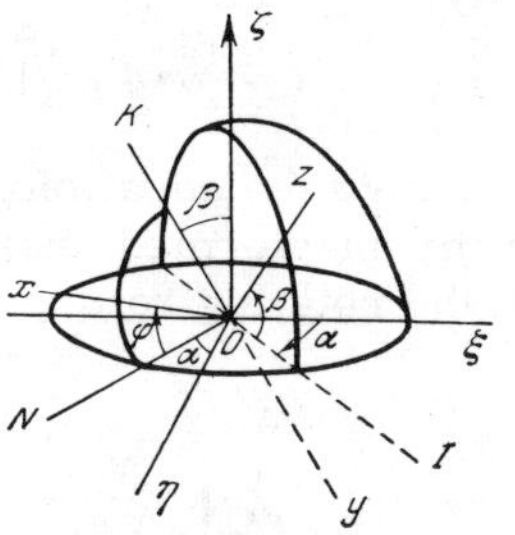

FIGURE 82

orthogonal trihedron, and the positive direction of the axis OK is chosen so that the system is a left-handed one. The plane NOK coincides with the plane NOy. We denote by I the line of intersection of the planes $O\xi\eta$ and $O\zeta z$. The bihedral angle between the planes $O\xi\zeta$ and $O\zeta z$ is denoted by α, and the angle between the axis Oz and the plane $O\zeta\eta$— by β. The position of the rigid body is then completely determined by the angles α, β and $\varphi = (N, x)$, the angles α and φ being positive if they are measured clockwise from the axes ξ and N, looking, correspondingly, from positive directions of the axes ζ and z; β is positive if an observer looking from the positive direction of the axis ON measures it counterclockwise from the direction of the axis OI.

The angles α and β are usually called the angle of precession and nutation, and the angle φ — the angle of normal rotation. The angles α and β are related to the Euler angles ϑ and ψ by the relations

$$\psi = \alpha + \frac{\pi}{2}, \quad \vartheta = \frac{\pi}{2} - \beta. \tag{5.70}$$

We shall now obtain the differential equations of motion, proceeding from Lagrange equations of the second kind. We must therefore determine the kinetic and potential energies.

According to König's theorem the kinetic energy of the body is given by the sum of the kinetic energy of its translational motion (together with the center of inertia) and the kinetic energy of rotational motion about the center of inertia. Let the fixed point O of the body coincide with the center of inertia, and the body axes xyz be directed along the principal axes of inertia. Denoting the mass of the body by m, the velocity of the center of inertia by v_c, and the moments of inertia of the body with respect to these axes by I_x, I_y, I_z, we obtain the following expression for the kinetic energy

$$T=\frac{1}{2}mv_c^2+\frac{1}{2}(I_xp^2+I_yq^2+I_zr^2), \tag{5.71}$$

where p, q, r are projections of the angular velocity vector Ω on the axes Ox, Oy, Oz. Expressing them in terms of the Résal angles, we obtain

$$p=-\dot\beta,\quad q=\dot\alpha\cos\beta,\quad r=\dot\varphi+\dot\alpha\sin\beta, \tag{5.72}$$

and consequently,

$$T=\frac{1}{2}mv_c^2+\frac{1}{2}[I_x\dot\beta^2+I_y\dot\alpha^2\cos^2\beta+I_z(\dot\varphi+\dot\alpha\sin\beta)^2]. \tag{5.73}$$

If the body under consideration rotates about the axis Oz, then $I_x=I_y=I$. Consequently,

$$T=\frac{1}{2}mv_c^2+\frac{1}{2}[I(\dot\beta^2+\dot\alpha^2\cos^2\beta)+I_z(\dot\varphi+\dot\alpha\sin\beta)^2]. \tag{5.74}$$

We now derive the equations of motion. We take the axis $O\xi$ along the axis OA of the undeformed shaft. To choose the coordinate system associated with the rotating rotor, we take its cross section M passing through

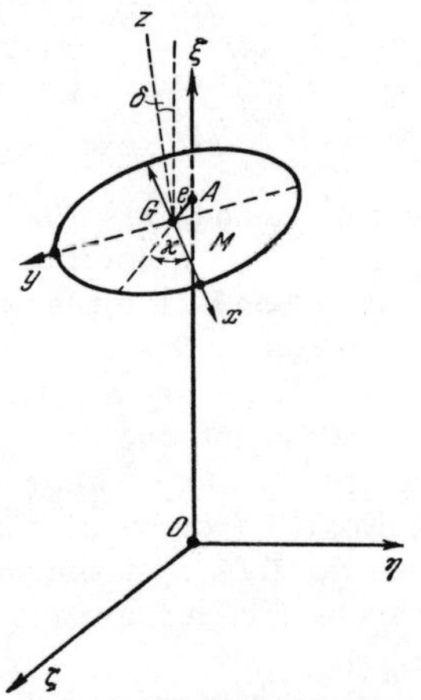

FIGURE 83

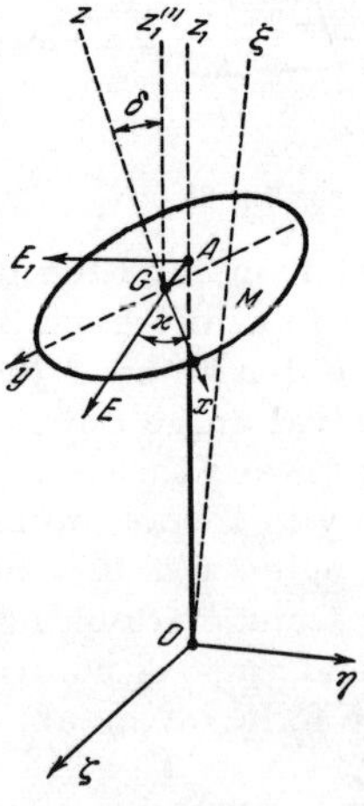

FIGURE 84

its center of gravity G perpendicular to its axis (Figure 83). (We recall that the point A lies in the plane of the disk M.) The origin of the moving coordinate system is taken at the center of gravity of the rotor, and the axis Gz is taken along its symmetry axis, i. e., perpendicular to the plane of the disk M. The axis Gy is taken along that diameter of the disk M which is its line of intersection with the plane containing the angle δ; the axis Gx is directed along the perpendicular diameter. The angle between AG and the axis Gx will be denoted by $\varkappa$.

Let us assume that the axis OA of our gyroscopic system is bent

(Figure 84). We denote the coordinates of the center of gravity of the rotor G in the fixed system by ξ_c, η_c, ζ_c, and characterize the position of the symmetry axis Gz of the rotor by the angles α and β (Figure 85). As the third Résal angle φ we take the angle formed by AG (i. e., the axis E associated with the rotor) and the line of nodes N.

Hence the position of the rotor is fully determined by the quantities $\xi_c, \eta_c, \zeta_c, \alpha, \beta, \varphi$.

Let us define the position of the rotor differently. We denote the coordinates of the point A in the fixed coordinate system by ξ, η, ζ. We draw at the point A the tangent z_1 to the axis of the bent shaft (Figure 84). The

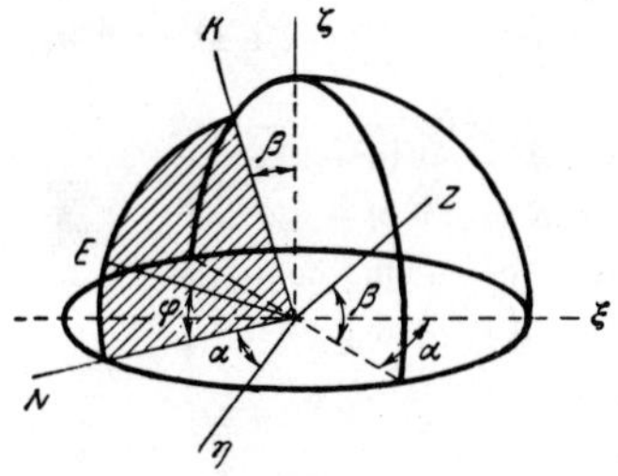

FIGURE 85

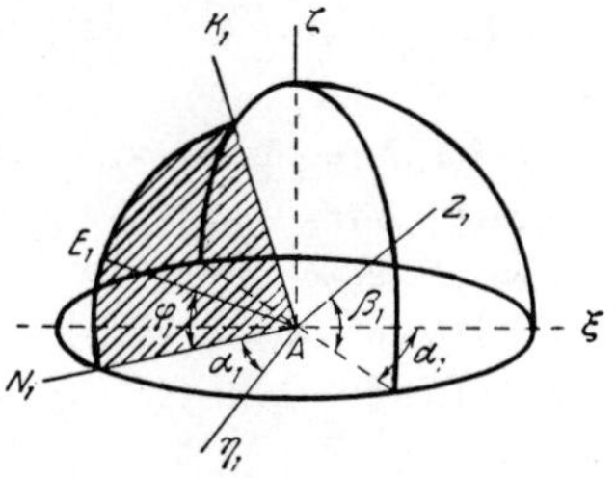

FIGURE 86

angles determining the position of this tangent will be denoted by α_1 and β_1. From the center of gravity of the rotor we draw a line $z_1^{(1)}$ parallel to z_1, and from the point A— the perpendicular E_1 on the axis $z_1^{(1)}$; this is the projection of E on the plane passing through A perpendicular to the axis z_1. As the third Résal angle we take the angle φ_1 formed by the axis E_1 and the line of nodes N_1; this is the intersection of this plane with the plane $O\xi\eta$.

Hence the position of the rotor is fully determined by the quantities $\xi, \eta, \zeta, \alpha_1, \beta_1, \varphi_1$ (Figure 86).

In order to establish the relation between the coordinates $\xi_c, \eta_c, \zeta_c, \alpha, \beta, \varphi$ and $\xi, \eta, \zeta, \alpha_1, \beta_1, \varphi_1$, we can apply the results given in the work of E. L. Nikolai /127/. For the coordinate ξ_c we have a geometric relation of the form

$$\xi_c = \xi(1 - e\sin\delta\sin\varphi) = l(1 - e\sin\delta\sin\varphi)\cos\alpha\cos\beta. \quad (5.75)$$

We proceed to establish the relation between the remaining five pairs of coordinates of the rotor.

We denote the projections of the segment $AG = e$ upon the axes ζ and η by e_ζ and e_η. Then

$$\eta_c = \eta + e_\eta, \quad \zeta_c = \zeta + e_\zeta. \quad (5.76)$$

We have, according to Figure 85,

$$\cos(E,N) = \cos\varphi, \quad \cos(E,K) = \sin\varphi,$$

and also

$$\left.\begin{aligned} \cos(E,\eta) &= \cos(E,N)\cos(N,\eta) + \cos(E,K)\cos(K,\eta) = \\ &= \cos\varphi\cos\alpha - \sin\varphi\sin\beta\sin\alpha, \\ \cos(E,\zeta) &= \cos(E,N)\cos(N,\zeta) + \cos(E,K)\cos(K,\zeta) = \\ &= \sin\varphi\cos\beta. \end{aligned}\right\} \quad (5.77)$$

Hence

$$\left.\begin{aligned} \eta_c &= \eta + l[\cos\varphi\cos\alpha - \sin\varphi\sin\beta\sin\alpha], \\ \zeta_c &= \zeta + l\sin\varphi\cos\beta. \end{aligned}\right\} \quad (5.78)$$

We shall now find the relation between the angles α, β and α_1, β_1, using for this the equalities

$$\cos(z_1, x) = 0, \quad \cos(z_1, y) = -\sin\delta. \tag{5.79}$$

We have the relations

$$\left.\begin{aligned} \cos(z_1, x) &= \cos(z_1, \xi)\cos(\xi, x) + \\ &+ \cos(z_1, \eta)\cos(\eta, x) + \cos(z_1, \zeta)\cos(\zeta, x), \\ \cos(z_1, y) &= \cos(z_1, \xi)\cos(\xi, y) + \\ &+ \cos(z_1, \eta)\cos(\eta, y) + \cos(z_1, \zeta)\cos(\zeta, y). \end{aligned}\right\} \tag{5.80}$$

According to Figure 86

$$\cos(z_1, \xi) = \cos\beta_1\cos\alpha_1, \quad \cos(z_1, \eta) = \cos\beta_1\sin\alpha_1, \quad \cos(z_1, \zeta) = \sin\beta_1; \tag{5.81}$$

we also have

$$\left.\begin{aligned} &\cos(x, N) = \cos(\varphi - \varkappa), && \cos(x, K) = \sin(\varphi - \varkappa), \\ &\cos(y, N) = -\sin(\varphi - \varkappa), && \cos(y, K) = \cos(\varphi - \varkappa), \\ &\cos(N, \xi) = -\sin\alpha, && \cos(K, \xi) = -\sin\beta\cos\alpha, \\ &\cos(N, \eta) = \cos\alpha, && \cos(K, \eta) = -\sin\beta\sin\alpha, \\ &\cos(N, \zeta) = 0, && \cos(K, \zeta) = \cos\beta. \end{aligned}\right\} \tag{5.82}$$

Deriving

$$\cos(\xi, x), \quad \cos(\eta, x), \quad \cos(\zeta, x),$$
$$\cos(\xi, y), \quad \cos(\eta, y), \quad \cos(\zeta, y)$$

and substituting their values and (5.81) into (5.80), we obtain

$$\begin{aligned} &\cos\beta_1\cos\alpha_1[\sin\alpha\cos(\varphi - \varkappa) + \sin\beta_1\cos\alpha\sin(\varphi - \varkappa)] = \\ &\qquad = \cos\beta_1\sin\alpha_1[\cos\alpha\cos(\varphi - \varkappa) - \sin\beta\sin\alpha\sin(\varphi - \varkappa)] + \\ &\qquad\qquad + \sin\beta_1\cos\beta\sin(\varphi - \varkappa), \end{aligned} \tag{5.83}$$

$$\begin{aligned} &\sin\delta + \cos\beta_1\cos\alpha_1[\sin\alpha\sin(\varphi - \varkappa) - \sin\beta\cos\alpha\cos(\varphi - \varkappa)] = \\ &\qquad = \cos\beta_1\sin\alpha_1[\cos\alpha\sin(\varphi - \varkappa) + \sin\beta\sin\alpha\cos(\varphi - \varkappa)] - \\ &\qquad\qquad - \sin\beta_1\cos\beta\cos(\varphi - \varkappa). \end{aligned} \tag{5.84}$$

Multiplying (5.83) by $\sin(\varphi - \varkappa)$, and (5.84) by $\cos(\varphi - \varkappa)$ and adding, we obtain

$$\cos\beta_1\sin(\alpha - \alpha_1) = -\sin\delta\sin(\varphi - \varkappa), \tag{5.85}$$

and by subtracting,

$$\cos\beta_1\sin\beta\cos(\alpha - \alpha_1) - \sin\beta_1\cos\beta = \sin\delta\cos(\varphi - \varkappa). \tag{5.86}$$

Before deriving direct expressions for α and β by means of α_1 and β_1, we shall obtain the relation between the angles φ and φ_1. To do this we

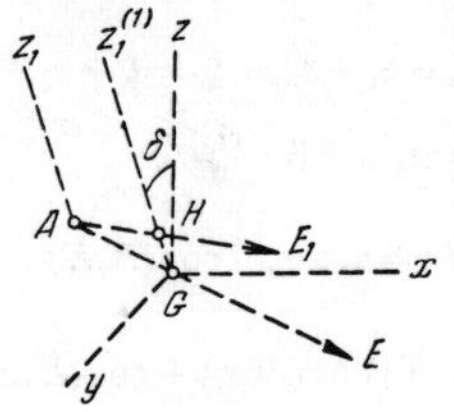

FIGURE 87

note, referring to Figure 87, that the axes y, z, and $z_1^{(1)}$ lie in the same plane, as do the axes E, z_1 and E_1.

Consider the equalities

$$\left.\begin{aligned} \cos(E, \eta) &= \cos(E, E_1)\cos(E_1, \eta) + \cos(E, z_1)\cos(z_1, \eta), \\ \cos(E, \zeta) &= \cos(E, E_1)\cos(E_1, \zeta) + \cos(E, z_1)\cos(z_1, \zeta). \end{aligned}\right\} \tag{5.87}$$

We can further write

$$\cos(E_1, N_1) = \cos\varphi_1, \quad \cos(E_1, K_1) = \sin\varphi_1, \tag{5.88}$$

and also

$$\left.\begin{aligned}\cos(E_1, \eta) &= \cos(E_1, N_1)\cos(N_1, \eta) + \cos(E_1, K_1)\cos(K_1, \eta) = \\ &= \cos\varphi_1\cos\alpha_1 - \sin\varphi_1\sin\beta_1\sin\alpha_1, \\ \cos(E_1, \zeta) &= \cos(E_1, N_1)\cos(N_1, \zeta) + \cos(E_1, K_1)\cos(K_1, \zeta) = \\ &= \sin\varphi_1\cos\beta_1.\end{aligned}\right\} \tag{5.89}$$

Let us now derive the quantities $\cos(E, E_1)$ and $\cos(E, z_1)$. We have from the right triangle AGH (see Figure 87),

$$\sin(E, E_1) = \frac{GH}{AG} = \frac{GH}{e}.$$

After a number of operations we find

$$GH = e\sin\varkappa\sin\delta.$$

Thus,

$$\sin(E, E_1) = \sin\varkappa\sin\delta,$$

and consequently,

$$\cos(E, E_1) = \sqrt{1 - \sin^2\varkappa\sin^2\delta}, \quad \cos(E, z_1) = -\sin(E, E_1) = -\sin\varkappa\sin\delta. \tag{5.90}$$

Substituting (5.88), (5.89), and (5.90) into the expressions (5.87), we obtain the following relation between the angles φ and φ_1:

$$\left.\begin{aligned}\cos\varphi\cos\alpha - \sin\varphi\sin\beta\sin\alpha &= \sqrt{1 - \sin^2\varkappa\sin^2\delta}\,[\cos\varphi_1\cos\alpha_1 - \\ &- \sin\varphi_1\sin\beta_1\sin\alpha_1] - \sin\varkappa\sin\delta\cos\beta_1\sin\alpha_1, \\ \sin\varphi\cos\beta &= \sqrt{1 - \sin^2\varkappa\sin^2\delta}\sin\varphi_1\cos\beta_1 - \sin\varkappa\sin\delta\sin\beta_1.\end{aligned}\right\} \tag{5.91}$$

It remains to express η_c, ζ_c, α, β, φ by η, ζ, α_1, β_1, φ_1 from the relations (5.78), (5.85), and (5.86).

We find from relations (5.91), up to fourth powers of η, ζ, α_1, and β_1, since δ and e are constant, that

$$\begin{aligned}\sin\varphi &= h\sin\varphi_1\left[1 + \frac{\beta^2}{2} - \frac{\beta_1^2}{2} + \frac{\beta_1^4}{24} + \frac{5\beta^4}{24} - \frac{\beta_1^2\beta^2}{4}\right] - \\ &\qquad - \sin\varkappa\sin\delta\left[\beta_1 - \frac{\beta_1^3}{6} + \frac{\beta_1\beta^2}{2}\right],\end{aligned} \tag{5.92}$$

$$\begin{aligned}\cos\varphi &= h\left\{\cos\varphi_1\left[1 + \frac{\alpha^2}{2} - \frac{\alpha_1^2}{2} + \frac{\alpha_1^4}{24} + \frac{5\alpha^4}{24} - \frac{\alpha^2\alpha_1^2}{4}\right] + \sin\varphi_1\left[\alpha\beta - \alpha_1\beta_1 + \right.\right. \\ &\qquad \left.\left. + \frac{\alpha^3\beta}{3} + \frac{\alpha_1^3\beta_1}{6} + \frac{\alpha\beta^3}{3} + \frac{\alpha_1\beta_1^3}{6} - \frac{\alpha\beta\beta_1^2}{2} - \frac{\alpha_1\beta_1\alpha^2}{2}\right]\right\} - \\ &\qquad - \sin\varkappa\sin\delta\left[\alpha_1 - \frac{\alpha_1\beta_1^2}{2} - \frac{\alpha_1^3}{6} + \frac{\alpha_1\alpha^2}{2} + \beta\beta_1\alpha\right],\end{aligned} \tag{5.93}$$

where

$$h = \sqrt{1 - \sin^2\varkappa\sin^2\delta}.$$

Putting these values of $\sin\varphi$ and $\cos\varphi$ in the right-hand sides of (5.85) and (5.86), we obtain after a number of manipulations

$$\left.\begin{aligned}\alpha &= \alpha_1 - h\sin\delta\sin(\varphi_1 - \varkappa) + \sin^2\delta H_1(\beta_1, \varphi_1), \\ \beta &= \beta_1 + h\sin\delta\cos(\varphi_1 - \varkappa) + \sin^2\delta H_2(\beta_1, \varphi_1),\end{aligned}\right\} \tag{5.94}$$

where

$$\begin{aligned}H_1(\beta_1, \varphi_1) &= -\frac{1}{2\sin\delta}\beta_1^2\sin(\varphi_1 - \varkappa) - \frac{1}{2}\beta_1\sin 2(\varphi_1 - \varkappa) - \\ &- \frac{h}{2}\sin\delta\left[\frac{1}{6}\sin 3(\varphi_1 - \varkappa) + \sin(\varphi_1 - \varkappa) - \frac{1}{2}\sin(\varphi_1 - 3\varkappa)\right],\end{aligned} \tag{5.95}$$

$$H_2(\beta_1, \varphi_1) = -\frac{\beta_1}{4} + \frac{\beta_1}{4}\cos 2(\varphi_1 - \varkappa) + \frac{h}{24}\sin\delta\cos 3(\varphi_1 - \varkappa) + \\ + \frac{3}{8} h \sin\delta\cos(\varphi_1 - \varkappa) - \frac{h}{4}\sin\delta\cos(\varphi_1 - 3\varkappa). \quad (5.95)$$

$H_1(\beta_1, \varphi_1)$ and $H_2(\beta_1, \varphi_1)$ contain first powers of β_1 and $\sin\delta$.

Relations (5.92) and (5.93) may now be written as

$$\left.\begin{aligned} \sin\varphi &= h\sin\varphi_1 + R_1(\beta_1, \varphi_1), \\ \cos\varphi &= h\cos\varphi_1 + R_2(\beta_1, \varphi_1), \end{aligned}\right\} \quad (5.96)$$

where $R_1(\beta_1, \varphi_1)$ and $R_2(\beta_1, \varphi_1)$ contain second and fourth order terms.

Substituting (5.94) and (5.96) into the right-hand sides of (5.78), we finally obtain the following expressions for η_c and ζ_c:

$$\left.\begin{aligned} \eta_c &= \eta + eh\cos\varphi_1 + Z_1(\alpha_1, \beta_1, \varphi_1), \\ \zeta_c &= \zeta + eh\sin\varphi_1 + Z_2(\beta_1, \varphi_1), \end{aligned}\right\} \quad (5.97)$$

where

$$Z_1(\alpha_1, \beta_1, \varphi_1) = -\frac{eh\alpha_1^2}{2}\cos\varphi_1 - e\alpha_1\sin\delta\sin\varkappa - eh\alpha_1\beta_1\sin\varphi_1,$$

$$Z_2(\beta_1, \varphi_1) = -\frac{e\beta_1^2 h}{2}\sin\varphi_1 - e\beta_1\sin\delta\sin\varkappa$$

contain terms of the third order with respect to the small quantities e, α_1, β_1, $\sin\delta$.

Differentiating (5.94), we obtain

$$\left.\begin{aligned} \dot{\alpha} &= \dot{\alpha}_1 - h\sin\delta\cos(\varphi_1 - \varkappa)\,\dot{\varphi}_1 + \sin^2\delta\,\dot{H}_1(\beta_1, \varphi_1), \\ \dot{\beta} &= \dot{\beta}_1 - h\sin\delta\sin(\varphi_1 - \varkappa)\,\dot{\varphi}_1 + \sin^2\delta\,\dot{H}_2(\beta_1, \varphi_1), \end{aligned}\right\} \quad (5.98)$$

where we have, according to (5.95),

$$\dot{H}_1(\beta_1, \varphi_1) = \left[-\frac{1}{\sin\delta}\beta_1\sin(\varphi_1 - \varkappa) - \frac{1}{2}\sin 2(\varphi_1 - \varkappa)\right]\dot{\beta}_1 + \\ + \left[-\frac{1}{2\sin\delta}\beta_1^2\cos(\varphi_1 - \varkappa) - \beta_1\cos 2(\varphi_1 - \varkappa) - \frac{h}{4}\sin\delta\cos 3(\varphi_1 - \varkappa) - \right. \\ \left. - \frac{h}{2}\sin\delta\cos(\varphi_1 - \varkappa) + \frac{h}{4}\sin\delta\cos(\varphi_1 - 3\varkappa)\right]\dot{\varphi}_1,$$

$$\dot{H}_2(\beta_1, \varphi_1) = -\frac{1}{2}\dot{\beta}_1\sin^2(\varphi_1 - \varkappa) + \left\{h\sin\delta\left[\frac{1}{4}\sin(\varphi_1 - 3\varkappa) - \right.\right. \\ \left.\left. - \frac{1}{8}\sin 3(\varphi_1 - \varkappa) - \frac{3}{8}\sin(\varphi_1 - \varkappa)\right] - \frac{\beta_1}{2}\sin 2(\varphi_1 - \varkappa)\right\}\dot{\varphi}_1.$$

Differentiating the expressions in (5.97), we obtain

$$\left.\begin{aligned} \dot{\eta}_c &= \dot{\eta} - eh\sin\varphi_1\dot{\varphi}_1 + \dot{Z}_1(\alpha_1, \beta_1, \varphi_1), \\ \dot{\zeta}_c &= \dot{\zeta} + eh\cos\varphi_1\dot{\varphi}_1 + \dot{Z}_2(\beta_1, \varphi_1), \end{aligned}\right\} \quad (5.99)$$

where

$$\dot{Z}_1(\alpha_1, \beta_1, \varphi_1) = e\left\{\dot{\alpha}_1[-h\alpha_1\cos\varphi_1 - \sin\delta\sin\varkappa - \beta_1 h\sin\varphi_1] - \right. \\ \left. - \dot{\beta}_1\alpha_1 h\sin\varphi_1 + \dot{\varphi}_1\left[\frac{h}{2}\alpha_1^2\sin\varphi_1 - \alpha_1\beta_1 h\cos\varphi_1\right]\right\},$$

$$\dot{Z}_2(\beta_1, \varphi_1) = e\left\{\dot{\beta}_1[-h\beta_1\sin\varphi_1 - \sin\delta\sin\varkappa] - \frac{h}{2}\beta_1^2\cos\varphi_1\dot{\varphi}_1\right\}.$$

From the expressions (5.99) we deduce after a few manipulations

$$\dot{\varphi}^2 = h^2\dot{\varphi}_1^2 + 2h\dot{\varphi}_1 G_2 + G_4, \quad (5.100)$$

where

$$G_2 = h^2\sin\delta\left[\dot{\beta}_1\cos(\varphi_1 - \varkappa)\,\dot{\varphi}_1 + \dot{\beta}_1\sin\varphi_1\cos\varkappa + \right. \\ \left. + \frac{1}{2}h\sin\delta\sin^2\varkappa\,\dot{\varphi}_1 + \frac{1}{2}h\sin\delta\cos 2(\varphi_1 - \varkappa)\,\dot{\varphi}_1\right],$$

$$G_4 = \dot{\beta}_1^2\sin^2\delta\sin^2\varphi_1 + 2h\dot{\beta}_1\dot{\varphi}_1\beta_1^2\sin\delta\sin(\varphi_1 - \varkappa) + \frac{2}{3}h\dot{\varphi}_1^2\beta_1^3\sin\delta\cos(\varphi_1 - \varkappa) +$$

$$+h\dot{\beta}_1\dot{\varphi}_1 \sin^3\delta \left[\sin\varphi_1 \left(\cos\varkappa - \frac{1}{2}\cos 3\varkappa \right) + \frac{3}{4}\sin 2\varkappa \cos(\varphi_1 - \varkappa) + \right.$$

$$\left. + \frac{1}{4}\sin 3(\varphi_1 - \varkappa) + \frac{5}{8}\sin(3\varphi_1 - \varkappa) + \frac{3}{8}\sin(5\varphi_1 - 3\varkappa) \right] +$$

$$+\dot{\beta}_1\dot{\varphi}_1\beta_1 \sin^2\delta\,[\sin 2\varphi_1 + 2\sin 2(\varphi_1 - \varkappa)] +$$

$$+\dot{\varphi}_1^2\beta_1^2 \sin^2\delta \left[\frac{1}{2} + \frac{5}{2}\cos 2(\varphi_1 - \varkappa) \right] + h\beta_1\dot{\varphi}_1^2 \sin^3\delta \left[\frac{5}{2}\cos(\varphi_1 - \varkappa) + \right.$$

$$\left. + 2\cos 3(\varphi_1 - \varkappa) - \cos(\varphi_1 - 3\varkappa) - \frac{1}{2}\cos(\varphi_1 + \varkappa) \right] + \dot{\varphi}_1^2 \sin^4\delta \left[\frac{7}{8} - \cos 2\varkappa + \right.$$

$$\left. + \frac{1}{4}\cos 4\varkappa - \frac{1}{4}\cos 2\varphi_1 + \frac{3}{2}\cos 2(\varphi_1 - \varkappa) - \frac{3}{4}\cos(2\varphi_1 - 4\varkappa) + \frac{3}{8}\cos 4(\varphi_1 - \varkappa) \right].$$

Substituting (5.98), (5.99), and (5.100) into the right-hand side of (5.74), we finally obtain for the kinetic energy

$$\begin{aligned} T = {} & \frac{1}{2} m\,[\dot{\eta}_0^2 + \dot{\zeta}_0^2] + \frac{1}{2} I\,[\dot{\beta}_{10}^2 + \dot{\alpha}_{10}^2] + \frac{1}{2} I_z\,[h^2\dot{\varphi}_1^2 + 2h\dot{\varphi}_1 G_2 + 2h\dot{\alpha}_{10}\dot{\varphi}_1\beta_{10}] + \\ & + \Big\{ m\,[\dot{\eta}_0\dot{Z}_1 + \dot{\zeta}_0\dot{Z}_2] + \frac{1}{2} I \left[2\dot{\beta}_{10}\dot{H}_2 \sin^2\delta + 2\dot{\alpha}_{10}\dot{H}_1 \sin^2\delta - \frac{1}{2}\dot{\alpha}_{10}^2\beta_{10}^2 \right] + \\ & + \frac{1}{2} I_z \left[2h\dot{\varphi}_1 (\dot{H}_1\beta_{10} + H_2\dot{\alpha}_{10}) \sin^2\delta - \frac{h}{3}\dot{\alpha}_{10}\beta_{10}^3\dot{\varphi}_1 + \dot{\alpha}_{10}^2\beta_{10}^2 + G_4 + 2\dot{\alpha}_{10}\beta_{10}G_2 \right] \Big\}, \end{aligned} \quad (5.101)$$

where we have introduced for brevity the following notation:

$$\eta_0 = \eta + eh\cos\varphi_1, \qquad \zeta_0 = \zeta + eh\sin\varphi_1,$$

$$\dot{\eta}_0 = \dot{\eta} - eh\sin\varphi_1\dot{\varphi}_1, \qquad \dot{\zeta}_0 = \dot{\zeta} + eh\cos\varphi_1\dot{\varphi}_1,$$

$$\alpha_{10} = \alpha_1 - h\sin\delta\sin(\varphi_1 - \varkappa), \qquad \beta_{10} = \beta_1 + h\sin\delta\cos(\varphi_1 - \varkappa),$$

$$\dot{\alpha}_{10} = \dot{\alpha}_1 - h\sin\delta\cos(\varphi_1 - \varkappa)\,\dot{\varphi}_1, \qquad \dot{\beta}_{10} = \dot{\beta}_1 - h\sin\delta\sin(\varphi_1 - \varkappa)\,\dot{\varphi}_1.$$

We now determine the potential energy of the system; it is the sum of the potential energy of the elastic forces of the deformed shaft and the potential energy of the spring.

The potential energy of the system may be written as the sum

$$V = V_{\rm I} + V_{\rm II}, \quad (5.102)$$

where the first term refers to the deflection in the plane $O\xi\eta$, and the second — in the plane $O\xi\zeta$.

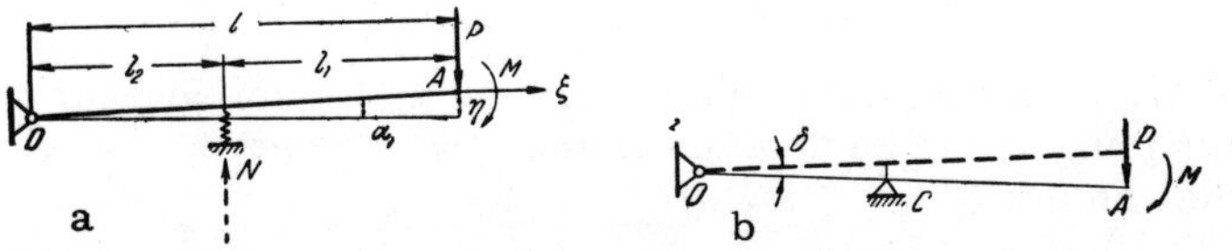

a b

FIGURE 88

To determine $V_{\rm I}$ we must find the potential energy of the beam (Figure 88a) under the condition that the tangent to the elastic line at point A forms together with the ξ-axis an angle α_1, and that the deflection at this point is η. We denote by P and M the force and the moment which should be impressed on the beam at the point A in order to yield the deformation of the beam and the spring (P and M, reactions of the rotor upon the shaft, N, reaction of the spring).

$V_{\rm I}$ may then be defined as the sum

$$V_{\rm I} = V_{\rm s} + V_{\rm b},$$

where $V_{\rm s}$ is the potential energy of the spring deformation, and $V_{\rm b}$ the potential energy of the bent beam.

We assume that the relation between the elastic force of the spring and the displacement η_{l_2} is expressed by a third degree polynomial

$$N = a_1\eta_{l_2} + a_3\eta_{l_2}^3, \tag{5.103}$$

where a_1 and a_3 are some constants.

V_S is then determined by the expression

$$V_S = \int_0^{\eta_{l_2}} N(\eta_{l_2})\, d\eta_{l_2},$$

from which we find, using (5.103),

$$V_S = \frac{1}{2} a_1 \eta_{l_2}^2 + \frac{1}{3} a_3 \eta_{l_2}^4. \tag{5.104}$$

We use the notation

$$M_{l_2} = P(l - l_2) + M. \tag{5.105}$$

Then using well-known formulas of the theory of elasticity, we obtain the expression

$$V_b = \frac{1}{2EI}\left\{\frac{M_{l_2} l_2}{2}\,\frac{2}{3} M_{l_2} + \frac{M_{l_2}(l-l_2)}{2}\left(\frac{2}{3} M_{l_2} + \frac{1}{3} M\right) + \frac{Ml}{2}\left(\frac{1}{3} M_{l_2} + \frac{2}{3} M\right)\right\}$$

for V_b or, inserting the value of M_{l_2},

$$V_b = \frac{1}{2EI}\left\{\frac{l}{3}\left[P^2 l_1^2 + 3PMl_1 + 3M^2\right] - \frac{l_2}{6}\left[MPl_1 + M^2\right]\right\}. \tag{5.106}$$

Hence we obtain for V_I the expression

$$V_I = \frac{1}{2} a_1 \eta_{l_2}^2 + \frac{1}{3} a_3 \eta_{l_2}^4 + \frac{1}{2EI}\left\{\frac{l}{3}\left[P^2 l_1^2 + 3PMl_1 + 3M^2\right] - \frac{l_2}{6}\left[MPl_1 + M^2\right]\right\}, \tag{5.107}$$

in which P, M, and η_{l_2} on the right-hand side should be expressed in terms of η and α_1.

From Figure 88b we have

$$\left.\begin{aligned} \eta &= \eta^* + \frac{\eta_{l_2}}{l_2} l, \\ \alpha_1 &= \alpha_1^* + \frac{\eta_{l_2}}{l_2}. \end{aligned}\right\} \tag{5.108}$$

Here η^* and α_1^* denote correspondingly the deflection and the angle of rotation at the end of the spring due to the force P and the moment M.

We further obtain

$$\left.\begin{aligned} \eta^* &= \vartheta(l - l_2) + \frac{P(l-l_2)^3}{3EI} + \frac{M(l-l_2)^2}{2EI}, \\ \alpha_1^* &= \vartheta + \frac{P(l-l_2)^2}{2EI} + \frac{M(l-l_2)}{EI}, \end{aligned}\right\} \tag{5.109}$$

where

$$\vartheta = \frac{1}{3}\,\frac{P(l-l_2)+M}{EI}\, l_2 \tag{5.110}$$

is the angle of rotation at the point C.

Substituting (5.109) and (5.110) into (5.108), and taking into account that according to (5.103)

$$a_1\eta_{l_2} + a_3\eta_{l_2}^3 = \frac{1}{l_2}\left[P(l - l_2) + M\right] \tag{5.111}$$

and consequently, that

$$\eta_{l_2} = \frac{Pl_1 + M}{l_2 a_1} - \frac{a_3\left[Pl_1 + M\right]^3}{l_2^3 a_1^4}, \tag{5.112}$$

up to terms of the third order, we find

$$\left.\begin{aligned}\eta &= \frac{[Pl_1+M]\,l}{l_2^2 a_1} - \frac{a_3\,[Pl_1+M]^3}{l_2^4 a_1^4}\,l + \frac{1}{3}\,\frac{Pl_1+M}{EI}\,l_1 l_2 + \frac{Pl_1^3}{3EI} + \frac{Ml_1^2}{2EI},\\ \alpha_1 &= \frac{[Pl_1+M]}{l_2^2 a_1} - \frac{a_3\,[Pl_1+M]^3}{l_2^4 a_1^4} + \frac{1}{3}\,\frac{Pl_1+M}{EI}\,l_2 + \frac{Pl_1^2}{2EI} + \frac{Ml_1}{EI}.\end{aligned}\right\} \tag{5.113}$$

Reversing these relations up to terms of the third order with respect to η and α_1, we find

$$\left.\begin{aligned}P &= p_1\eta + p_2\alpha_1 + p_3\eta\alpha_1^2 + p_4\alpha_1\eta^2 + p_5\eta^3 + p_6\alpha_1^3,\\ M &= q_1\eta + q_2\alpha_1 + q_3\eta\alpha_1^2 + q_4\alpha_1\eta^2 + q_5\eta^3 + q_6\alpha_1^3,\end{aligned}\right\} \tag{5.114}$$

where p_i, q_i $(i = 1, 2, \ldots, 6)$ are some constants which we do not write explicitly.

Substituting the values of P and M into the right-hand side of (5.107), taking into account (5.112), we obtain the expression

$$V_{\rm I} = V_{\rm I}(\eta, \alpha_1), \tag{5.115}$$

for $V_{\rm I}$ in which the right-hand side is a certain fourth degree polynomial in η and α_1.

Replacing η and α_1 correspondingly by ζ and β_1, we obtain $V_{\rm II}$.

We finally obtain the expression

$$\begin{aligned}V = \frac{1}{2}a_{02}[\eta^2+\zeta^2] + \frac{1}{2}a_{20}[\alpha_1^2+\beta_1^2] + a_{11}[\alpha_1\eta+\beta_1\zeta] + a_{12}[\alpha_1\eta^2+\beta_1\zeta^2] +\\ + a_{21}[\alpha_1^2\eta+\beta_1^2\zeta] + a_{04}[\eta^4+\zeta^4] + a_{40}[\alpha_1^4+\beta_1^4] + a_{22}[\alpha_1^2\eta^2+\beta_1^2\zeta^2] +\\ + a_{13}[\alpha_1\eta^3+\beta_1\zeta^3] + a_{31}[\alpha_1^3\eta+\beta_1^3\zeta],\end{aligned} \tag{5.116}$$

for the potential energy where a_{ij} are known constants.

Having derived the expressions for the kinetic and potential energies, we can construct without difficulty the Lagrangian

$$\mathscr{L} = T - V \tag{5.117}$$

and obtain the differential equations of motion by means of the Lagrange equation

$$\frac{d}{dt}\left\{\frac{\partial\overline{\mathscr{L}}}{\partial\dot{q}}\right\} - \frac{\partial\overline{\mathscr{L}}}{\partial q} = Q, \tag{5.118}$$

where $\overline{\mathscr{L}}$ is a homogeneous quadratic form with slowly varying coefficients, and Q, perturbing forces whose explicit forms are found after differentiating in $\overline{\mathscr{L}}$ terms of order higher than the second with respect to the quantities η, ζ, α_1, β_1, δ, and their derivatives, and also terms of the first order that do not depend on the coordinates η, ζ and the angles α_1, β_1.

We now construct the equations of motion. Differentiating the expression for $\overline{\mathscr{L}}$, we obtain

$$\left.\begin{aligned}&\frac{\partial\overline{\mathscr{L}}}{\partial\dot{\eta}} = m\dot{\eta} - meh\sin\varphi_1\dot{\varphi}_1 + m\dot{Z}_1, \quad \frac{\partial\overline{\mathscr{L}}}{\partial\dot{\zeta}} = m\dot{\zeta} + meh\cos\varphi_1\dot{\varphi}_1 + m\dot{Z}_2,\\ &\frac{\partial\overline{\mathscr{L}}}{\partial\dot{\alpha}_1} = I\dot{\alpha}_1 + I_z h\dot{\varphi}_1\beta_1 - Ih\sin\delta\cos(\varphi_1-\varkappa)\,\dot{\varphi}_1 +\\ &+ I_z\dot{h}\varphi_1\sin\delta\cos(\varphi_1-\varkappa) + W_1(\alpha_1, \beta_1, \varphi_1, \dot{\alpha}_1, \dot{\beta}_1, \dot{\varphi}_1, \dot{\eta}),\\ &\frac{\partial\overline{\mathscr{L}}}{\partial\dot{\beta}_1} = I\dot{\beta}_1 - Ih\sin\delta\dot{\varphi}_1\sin(\varphi_1-\varkappa) + W_2(\alpha_1, \beta_1, \varphi_1, \dot{\alpha}_1, \dot{\beta}_1, \dot{\varphi}_1, \dot{\eta}, \dot{\zeta}),\end{aligned}\right\} \tag{5.119}$$

where

$$\left.\begin{aligned}
&W_1(\alpha_1,\ \beta_1,\ \varphi_1,\ \dot\alpha_1,\ \dot\beta_1,\ \dot\varphi_1,\ \dot\eta)=I\dot H_1\sin^2\delta-\frac{1}{2}I\dot\alpha_{10}\beta_{10}+\\
&\qquad+I_zh\dot\varphi_1H_2\sin^2\delta-\frac{1}{6}Ih\beta_{10}^3\dot\varphi_1+I_z\dot\alpha_{10}\beta_{10}^2+I_z\beta_{10}G_2+m\dot\eta_0\ddot Z_{1\dot\alpha_1},\\
&W_2(\alpha_1,\ \beta_1,\ \varphi_1,\ \dot\alpha_1,\ \dot\beta_1,\ \dot\varphi_1,\ \dot\eta,\ \dot\zeta)=\\
&=-e\alpha_1hm\dot\eta_0\sin\varphi_1-me\dot\zeta_0(h\beta_1\sin\varphi_1-\sin\delta\sin\varkappa)+I\dot H_2\sin^2\delta-\\
&-\frac{1}{2}I\dot\beta_{10}\sin^2\delta\sin^2(\varphi_1-\varkappa)+I_2[\beta_1\sin^2\delta\sin^2\varphi_1+h\dot\varphi_1\beta_1^2\sin\delta\sin(\varphi_1-\varkappa)]-\\
&\qquad-[I\dot\alpha_{10}\sin^2\delta+I_zh\dot\varphi_1\sin^2\delta]\left[\frac{\beta_1}{\sin\delta}\sin(\varphi_1-\varkappa)+\frac{1}{2}\sin 2(\varphi_1-\varkappa)\right].
\end{aligned}\right\}\qquad(5.120)$$

Here W_1 and W_2 contain terms of the third order with respect to $\alpha_1,\ \beta_1,\ \dot\alpha_1,\ \dot\beta_1,\ e,\ \sin\delta,\ \eta,\ \zeta$.

We also derive

$$\frac{\partial\overline{\mathscr{L}}}{\partial\dot\varphi_1}=I_zh\dot\varphi_1+\Phi(\alpha_1,\ \beta_1,\ \eta,\ \zeta,\ \varphi_1,\ \dot\alpha_1,\ \dot\beta_1,\ \dot\eta,\ \dot\zeta,\ \dot\varphi_1),\qquad(5.121)$$

where Φ contains second order terms.

We further derive

$$\left.\begin{aligned}
&\frac{\partial\overline{\mathscr{L}}}{\partial\eta}=-(a_{02}\eta+a_{11}\alpha_1+2a_{12}\alpha_1\eta+a_{21}\alpha_1^2+\\
&\qquad+4a_{04}\eta^3+2a_{22}\alpha_1^2\eta+3a_{13}\alpha_1\eta^2+a_{31}\alpha_1^3),\\
&\frac{\partial\overline{\mathscr{L}}}{\partial\zeta}=-(a_{02}\zeta+a_{11}\beta_1+2a_{12}\beta_1\zeta+a_{21}\beta_1^2+\\
&\qquad+4a_{04}\zeta^3+2a_{22}\beta_1^2\zeta+3a_{13}\beta_1\zeta^2+a_{31}\beta_1^3).
\end{aligned}\right\}\qquad(5.122),$$

$$\left.\begin{aligned}
&\frac{\partial\overline{\mathscr{L}}}{\partial\alpha_1}=-[a_{20}\alpha_1+a_{11}\eta+a_{12}\eta^2+2a_{21}\alpha_1\eta+4a_{40}\alpha_1^3+\\
&+2a_{22}\alpha_1\eta^2+a_{13}\eta^3+3a_{31}\alpha_1^2\eta-R_1(\alpha_1,\ \beta_1,\ \eta,\ \varphi_1,\ \dot\alpha_1,\ \dot\beta_1,\ \dot\eta,\ \dot\varphi_1)],\\
&\frac{\partial\overline{\mathscr{L}}}{\partial\beta_1}=-[-I_zh\dot\varphi_1\dot\alpha_1+a_{20}\beta_1+a_{11}\zeta+a_{12}\zeta^2+2a_{21}\beta_1\zeta+4a_{40}\beta_1^3+2a_{22}\beta_1\zeta^2+\\
&+a_{13}\zeta^3+3a_{31}\beta_1^2\zeta+I_zh\sin\delta\dot\varphi_1^2\cos(\varphi_1-\varkappa)-R_2(\alpha_1,\ \beta_1,\ \zeta,\ \varphi_1,\ \dot\alpha_1,\ \dot\beta_1,\ \dot\zeta,\ \dot\varphi_1)],
\end{aligned}\right\}\qquad(5.123)$$

where

$$\begin{aligned}
&R_1(\alpha_1,\ \beta_1,\ \eta,\ \varphi_1,\ \dot\alpha_1,\ \dot\beta_1,\ \dot\eta,\ \dot\varphi_1)=m\left\{\frac{1}{2}\dot\varphi_1e^2h^2(\dot\alpha_1+\beta_1\dot\varphi_1)\sin 2\varphi_1+\right.\\
&\qquad\left.+e^2h^2\dot\varphi_1(\dot\beta_1-\alpha_1\dot\varphi_1)\sin^2\varphi_1-eh\dot\eta[(\dot\beta_1-\alpha_1\dot\varphi_1)\sin\varphi_1+(\dot\alpha_1+\beta_1\dot\varphi_1)\cos\varphi_1]\right\},
\end{aligned}$$

and

$$R_2(\alpha_1,\ \beta_1,\ \zeta,\ \varphi_1,\ \dot\alpha_1,\ \dot\beta_1,\ \dot\zeta,\ \dot\varphi_1)=\left(\frac{\partial T}{\partial\beta_1}\right)_2$$

denote the sum of terms, beginning with the second order term, which are obtained on differentiating the kinetic energy (5.101) with respect to β_1.

$$\frac{\partial\overline{\mathscr{L}}}{\partial\varphi_1}=\frac{\partial T}{\partial\varphi_1}\qquad(5.124)$$

is a form of the same order of smallness as T, i.e., it begins with terms of the second order. We shall not write down the explicit expression for $\frac{\partial T}{\partial\varphi_1}$.

Differentiating the expressions in (5.119) with respect to time, we find

$$\left.\begin{aligned}
&\frac{d}{dt}\left\{\frac{\partial\overline{\mathscr{L}}}{\partial\dot\eta}\right\}=m\ddot\eta-meh\dot\varphi_1^2\cos\varphi_1-meh\ddot\varphi_1\sin\varphi_1+m\ddot Z_1,\\
&\frac{d}{dt}\left\{\frac{\partial\overline{\mathscr{L}}}{\partial\dot\zeta}\right\}=m\ddot\zeta-meh\dot\varphi_1^2\sin\varphi_1+meh\ddot\varphi_1\cos\varphi_1+m\ddot Z_2,
\end{aligned}\right\}\qquad(5.125)$$

$$\left.\begin{aligned}
&\frac{d}{dt}\left\{\frac{\partial \overline{\mathscr{L}}}{\partial \dot{\alpha}_1}\right\} = I\ddot{\alpha}_1 + I_z h\dot{\varphi}_1\dot{\beta}_1 + I_z h\ddot{\varphi}_1\beta_1 - Ih\sin\delta\ddot{\varphi}_1\cos(\varphi_1-\varkappa) + \\
&+ Ih\dot{\varphi}_1^2\sin\delta\sin(\varphi_1-\varkappa) + I_z h^2\ddot{\varphi}_1\sin\delta\cos(\varphi_1-\varkappa) - I_z h^2\dot{\varphi}_1^2\sin\delta\sin(\varphi_1-\varkappa) + \dot{W}_1, \\
&\frac{d}{dt}\left\{\frac{\partial \overline{\mathscr{L}}}{\partial \dot{\beta}_1}\right\} = I\ddot{\beta}_1 - Ih\ddot{\varphi}_1\sin\delta\sin(\varphi_1-\varkappa) - Ih\dot{\varphi}_1^2\sin\delta\cos(\varphi_1-\varkappa) + \dot{W}_2.
\end{aligned}\right\} \tag{5.125}$$

Differentiating the right-hand side of (5.121), we obtain

$$\frac{d}{dt}\left\{\frac{\partial \overline{\mathscr{L}}}{\partial \dot{\varphi}_1}\right\} = I_z h^2\ddot{\varphi}_1 + \dot{\Phi}, \tag{5.126}$$

where $\dot{\Phi}$ contains terms starting with the second order of smallness.

Substituting (5.122), (5.123), (5.124), (5.125), and (5.126) into (5.118), we finally obtain the following system of differential equations:

$$\left.\begin{aligned}
&m\ddot{\eta} + a_{02}\eta + a_{11}\alpha = meh\dot{\varphi}_1^2\cos\varphi_1 + Q_\eta, \\
&m\ddot{\zeta} + a_{02}\zeta + a_{11}\beta = meh\dot{\varphi}_1^2\sin\varphi_1 + Q_\zeta, \\
&I\ddot{\alpha}_1 + I_z h\dot{\varphi}_1\dot{\beta}_1 + a_{11}\eta + a_{20}\alpha_1 = \\
&\qquad = (I_z - I)h\dot{\varphi}_1^2\sin\delta\sin(\varphi_1-\varkappa) + Q_{\alpha_1}, \\
&I\ddot{\beta}_1 - I_z h\dot{\varphi}_1\dot{\alpha}_1 + a_{11}\zeta + a_{20}\beta_1 = \\
&\qquad = -(I_z - I)h\dot{\varphi}_1^2\sin\delta\cos(\varphi_1-\varkappa) + Q_{\beta_1}, \\
&\qquad I_z h^2\ddot{\varphi}_1 = Q_{\varphi_1},
\end{aligned}\right\} \tag{5.127}$$

where

$$\left.\begin{aligned}
Q_\eta &= meh\ddot{\varphi}_1\sin\varphi_1 - m\ddot{Z}_1 - [2a_{12}\alpha_1\eta + a_{21}\alpha_1^2 + 4a_{04}\eta^3 + \\
&\qquad + 2a_{22}\alpha_1^2\eta + 3a_{13}\alpha_1\eta^2 + a_{31}\alpha_1^3], \\
Q_\zeta &= -meh\ddot{\varphi}_1\cos\varphi_1 - m\ddot{Z}_2 - [2a_{12}\beta_1\zeta + a_{21}\beta_1^2 + 4a_{04}\zeta^3 + \\
&\qquad + 2a_{22}\beta_1^2\zeta + 3a_{13}\beta_1\zeta^2 + a_{31}\beta_1^3], \\
Q_{\alpha_1} &= -I_z h\ddot{\varphi}_1\beta_1 + (Ih - I_z h^2)\ddot{\varphi}_1\sin\delta\cos(\varphi_1-\varkappa) - \\
&\quad - W_1 + R_1 - [a_{12}\eta^2 + 2a_{21}\alpha_1\eta + 4a_{40}\alpha_1^3 + 2a_{22}\alpha_1\eta^2 + \\
&\qquad + a_{13}\eta^3 + 3a_{31}\alpha_1^2\eta], \\
Q_\beta &= Ih\ddot{\varphi}_1\sin\delta\sin(\varphi_1-\varkappa) - \dot{W}_2 + R_2 - [a_{12}\zeta^2 + \\
&\qquad + 2a_{21}\beta_1\zeta + 4a_{40}\beta_1^3 + 2a_{22}\beta_1\zeta^2 + a_{13}\zeta^3 + 3a_{31}\beta_1^2\zeta], \\
Q_{\varphi_1} &= -\dot{\Phi}_1 + \frac{\partial T}{\partial \varphi_1}.
\end{aligned}\right\} \tag{5.128}$$

Once the explicit expressions of the right-hand sides of equations (5.127) are known, it is not difficult to construct with the help of formulas (5.55) the equations of the first (or higher) approximation, and to analyze vibrational processes in the specified mechanical system.

It is clear that only the fundamental resonance may occur in the first approximation (i. e., when the terms (5.128) are neglected in the right-hand sides of (5.127)).

When the terms (5.128) are taken into account a number of phenomena can be established which are specific for nonlinear vibrational systems. Thus a parametric resonance may be observed due to the presence of terms of the form $\eta\sin\varphi_1$, $\zeta\sin\varphi_1$, $\alpha_1\sin\varphi_1$, $\beta_1\sin\varphi_1$, $\dot{\alpha}_1\sin\varphi_1$, $\dot{\beta}_1\sin\varphi_1$ etc. In addition an n-th kind resonance as well as various combination resonances may occur in the system due to the presence of nonlinear terms of the type η^3, $\eta^2\alpha_1$, $\eta\alpha_1$, α_1^2, etc.; the presence of nonlinear terms deforms the resonance curves, gives rise to hysteresis effects, etc. A resonance may also appear in the system due to the presence of external perturbing forces with multiple frequencies of the type $\sin 2\varphi_1$, $\sin 3\varphi_1$, $\sin 4\varphi_1$, etc.

It should however be stressed that all these additional resonances are very sharp and may be detected only in a narrow frequency band and when the friction is small.

We note in conclusion that in the second approximation φ_1 ceases to be a linear function of time, and should be determined from the system of equations (5.127).

§ 3. Forced vibrations during a transition through critical numbers of revolutions in a centrifuge taking account of the gyroscopic effect of the rotor

We shall analyze briefly the nonstationary process of transition through critical numbers of revolutions for a system described by equations (5.127) under a number of simplifications.

We first of all neglect the influence of the shaft's deformation upon the law of variation of the angular velocity $\dot{\varphi}_1$. The variation of the angular velocity may in this case be derived by integrating the last of equations (5.127) independently of the first four.

Discarding nonlinear terms also in the right-hand sides of (5.127), we obtain the following system of equations:

$$\left.\begin{aligned}
&m\ddot{\eta}+a_{02}\eta+a_{11}\alpha_1=meh\,[\dot{\varphi}_1^2\cos\varphi_1+\ddot{\varphi}_1\sin\varphi_1],\\
&m\ddot{\zeta}+a_{02}\zeta+a_{11}\beta_1=meh\,[\dot{\varphi}_1^2\sin\varphi_1-\ddot{\varphi}_1\cos\varphi_1],\\
&I\ddot{\alpha}_1+I_zh\dot{\varphi}_1\dot{\beta}_1+a_{11}\eta+a_{20}\alpha_1=\\
&\qquad=(I_zh-I)\,h\sin\delta\,[\dot{\varphi}_1^2\sin(\varphi_1-\varkappa)-\\
&\qquad\qquad-\ddot{\varphi}_1\cos(\varphi_1-\varkappa)]-I_zh\ddot{\varphi}_1\beta_1,\\
&I\ddot{\beta}_1-I_zh\dot{\varphi}_1\dot{\alpha}_1+a_{11}\zeta+a_{20}\beta_1=\\
&\qquad=-(I_zh-I)\,h\sin\delta\dot{\varphi}_1^2\cos(\varphi_1-\varkappa)+\\
&\qquad\qquad+Ih\sin\delta\ddot{\varphi}_1\sin(\varphi_1-\varkappa),
\end{aligned}\right\}\qquad(5.129)$$

which describe deformation vibrations of the shaft conditioned by the static and dynamic nonequilibrium of the rotor, under the assumption of nondeformability of the shaft and of a linear dependence between the elastic force of the spring in the upper bearing and the displacement.

While investigating the vibrations it is generally necessary to take into account friction forces (internal and external) which affect considerably the development of vibration amplitudes in a stationary mode within a resonance zone, as well as during a transition through the resonance zone.

We assume that the forces of friction caused by an external agent are proportional to the deformation velocity of the shaft, and are described by the dissipative function

$$\begin{aligned}\Phi_1=\frac{1}{2}(&\varkappa_{11}\dot{\eta}^2+2\varkappa_{12}\dot{\eta}\dot{\zeta}+2\varkappa_{13}\dot{\eta}\dot{\alpha}_1+2\varkappa_{14}\dot{\eta}\dot{\beta}_1+\varkappa_{22}\dot{\zeta}^2+\\
&+2\varkappa_{23}\dot{\zeta}\dot{\alpha}_1+2\varkappa_{24}\dot{\zeta}\dot{\beta}_1+\varkappa_{33}\dot{\alpha}_1^2+2\varkappa_{34}\dot{\alpha}_1\dot{\beta}_1+\varkappa_{44}\dot{\beta}_1^2),\end{aligned}\qquad(5.130)$$

where $\varkappa_{ij}$ $(i,\ j=1,\ 2,\ 3,\ 4)$ are coefficients of external friction, and the forces of internal friction are described by the dissipative function

$$\Phi_2 = \frac{1}{2}\{k_1[\dot{\eta} - \nu(\tau)\eta]^2 + k_2[\dot{\zeta} + \nu(\tau)\zeta]^2 + \\ + k_3[\dot{\alpha}_1 - \nu(\tau)\beta_1]^2 + k_4[\dot{\beta}_1 + \nu(\tau)\alpha_1]^2\}, \qquad (5.131)$$

where k_1, k_2, k_3, k_4 are coefficients of internal friction.

We discard in the right-hand sides of equations (5.129) terms of linear perturbations containing as a factor the angular acceleration $\ddot{\varphi}_1 = \frac{d\nu}{dt}$, since in the resonance zone $\frac{d\nu}{dt} \ll \nu^2$.

It is much easier to construct approximate solutions for the system (5.129) corresponding to a monofrequency mode under the above simplifications by introducing the complex variables

$$z_1 = \eta + i\zeta, \quad z_2 = \alpha_1 + i\beta_1, \qquad (5.132)$$

and by excluding from consideration the state of inverse precession; this will be discussed in greater detail below.

The system (5.129) may be written in terms of the new variables z_1 and z_2, the forces of internal and external friction being taken into account, in the form

$$\left.\begin{aligned} m\ddot{z}_1 + a_{02}z_1 + a_{11}z_2 = & -\varkappa_{11}\dot{z}_1 - \varkappa_{12}\dot{z}_2 - k_{11}[\dot{z}_1 - i\nu(\tau)z_1] - \\ & - k_{12}[\dot{z}_2 - i\nu(\tau)z_2] + meh\nu^2(\tau)e^{i\varphi(t)}, \\ I\ddot{z}_2 - iI_zh\nu(\tau)\dot{z}_2 + a_{20}z_2 + a_{11}z_1 = & -\varkappa_{21}\dot{z}_1 - \varkappa_{22}\dot{z}_2 - \\ - k_{21}[\dot{z}_1 - i\nu(\tau)z_1] - k_{22}&[\dot{z}_2 - i\nu(\tau)z_2] - \\ & - i(I_z - I)h\sin\delta\,\nu^2(\tau)e^{i[\varphi(t) - \varkappa]}, \end{aligned}\right\} \qquad (5.133)$$

where k_{ij} $(i, j = 1, 2)$ are constant coefficients.

The unperturbed system corresponding to the system of equations (5.133) will be

$$\left.\begin{aligned} m\ddot{z}_1 + a_{02}z_1 + a_{11}z_2 = 0, \\ I\ddot{z}_2 - iI_zh\nu(\tau)\dot{z}_2 + a_{20}z_2 + a_{11}z_1 = 0. \end{aligned}\right\} \qquad (5.134)$$

Since the vibrations of the rotor conditioned by its static and dynamic nonequilibrium bear the character of direct precession, we look for the particular solution of the system (5.134) corresponding to principal vibrations at the frequency $\omega_k(\tau)$ $(k = 1, 2)$ in the form

$$z_j = a_k\varphi_j^{(k)}e^{i(\omega_k(\tau)t + \psi_k)} \quad (j = 1, 2), \qquad (5.135)$$

where a_k and ψ_k are real constants; τ is regarded as a constant parameter.

Substituting (5.135) into (5.134), we obtain a system of equations for determining the basic functions $\varphi_j^{(k)}(\tau)$:

$$\left.\begin{aligned} (-m\omega_k^2 + a_{02})\varphi_1^{(k)}(\tau) + a_{11}\varphi_2^{(k)}(\tau) = 0, \\ a_{11}\varphi_1^{(k)}(\tau) + (-I\omega_k^2 + I_zh\nu(\tau)\omega_k + a_{20})\varphi_2^{(k)}(\tau) = 0, \end{aligned}\right\} \qquad (5.136)$$

and the following characteristic equation for the eigenfrequencies ω_k:

$$D(\omega) = (a_{02} - m\omega^2)[(a_{20} - I\omega^2) + I_zh\nu(\tau)\omega] - a_{11}^2 = 0. \qquad (5.137)$$

The particular solution of the perturbed system (5.133) describing forced nonstationary vibrations corresponding to a transition through the lowest eigenfrequency in the state of direct precession is sought for in the form

$$z_j = a\varphi_j^{(1)}(\tau)e^{i[\varphi(t) + \psi]} \quad (j = 1, 2, \ldots, N), \qquad (5.138)$$

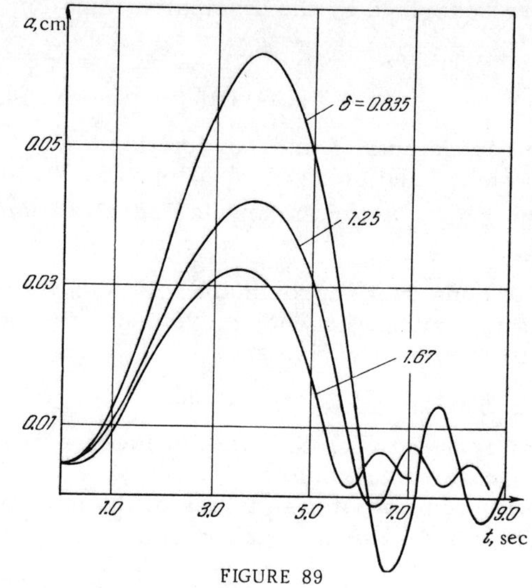

FIGURE 89

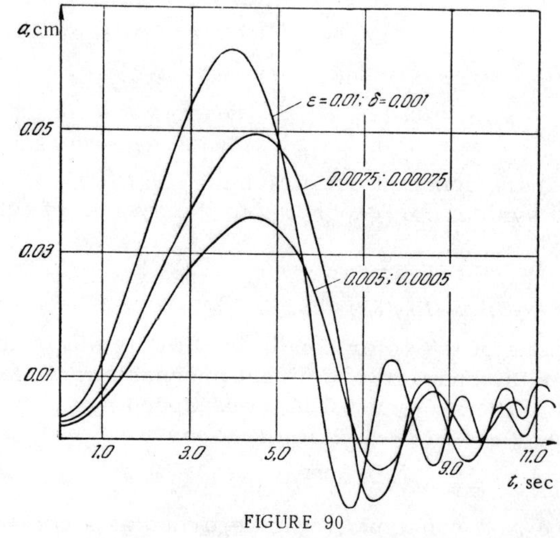

FIGURE 90

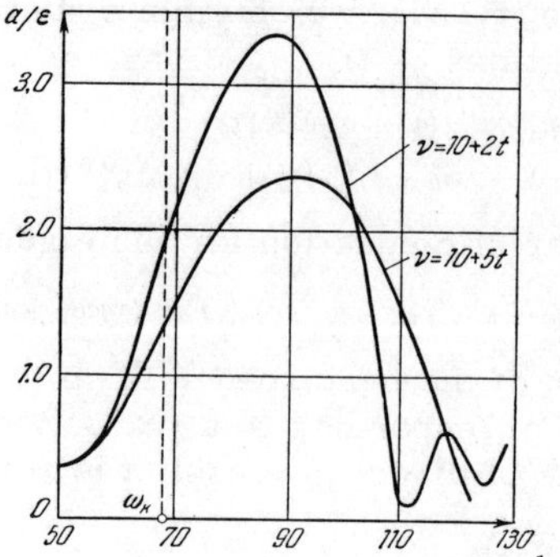

FIGURE 91

where a and ψ are to be determined from the system of equations of the first approximation

$$\left.\begin{aligned}\frac{da}{dt} &= \varepsilon A_1(\tau, a, \psi),\\ \frac{d\psi}{dt} &= \omega_1(\tau) - \nu(\tau) + \varepsilon B_1(\tau, a, \psi),\end{aligned}\right\} \quad (5.139)$$

in which the functions $A_1(\tau, a, \psi)$ and $B_1(\tau, a, \psi)$ should be determined from equations of the type (5.55); we now construct these equations.

Substituting the expressions in the right-hand sides of equations (5.133) into (5.55), and using (5.138), we obtain after a few manipulations the system

$$\left.\begin{aligned}&\varepsilon\left\{m_1(\tau)[\omega_1(\tau) - \nu(\tau)]\frac{\partial A_1}{\partial\psi} - [2m_1(\tau)\omega_1(\tau) + m_2(\tau)]aB_1\right\} =\\ &= meh\nu^2(\tau)\varphi_1^{(1)}(\tau)\cos\psi - (I_z h - I)\varphi_2^{(1)}(\tau)\sin\delta\,\nu^2(\tau)\sin(\psi - \varkappa),\\ &\varepsilon\left\{m_1(\tau)[\omega_1(\tau) - \nu(\tau)]a\frac{\partial B_1}{\partial\psi} + [2m_1(\tau)\omega_1(\tau) + m_2(\tau)]A_1\right\} =\\ &= m_3(\tau)a - \frac{d}{d\tau}[m_1(\tau)\omega_1(\tau)]a -\\ &- \sum_{i,j=1}^{2}\{\varkappa_{ij}\omega_1(\tau) + k_{ij}[\omega_1(\tau) - \nu(\tau)]\varphi_i^{(1)}(\tau)\varphi_j^{(1)}(\tau)a -\\ &- meh\nu^2(\tau)\varphi_1^{(1)}(\tau)\sin\psi - (I_z h - I)\varphi_2^{(1)}(\tau)\sin\delta\,\nu^2(\tau)\ \mathrm{os}(\psi + \varkappa)\},\end{aligned}\right\} \quad (5.140)$$

where

$$\left.\begin{aligned}m_1(\tau) &= m\varphi_1^{(1)^2}(\tau) + I\varphi_2^{(1)^2}(\tau),\\ m_2(\tau) &= I_z h\nu(\tau)\varphi_2^{(1)^2}(\tau),\\ m_3(\tau) &= I_z h\nu(\tau)\varphi_2^{(1)}(\tau)\frac{d\varphi_2^{(1)}(\tau)}{d\tau}.\end{aligned}\right\} \quad (5.141)$$

Determining $A_1(\tau, a, \psi)$ and $B_1(\tau, a, \psi)$ from the system (5.140) as particular solutions periodic in ψ, and putting these in the right-hand sides of the equations of the first approximation (5.139), we obtain the following system of equations in an explicit form for determining a and ψ:

$$\left.\begin{aligned}\frac{da}{dt} &= \frac{m_3(\tau) - \frac{d}{d\tau}[m_1(\tau)\omega_1(\tau)] - \sum\limits_{i,j=1}^{2}\{\varkappa_{ij}\omega_1(\tau) + k_{ij}[\omega_1(\tau) - \nu(\tau)]\varphi_i^{(1)}(\tau)\varphi_j^{(1)}(\tau)\}}{2m_1(\tau)\omega_1(\tau) + m_2(\tau)}a -\\ &- \frac{\nu^2(\tau)}{m_1(\tau)[\omega_1(\tau) + \nu(\tau)] + m_2(\tau)}\{meh\,\varphi_2^{(1)}(\tau)\sin\psi +\\ &\qquad + (I_z h - I)\sin\delta\,\varphi_2^{(1)}(\tau)\cos(\psi + \varkappa)\},\\ \frac{d\psi}{dt} &= \omega_1(\tau) - \nu(\tau) - \frac{1}{a}\,\frac{\nu^2(\tau)}{m_1(\tau)[\omega_1(\tau) + \nu(\tau)] + m_2(\tau)}\times\\ &\times\{meh\psi_1^{(1)}(\tau)\cos\psi - (I_z h - I)\sin\delta\,\varphi_2^{(1)}(\tau)\sin(\psi - \varkappa)\}.\end{aligned}\right\} \quad (5.142)$$

Equating the right-hand sides of the obtained system (5.142) to zero, and eliminating the phase ψ, we obtain, as usual, a formula for the resonance curves in stationary mode.

Specifying the numerical values of the parameters entering into the system of equations (5.133) and the law of variation of the rotor's state of rotation, and integrating numerically (or with the help of an analog computor) the system (5.142), we can easily obtain amplitude curves characterizing the nonstationary mode — a transition of the rotor through critical numbers of revolutions. These curves are of the type shown in Figures 89–91, which are taken from /42/.

§ 4: Reduction of a system of equations of the type (5.4) to "normal" coordinates

In many cases it is advisable to reduce the system of N second order equations (5.4) to a system of $2N$ first order equations.

Let, as before, the vibrations of the system under consideration be characterized by a Lagrangian of the type

$$\mathcal{L}=\frac{1}{2}\left\{\sum_{i,j=1}^{N} a_{ij}(\tau)\dot{q}_i\dot{q}_j+2\sum_{i,j=1}^{N} g_{ij}(\tau)q_i\dot{q}_j-\sum_{i,j=1}^{N} b_{ij}(\tau)q_iq_j\right\}, \tag{5.143}$$

where $\tau=\varepsilon t$, $a_{ij}(\tau)=a_{ji}(\tau)$, $b_{ij}(\tau)=b_{ji}(\tau)$, $g_{ij}(\tau)\neq g_{ji}(\tau)$, and the full energy of the dynamical system considered is a positive definite quadratic form, i. e., the discriminant

$$\Delta=|a_{ij}(\tau)|\neq 0, \tag{5.144}$$

for any τ.

The original system of differential equations

$$\frac{d}{dt}\left[\frac{\partial\mathcal{L}}{\partial\dot{q}_i}\right]-\frac{\partial\mathcal{L}}{\partial q_i}=\varepsilon Q_i(\tau,\theta,q_1,q_2,\ldots q_N,\dot{q}_1,\ldots,\dot{q}_N,\varepsilon)\ (i=1,2,\ldots,N), \tag{5.145}$$

where $\varepsilon Q_i(\tau,\theta,q_1,q_2,\ldots,q_N,\dot{q}_1,\ldots,\dot{q}_N,\varepsilon)$ are small perturbing forces satisfying the conditions indicated above, may now be easily reduced to a system of $2N$ first order equations. To do this we introduce, as usual, the new variables

$$q_{N+i}=\frac{\partial\mathcal{L}}{\partial\dot{q}_i}\qquad(i=1,2,\ldots,N) \tag{5.146}$$

or, explicitly,

$$q_{N+i}=\sum_{j=1}^{N} a_{ij}(\tau)\dot{q}_j+\sum_{j=1}^{N} g_{ji}(\tau)q_j\qquad(i=1,2,\ldots,N). \tag{5.147}$$

Solving the system (5.147) for $\dot{q}_j$ $(j=1,2,\ldots,N)$ we obtain

$$\dot{q}_j=\sum_{i=1}^{N}\bar{a}_{ij}(\tau)q_{N+i}+\sum_{i=1}^{N}\bar{g}_{ij}(\tau)q_i\qquad(j=1,2,\ldots,N), \tag{5.148}$$

where the coefficients $a_{ij}(\tau)$, $g_{ji}(\tau)$, $\bar{a}_{ij}(\tau)$, $\bar{g}_{ij}(\tau)$ are related by

$$\sum_{j=1}^{N}\bar{a}_{\sigma j}(\tau)a_{j\varrho}(\tau)=\begin{cases}1, & \sigma=\varrho,\\ 0, & \sigma\neq\varrho,\end{cases} \tag{5.149}$$

$$\sum_{j=1}^{N}\bar{a}_{\varrho j}(\tau)g_{\sigma j}(\tau)-\bar{g}_{\varrho\sigma}(\tau)=0\qquad(\varrho,\sigma=1,2,\ldots,N). \tag{5.150}$$

Since $\Delta\neq 0$ for any τ, the coefficients $\bar{a}_{ij}(\tau)$ and $\bar{g}_{ij}(\tau)$ are also unlimitedly differentiable functions for any finite values of τ.

Using the notation of (5.146), we have

$$H=\sum_{i=1}^{N}\dot{q}_iq_{N+i}-\mathcal{L}. \tag{5.151}$$

The system of equations (5.4) may, according to well-known principles of mechanics, then be replaced by the system of equations

$$\frac{dq_i}{dt}=\frac{\partial\widehat{H}}{\partial q_{N+i}},\quad\frac{dq_{N+i}}{dt}=-\frac{\partial\widehat{H}}{\partial q_i}+\varepsilon\widehat{Q}_i\qquad(i=1,2,\ldots,N), \tag{5.152}$$

where $\widehat{H}$ and $\widehat{Q}_i$ are expressions associated with the functions H and Q_i, obtained by replacing in the latter the velocities $\dot{q}_i$ $(i=1, 2, \ldots, N)$ by their values from (5.148). Substituting the $\dot{q}_i$ $(i=1, 2, \ldots, N)$ into (5.151), we obtain

$$\widehat{H}=\frac{1}{2}\sum_{i,\,j=1}^{N}\alpha_{ij}(\tau)q_iq_j+\sum_{i,\,j=1}^{N}\gamma_{ij}(\tau)q_iq_{N+j}+\frac{1}{2}\sum_{i,\,j=1}^{N}\beta_{ij}(\tau)q_{N+i}q_{N+j}, \qquad (5.153)$$

where the coefficients $\alpha_{ij}(\tau)$, $\beta_{ij}(\tau)$, $\gamma_{ij}(\tau)$ are according to the above unlimitedly differentiable functions for any finite values of τ; $\alpha_{ij}(\tau)=\alpha_{ji}(\tau)$; $\beta_{ij}(\tau)=\beta_{ji}(\tau)$; and $\gamma_{ij}(\tau)\neq\gamma_{ji}(\tau)$.

Using the expression for $\widehat{H}$, (5.152) may be written explicitly in the form

$$\left.\begin{aligned}&\sum_{j=1}^{N}\alpha_{ij}(\tau)q_i+\sum_{j=1}^{N}\left[\gamma_{ij}(\tau)q_{N+j}+\frac{dq_{N+i}}{dt}\right]=\varepsilon\widehat{Q}_i,\\&\sum_{j=1}^{N}\left[\gamma_{ji}(\tau)q_j-\frac{dq_i}{dt}\right]+\sum_{j=1}^{N}\beta_{ij}(\tau)q_{N+j}=0 \qquad (i=1, 2, \ldots, N).\end{aligned}\right\} \qquad (5.154)$$

These equations can be easily reduced to "normal" coordinates using B. V. Bulgakov's results /18/.

Let us construct for the unperturbed equations, i. e., equations (5.154) with $\varepsilon=0$, the functional matrix

$$f(\omega)=\left\|\begin{matrix}\alpha_{11}(\tau) & \ldots & \alpha_{1N}(\tau) & \gamma_{11}(\tau)+\omega & \ldots & \gamma_{1N}(\tau)\\ \cdots & \cdots & \cdots & \cdots & \cdots & \cdots\\ \alpha_{N1}(\tau) & \ldots & \alpha_{NN}(\tau) & \gamma_{N1}(\tau) & \ldots & \gamma_{NN}(\tau)+\omega\\ \gamma_{11}(\tau)-\omega & \ldots & \gamma_{N1}(\tau) & \beta_{11}(\tau) & \ldots & \beta_{1N}(\tau)\\ \cdots & \cdots & \cdots & \cdots & \cdots & \cdots\\ \gamma_{1N}(\tau) & \ldots & \gamma_{NN}(\tau)-\omega & \beta_{N1}(\tau) & \ldots & \beta_{NN}(\tau)\end{matrix}\right\| \qquad (5.155)$$

This matrix is quasi-hermitian since $f(-\omega)$ coincides with the transposed matrix; hence it follows that

$$D_1(\omega)=D_1(-\omega), \qquad (5.156)$$

i. e., that the characteristic determinant $D_1(\omega)$ contains only odd powers of ω.

As $\widehat{H}$ is a positive definite quadratic form, all the roots of the equation $D_1(\omega)=0$ are of the form $\omega=\pm i\omega_h(\tau)$.

We assume that the equation $D_1(\omega)=0$ does not admit multiple roots in the interval $0\leqslant\tau\leqslant L$. Then for any h there exists at least one diagonal nonvanishing element of the associated matrix

$$F(i\omega_h)=\|F_{ij}(i\omega_h)\| \qquad (5.157)$$

we denote it by $F_{l(h)\,l(h)}(i\omega_h)$. We note that

$$\Delta_h(\omega)=\frac{D_1(\omega)}{\omega^2+\omega_h^2}, \qquad (5.158)$$

and also that the $F_{l(h)l(h)}(i\omega_h)$ are real, and $\Delta_h(i\omega_h)$, like $F_{l(h)l(h)}(i\omega_h)$, do not vanish in the interval $0\leqslant\tau\leqslant L$.

We then transform the dependent variables q_j, q_{N+j} $(j=1, 2, \ldots, N)$ to the new variables x_j, y_j $(j=1, 2, \ldots, N)$ according to the formulas

$$\left.\begin{aligned}q_j&=\sum_{h=1}^{N}[X_j^{(h)}(\tau)x_h+Y_j^{(h)}(\tau)y_h],\\ q_{N+j}&=\sum_{h=1}^{N}[X_{N+j}^{(h)}(\tau)x_h+Y_{N+j}^{(h)}(\tau)y_h] \quad (j=1, 2, \ldots, N),\end{aligned}\right\} \qquad (5.159)$$

where $X_k^{(h)}(\tau)$, $Y_k^{(h)}(\tau)$ $(k=1, 2, \ldots 2N)$, defined by

$$X_k^{(h)}(\tau)-iY_k^{(h)}(\tau)=\frac{F_{kl}(i\omega_h)}{\sqrt{\omega_h|\Delta_h(i\omega_h)F_{ll}(i\omega_h)|}} \qquad (k=1,2, \ldots, 2N), \tag{5.160}$$

are unlimitedly differentiable for any finite values of τ.

We shall write down relations satisfied by the quantities $X_k^{(h)}(\tau)$, $Y_k^{(h)}(\tau)$ $(k=1, 2, \ldots, 2N)$. To do this we label the elements of the matrix $f(\omega)$ as follows:

$$f_{ij}(\tau)=\begin{cases}\alpha_{ij}(\tau), & i, j=1, 2, \ldots, N,\\ \gamma_{ij}(\tau), & i=1, 2, \ldots, N;\ j=N+1, \ldots, 2N,\\ \gamma_{ji}(\tau), & i=N+1, \ldots, 2N;\ j=1,2, \ldots, N,\\ \beta_{ij}(\tau), & i, j=N+1, \ldots, 2N.\end{cases}$$

Keeping in mind the obvious relation

$$F_{ll}(i\omega_h)F_{kj}(i\omega_h)-F_{lj}(i\omega_h)F_{kl}(i\omega_h)=0, \tag{5.161}$$

and the identity

$$\mp i\omega_h F_{N+j,\,j}(i\omega_h)+\sum_{k=1}^{2N} f_{jk}(\tau)F_{kj}(i\omega_h)=0 \tag{5.162}$$

(minus for $j\leqslant N$, and plus for $j>N$), we obtain the following relations for $X_k^{(h)}(\tau)$ and $Y_k^{(h)}(\tau)$:

$$\left.\begin{aligned}&\sum_{j,\,k=1}^{2N} f_{jk}(\tau)Y_j^{(l)}Y_j^{(h)}=\sum_{j,\,k=1}^{2N} f_{jk}(\tau)X_j^{(h)}X_k^{(l)}=\begin{cases}\omega_l(\tau), & l=h,\\ 0, & l\neq h,\end{cases}\\ &\sum_{j,\,k=1}^{2N} f_{jk}(\tau)Y_k^{(h)}X_j^{(l)}=0 \quad \text{(for any } h \text{ and } l)\end{aligned}\right\} \tag{5.163}$$

Substituting now the values of q_j, q_{N+j} $(j=1, 2, \ldots, N)$ from (5.159) into equations (5.154), we obtain

$$\left.\begin{aligned}&\sum_{j=1}^{N}\alpha_{ij}(\tau)\sum_{h=1}^{N}[X_j^{(h)}(\tau)x_h+Y_j^{(h)}(\tau)y_h]+\\ &\quad+\sum_{j=1}^{N}\gamma_{ij}(\tau)\sum_{h=1}^{N}[X_{N+j}^{(h)}(\tau)x_h+Y_{N+j}^{(h)}(\tau)y_h]+\\ &\quad+\sum_{h=1}^{N}\left[X_{N+i}^{(h)}(\tau)\frac{dx_h}{dt}+Y_{N+i}^{(h)}(\tau)\frac{dy_h}{dt}\right]=\\ &\qquad=\varepsilon\widehat{Q}_i-\varepsilon\sum_{h=1}^{N}\left[\frac{dX_{N+i}^{(h)}(\tau)}{d\tau}x_h+\frac{dY_{N+i}^{(h)}(\tau)}{d\tau}y_h\right].\\ &\sum_{j=1}^{N}\gamma_{ji}(\tau)\sum_{h=1}^{N}[X_j^{(h)}(\tau)x_h+Y_j^{(h)}(\tau)y_h]+\\ &\quad+\sum_{j=1}^{N}\beta_{ij}(\tau)\sum_{h=1}^{N}[X_{N+j}^{(h)}(\tau)x_h+Y_{N+j}^{(h)}(\tau)y_h]-\\ &\quad-\sum_{h=1}^{N}\left[X_i^{(h)}(\tau)\frac{dx_h}{dt}+Y_i^{(h)}(\tau)\frac{dy_h}{dt}\right]=\\ &\quad=\varepsilon\sum_{h=1}^{N}\left[\frac{dX_i^{(h)}(\tau)}{d\tau}x_h+\frac{dY_i^{(h)}(\tau)}{d\tau}y_h\right] \qquad (i=1, 2, \ldots, N).\end{aligned}\right\} \tag{5.164}$$

Multiplying the first equation by $Y_i^{(h)}(\tau)$, the second by $Y_{N+i}^{(h)}(\tau)$, and then adding them to each other and summing over i, we obtain, using relations (5.162) and (5.163),

$$\frac{dx_h}{dt}-\omega_h(\tau)y_h=\varepsilon\Big\{\sum_{i=1}^{N}\Big[\widehat{Q}_iY_i^{(h)}(\tau)+\sum_{h_1=1}^{N}\Big(\frac{dX_i^{(h_1)}(\tau)}{dt}Y_{N+i}^{(h)}(\tau)-$$
$$-\frac{dX_{N+i}^{(h_1)}(\tau)}{dt}Y_i^{(h)}(\tau)\Big)x_{h_1}+\sum_{h_1=1}^{N}\Big(\frac{dY_i^{(h_1)}(\tau)}{d\tau}Y_{N+i}^{(h)}(\tau)-$$
$$-\frac{dY_{N+i}^{(h_1)}(\tau)}{d\tau}Y_i^{(h)}(\tau)\Big)y_{h_1}\Big]\Big\}\qquad(h=1,\,2,\,\ldots,\,N).\tag{5.165}$$

Multiplying the first equation of (5.164) by $X_i^{(h)}(\tau)$, the second by $X_{N+i}^{(h)}(\tau)$, and performing similar operations, we obtain

$$\frac{dy_h}{dt}+\omega_h(\tau)x_h=\varepsilon\Big\{\sum_{i=1}^{N}\Big[\widehat{Q}_iX_i^{(h)}(\tau)+\sum_{h_1=1}^{N}\Big(\frac{dX_i^{(h_1)}(\tau)}{d\tau}X_{N+i}^{(h)}(\tau)-$$
$$-\frac{dX_{N+i}^{(h_1)}(\tau)}{d\tau}X_i^{(h)}(\tau)\Big)x_{h_1}+\sum_{h_1=1}^{N}\Big(\frac{dY_i^{(h_1)}(\tau)}{d\tau}X_{N+i}^{(h)}(\tau)-$$
$$-\frac{dY_{N+i}^{(h_1)}(\tau)}{d\tau}X_i^{(h)}(\tau)y_{h_1}\Big)\Big]\Big\}\qquad(h=1,\,2,\,\ldots,\,N),\tag{5.166}$$

where the values of q_i, q_{N+i} $(i=1,\,2,\,\ldots,\,N)$ from (5.159) should also be substituted into the functions $\widehat{Q}_i$ $(i=1,\,2,\,\ldots,\,N)$. Denoting the right-hand sides of equations (5.165) and (5.166) correspondingly by $\varepsilon Q_h^{(1)}(\tau,\,\theta,\,x,\,y,\,\varepsilon)$ and $\varepsilon Q_h^{(2)}(\tau,\,\theta,\,x,\,y,\,\varepsilon)$ (the set of quantities $x_1,\,x_2,\,\ldots,\,x_N$ is in what follows represented by the single letter x, etc.), we finally obtain the following system of $2N$ first order equations

$$\left.\begin{aligned}&\frac{dx_h}{dt}-\omega_h(\tau)y_h=\varepsilon Q_h^{(1)}(\tau,\,\theta,\,x,\,y,\,\varepsilon),\\&\frac{dy_h}{dt}+\omega_h(\tau)x_h=\varepsilon Q_h^{(2)}(\tau,\,\theta,\,x,\,y,\,\varepsilon)\qquad(h=1,\,2,\,\ldots,\,N).\end{aligned}\right\}\tag{5.167}$$

§ 5. Example of reducing a system of equations with gyroscopic terms to "normal" coordinates

As the simplest example we consider a vibrational system characterized by a Lagrangian of the form

$$\mathcal{L}=\frac{1}{2}\,[I\,(\dot{q}_1^2+\dot{q}_2^2)+2I_z\dot{\varphi}(\tau)\,\dot{q}_1q_2-a_{11}\,(q_1^2+q_2^2)],\tag{5.168}$$

encountered in the investigation of certain gyroscopic systems as, for example, a shaft with a disk fixed onto it, etc.

We assume that I, I_z, and a_{11} are constants, $\varphi(\tau)$ is some slowly varying function of time, and that the right-hand side of (5.168) satisfies condition (5.121). It is also assumed, for simplicity, that there are no external perturbing forces.

Differentiating (5.168), we obtain

$$\frac{\partial\mathcal{L}}{\partial\dot{q}_1}=I\dot{q}_1+I_z\dot{\varphi}(\tau)\,q_2,\tag{5.169}$$

$$\left.\begin{aligned}&\frac{\partial \mathcal{L}}{\partial \dot{q}_2}=I\dot{q}_2,\\&\frac{\partial \mathcal{L}}{\partial q_1}=-a_{11}q_1, \quad \frac{\partial \mathcal{L}}{\partial q_2}=I_z\dot{\varphi}(\tau)\dot{q}_1-a_{11}q_2,\end{aligned}\right\} \tag{5.169}$$

and also

$$\left.\begin{aligned}&\frac{d}{dt}\left[\frac{\partial \mathcal{L}}{\partial \dot{q}_1}\right]=I\ddot{q}_1+I_z\dot{\varphi}(\tau)\dot{q}_2+\varepsilon I_z\ddot{\varphi}(\tau)q_2,\\&\frac{d}{dt}\left[\frac{\partial \mathcal{L}}{\partial \dot{q}_2}\right]=I\ddot{q}_2.\end{aligned}\right\} \tag{5.170}$$

We then obtain a system of second order differential equations

$$\left.\begin{aligned}&I\ddot{q}_1+I_z\dot{\varphi}(\tau)\dot{q}_2+a_{11}q_1=\varepsilon I\ddot{\varphi}(\tau)q_2,\\&I\ddot{q}_2-I_z\dot{\varphi}(\tau)\dot{q}_1+a_{11}q_2=0.\end{aligned}\right\} \tag{5.171}$$

In order to reduce this system to a system of four first order equations, we introduce, as indicated above, new variables

$$q_3=I\dot{q}_1+I_z\dot{\varphi}(\tau)q_2, \quad q_4=I\dot{q}_2. \tag{5.172}$$

We then derive according to (5.151) an expression for the Hamiltonian

$$H=\dot{q}_1\frac{\partial \mathcal{L}}{\partial \dot{q}_1}+\dot{q}_2\frac{\partial \mathcal{L}}{\partial \dot{q}_2}-\mathcal{L}. \tag{5.173}$$

Substituting the value of $\mathcal{L}$ into the right-hand side, and eliminating $\dot{q}_1$ and $\dot{q}_2$, we obtain

$$\widehat{H}=\frac{1}{2}\left\{\frac{1}{I}\left[q_3^2+q_4^2-2I_z\dot{\varphi}(\tau)q_3q_2+I_z^2\dot{\varphi}^2(\tau)q_2^2\right]+a_{11}(q_1^2+q_2^2)\right\}. \tag{5.174}$$

It is then possible, according to (5.152), to consider instead of the system of equations (5.171) the following system of four first order equations:

$$\left.\begin{aligned}&a_{11}q_1+\frac{dq_3}{dt}=\varepsilon I_z\ddot{\varphi}(\tau)q_2,\\&\left(a_{11}+\frac{I_z^2}{I}\dot{\varphi}^2(\tau)\right)q_2-\frac{I_z}{I}\dot{\varphi}(\tau)q_3+\frac{dq_4}{dt}=0,\\&-\frac{dq_1}{dt}-\frac{I_z}{I}\dot{\varphi}(\tau)q_2+\frac{1}{I}q_3=0,\\&-\frac{dq_2}{dt}+\frac{1}{I}q_4=0.\end{aligned}\right\} \tag{5.175}$$

Let us transform this system to "normal" coordinates.

We first of all construct the functional matrix for the unperturbed equations. According to (5.155)

$$f(\omega)=\left\|\begin{matrix}a_{11} & 0 & \omega & 0\\ 0 & a_{11}+\frac{I_z^2}{I}\dot{\varphi}^2(\tau) & -\frac{I_z}{I}\dot{\varphi}(\tau) & \omega\\ -\omega & -\frac{I_z}{I}\dot{\varphi}(\tau) & \frac{1}{I} & 0\\ 0 & -\omega & 0 & \frac{1}{I}\end{matrix}\right\| \tag{5.176}$$

We then derive the characteristic equation

$$D_1(\omega)=\omega^4+\left[\frac{2a_{11}}{I}+\frac{I_z^2}{I}\dot{\varphi}^2(\tau)\right]\omega^2+\frac{a_{11}^2}{I^2}=0. \tag{5.177}$$

Denoting the roots of this equation by

$$\omega = \pm i\omega_h(\tau) \qquad (h = 1, 2),$$

we find

$$\left.\begin{aligned} \omega_1(\tau) &= \frac{I_z}{2I}\dot{\varphi}(\tau) - \sqrt{\frac{a_{11}}{I} + \frac{I_z^2}{4I^2}\dot{\varphi}^2(\tau)}, \\ \omega_2(\tau) &= \frac{I_z}{2I}\dot{\varphi}(\tau) + \sqrt{\frac{a_{11}}{I} + \frac{I_z^2}{4I^2}\dot{\varphi}^2(\tau)}. \end{aligned}\right\} \tag{5.178}$$

We construct the matrix associated to the matrix (5.176)

$$F(\omega) = \left\| \begin{matrix} \frac{a_{11}}{I^2} + \frac{\omega^2}{I} & -\frac{I_z}{I^2}\dot{\varphi}(\tau)\omega & -\omega^3 - \frac{\omega a_{11}}{I} - \frac{I_z^2}{I^2}\dot{\varphi}^2(\tau)\,\omega & \frac{I_z}{I}\dot{\varphi}(\tau)\,\omega^2 \\ \frac{I_z}{I^2}\dot{\varphi}(\tau)\omega & \frac{a_{11}}{I^2} + \frac{\omega^2}{I} & \frac{I_z}{I^2}\dot{\varphi}(\tau)\,a_{11} & -\omega^3 - \frac{a_{11}\omega}{I} \\ \omega^3 + \frac{\omega a_{11}}{I} + \frac{I_z^2}{I^2}\dot{\varphi}(\tau)\,\omega & \frac{I_z}{I^2}\dot{\varphi}(\tau)\,a_{11} & \frac{a_{11}^2}{I} + \frac{I_z^2}{I^2}\dot{\varphi}^2(\tau)\,a_{11} + a_{11}\omega^2 & -\frac{I_z}{I}\dot{\varphi}(\tau)\,\omega a_{11} \\ \frac{I_z}{I}\dot{\varphi}(\tau)\,\omega^2 & \omega^3 + \frac{a_{11}\omega}{I} & \frac{I_z}{I}\dot{\varphi}(\tau)\,a_{11}\omega & \frac{a_{11}^2}{I} + \frac{I_z^2}{I}\dot{\varphi}^2(\tau)\,\omega^2 + a_{11}\omega^2 \end{matrix} \right\|. \tag{5.179}$$

We derive

$$\left.\begin{aligned} F_{2(1)2(1)}(i\omega_1) &= -\frac{I_z}{I^2}\dot{\varphi}(\tau)\,\omega_1, \\ F_{2(2)2(2)}(i\omega_2) &= -\frac{I_z}{I^2}\dot{\varphi}(\tau)\,\omega_2. \end{aligned}\right\} \tag{5.180}$$

and obtain, according to (5.158),

$$\Delta_h(\omega)\Big|_{\omega=i\omega_h} = \frac{D_1(\omega)}{\omega^2+\omega_h^2}\Big|_{\omega=i\omega_h} = \frac{\omega^4 + \left(\frac{2a_{11}}{I} + \frac{I_z^2}{I^2}\dot{\varphi}^2(\tau)\right)\omega^2 + \frac{a_{11}^2}{I^2}}{\omega^2+\omega_h^2}\Bigg|_{\omega=i\omega_h}. \tag{5.181}$$

Removing the indefiniteness, we obtain

$$\Delta_h(\omega)\Big|_{\omega=i\omega_h} = 2\omega^2 + \frac{2a_{11}}{I} + \frac{I_z^2}{I^2}\dot{\varphi}^2(\tau)\Big|_{\omega=i\omega_h} \tag{5.182}$$

and consequently,

$$\left.\begin{aligned} \Delta_1(i\omega_1) &= 2\frac{I_z}{I}\dot{\varphi}(\tau)\sqrt{\frac{a_{11}}{I} + \frac{I_z^2}{4I^2}\dot{\varphi}^2(\tau)}, \\ \Delta_2(i\omega_2) &= -2\frac{I_z}{I}\dot{\varphi}(\tau)\sqrt{\frac{a_{11}}{I} + \frac{I_z^2}{4I^2}\dot{\varphi}^2(\tau)}. \end{aligned}\right\} \tag{5.183}$$

For simplicity we use the notation

$$\sqrt{\frac{a_{11}}{I} + \frac{I_z^2}{4I^2}\dot{\varphi}^2(\tau)} = m(\tau). \tag{5.184}$$

The denominator in formula (5.160) then assumes, on the grounds of (5.178), (5.180), and (5.183), the value

$$\sqrt{\omega_h\,|\Delta_h(i\omega_h)\,F_{2(h)2(h)}(i\omega_h)|} = \omega_h\frac{I_z}{I}\dot{\varphi}(\tau)\sqrt{\frac{2m(\tau)}{I}}. \tag{5.185}$$

From (5.160) we have

$$\left.\begin{aligned} X_1^{(1)}(\tau) &= 0, & Y_1^{(1)}(\tau) &= \frac{1}{\sqrt{2Im(\tau)}}, \\ X_1^{(2)}(\tau) &= 0, & Y_1^{(2)}(\tau) &= \frac{1}{\sqrt{2Im(\tau)}}, \\ X_2^{(1)}(\tau) &= -\frac{1}{\sqrt{2Im(\tau)}}, & Y_2^{(1)}(\tau) &= 0, \\ X_2^{(2)}(\tau) &= -\frac{1}{\sqrt{2Im(\tau)}}, & Y_2^{(2)}(\tau) &= 0, \end{aligned}\right\} \tag{5.186}$$

$$
\left.\begin{aligned}
X_3^{(1)}(\tau) &= \frac{a_{11}}{\omega_1(\tau)\sqrt{2Im(\tau)}}, & Y_3^{(1)}(\tau) &= 0,\\
X_3^{(2)}(\tau) &= \frac{a_{11}}{\omega_2(\tau)\sqrt{2Im(\tau)}}, & Y_3^{(2)}(\tau) &= 0,\\
X_4^{(1)}(\tau) &= 0, & Y_4^{(1)}(\tau) &= \frac{I\omega_1(\tau)}{\sqrt{2Im(\tau)}},\\
X_4^{(2)}(\tau) &= 0, & Y_4^{(2)}(\tau) &= \frac{I\omega_2(\tau)}{\sqrt{2Im(\tau)}}.
\end{aligned}\right\} \tag{5.186}
$$

Consequently the transition to "normal" coordinates is realized by the formulas

$$
\left.\begin{aligned}
q_1 &= \frac{1}{\sqrt{2Im(\tau)}}[y_1+y_2],\\
q_2 &= -\frac{1}{\sqrt{2Im(\tau)}}[x_1+x_2],\\
q_3 &= \frac{a_{11}}{\sqrt{2Im(\tau)}}\left[\frac{1}{\omega_1(\tau)}x_1+\frac{1}{\omega_2(\tau)}x_2\right],\\
q_4 &= \frac{I}{\sqrt{2Im(\tau)}}[\omega_1(\tau)y_1+\omega_2(\tau)y_2].
\end{aligned}\right\} \tag{5.187}
$$

Substituting the values of q_i $(i=1, 2, 3, 4)$ into the right-hand side of (5.174), we obtain an expression for the Hamiltonian in terms of the "normal" coordinates

$$
H=\frac{1}{2}[\omega_1(\tau)x_1^2+\omega_2(\tau)x_2^2+\omega_1(\tau)y_1^2+\omega_2(\tau)y_2^2]. \tag{5.188}
$$

Using equations (5.175) and (5.187) directly, we finally obtain the following equations for the variables x_1, x_2, y_1, y_2:

$$
\left.\begin{aligned}
\frac{dx_1}{dt}-\omega_1(\tau)y_1 &= \varepsilon[f_1(\tau)x_1+f_2(\tau)x_2],\\
\frac{dx_2}{dt}-\omega_2(\tau)y_2 &= \varepsilon[f_3(\tau)x_1+f_4(\tau)x_2],\\
\frac{dy_1}{dt}+\omega_1(\tau)x_1 &= \varepsilon[f_5(\tau)y_1+f_6(\tau)y_2],\\
\frac{dy_2}{dt}+\omega_2(\tau)x_2 &= \varepsilon[f_7(\tau)y_1+f_8(\tau)y_2],
\end{aligned}\right\} \tag{5.189}
$$

where

$$
\left.\begin{aligned}
f_1(\tau) &= \left[-\frac{I_z\dot{\varphi}_\tau(\tau)}{2Im(\tau)}+\frac{3\omega_1(\tau)m'_\tau(\tau)}{4m^2(\tau)}+\frac{a_{11}\omega'_1(\tau)}{2I\omega_1^2(\tau)m(\tau)}+\frac{a_{11}m'_\tau(\tau)}{4I\omega_1(\tau)m^2(\tau)}\right],\\
f_2(\tau) &= \left[-\frac{I_z\dot{\varphi}'_\tau(\tau)}{2Im(\tau)}+\frac{3\omega_1(\tau)m'_\tau(\tau)}{4m^2(\tau)}+\frac{a_{11}\omega'_2(\tau)}{2I\omega_2^2(\tau)m(\tau)}+\frac{a_{11}m'_\tau(\tau)}{4I\omega_2(\tau)m^2(\tau)}\right],\\
f_3(\tau) &= \left[-\frac{I_z\dot{\varphi}'_\tau(\tau)}{2Im(\tau)}+\frac{3\omega_2(\tau)m'_\tau(\tau)}{4m^2(\tau)}+\frac{a_{11}\omega'_1(\tau)}{2I\omega_1^2(\tau)m(\tau)}+\frac{a_{11}m'_\tau(\tau)}{4I\omega_1(\tau)m^2(\tau)}\right],\\
f_4(\tau) &= \left[-\frac{I_z\dot{\varphi}'_\tau(\tau)}{2Im(\tau)}+\frac{3\omega_2(\tau)m'_\tau(\tau)}{4m^2(\tau)}+\frac{a_{11}\omega'_2(\tau)}{2I\omega_2^2(\tau)m(\tau)}+\frac{a_{11}m'_\tau(\tau)}{4I\omega_2(\tau)m^2(\tau)}\right],\\
f_5(\tau) &= \left[-\frac{3m'_\tau(\tau)a_{11}}{2Im^2(\tau)\omega_1(\tau)}+\frac{\omega'_1(\tau)}{m(\tau)}-\frac{\omega_1(\tau)m'_\tau(\tau)}{2m^2(\tau)}\right],\\
f_6(\tau) &= \left[-\frac{3m'_\tau(\tau)a_{11}}{2Im^2(\tau)\omega_1(\tau)}+\frac{\omega'_2(\tau)}{m(\tau)}-\frac{\omega_2(\tau)m'_\tau(\tau)}{2m^2(\tau)}\right],\\
f_7(\tau) &= \left[-\frac{3m'_\tau(\tau)a_{11}}{2Im^2(\tau)\omega_2(\tau)}+\frac{\omega'_1(\tau)}{m(\tau)}-\frac{\omega_1(\tau)m'_\tau(\tau)}{2m^2(\tau)}\right],\\
f_8(\tau) &= \left[-\frac{3m'_\tau(\tau)a_{11}}{2Im^2(\tau)\omega_2(\tau)}+\frac{\omega'_2(\tau)}{m(\tau)}-\frac{\omega_2(\tau)m'_\tau(\tau)}{2m^2(\tau)}\right].
\end{aligned}\right\} \tag{5.190}
$$

§ 6. Construction of the general solution for a system of nonlinear equations of the type (5.167)

We now present a method for constructing an approximate asymptotic solution for a system of nonlinear differential equations of the type

$$\left.\begin{aligned} &\frac{dx_h}{dt} - \omega_h(\tau) y_h = \varepsilon Q_h^{(1)}(\tau, \theta, x, y, \varepsilon), \\ &\frac{dy_h}{dt} + \omega_h(\tau) x_h = \varepsilon Q_h^{(2)}(\tau, \theta, x, y, \varepsilon) \\ &\qquad (h = 1, 2, \ldots, N), \end{aligned}\right\} \tag{5.191}$$

where, as before, $\tau = \varepsilon t$; $\frac{d\theta}{dt} = \nu(\tau)$; $Q_h^{(1)}(\tau, \theta, x, y, \varepsilon)$ and $Q_h^{(2)}(\tau, \theta, x, y, \varepsilon)$ $(h = 1, 2, \ldots, N)$ are periodic functions of θ with period 2π, and $Q_h^{(1)}(\tau, \theta, x, y, \varepsilon)$, $Q_h^{(2)}(\tau, \theta, x, y, \varepsilon)$, $\omega_h(\tau)$, and $\nu(\tau)$ have the desired number of derivatives for all finite values of τ, θ, x, y and sufficiently small ε (the set $x_1, x_2, \ldots, x_N$ or $y_1, y_2, \ldots, y_N$ is, as mentioned before, denoted correspondingly by the single letter x or y).

To construct the approximate solution we apply the method expounded in § 1 directly, but now look for an approximate solution corresponding to the monofrequency mode. To apply this method without essential modification and to preserve the conformity of presentation we must transform the system of $2N$ first order equation to a system of N second order equations of the "quasi-normal" type; this is easily done as the right-hand sides are proportional to the small parameter ε.

Differentiating the first equation with respect to time we obtain

$$\frac{d^2x_h}{dt^2} - \omega_h(\tau)\frac{dy_h}{dt} = \varepsilon\left\{y_h \frac{d\omega_h}{d\tau} + \varepsilon\frac{\partial Q_h^{(1)}}{\partial\tau} + \nu(\tau)\frac{\partial Q_h^{(1)}}{\partial\theta} + \right.$$
$$\left. + \sum_{i=1}^{N} \frac{\partial Q_h}{\partial x_i}\frac{dx_i}{dt} + \sum_{i=1}^{N} \frac{\partial Q_h^{(1)}}{\partial y_i}\frac{dy_i}{dt}\right\} \quad (h = 1, 2, \ldots, N). \tag{5.192}$$

Inserting (to an accuracy of terms of the second order) the values

$$\frac{dy_h}{dt} = -\omega_h(\tau) x_h + \varepsilon Q_h^{(2)}(\tau, \theta, x, y, \varepsilon),$$

we obtain the following system of N second order equations which up to terms of the second order is equivalent to the system (5.191):

$$\frac{d^2x_h}{dt^2} + \omega_h^2(\tau) x_h = \varepsilon X_h(\tau, \theta, x, \dot{x}, \varepsilon) \qquad (h = 1, 2, \ldots, N), \tag{5.193}$$

where

$$X_h(\tau, \theta, x, \dot{x}, \varepsilon) = \frac{1}{\omega_h(\tau)}\frac{d\omega_h(\tau)}{d\tau}\frac{dx_h}{dt} + \nu(\tau)\frac{\partial Q_h^{(1)}(\tau, \theta, x, \dot{x}, \varepsilon)}{\partial\theta} +$$
$$+ \sum_{i=1}^{N} \frac{\partial Q_h^{(1)}(\tau, \theta, x, \dot{x}, \varepsilon)}{\partial x_i}\frac{dx_i}{dt} - \sum_{i=1}^{N} \frac{\partial Q_h^{(1)}(\tau, \theta, x, \dot{x}, \varepsilon)}{\partial \dot{x}_i}\omega_i^2(\tau) x_i +$$
$$+ \omega_h(\tau) Q_h^{(2)}(\tau, \theta, x, \dot{x}, \varepsilon). \tag{5.194}$$

Such an elimination of variables allows us to increase the degree of accuracy without difficulty. However, to make the elementary calculations shorter, we confine ourselves to the specified order; this is sufficient for constructing the first and the first improved approximation.

Let us introduce several notations and auxiliary definitions simplifying the calculations and the final form of approximate solutions. We denote by

m the combined index representing the set of N indexes $m_1, m_2, \ldots, m_N$, each of which may assume any integral values; by a, the set of quantities $a_1, a_2, \ldots, a_N$; by p and q, the sets of numbers $p_1, p_2, \ldots, p_N$ and $q_1, q_2, \ldots, q_N$, respectively, where p_i and q_i are mutually prime for all $i = 1, 2, \ldots, N$. by $p\varphi + \psi$, the set $p_1\varphi_1 + \psi_1,\ p_2\varphi_2 + \psi_2, \ldots, p_N\varphi_N + \psi_N$, where $\varphi_h = \frac{1}{q_h}\theta$; by $\omega(\tau)$, the set $\omega_1(\tau), \omega_2(\tau), \ldots, \omega_N(\tau)$; by $\omega^2(\tau)$, the set $\omega_1^2(\tau), \omega_2^2(\tau), \ldots, \omega_N^2(\tau)$; $m(p\varphi+\psi) = m_1(p_1\varphi_1+\psi_1) + \ldots + m_N(p_N\varphi_N+\psi_N)$; and $m^2 = m_1^2 + m_2^2 + \ldots + m_N^2$. We introduce analogous notations to designate the differentiation and integration of functions depending on the sets of variables a, $p\varphi+\psi$, etc., i. e.,

$$\sum_{i=1}^{N} \frac{\partial}{\partial a_i} = \frac{\partial}{\partial a}, \quad \sum_{i=1}^{N} \frac{\partial}{\partial (p_i\varphi_i+\psi_i)} = \frac{\partial}{\partial (p\varphi+\psi)}, \tag{5.195}$$

$$\underbrace{\int \ldots \int}_{N} f(p_1\varphi_1+\psi_1, \ldots, p_N\varphi_N+\psi_N)\, d(p_1\varphi_1+\psi_1) \ldots d(p_N\varphi_N+\psi_N) =$$

$$= \int f(p\varphi+\psi)\, d(p\varphi+\psi). \tag{5.196}$$

We now derive approximate solutions for the system of equations (5.193). The conditions listed in Chapter IV, § 1 (p. 158) are satisfied in the entire interval $0 \leqslant \tau \leqslant L$ because of the restrictions imposed on the Lagrangian.

We look for an approximate solution of equations (5.193), as in § 1, in the form of asymptotic series

$$x_h = a_h \cos(p_h\varphi_h + \psi_h) + \varepsilon u_h^{(1)}(\tau, \theta, a, p\varphi+\psi) + \\ + \varepsilon^2 u_h^{(2)}(\tau, \theta, a, p\varphi+\psi) + \ldots \quad (h = 1, 2, \ldots, N), \tag{5.197}$$

where $\tau = \varepsilon t$, $u_h^{(i)}(\tau, \theta, a, p\varphi+\psi)$ $(h = 1, 2, \ldots, N;\ i = 1, 2, \ldots)$ are periodic functions in θ and $p\varphi+\psi$ with period 2π, and the quantities a_h, ψ_h $(h = 1, 2, \ldots, N)$ are determined as functions of time by the equations

$$\left.\begin{aligned} \frac{da_h}{dt} &= \varepsilon A_1^{(h)}(\tau, a, \psi) + \varepsilon^2 A_2^{(h)}(\tau, a, \psi) + \ldots, \\ \frac{d\psi_h}{dt} &= \omega_h(\tau) - \frac{p_h}{q_h}\nu(\tau) + \varepsilon B_1^{(h)}(\tau, a, \psi) + \varepsilon^2 B_2^{(h)}(\tau, a, \psi) + \ldots \\ &(h = 1, 2, \ldots, N), \end{aligned}\right\} \tag{5.198}$$

where the choice of p_h and q_h depends on the resonance under consideration.

To construct the asymptotic series (5.197) we must find the functions $u_h^{(i)}(\tau, \theta, a, p\varphi+\psi)$ $(h = 1, 2, \ldots, N;\ i = 1, 2, \ldots)$ and also expressions for $A_1^{(h)}(\tau, a, \psi), A_2^{(h)}(\tau, a, \psi), \ldots, B_1^{(h)}(\tau, a, \psi), B_2^{(h)}(\tau, a, \psi), \ldots$ $(h = 1, 2, \ldots, N)$; the functions $u_h^{(i)}(\tau, \theta, a, p\varphi+\psi)$ $(h = 1, 2, \ldots, N;\ i = 1, 2, \ldots)$ should not contain terms with denominators which can vanish in the interval $0 \leqslant \tau \leqslant L$. This condition yields additional relations from which we derive the quantities $A_1^{(h)}(\tau, a, \psi)$, $B_1^{(h)}(\tau, a, \psi), \ldots$ $(h = 1, 2, \ldots, N)$.

Using (5.197) and (5.198) we obtain the following equations for determining the required functions:

$$\frac{dx_h}{dt} = -a_h\omega_h(\tau)\sin(p_h\varphi_h+\psi_h) + \varepsilon\Big\{A_1^{(h)}(\tau, a, \psi)\cos(p_h\varphi_h+\psi_h) - \\ \text{'n}\,(p_h\varphi_h+\psi_h)\, B_1^{(h)}(\tau, a, \psi) + \frac{\partial u_h^{(1)}}{\partial(p\varphi+\psi)}\omega(\tau) + \\ + \frac{\partial u_h^{(1)}}{\partial\theta}\nu(\tau)\Big\} + \varepsilon^3\{\ldots\} + \ldots \tag{5.199}$$

$$\frac{d^2x_h}{dt^2} = -a_h\omega_h^2(\tau)\cos(p_h\varphi_h+\psi_h) + \varepsilon\Big\{-\Big[a_h\frac{\partial B_1^{(h)}}{\partial\psi}\Big[\omega(\tau)-\frac{p}{q}\nu(\tau)\Big] + 2\omega_h(\tau)A_1^{(h)} + a_h\omega_h'(\tau)\Big]\sin(p_h\varphi_h+\psi_h) + \Big[\frac{\partial A_1^{(h)}}{\partial\psi}\Big[\omega(\tau)-\frac{p}{q}\nu(\tau)\Big] - 2a_h\omega_h(\tau)B_1^{(h)}\Big]\cos(p_h\varphi_h+\psi_h) + \frac{\partial^2 u_h^{(1)}}{\partial\theta^2}\nu^2(\tau) + 2\frac{\partial^2 u_h^{(1)}}{\partial\theta\,\partial(p\varphi+\psi)}\nu(\tau)\omega(\tau) + \frac{\partial^2 u_h^{(1)}}{\partial(p\varphi+\psi)^2}\omega^2(\tau)\Big\} + \varepsilon^2\{\ldots\} + \ldots \tag{5.200}$$

We substitute these expressions into the left-hand sides of equations (5.193), expand the right-hand sides in a Taylor series after inserting the values of x_h from (5.197) and $\frac{dx_h}{dt}$ from (5.199), and then equate the coefficients of ε on both sides; this yields the following equations for the determination of $u_h^{(1)}(\tau, a, \theta, p\varphi+\psi)$, $A_1^{(h)}(\tau, a, \psi)$, $B_1^{(h)}(\tau, a, \psi)$ $(h=1, 2, \ldots, N)$:

$$\frac{\partial^2 u_h^{(1)}}{\partial\theta^2}\nu^2(\tau) + 2\frac{\partial^2 u_h^{(1)}}{\partial\theta\,\partial(p\varphi+\psi)}\nu(\tau)\omega(\tau) + \frac{\partial^2 u_h^{(1)}}{\partial(p\varphi+\psi)^2}\omega^2(\tau) + \omega_h^2(\tau)u_1^{(h)} = X_h^{(0)}(\tau, a, \theta, p\varphi+\psi) + \Big[a_h\frac{\partial B_1^{(h)}}{\partial\psi}\Big[\omega(\tau)-\frac{p}{q}\nu(\tau)\Big] + 2\omega_h(\tau)A_1^{(h)} + a_h\frac{d\omega_h(\tau)}{d\tau}\Big]\sin(p_h\varphi_h+\psi_h) - \Big[\frac{\partial A_1}{\partial\psi}\Big[\omega(\tau)-\frac{p}{q}\nu(\tau)\Big] - 2a_h\omega_h(\tau)B_1^{(h)}\Big]\cos(p_h\varphi_h+\psi_h) = \overline{X}_h^{(0)}(\tau, a, \theta, p\varphi+\psi) \quad (h=1, 2, \ldots, N). \tag{5.201}$$

where

$$\overline{X}_h^{(0)}(\tau, \theta, a, p\varphi+\psi) = X_h(\tau, \theta, a\cos(p\varphi+\psi), -a\omega\sin(p\varphi+\psi), 0) \quad (h=1, 2, \ldots, N). \tag{5.202}$$

To determine $u_h^{(1)}(\tau, a, \theta, p\varphi+\psi)$ we represent $\overline{X}_h^{(0)}(\tau, \theta, a, p\varphi+\psi)$ $(h=1, 2, \ldots, N)$ in the form of series

$$\overline{X}_h^{(0)}(\tau, \theta, a, p\varphi+\psi) = \sum_{n,m}\overline{X}_{hnm}^{(0)}(\tau, a)\,e^{i\{n\theta+m(p\varphi+\psi)\}} \quad (h=1, 2, \ldots, N), \tag{5.203}$$

where the summation runs over all the values of the combined index m, and the $\overline{X}_{hnm}^{(0)}(\tau, a)$ are determined by

$$\overline{X}_{hnm}^{(0)}(\tau, a) = \frac{2}{(2\pi)^{N+1}}\int_0^{2\pi}\int_0^{2\pi}\overline{X}_h^{(0)}(\tau, a)\,e^{-i\{n\theta+m(p\varphi+\psi)\}}\,d\theta\,d(p\varphi+\psi). \tag{5.204}$$

The function $u_h^{(1)}(\tau, \theta, a, p\varphi+\psi)$ is also sought for in the form of a series

$$u_h^{(1)}(\tau, \theta, a, p\varphi+\psi) = \sum_{n,m}k_{nm}^{(h1)}(\tau, a)\,e^{i\{n\theta+m(p\varphi+\psi)\}}. \tag{5.205}$$

Substituting these expressions of $\overline{X}_h^{(0)}(\tau, \theta, a, p\varphi+\psi)$ and $u_h^{(1)}(\tau, \theta, a, p\varphi+\psi)$ into equations (5.201) and equating the coefficients of equal harmonics, we obtain the following relations for the unknown $k_{nm}^{(h1)}(\tau, a)$:

$$\{\omega_h^2(\tau) - [n^2\nu^2(\tau) + 2n\nu(\tau)m\omega(\tau) + m^2\omega^2(\tau)]\}\,k_{nm}^{(h1)}(\tau, a) = \overline{X}_{hnm}^{(0)}(\tau, a), \tag{5.206}$$

from which we obtain

$$k_{nm}^{(h1)}(\tau, a) = \frac{\overline{X}_{hnm}^{(0)}(\tau, a)}{\omega_h^2(\tau) - [n\nu(\tau) + m(\tau)]^2} \quad (h=1, 2, \ldots, N). \tag{5.207}$$

We thus obtain for $u_h^{(1)}(\tau, \theta, a, p\varphi+\psi)$ the expression

$$u_h^{(1)}(\tau, \theta, a, p\varphi+\psi) = \sum_{n,m}\frac{\overline{X}_{hnm}^{(0)}(\tau, a)}{\omega_h^2(\tau) - [n\nu(\tau) + m\omega(\tau)]^2} \quad (h=1, 2, \ldots, N). \tag{5.208}$$

The right-hand sides of these series should not contain terms in which the denominators $\omega_h^2(\tau) - [n\nu(\tau) + m\omega(\tau)]^2$ may vanish. Hence the terms for which n and m satisfy the relation $n + m\frac{p}{q} = \pm\frac{p_h}{q_h}$ $(h = 1, 2, \ldots, N)$ should not appear in the summation, which requires that $A_1^{(h)}(\tau, a, \psi)$ and $B_1^{(h)}(\tau, a, \psi)$ should satisfy the relations

$$\sum_{n,m}{}' X_{hnm}^{(0)}(\tau, \theta, a, p\varphi+\psi)\, e^{i\{n\theta+m(p\varphi+\psi)\}} = 0, \tag{5.209}$$

where the summation $\sum\limits_{n,m}{}'$ extends over those values of $n, m_1, m_2, \ldots, m_N$, for which $n + m\frac{p}{q} = \pm\frac{p_h}{q_h}$ $(h = 1, 2, \ldots, N)$.

Conditions (5.209) yield $2N$ relations for the determination of $A_1^{(h)}(\tau, a, \psi)$ and $B_1^{(h)}(\tau, a, \psi)$

$$\left.\begin{aligned}
&\left(\omega(\tau) - \frac{p}{q}\nu(\tau)\right)\frac{\partial A_1^{(h)}}{\partial\psi} - 2a_h\omega_h(\tau)B_1^{(h)} = \\
&\quad = \frac{2}{(2\pi)^{N+1}}\sum_\sigma e^{i\sigma q\psi}\int_0^{2\pi}\int_0^{2\pi} X_h^{(0)}(\tau, a, \theta, p\varphi+\psi)\times \\
&\qquad\times\cos(p_h\varphi_h+\psi_h)\, e^{-i\sigma q\psi}\, d\theta\, d(p\varphi+\psi), \\
&\left(\omega(\tau) - \frac{p}{q}\nu(\tau)\right)a_h\frac{\partial B_1^{(h)}}{\partial\psi} + 2\omega_h(\tau)A_1^{(h)} = \\
&\quad = -a_h\frac{d\omega_h(\tau)}{d\tau} - \frac{2}{(2\pi)^{N+1}}\sum_\sigma e^{i\sigma q\psi}\int_0^{2\pi}\int_0^{2\pi} X_h^{(0)}(\tau, a, \theta, p\varphi+\psi)\times \\
&\qquad\times\sin(p_h\varphi_h+\psi_h)\, e^{-i\sigma q\psi}\, d\theta\, d(p\varphi+\psi) \\
&\qquad (h = 1, 2, \ldots, N),
\end{aligned}\right\} \tag{5.210}$$

where σ is the combined index $\sigma_1, \sigma_2, \ldots, \sigma_N$, and $u_h^{(1)}(\tau, \theta, a, p\varphi+\psi)$ are determined by the expressions

$$u_h^{(1)}(\tau, \theta, a, p\varphi+\psi) = \sum_{\substack{n,m \\ \left(n+m\frac{p}{q} \neq \pm\frac{p_h}{q_h}\right)}} \frac{X_{hnm}^{(0)}(\tau, a)\, e^{i[n\theta+m(p\varphi+\psi)]}}{\omega_h^2(\tau) - [n\nu(\tau)+m\omega(\tau)]^2} \quad (h = 1, 2, \ldots, N). \tag{5.211}$$

From equations (5.210) we obtain

$$\left.\begin{aligned}
A_1^{(h)}(\tau, a, \psi) &= -\frac{a_h}{2\omega_h(\tau)}\frac{d\omega_h(\tau)}{d\tau} + \frac{2}{(2\pi)^{N+1}}\sum_\sigma e^{i\sigma q\psi}\times \\
&\times\left[\frac{[q\omega(\tau) - p\nu(\tau)]\sigma i\int_0^{2\pi}\int_0^{2\pi} X_h^{(0)}\cos(p_h\varphi_h+\psi_h)\, e^{-i\sigma q\psi}\, d\theta\, d(p\varphi+\psi)}{4\omega_h^2(\tau) - \sigma^2[\omega(\tau)q - p\nu(\tau)]^2}\right. \\
&\left.\quad - \frac{2\omega_h(\tau)\int_0^{2\pi}\int_0^{2\pi} X_h^{(0)} e^{-i\sigma q\psi}\sin(p_h\varphi_h+\psi_h)\, d\theta\, d(p\varphi+\psi)}{4\omega_h^2(\tau) - \sigma^2[\omega(\tau)q - p\nu(\tau)]^2}\right], \\
B_1^{(h)}(\tau, a, \psi) &= -\frac{2}{(2\pi)^{N+1}}\sum_\sigma e^{i\sigma q\psi}\times \\
&\times\left\{\frac{[q\omega(\tau) - p\nu(\tau)]\sigma i\int_0^{2\pi}\int_0^{2\pi} X_h^{(0)}\sin(p_h\varphi_h+\psi_h)\, e^{-i\sigma q\psi}\, d\theta\, d(p\varphi+\psi)}{4\omega_h^2(\tau) - \sigma^2[\omega(\tau)q - p\nu(\tau)]^2} + \right. \\
&\left.\quad + \frac{2\omega_h(\tau)\int_0^{2\pi}\int_0^{2\pi} X_h^{(0)} e^{-i\sigma q\psi}\cos(p_h\varphi_h+\psi_h)\, d\theta\, d(p\varphi+\psi)}{4\omega_h^2(\tau) - \sigma^2[\omega(\tau)q - p\nu(\tau)]^2}\right\}.
\end{aligned}\right\} \tag{5.212}$$

Replacing the symbolic notation we have

$$A_1^{(h)}(\tau, a, \psi) = -\frac{a_h}{2\omega_h(\tau)}\frac{d\omega_h(\tau)}{d\tau} + \frac{2}{(2\pi)^{N+1}}\sum_{\sigma_1, \ldots, \sigma_N} e^{i\{\sigma_1 q_1\psi_1 + \ldots + \sigma_N q_N \psi_N\}} \times$$

$$\times \frac{i\{\sigma_1[q_1\omega_1(\tau) - p_1\nu(\tau)] + \ldots + \sigma_N[q_N\omega_N(\tau) - p_N\nu(\tau)]\}}{4\omega_h^2(\tau) - \{\sigma_1^2[q_1\omega_1(\tau) - p_1\nu(\tau)]^2 + \ldots + \sigma_N^2[q_N\omega_N(\tau) - p_N\nu(\tau)]^2\}} \int_0^{2\pi}\int_0^{2\pi}\cdots$$

$$\cdots\int_0^{2\pi} X_h^{(0)}(\tau, \theta, a_1, \ldots, a_N, p_1\varphi_1 + \psi_1, \ldots, p_N\varphi_N + \psi_N)\cos(p_h\varphi_h + \psi_h) \times$$

$$\times e^{-i\{\sigma_1 q_1\psi_1 + \ldots + \sigma_N q_N\psi_N\}}\, d\theta\, d(p_1\varphi_1 + \psi_1) \ldots d(p_N\varphi_N + \psi_N) -$$

$$-\frac{4}{(2\pi)^{N+1}}\sum_{\sigma_1, \ldots, \sigma_N}\frac{e^{i\{\sigma_1 q_1\psi_1 + \ldots + \sigma_N q_N\psi_N\}\omega_h(\tau)}}{4\omega_h^2(\tau) - \{\sigma_1^2[q_1\omega_1(\tau) - p_1\nu(\tau)]^2 + \ldots + \sigma_N^2[q_N\omega_N(\tau) - p_N\nu(\tau)]^2\}} \times$$

$$\times \int_0^{2\pi}\int_0^{2\pi}\cdots\int_0^{2\pi} X_h^{(0)}(\tau, \theta, a_1, \ldots, a_N, p_1\varphi_1 + \psi_1, \ldots, p_N\varphi_N + \psi_N) \times$$

$$\times \sin(p_h\varphi_h + \psi_h)\, e^{-i\{\sigma_1 q_1\psi_1 + \ldots + \sigma_N q_N\psi_N\}}\, d\theta\, d(p_1\varphi_1 + \psi_1) \ldots d(p_N\varphi_N + \psi_N). \quad (5.213)$$

A similar expression is obtained for $B_1^{(h)}(\tau, a, \psi)$.

We have thus derived a solution of the system of equations (5.193) in the first, and the first improved approximation. We shall not consider the second approximation which may be carried out according to the method given in § 1 of this chapter. We only remark that it requires an improvement in the accuracy of the system (5.199) obtained from the system (5.191) by successive elimination of y_h and $\frac{dy_h}{dt}$ with a specified degree of accuracy.

Returning to the former variables $q_1, \ldots, q_N$, we obtain in the first approximation (using formulas (5.159) and (5.197)) the expressions

$$q_j = \sum_{h=1}^{N}\left\{X_j^{(h)}(\tau)\, a_h\cos(p_h\varphi_h + \psi_h) - \right.$$

$$\left. - Y_j^{(h)}(\tau)\frac{a_h}{\omega_h(\tau)}\sin(p_h\varphi_h + \psi_h)\right\} \quad (j = 1, 2, \ldots, N), \quad (5.214)$$

where $X_j^{(h)}(\tau)$ and $Y_j^{(h)}(\tau)$ $(j, h = 1, 2, \ldots, N)$ are determined by formulas (5.160), and a_h and ψ_h $(h = 1, 2, \ldots, N)$— from the equations of the first approximation.

As an example we consider the system of equations (5.189) given in the preceding section.

To construct the approximate solution in the first approximation we first derive a system of the second order equivalent (with an accuracy of second order terms) to the system (5.189). After a few operations we obtain

$$\left.\begin{aligned} \frac{d^2x_1}{dt^2} + \omega_1^2(\tau)x_1 &= \varepsilon\left\{\Phi_1(\tau)\frac{dx_1}{dt} + \Phi_2(\tau)\frac{dx_2}{dt}\right\}, \\ \frac{d^2x_2}{dt^2} + \omega_2^2(\tau)x_2 &= \varepsilon\left\{\Phi_3(\tau)\frac{dx_1}{dt} + \Phi_4(\tau)\frac{dx_2}{dt}\right\}, \end{aligned}\right\} \quad (5.215)$$

where

$$\left.\begin{aligned} \Phi_1(\tau) &= \frac{\omega_1'(\tau)}{\omega_1(\tau)} + f_1(\tau) + f_5(\tau), \\ \Phi_2(\tau) &= \frac{\omega_1(\tau)}{\omega_2(\tau)}f_6(\tau) + f_2(\tau), \\ \Phi_3(\tau) &= \frac{\omega_2(\tau)}{\omega_1(\tau)}f_7(\tau) + f_3(\tau), \\ \Phi_4(\tau) &= \frac{\omega_2(\tau)}{\omega_2(\tau)} + f_4(\tau) + f_8(\tau). \end{aligned}\right\} \quad (5.216)$$

For the improved first approximation of the system (5.215) we obtain

$$\left.\begin{aligned}x_1 &= a_1\cos\psi_1 + \varepsilon\frac{\Phi_2(\tau)\,a_2\omega_2(\tau)}{\omega_2^2(\tau)-\omega_1^2(\tau)}\sin\psi_2,\\ x_2 &= a_2\cos\psi_2 + \varepsilon\frac{\Phi_3(\tau)\,a_1\omega_1(\tau)}{\omega_1^2(\tau)-\omega_2^2(\tau)}\sin\psi_1,\end{aligned}\right\}\tag{5.217}$$

where a_1, a_2, ψ_1, ψ_2 are to be determined from the system of equations of the first approximation

$$\left.\begin{aligned}\frac{da_1}{dt} &= \varepsilon\left[\frac{\Phi_1(\tau)}{2}-\frac{\omega_1'(\tau)}{2\omega_1(\tau)}\right]a_1,\\ \frac{da_2}{dt} &= \varepsilon\left[\frac{\Phi_4(\tau)}{2}-\frac{\omega_2'(\tau)}{2\omega_2(\tau)}\right]a_2,\\ \frac{d\psi_1}{dt} &= \omega_1(\tau),\\ \frac{d\psi_2}{dt} &= \omega_2(\tau),\end{aligned}\right\}\tag{5.218}$$

whose integration yields

$$\left.\begin{aligned}a_1 &= a_{10}e^{-\delta_1(\tau,\,t)},\\ a_2 &= a_{20}e^{-\delta_2(\tau,\,t)},\\ \psi_1 &= \bar{\omega}_1(\tau,\,t)+\theta_1,\\ \psi_2 &= \bar{\omega}_2(\tau,\,t)+\theta_2.\end{aligned}\right\}\tag{5.219}$$

Here

$$\left.\begin{aligned}\delta_1(\tau,\,t) &= -\int\left[\frac{\Phi_1(\tau)}{2}-\frac{\omega_1'(\tau)}{2\omega_1(\tau)}\right]d\tau,\\ \delta_2(\tau,\,t) &= -\int\left[\frac{\Phi_4(\tau)}{2}-\frac{\omega_2'(\tau)}{2\omega_2(\tau)}\right]d\tau,\\ \bar{\omega}_1(\tau,\,t) &= \int\omega_1(\tau)\,dt,\quad \bar{\omega}_2(\tau,\,t)=\int\omega_2(\tau)\,dt,\end{aligned}\right\}\tag{5.220}$$

and a_{10}, a_{20}, θ_1, θ_2 are integration constants.

Substituting the values of (5.219) into the right-hand sides of the expressions in (5.217), we obtain

$$\left.\begin{aligned}x_1 &= a_{10}e^{-\delta_1(\tau,\,t)}\cos[\bar{\omega}_1(\tau,\,t)+\theta_1]+\\ &\quad+\varepsilon\frac{\Phi_2(\tau)\,\omega_2(\tau)\,a_{20}e^{-\delta_2(\tau,\,t)}}{\omega_2^2(\tau)-\omega_1^2(\tau)}\sin[\bar{\omega}_2(\tau,\,t)+\theta_2],\\ x_2 &= a_{20}e^{-\delta_2(\tau,\,t)}\cos[\bar{\omega}_2(\tau,\,t)+\theta_2]+\\ &\quad+\varepsilon\frac{\Phi_3(\tau)\,\omega_1(\tau)\,a_{10}e^{-\delta_1(\tau,\,t)}}{\omega_1^2(\tau)-\omega_2^2(\tau)}\sin[\bar{\omega}_1(\tau,\,t)+\theta_1].\end{aligned}\right\}\tag{5.221}$$

Transforming to the former variables q_1 and q_2, we obtain in the first approximation

$$\left.\begin{aligned}q_1 &= -\frac{1}{\sqrt{2Im(\tau)}}\{a_{10}e^{-\delta_1(\tau,\,t)}\sin[\bar{\omega}_1(\tau,\,t)+\theta_1]+\\ &\quad+a_{20}e^{-\delta_2(\tau,\,t)}\sin[\bar{\omega}_2(\tau,\,t)+\theta_2]\},\\ q_2 &= -\frac{1}{\sqrt{2Im(\tau)}}\{a_{10}e^{-\delta_1(\tau,\,t)}\cos[\bar{\omega}_1(\tau,\,t)+\theta_1]+\\ &\quad+a_{20}e^{-\delta_2(\tau,\,t)}\cos[\bar{\omega}_2(\tau,\,t)+\theta_2]\}.\end{aligned}\right\}\tag{5.222}$$

Finally we give another relatively simple method of constructing the first approximation for the general solution of the system (5.191) based on applying the averaging principle.

To simplify the calculations we consider separately the two cases: the case of an autonomous gyroscopic system when the right-hand sides of the system of equations (5.145) do not depend explicitly on time, and the case of a nonautonomous gyroscopic system.

In the case of an autonomous gyroscopic system (rather "quasi-autonomous" since we assume that the dependence on τ — the slowing time — is retained) when the right-hand sides of (5.145) do not depend explicitly on time being only functions of the variables $q\ (q_1,\ q_2,\ \ldots,\ q_N)$ and τ, we obtain after the substitution of variables according to (5.159) a system of differential equations in the quasi-normal coordinates $x\ (x_1,\ x_2,\ \ldots,\ x_N)$, $y\ (y_1,\ y_2,\ \ldots,\ y_N)$ of the form

$$\left.\begin{aligned} &\frac{dx_h}{dt} - \omega_h(\tau)\, y_h = \varepsilon Q_h^{(1)}(\tau,\ x_1,\ x_2,\ \ldots,\ x_N,\ y_1,\ y_2,\ \ldots,\ y_N,\ \varepsilon),\\ &\frac{dy_h}{dt} + \omega_h(\tau)\, x_h = \varepsilon Q_h^{(2)}(\tau,\ x_1,\ x_2,\ \ldots,\ x_N,\ y_1,\ y_2,\ \ldots,\ y_N,\ \varepsilon)\\ &\qquad\qquad (h = 1,\ 2,\ \ldots,\ N). \end{aligned}\right\} \tag{5.223}$$

The nonattenuating vibrations

$$x_h = a_h \cos(\omega_h(\tau)\, t + \psi_h), \quad y_h = -a_h \sin(\omega_h(\tau)\, t + \psi_h) \qquad (h = 1,\ 2,\ \ldots,\ N), \tag{5.224}$$

are possible in the unperturbed system (i. e., in the system (5.223) for $\varepsilon = 0$ and $\tau = \text{const}$) under the previously indicated assumptions. As the system (5.223) is close ($\varepsilon \ll 1$) to a linear conservative one, we look for its solution in the first approximation in the form

$$\left.\begin{aligned} x_h &= a_h \cos(\vartheta_h + \psi_h),\\ y_h &= -a_h \sin(\vartheta_h + \psi_h) \qquad (h = 1,\ 2,\ \ldots,\ N), \end{aligned}\right\} \tag{5.225}$$

where a_h, ψ_h are slowly varying functions, $\frac{d\vartheta}{dt} = \omega_h(\tau)$, $\tau = \varepsilon t$.

Regarding (5.225) as transformation formulas for the system of equations (5.223) to the new independent variables ψ_h, we obtain instead of (5.223) the following system of equations in the standard form:

$$\left.\begin{aligned} \frac{da_h}{dt} &= \varepsilon\, [Q_{h0}^{(1)}(\tau,\ a,\ \vartheta,\ \psi,\ \varepsilon) \cos(\vartheta_h + \psi_h) -\\ &\qquad - Q_{h0}^{(2)}(\tau,\ a,\ \vartheta,\ \psi,\ \varepsilon) \sin(\vartheta_h + \psi_h)],\\ \frac{d\psi_h}{dt} &= -\frac{\varepsilon}{a_h}\, [Q_{h0}^{(1)}(\tau,\ a,\ \vartheta,\ \psi,\ \varepsilon) \sin(\vartheta_h + \psi_h) +\\ &\qquad + Q_{h0}^{(2)}(\tau,\ a,\ \vartheta,\ \psi,\ \varepsilon) \cos(\vartheta_h + \psi_h)], \end{aligned}\right\} \tag{5.226}$$

where

$$Q_{h0}^{(i)}(\tau,\ a,\ \vartheta,\ \psi,\ \varepsilon) = Q_h^{(i)}(\tau,\ a_1 \cos(\vartheta_1 + \psi_1),\ \ldots,\ -a_1 \sin(\vartheta_1 + \psi_1),\ \ldots,\ \varepsilon)$$
$$(h = 1,\ 2,\ \ldots,\ N;\ i = 1,\ 2).$$

Averaging this system of equations over the angular variables $\vartheta_h + \psi_h$ ($h = 1$, $2,\ \ldots,\ N$), we obtain the equations of the first approximation for a_h, ψ_h:

$$\left.\begin{aligned} \frac{da_h}{dt} &= -\frac{\varepsilon}{(2\pi)^N} \int_0^{2\pi} \cdots \int_0^{2\pi} [Q_{h0}^{(1)}(\tau,\ a,\ \vartheta,\ \psi,\ \varepsilon) \cos(\vartheta_h + \psi) -\\ &\qquad - Q_{h0}^{(2)}(\tau,\ a,\ \vartheta,\ \psi,\ \varepsilon) \sin(\vartheta_h + \psi_h)]\, d(\vartheta + \psi),\\ \frac{d\psi_h}{dt} &= -\frac{\varepsilon}{(2\pi)^N a_h} \int_0^{2\pi} \cdots \int_0^{2\pi} [Q_{h0}^{(1)}(\tau,\ a,\ \vartheta,\ \psi,\ \varepsilon) \sin(\vartheta_h + \psi_h) +\\ &\qquad + Q_{h0}^{(2)}(\tau,\ a,\ \vartheta,\ \psi,\ \varepsilon) \cos(\vartheta_h + \psi_h)]\, d(\vartheta + \psi)\\ &\qquad (h = 1,\ 2,\ \ldots,\ N), \end{aligned}\right\} \tag{5.227}$$

where

$$d(\vartheta + \psi) = d(\vartheta_1 + \psi_1) \ldots d(\vartheta_N + \psi_N).$$

We now consider the nonautonomous case for which the system (5.145) is reduced by means of the substitution of variables (5.159) to the form

$$\left.\begin{aligned} &\frac{dx_h}{dt}-\omega_h(\tau)\,y_h=\varepsilon Q_h^{(1)}(\tau,\,\theta,\,x_1,\,\ldots,\,x_N,\,y_1,\,\ldots,\,y_N,\,\varepsilon),\\ &\frac{dy_h}{dt}+\omega_h(\tau)\,x_h=\varepsilon Q_h^{(2)}(\tau,\,\theta,\,x_1,\,\ldots,\,x_N,\,y_1,\,\ldots,\,y_N,\,\varepsilon)\\ &\qquad (h=1,\,2,\,\ldots,\,N), \end{aligned}\right\} \tag{5.228}$$

where the right-hand sides are periodic in θ with period 2π, $\frac{d\theta}{dt}=\nu(\tau)$ and $\tau=\varepsilon t$.

We introduce into the system (5.228) the new variables a_h, ψ_h $(h=1,\,2,\,\ldots,\,N)$ defined by

$$\left.\begin{aligned} &x_h=a_h\cos(\theta+\psi_h),\\ &y_h=-\frac{a_h\nu(\tau)}{\omega_h(\tau)}\sin(\theta+\psi_h)\\ &(h=1,\,2,\ldots,\,N), \end{aligned}\right\} \tag{5.229}$$

where $\tau=\varepsilon t$.

Then after a number of operations we obtain instead of (5.228) the system

$$\left.\begin{aligned} \frac{da_h}{dt}=\varepsilon&\left(\frac{\omega'_{h\tau}(\tau)}{\omega_h(\tau)}-\frac{\nu'_\tau(\tau)}{\nu(\tau)}\right)a_h\sin^2(\theta+\psi_h)+\\ &+\frac{\omega_h^2(\tau)-\nu^2(\tau)}{\nu(\tau)}\,a_h\cos(\theta+\psi_h)\sin(\theta+\psi_h)+\\ &+\varepsilon\Big[Q_{h0}^{(1)}(\tau,\,a,\,\theta,\,\theta+\psi,\,\varepsilon)\cos(\theta+\psi_h)-\\ &\qquad-Q_{h0}^{(2)}(\tau,\,a,\,\theta,\,\theta+\psi,\,\varepsilon)\frac{\omega_h(\tau)}{\nu(\tau)}\sin(\theta+\psi_h)\Big],\\ \frac{d\psi_h}{dt}=\varepsilon&\left(\frac{\omega'_{h\tau}(\tau)}{\omega_h(\tau)}-\frac{\nu'_\tau(\tau)}{\nu(\tau)}\right)\sin(\theta+\psi_h)\cos(\theta+\psi_h)+\\ &+\frac{\omega_h^2(\tau)-\nu^2(\tau)}{\nu(\tau)}\cos^2(\theta+\psi_h)-\\ &-\frac{\varepsilon}{a_h}\Big[Q_{h0}^{(1)}(\tau,\,a,\,\theta,\,\theta+\psi,\,\varepsilon)\sin(\theta+\psi_h)+\\ &\qquad+Q_{h0}^{(2)}(\tau,\,a,\,\theta,\,\theta+\psi,\,\varepsilon)\frac{\omega_h(\tau)}{\nu(\tau)}\cos(\theta+\psi_h)\Big]\\ &\qquad(h=1,\,2,\,\ldots,\,N), \end{aligned}\right\} \tag{5.230}$$

where

$$\begin{aligned} &Q_{h0}^{(i)}(\tau,\,a,\,\theta,\,\theta+\psi,\,\varepsilon)=\\ &\qquad=Q_h^{(i)}\left(\tau,\,\theta,\,a,\cos(\theta+\psi),\,\ldots,\,-\frac{a_1\nu(\tau)}{\omega_1(\tau)}\sin(\theta+\psi),\,\ldots,\,\varepsilon\right)\\ &\qquad\qquad(h=1,\,2,\,\ldots,\,N;\;i=1,\,2). \end{aligned}$$

Considering vibrations in a region near the resonance at the k-th eigenfrequency, and putting in that case

$$\omega_k(\tau)-\nu(\tau)=\varepsilon\Delta(\tau), \tag{5.231}$$

we can apply the averaging principle to the k-th pair of equations of the system (5.230) and obtain

$$\left.\begin{aligned} \frac{da_k}{dt}=\varepsilon&\left(\frac{\omega'_{k\tau}(\tau)}{\omega_k(\tau)}-\frac{\nu'_\tau(\tau)}{\nu(\tau)}\right)a_k+\frac{\varepsilon}{(2\pi)^N}\int_0^{2\pi}\!\!\cdots\!\int_0^{2\pi}\Big[Q_{k0}^{(1)}(\tau,\,a,\,\theta,\,\theta+\psi,\,\varepsilon)\cos(\theta+\psi_k)-\\ &-Q_{k0}^{(2)}(\tau,\,a,\,\theta,\,\theta+\psi,\,\varepsilon)\frac{\omega_k(\tau)}{\nu(\tau)}\sin(\theta+\psi_k)\Big]\,d(\theta+\psi),\\ \frac{d\psi_k}{dt}=\omega_k&(\tau)-\nu(\tau)-\frac{\varepsilon}{(2\pi)^N a_k}\int_0^{2\pi}\!\!\cdots\!\int_0^{2\pi}\Big[Q_{k0}^{(1)}(\tau,\,a,\,\theta,\,\theta+\psi,\,\varepsilon)\sin(\theta+\psi_k)+\\ &+Q_0^{(2)}(\tau,\,a,\,\theta,\,\theta+\psi,\,\varepsilon)\frac{\omega_k(\tau)}{\nu(\tau)}\cos(\theta+\psi_k)\Big]\,d(\theta+\psi). \end{aligned}\right\} \tag{5.232}$$

§ 7. Construction of asymptotic solutions for a nonlinear equation with gyroscopic terms in the presence of internal resonance

While examining nonlinear systems with many degrees of freedom we have imposed a number of restrictions among which the absence of any kind of internal resonances in the considered system seemed to be an essential condition. This restriction, which was introduced for simplification of calculations and of final formulas, is however not necessary when applying asymptotic methods. Approximate solutions may be constructed in an analogous way in the presence of an internal resonance.

Let us for simplicity consider the particular case when the problem of vibrations reduces to the study of a system of nonlinear differential equations of the type

$$\left.\begin{aligned} a_{11}(\tau)\frac{d^2q_1}{dt^2}+\gamma(\tau)\frac{dq_2}{dt}+b_{11}(\tau)q_1=\varepsilon f_1(\tau,\theta,q_1,\dot{q}_1,q_2,\dot{q}_2),\\ a_{22}(\tau)\frac{d^2q_2}{dt^2}-\gamma(\tau)\frac{dq_1}{dt}+b_{22}(\tau)q_2=\varepsilon f_2(\tau,\theta,q_1,\dot{q}_1,q_2,\dot{q}_2),\end{aligned}\right\} \tag{5.233}$$

where, as always, q_1, q_2 are generalized coordinates; t is the time; $\tau=\varepsilon t$; and the coefficients $a_{11}(\tau), a_{22}(\tau), b_{11}(\tau), b_{22}(\tau), \gamma(\tau)$ and the functions on the right-hand sides of these equations satisfy the conditions listed in Chapter IV, § 1 (p. 158) for all $0\leqslant\tau\leqslant L$.

We consider together with the system (5.233) the equations of "unperturbed motion"

$$\left.\begin{aligned} a_{11}(\tau)\frac{d^2q_1}{dt^2}+\gamma(\tau)\frac{dq_2}{dt}+b_{11}(\tau)q_1=0,\\ a_{22}(\tau)\frac{d^2q_2}{dt^2}-\gamma(\tau)\frac{dq_1}{dt}+b_{22}(\tau)q_2=0,\end{aligned}\right\} \tag{5.234}$$

whose coefficients depend on τ as a parameter. The general solution of this system has the form

$$q_1=a_1\cos\psi_1+a_2\cos\psi_2,$$
$$q_2=c_1(\tau)a_1\sin\psi_1+c_2(\tau)a_2\sin\psi_2,$$

where $\psi_1=\omega_1 t+\varphi_1$, $\psi_2=\omega_2 t+\varphi_2$, a_1, a_2, φ_1, φ_2 are arbitrary constants depending on the initial conditions, ω_1 and ω_2 are roots of the characteristic equation

$$D(\omega)=\begin{vmatrix} b_{11}(\tau)-\omega^2a_{11}(\tau) & \gamma(\tau)\omega\\ \gamma(\tau)\omega & b_{22}(\tau)-\omega^2a_{22}(\tau)\end{vmatrix}=0, \tag{5.235}$$

and $c_1(\tau)$ and $c_2(\tau)$ are basic functions determined by

$$c_1(\tau)=\frac{a_{11}(\tau)\omega_1^2(\tau)-b_{11}(\tau)}{\gamma(\tau)\omega_1(\tau)},\quad c_2(\tau)=\frac{a_{11}(\tau)\omega_2^2(\tau)-b_{11}(\tau)}{\gamma(\tau)\omega_2(\tau)}.$$

It is now assumed that an internal resonance may occur in the vibrating system under investigation for some value $\tau=\tau^*$ $(0\leqslant\tau^*\leqslant L)$, i. e., the equality

$$\omega_1(\tau^*)=\frac{p}{q}\omega_2(\tau^*),$$

where p and q are some small, mutually prime integers, is possible. It is also assumed that the frequency of the external force $\nu(\tau)$ may coincide for some values of τ with one of the eigenfrequencies $\omega_1(\tau)$ or $\omega_2(\tau)$, i. e., the equalities

$$\nu(\tau_1)=\omega_1(\tau_1)\quad\text{or}\quad\nu(\tau_2)=\omega_2(\tau_2),$$

where τ_1, τ_2 belong to the interval $0\leqslant\tau\leqslant L$, are possible.

The asymptotic solution of the system of differential equations (5.233) is then looked for in the form of asymptotic series

$$q_1 = a_1 \cos(\theta+\psi_1) + a_2 \cos(\theta+\psi_2) + \varepsilon u_1(\tau, a_1, a_2, \theta, \theta+\psi_1, \theta+\psi_2) + \varepsilon^2 \ldots, \qquad (5.236)$$
$$q_2 = c_1(\tau) a_1 \sin(\theta+\psi_1) + c_2(\tau) a_2 \sin(\theta+\psi_2) + \varepsilon v_1(\tau, a_1, a_2, \theta, \theta+\psi_1, \theta+\psi_2) + \varepsilon^2 \ldots,$$

where $\tau = \varepsilon t$, $u_1(\tau, a_1, a_2, \theta, \theta+\psi_1, \theta+\psi_2, \ldots)$, $v_1(\tau, a_1, a_2, \theta, \theta+\psi_1, \theta+\psi_2)$, ... are certain functions, periodic in θ, $\theta+\psi_1$, $\theta+\psi_2$ with period 2π, and the quantities a_1, a_2, ψ_1, ψ_2 are to be determined from the system of differential equations

$$\left.\begin{aligned}
\frac{da_1}{dt} &= \varepsilon A_1^{(1)}(\tau, a_1, a_2, \psi_1, \psi_2) + \varepsilon^2 A_2^{(1)}(\tau, a_1, a_2, \psi_1, \psi_2) + \ldots, \\
\frac{da_2}{dt} &= \varepsilon A_1^{(2)}(\tau, a_1, a_2, \psi_1, \psi_2) + \varepsilon^2 A_2^{(2)}(\tau, a_1, a_2, \psi_1, \psi_2) + \ldots, \\
\frac{d\psi_1}{dt} &= \omega_1(\tau) - \nu(\tau) + \varepsilon B_1^{(1)}(\tau, a_1, a_2, \psi_1, \psi_2) + \varepsilon^2 B_2^{(1)}(\tau, a_1, a_2, \psi_1, \psi_2) + \ldots, \\
\frac{d\psi_2}{dt} &= \omega_2(\tau) - \nu(\tau) + \varepsilon B_1^{(2)}(\tau, a_1, a_2, \psi_1, \psi_2) + \varepsilon^2 B_2^{(2)}(\tau, a_1, a_2, \psi_1, \psi_2) + \ldots
\end{aligned}\right\} \qquad (5.237)$$

Confining ourselves to the first approximation, which is absolutely sufficient in practice, we look for a solution of the system of differential equations (5.233) in the form

$$\left.\begin{aligned}
q_1 &= a_1 \cos\psi_1 + a_2 \cos\psi_2, \\
q_2 &= c_1(\tau) a_1 \sin\psi_1 + c_2(\tau) a_2 \sin\psi_2,
\end{aligned}\right\} \qquad (5.238)$$

where $\tau = \varepsilon t$, and the quantities a_1, a_2, ψ_1, ψ_2 are to be determined from the system of equations of the first approximation

$$\left.\begin{aligned}
\frac{da_1}{dt} &= \varepsilon A_1^{(1)}(\tau, a_1, a_2, \psi_1, \psi_2), \\
\frac{da_2}{dt} &= \varepsilon A_1^{(2)}(\tau, a_1, a_2, \psi_1, \psi_2), \\
\frac{d\psi_1}{dt} &= \omega_1(\tau) - \nu(\tau) + \varepsilon B_1^{(1)}(\tau, a_1, a_2, \psi_1, \psi_2), \\
\frac{d\psi_2}{dt} &= \omega_2(\tau) - \nu(\tau) + \varepsilon B_1^{(2)}(\tau, a_1, a_2, \psi_1, \psi_2).
\end{aligned}\right\} \qquad (5.239)$$

The problem in the first approximation is thus reduced to deriving $A_1^{(1)}(\tau, a_1, a_2, \psi_1, \psi_2)$, $A_1^{(2)}(\tau, a_1, a_2, \psi_1, \psi_2)$, $B_1^{(1)}(\tau, a_1, a_2, \psi_1, \psi_2)$, $B_1^{(2)}(\tau, a_1, a_2, \psi_1, \psi_2)$. For a unique determination of these quantities the functions $u_1(\tau, a_1, a_2, \theta, \theta+\psi_1, \theta+\psi_2)$, $v_1(\tau, a_1, a_2, \theta, \theta+\psi_1, \theta+\psi_2)$ must not contain terms with denominators which can vanish.

We now derive these quantities. Differentiating the right-hand sides of the series (5.236) twice, keeping in mind the relations (5.237), and substituting the values of q_1, q_2 and their derivatives into equations (5.233), we equate the coefficients of ε, obtaining the following equations for $u_1(\tau, a_1, a_2, \theta, \theta+\psi_1, \theta+\psi_2)$, $v_1(\tau, a_1, a_2, \theta, \theta+\psi_1, \theta+\psi_2)$:

$$\left.\begin{aligned}
a_{11}(\tau)\frac{d^2u_1}{dt^2} + \gamma(\tau)\frac{dv_1}{dt} + b_{11}u_1 &= \bar{f}_1(\tau, \theta, x_0, \dot{x}_0, y_0, \dot{y}_0), \\
a_{22}(\tau)\frac{d^2v_1}{dt^2} - \gamma(\tau)\frac{du_1}{dt} + b_{22}v_1 &= \bar{f}_2(\tau, \theta, x_0, \dot{x}_0, y_0, \dot{y}_0),
\end{aligned}\right\} \qquad (5.240)$$

where

$$\begin{aligned}
\bar{f}_1(\tau, \theta, x_0, \dot{x}_0, y_0, \dot{y}_0) &= f_1(\tau, \theta, x_0, \dot{x}_0, y_0, \dot{y}_0) - \\
&- \left\{a_{11}\frac{dA_1^{(1)}}{dt} - 2a_{11}a_1\omega_1 B_1^{(1)} + \gamma c_1 a_1 B_1^{(1)}\right\}\cos(\theta+\psi_1) - \\
&- \left\{a_{11}\frac{dA_1^{(2)}}{dt} - (2a_{11}\omega_2 - \gamma c_2) a_2 B_1^{(2)}\right\}\cos(\theta+\psi_2) + \\
&+ \left\{a_{11}a_1\frac{dB_1^{(1)}}{dt} + (2a_{11}\omega_1 - \gamma c_1) A_1^{(1)} + a_{11}a_1\frac{d\omega_1}{d\tau} - \gamma a_1\frac{dc_1}{d\tau}\right\} \times \\
\times \sin(\theta+\psi_1) &+ \left\{a_{11}a_2\frac{dB_1^{(2)}}{dt} + (2a_{11}\omega_2 - \gamma c_2) A_1^{(2)} + a_{11}a_2\frac{d\omega_2}{d\tau} - \gamma a_2\frac{dc_2}{d\tau}\right\}\sin(\theta+\psi_2),
\end{aligned} \qquad (5.241)$$

$$\bar{f}_2(\tau, \theta, x_0, \dot{x}_0, y_0, \dot{y}_0) = f_2(\tau, \theta, x_0, \dot{x}_0, y_0, \dot{y}_0) -$$
$$- \left\{ a_{22}c_1 \frac{dA_1^{(1)}}{dt} - (2a_{22}c_1\omega_1 - \gamma a_1) B_1^{(1)} \right\} \sin(\theta + \psi_1) -$$
$$- \left\{ a_{22}c_2 \frac{dA_1^{(2)}}{dt} - (2a_{22}c_2\omega_2 + \gamma a_2) B_1^{(2)} \right\} \sin(\theta + \psi_2) -$$
$$- \left\{ a_{22}c_1a_1 \frac{dB_1^{(1)}}{dt} + (2a_{22}c_1\omega_1 - \gamma) A_1^{(1)} + 2a_{22}a_1\omega_1 \frac{dc_1}{d\tau} + \right.$$
$$\left. + a_{22}c_1a_1 \frac{d\omega_1}{d\tau} \right\} \cos(\theta + \psi_1) - \left\{ a_{22}c_2a_2 \frac{dB_1^{(2)}}{dt} + \right.$$
$$\left. + (2a_{22}c_2\omega_2 - \gamma) A_1^{(2)} + 2a_{22}a_2\omega_2 \frac{dc_2}{d\tau} + a_{22}c_2a_2 \frac{d\omega_2}{d\tau} \right\} \cos(\theta + \psi_2), \qquad (5.241)$$

and

$$\left. \begin{aligned} x_0 &= a_1 \cos(\theta + \psi_1) + a_2 \cos(\theta + \psi_2), \\ y_0 &= c_1a_1 \sin(\theta + \psi_1) + c_2a_2 \sin(\theta + \psi_2), \\ \dot{x}_0 &= -a_1\omega_1 \sin(\theta + \psi_1) - a_2\omega_2 \sin(\theta + \psi_2), \\ \dot{y}_0 &= c_1a_1\omega_1 \cos(\theta + \psi_1) + c_2a_2\omega_2 \cos(\theta + \psi_2). \end{aligned} \right\} \qquad (5.242)$$

Representing the expressions in (5.241) as Fourier sums

$$\bar{f}_i(\tau, \theta, x_0, \dot{x}_0, y_0, \dot{y}_0) = \sum_k \sum_n \sum_m \bar{f}_{iknm}(\tau, a_1, a_2)\, e^{i\{k\theta + n(\theta+\psi_1) + m(\theta+\psi_2)\}}, \qquad (5.243)$$

where

$$\bar{f}_{iknm}(\tau, a_1, a_2) = \frac{1}{8\pi^3} \int_0^{2\pi}\int_0^{2\pi}\int_0^{2\pi} \bar{f}_i(\tau, \theta, x_0, \dot{x}_0, y_0, \dot{y}_0) \times$$
$$\times e^{-i\{k\theta + n(\theta+\psi_1) + m(\theta+\psi_2)\}}\, d\theta\, d(\theta + \psi_1)\, d(\theta + \psi_2) \qquad (i = 1, 2), \qquad (5.244)$$

we find from equations (5.240) the expressions (with an accuracy of the order of ε)

$$\left. \begin{aligned} & u_1(\tau, a_1, a_2, \theta, \theta + \psi_1, \theta + \psi_2) = \\ & \qquad = \sum_k \sum_n \sum_m u_{1knm}(\tau, a_1, a_2)\, e^{i\{k\theta + n(\theta+\psi_1) + m(\theta+\psi_2)\}}, \\ & v_1(\tau, a_1, a_2, \theta, \theta + \psi_1, \theta + \psi_2) = \\ & \qquad = \sum_k \sum_n \sum_m v_{1knm}(\tau, a_1, a_2)\, e^{i\{k\theta + n(\theta+\psi_1) + m(\theta+\psi_2)\}}, \end{aligned} \right\} \qquad (5.245)$$

for $u_1(\tau, a_1, a_2, \theta, \theta + \psi_1, \theta + \psi_2)$ and $v_1(\tau, a_1, a_2, \theta, \theta + \psi_1, \theta + \psi_2)$. Here the coefficients on the right-hand sides have the form

$$\left. \begin{aligned} u_{1knm}(\tau, a_1, a_2) &= \frac{[b_{22}(\tau) - a_{22}(\tau)(k\nu + n\omega_1 + m\omega_2)^2]\, \bar{f}_{1knm}(\tau, a_1, a_2)}{D(k\nu + n\omega_1 + m_1\omega_2)} - \\ & \quad - \frac{i\gamma(\tau)(k\nu + n\omega_1 + m\omega_2)\, \bar{f}_{2knm}(\tau, a_1, a_2)}{D(k\nu + n\omega_1 + m\omega_2)}, \\ v_{1knm}(\tau, a_1, a_2) &= \frac{-[b_{11}(\tau) - a_{11}(\tau)(k\nu + n\omega_1 + m\omega_2)^2]\, \bar{f}_{2knm}(\tau, a_1, a_2)}{D(k\nu + n\omega_1 + m\omega_2)} + \\ & \quad + \frac{i\gamma(\tau)(k\nu + n\omega_1 + m\omega_2)\, \bar{f}_{1knm}(\tau, a_1, a_2)}{D(k\nu + n\omega_1 + m_1\omega_2)}. \end{aligned} \right\} \qquad (5.246)$$

To ensure that the functions $u_1(\tau, a_1, a_2, \theta, \theta + \psi_1, \theta + \psi_2)$ and $v_1(\tau, a_1, a_2, \theta, \theta + \psi_1, \theta + \psi_2)$ do not contain terms with denominators which can vanish for some values of τ, the Fourier series (5.243) must not contain terms with those indexes k, n, m, for which $D(k\nu + n\omega_1 + m\omega_2) = 0$, i.e., for which

$$k\nu + n\omega_1 + m\omega_2 = \pm\, \omega_1$$

or

$$k\nu + n\omega_1 + m\omega_2 = \pm\, \omega_2.$$

Equating the corresponding Fourier coefficients to zero in (5.245), we obtain relations from which we can determine the quantities $A_1^{(1)}(\tau, a_1, a_2, \psi_1, \psi_2)$,

$A_1^{(2)}(\tau, a_1, a_2, \psi_1, \psi_2)$, $B_1^{(1)}(\tau, a_1, a_2, \psi_1, \psi_2)$, $B_1^{(2)}(\tau, a_1, a_2\ \psi_1, \psi_2)$ depending upon the choice of the resonance under consideration or, more accurately, upon the relation between $\nu(\tau)$, $\omega_1(\tau)$, and $\omega_2(\tau)$ which may take place during vibrations.

Let us consider the following possibilities.

First case: for certain values of $\tau\,(0\leqslant\tau\leqslant L)$ the following relations are possible between $\nu(\tau)$, $\omega_1(\tau)$, and $\omega_2(\tau)$:

$$\nu(\tau_1)=\omega_1(\tau_1),\ \nu(\tau_2)=\omega_2(\tau_2),\ \omega_1(\tau_1)\neq\frac{p}{q}\omega_2(\tau).$$

Second case:

$$\nu(\tau_1)=\omega_1(\tau_1),\ \omega_1(\tau_2)=\frac{p}{q}\omega_2(\tau_2),\ \nu(\tau)\neq\omega_1(\tau).$$

Third case:

$$\nu(\tau_1)=\omega_2(\tau_1),\ \omega_1(\tau_2)=\frac{p}{q}\omega_2(\tau_2),\ \nu(\tau)\neq\omega_1(\tau),$$

where p and q are some mutually prime integers, and $\tau_1\neq\tau_2$, $0\leqslant\tau_1\leqslant L$, $0\leqslant\tau_2\leqslant L$.

Corresponding to these cases, we derive the quantities

$$A_1^{(1)}(\tau, a_1, a_2, \psi_1, \psi_2),\quad A_1^{(2)}(\tau, a_1, a_2, \psi_1, \psi_2),$$
$$B_1^{(1)}(\tau, a_1, a_2, \psi_1, \psi_2),\quad B_1^{(2)}(\tau, a_1, a_2, \psi_1, \psi_2).$$

In the first case we obtain after a series of operations

$$\left.\begin{aligned}&A_1^{(1)}(\tau, a_1, a_2, \psi_1, \psi_2)=\sum_\sigma A_{1\sigma}^{(1)},\quad A_1^{(2)}(\tau, a_1, a_2, \psi_1, \psi_2)=\sum_\sigma A_{1\sigma}^{(2)},\\&B_1^{(1)}(\tau, a_1, a_2, \psi_1, \psi_2)=\sum_\sigma B_{1\sigma}^{(1)},\quad B_1^{(2)}(\tau, a_1, a_2, \psi_1, \psi_2)=\sum_\sigma B_{1\sigma}^{(2)},\end{aligned}\right\}\tag{5.247}$$

where $A_{1\sigma}^{(1)}$, $A_{1\sigma}^{(2)}$, $B_{1\sigma}^{(1)}$ and $B_{1\sigma}^{(2)}$, periodic in ψ_1 and ψ_2, are certain solutions of the system of equations

$$\left.\begin{aligned}&R_{\sigma j}\frac{dA_{1\sigma}^{(j)}}{dt}-L_{\sigma j}a_jB_{1\sigma}^{(j)}=\\&\quad=\frac{e^{-i\sigma\psi_1}}{4\pi^3}\int_0^{2\pi}\int_0^{2\pi}\int_0^{2\pi}\{h_{1j}^\sigma(\tau)f_1(\tau,\theta,x_0,\dot{x}_0,y_0,\dot{y}_0)\cos(\theta+\psi_j)-\\&-g_{1j}^\sigma(\tau)f_2(\tau,\theta,x_0,\dot{x}_0,y_0,\dot{y}_0)\sin(\theta+\psi_j)\}e^{i\sigma\psi_1}d\theta\,d(\theta+\psi_1)\,d(\theta+\psi_2)+\\&\quad+\frac{e^{-i\sigma\psi_2}}{4\pi^3}\int_0^{2\pi}\int_0^{2\pi}\int_0^{2\pi}\{h_{2j}^\sigma(\tau)f_1(\tau,\theta,x_0,\dot{x}_0,y_0,\dot{y}_0)\cos(\theta+\psi_j)-\\&-g_{2j}^\sigma(\tau)f_2(\tau,\theta,x_0,\dot{x}_0,y_0,\dot{y}_0)\sin(\theta+\psi_j)\}e^{i\sigma\psi_2}d\theta\,d(\theta+\psi_1)\,d(\theta+\psi_2),\\&R_{\sigma j}a_j\frac{dB_{1\sigma}^{(j)}}{dt}+L_{\sigma j}A_{1\sigma}^{(j)}=\\&\quad=-\frac{e^{-i\sigma\psi_1}}{4\pi^3}\int_0^{2\pi}\int_0^{2\pi}\int_0^{2\pi}\{h_{1j}^\sigma(\tau)f_1(\tau,\theta,x_0,\dot{x}_0,y_0,\dot{y}_0)\sin(\theta+\psi_j)-\\&-g_{1j}^\sigma(\tau)f_2(\tau,\theta,x_0,\dot{x}_0,y_0,\dot{y}_0)\cos(\theta+\psi_i)\}e^{i\sigma\psi_1}d\theta\,d(\theta+\psi_1)\,d(\theta+\psi_2)-\\&\quad-\frac{e^{-i\sigma\psi_2}}{4\pi^3}\int_0^{2\pi}\int_0^{2\pi}\int_0^{2\pi}\{h_{2j}^\sigma(\tau)f_1(\tau,\theta,x_0,\dot{x}_0,y_0,\dot{y}_0)\sin(\theta+\psi_j)-\\&\quad-g_{2j}^\sigma(\tau)f_2(\tau,\theta,x_0,\dot{x}_0,y_0,\dot{y}_0)\cos(\theta+\psi_j)\}\times\\&\quad\times e^{i\sigma\psi_2}d\theta\,d(\theta+\psi_1)\,d(\theta+\psi_2)-a_j\left(a_{11}\frac{d\omega_j}{d\tau}-\gamma\frac{dc_j}{d\tau}\right)\times\\&\times(h_{1j}^\sigma(\tau)+h_{2j}^\sigma(\tau))+a_j\left(a_{22}c_j\frac{d\omega_j}{d\tau}+2a_{22}\omega_j\frac{dc_j}{d\tau}\right)(g_{1j}^\sigma(\tau)+g_{2j}^\sigma(\tau))\\&\qquad\qquad(j=1,2).\end{aligned}\right\}\tag{5.248}$$

We have introduced here the notation

$$\left.\begin{aligned} h_{1j}^{\sigma}(\tau) &= b_{22}(\tau) - a_{22}(\tau)\,[\sigma(\nu-\omega_1)+\omega_j]^2, \\ h_{2j}^{\sigma}(\tau) &= b_{22}(\tau) - a_{22}(\tau)\,[\sigma(\nu-\omega_2)+\omega_j]^2, \\ g_{1j}^{\sigma}(\tau) &= \gamma(\tau)\,[\sigma(\nu-\omega_1)+\omega_j], \\ g_{2j}^{\sigma}(\tau) &= \gamma(\tau)\,[\sigma(\nu-\omega_2)+\omega_j], \end{aligned}\right\} \tag{5.249}$$

$$\left.\begin{aligned} R_{\sigma j} &= a_{11}(\tau)\,\{h_{1j}^{\sigma}(\tau)+h_{2j}^{\sigma}(\tau)\} - a_{22}(\tau)\,[g_{1j}^{\sigma}(\tau)+g_{2j}^{\sigma}(\tau)], \\ L_{\sigma j} &= (2a_{11}(\tau)\,\omega_1 - \gamma(\tau)\,c_j(\tau)\,[h_{1j}^{\sigma}(\tau)+h_{2j}^{\sigma}(\tau)] - \\ &\quad - (2a_{22}(\tau)\,c_j(\tau)\,\omega_1 - \gamma(\tau))\,(g_{1j}^{\sigma}(\tau)+g_{2j}^{\sigma}(\tau)) \quad (j=1,2), \end{aligned}\right\} \tag{5.250}$$

and the differentiation and integration is performed everywhere with respect to t (contained explicitly) since τ is considered as a parameter.

In the second and third cases we obtain

$$\left.\begin{aligned} A_1^{(j)}(\tau, a_1, a_2, \psi_1, \psi_2) &= \sum_{\sigma}\sum_{k} A_{1\sigma k}^{(j)}, \\ B_1^{(j)}(\tau, a_1, a_2, \psi_1, \psi_2) &= \sum_{\sigma}\sum_{k} B_{1\sigma k}^{(j)} \quad (j=1,2), \end{aligned}\right\} \tag{5.251}$$

where $A_{1\sigma k}^{(j)}$ and $B_{1\sigma k}^{(j)}$ $(j=1,2)$ are solutions of the system of equations

$$\begin{aligned} R_{\sigma kj}\frac{dA_{1\sigma k}^{(j)}}{dt} - L_{\sigma kj}a_jB_{1\sigma k}^{(j)} &= \\ = \frac{1}{4\pi^3} e^{i\{\sigma(q\psi_1-p\psi_2)-k\psi_i\}} &\int_0^{2\pi}\int_0^{2\pi}\int_0^{2\pi} h_{ij}^{\sigma k}(\tau)\, f_1(\tau, \theta, x_0, \dot{x}_0, y_0, \dot{y}_0) \times \\ &\times e^{-i\{\sigma(q\psi_1-p\psi_2)-k\psi_i\}} \cos(\theta+\psi_i)\, d\theta\, d(\theta+\psi_1)\, d(\theta+\psi_2) - \\ - \frac{1}{4\pi^3} e^{i\{\sigma(q\psi_1-p\psi_2)-k\psi_i\}} &\int_0^{2\pi}\int_0^{2\pi}\int_0^{2\pi} g_{ij}^{\sigma k}(\tau)\, f_2(\tau, \theta, x_0, \dot{x}_0, y_0, \dot{y}_0) \times \\ &\times e^{-i\{\sigma(q\psi_1-p\psi_2)-k\psi_i\}} \sin(\theta+\psi_j)\, d\theta\, d(\theta+\psi_1)\, d(\theta+\psi_2), \\ R_{\sigma kj}a_j\frac{dB_{1\sigma k}^{(j)}}{dt} + L_{\sigma kj}A_{1\sigma k}^{(j)} &= -\frac{1}{4\pi} e^{i\{\sigma(q\psi_1-p\psi_2)-k\psi_i\}} \int_0^{2\pi}\int_0^{2\pi}\int_0^{2\pi} h_{ij}^{\sigma k}(\tau) \times \\ \times f_1(\tau, \theta, x_0, \dot{x}_0, y_0, \dot{y}_0)\, &e^{-i\{\sigma(q\psi_1-p\psi_2)-k\psi_i\}} \sin(\theta+\psi_j)\, d\theta\, d(\theta+\psi_1) \times \\ \times d(\theta+\psi_2) - \frac{1}{4\pi^3} &e^{-i\{\sigma(q\psi_1-p\psi_2)-k\psi_i\}} \int_0^{2\pi}\int_0^{2\pi}\int_0^{2\pi} g_{ij}^{\sigma}(\tau) \times \\ &\times f_2(\tau, \theta, x_0, \dot{x}_0, y_0, \dot{y}_0)\, e^{-i\{\sigma(q\psi_1-p\psi_2)-k\psi_i\}} \times \\ &\times \cos(\theta+\psi_j)\, d\theta\, d(\theta+\psi_1)\, d(\theta+\psi_2) - \\ - a_j\left(a_{11}\frac{d\omega_j}{d\tau} - \gamma\frac{dc_j}{d\tau}\right) h_{ij}^{\sigma k}(\tau) &+ a_j\left(a_{22}c_j\frac{d\omega_j}{d\tau} + 2a_{22}\omega_j\frac{dc_j}{d\tau}\right) g_{ij}^{\sigma k}(\tau), \end{aligned} \tag{5.252}$$

where

$$\left.\begin{aligned} h_{ij}^{\sigma k}(\tau) &= b_{22}(\tau) - a_{22}(\tau)\,[k(\nu-\omega_i)+\sigma(q\omega_1-p\omega_2)+\omega_j]^2, \\ g_{ij}^{\sigma k}(\tau) &= \gamma(\tau)\,[k(\nu-\omega_i)+\sigma(q\omega_1-p\omega_2)+\omega_j] \quad (j=1,2). \end{aligned}\right\} \tag{5.253}$$

Here $i=1$ corresponds to the second case, and $i=2$ — to the third.

In the system (5.248) and in equations (5.252) the total time derivative should be taken as

$$\frac{dA_{1\sigma}^{(j)}}{dt} = \frac{\partial A_{1\sigma}^{(j)}}{\partial\psi_1}(\omega_1-\nu) + \frac{\partial A_{1\sigma}^{(j)}}{\partial\psi_2}(\omega_2-\nu). \tag{5.254}$$

After deriving explicit expressions for $A_1^{(i)}(\tau, a_1, a_2, \psi_1, \psi_2)$, $B_1^{(i)}(\tau, a_1, a_2, \psi_1, \psi_2)$ and also the functions $u_1(\tau, \theta, a_1, a_2, \theta+\psi_1, \theta+\psi_2)$ and $v_1(\tau, \theta, a_1, a_2, \theta+\psi_1, \theta+\psi_2)$, the solution of the system of equations (5.233) can be written in the first approximation, as well as in the "improved" first approximation, i. e., taking into account higher harmonics of vibrations.

§ 8. Investigation of the nonstationary mode in a gyroscopic system of centrifugal type in the presence of internal resonance

Let us apply the results of the preceding section to the study of vibrations in the nonstationary mode of an actual gyroscopic system. We consider the scheme of a gyroscope applicable in some centrifugal devices (Figure 92).

Let the material axis of the gyroscope OA be fixed vertically by means of two bearings C and O, the upper bearing C having equal rigidity at all

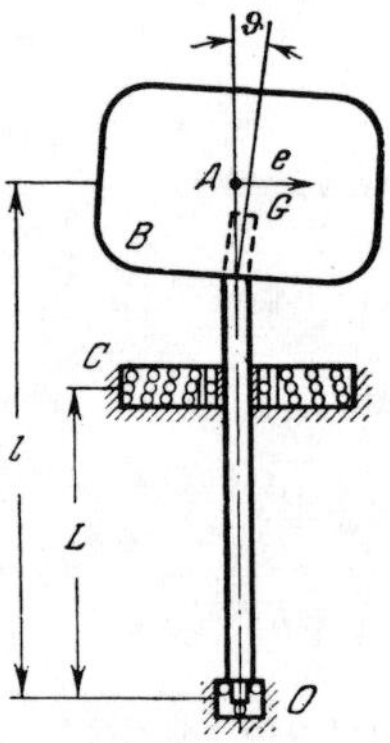

FIGURE 92

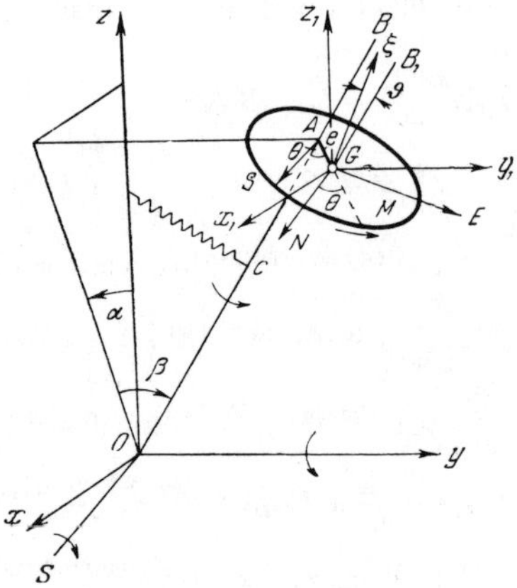

FIGURE 93

radii. (One could consider again the case when the rigidity of the upper bearing depends nonlinearly on the displacement of the axis; this would, however, complicate the problem.) We neglect the elasticity and mass of the gyroscope axis*, and consider the rotor of the gyroscope B to be unbalanced statically, as well as dynamically.

We introduce the following notation: m, mass of the gyroscope rotor; I, moment of inertia of the rotor with respect to the symmetry axis; J, moment of inertia with respect to a transversal axis passing through the rotor's center of gravity perpendicular to the symmetry axis; P, rotor's weight; l, distance from the step bearing to the rotor's center of gravity; L, distance from the step bearing to the upper bearing; e, displacement of the rotor's center of gravity G with respect to the rotation axis (linear eccentricity of the rotor); ϑ, angle between the symmetry axis of the rotor and the rotation axis (angular eccentricity of the rotor).

We denote by α and β the angles defining the position of the gyroscope axis in the fixed coordinate system $Oxyz$, whose axis Oz points vertically

* The nonstationary mode for a system of this type was studied under some restrictions, in particular for large rotation velocities, by A. A. Smelkov /152/.

upwards (Figure 93). The three Euler angles α, β, and θ (the rotation angle of the rotor) and the line of nodes S determine completely the position of the rotor in space.

Let us introduce a second coordinate system x_1, y_1, z_1 with axes parallel to the system x, y, z and origin at the intersection of the rotor's symmetry axis $G\xi$ with the plane Oxy. The position of the rotor's symmetry axis is then fully specified by the Euler angles α_1, β_1, θ and the line of nodes N (cf. § 2 and Figure 86).

We denote the coordinates of the center of gravity by x_G, y_G, z_G.

We limit the consideration to small deviations of the gyroscope axis from the vertical, regarding the linear and angular eccentricities as small.

Moreover, we assume that the segment $AG = e$ lies in the plane of the angle $B_1G\xi = \vartheta$. Since we intend to construct only the first approximation when deriving the equations of motion, we need only take into account linear terms in the small quantities

$$\alpha,\ \beta,\ \alpha_1,\ \beta_1,\ x_G,\ y_G.$$

One can establish, according to the notation introduced, the relations

$$\alpha - \alpha_1 = \vartheta\cos\theta, \quad \beta - \beta_1 = \vartheta\sin\theta, \tag{5.255}$$

$$x_G = l\alpha + e\cos\theta, \quad y_G = l\beta + e\sin\theta, \quad z_G = l. \tag{5.256}$$

To obtain the equations of motion we apply the law of moments for the projections on the fixed axes x, y, z to the motion of the rotor with respect to the fixed bearing.

Denoting the projections of the rotor's angular momentum K (with respect to the point O) on these axes, correspondingly, by K_x, K_y, K_z, we obtain

$$\frac{dK_x}{dt} = L_x, \quad \frac{dK_y}{dt} = L_y, \quad \frac{dK_z}{dt} = L_z, \tag{5.257}$$

where L_x, L_y, L_z are projections of the total moment of the external forces acting on the rotor at the point O_1 taken with respect to the fixed bearing O, and

$$K = K_G + r_G \times m r_G,$$

where K_G is the angular momentum of the rotor with respect to the center of inertia, and r_G, the radius vector of the center of inertia.

Let us calculate K_x, K_y, K_z. Since the axes N, E, ξ are the principal axes of inertia of the rotor, we have

$$\left.\begin{aligned} K_{GN} &= J\Omega_N = -J\dot{\beta}_1, \\ K_{GE} &= J\Omega_E = J\dot{\alpha}_1, \\ K_{G\xi} &= I\Omega_\xi = I\dot{\theta}, \end{aligned}\right\} \tag{5.258}$$

where $K_{GN}, K_{GE}, K_{G\xi}$ are projections of the momentum K_G on the axes N, E, ξ.

Projecting the vector K_G on the axes x, y, z, we obtain

$$\left.\begin{aligned} K_{Gx} &= K_{GN} + \alpha_1 K_{G\xi} = -J\dot{\beta}_1 + I\dot{\theta}\alpha_1, \\ K_{Gy} &= K_{GE} + \beta_1 K_{G\xi} = J\dot{\alpha}_1 + I\dot{\theta}\beta_1, \\ K_{Gz} &= K_{G\xi} = I\dot{\theta}. \end{aligned}\right\} \tag{5.259}$$

We can then write, using the coordinates of the center of gravity (5.256),

$$\left.\begin{aligned} K_x &= -J\dot{\beta}_1 + I\dot{\theta}\alpha_1 - ml^2\dot{\beta} - mle\dot{\theta}\cos\theta, \\ K_y &= J\dot{\alpha}_1 + I\dot{\theta}\beta_1 + ml^2\dot{\alpha} - mle\dot{\theta}\sin\theta, \\ K_z &= I\dot{\theta}. \end{aligned}\right\}$$

Since $r_G = i(l\dot{\alpha} - \varepsilon\dot{\theta}\sin\theta) + j(l\dot{\beta} + e\dot{\theta}\cos\theta)$, we have, with the specified accuracy,

$$r_G \times m\dot{r}_G = -im(l^2\dot{\beta} + le\dot{\theta}\cos\theta) + jm(l^2\dot{\alpha} - le\dot{\theta}\sin\theta). \qquad (5.261)$$

Substituting the values α_1 and β_1 from (5.255) into the right-hand sides of (5.260), we obtain

$$\left.\begin{aligned} K_x &= I\dot{\theta}\alpha + (J - I)\dot{\theta}\vartheta\cos\theta - (J + ml^2)\dot{\beta} - mle\dot{\theta}\cos\theta, \\ K_y &= I\dot{\theta}\beta + (J - I)\dot{\theta}\vartheta\sin\theta + (J + ml^2)\dot{\alpha} - mle\dot{\theta}\sin\theta, \\ K_z &= I\dot{\theta}. \end{aligned}\right\} \qquad (5.262)$$

Calculating the moment of the external forces acting on the rotor, we find

$$\left.\begin{aligned} L_x &= kL^2\beta - (l\beta + e\sin\theta)P, \\ L_y &= -kL^2\alpha + (l\alpha + e\cos\theta)P, \\ L_z &= I\gamma, \end{aligned}\right\} \qquad (5.263)$$

where $\gamma = \ddot{\theta}$ is a constant angular acceleration of the normal rotation of the rotor caused by a constant moment suddenly applied to the gyroscope axis.

Inserting the values of (5.262) and (5.263) into (5.257), we obtain the differential equations

$$\left.\begin{aligned} I_1\ddot{\beta} - I\dot{\theta}\dot{\alpha} + b\beta &= A[\dot{\theta}^2\sin\theta - \ddot{\theta}\cos\theta] - eP\sin\theta + I\ddot{\theta}\alpha, \\ I_1\ddot{\alpha} + I\dot{\theta}\dot{\beta} + b\alpha &= A[\dot{\theta}^2\cos\theta + \ddot{\theta}\sin\theta] + eP\cos\theta - I\ddot{\theta}\beta, \quad \ddot{\theta} = \ddot{\theta}(t), \end{aligned}\right\} \qquad (5.264)$$

where

$$(I - J)\vartheta + mle = A, \quad kL^2 - Pl = b, \quad J + ml^2 = I_1. \qquad (5.265)$$

I_1 is the moment of inertia of the rotor with respect to an axis passing through the lower bearing perpendicular to the symmetry axis.

We now construct an approximate solution for the system of equations (5.264) with the help of the formulas derived in the preceding section. Since these equations are linear (their variable coefficients being only the coefficients of the gyroscopic terms and also the amplitudes and frequencies of the external periodic forces), we obtain very simple equations of the first approximation whose numerical integration involves no additional difficulties. (These equations may even be reduced to quadratures.)

When considering a nonstationary process in a gyroscopic system described by the differential equations (5.264), we assume that we are dealing with the first case mentioned in the preceding section (p. 254), i. e., that the following relations between $\nu(\tau)$, $\omega_1(\tau)$, and $\omega_2(\tau)$ are possible for certain values of τ in the specified interval $0 \leqslant \tau \leqslant L$:

$$\nu(\tau_1) = \omega_1(\tau_1), \quad \nu(\tau_2) = \omega_2(\tau_2), \quad \omega_1(\tau) \neq \frac{p}{q}\omega_2(\tau).$$

To construct the solution, we must use the formulas (5.235), (5.238), (5.247), and the equations (5.239), (5.248).

Comparing equations (5.264) with equations (5.233), we have

$$a_{11}(\tau) = a_{22}(\tau) = I_1, \quad \gamma(\tau) = I\nu(\tau), \quad b_{11}(\tau) = b_{22}(\tau) = b.$$

We further set up, according to (5.235), the characteristic equations for the "eigen"-frequencies

$$D(\omega) = \begin{vmatrix} b - I_1\omega^2 & I\nu(\tau)\omega \\ I\nu(\tau)\omega & b - I_1\omega^2 \end{vmatrix} = 0, \qquad (5.266)$$

from which we obtain

$$\omega_1(\tau)=\frac{\sqrt{I^2\nu^2(\tau)+4bI_1}-I\nu(\tau)}{2I_1},\quad \omega_2(\tau)=\frac{\sqrt{I^2\nu^2(\tau)+4bI_1}+I\nu(\tau)}{2I_1}. \tag{5.267}$$

We derive the basic functions

$$\left.\begin{aligned} c_1&=\frac{I_1\omega_1^2(\tau)-b}{I\nu(\tau)\omega_1(\tau)}=\frac{I_1\frac{1}{I_1}[b+I\nu(\tau)\omega_1(\tau)]-b}{I\nu(\tau)\omega_1(\tau)}=1,\\ c_2&=\frac{I_1\omega_2^2(\tau)-b}{I\nu(\tau)\omega_2(\tau)}=\frac{I_1\frac{1}{I_1}[b+I\nu(\tau)\omega_2(\tau)]-b}{I\nu(\tau)\omega_2(\tau)}=1.\end{aligned}\right\} \tag{5.268}$$

Thus, the solution of the system of differential equations (5.264) may be written in the form

$$\left.\begin{aligned}\alpha&=a_1\cos(\theta+\psi_1)+a_2\cos(\theta+\psi_2),\\ \beta&=a_1\sin(\theta+\psi_1)+a_2\sin(\theta+\psi_2),\end{aligned}\right\} \tag{5.269}$$

where $\frac{d\theta}{dt}=\nu_0+\gamma t=\nu(\tau)$ (we write $\nu(\tau)$ since γ is assumed to be of the order of ε), and the quantities a_1, a_2, ψ_1, and ψ_2 should be determined from the equations of the first approximation.

Friction was not taken into account in the differential equations describing the motion of gyroscope axes. To account for friction we consider, for example, Rayleigh's dissipative function

$$R=\frac{1}{2}[b_1\dot{\alpha}^2+2h\dot{\alpha}\dot{\beta}+b_2\dot{\beta}^2], \tag{5.270}$$

where b_1, b_2, h are some constants, and obtain for the perturbing forces

$$\left.\begin{aligned}\varepsilon f_1(\tau,\theta,\alpha,\dot{\alpha},\beta,\dot{\beta})&=A\left[\nu^2(\tau)\cos\theta+\frac{d\nu(\tau)}{d\tau}\sin\theta\right]-\\ &-I\frac{d\nu(\tau)}{d\tau}\beta-b_1\dot{\alpha}-h\dot{\beta}+eP\cos\theta,\\ \varepsilon f_2(\tau,\theta,\alpha,\dot{\alpha},\beta,\dot{\beta})&=A\left[\nu^2(\tau)\sin\theta-\frac{d\nu(\tau)}{d\tau}\cos\theta\right]+\\ &+I\frac{d\nu(\tau)}{d\tau}\alpha-h\dot{\alpha}-b_2\dot{\beta}-eP\sin\theta.\end{aligned}\right\} \tag{5.271}$$

Substituting the values of α, β, $\dot{\alpha}$, $\dot{\beta}$ found from (5.269) into the right-hand sides of these expressions, we obtain

$$\left.\begin{aligned}&\varepsilon f_1(\tau,\theta,\alpha_0,\dot{\alpha}_0,\beta_0,\dot{\beta}_0)=\left[(A\nu^2(\tau)+eP)\cos\theta+A\frac{d\nu(\tau)}{d\tau}\sin\theta\right]-\\ &-I\frac{d\nu(\tau)}{d\tau}[a_1\sin(\theta+\psi_1)+a_2\sin(\theta+\psi_2)]+b_1[a_1\omega_1(\tau)\sin(\theta+\psi_1)+\\ &+a_2\omega_2(\tau)\sin(\theta+\psi_2)]-h[a_1\omega_1(\tau)\cos(\theta+\psi_1)+a_2\omega_2(\tau)\cos(\theta+\psi_2)],\\ &\varepsilon f_2(\tau,\theta,\alpha_0,\dot{\alpha}_0,\beta_0,\dot{\beta}_0)=\left[(A\nu^2(\tau)-eP)\sin\theta-A\frac{d\nu(\tau)}{d\tau}\cos\theta\right]+\\ &+I\frac{d\nu(\tau)}{d\tau}[a_1\cos(\theta+\psi_1)+a_2\cos(\theta+\psi_2)]+h[a_1\omega_1(\tau)\sin(\theta+\psi_1)+\\ &+a_2\omega_2(\tau)\sin(\theta+\psi_2)]-b_2[a_1\omega_1(\tau)\cos(\theta+\psi_1)+a_2\omega_2(\tau)\cos(\theta+\psi_2)].\end{aligned}\right\} \tag{5.272}$$

We now construct the equations of the first approximation. We need to calculate the integrals in the right-hand sides of equations (5.248), substituting first the values of $f_1(\tau,\theta,\alpha_0,\dot{\alpha}_0,\beta_0,\dot{\beta}_0)$ and $f_2(\tau,\theta,\alpha_0,\dot{\alpha}_0,\beta_0,\dot{\beta}_0)$.

These integrals are different from zero if $\sigma=0,\ \pm 1$.

After a number of calculations we obtain equations for

$$A_{1\sigma}^{(j)}(\tau,a_1,a_2,\psi_1,\psi_2),\quad B_{1\sigma}^{(j)}(\tau,a_1,a_2,\psi_1,\psi_2)\qquad (j=1,\ 2;\ \sigma=0,\ \pm 1).$$

For $\sigma=0$

$$\left.\begin{aligned}
&\varepsilon\left\{3I_1\frac{dA_{10}^{(j)}}{dt}-3\,[2I_1\omega_j(\tau)-I\nu(\tau)]\,a_jB_{10}^{(j)}\right\}=ha_j\omega_j(\tau),\\
&\varepsilon\left\{3I_1\frac{dB_{10}^{(j)}}{dt}+3\,[2I_1\omega_j(\tau)-I\nu(\tau)\,]\,A_{10}^{(j)}\right\}=\\
&=-\left[(b_1+2b_2)\,\omega_j(\tau)-3I\frac{d\nu(\tau)}{d\tau}\right]a_j-3I_1a_j\frac{d\omega_j(\tau)}{d\tau}\qquad(j=1,\ 2).
\end{aligned}\right\}\tag{5.273}$$

For $\sigma=1$

$$\left.\begin{aligned}
&\varepsilon\left\{I_1\frac{dA_{11}^{(j)}}{dt}-[2I_1\omega_j(\tau)-I\nu(\tau)]\,a_j\,B_{11}^{(j)}\right\}=\\
&\qquad=\frac{1}{2}\,[A\nu^2(\tau)+eP]\,e^{i\psi_j}-\frac{1}{2i}\,A\,\frac{d\nu(\tau)}{d\tau}\,e^{i\psi_j},\\
&\varepsilon\left\{I_1a_j\frac{dB_{11}^{(j)}}{dt}+[2I_1\omega_j(\tau)-I\nu(\tau)]\,A_{11}^{(j)}\right\}=\\
&\qquad=-\frac{1}{2i}\,[A\nu^2(\tau)+eP]\,e^{i\psi_j}-\frac{1}{2}\,A\,\frac{d\nu(\tau)}{d\tau}\,e^{i\psi_j}\qquad(j=1,\ 2).
\end{aligned}\right\}\tag{5.274}$$

For $\sigma=-1$

$$\left.\begin{aligned}
&\varepsilon\left\{I_1\frac{dA_{1-1}^{(j)}}{dt}-[2I_1\omega_j(\tau)-I\nu(\tau)]\,a_j\,B_{1-1}^{(j)}\right\}=\\
&\qquad=\frac{1}{2}\,[A\nu^2(\tau)+eP]\,e^{-i\psi_j}+\frac{1}{2i}\,A\,\frac{d\nu(\tau)}{d\tau}\,e^{-i\psi_j},\\
&\varepsilon\left\{I_1a_j\frac{dB_{1-1}^{(j)}}{dt}+[2I_1\omega_j(\tau)-I\nu(\tau)]a_jA_{1-1}^{(j)}\right\}=\\
&\qquad=\frac{1}{2i}\,[A\nu^2(\tau)+eP]\,e^{-i\psi_j}-\frac{1}{2}\,A\,\frac{d\nu(\tau)}{d\tau}\,e^{-i\psi_j}\qquad(j=1,\ 2).
\end{aligned}\right\}\tag{5.275}$$

Solving these equations, we find

$$\left.\begin{aligned}
&\varepsilon A_{10}^{(j)}=-\frac{[(b_1+2b_2)]\,\omega_j(\tau)-3I\dfrac{d\nu(\tau)}{d\tau}}{3\,[2I_1\omega_j(\tau)-I\nu(\tau)]}\,a_j-\frac{I_1\dfrac{d\omega_j(\tau)}{d\tau}}{2I_1\omega_j(\tau)-I\nu(\tau)}\,a_j,\\
&\varepsilon B_{10}^{(j)}=-\frac{h\omega_j(\tau)}{3[2I_1\omega_j(\tau)-I\nu(\tau)]},\\
&\varepsilon A_{11}^{(j)}=-\frac{\dfrac{1}{2i}\,[A\nu^2(\tau)-eP]\,e^{i\psi_j}+\dfrac{1}{2}\,A\dfrac{d\nu(\tau)}{d\tau}\,e^{i\psi_j}}{I_1\,[\omega_j(\tau)+\nu(\tau)]-I\nu(\tau)},\\
&\varepsilon B_{11}^{(j)}=\frac{-\dfrac{1}{2}\,[A\nu^2(\tau)+eP]\,e^{i\psi_j}+\dfrac{1}{2i}\,A\,\dfrac{d\nu(\tau)}{d\tau}\,e^{i\psi_j}}{a_j\,\{I_1\,[\omega_j(\tau)+\nu(\tau)]-I\nu(\tau)\}},\\
&\varepsilon A_{1-1}^{(j)}=\frac{\dfrac{1}{2i}\,[A\nu^2(\tau)-eP]\,e^{-i\psi_j}-\dfrac{1}{2}\,A\,\dfrac{d\nu(\tau)}{d\tau}\,e^{-i\psi_j}}{I_1\,[\omega_j(\tau)+\nu(\tau)]-I\nu(\tau)},\\
&\varepsilon B_{1-1}^{(j)}=\frac{-\dfrac{1}{2}\,[A\nu^2(\tau)+eP]\,e^{-i\psi_j}-\dfrac{1}{2i}\,A\,\dfrac{d\nu(\tau)}{d\tau}\,e^{-i\psi_j}}{a_j\,\{I_1\,[\omega_j(\tau)+\nu(\tau)]-I\nu(\tau)\}}\qquad(j=1,\ 2).
\end{aligned}\right\}\tag{5.276}$$

We can write with the help of formulas (5.239) the equations of the first approximation

$$\left.\begin{aligned}
\frac{da_1}{dt}=&\frac{-\left[(b_1+2b_2)\,\omega_1(\tau)-3I\dfrac{d\nu(\tau)}{d\tau}\right]}{3\,[2I_1\omega_1(\tau)-I\nu(\tau)]}\,a_1-\frac{I_1\dfrac{d\omega_1(\tau)}{d\tau}}{2I_1\omega_1(\tau)-I\nu(\tau)}\,a_1-\\
&-\frac{[A\nu^2(\tau)-eP]\sin\psi_1-A\dfrac{d\nu(\tau)}{d\tau}\cos\psi_1}{I_1\,[\omega_1(\tau)+\nu(\tau)]-I\nu(\tau)},
\end{aligned}\right\}\tag{5.277}$$

$$
\left.
\begin{aligned}
\frac{da_2}{dt} &= -\frac{\left[(b_1+2b_2)\,\omega_2(\tau) - 3I\frac{d\nu(\tau)}{d\tau}\right]}{3\,[2I_1\omega_2(\tau) - I\nu(\tau)]}\,a_2 - \frac{I_1\frac{d\omega_2(\tau)}{d\tau}}{2I_1\omega_2(\tau) - I\nu(\tau)}\,a_2 - \\
&\qquad - \frac{[A\nu^2(\tau) - eP]\sin\psi_2 - A\frac{d\nu(\tau)}{d\tau}\cos\psi_2}{I_1\,[\omega_2(\tau)+\nu(\tau)] - I\nu(\tau)}, \\
\frac{d\psi_1}{dt} &= \omega_1(\tau) - \nu(\tau) - \frac{h\omega_1(\tau)}{3\,[2I_1\omega_1(\tau) - I\nu(\tau)]} - \\
&\qquad - \frac{[A\nu^2(\tau) + eP]\cos {}_1 - A\frac{d\nu(\tau)}{d\tau}\sin\psi_1}{a_1\,[I_1(\omega_1(\tau)+\nu(\tau)) - I\nu(\tau)]}, \\
\frac{d\psi_2}{dt} &= \omega_2(\tau) - \nu(\tau) - \frac{h\omega_2(\tau)}{3\,[2I_1\omega_2(\tau) - I\nu(\tau)]} - \\
&\qquad - \frac{[A\nu^2(\tau) + eP]\cos\psi_2 - A\frac{d\nu(\tau)}{d\tau}\sin\psi_2}{a_2\,[I_1(\omega_2(\tau)+\nu(\tau)) - I\nu(\tau)]}.
\end{aligned}
\right\}
\tag{5.277}
$$

The inequalities $\left|I\frac{d\nu(\tau)}{d\tau}\right| \ll b_1\omega_j(\tau)$, $\left|I_1\frac{d\omega_j(\tau)}{d\tau}\right| \ll b_1\omega_j(\tau)$, as well as $\left|\frac{d\nu(\tau)}{d\tau}\right| \ll \nu^2(\tau)$ and $eP \ll A\nu^2(\tau)$ hold even for quite large velocities of transition through resonance. Thus the quantities $I\frac{d\nu(\tau)}{d\tau}$, $I_1\frac{d\omega_j(\tau)}{d\tau}$, $\frac{d\nu(\tau)}{d\tau}$, eP may be neglected in the first approximation without affecting the accuracy of calculations, and may be referred to the second approximation. We may also assume for simplicity that $h=0$, $b_1=b_2=\delta$ in Rayleigh's dissipative function; the equations of the first approximation then become

$$
\left.
\begin{aligned}
\frac{da_1}{dt} &= -\frac{\delta\omega_1(\tau)\,a_1}{2I_1\omega_1(\tau) - I\nu(\tau)} - \frac{A\nu^2(\tau)\sin\psi_1}{I_1(\omega_1(\tau)+\nu(\tau)) - I\nu(\tau)}, \\
\frac{da_2}{dt} &= -\frac{\delta\omega_2(\tau)\,a_2}{2I_1\omega_2(\tau) - I\nu(\tau)} - \frac{A\nu^2(\tau)\sin\psi_2}{I_1(\omega_2(\tau)+\nu(\tau)) - I\nu(\tau)}, \\
\frac{d\psi_1}{dt} &= \omega_1(\tau) - \nu(\tau) - \frac{A\nu^2(\tau)\cos\psi_1}{a_1\,[I_1(\omega_1(\tau)+\nu(\tau)) - I\nu(\tau)]}, \\
\frac{d\psi_2}{dt} &= \omega_2(\tau) - \nu(\tau) - \frac{A\nu^2(\tau)\cos\psi_2}{a_2\,[I_1(\omega_2(\tau)+\nu(\tau)) - I\nu(\tau)]}.
\end{aligned}
\right\}
\tag{5.278}
$$

This system of equations of the first approximation may be integrated with even less difficulty since it separates into two independent systems of two equations.

We now derive the relations determining the dependence of the vibration amplitude on the frequency of the external force (the resonance curve) in a stationary mode. We must equate the right-hand sides of equations (5.278) to zero and eliminate from the obtained expressions the phases of vibration ψ_1 and ψ_2; this yields the relations

$$
\left.
\begin{aligned}
a_1^2 &= \frac{(A\nu^2)^2}{[I_1(\omega_1+\nu) - I\nu]^2\left\{(\omega_1-\nu)^2 + \frac{\delta^2\omega_1^2}{(2I_1\omega_1 - I\nu)^2}\right\}}, \\
a_2^2 &= \frac{(A\nu^2)^2}{[I_1(\omega_2+\nu) - I\nu]^2\left\{(\omega_2-\nu)^2 + \frac{\delta^2\omega_2^2}{(2I_1\omega_2 - I\nu)^2}\right\}},
\end{aligned}
\right\}
\tag{5.279}
$$

allowing us to construct the resonance curves for the separate harmonic components.

In these formulas ω_1 and ω_2 are not constants, but depend on ν according to the relations (5.207). The expressions (5.200) may be expressed in the form

$$
\left.
\begin{aligned}
\alpha &= U\cos(\theta+\varkappa), \\
\beta &= U\sin(\theta+\varkappa),
\end{aligned}
\right\}
\tag{5.280}
$$

where the amplitude of the resultant vibration U and the phase shift $\varkappa$ are determined by

$$U=\sqrt{(a_1\cos\psi_1+a_2\cos\psi_2)^2+(a_1\sin\psi_1+a_2\sin\psi_2)^2}, \qquad (5.281)$$

$$\operatorname{tg}\varkappa=\frac{a_1\sin\psi_1+a_2\sin\psi_2}{a_1\cos\psi_1+a_2\cos\psi_2}. \qquad (5.282)$$

We use these formulas with the following numerical values of the parameters of the considered system:

$$P=5\,\text{kg},\ l=25\,\text{cm},\ L=15\,\text{cm},$$

$$k=5\,\text{kg/cm},\ I=0.1\,\text{kg}\cdot\text{cm}\cdot\text{sec}^2,\ I_1=0.2\,\text{kg}\cdot\text{cm}\cdot\text{sec}^2.$$

To solve this problem we confine ourselves to constructing resonance curves for stationary modes as well as for different modes of transition through resonance. We do not then need to find the absolute amplitude values a_1 and

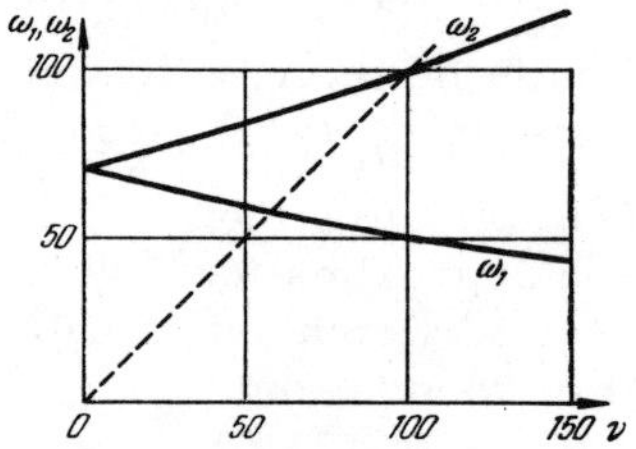

FIGURE 94

a_2 as it is sufficient to plot the graphs for the ratios $\frac{a_1}{A}$ and $\frac{a_2}{A}$ (which can be done because of the linearity of equation (5.264)). We do not therefore need to give the numerical values of the gyroscope parameters which characterize A, i.e., the numerical values of e and ϑ. The friction coefficient δ is taken equal to 0.02.

Inserting these numerical values into (5.267), we find

$$\left.\begin{aligned}\omega_1(\tau)&=\frac{\sqrt{800+0.1\nu^2(\tau)}-0.2\nu(\tau)}{0.4},\\ \omega_2(\tau)&=\frac{\sqrt{800+0.1\nu^2(\tau)}+0.2\nu(\tau)}{0.4}\end{aligned}\right\} \qquad (5.283)$$

(when $\nu=0$ $\omega_1=\omega_2=70.7\,\text{sec}^{-1}$). The dependence of ω_1 and ω_2 on ν is shown in Figure 94, from which it is clear that as the frequency of the external force ν varies from 0 to $100-110\,\text{sec}^{-1}$, a resonance may occur twice in the system, the first time when $\nu=59\,\text{sec}^{-1}$, and the second — when $\nu=101\,\text{sec}^{-1}$. Using formulas (5.279), we obtain the relations

$$\frac{a_j}{A_1}=\frac{\nu^2}{(\omega_j+0{,}5\nu)\sqrt{(\omega_j-\nu)^2+\dfrac{500\omega_j^2}{(2\omega_j-0.5\nu)^2}}},\quad A_1=10A\quad (j=1,\ 2), \qquad (5.284)$$

which we use for constructing the stationary resonance curves characterizing the amplitude variation of the separate harmonic components entering in the expressions for α and β (cf. Figures 95 and 96).

To investigate the gyroscopic system during various modes of transition through resonance we must integrate numerically the equations of the first approximation (5.278) for various rates of change of the external force frequency $\nu(\tau)$. We assume for simplicity that $\nu(\tau)$ varies linearly, $\nu(\tau)=\nu_0+\gamma t$. We examine the behavior of the system for the values $\gamma=8\,\text{sec}^{-2}$, $20\,\text{sec}^{-2}$, $50\,\text{sec}^{-2}$.

Substituting the numerical values into equations (5.278), we obtain

$$\left.\begin{aligned}\frac{d\left(\frac{a_j}{A}\right)}{dt}&=-\frac{14.14\omega_j(\tau)}{4\omega_j(\tau)-\nu(\tau)}\left(\frac{a_j}{A}\right)-\frac{10\nu^2(\tau)}{2\omega_j(\tau)+\nu(\tau)}\sin\psi_j,\\ \frac{d\psi_j}{dt}&=\omega_j(\tau)-\nu(\tau)-\frac{10\nu^2(\tau)}{\left(\frac{a_j}{A}\right)[2\omega_j(\tau)+\nu(\tau)]}\cos\psi_j,\end{aligned}\right\} \tag{5.285}$$

where

$$j=1,\ 2,\qquad \nu(\tau)=\nu_0+\gamma t,\qquad \omega_j(\tau)=\frac{\sqrt{800+0.1\nu^2(\tau)\mp 0.2\nu(\tau)}}{0.4}.$$

Integrating this system of equations numerically, we obtain the curves of transition through resonance for the accelerations $\gamma=8\sec^{-2}$, $20\sec^{-2}$,

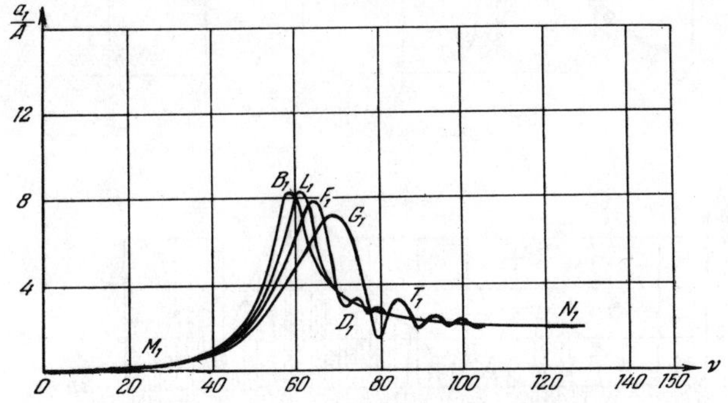

FIGURE 95

$50\sec^{-2}$. The curves corresponding to these accelerations in Figure 95 are $M_1L_1N_1$, $M_1F_1D_1$, and $M_1G_1T_1$, in Figure 96 — the curves $M_2L_2N_2$, $M_2F_2D_2$, and $M_2G_2T_2$.

In Figure 97 we show, according to formula (5.281), the curves of U as a function of the frequency for the stationary mode (the curve $ABCDE$), and for various accelerations (the curve $A_1B_1C_1D_1E_1$ for $\gamma=20\sec^{-2}$, the curve $A_2B_2C_2D_2E_2$ for $\gamma=50\sec^{-2}$).

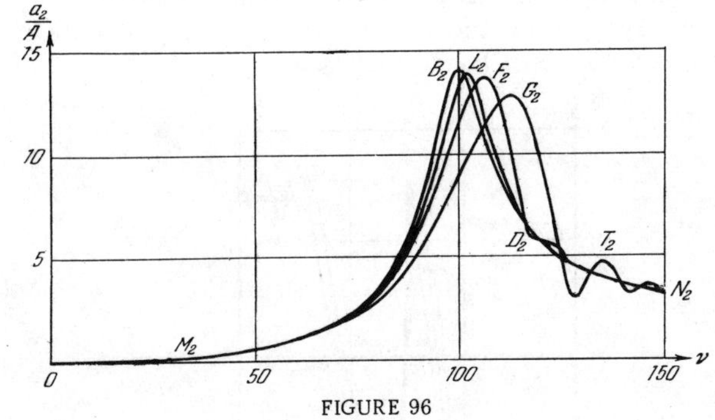

FIGURE 96

In Figure 98 we show for clarity the dependence of $\frac{a}{A_1}=U\cos(\theta+\varkappa)$ on the time during which the vibrating system under investigation passes through both resonances when $\gamma=50\sec^{-2}$.

The curves in Figure 99 describe the dependence of $\varkappa$ and ν in the stationary mode and for different velocities of transition through resonance.

For practical purposes it is sufficient to construct only the last two graphs which characterize completely the behavior of the system during a transition through resonance, namely, the displacement and lowering of the resonance curve maxima during a transition through resonant values of the

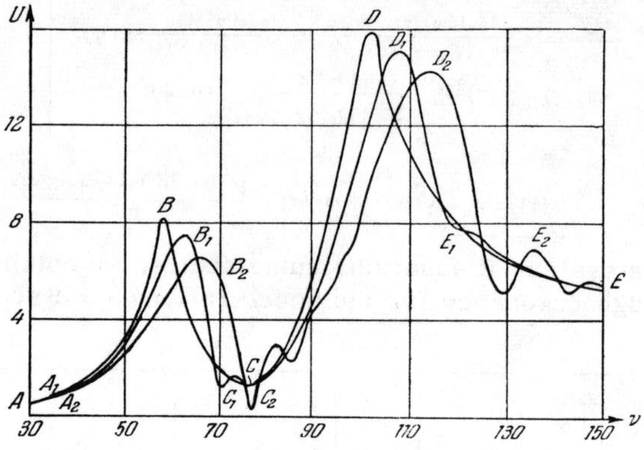

FIGURE 97

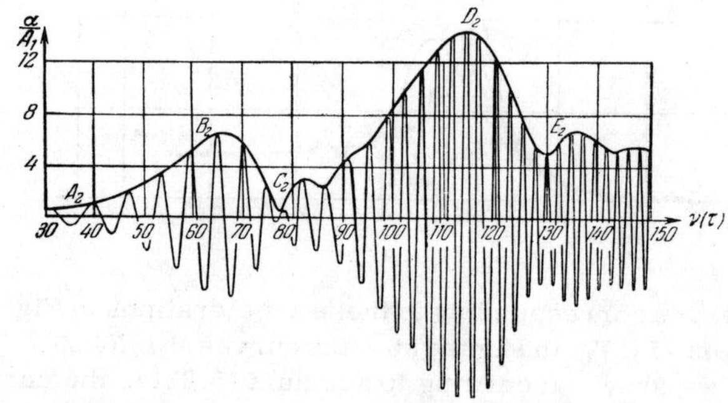

FIGURE 98

frequency, the appearance of beats, the variation in the shift between the phases of the perturbing force and the vibrations, etc.

We shall not analyze here the resonance curves just obtained, since we did this in great detail in Chapter III, § 6.

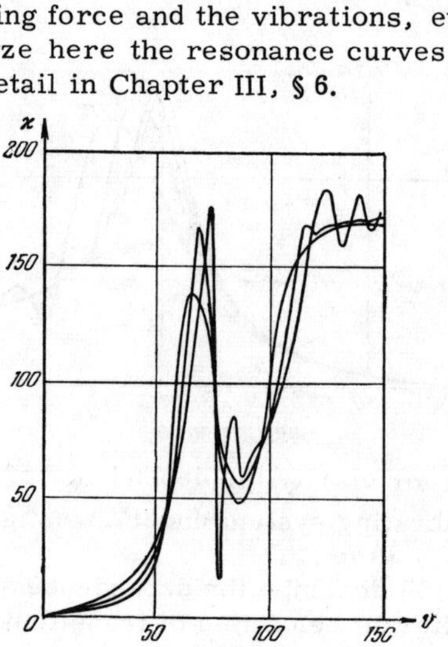

FIGURE 99

§ 9. Nonstationary vibrations of coaxial rotors

We consider the complex nonstationary process associating a double-frequency perturbation of the k-th resonating eigenfrequency in coaxial rotors (see V. A. Grobov /42/). The study of vibrations of coaxial rotors is very important in practice, since in recent years new types of gas-turbine devices with coaxial rotors have appeared in the airforce and the navy; the method of calculation is still insufficiently illustrated in the literature.

The dynamics of coaxial rotors has a number of peculiarities. For example, each of the shafts is under the action of perturbations of different frequencies determined by the nonequilibrium of the disks, i. e., the perturbation bears a double-frequency character which may consequently, as indicated in Chapter IV, § 8, give rise to a number of specific effects: transfer of vibrational energy from the external shaft to the internal one (and in the opposite direction), fractional resonances and self-vibrating modes, beats, simultaneous modes of direct and inverse precession, etc.

The angular velocities of the rotors during the start in double-rotor coaxial motors are determined by the type of the regulating system. We consider the case when $\frac{\nu_1(t)}{\nu_2(t)} = k(t)$ is some function of time determined by the structure of the regulating system ($\nu_1(t)$ and $\nu_2(t)$ are the angular velocities of the internal and external shafts, respectively).

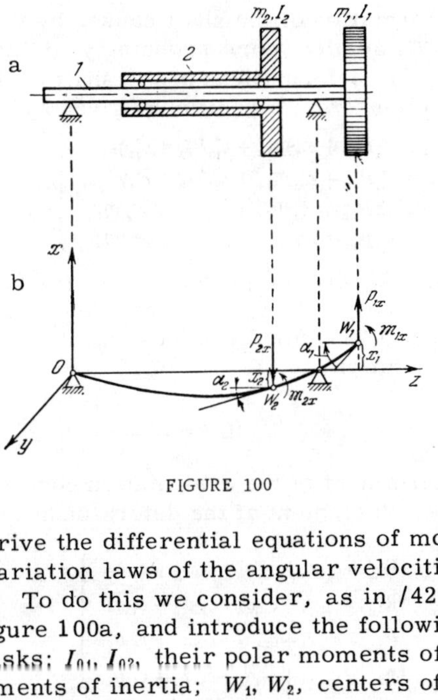

FIGURE 100

Let us briefly derive the differential equations of motion of coaxial rotors for arbitrary variation laws of the angular velocities of the internal and external shafts. To do this we consider, as in /42/, the coaxial rotors 1 and 2 shown in Figure 100a, and introduce the following notation: m_1, m_2, the masses of the disks; I_{01}, I_{02}, their polar moments of inertia; I_1, I_2, their equatorial moments of inertia; W_1, W_2, centers of the elastic line of the shafts in the plane of the disks m_1, m_2; e_1, e_2, the eccentricities of the disks; δ_1, δ_2, the angular eccentricities characterizing the dynamical nonequilibrium of the disks; ν_1, ν_2, the angular velocities of the internal and external disks.

We assume when deriving the differential equations describing the deformation vibrations of the rotors: 1) that the masses of the two shafts are small compared to the masses of the disks, and may therefore be neglected; 2) that the external shaft does not increase the rigidity of the internal shaft considerably, and that the disk placed on it acts as an "associated" mass for the internal shaft; 3) the angular velocities $v_1(t)$, $v_2(t)$ are given functions of time (otherwise we would get a more complicated system of equations of the type (5.127)).

Before deriving the equations of motion we shall first obtain expressions for the kinetic and potential energies.

For the kinetic energy of a system of two rotating disks, we obtain after a series of calculations (cf., for example, /42/) the expression

$$T = \frac{1}{2} \sum_{j=1}^{2} \{m_j(\dot{x}_j^2 + \dot{y}_j^2) + I_j(\dot{\alpha}_j^2 + \dot{\beta}_j^2) + I_{0j}\dot{\alpha}_j\beta_j\omega_j(t) + I_{0j}\omega_j^2(t) +$$
$$+ 2m_j e_j \omega_j(t)(\dot{y}_j \cos\varphi_j - \dot{x}_j \sin\varphi_j) - 2I_{0j}\delta_j(\dot{\alpha}_j - \omega_j\beta_j)\cos(\varphi_j - \chi_j) -$$
$$- 2I\delta_j\omega_j(t)[\dot{\alpha}_j\cos(\varphi_j - \chi_j) + \dot{\beta}_j\sin(\varphi_j - \chi_j)], \quad (5.286)$$

where x_j, y_j ($j = 1, 2$) are the coordinates of the points W_j ($j = 1, 2$); α_j, β_j ($j = 1, 2$), the angles defining the position of the tangent to the elastic line of the shaft at the point W_j; φ_j ($j = 1, 2$), rotation angles of the rotors; and χ_j, the phase difference between these angles.

Let us now derive the expression for the potential energy of the deformed internal shaft.

We consider the deformation of the shaft caused by the forces P_1 and P_2, and the moments $\mathfrak{M}_1$, $\mathfrak{M}_2$ applied correspondingly at the points W_1 and W_2. The linear and angular deformations of the shaft at the points W_1 and W_2 are, according to well-known propositions, given by the formulas

$$\left. \begin{array}{l} x_1 = c_{11}^* P_{1x} + c_{12}^* \mathfrak{M}_{1x} + c_{13}^* P_{2x} + c_{14}^* \mathfrak{M}_{2x}, \\ \alpha_1 + c_{21}^* P_{1x} + c_{22}^* \mathfrak{M}_{1x} + c_{23}^* P_{2x} + c_{24}^* \mathfrak{M}_{2x}, \\ x_2 = c_{31}^* P_{1x} + c_{32}^* \mathfrak{M}_{1x} + c_{33}^* P_{2x} + c_{34}^* \mathfrak{M}_{2x}, \\ \alpha_2 = c_{41}^* P_{1x} + c_{42}^* \mathfrak{M}_{1x} + c_{43}^* P_{2x} + c_{44}^* \mathfrak{M}_{2x}, \end{array} \right\} \quad (5.287)$$

where P_{1x}, P_{2x} and $\mathfrak{M}_{1x}$, $\mathfrak{M}_{2x}$ are the projections of the forces and the moments on the axis Ox; $c_{jk}^* = c_{kj}^*$ ($j, k = 1, 2, 3, 4$) are deformation coefficients.

Solving the equations (5.287) for the forces P_{1x}, P_{2x} and the moments $\mathfrak{M}_{1x}$, $\mathfrak{M}_{2x}$ and using the notation

$$c_{jk} = \frac{\Delta_{jk}(c_{jk}^*)}{\Delta(c_{jk}^*)} \quad (j, k = 1, 2, 3, 4),$$

where $\Delta(c_{jk}^*)$ is the determinant of the deformation coefficients, and $\Delta_{jk}(c_{jk}^*)$ the cofactor of the (j, k)-th element of the determinant Δ, we obtain

$$\left. \begin{array}{l} P_{1x} = c_{11}x_1 + c_{21}\alpha_1 + c_{31}x_2 + c_{41}\alpha_2, \\ \mathfrak{M}_{1x} = c_{12}x_1 + c_{22}\alpha_1 + c_{32}x_2 + c_{42}\alpha_2, \\ P_{2x} = c_{13}x_1 + c_{23}\alpha_1 + c_{33}x_2 + c_{43}\alpha_2, \\ \mathfrak{M}_{2x} = c_{14}x_1 + c_{24}\alpha_1 + c_{34}x_2 + c_{44}\alpha_2. \end{array} \right\} \quad (5.288)$$

Similar expressions can be obtained for P_{1y}, P_{2y}, $\mathfrak{M}_{1y}$, $\mathfrak{M}_{2y}$ if one considers the deformation of the shaft in the yz plane; these differ from the expressions (5.288) only by x_1, x_2, α_1, α_2 being replaced correspondingly by y_1, y_2, β_1, β_2.

The total potential energy of deformation, consisting of the sum of potential energies of deformation in the planes xz and yz, is

$$V = V_x + V_y = \frac{1}{2}(P_{1x}x_1 + \mathfrak{M}_{1x}\alpha_1 + P_{2x}x_2 + \mathfrak{M}_{2x}\alpha_2) + \\ + \frac{1}{2}(P_{1y}y_1 + \mathfrak{M}_{1y}\beta_1 + P_{2y}y_2 + \mathfrak{M}_{2y}\beta_2). \qquad (5.289)$$

Substituting into (5.289) the values of the forces and the moments given by (5.288), we obtain

$$V_x = \frac{1}{2}(c_{11}x_1^2 + 2c_{13}x_1x_2 + c_{33}x_2^2) + \frac{1}{2}(c_{22}\alpha_1^2 + 2c_{24}\alpha_1\alpha_2 + c_{44}\alpha_2^2) + \\ + c_{12}x_1\alpha_1 + c_{14}x_1\alpha_2 + c_{23}x_2\alpha_1 + c_{34}x_2\alpha_2 \qquad (5.290)$$

and a similar expression for V_y.

Substituting the values of V (5.289) and T (5.286) into the Lagrange equations, introducing the complex quantities $z_1 = x_1 + iy_1$, $z_2 = \alpha_1 + i\beta_1$, $z_3 = x_2 + iy_2$, $z_4 = \alpha_2 + i\beta_2$, taking into account friction forces, regarding them as being proportional to the deformation velocity, and taking into consideration the friction by means of the dissipative function

$$\Phi = \frac{1}{2}\sum_{j,k=1}^{4} \varkappa_{jk}\dot{z}_j\dot{z}_k, \qquad (5.291)$$

we obtain the following system of differential equations for the transversal vibrations of the coaxial rotors:

$$\left.\begin{aligned}
&m_1\ddot{z}_1 + \sum_{j=1}^{4} c_{1j}z_j = -\sum_{k=1}^{4}\varkappa_{1k}\dot{z}_k + m_1e_1\nu_1^2(t)\,e^{i\varphi_1(t)},\\
&I_1\ddot{z}_2 - iI_{01}\nu_1(t)\dot{z}_2 + \sum_{j=1}^{4} c_{2j}z_j = -\sum_{k=1}^{4}\varkappa_{2k}\dot{z}_k -\\
&\qquad\qquad - i(I_{01} - I_1)\,\delta_1\nu_1^2(t)\,e^{i[\varphi_1(t)-\chi_1]},\\
&m_2\ddot{z}_3 + \sum_{j=1}^{4} c_{3j}z_j = -\sum_{k=1}^{4}\varkappa_{3k}\dot{z}_k + m_2e_2\nu_2^2(t)\,e^{i\varphi_2(t)},\\
&I_2\ddot{z}_4 - iI_{01}\nu_1(t)\dot{z}_4 + \sum_{j=1}^{4} c_{4j}z_j = -\sum_{k=1}^{4}\varkappa_{4k}\dot{z}_k -\\
&\qquad\qquad - (I_{02} - I_2)\,\delta_2\nu_2^2(t)\,e^{i[\varphi_1(t)-\chi_2]},
\end{aligned}\right\} \qquad (5.292)$$

where $\frac{d\varphi_i(t)}{dt} = \nu_i(t)$ $(i = 1,\ 2)$.

We shall consider together with the system (5.292) also the unperturbed system which is obtained by putting $e_1 = e_2 = 0$, $\delta_1 = \delta_2 = 0$, $\varkappa_{jk} = 0$ $(j,\ k = 1,\ 2,\ 3,\ 4)$ in (5.292):

$$\left.\begin{aligned}
&m_1\ddot{z}_1 + \sum_{j=1}^{4} c_{1j}z_j = 0, \quad I_1\ddot{z}_2 - iI_{01}\nu_1(t)\dot{z}_2 + \sum_{j=1}^{4} c_{2j}z_j = 0,\\
&m_2\ddot{z}_3 + \sum_{j=1}^{4} c_{3j}z_j = 0, \quad I_2\ddot{z}_4 - iI_{01}\nu_2(t)\dot{z}_4 + \sum_{j=1}^{4} c_{4j}z_j = 0.
\end{aligned}\right\} \qquad (5.293)$$

When studying forced vibrations of coaxial rotors determined by the non-equilibrium of the disks m_1 and m_2, it is possible, because of the linearity of the system of differential equations (5.292), to use successfully the idea of the monofrequency method. We regard in turn vibrations at frequencies close to ν_1 and ν_2, and then calculate the total amplitude of vibration by adding the derived harmonics (since the superposition principle is valid in linear systems).

According to the results of Chapter IV, § 8, the development of large amplitudes for each harmonic during their combined action almost coincides in practice with the development of the amplitudes of the resonating harmonic in the case of its separate action. This is so if the regions of large amplitudes of both harmonics do not overlap, even in nonlinear systems for which the superposition principle is, generally speaking, not valid. If the resonance zones overlap, then the mutual influence of harmonics in a nonlinear vibrating system is considerable. The mutual influence of harmonics during a transition through resonance decreases with increasing rate of transition. Thus even in nonlinear systems it is possible in many cases to treat the harmonics separately, and then, if desired, to take into account their mutual influence.

While investigating forced vibrations of coaxial rotors we shall consider only the case when the frequencies of both harmonics ν_1 and ν_2, while changing with time, cross successively the values of one of the eigenfrequencies ω_k $(k = 1, 2, 3, 4)$ of the system under consideration; this is the most interesting one. This may be the case in double-rotor turbo-jet and turbo-propeller engines, in which the power is regulated by a change in the rotation velocity of the propeller turbine. The problem may in this case be solved by considering separately the harmonics caused by perturbations of the frequencies ν_1 and ν_2, and then adding the vibration components for deriving the total amplitude of combined vibrations.

We now construct the asymptotic solution in the first approximation of the system of differential equations (5.292) describing the forced vibrations caused by a double-frequency perturbation in the neighborhood of the k-th resonating eigenfrequency ω_k. We have

$$z_j = a_1 c_{jk}(\tau) e^{i(\varphi_1+\psi_1)} + a_2 c_{jk}(\tau) e^{i(\varphi_2+\psi_2)} \qquad (j = 1, 2, 3, 4) \tag{5.294}$$

or, using the previously defined notation,

$$\left.\begin{aligned} x_1 &= a_1 c_{1k}(\tau)\cos(\varphi_1+\psi_1) + a_2 c_{1k}(\tau)\cos(\varphi_2+\psi_2),\\ y_1 &= a_1 c_{1k}(\tau)\sin(\varphi_1+\psi_1) + a_2 c_{1k}(\tau)\sin(\varphi_2+\psi_2),\\ \alpha_1 &= a_1 c_{2k}(\tau)\cos(\varphi_1+\psi_1) + a_2 c_{2k}(\tau)\cos(\varphi_2+\psi_2),\\ \beta_1 &= a_1 c_{2k}(\tau)\sin(\varphi_1+\psi_1) + a_2 c_{2k}(\tau)\sin(\varphi_2+\psi_2), \end{aligned}\right\} \tag{5.295}$$

and also

$$\left.\begin{aligned} x_2 &= a_1 c_{3k}(\tau)\cos(\varphi_1+\psi_1) + a_2 c_{3k}(\tau)\cos(\varphi_2+\psi_2),\\ y_2 &= a_1 c_{3k}(\tau)\sin(\varphi_1+\psi_1) + a_2 c_{3k}(\tau)\sin(\varphi_2+\psi_2),\\ \alpha_2 &= a_1 c_{4k}(\tau)\cos(\varphi_1+\psi_1) + a_2 c_{4k}(\tau)\cos(\varphi_2+\psi_2),\\ \beta_2 &= a_1 c_{4k}(\tau)\sin(\varphi_1+\psi_1) + a_2 c_{4k}(\tau)\sin(\varphi_2+\psi_2), \end{aligned}\right\} \tag{5.296}$$

where $\frac{d\varphi_1}{dt} = \nu_1(\tau)$, $\frac{d\varphi_2}{dt} = \nu_2(\tau)$ (we write here and henceforth $\nu_i(\tau)$ $(i = 1, 2)$ in order to stress the slow change of these quantities, which happens in practice); a_1, ψ_1, the amplitude and phase of the first harmonic of frequency $\nu_1(\tau)$; a_2, ψ_2, the amplitude and phase of the second harmonic of frequency $\nu_2(\tau)$; $c_{jk}(\tau)$ $(j = 1, 2, 3, 4)$ normal functions corresponding to the k-th eigenfrequency $\omega_k(\tau)$ which are derived from the system of unperturbed equations and depend on the slowing time (the unperturbed system contains gyroscopic terms which include the slowing time through the frequencies $\nu_1(\tau)$ and $\nu_2(\tau)$).

We obtain for the amplitudes a_1, a_2 and the phases ψ_1, ψ_2, according to the general methods expounded in the present chapter, the following system of equations of the first approximation:

$$\left.\begin{aligned}\frac{da_1}{dt}&=-\delta_e^{(k)}(\tau)\,a_1-E_1^{(k)}(\tau)\sin\psi_1-E_2^{(k)}(\tau)\cos(\psi_1-\chi_1),\\ \frac{d\psi_1}{dt}&=\omega_k(\tau)-\nu_1(\tau)-\frac{E_1^{(k)}(\tau)}{a_1}\cos\psi_1-\frac{E_2^{(k)}(\tau)}{a_1}\sin(\psi_1-\chi_1),\end{aligned}\right\}\tag{5.297}$$

$$\left.\begin{aligned}\frac{da_2}{dt}&=-\delta_e^{(k)}(\tau)\,a_2\mp E_3^{(k)}(\tau)\sin\psi_2\mp E_4^{(k)}(\tau)\cos(\psi_2-\chi_2),\\ \frac{d\psi_2}{dt}&=\omega_k(\tau)-\nu_2(\tau)-\frac{E_3^{(k)}(\tau)}{a_2}\cos\psi_2-\frac{E_4^{(k)}(\tau)}{a_2}\sin(\psi_2-\chi_2),\end{aligned}\right\}\tag{5.298}$$

where for brevity we have used the notation

$$\delta_e^{(k)}(\tau)=\frac{\bar{m}_3^{(k)}(\tau)+\frac{d}{d\tau}\left[\bar{m}_1^{(k)}(\tau)\,\omega_k(\tau)\right]+\sum_{i,\,j=1}^{4}\varkappa_{ij}c_{ik}(\tau)\,c_{jk}(\tau)\,\omega_k(\tau)}{2\bar{m}_1^{(k)}(\tau)\,\omega_k(\tau)+\bar{m}_2^{(k)}(\tau)},\tag{5.299}$$

$$\left.\begin{aligned}\bar{m}_1^{(k)}(\tau)&=m_1c_{1k}^2(\tau)+I_1c_{2k}^2(\tau)+m_2c_{3k}^2(\tau)+I_2c_{4k}^2(\tau),\\ \bar{m}_2^{(k)}(\tau)&=I_{01}\nu_1(\tau)\,c_{2k}^2(\tau)+I_{02}\nu_2(\tau)\,c_{4k}^2(\tau),\\ \bar{m}_3^{(k)}(\tau)&=I_{01}\nu_1(\tau)\,c_{2k}(\tau)\frac{dc_{3k}(\tau)}{d\tau}+I_{02}\nu_2(\tau)\,c_{4k}(\tau)\frac{dc_{4k}(\tau)}{d\tau},\end{aligned}\right\}\tag{5.300}$$

$$\left.\begin{aligned}E_1^{(k)}(\tau)&=\frac{m_1e_1c_{1k}(\tau)\,\nu_1^2(\tau)}{\bar{m}_1^{(k)}(\tau)\left[\omega_k(\tau)+\nu_1(\tau)\right]+\bar{m}_2^{(k)}(\tau)},\\ E_2^{(k)}(\tau)&=\frac{(I_{01}-I_2)\,\delta_1c_{2k}(\tau)\,\nu_1^2(\tau)}{\bar{m}_1^{(k)}(\tau)\left[\omega_k(\tau)+\nu_1(\tau)\right]+\bar{m}_2^{(k)}(\tau)}.\end{aligned}\right\}\tag{5.301}$$

When the signs of ν_1 and ν_2 are equal (i. e., equal directions of rotation of the shafts), then

$$\left.\begin{aligned}E_3^{(k)}(\tau)&=\frac{m_2e_2c_{3k}(\tau)\,\nu_2^2(\tau)}{\bar{m}_1^{(k)}(\tau)\left[\omega_k(\tau)+\nu_2(\tau)\right]+\bar{m}_2^{(k)}(\tau)},\\ E_4^{(k)}(\tau)&=\frac{(I_{01}-I_2)\,\delta_2c_{4k}(\tau)\,\nu_2^2(\tau)}{\bar{m}_1^{(k)}(\tau)\left[\omega_k(\tau)+\nu_2(\tau)\right]+\bar{m}_2^{(k)}(\tau)},\end{aligned}\right\}\tag{5.302}$$

and the upper sign (minus) should be taken in the first of equations (5.298).

When the directions of rotations of the shafts are different, the amplitudes of perturbations $E_3^{(k)}(\tau)$ and $E_4^{(k)}(\tau)$ are determined by the formulas

$$\left.\begin{aligned}E_3^{(k)}(\tau)&=\frac{m_2e_2c_{3k}(\tau)\,\nu_2^2(\tau)}{\bar{m}_1^{(k)}(\tau)\left[3\omega_k(\tau)-\nu_2(\tau)\right]+\bar{m}_2^{(k)}(\tau)},\\ E_4^{(k)}(\tau)&=\frac{(I_{01}-I_2)\,\delta_ec_{4k}(\tau)\,\nu_2^2(\tau)}{\bar{m}_1^{(k)}(\tau)\left[3\omega_k(\tau)-\nu_2(\tau)\right]+\bar{m}_2^{(k)}(\tau)},\end{aligned}\right\}\tag{5.303}$$

and the lower sign (plus) is taken in the first of equations (5.298).

To obtain formulas for the stationary amplitude values of the first and second terms in (5.294), we equate the right-hand sides of equations (5.297) and (5.298) to zero and eliminate ψ_1 and ψ_2.

We obtain

$$a_1=\sqrt{\frac{E_1^{(k)2}+E_2^{(k)2}+2E_1^{(k)}E_2^{(k)}\cos\chi_1}{\delta_e^{(k)2}+(\omega_k-\nu_1)^2}},\tag{5.304}$$

$$a_2=\sqrt{\frac{E_3^{(k)2}+E_4^{(k)2}+2E_3^{(k)}E_4^{(k)}\cos\chi_2}{\delta_0^{(k)2}+(\omega_k-\nu_2)^2}}.\tag{5.305}$$

For the phase angles ψ_1 and ψ_2 in a stationary mode of vibration, we obtain the formulas

$$\cos\psi_1=\frac{(\omega_k-\nu_1)\left(E_1^{(k)}\sin\delta_1+E_2^{(k)}\sin\delta_2\right)+\delta_e^{(k)}\left(E_1^{(k)}\cos\delta_1+E_2^{(k)}\cos\delta_2\right)}{E_1^{(k)2}+E_2^{(k)2}+2E_1^{(k)}E_2^{(k)}\cos\chi_1}\,a_1,\tag{5.306}$$

$$\cos\psi_2=\frac{(\omega_k-\nu_2)\left(E_3^{(k)}\cos\delta_3+E_4^{(k)}\cos\delta_4\right)\pm\delta_e^{(k)}\left(E_3^{(k)}\sin\delta_3+E_4^{(k)}\sin\delta_4\right)}{E_3^{(k)2}+E_4^{(k)2}+2E_3^{(k)}E_4^{(k)}\cos\chi_2}\,a_2,\tag{5.307}$$

where in the last formula the plus sign is taken when the rotors rotate in the same direction, and the minus sign when they rotate in different directions.

Given the numerical values, we can construct the stationary resonance curves characterizing the amplitudes a_1 and a_2 as functions of the external frequencies ν_1 and ν_2 with the help of formulas (5.304) and (5.305) (as was done in the preceding section).

Numerical integration (or integration with the help of analogue computors) of the systems of equations (5.297) and (5.298) allows us to construct the curves of transition through resonance frequency values ω_k. We can also discuss the results, as was done, for example in Chapter IV, § 8, now obviously taking into account that the initial system of differential equations contains gyroscopic terms, which gives rise to additional phenomena characteristic of such systems during nonstationary processes.

§ 10. Construction of asymptotic approximations for systems of differential equations of the type (1.18)

The investigation of a number of problems concerning nonstationary vibrations in nonlinear systems with gyroscopic effects leads to the following system of differential equations

$$\left.\begin{aligned} &\frac{d}{dt}\left[\frac{\partial \mathcal{L}}{\partial \dot{q}_j}\right]-\frac{\partial \mathcal{L}}{\partial q_j}=\varepsilon Q_j(y, \dot{y}, q_1, \ldots, q_N, \dot{q}_1, \ldots, \dot{q}_N, \varepsilon), \\ &\frac{d^2 y}{dt^2}=\varepsilon Y(y, \dot{y}, q_1, \ldots, q_N, \dot{q}_1, \ldots, \dot{q}_N, \varepsilon) \qquad (j=1, 2, \ldots, N), \end{aligned}\right\} \tag{5.308}$$

where

$$\mathcal{L}=\frac{1}{2}\left\{\sum_{i, j=1}^{N} a_{ij}(\dot{y}) \dot{q}_i \dot{q}_j+2 \sum_{i, j=1}^{N} g_{ij}(\dot{y}) q_i \dot{q}_j-\sum_{i, j=1}^{N} b_{ij}(\dot{y}) q_i q_j\right\}, \tag{5.309}$$

and the functions $Q_j(y, \dot{y}, q_1, \ldots, q_N, \dot{q}_1, \ldots, \dot{q}_N, \varepsilon)$ $(j=1, 2, \ldots, N)$ and $Y(y, \dot{y}, q_1, \ldots, q_N, \dot{q}_1, \ldots, \dot{q}_N, \varepsilon)$ are periodic in y with period 2π, and satisfy the same conditions as the functions (5.3).

We obtained an actual system of equations of the type (5.308) in § 2 of the present chapter when studying the vibrational process of a centrifuge. Such systems occur while investigating the nonstationary vibrational process in various systems coupled to a low power energy source, while constructing higher approximations for systems of equations of the type (5.129), etc.

Approximate solutions of the system (5.308) corresponding to the nonstationary monofrequency mode can be constructed with the help of the results derived in § 1 of the present chapter, and the arguments suggested in connection with the system (1.16) in Chapter III, § 13.

By the unperturbed system corresponding to the perturbed equations (5.308) we mean the system of equations

$$\sum_{j=1}^{N} a_{ij}(\dot{y}) \ddot{q}_{ij}+\sum_{j=1}^{N} c_{ij}(\dot{y}) \dot{q}_j+\sum_{j=1}^{N} b_{ij}(\dot{y}) q_j=0 \qquad (i=1, 2, \ldots, N), \tag{5.310}$$

where $\dot{y}=\text{const}$.

We assume that the Lagrangian $\mathscr{L}$ (5.309) is such that the total energy of the unperturbed vibrating system under consideration is for any constant $\dot{y}$ a positive definite quadratic form, and that the conditions listed in the first section of Chapter IV (p. 158) are satisfied. These conditions guarantee the existence of a stable monofrequency mode in the unperturbed system (5.310)

$$q_j = a\varphi_j^{(1)}(\dot{y})\, e^{i(\omega_1 t+\varphi)} + a\overset{*}{\varphi}{}_j^{(1)}(\dot{y})\, e^{-i(\omega_1 t+\varphi)} \qquad (j = 1, 2, \ldots, N), \tag{5.311}$$

where ω_1 is an eigenfrequency determined by the equation

$$D(\omega) = \| -a_{ij}(\dot{y})\,\omega^2 + ic_{ij}(\dot{y})\,\omega + b_{ij}(\dot{y}) \| = 0, \tag{5.312}$$

and $\varphi_j^{(1)}(\dot{y})$ $(j = 1, 2, \ldots, N)$ are basic functions which are nontrivial solutions of the system of algebraic equations

$$\sum_{j=1}^{N} \{-a_{ij}(\dot{y})\,\omega_1^2 + ic_{ij}(\dot{y})\,\omega_1 + b_{ij}(\dot{y})\}\, \varphi_j^{(1)} = 0 \qquad (i = 1, 2, \ldots, N). \tag{5.313}$$

Under these assumptions, the solution of the perturbed equations (5.308) close to the two-parametric family of the particular solutions (5.311) of the unperturbed system in the resonance case $\left(\omega_1 - \frac{p}{q}\dot{y} = \varepsilon\Delta(\dot{y})\right)$, is looked for in the form of asymptotic series

$$q_j = u_j^{(0)}(\dot{y}, a, p\varphi+\psi) + \varepsilon u_j^{(1)}(\dot{y}, a, y, p\varphi+\psi) + \\ + \varepsilon^2 u_j^{(2)}(\dot{y}, a, y, p\varphi+\psi) + \ldots \qquad (j = 1, 2, \ldots, N), \tag{5.314}$$

where

$$u_j^{(0)}(\dot{y}, a, p\varphi+\psi) = a\varphi_j^{(1)}(\dot{y})\, e^{i(p\varphi+\psi)} + a\overset{*}{\varphi}{}_j^{(1)}(\dot{y})\, e^{-i(p\varphi+\psi)} \\ (j = 1, 2, \ldots, N), \tag{5.315}$$

$u_j^{(l)}(\dot{y}, a, y, p\varphi+\psi)$ $(j = 1, 2, \ldots, N;\ l = 1, 2, \ldots)$ are periodic functions in y and $p\varphi+\psi$ with period 2π; $\varphi = \frac{1}{q}y$; p and q are mutually prime integers depending on the resonance near which we intend to examine the behavior of the specified system; and the quantities a, ψ and $\dot{y}$ are to be determined from the system of equations

$$\left.\begin{aligned} \frac{da}{dt} &= \varepsilon A_1(\dot{y}, a, \psi) + \varepsilon^2 A_2(\dot{y}, a, \psi) + \ldots, \\ \frac{d\psi}{dt} &= \omega_1(\dot{y}) - \frac{p}{q}\dot{y} + \varepsilon B_1(\dot{y}, a, \psi) + \varepsilon^2 B_2(\dot{y}, a, \psi) + \ldots, \\ \frac{d\dot{y}}{dt} &= \varepsilon D_1(\dot{y}, a, \psi) + \varepsilon^2 D_2(\dot{y}, a, \psi) + \ldots, \end{aligned}\right\} \tag{5.316}$$

where the functions $A_1(\dot{y}, a, \psi)$, $B_1(\dot{y}, a, \psi)$, $D_1(\dot{y}, a, \psi)$, $A_2(\dot{y}, a, \psi)$, $B_2(\dot{y}, a, \psi)$, $D_2(\dot{y}, a, \psi)$, ... are derived by the method given in § 1 of the present chapter.

In order to obtain the solutions in the first approximation we must construct the system (5.316) up to and including terms of the first order. This may be achieved, as was shown more than once, by the method of harmonic balance.

If a solution is to be derived so that the behavior of the system not only in the resonance zone, but also during an approach to the latter (the difference $\omega_1(\dot{y}) - \frac{p}{q}\dot{y}$ is not small) is to be studied, then it already becomes necessary to apply the method of successive substitutions (see Chapter III, § 10), first transforming the system of equations (5.308) to a more convenient form by means of the changes of variables given in Chapter VII.

Let us now consider an actual example.

We consider nonstationary deformation vibrations of a shaft with an unbalanced disk placed asymmetrically with respect to the bearings, taking into account the coupling to a motor /42/.

The equations of motion of the disk have the form

$$\left.\begin{aligned}
&m\ddot{x} + c_{11}x - c_{12}\alpha = m\varepsilon\,(\dot{y}^2\cos y + \ddot{y}\sin y),\\
&m\ddot{z} + c_{11}z - c_{12}\beta = m\varepsilon\,(\dot{y}^2\sin y - \ddot{y}\cos y),\\
&I\ddot{\alpha} + I_0\dot{y}\dot{\beta} + c_{22}\alpha - c_{12}x = -I_0\ddot{y}\beta + (I_0 - I)\,\delta\,[\dot{y}^2\sin(y-\chi) - \ddot{y}\cos(y-\chi)],\\
&I\ddot{\beta} - I_0\dot{y}\dot{\alpha} + c_{22}\beta - c_{12}z = -(I_0 - I)\,\delta\dot{y}^2\cos(y-\chi) + I\delta\ddot{y}\sin(y-\chi),\\
&I_0\ddot{y} = M(\dot{y}) - R(\dot{y}) - m\varepsilon\,(\ddot{z}\cos y - \ddot{x}\sin y) - I_0\ddot{\alpha}\beta - I_0\dot{\alpha}\dot{\beta} +\\
&\qquad\qquad + 2I_0\delta\dot{y}\dot{\beta}\cos(y-\chi) - I_0\delta\dot{y}^2\beta\sin(y-\chi),
\end{aligned}\right\}\qquad (5.317)$$

where m is the mass of the disk; I, the moment of inertia of the disk with respect to a transverse axis passing through the center of gravity perpendicular to the symmetry axis; I_0, the moment of inertia of the disk with respect to the symmetry axis; ε, the displacement of the disk's center of gravity from the symmetry axis (linear eccentricity); δ, the angle between the symmetry axis of the rotor and the axis of the shaft on which it is fastened (angular eccentricity of the disk); $M(\dot{y})$, the torsional moment of the motor; $R(\dot{y})$, the moment of resistance to rotational motion; x, z, coordinates defining the disk's position in a system of coordinates attached to it; α, β, y, Résal angles; c_{11}, c_{12}, c_{22}, coefficients characterizing the geometry and the conditions of fastening of the shaft and the disk.

Neglecting second order terms in the right-hand sides of the system (5.317), and also the perturbations conditioned by the dynamical nonequilibrium of the disk, one can consider instead of (5.317) the system

$$\left.\begin{aligned}
&m\ddot{x} + c_{11}x - c_{12}\alpha = m\varepsilon\dot{y}^2\cos y,\\
&m\ddot{z} + c_{11}z - c_{12}\beta = m\varepsilon\dot{y}^2\sin y,\\
&I\ddot{\alpha} + I_0\dot{y}\beta - c_{22}\alpha - c_{11}x = -I_0\ddot{y}\beta,\\
&I\ddot{\beta} - I_0\dot{y}\alpha + c_{22}\beta - c_{12}z = 0,\\
&I_0\ddot{y} = M(\dot{y}) - R(\dot{y}) - m\varepsilon\,(\ddot{z}\cos y - \ddot{x}\sin y) - I_0\ddot{\alpha}\beta - I_0\dot{\alpha}\dot{\beta}.
\end{aligned}\right\}\qquad (5.318)$$

We assume that the specified vibrating system is also subject to the action of external friction forces described by the dissipative function

$$\Phi_1 = \frac{1}{2}\left[\varkappa_1\dot{x}^2 + \varkappa_2\dot{z}^2 + \varkappa_3\dot{\alpha}^2 + \varkappa_4\dot{\beta}^2\right], \qquad (5.319)$$

where $\varkappa_i$ $(i = 1, 2, 3, 4)$ are constant coefficients of the external friction forces, and to the action of internal friction forces described by the dissipative function

$$\Phi_2 = \frac{1}{2}\left[k_1(\dot{x} - \dot{y}z)^2 + k_2(\dot{z} + \dot{y}z)^2 + k_3(\dot{\alpha} - \dot{y}\beta)^2 + k_4(\dot{\beta} + \dot{y}\alpha)^2\right], \qquad (5.320)$$

where k_i $(i = 1, 2, 3, 4)$ are constant coefficients of the internal friction forces.

The solution of the system (5.318) corresponding to the resonant monofrequency mode close to the first eigenfrequency has in the first approximation the form

$$q_j = a\varphi_j^{(1)}(\dot{y})\, e^{i(y+\psi)} + a\overset{*}{\varphi}_j^{(1)}(\dot{y})\, e^{-i(y+\psi)} \qquad (j = 1, 2, 3, 4), \tag{5.321}$$

where $q_1 = x$; $q_2 = z$; $q_3 = \alpha$; $q_4 = \beta$ $(p = q = 1)$; and $\varphi_j^{(1)}(\dot{y})$ $(j = 1, 2, 3, 4)$ are nontrivial solutions of the system of algebraic equations

$$\left.\begin{aligned}
(c_{11} - m\omega_1^2)\,\varphi_1 - c_{12}\varphi_3 &= 0,\\
(c_{11} - m\omega_1^2)\,\varphi_2 - c_{12}\varphi_4 &= 0,\\
c_{11}\varphi_1 + (I\omega_1^2 + c_{22})\,\varphi_3 - I_0\dot{y}\varphi_4 &= 0,\\
c_{12}\varphi_2 + I_0\dot{y}\varphi_3 + (I\omega_1^2 - c_{22})\,\varphi_4 &= 0.
\end{aligned}\right\} \tag{5.322}$$

Here $\omega_1 = \omega_1(\dot{y})$ is an eigenfrequency determined by the equation

$$(c_{11} - m\omega_1^2)^2 (I^2\omega_1^4 - c_{22}^2 + I_0^2\dot{y}^2) + (c_{11} - m\omega_1^2)\,[c_{12}^2 (I\omega_1^2 + c_{22}) + \\ + c_{11}c_{12}(I\omega_1^2 - c_{22})] + c_{11}c_{12}^2 = 0, \tag{5.323}$$

and the quantities a, ψ, and $\dot{y}$ are to be determined from the system of equations of the first approximation

$$\left.\begin{aligned}
\frac{da}{dt} &= -\delta_e(\dot{y}, \psi)\, a - E_1(\dot{y}) \cos\left(\psi - \frac{\pi}{4}\right),\\
\frac{d\psi}{dt} &= \omega_1(\dot{y}) - \dot{y} + \frac{1}{a} E_1(\dot{y}) \sin\left(\psi - \frac{\pi}{4}\right),\\
\frac{d\dot{y}}{dt} &= \frac{1}{I_0}\Big\{M(\dot{y}) - R(\dot{y}) + \frac{\sqrt{2}}{2} m\varepsilon\omega_1^2(\dot{y})\big(1 - \varphi_2^{(1)}(\dot{y})\big) a \cos\left(\psi - \frac{\pi}{4}\right) +\\
&+ a^2\dot{y}\,[k_1\varphi_2^{(1)2}(\dot{y}) + k_2 + k_3\varphi_4^{(1)2}(\dot{y}) + k_4\varphi_3^{(1)2}(\dot{y})]\Big\} = f(\dot{y}, a, \psi),
\end{aligned}\right\} \tag{5.324}$$

where we have used the notation

$$\delta_e(\dot{y}, \psi) = \frac{\sum\limits_{j=1}^{4} [\varkappa_j\omega_1(\dot{y}) + k_j(\omega_1(\dot{y}) - \dot{y})]\, \varphi_j^{(1)}(\dot{y})\, \overset{*}{\varphi}_j^{(1)}(\dot{y}) + m_3^{(1)}(\dot{y}) + \frac{d}{d\dot{y}}[m_1^{(1)}(\dot{y})\,\omega_1(\dot{y})]\, f(\dot{y}, a, \psi)}{2m_1^{(1)}(\dot{y})\,\omega_1(\dot{y}) + m_2^{(1)}(\dot{y})}, \tag{5.325}$$

$$E_1(\dot{y}) = \frac{\sqrt{2}}{2}\, \frac{m\varepsilon\dot{y}^2\,[1 + \varphi_2^{(1)}(\dot{y})]}{m_1^{(1)}(\dot{y})\,(\omega_1(\dot{y}) + \dot{y}) + m_2^{(1)}(\dot{y})}, \tag{5.326}$$

$$\left.\begin{aligned}
m_1^{(1)}(\dot{y}) &= m\,[\varphi_1^{(1)}(\dot{y})\,\overset{*}{\varphi}_1^{(1)}(\dot{y}) + \varphi_1^{(1)}(\dot{y})\overset{*}{\varphi}_2^{(1)}(\dot{y}) + I\,(\varphi_3^{(1)}(\dot{y})\,\overset{*}{\varphi}_3^{(1)}(\dot{y}) + \varphi_4^{(1)}(\dot{y})\,\overset{*}{\varphi}_4^{(1)}(\dot{y}))],\\
m_2^{(1)}(\dot{y}) &= I_0\dot{y}\,[\varphi_4^{(1)}(\dot{y})\,\overset{*}{\varphi}_4^{(1)}(\dot{y}) - \varphi_3^{(1)}(\dot{y})\,\overset{*}{\varphi}_3^{(1)}(\dot{y})],\\
m_3^{(1)}(\dot{y}) &= I_0\dot{y}\left[\overset{*}{\varphi}_4^{(1)}(\dot{y})\,\frac{d\varphi_3^{(1)}(\dot{y})}{d\dot{y}} - \overset{*}{\varphi}_3^{(1)}(\dot{y})\,\frac{d\varphi_4^{(1)}(\dot{y})}{d\dot{y}}\right] f(\dot{y}, a, \psi).
\end{aligned}\right\} \tag{5.327}$$

Chapter VI

MONOFREQUENCY VIBRATIONS OF SYSTEMS WITH DISTRIBUTED PARAMETERS

§ 1. The construction of approximate solutions without a derivation of the exact differential equations of the problem

We consider first a simple energy interpretation of the formulas and equations of the first and higher approximations obtained in the fourth chapter. We will then be able to formulate a few convenient rules with which we can derive the equations of the first and higher approximations (equations which we use when analyzing the specified vibrational process) without constructing beforehand the exact equations describing the vibrational process in the given system.

Consider the expression of the virtual work of the perturbing forces $\varepsilon Q_j(\tau, \theta, q_1, \ldots, q_N, \dot{q}_1, \ldots, \dot{q}_N, \varepsilon)$ $(j = 1, 2, \ldots, N)$ in a mode of sinusoidal vibrations

$$q_{i0} = \varphi_i^{(1)}(\tau) a \cos(p\varphi + \psi),$$

$$\dot{q}_{i0} = -\varphi_i^{(1)}(\tau) a\omega_1(\tau) \sin(p\varphi + \psi) \qquad (i = 1, 2, \ldots, N) \tag{6.1}$$

for the virtual displacement

$$\delta q_{j0} = \varphi_j^{(1)}(\tau) \cos(p\varphi + \psi)\, \delta a - \varphi_j^{(1)}(\tau)\, a \sin(p\varphi + \psi)\, \delta\psi \qquad (j = 1, 2, \ldots, N), \tag{6.2}$$

corresponding to the variations of amplitude and total phase of the first "normal" mode. As before, τ is taken to be a constant parameter and it is assumed in the final formulas and in the equations of the first and higher approximations that $\tau = \varepsilon t$.

We then have, with an accuracy of second order terms,

$$\delta W = \varepsilon \sum_{j=1}^{N} Q_{j0}^{(1)}(\tau, \theta, a, p\varphi + \psi)\,[\varphi_j^{(1)}(\tau) \cos(p\varphi + \psi)\, \delta a - \varphi_j^{(1)}(\tau)\, a \sin(p\varphi + \psi)\, \delta\psi]\,^{*} \tag{6.3}$$

Denoting the average value of this work over a complete period of vibration by $\overline{\delta W}$, and expanding it in a double Fourier sum, we obtain

$$\overline{\delta W} = \frac{\varepsilon\delta a}{4\pi^2} \sum_{\sigma} e^{i\sigma q\psi} \int_0^{2\pi}\int_0^{2\pi} \sum_{j=1}^{N} Q_{j0}^{(1)}(\tau, \theta, a, p\varphi + \psi)\, \varphi_j^{(1)}(\tau) \times$$
$$\times e^{-i\sigma q\psi} \cos(p\varphi + \psi)\, d\theta\, d(p\varphi + \psi) - \frac{\varepsilon a\delta\psi}{4\pi^2} \sum_{\sigma} e^{i\sigma q\psi} \int_0^{2\pi}\int_0^{2\pi} \sum_{j=1}^{N} Q_{j0}^{(1)}(\tau, \theta, a, p\varphi + \psi) \times$$
$$\times \varphi_j^{(1)}(\tau)\, e^{-i\sigma q\psi} \sin(p\varphi + \psi)\, d\theta\, d(p\varphi + \psi). \tag{6.4}$$

* δW – the differential form of the virtual work of the perturbing forces – is generally speaking not a total differential.

The coefficients of the variations δa and $\delta\psi$ will be denoted symbolically by $\frac{\overline{\delta W}}{\delta a}$ and $\frac{\overline{\delta W}}{\delta\psi}$; the system of equations (4.45) (see Chapter IV), may now be written in the form

$$\left.\begin{aligned}&\varepsilon\left[\omega_1(\tau)-\frac{p}{q}\nu(\tau)\right]\frac{\partial A_1}{\partial\psi}-2\varepsilon\omega_1(\tau)aB_1=\frac{2}{m_1(\tau)}\frac{\overline{\delta W}}{\delta a},\\&\varepsilon\left[\omega_1(\tau)-\frac{p}{q}\nu(\tau)\right]a\frac{\partial B_1}{\partial\psi}+2\varepsilon\omega_1(\tau)A_1=\\&\qquad=-\frac{\varepsilon a}{m_1(\tau)}\frac{d[m_1(\tau)\omega_1(\tau)]}{d\tau}+\frac{2}{m_1(\tau)a}\frac{\overline{\delta W}}{\delta\psi}.\end{aligned}\right\}\qquad(6.5)$$

Thus, to construct the system of partial differential equations for $A_1(\tau, a, \psi)$ and $B_1(\tau, a, \psi)$, we must find the average value of the virtual work of the perturbing forces during a vibration period in a sinusoidal mode for virtual displacements corresponding to the variations in their amplitude a and phase ψ, and then substitute the "partial derivatives" $\frac{\overline{\delta W}}{\delta a}$, $\frac{\overline{\delta W}}{\delta\psi}$ of the virtual work thus obtained into equation (6.5).

The equations of the first approximation (4.50) may be easily given a direct energy interpretation. To do this we express the average value of the virtual work (6.4) in the form

$$\overline{\delta W}=\sum_{\delta}\overline{\delta W_\sigma}=\sum_{\sigma}\left\{\frac{\overline{\delta W_\sigma}}{\delta a}\delta a+\frac{\overline{\delta W}}{\delta\psi}\delta\psi\right\},\qquad(6.6)$$

where $\overline{\delta W}$ denotes the average value of the virtual work over a vibration period performed by the σ-th term in the Fourier expansion of the perturbing force in a sinusoidal mode for virtual displacements corresponding to the variations of the vibration amplitude and phase.

The equations of the first approximation (4.50) assume in this notation the form

$$\left.\begin{aligned}\frac{da}{dt}&=-\frac{\varepsilon a}{2m_1(\tau)\omega_1(\tau)}\frac{d[m_1(\tau)\omega_1(\tau)]}{d\tau}+\\&\quad+\frac{2}{m_1(\tau)}\sum_{\sigma}\frac{[p\omega_1(\tau)-q\nu(\tau)]\sigma i\frac{\overline{\delta W_\sigma}}{\delta a}+2\omega_1(\tau)\frac{\overline{\delta W_\sigma}}{\delta\psi}\frac{1}{a}}{4\omega_1^2(\tau)-[p\omega_1(\tau)-q\nu(\tau)]^2\sigma^2},\\\frac{d\psi}{dt}&=\omega_1(\tau)-\frac{p}{q}\nu(\tau)+\\&\quad+\frac{2}{m_1(\tau)}\sum_{\sigma}\frac{[p\omega_1(\tau)-q\nu(\tau)]\sigma i\frac{\overline{\delta W_\sigma}}{\delta\psi}\frac{1}{a}-2\omega_1(\tau)\frac{\overline{\delta W}}{\delta a}}{4\omega_1^2(\tau)-[p\omega_1(\tau)-q\nu(\tau)]^2\sigma^2}.\end{aligned}\right\}\qquad(6.7)$$

Thus, to obtain the equations of the first approximation, we must find the average value of the virtual work of the perturbing forces over a complete vibration period in a sinusoidal mode for virtual displacements corresponding to the variations in the amplitude and phase, expand the result in a Fourier series, and then substitute the "partial derivatives" of the σ-th term into equation (6.7).

To express $u_i^{(1)}(\tau, a, \theta, p\varphi+\psi)$ given by formula (4.42) in a simpler form, we consider the expression of the virtual work of the perturbing forces $\varepsilon Q_j(\tau, \theta, q_1, \ldots, q_N, \dot q_1, \ldots, \dot q_N, \varepsilon)$ $(j=1, 2, \ldots, N)$in a sinusoidal mode of vibrations

$$q_{i_0}=\varphi_i^{(1)}(\tau)a\cos(p\varphi+\psi),\quad \dot q_{i0}=-\omega_1(\tau)\varphi_i^{(1)}(\tau)a\sin(p\varphi+\psi)$$
$$(i=1, 2, \ldots, N)\qquad(6.8)$$

for the virtual displacements

$$\delta q_{j0}^{(k)} = \varphi_j^{(k)}(\tau)\cos(p\varphi+\psi)\,\delta a - \varphi_j^{(k)}(\tau)\sin(p\varphi+\psi)\,\delta\psi =$$
$$= \varphi_j^{(k)}(\tau)\,\delta\,[a\cos(p\varphi+\psi)] \qquad (j=1,\ 2,\ \ldots,\ N), \tag{6.9}$$

corresponding to the variations of the amplitude and phase of the k-th normal mode.

We can write with an accuracy of first order terms

$$\delta W^{(k)} = \varepsilon\sum_{j=1}^{N} Q_{j0}^{(1)}(\tau,\ \theta,\ a,\ p\varphi+\psi)\,\delta q_{j0}^{(k)} =$$
$$= \varepsilon\sum_{j=1}^{N} Q_{j0}^{(1)}(\tau,\ \theta,\ a,\ p\varphi+\psi)\,\varphi_j^{(k)}(\tau)\,\delta\,[a\cos(p\varphi+\psi)], \tag{6.10}$$

from which we obtain for the generalized force acting on the k-th normal coordinate, the expression

$$\varepsilon\sum_{j=1}^{N} Q_{j0}^{(1)}(\tau,\ \theta,\ a,\ p\varphi+\psi)\,\varphi_j^{(k)}(\tau) = \left(\frac{\delta W^{(k)}}{\delta y}\right)_{y=a\cos(p\varphi+\psi)}. \tag{6.11}$$

We express it formally as a Fourier double sum

$$\left(\frac{\delta W^{(k)}}{\delta y}\right)_{y=a\cos(p\varphi+\psi)} = \sum_{n,\ m}\left(\frac{\delta W^{(k)}}{\delta y}\right)_{n,\ m\ y=a\cos(p\varphi+\psi)}.$$

Thus $u_i^{(1)}(\tau,\ a,\ \theta,\ p\varphi+\psi)$ may be represented in the form

$$\varepsilon u_i^{(1)}(\tau,\ a,\ \theta,\ p\varphi+\psi) = \frac{1}{4\pi^2}\sum_{n,\ m}\sum_{k=1}^{N}\frac{\varphi_i^{(k)}(\tau)\left(\frac{\delta W^{(k)}}{\delta y}\right)_{n,\ m\ y=a\cos(p\varphi+\psi)}}{m_k(\tau)\{\omega_k^2(\tau)-[\omega_1(\tau)m+\nu(\tau)n]^2\}} -$$
$$-2\omega_1(\tau)\,a\sum_{k=2}^{N}\frac{\left(\frac{\delta W^{T(k)}}{\delta y}\right)_{y=a\cos(p\varphi+\psi)}}{m_k(\tau)\,[\omega_k^2(\tau)-\omega_1^2(\tau)]}\sin(p\varphi+\psi) \quad (i=1,\ 2,\ \ldots,\ N), \tag{6.12}$$

where $\delta W^{T(k)}$ denotes the virtual work of the "forces" $\frac{d}{d\tau}\left[\frac{\partial T}{\partial \dot{q}_j}\right]_{\dot{q}_i=\varphi_i^{(1)}}$ for virtual displacements $\delta q^{(k)}$ corresponding to variations of the amplitude and phase of the k-th normal mode. Hence, to determine $u_i^{(1)}(\tau,\ a,\ \theta,\ p\varphi+\psi)$ we must find the "derivatives" $\left(\frac{\delta W^{(k)}}{\delta y}\right)_{y=a\cos(p\varphi+\psi)}$, $\left(\frac{\delta W^{T(k)}}{\delta y}\right)_{y=a\cos(p\varphi+\psi)}$ of the virtual work performed by the generalized forces $\varepsilon Q_j(\tau,\ \theta,\ q_1,\ \ldots,\ q_N,\ \dot{q}_1,\ \ldots,\ \dot{q}_N,\ \varepsilon)$ $(j=1,\ 2,\ \ldots,\ N)$ and the "forces" $\frac{d}{d\tau}\left[\frac{\partial T}{\partial \dot{q}_j}\right]_{\dot{q}_i=\varphi_i^{(1)}(\tau)}$ in a sinusoidal mode for the virtual displacements $\delta q_{j0}^{(k)}$ corresponding to the variations of the amplitude and phase of the k-th normal mode, expand the first expression in a Fourier sum, and then substitute the (n, m)-th terms of this sum and $\left(\frac{\delta W^{T(k)}}{\delta y}\right)_{y=a\cos(p\varphi+\psi)}$ into (6.12).

Equations (4.48) defining $A_2(\tau,\ a,\ \psi)$ and $B_2(\tau,\ a,\ \psi)$ may also be written in a simpler form. As above, we can express the sums on the right-hand sides by "partial derivatives" of the virtual work averaged over a complete period. These equations then take the form

$$\left[\omega_1(\tau)-\frac{p}{q}\nu(\tau)\right]\frac{\partial A_2}{\partial\psi} - 2\omega_1(\tau)\,aB_2 =$$
$$= -\left[\frac{\partial A_1}{\partial a}A_1 + \frac{\partial A_1}{\partial\psi}B_1 + \frac{\partial A_1}{\partial\tau} - aB_1^2 + \frac{dm_1(\tau)}{d\tau}\frac{A_1}{m_1(\tau)} -\right.$$
$$\left.-\gamma(\tau)\frac{a}{m_1(\tau)}\right] + \frac{2}{m_1(\tau)}\frac{\overline{\delta W_2}}{\delta a} + \frac{2}{m_1(\tau)}\left(\frac{\overline{\delta\delta W}}{\delta a}\right)_{\substack{\delta q=\Delta q\\ \delta q'=\Delta q'}} - \frac{2}{m_1(\tau)}\frac{\overline{\delta W_T}}{\delta a}, \tag{6.13}$$

$$\left[\omega_1(\tau)-\frac{p}{q}\nu(\tau)\right]a\frac{\partial B_2}{\partial\psi}+2\omega_1(\tau)A_2=-\left\{a\frac{\partial B_1}{\partial a}A_1+a\frac{\partial B_1}{\partial\psi}B_1+\right.$$

$$\left.+a\frac{\partial B_1}{\partial\tau}+2A_1B_1+\frac{dm_1(\tau)}{d\tau}\frac{a}{m_1(\tau)}B_1\right\}+\frac{2}{m_1(\tau)a}\frac{\overline{\delta W_2}}{\delta\psi}+$$

$$+\frac{2}{m_1(\tau)a}\left(\frac{\overline{\delta\delta W}}{\delta\psi}\right)_{\substack{\delta q=\Delta q\\ \delta q'=\Delta q'}}-\frac{2}{m_1(\tau)a}\frac{\overline{\delta W_T}}{\delta\psi}, \tag{6.13}$$

where $\overline{\delta W_2}$, $\overline{\delta W_T}$, and $\overline{(\delta\delta W)}_{\substack{\delta q=\Delta q\\ \delta q'=\Delta q'}}$ denote the expression (6.4), in which $\varepsilon Q_{j0}^{(1)}(\tau,\theta,a,p\varphi+\psi)$ is replaced correspondingly by $Q_{j0}^{(2)}(\tau,\theta,a,p\varphi+\psi)$, $\frac{d}{dt}\left[\frac{\partial T(\dot q)}{\partial\dot q_i}\right]_{\dot q_j=\dot u_i^{(1)}}$, and $(\delta W)_{\substack{\delta q=\Delta q\\ \delta q'=\Delta q'}}$, where $(\delta W)_{\substack{\delta q=\Delta q\\ \delta q'=\Delta q'}}$ is the expression for the virtual work, in which

$$\Delta q_i=u_i^{(1)}\ \text{and}\ \Delta q_i'=\frac{d\varphi_i^{(1)}}{d\tau}a\cos(p\varphi+\psi)+\varphi_i^{(1)}(\tau)\cos(p\varphi+\psi)A_1-$$

$$-\varphi_i^{(1)}(\tau)a\sin(p\varphi+\psi)B_1+\frac{\partial u_i^{(1)}}{\partial\theta}\nu(\tau)+\frac{\partial u_i^{(1)}}{\partial(p\varphi+\psi)}\omega_1(\tau).$$

Thus, to construct the system of partial differential equations for $A_2(\tau,a,\psi)$ and $B_2(\tau,a,\psi)$, we must find the average (over a period of vibration) of the virtual work of the perturbing forces $Q_j^{(2)}(\tau,\theta,q_1,\ldots,q_N,\dot q_1,\ldots,\dot q_N)$ $(j=1,2,\ldots,N)$ and the "forces" $\frac{d}{dt}\left[\frac{\partial T(\dot q)}{\partial\dot q_j}\right]_{\dot q_j=\dot u_i}$, $(\delta W)_{\substack{\delta q=\Delta q\\ \delta q'=\Delta q'}}$ in a mode of sinusoidal vibrations for virtual displacements corresponding to the variations of the amplitude and phase of normal vibrations, and then substitute the "partial derivatives" of the virtual works obtained into equations (6.13).

The rules just quoted become considerably simpler if the system under investigation is subjected only to the action of potential forces. Then

$$\varepsilon\sum_{j=1}^{N}Q_j\delta q_j=\delta W=-\delta V, \tag{6.14}$$

where V is that part of the potential energy due to the perturbation. We call it the "perturbed" potential energy. To construct the system of partial differential equations for $A_1(\tau,a,\psi)$ and $B_1(\tau,a,\psi)$ in such a case, we must find the variation of the "perturbed" potential energy corresponding to the variations of the amplitude and phase of vibration, average over a period of vibration, and instead of $\frac{\overline{\delta W}}{\delta a}$ and $\frac{\overline{\delta W}}{\delta\psi}$, substitute the "partial derivatives" with opposite signs into equations (6.5).

To set up the equations of the first approximation we must in addition express the "partial derivatives" as Fourier sums and in place of $\frac{\delta W_\sigma}{\delta a}$ and $\frac{\delta W_\sigma}{\delta\psi}$ substitute the σ-th terms of these sums with opposite signs into equations (6.7).

For deriving $u_i^{(1)}(\tau,a,\theta,p\varphi+\psi)$, we must replace δq_j in the variation of the "perturbed" potential energy by $\varphi_j^{(k)}(\tau)$, expand the result in Fourier double sums, and substitute the (n,m)-th terms of these sums into the first sum in (6.12).

The second sum is derived according to the previously stated rule.

The construction of equations (6.13) for determining $A_2(\tau,a,\psi)$ and $B_2(\tau,a,\psi)$ becomes much simpler: we must find the variation of the "perturbed" potential energy δV, replace δq_j in the latter by $u_j^{(1)}$, derive again the variation of the result, corresponding to the variations of the amplitude and phase of vibration, take its average over a period of vibration, and substitute the "partial derivatives" with opposite signs into equation (6.13) in place of $\left(\frac{\delta\delta W}{\delta a}\right)_{\substack{\delta q=\Delta q\\ \delta q'=\Delta q'}}$ and $\left(\frac{\delta\delta W}{\delta\psi}\right)_{\substack{\delta q=\Delta q\\ \delta q'=\Delta q'}}$, replacing δW_2 in the expression of δV by a quantity of the order of ε^2.

Let us interpret equations (4.82) which are obtained when investigating monofrequency vibrations in a system described by the linear differential equations with slowly varying coefficients (4.79). In the given case the expression of the perturbing forces in a sinusoidal mode of vibrations has the form

$$\varepsilon Q_{j0}^{(1)}(\tau,\ a,\ \theta,\ \theta+\psi)=\varepsilon E_j(\tau)\sin\theta-a\omega_1(\tau)\sin(\theta+\psi)\sum_{i=1}^{N}\lambda_{ij}(\tau)\varphi_i^{(1)}(\tau). \tag{6.15}$$

Substituting the right-hand side of (6.15) into the expression of the average virtual work (6.4), we obtain after a number of operations the equations of the first approximation in the form

$$\left.\begin{aligned}\frac{da}{dt}&=-\frac{\varepsilon a}{2m_1(\tau)\,\omega_1(\tau)}\,\frac{d\,[m_1(\tau)\,\omega_1(\tau)]}{d\tau}+\frac{1}{a\omega_1(\tau)\,m_1(\tau)}\,\frac{\overline{\delta W_0}}{\delta\psi}+\\&\quad+\frac{2i}{m_1(\tau)\,[\omega_1(\tau)+\nu(\tau)]}\left[\frac{\overline{\delta W_1}}{\delta a}+\frac{\overline{\delta W_{-1}}}{\delta a}\right],\\\frac{d\psi}{dt}&=\omega_1(\tau)-\nu(\tau)-\frac{2}{am_1(\tau)\,[\omega_1(\tau)+\nu(\tau)]}\left[\frac{\overline{\delta W_1}}{\delta a}+\frac{\overline{\delta W_{-1}}}{\delta a}\right].\end{aligned}\right\} \tag{6.16}$$

We can write

$$\frac{\overline{\delta W_k}}{\delta a}=e^{ik\psi}\,\frac{\delta w_k(\tau)}{\delta a},\quad \frac{\overline{\delta W_0}}{\delta\psi}=a^2\,\frac{\delta w_0(\tau)}{\delta\psi}, \tag{6.17}$$

and introducing also the complex function

$$z=ae^{i\psi}, \tag{6.18}$$

the system (6.16) can be reduced to a linear inhomogeneous equation of the first order

$$\frac{dz}{dt}=\left\{\frac{2\,\dfrac{\delta w_0(\tau)}{\delta\psi}-\varepsilon\,\dfrac{d\,[m_1(\tau)\,\omega_1(\tau)]}{d\tau}}{2m_1(\tau)\,\omega_1(\tau)}+i\,[\omega_1(\tau)-\nu(\tau)]\right\}z-\frac{4i}{m_1(\tau)\,[\omega_1(\tau)+\nu(\tau)]}\,\frac{\delta w_{-1}(\tau)}{\delta a}. \tag{6.19}$$

The energy method presented above permits us to obtain approximate solutions without constructing beforehand the exact differential equations of problems of the type (4.3). The approximate solutions are according to this method obtained by setting up the linear system describing the unperturbed motion (4.4), checking whether it satisfies the required conditions, calculating its eigenfrequencies and "normal" functions, and then constructing the equations of the first (second) approximation using the rules quoted above.

On the grounds of the energy interpretation, we construct the equations of the first (second) approximation using directly the form of the work (or potential energy) and the kinetic energy. This suggests the idea of applying this method formally for constructing approximate solutions of partial equations describing vibrational processes in systems with distributed parameters.

Indeed, vibrational systems with distributed parameters (described by nonlinear partial differential equations close to linear hyperbolic equations) may, exactly as systems with a finite number of degrees of freedom, sustain monofrequency vibrational modes under certain conditions; if these modes are stable their investigation is of physical interest. It is thus sensible to try and extend the method of treating monofrequency vibrations (developed by us for vibrating systems with N degrees of freedom), to vibrating systems with distributed parameters.

In the following sections we shall illustrate by actual examples of vibrating systems with distributed parameters the effectiveness of the energy method for investigating the monofrequency vibrational process.

§ 2. Transversal vibrations of a rod subjected to the action of a longitudinal sinusoidal force of variable frequency

As the first example illustrating the application of the formal method described above, we consider the transversal vibrations of a rod of length l, whose free end is subjected to the axial force

$$S = S_0 + \varepsilon F(t), \tag{6.20}$$

where

$$F(t) = A + B \sin \alpha t \tag{6.21}$$

(Figure 101). Let EI be the rigidity of the rod; γ, the weight of a unit volume; g, the acceleration due to gravity; and Ω, the area of transversal cross section. Neglecting rotational inertia and the shearing force, we obtain for the potential energy of bending the expression

$$V = \frac{1}{2}\left(E - \frac{S}{\Omega}\right) I \int_0^l \left(\frac{\partial^2 y}{\partial x^2}\right)^2 dx - \frac{1}{2} S \int_0^l \left(\frac{\partial y}{\partial x}\right)^2 dx. \tag{6.22}$$

We use the notation

$$V = V_0 + \varepsilon V_1, \tag{6.23}$$

where

$$V_0 = \frac{1}{2}\left(E - \frac{S_0}{\Omega}\right) I \int_0^l \left(\frac{\partial^2 y}{\partial x^2}\right)^2 dx - \frac{1}{2} S_0 \int_0^l \left(\frac{\partial y}{\partial x}\right)^2 dx \tag{6.24}$$

is the potential energy of the "unperturbed" system, and

$$\varepsilon V_1 = -\frac{1}{2} \frac{\varepsilon F(t)}{\Omega} I \int_0^l \left(\frac{\partial^2 y}{\partial x^2}\right)^2 dx - \frac{1}{2} \varepsilon F(t) \int_0^l \left(\frac{\partial y}{\partial x}\right)^2 dx, \tag{6.25}$$

the "perturbed" potential energy appearing on account of the force $\varepsilon F(t)$.

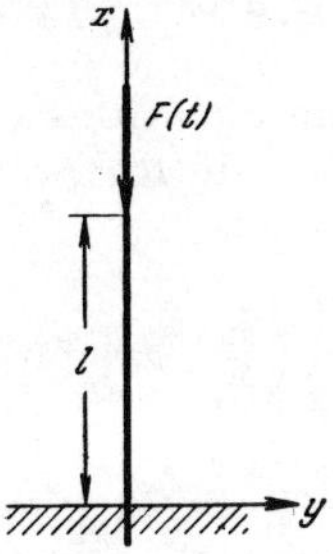

FIGURE 101

For the kinetic energy we have the expression

$$T = \frac{1}{2} \frac{\gamma \Omega}{g} \int_0^l \left(\frac{\partial y}{\partial t}\right)^2 dx. \tag{6.26}$$

The differential equation of the "unperturbed" motion has the form

$$\left(E-\frac{S_0}{\Omega}\right)I\frac{\partial^4 y}{\partial x^4}+S_0\frac{\partial^2 y}{\partial x^2}+\frac{\gamma\Omega}{g}\frac{\partial^2 y}{\partial t^2}=0, \tag{6.27}$$

and the boundary conditions at the fixed end are

$$y\,|_{x=0}=0,\quad \left.\frac{\partial y}{\partial x}\right|_{x=0}=0, \tag{6.28}$$

and at the free end

$$\left.\frac{\partial^2 y}{\partial x^2}\right|_{x=l}=0,\ \left(E-\frac{S_0}{\Omega}\right)I\left.\frac{\partial^3 y}{\partial x^3}\right|_{x=l}=-S_0\left.\frac{\partial y}{\partial x}\right|_{x=l}.$$

We solve equations (6.27) by separating the variables, looking for a solution in the form $e^{i\lambda t}\varphi(x)$, where $\varphi(x)$ is a normal function. After a few operations we obtain the frequencies and the normal functions

$$\varphi^{(k)}(x)=(\lambda_1^2\sin\lambda_1 l+\lambda_1\lambda_2\operatorname{sh}\lambda_2 l)(\cos\lambda_1 x-\operatorname{ch}\lambda_2 x)-$$
$$-(\lambda_1^2\cos\lambda_1 l+\lambda_2^2\operatorname{ch}\lambda_2 l)\left(\sin\lambda_1 x-\frac{\lambda_1}{\lambda_2}\operatorname{sh}\lambda_2 x\right), \tag{6.29}$$

where $i\lambda_1$ and $i\lambda_2$ are roots of the equation

$$\left(E-\frac{S_0}{\Omega}\right)I\lambda^4+S_0\lambda^2-\frac{\omega_k^2\gamma\Omega}{g}=0,$$

and ω_k is the frequency of the k-th normal mode.

This problem may be solved simply for several other boundary conditions. For example, in the case of hinged ends the equation of the "perturbed" motion may be reduced to an ordinary differential equation by a suitable change of variables.

The boundary conditions (6.28) do not permit such a change of variables, and the given problem can therefore not be solved in the usual way. However, the method of constructing the amplitude and phase equations proceeding directly from the expression for the energy (see above) allows us to solve this problem without difficulty.

Intensive vibrations may occur in the rod under the action of an axial force of a frequency approximately twice as large as the fundamental eigenfrequency (parametric resonance). To study these vibrations, we look for the solution of the "perturbed" motion in the form

$$y=\varphi^{(1)}(x)\,a\cos\left(\frac{1}{2}\theta+\psi\right), \tag{6.30}$$

where a and ψ should satisfy the equations of the first approximation which we construct with the help of the rule formulated above (obviously, $p=1$, $q=2$). We have

$$\delta V=-\delta\left\{\frac{1}{2}\frac{I}{\Omega}\int_0^l\left(\frac{\partial^2 y}{\partial x^2}\right)^2dx+\int_0^l\left(\frac{\partial y}{\partial x}\right)^2dx\right\}F(t)=$$
$$=\left\{\frac{I}{\Omega}\int_0^l\left[\frac{\partial^2\varphi^{(1)}(x)}{\partial x^2}\right]^2dx+\int_0^l\left[\frac{\partial\varphi^{(1)}(x)}{\partial x}\right]^2dx\right\}\times$$
$$\times(A+B\sin\theta)\left[\cos\left(\frac{1}{2}\theta+\psi\right)\delta a-a\sin\left(\frac{1}{2}\theta+\psi\right)\delta\psi\right]a\cos\left(\frac{1}{2}\theta+\psi\right). \tag{6.31}$$

We find the average value of δV over a complete period

$$\overline{\delta V} = -\frac{1}{4\pi}\int_0^{4\pi} \Phi^{(1)} (A + B\sin\theta)\left[\cos\left(\frac{1}{2}\theta + \psi\right)\delta a - a\sin\left(\frac{1}{2}\theta + \psi\right)\delta\psi\right] \times$$
$$\times a\cos\left(\frac{1}{2}\theta + \psi\right) d\theta = \Phi^{(1)}\frac{aA}{2}\delta a - \Phi^{(1)}\frac{aB}{8i}e^{-2i\psi}\delta a + \Phi^{(1)}\frac{aB}{8i}e^{2i\psi}\delta a +$$
$$+ \Phi^{(1)}\frac{a^2B}{8}e^{-2i\psi}\delta\psi + \Phi^{(1)}\frac{a^2B}{8}c^{2i\psi}\delta\psi, \qquad (6.32)$$

where

$$\Phi^{(1)} = \frac{I}{\Omega}\int_0^l \left(\frac{\partial\varphi^{(1)}(x)}{\partial x^2}\right)^2 dx + \int_0^l \left(\frac{\partial\varphi^{(1)}(x)}{\partial x}\right)^2 dx.$$

Substituting the σ-th "partial derivatives" into (6.7), we obtain the equations of the first approximation

$$\left.\begin{aligned} \frac{da}{dt} &= -\frac{\varepsilon\Phi^{(1)}aB}{2m_1\nu}\cos 2\psi, \\ \frac{d\psi}{dt} &= \omega_1 - \frac{\nu}{2} + \frac{\varepsilon\Phi^{(1)}A}{2m_1\nu} + \frac{\varepsilon\Phi^{(1)}B}{2m_1\nu}\sin 2\psi, \end{aligned}\right\} \qquad (6.33)$$

where

$$m_1 = \frac{\gamma\Omega}{g}\int_0^l [\varphi^{(1)}(x)]^2 dx.$$

We proceed to construct the second approximation. To do this we must find $u_i^{(1)}(\tau, a, \theta, p\varphi + \psi)$ by means of the corresponding rule. (In what follows, we assume for simplicity that $A = 0$.) We have

$$\left(\frac{\delta V^{(k)}}{\delta y}\right)_{y=a\cos\left(\frac{1}{2}\theta+\psi\right)} = \left\{\frac{I}{\Omega}\int_0^l \frac{\partial^2\varphi^{(1)}(x)}{\partial x^2}\frac{\partial^2\varphi^{(k)}(x)}{\partial x^2}dx +\right.$$
$$\left. + \int_0^l \frac{\partial\varphi^{(1)}(x)}{\partial x}\frac{\partial\varphi^{(k)}(x)}{\partial x}dx\right\} F(t)\, a\cos\left(\frac{1}{2}\theta + \psi\right). \qquad (6.34)$$

Expanding the above equation in a series, we obtain

$$\left(\frac{\delta V^{(k)}}{\delta y}\right)_{y=a\cos\left(\frac{1}{2}\theta+\psi\right)} = \sum_{n,m}\left(\frac{\delta V^{(k)}_{n,m}}{\delta y}\right)_{y=a\cos\left(\frac{1}{2}\theta+\psi\right)} =$$
$$= -\frac{a\Phi^{(k)}B}{4i}e^{i\left[\theta+\left(\frac{1}{2}\theta+\psi\right)\right]} + \frac{a\Phi^{(k)}B}{4i}e^{-i\left[\theta+\left(\frac{1}{2}\theta+\psi\right)\right]} -$$
$$- \frac{a\Phi^{(k)}B}{4i}e^{i\left[\theta-\left(\frac{1}{2}\theta+\psi\right)\right]} + \frac{a\Phi^{(k)}B}{4i}e^{-i\left[\theta-\left(\frac{1}{2}\theta+\psi\right)\right]}, \qquad (6.35)$$

where

$$\Phi^{(k)} = \left\{\frac{I}{\Omega}\int_0^l \frac{\partial^2\varphi^{(1)}(x)}{\partial x^2}\frac{\partial^2\varphi^{(k)}(x)}{\partial x^2}dx + \int_0^l \frac{\partial\varphi^{(1)}(x)}{\partial x}\frac{\partial\varphi^{(k)}(x)}{\partial x}dx\right\}.$$

Substituting the (n, m)-th terms of the Fourier sum (6.35) into (6.12), we obtain

$$\varepsilon u_x^{(1)}\left(\tau, a, \theta, \frac{1}{2}\theta + \psi\right) =$$
$$= \sum_{k=1,2,3,\ldots} \frac{a\Phi^{(k)}\varphi^{(k)}(x)}{m_k[\omega_k^2 - (\omega_1+\nu)^2]}\frac{B}{2}\sin\left[\theta + \left(\frac{1}{2}\theta + \psi\right)\right] +$$
$$+ \sum_{k=2,3,4,\ldots} \frac{a\Phi^{(k)}\varphi^{(k)}(x)}{m_k[\omega_k^2 - (\omega_1-\nu)^2]}\frac{B}{2}\sin\left[\theta - \left(\frac{1}{2}\theta + \psi\right)\right]. \qquad (6.36)$$

Thus, as the second approximation we take

$$y = \varphi^{(1)}(x)\, a\cos\left(\frac{1}{2}\theta + \psi\right) + \varepsilon u_x^{(1)}\left(\tau, a, \theta, \frac{1}{2}\theta + \psi\right), \qquad (6.37)$$

where a and ψ are to be determined from the equations of the second approximation whose construction requires the derivation of $A_2(\tau, a, \psi)$ and $B_2(\tau, a, \psi)$.

According to the preceding section

$$-\delta V_{\delta q=u_x^{(1)}}=\left\{\frac{I}{\Omega}\int_0^l \frac{\partial^2 y}{\partial x^2}\frac{\partial^2 u_i^{(1)}}{\partial x^2}dx+\int_0^l \frac{\partial y}{\partial x}\frac{\partial u_i^{(1)}}{\partial x}dx\right\}F(t);$$

we further obtain

$$-\delta\delta V_{\delta q=u_x^{(1)}}=\frac{1}{2}\left\{\sum_{k=1,2,\ldots}\frac{\Phi^{(k)^2}aB^2}{m_k[\omega_k^2-(\omega_1+\nu)^2]}\sin\left[\theta+\left(\frac{1}{2}\theta+\psi\right)\right]+\right.$$
$$\left.+\sum_{k=2,3,\ldots}\frac{\Phi^{(k)^2}aB^2}{m_k[\omega_k^2-(\omega_1-\nu)^2]}\sin\left[\theta-\left(\frac{1}{2}\theta+\psi\right)\right]\right\}\times$$
$$\times\left[\cos\left(\frac{1}{2}\theta+\psi\right)\delta a-a\sin\left(\frac{1}{2}\theta+\psi\right)\delta\psi\right],$$

from which we find

$$\frac{\overline{\delta\delta V}_{\delta q=u_x^{(1)}}}{\delta a}=\frac{aB^2}{8}\left\{\sum_{k=1,2,\ldots}\frac{\Phi^{(k)^2}}{m_k[\omega_k^2-(\omega_1+\nu)^2]}+\sum_{k=2,3,\ldots}\frac{\Phi^{(k)^2}}{m_k[\omega_k^2-(\omega_1-\nu)^2]}\right\}, \tag{6.38}$$

$$\frac{\overline{\delta\delta V}_{\delta q=u_x^{(1)}}}{\delta\psi}=0.$$

Applying the appropriate rule, we obtain

$$\left.\begin{aligned}\frac{\overline{\delta W_T}}{\delta a}&=-\frac{aB^2\Phi^{(1)}}{8m_1}\sum_{k=2,3,\ldots}\frac{\Phi^{(k)}m_{1k}}{m_k[\omega_k^2-(\omega_1-\nu)^2]}+D\cos 2\psi,\\ \frac{\overline{\delta W_T}}{\delta\psi}&=-Da\sin 2\psi,\end{aligned}\right\} \tag{6.39}$$

where

$$m_{1k}=\frac{\gamma\Omega}{g}\int_0^l\varphi^{(1)}(x)\,\varphi^{(k)}(x)\,dx,$$

$$D=\left[\frac{1}{4}-(\omega_1-\nu)^2\right]aB\frac{d\nu}{d\tau}\sum_{k=2,3,\ldots}\frac{\Phi^{(k)}m_{1k}}{m_k[\omega_k^2-(\omega_1-\nu)^2]}.$$

Substituting (6.38) and (6.39) into the system (6.13), we obtain $A_2(\tau, a, \psi)$ and $B_2(\tau, a, \psi)$.

We finally obtain the equations of the second approximation

$$\left.\begin{aligned}\frac{da}{dt}&=-\frac{\varepsilon\Phi^{(1)}aB}{2m_1\nu}\cos 2\psi+\frac{\varepsilon^2 D}{2m_1\nu}\sin 2\psi,\\ \frac{d\psi}{dt}&=\omega_1-\frac{\nu}{2}+\frac{\varepsilon\Phi^{(1)}B}{2m_1\nu}\sin 2\psi+\frac{\varepsilon^2 D}{2m_1\nu a}\cos 2\psi+\\ &+\frac{\Phi^{(1)^2}\varepsilon^2B^2}{8m_1^2\omega_1\nu^2}-\frac{\varepsilon^2B^2\Phi^{(1)}}{16m_1\omega_1}\sum_{k=2,3,\ldots}\frac{\Phi^{(k)}m_{1k}}{m_k[\omega_k^2-(\omega_1-\nu)^2]}-\\ &-\frac{\varepsilon^2B^2}{8m_1\nu_1}\left\{\sum_{k=1,2,\ldots}\frac{\Phi^{(k)^2}}{m_k[\omega_k^2-(\omega_1+\nu)^2]}+\sum_{k=2,3,\ldots}\frac{\Phi^{(k)^2}}{m_k[\omega_k^2-(\omega_1-\nu)^2]}\right\}.\end{aligned}\right\} \tag{6.40}$$

We illustrate this procedure by performing the calculation for the following data:

$$\left.\begin{aligned}&E=2\cdot10^6\ \text{kg/cm}^2, &&\Omega=79\ \text{cm}^2, &&I=5000\ \text{cm}^4,\\ &\gamma=7.8\ \text{g/cm}^3, &&g=981\ \text{cm/sec}^2, &&S_0=4\cdot10^4\ \text{kg},\\ &\varepsilon B=4\cdot10^3\ \text{kg}, &&l=500\ \text{cm},\end{aligned}\right\} \tag{6.41}$$

For clarity, we give in Table 9 the values of some of the quantities appearing in the equations of the first and second approximations.

TABLE 9

k	ω_k	m_k	m_{1k}	$\Phi^{(k)}$
1	43.947	20598.607		0.6455
2	339.688	$1434006\cdot 10^5$	—32.9930	—114.3217
3	945.5	$22808\cdot 10^{13}$	$46757\cdot 10^4$	—8686.152

The equations of the first approximation are

$$\left.\begin{aligned}\frac{da}{dt}&=-\frac{62.7015a}{\nu}\cos 2\psi,\\ \frac{d\psi}{dt}&=43.947-\frac{\nu}{2}+\frac{62.7015}{\nu}\sin 2\psi.\end{aligned}\right\}\tag{6.42}$$

Let us calculate the sums entering into the expression for $u_x^{(1)}\left(\tau,\,a,\,\theta,\,\frac{1}{2}\theta+\psi\right)$ and into the equations of the second approximation. This is easily done by virtue of the very rapid convergence of these sums.

We obtain for $u_x^{(1)}\left(\tau,\,a,\,\theta,\,\frac{1}{2}\theta+\psi\right)$:

$$\begin{aligned}u_x^{(1)}\left(\tau,\,a,\,\theta,\,\frac{1}{2}\theta+\psi\right)&=\left\{-\frac{0.0627\varphi^{(1)}(x)}{(2\omega_1+\nu)\nu}+\frac{1.5944\cdot 10^{-6}\varphi^{(2)}(x)}{[\omega_2^2-(\omega_1+\nu)^2]}+\right.\\&\left.+\frac{7.6168\cdot 10^{-11}\varphi^{(3)}(x)}{[\omega_3^2-(\omega_1+\nu)^2]}+\ldots\right\}a\sin\left[\theta+\left(\frac{1}{2}\theta+\psi\right)\right]+\\&+\left\{\frac{1.5944\cdot 10^{-6}\varphi^{(2)}(x)}{[\omega_2^2-(\omega_1-\nu)^2]}+\frac{7.6168\cdot 10^{-11}\varphi^{(3)}(x)}{[\omega_3^2-(\omega_1-\nu)^2]}+\ldots\right\}a\sin\left[\theta-\left(\frac{1}{2}\theta+\psi\right)\right].\end{aligned}\tag{6.43}$$

Assuming that $\nu\cong 2\omega_1$, we obtain

$$\begin{aligned}&u_x^{(1)}\left(\tau,\,a,\,\theta,\,\frac{1}{2}\theta+\psi\right)=\\&=\{-4.0581\cdot 10^{-6}\varphi^{(1)}(x)+1.6269\cdot 10^{-11}\varphi^{(2)}(x)+8.1705\cdot 10^{-17}\varphi^{(3)}(x)+\ldots\}\times\\&\times a\sin\left[\theta+\left(\frac{1}{2}\theta+\psi\right)\right]+\{1.4053\cdot 10^{-11}\varphi^{(2)}(x)+\\&+8.0373\cdot 10^{-17}\varphi^{(3)}(x)+\ldots\}a\sin\left[\theta-\left(\frac{1}{2}\theta+\psi\right)\right].\end{aligned}\tag{6.44}$$

The order of magnitude of the coefficients of the basic functions indicates that it is practically meaningless to compute $u_x^{(1)}\left(\tau,\,a,\,\theta,\,\frac{1}{2}\theta+\psi\right)$ in the given case.

The equations of the second approximation assume after the substitution of the numerical values the form

$$\left.\begin{aligned}\frac{da}{dt}&=-\frac{62.7015}{\nu}\cos 2\psi+\left\{\frac{(\omega_1-\nu)^2}{\nu}\frac{d\nu}{dt}\left[\frac{3.2097\cdot 10^{-1}}{[\omega_2^2-(\omega_1-\nu)^2]}-\frac{2.1730\cdot 10^{-8}}{[\omega_3^2-(\omega_1-\nu)^2]}+\ldots\right]+\right.\\&\left.+\frac{1}{\nu}\frac{d\nu}{dt}\left[\frac{8.0242\cdot 10^{-2}}{[\omega_2^2-(\omega_1-\nu)^2]}-\frac{5.4324\cdot 10^{-8}}{[\omega_3^2-(\omega_1-\nu)^2]}\right]\right\}a\sin 2\psi,\\ \frac{d\psi}{dt}&=43.947-\frac{\nu}{2}+\frac{62.7015}{\nu}\sin 2\psi+\\&+\left\{\frac{(\omega_1-\nu)^2}{\nu}\frac{d\nu}{dt}\left[\frac{3.2097\cdot 10^{-1}}{[\omega_2^2-(\omega_1-\nu)^2]}-\frac{2.1730\cdot 10^{-8}}{[\omega_3^2-(\omega_1-\nu)^2]}+\ldots\right]+\right.\\&\left.+\frac{1}{\nu}\frac{d\nu}{dt}\left[\frac{8.0242\cdot 10^{-2}}{[\omega_2^2-(\omega_1-\nu)^2]}-\frac{5.4324\cdot 10^{-8}}{[\omega_3^2-(\omega_1-\nu)^2]}+\ldots\right]\right\}\cos 2\psi+\frac{44.7298}{(2\omega_1+\nu)\nu}-\\&-\frac{0.2014}{[\omega_2^2-(\omega_1+\nu)^2]}-\frac{7.312\cdot 10^{-4}}{[\omega_3^2-(\omega_1+\nu)^2]}+\ldots-\frac{0.2014}{[\omega_2^2-(\omega_1-\nu)^2]}-\frac{7.312\cdot 10^{-4}}{[\omega_3^2-(\omega_1-\nu)^2]}+\ldots\end{aligned}\right\}\tag{6.45}$$

We calculate for clarity the coefficients in the equations of the second approximation with the assumptions $\nu=\nu(\tau)\cong 2\omega_1$, and $\frac{d\nu(\tau)}{d\tau}\cong\frac{1}{10}2\omega_1$; in that case

$$\left.\begin{aligned}
&\frac{da}{dt}=-0.7134a\cos 2\psi+\{5.4638\cdot 10^{-3}+7.0725\cdot 10^{-8}+\ldots\}\,a\sin 2\psi,\\
&\frac{d\psi}{dt}=43.947-\frac{\nu}{2}+0.7134\sin 2\psi+\\
&\quad+\{5.4638\cdot 10^{-3}+7.0725\cdot 10^{-8}+\ldots\}\cos 2\psi+5.7900\cdot 10^{-3}-\\
&-1.0450\cdot 10^{-3}+8.4328\cdot 10^{-6}+\ldots+2.8950\cdot 10^{-3}+2.0554\cdot 10^{-6}-\\
&\qquad-7.8433\cdot 10^{-10}-\ldots-1.7755\cdot 10^{-6}-7.7155\cdot 10^{-10}-\ldots
\end{aligned}\right\}\quad(6.46)$$

Analyzing the last equations, we notice that the correction to the eigenfrequency in the second approximation, being approximately equal to $8.6\cdot 10^{-3}$, is only 0.02 % of the eigenfrequency of the unperturbed system. The same conclusion may be made about the correction for the resonance zone. It is therefore sufficient to confine oneself in practice to the equations of the first approximation (6.42), which enable us to study the nonstationary process, determine the zone of stability, etc. If $\nu=\mathrm{const}$, these equations yield themselves to elementary integration in quadratures. However, the nature of the solutions of the equations of the first approximation (6.33) (and also of the second approximation) may also be demonstrated directly without integration. For example, let

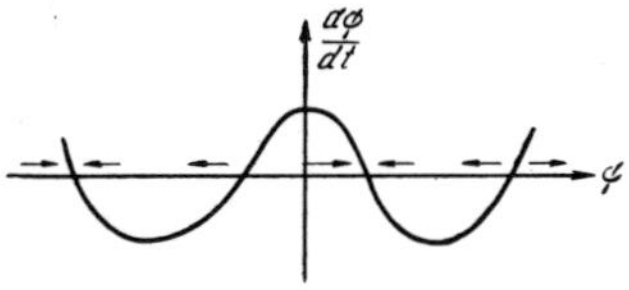

FIGURE 102

$$\left|\omega_1-\frac{\nu}{2}+\frac{\varepsilon\Phi^{(1)}A}{2m_1\nu}\right|<\frac{\varepsilon\Phi^{(1)}B}{2m_1\nu}. \quad (6.47)$$

Then according to the second equation of the system (6.33) the derivative $\frac{d\psi}{dt}$ is a sign-reversing function of ψ, shown graphically in Figure 102, and there obviously exist constant solutions which are roots of the equation

$$\omega_1-\frac{\nu}{2}+\frac{\varepsilon\Phi^{(1)}A}{2m_1\nu}+\frac{\varepsilon\Phi^{(1)}B}{2m_1\nu}\sin 2\psi=0. \quad (6.48)$$

These solutions are known to be stable if

$$\frac{\Phi^{(1)}B}{m_1\nu}\cos 2\psi<0,$$

and unstable if

$$\frac{\Phi^{(1)}B}{m_1\nu}\cos 2\psi>0.$$

If the value of ψ at the initial moment of time does not coincide with the stable stationary value given by equation (6.48), then it will approach the latter, increasing or decreasing monotonically. The system will sustain vibrations at a frequency equal to half the frequency of the external force $\left(\frac{\nu}{2}\cong\omega_1\right)$.

The inequality (6.47) is the instability condition for the axis of the undeformed rod under the action of a sinusoidal axial load (with the frequency $\nu\cong 2\omega_1$). The resonance zone, which is the zone of instability, is according to this inequality defined (up to and including terms of the first order) by the inequality

$$\omega_1+\frac{\varepsilon\Phi^{(1)}A}{4m_1\omega_1}-\left|\frac{\varepsilon\Phi^{(1)}B}{4m_1\omega_1}\right|<\frac{\nu}{2}<\omega_1+\frac{\varepsilon\Phi^{(1)}A}{4m_1\omega_1}+\left|\frac{\varepsilon\Phi^{(1)}B}{4m_1\omega_1}\right|. \quad (6.49)$$

To study the nonstationary vibrating process of the beam when subjected to the action of an external force with variable frequency, we assume that $\nu=\nu(\tau)$ varies according to the law

$$\nu(\tau)=85+t,$$

and integrate the system

$$\left.\begin{aligned}\frac{da}{dt}&=-\frac{62.7015}{85+t}a\cos 2\psi,\\ \frac{d\psi}{dt}&=43.947-\frac{85+t}{2}+\frac{62.7015}{85+t}\sin 2\psi\end{aligned}\right\} \quad (6.50)$$

numerically with the initial values $t=0,\ a=a_0,\ \psi=\frac{\pi}{4}$.

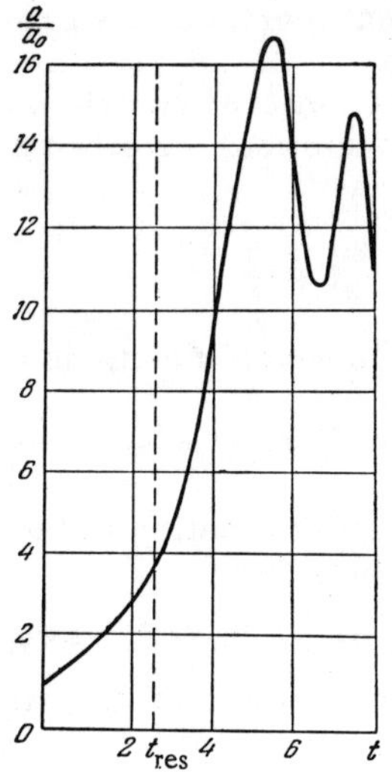

FIGURE 103

We obtain the amplitude curve shown in Figure 103, which characterizes the nonstationary mode – transition through the parametric resonance.

§ 3. Transversal vibrations of a beam subjected to the action of a pulsating force with a mobile point of application

As the second example* we consider the transversal vibrations of a beam subjected to the action of a slowly moving additional load and a perturbing force whose point of application moves together with the load (cf. Figure 104).

We denote the area of the beam cross section by A, and assume that it is small compared to its length l.

* This example is studied in detail in /120/.

Let a certain small (compared to the mass of the beam) mass εM be moving along the beam with velocity εv. In addition, let the beam be subjected to the action of a vertical periodic force $\varepsilon F(\theta, \tau)$, whose point of application coincides in the considered time interval with the center of gravity of the mass εM. We introduce the notation: ϱ, the density of the beam's material, E, Young's modulus, I, the moment of inertia of a cross section of the beam with respect to the axis perpendicular to the plane of deformation.

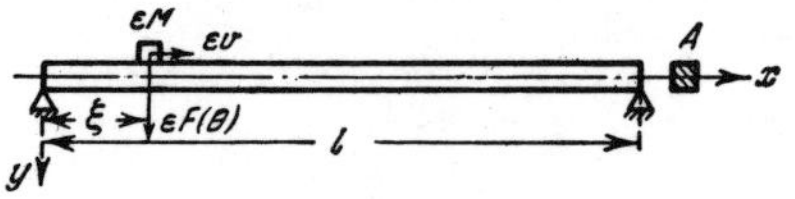

FIGURE 104

While considering transversal vibrations of the beam in the vertical plane, we neglect the rotational inertia of the cross sections and the shearing forces.

The equation describing free lateral vibrations in the plane of bending (equation of the "unperturbed" motion) for a beam freely supported at the ends is known to be of the form

$$\frac{\partial^2 y}{\partial t^2}+\frac{EI}{\varrho A}\frac{\partial^4 y}{\partial x^4}=0. \tag{6.51}$$

The eigenfrequencies and "normal" functions are

$$\omega_k=\frac{ak\pi}{l}, \quad \varphi_x^{(k)}=\sin\frac{k\pi}{l}x, \quad a^2=\frac{EI}{\varrho A} \qquad (k=1, 2, 3, \ldots). \tag{6.52}$$

Let us construct the first approximation in the case when the beam is under the action of the perturbing force

$$\varepsilon\Phi(\tau, \theta, \ddot{y})=[\varepsilon F(\theta)+\varepsilon Mg-\varepsilon M\ddot{y}]_{x=\xi}, \tag{6.53}$$

where $\xi=\varepsilon v_1 t$ is the current coordinate of its point of application. (As the generalized coordinate we take the beam's deflection y.) For definiteness we put $\varepsilon F(\theta)=F\sin\theta$.

The solution corresponding to a monofrequency vibration of the beam, close to the first normal mode, is, according to (4.81) and (6.16),

$$y^{(1)}=\sin\frac{\pi x}{l}a\cos(\theta+\psi), \tag{6.54}$$

where a and ψ are to be determined from the equations of the first approximation. To construct these equations we shall first derive the expression of the perturbing force $\varepsilon\Phi(\tau, \theta, \ddot{y})$ in the mode of sinusoidal vibrations

$$\left.\begin{aligned} y^{(1)}=\sin\frac{\pi x}{l}a\cos(\theta+\psi), \quad \dot{y}^{(1)}=-\omega_1\sin\frac{\pi x}{l}a\sin(\theta+\psi), \\ \ddot{y}=-\omega_1^2\sin\frac{\pi x}{l}a\cos(\theta+\psi). \end{aligned}\right\} \tag{6.55}$$

We have

$$\varepsilon\Phi_0^{(1)}=\varepsilon\left[F_0\sin\theta+Mg+M\omega_1^2\sin\frac{\pi}{l}\xi a\cos(\theta+\psi)\right]. \tag{6.56}$$

The expression of the virtual work corresponding to variations in the amplitude and phase of vibration is

$$\varepsilon\Phi_0^{(1)}\delta y^{(1)}=\varepsilon\left[F_0\sin\theta+Mg+M\omega_1^2\sin\frac{\pi}{l}\xi a\cos(\theta+\psi)\right]\times$$
$$\times\left[\sin\frac{\pi}{l}\xi\cos(\theta+\psi)\,\delta a-a\sin\frac{\pi}{l}\sin(\theta+\psi)\,\delta\psi\right] \tag{6.57}$$

We then derive the value of the averaged virtual work, and obtain for its "partial" derivatives the expressions

$$\left.\begin{aligned}
&\frac{\overline{\delta W_1}}{\delta a}=\frac{i}{4}e^{i\psi}F_0\sin\frac{\pi}{l}\xi, \qquad \frac{\overline{\delta W_{-1}}}{\delta a}=-\frac{i}{4}e^{-i\psi}F_0\sin\frac{\pi}{l}\xi,\\
&\frac{\overline{\delta W_1}}{\delta\psi}=-\frac{a}{4}e^{i\psi}F_0\sin\frac{\pi}{l}\xi, \qquad \frac{\overline{\delta W_{-1}}}{\delta\psi}=-\frac{a}{4}e^{-i\psi}F_0\sin\frac{\pi}{l}\xi,\\
&\frac{\overline{\delta W_0}}{\delta a}=\frac{a}{2}M\omega_1^2\sin\frac{\pi}{l}\xi.
\end{aligned}\right\} \tag{6.58}$$

We also derive

$$m_1=\int_0^l \varrho A\sin^2\frac{\pi}{l}x\,dx=\varrho\frac{Al}{2}.$$

Then, using (6.16), one obtains the equations of the first approximation

$$\left.\begin{aligned}
&\frac{da}{dt}=-\frac{2\varepsilon F_0\sin\frac{\pi}{l}\xi}{\varrho Al(\omega_1+\nu)}\cos\psi,\\
&\frac{d\psi}{dt}=\omega_1-\nu-\frac{\varepsilon\omega_1 M\sin^2\frac{\pi}{l}\xi}{\varrho Al}-\frac{2\varepsilon F_0\sin\frac{\pi}{l}\xi}{a\varrho Al(\omega_1+\nu)}\sin\psi
\end{aligned}\right\} \tag{6.59}$$

or, substituting $z=ae^{i\psi}$,

$$\frac{dz}{dt}=i\left(\omega_1-\nu-\frac{\varepsilon\omega_1 M\sin^2\frac{\pi}{l}\xi}{\varrho Al}\right)z+\frac{4\varepsilon F_0\sin\frac{\pi}{l}\xi}{\varrho Al(\omega_1+\nu)}. \tag{6.60}$$

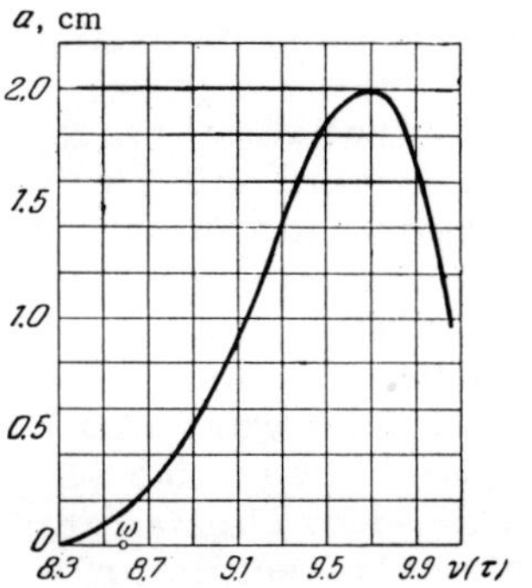

FIGURE 105

Introducing the notation

$$p(\tau)=\omega_1-\nu-\frac{\varepsilon\omega_1 M\sin^2\frac{\pi}{l}\xi}{\varrho Al}, \qquad q(\tau)=\frac{2\varepsilon F_0\sin\frac{\pi}{l}\xi}{\varrho Al(\omega_1+\nu)}$$

and integrating equation (6.60) with the condition $a(0)=0$, we obtain

$$z=-\frac{2\varepsilon F_0}{\varrho Al}e^{i\int_0^t p(\tau)dt}\left[\int^t\frac{\sin\frac{\pi}{l}\xi}{\omega_1-\nu(\tau)}e^{-i\int_0^t p(\tau)dt}dt\right]. \tag{6.61}$$

Separating the real and imaginary parts according to the first formula of (4.89), we find*

$$a^2(t)=\frac{4\varepsilon^2 F_0^2}{\varrho^2 A^2 l^2}\left\{\left[\int_0^t \frac{\sin\frac{\pi\varepsilon\nu_1}{l}t}{\omega_1+\nu(\tau)}\sin\left[\int_0^t p(\tau)\,dt\right]dt\right]^2+\right.$$

$$\left.+\left[\int_0^t \frac{\sin\frac{\pi\varepsilon\nu_1}{l}t}{\omega_1+\nu(\tau)}\cos\left[\int_0^t p(\tau)\,dt\right]dt\right]^2\right\}. \tag{6.62}$$

Using the second formula of (4.89), we can also find the expression for ψ. It is not difficult to derive with the help of (6.62) the time dependence of the amplitude of the fundamental vibration tone in a nonstationary mode. This dependence is shown for the numerical values

$$\left.\begin{array}{c} l=1000\text{ cm},\quad A=30\text{ cm}^2,\\ \varrho=7.8\text{g}/\text{cm}^3, I_z=\frac{bh^3}{12}=90\text{cm}^4,\\ E=2\cdot 10^9\cdot 981\text{g}/\text{cm}\cdot\text{sec}^2, \varepsilon=0.1,\\ F_0=10^6\cdot 981\text{ g}/\text{cm}\cdot\text{sec}^2,\\ \nu_1=10\text{cm}/\text{sec}, M=5\cdot 10^4\text{ g},\\ \nu=\nu_0+\alpha t,\quad \alpha=\frac{0{,}01}{\pi}\nu_0^2,\\ \nu_0=8.3\text{ sec}^{-1} \end{array}\right\} \tag{6.63}$$

in Figure 105, where it is assumed that the point of application of the perturbing force and the mass lie initially on the left bearing.

§ 4. Transversal vibrations of a rod of double-valued rigidity in a transient rotational mode**

We shall consider the case of spatial transversal vibrations of a rod or shaft of double-valued rigidity, rotating with a variable angular velocity $\omega(\tau)$, where $\tau=\varepsilon t$, and t is the time in units of the order of a period of the shaft's normal transversal vibrations.

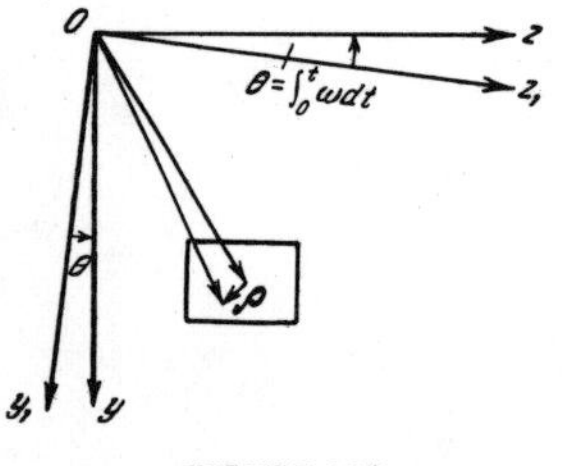

FIGURE 106

We shall first derive briefly the differential equation of motion. A detailed derivation may be found in the work of B. I. Moseenkov /121/.

In addition to the assumptions usually made in the study of small transversal vibrations of thin rods (rectilinearity in the unstressed state, neglect of shearing forces and of rotational inertia of the cross sections), we assume that the rod is not twisted in the course of vibrations. We also assume that the rod is not in static equilibrium, i. e., that the line connecting the centers of gravity of the elements of the nonrotating rod is displaced with respect to the rectilinear axis and constitutes a planar curve, and that the variation law of nonequilibrium is given. We neglect longitudinal vibrations while studying transversal vibrations.

* The approximate solution (6.54) obtained here coincides in the particular case $\nu=\text{const}$, $\xi=\text{const}$, $\varepsilon M=0$, as should have been expected, with the exact solution derived by S. P. Timoshenko /154/.

** A more detailed presentation of the examples considered in this and the next two sections may be found in the monograph of Yu. A. Mitropol'skii and B. I. Moseenkov /116/.

To construct the equations of motion we introduce two systems of coordinates $x_1y_1z_1$ and xyz, the first of which is fixed and the second rotates with the angular velocity ω; the axes x_1 and x are directed along the rotation axis which coincides with the rectilinear axis of the nonrotating rod, and the origin O is chosen at the left bearing. The other axes of the rotating system of coordinates, y and z, are directed along the principal directions of the rod's cross section, and the y_1-axis of the fixed system is directed vertically downwards (Figure 106).

We denote by l the length of the rod, by m, its linear density, and by $\boldsymbol{\varrho}(x)$, the eccentricity vector of the center of gravity which, in view of our choice of the coordinate axes of the rotating system, is only a function of x.

The kinetic energy of deformation vibrations of the rod under the above-mentioned restrictions and with the introduced notation obviously has the following form for a rotational transient **mode**,

$$T=\frac{m}{2}\int_0^l\left\{\left[\frac{\partial y}{\partial t}-\omega(\tau)(z-\varrho_z)\right]^2+\left[\frac{\partial z}{\partial t}+\omega(\tau)(y-\varrho_y)\right]^2\right\}dx, \tag{6.64}$$

where ϱ_z and ϱ_y are projections of the vector $\boldsymbol{\varrho}$ on the moving axes z and y.

The potential energy of deformation of the rod, taking its weight into account, is expressed by

$$V=\int_0^l\left\{\frac{E}{2}\left[I_z\left(\frac{\partial^2 y}{\partial x^2}\right)^2+I_y\left(\frac{\partial^2 z}{\partial x^2}\right)^2\right]-mg(y\cos\theta-z\sin\theta)\right\}dx, \tag{6.65}$$

where I_z and I_y are the principal moments of inertia of a cross section; E, Young's modulus; g, the acceleration due to gravity; $\theta=\int_0^t\omega(\tau)\,dt$, the angle of rotation of the moving coordinate system with respect to the fixed one.

Since the asymmetry of a cross section is small, and the influence of the weight and the static nonequilibrium is also small, the expressions of the kinetic and potential energies may be represented as sums of perturbed and unperturbed energies.

We have

$$T=T_0+\varepsilon T_1, \tag{6.66}$$

where

$$T_0=\frac{m}{2}\int_0^l\left\{\left[\frac{\partial y}{\partial t}-\omega(\tau)z\right]^2+\left[\frac{\partial z}{\partial t}+\omega(\tau)y\right]^2\right\}dx; \tag{6.67}$$

$$\varepsilon T_1=\frac{m}{2}\int_0^l\left\{\varrho_z^2\omega^2(\tau)-2\varrho_z\omega(\tau)\left[\frac{\partial y}{\partial t}-\omega(\tau)z\right]+\varrho_y^2\omega^2(\tau)+\right.$$
$$\left.+2\varrho_y\omega(\tau)\left[\frac{\partial z}{\partial t}+\omega(\tau)y\right]\right\}dx, \tag{6.68}$$

and

$$V=V_0+\varepsilon V_1, \tag{6.69}$$

where

$$V_0=\frac{E(I_z+I_y)}{2}\int_0^l\left[\left(\frac{\partial^2 y}{\partial x^2}\right)^2+\left(\frac{\partial^2 z}{\partial x^2}\right)^2\right]dx, \tag{6.70}$$

$$\varepsilon V_1=\int_0^l\left\{\frac{E(I_z-I_y)}{4}\left[\left(\frac{\partial^2 y}{\partial x^2}\right)^2-\left(\frac{\partial^2 z}{\partial x^2}\right)^2\right]-mg(y\cos\theta-z\sin\theta)\right\}dx. \tag{6.71}$$

Applying the Ostrogradskii-Hamilton principle, we can easily derive from the expressions (6.67) and (6.70) the equations of unperturbed motion. We have

$$\delta \int_{t_1}^{t_2} (T_0 - V_0)\, dt = 0. \tag{6.72}$$

Substituting in (6.72) the values of T_0 and V_0 from (6.67) and (6.70), and performing the variation, we obtain the following system of equations of unperturbed motion*

$$\left.\begin{aligned} &C^2 \frac{\partial^4 y}{\partial x^4} + \frac{\partial^2 y}{\partial t^2} - 2\omega(\tau)\frac{\partial z}{\partial t} - \varepsilon \frac{d\omega(\tau)}{d\tau} z - \omega^2(\tau)\, y = 0, \\ &C^2 \frac{\partial^4 z}{\partial x^4} + \frac{\partial^2 z}{\partial t^2} + 2\omega(\tau)\frac{\partial y}{\partial t} + \varepsilon \frac{d\omega(\tau)}{d\tau} y - \omega^2(\tau)\, z = 0, \end{aligned}\right\} \tag{6.73}$$

where

$$C^2 = \frac{E(I_z + I_y)}{2m}. \tag{6.74}$$

The boundary conditions for the system (6.73), in the case of isotropic hinged-fastening, are

$$\left.\begin{aligned} &y(0, t) = y(l, t) = z(0, t) = z(l, t) = 0, \\ &\left.\frac{\partial^2 y}{\partial x^2}\right|_{x=0} = \left.\frac{\partial^2 y}{\partial x^2}\right|_{x=l} = \left.\frac{\partial^2 z}{\partial x^2}\right|_{x=0} = \left.\frac{\partial^2 z}{\partial x^2}\right|_{x=l} = 0. \end{aligned}\right\} \tag{6.75}$$

Multiplying the second equation of the system (6.73) by $i = \sqrt{-1}$ and adding it to the first, we obtain a complex differential equation with respect to the moving coordinate system

$$C^2 \frac{\partial^4 \xi}{\partial x^4} + \frac{\partial^2 \xi}{\partial t^2} + 2i\omega(\tau)\frac{\partial \xi}{\partial t} + i\varepsilon \frac{d\omega(\tau)}{d\tau} \xi - \omega^2(\tau)\, \xi = 0, \tag{6.76}$$

where

$$\xi(x, t) = y(x, t) + iz(x, t). \tag{6.77}$$

Using the transition formula from the moving to the fixed coordinate system

$$\xi = \eta e^{-i\theta}, \tag{6.78}$$

we obtain without difficulty the differential equation of the unperturbed motion in the fixed coordinate system

$$C^2 \frac{\partial^4 \eta}{\partial x^4} + \frac{\partial^2 \eta}{\partial t^2} = 0, \tag{6.79}$$

where

$$\eta(x, t) = y_1(x, t) + iz_1(x, t). \tag{6.80}$$

The boundary conditions (6.75) assume for equation (6.79) the form

$$\eta(0, t) = \eta(l, t) = 0, \quad \left.\frac{\partial^2 \eta}{\partial x^2}\right|_{x=0} = \left.\frac{\partial^2 \eta}{\partial x^2}\right|_{x=l} = 0. \tag{6.81}$$

Solving equation (6.79) in the usual way, we obtain the frequencies and the normal functions (the functions of deformation)

$$k_n = \frac{n^2\pi^2}{l^2} \sqrt{\frac{E(I_z + I_y)}{2m}}, \tag{6.82}$$

* It is not difficult to derive the equations of perturbed motion directly and to solve them. However, we intend here to construct an approximate solution according to the procedure given in § 1 of this chapter.

$$\varphi^{(n)}(x) = \sin\frac{n\pi}{l}x \qquad (n = 1,\ 2,\ 3,\ \ldots). \tag{6.83}$$

Hence the particular solution of equation (6.79) corresponding to the first normal mode is written in the form

$$\eta^{(1)}(x,\ t) = \sin\frac{\pi}{l}x\,[a_1 e^{i(k_1 t+\psi_1)} + a_2 e^{-i(k_1 t+\psi_2)}], \tag{6.84}$$

where a_1, a_2, ψ_1, ψ_2 are real arbitrary constants.

Passing from the complex function $\eta(x,\ t)$ to the real functions $y_1(x,\ t)$ and $z_1(x,\ t)$ in (6.84), we find

$$\left.\begin{aligned} y_1^{(1)}(x,\ t) &= \sin\frac{\pi}{l}x\,[a_1\cos(k_1 t+\psi_1) + a_2\cos(k_1 t+\psi_2)],\\ z_1^{(1)}(x,\ t) &= \sin\frac{\pi}{l}x\,[a_1\sin(k_1 t+\psi_1) - a_2\sin(k_1 t+\psi_2)]. \end{aligned}\right\} \tag{6.85}$$

After these preliminary remarks concerning the solution of the unperturbed equation (6.79) we construct an approximate solution for the perturbed vibrations (taking into account the neglected expressions (6.68) and (6.71)), close to the first normal vibration (6.85).

Let $\frac{d\theta}{dt} = \omega(\tau) \approx k_1$ for some $\tau \in [0,\ L]$.

Then, taking into account the expressions (6.85), we look for an asymptotic solution of the specified problem in the first approximation in the form

$$\left.\begin{aligned} y_1^{(1)}(x,\ t) &= \sin\frac{\pi}{l}x\,[a_1\cos(\theta+\psi_1) + a_2\cos(k_1 t+\psi_2)],\\ z_1^{(1)}(x,\ t) &= \sin\frac{\pi}{l}x\,[a_1\sin(\theta+\psi_1) - a_2\sin(k_1 t+\psi_2], \end{aligned}\right\} \tag{6.86}$$

where a_1, a_2, ψ_1, ψ_2 are to be determined from the system of equations of the first approximation, which we shall now construct, using the energy method.

We first derive the perturbing forces corresponding to the "perturbed" potential energy (6.65) (for simplicity, we put $g = 0$ in formula (6.65), i.e., while examining the affect of the static nonequilibrium we neglect the perturbations due to the rod's weight) and the "perturbed" kinetic energy (6.68), which perturb the rod on account of the cross section asymmetry, the static nonequilibrium, and the variation of the angular rotation velocity.

We shall take into account in addition to these perturbing forces also small dissipative forces constituting the friction force, which is proportional to the displacement velocity of an element of the rod and oriented in the opposite direction to the latter.

The external friction force for an element of the rod is written in the form

$$F_{\text{т}} = -\varkappa\frac{\partial\eta}{\partial t}dx, \tag{6.87}$$

where $\varkappa$ is the coefficient of external friction referred to unit length of the rod.

To obtain explicit expressions for the perturbing forces defined in the fixed coordinate system, with respect to which we have considered the unperturbed system (6.79), we first derive the distributed forces in the moving coordinate system.

Replacing T_0 and V_0 by the values of εT_1 and εV_1 given by (6.68) and (6.71) in the integrand of (6.72), and performing the variation, we obtain the distributed load referred to the principal axes of bending

$$\varepsilon q_y(x,t) = m\varrho_y\omega^2(\tau) + \varepsilon m\varrho_z\frac{d\omega(\tau)}{d\tau} + \frac{E(I_y-I_z)}{2}\frac{\partial^4 y}{\partial x^4}, \tag{6.88}$$

$$\varepsilon q_z(x,t) = m\varrho_z\omega^2(\tau) - \varepsilon m\varrho_y\frac{d\omega(\tau)}{d\tau} - \frac{E(I_y-I_z)}{2}\frac{\partial^4 z}{\partial x^4}. \tag{6.89}$$

Multiplying (6.89) by $i=\sqrt{-1}$ and adding to (6.88), we obtain a complex expression for the distributed load implied by the perturbing forces in the moving coordinate system

$$\varepsilon q(x,\ t) = m\varrho\left[\ \omega^2(\tau) - i\varepsilon\frac{d\omega(\tau)}{d\tau}\right] + \frac{E(I_y-I_z)}{2}\frac{\partial^4\xi}{\partial x^4}, \tag{6.90}$$

where

$$q(x,\ t) = q_y(x,\ t) + iq_z(x,\ t),\quad \varrho = \varrho_y + i\varrho_z,$$
$$\xi(x,\ t) = y(x,\ t) - iz(x,\ t).$$

Applying the transition formula (6.78) to the fixed coordinate system, taking account of the friction introduced through (6.87), and neglecting in (6.90) the second order term $i\varepsilon\varrho m\,\frac{d\omega(\tau)}{d\tau}$, we obtain the following expression for the perturbing force acting on an element dx of the rod:

$$\varepsilon Q(x,\ t) = \left[\ m\varrho\omega^2(\tau)\,e^{i\theta} + \frac{E(I_y-I_z)}{2}\frac{\partial^4\bar\eta}{\partial x^4}e^{2i\theta} - \varkappa\frac{\partial\eta}{\partial t}\right]dx, \tag{6.91}$$

where

$$\bar\eta(x,\ t) = y_1(x,\ t) - iz_1(x,\ t).$$

Separating the real and imaginary parts in the last expression, we find the perturbation forces acting on an element dx of the rod along the axes y_1 and z_1 of the fixed coordinate system

$$\left.\begin{aligned}\varepsilon Q_{y_1}(x,\ t) &= \Big[\ m\omega^2(\tau)(\varrho_y\cos\theta - \varrho_z\sin\theta) + \\ &+ \frac{E(I_y-I_z)}{2}\left(\frac{\partial^4 y_1}{\partial x^4}\cos 2\theta + \frac{\partial^4 z_1}{\partial x^4}\sin 2\theta\right) - \varkappa\frac{\partial y_1}{\partial t}\Big]\,dx,\\ \varepsilon Q_{z_1}(x,\ t) &= \Big[\ m\omega^2(\tau)(\varrho_z\cos\theta + \varrho_y\sin\theta) + \\ &+ \frac{E(I_y-I_z)}{2}\left(\frac{\partial^4 y_1}{\partial x^4}\sin 2\theta - \frac{\partial^4 z_1}{\partial x^4}\cos 2\theta\right) - \varkappa\frac{\partial z_1}{\partial t}\Big]\ dx.\end{aligned}\right\} \tag{6.92}$$

We derive expressions for the perturbing forces (6.92) in the mode of sinusoidal vibrations (6.86) of the direct and inverse precession of the rod in the first form of dynamic equilibrium.

Substituting (6.86) into (6.92), we obtain up to an accuracy of the first order

$$\left.\begin{aligned}\varepsilon Q^{(1)}_{y_1 0}(x,t) &= \Big\{m\omega^2(\tau)(\varrho_y\cos\theta - \varrho_z\sin\theta) + \sin\frac{\pi}{l}x\,\frac{E(I_y-I_z)\pi^4}{2l^4}\times\\ &\times[a_1\cos(\theta+\psi_1) + a_2\cos(2\theta + k_1t + \psi_2)] + \\ &+ \varkappa k_1\sin\frac{\pi}{l}x\,[a_1\sin(\theta+\psi_1) + a_2\sin(k_1t+\psi_2)]\Big\}\,dx,\\ \varepsilon Q^{(1)}_{z_1 0}(x,t) &= \Big\{m\omega^2(\tau)(\varrho_z\cos\theta + \varrho_y\sin\theta) + \sin\frac{\pi}{l}x\,\frac{E(I_y-I_z)\pi^4}{2l^4}\times\\ &\times[a_1\sin(\theta-\psi_1) + a_2\sin(2\theta + k_1t + \psi_2)] + \\ &+ \varkappa k_1\sin\frac{\pi}{l}x\,[a_2\cos(k_1t+\psi_2) - a_1\cos(\theta+\psi_1)]\Big\}\,dx.\end{aligned}\right\} \tag{6.93}$$

Let us write the expressions of the virtual work done by the perturbing forces in the mode of sinusoidal vibrations, i.e., the forces given by (6.93) acting along the virtual displacements

$$\left.\begin{aligned}\delta y_1^{(1)}(x,\ t) &= \sin\frac{\pi}{l}x\,[\cos(\theta+\psi_1)\,\delta a_1 - a_1\sin(\theta+\psi_1)\,\delta\psi_1 + \\ &\quad + \cos(k_1t+\psi_2)\,\delta a_2 - a_2\sin(k_1t+\psi_2)\,\delta\psi_2],\\ \delta z_1^{(1)}(x,\ t) &= \sin\frac{\pi}{l}x\,[\sin(\theta+\psi_1)\,\delta a_1 + a_1\cos(\theta+\psi_1)\,\delta\psi_1 - \\ &\quad - \sin(k_1t+\psi_2)\,\delta a_2 - a_2\cos(k_1t+\psi_2)\,\delta\psi_2],\end{aligned}\right\} \tag{6.94}$$

corresponding to variations of the amplitudes a_1, a_2 and the phases ψ_1, ψ_2 for direct and inverse precession in the first form of dynamic equilibrium.

With an accuracy of first order terms, we obtain

$$\left.\begin{aligned}\delta W_{y_1} &= \int_0^l \varepsilon Q_{y10}^{(1)}(x,\ t)\,\delta y_1^{(1)}(x,\ t),\\ \delta W_{z_1} &= \int_0^l \varepsilon Q_{z10}^{(1)}(x,\ t)\,\delta z_1^{(1)}(x,\ t).\end{aligned}\right\} \tag{6.95}$$

To simplify the calculation of the integrals (6.95) we assume that the static nonequilibrium is given by

$$\left.\begin{aligned}\varrho_y(x) &= \varrho_1\sin\frac{\pi}{l}x,\\ \varrho_z(x) &= -\varrho_2\sin\frac{\pi}{l}x,\end{aligned}\right\} \tag{6.96}$$

where ϱ_1, ϱ_2 are constants.

Substituting (6.93), (6.94), and (6.96) into the right-hand sides of (6.95) and integrating, we obtain

$$\left.\begin{aligned}\delta W_{y_1} &= \frac{l}{2}\Big\{m\omega^2(\tau)(\varrho_1\cos\theta+\varrho_2\sin\theta) + \frac{E(I_y-I_z)\pi^4}{2l^4}\times\\ &\times[a_1\cos(\theta-\psi_1)+a_2\cos(2\theta+k_1t+\psi_2)] + \varkappa k_1[a_1\sin(\theta+\psi_1)+\\ &+a_2\sin(k_1t+\psi_2)]\Big\}[\cos(\theta+\psi_1)\,\delta a_1 - a_1\sin(\theta+\psi_1)\,\delta\psi_1+\\ &+\cos(k_1t+\psi_2)\,\delta a_2 - a_2\sin(k_1t+\psi_2)\,\delta\psi_2],\\ \delta W_{z_1} &= \frac{l}{2}\Big\{m\omega^2(\tau)(\varrho_1\sin\theta-\varrho_2\cos\theta) + \frac{E(I_y-I_z)\pi^4}{2l^4}\times\\ &\times[a_1\sin(\theta-\psi_1)+a_2\sin(2\theta+k_1t+\psi_2)] + \varkappa k_1[a_2\cos(k_1t+\psi_2)-\\ &-a_1\cos(\theta+\psi_1)]\Big\}[\sin(\theta+\psi_1)\,\delta a_1 + a_1\cos(\theta+\psi_1)\,\delta\psi_1-\\ &-\sin(k_1t+\psi_2)\,\delta a_2 - a_2\cos(k_1t+\psi_2)\,\delta\psi_2].\end{aligned}\right\} \tag{6.97}$$

Let us now compute the average values of these virtual works over a complete period from the formulas

$$\left.\begin{aligned}\overline{\delta W}_{y_1} &= \frac{1}{4\pi^2}\int_0^{2\pi}\int_0^{2\pi}\delta W_{y_1}\,d(\theta+\psi_1)\,d(k_1t+\psi_2),\\ \overline{\delta W}_{z_1} &= \frac{1}{4\pi^2}\int_0^{2\pi}\int_0^{2\pi}\delta W_{z_1}\,d(\theta+\psi_1)\,d(k_1t+\psi_2).\end{aligned}\right\} \tag{6.98}$$

Substituting δW_{y_1} and δW_{z_1} from (6.97) into the right-hand sides of (6.98) and integrating, we find

$$\begin{aligned}\overline{\delta W}_{y_1} = \overline{\delta W}_{z_1} &= \frac{l}{4}\Big\{\Big[m\omega^2(\tau)(\varrho_1\cos\psi_1-\varrho_2\sin\psi_1)+\frac{E(I_y-I_z)\pi^4}{2l^4}a_1\cos 2\psi_1\Big]\times\\ &\times\delta a_1 - [m\omega^2(\tau)(a_1\varrho_2\cos\psi_1+a_1\varrho_1\sin\psi_1)+\\ &+\frac{E(I_y-I_z)\pi^4}{2l^4}a_1^2\sin 2\psi_1+\varkappa k_1a_1^2\Big]\delta\psi_1-\varkappa k_1a_2^2\delta\psi_2\Big\}.\end{aligned} \tag{6.99}$$

We also obviously have

$$m_1 = m \int_0^l \sin^2 \frac{\pi}{l} x\, dx = \frac{ml}{2}. \tag{6.100}$$

The relationships

$$\overline{\delta W} = \overline{\delta W}_{y_1} = \overline{\delta W}_{z_1} = \overline{\delta W}_1 (a_1, \psi_1) + \overline{\delta W}_2 (a_2, \psi_2), \tag{6.101}$$

hold for the considered problem, described by a linear partial differential equation.

Thus, computing the σ-th "partial derivatives" of $\overline{\delta W}_1$ and $\overline{\delta W}_2$ and substituting them into the system (6.7), we obtain two systems of equations of the first approximation for a_1, ψ_1 and a_2, ψ_2, correspondingly,

$$\left.\begin{aligned} \frac{da_1}{dt} &= -\alpha a_1 - \gamma_1(\tau)\sin\psi_1 - \gamma_2(\tau)\cos\psi_1 - \beta(\tau)\, a_1 \sin 2\psi_1, \\ \frac{d\psi_1}{dt} &= 1 - \Delta(\tau) - \gamma(\tau)\frac{1}{a_1}\cos\psi_1 + \gamma_2(\tau)\frac{1}{a_1}\sin\psi_1 - \beta(\tau)\cos 2\psi_1, \end{aligned}\right\} \tag{6.102}$$

$$\left.\begin{aligned} \frac{da_2}{dt} &= -\alpha a_2, \\ \frac{d\psi_2}{dt} &= 0, \end{aligned}\right\} \tag{6.103}$$

where

$$\left.\begin{aligned} &\Delta(\tau) = \frac{\omega(\tau)}{k_1}, \quad \beta(\tau) = \frac{I_y - I_z}{2\Delta(\tau)(I_y - I_z)}, \quad \alpha = \frac{\varkappa}{2mk_1}, \\ &\gamma_1(\tau) = \frac{\varrho_1 \Delta^2(\tau)}{1+\Delta(\tau)}, \quad \gamma_2(\tau) = \frac{\varrho_2 \Delta^2(\tau)}{1+\Delta(\tau)}. \end{aligned}\right\} \tag{6.104}$$

The expression of the average virtual work $\overline{\delta W}$ (for $\varkappa = 0$) could be derived indirectly by performing the variation (and then averaging) the expression $\frac{1}{2}(\varepsilon T_1 - \varepsilon V_1^*)$ where εV_1^* is defined by (6.71) with $g = 0$.

We consider now the problem of integrating the system (6.102) describing the variation of the amplitude a_1 and the phase ψ_1 in a direct precessional motion.

We introduce new variables u and v into the system (6.102) defined by

$$u = a_1 \cos\psi_1, \quad v = a_1 \sin\psi_1. \tag{6.105}$$

We obtain the system of equations of the first approximation in the form

$$\left.\begin{aligned} \frac{du}{dt} &= -\alpha u - [1 - \Delta(\tau) + \beta(\tau)]\, v - \gamma_2(\tau), \\ \frac{dv}{dt} &= [1 - \Delta(\tau) - \beta(\tau)]\, u - \alpha v - \gamma_1(\tau), \end{aligned}\right\} \tag{6.106}$$

where $\tau = \varepsilon t$.

In order to obtain the stationary values of u and v, we equate the right-hand sides of the system (6.106) to zero for a fixed τ.

We obtain

$$\left.\begin{aligned} u(\tau) &= \frac{-\alpha\gamma_2(\tau) + \gamma_1(\tau)[1 - \Delta(\tau) + \beta(\tau)]}{\alpha^2 + (1 - \Delta(\tau))^2 - \beta^2(\tau)}, \\ v(\tau) &= \frac{-\alpha\gamma_1(\tau) - \gamma_2(\tau)[1 - \Delta(\tau) - \beta(\tau)]}{\alpha^2 + (1 - \Delta(\tau))^2 - \beta^2(\tau)}. \end{aligned}\right\} \tag{6.107}$$

From (6.105) we obtain

$$a(\tau) = \sqrt{u^2(\tau) + v^2(\tau)}. \tag{6.108}$$

We now give some results of B. I. Moseenkov's research (using (6.106) and (6.107)) into the stationary mode and the transition through resonance for various rotation modes of a rod of double rigidity /121/. His results are of interest since they also supplement and demonstrate more completely the phenomena observed during a transition through resonance of vibrating systems. We have considered and discussed these topics more than once.

In the specified case of a rod of double rigidity, we can detect a number of new phenomena observed during a transition through a parametric resonance. While constructing the resonance curves corresponding to the stationary modes, or to a transition through resonance, we shall not go into the details of the method of integration of the system (6.106) as this has been discussed a number of times. We shall also omit the derivation of the formulas defining the zones of the parametric resonance.

We turn immediately to actual examples.

Let us consider transversal vibrations of a steel rod of rectangular cross section $7.28 \cdot 7.7\,\mathrm{cm}^2$, $l = 200\,\mathrm{cm}$, $D = 0.0079\,\mathrm{kg/cm}^3$, $E = 2.1 \cdot 10^6\,\mathrm{kg/cm}^2$. Let $\varkappa = 2.5 \cdot 10^{-3}\,\mathrm{kg \cdot sec/cm}^2$, $\varrho = \varrho_1 = \varrho_2 = 0.05\,\mathrm{mm}$, and let the angular velocity variation be given by $\Delta(\tau) = 0.9 + 0.003t_1$.

With these values we find according to (6.108) (where $u(\tau)$ and $v(\tau)$ are determined from (6.107) with account being taken of (6.104)) the stationary values of the amplitude a_1. We then construct the resonance curve of the stationary amplitudes for the stationary rotation mode of the rod in the interval $0.85 \leqslant \Delta(\tau) \leqslant 1.3$, representing the stable branches of the resonance curve by continuous lines, and the unstable — by dotted lines (Figure 107).

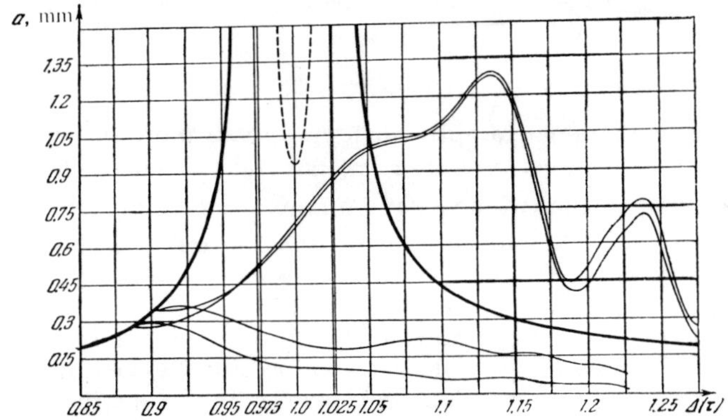

FIGURE 107

Integrating the system (6.106) for the given ϱ and for $\varrho = 0$, we obtain the curves of the vibration amplitudes during a transition through resonance of a statically unbalanced ($\varrho \neq 0$) and a statically balanced ($\varrho = 0$) rod for the same initial conditions (cf. the curves beginning at $\Delta(0) = 0.9$ in Figure 107).

Changing the initial conditions because of the change in the initial rotation velocity, i.e., taking $\Delta(\tau) = 0.884 + 0.001t$, we again construct the resonance curves in the transient mode of rotation. These curves begin at $\Delta(0) = 0.884$ (cf. Figure 107).

The considered case corresponds to a negligible difference of the rigidities $\left(\frac{I_z}{I_y} = 0.894,\ \delta = 0.056\right)$ and, consequently, to a relatively narrow zone of parametric resonance $(0.973 \leqslant \Delta(\tau) \leqslant 1.025)$.

Due to the narrowness of the parametric resonance zone and of the presence of friction, the initial amplitudes of the statically balanced rod ($\varrho = 0$) damp down in a transient mode with the given transition rate (0.003). In other words, the affect of the parametric resonance on the development of amplitudes lasts, for a narrow zone, for a short time and is therefore negligible.

The amplitudes of the statically unbalanced rod ($\varrho \neq 0$) develop mainly because of the presence of the fundamental resonance, and are stable under a change of the initial values with respect to the form of the resonance curve as well as with respect to the values accepted.

Thus, the radius a_1 of the circular motion in a transient mode depends in the given case more on the static nonequilibrium than on th: cross section asymmetry.

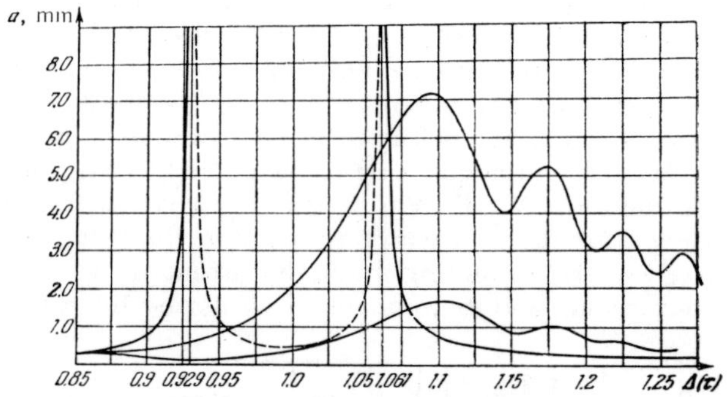

FIGURE 108

Let us consider another example, differing from the first by the dimensions of the rod's cross section which are now taken to be $7 \cdot 8\ \text{cm}^2$, ($\frac{I_z}{I_y} = 0.765$, $\delta = 0.133$), so that we now have a considerable difference in the rigidities. The zone of the parametric resonance is due to this difference being considerably wider than in the preceding case, and is defined by the inequality $0.929 \leqslant \Delta(\tau) \leqslant 1.061$ (Figures 108 and 109).

We construct the resonance curve of the stationary amplitudes in the stationary rotation mode of the rod in the interval $0.85 \leqslant \Delta(\tau) \leqslant 1.25$, designating, as above, the stable sections by continuous lines, and the unstable — by dotted lines.

The unstable branch dropped down almost to the $\Delta(\tau)$-axis, which is a result of the considerable increase in the parametric resonance zone.

We next construct the curves of transition through resonance for the following modes of rotation of the rod: 1) $\Delta(\tau) = 0.85 + 0.003t$, $\varrho \neq 0$— the upper curve, $\varrho = 0$— the lower curve (Figure 108); 2) $\Delta(\tau) = 0.85 + 0.0015t$, $\varrho \neq 0$ — the upper curve, $\varrho = 0$— the lower curve (Figure 109).

As is evident from the graph (Figure 108), the static nonequilibrium which generates the fundamental resonance at the boundary of the parametric resonance in the initial stage (i. e., before the action of the parametric resonance) is for the given transition rate the main cause of the increase in the vibration amplitude. The amplitude of vibrations of a statically unbalanced rod increases in the initial stage despite the presence of external friction, while the amplitude of a statically balanced rod attenuates due to the presence of friction forces, and drops to the minimum value up to the beginning of the action of the parametric resonance.

In the second stage of the transition process, i. e., in the stage of transition under the action of the parametric resonance, the main factor in the specified example becomes the difference in the rigidities caused by the cross section asymmetry. The increase in the amplitudes under the action of the parametric resonance is almost the same for the statically balanced and the statically unbalanced rods (15.8 and 14.1), which, naturally, leads to the conclusion of the dominant influence of the difference in the rigidities.

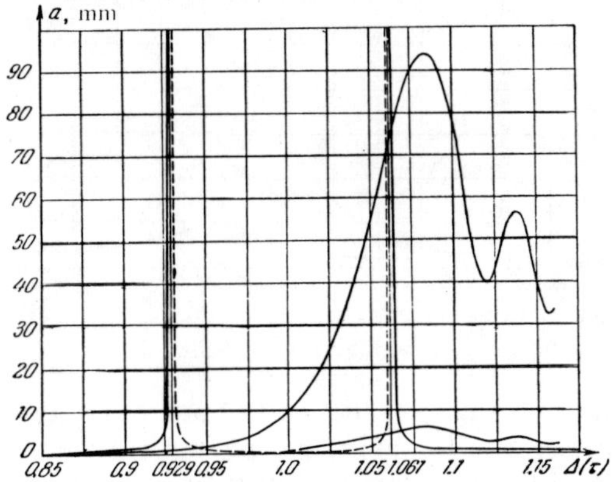

FIGURE 109

Thus, the static nonequilibrium develops the vibration amplitudes during the first stage of the transition mode, and produces the considerably increased initial values (of a_1) before the beginning of the action of the parametric resonance. This is the basic role of static nonequilibrium.

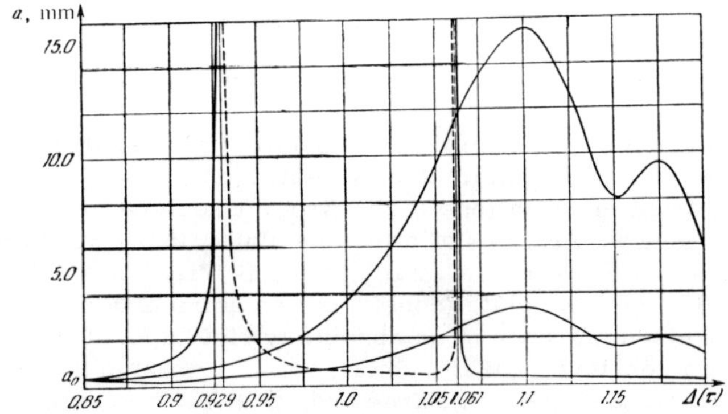

FIGURE 110

Comparing the graphs of the resonance amplitudes for different rates of transitions (cf. Figures 108 and 109), we notice that the maximum of the amplitude is shifted to the side of lower rotation velocities when the rate of transition is smaller, and that the amplitudes increase very much faster; this is especially clearly demonstrated for the statically unbalanced rod.

Let us still consider the case when the eccentricities of the center of gravity of each cross section lie on one of the principal planes of bending.

Leaving all the parameters of the problem unchanged, we construct the resonance curves of transition through the parametric resonance, first for the case when $\varrho_1 = \varrho\sqrt{2}$, and $\varrho_2 = 0$, i. e., for the case when the eccentricity lies in the principal plane of bending of the smallest rigidity (cf. Figure 110), and then for the case when $\varrho_1 = 0$, $\varrho_2 = \varrho\sqrt{2}$, i. e., for the case when the eccentricity lies in the principal plane of bending of the highest rigidity (cf. Figure 111).

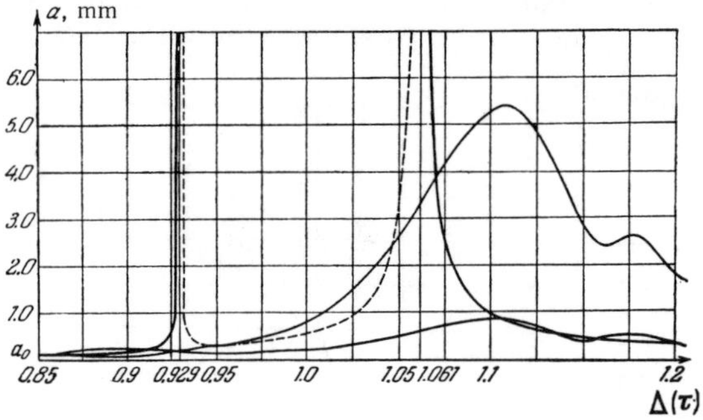

FIGURE 111

As is evident from the graphs (Figures 110 and 111), the position of the eccentricities with respect to the principal planes of bending considerably affects the magnitude of the maximum amplitude in a transient rotation mode for one and the same rate of transition. The largest amplitudes in a transient mode are attained when the plane of the eccentricities coincides with the principal plane of bending of the smallest rigidity (Figure 110), and the lowest— when it coincides with the principal plane of bending of the highest rigidity (Figure 111).

§ 5. Nonstationary vibrations of a turbine plate

The energy method can be applied successfully to the study of nonstationary processes in more complicated systems.

We shall construct the equations of the first approximation for nonstationary monofrequency vibrations of a turbine plate; these were investigated in detail, analytically and experimentally, by G. S. Pisarenko /132/.

Let l be the length of the plate (considered as a prismatic rod) fastened at one end to the circumference of an absolutely rigid disk of radius r_0, which rotates at a constant angular velocity ω. We assume that the plate bears a concentrated mass m at its free end, and that it is subjected to the action of a bending moment proportional to the rotation angle of the plate's end.

Let the plate be subjected to the action of an external perturbing uniformly distributed force

$$q = \varepsilon q_0 \sin\theta, \qquad (6.109)$$

where ε is a small positive parameter,

$$\frac{d\theta}{dt} = \nu(\tau), \qquad \tau = \varepsilon t.$$

The transversal vibrations of the plate in the field of the centrifugal forces, assuming that the vibrations occur in one of the principal planes of bending and that the dimensions of the plate cross section are small compared with its length, are described by the partial differential equation*

$$\frac{\partial^2}{\partial x^2}\left[EI\frac{\partial^2 u}{\partial x^2}\right]+\frac{\partial^2}{\partial x^2}\left[\varepsilon\overleftrightarrow{\Phi}\left(\frac{\partial^2 u}{\partial x^2}\right)\right]+\mu\omega^2(r_0+x)\frac{\partial u}{\partial x}-$$
$$-\omega^2\left[m(r_0+l)+\mu\left(r_0 l+\frac{l^2}{2}-r_0 x-\frac{x^2}{2}\right)\right]\frac{\partial^2 u}{\partial x^2}+\mu\frac{\partial^2 u}{\partial t^2}-\varepsilon q_0\sin\theta=0, \quad (6.110)$$

where μ is the mass of unit length of the plate, x, the coordinate defining the position of an element of the rod measured from the fastened end, $\varepsilon\overleftrightarrow{\Phi}\left(\frac{\partial^2 u}{\partial x^2}\right)$, a function which takes into account the imperfect elastic properties of the plate's material and, therefore, the energy dispersion.

To ease integration we introduce the relative coordinate $\zeta=\frac{x}{l}$ and the relative deflection $v=\frac{u}{l}$, where l is the length of the plate. Equation (6.110) then becomes

$$\frac{\partial^4 v}{\partial\zeta^4}-\frac{l^2}{EI}\left(p_0-\alpha r_0 l\zeta-\frac{\alpha l^2}{2}\zeta^2\right)\frac{\partial^2 v}{\partial\zeta^2}+\frac{\alpha l^3}{EI}(r_0+l\zeta)\frac{\partial v}{\partial\zeta}+$$
$$+\frac{\mu l^4}{EI}\frac{\partial^2 v}{\partial t^2}=\varepsilon\left\{\frac{q_0 l^3}{EI}\sin\theta-\frac{l}{EI}\frac{\partial^2}{\partial\zeta^2}\left[\overleftrightarrow{\Phi}\left(\frac{\partial^2 v}{\partial\zeta^2}\right)\right]\right\}, \quad (6.111)$$

where

$$\alpha=\mu\omega^2,\ p_0=m\omega^2(r_0+l)+\mu\omega^2\left(r_0 l+\frac{l^2}{2}\right). \quad (6.112)$$

The boundary conditions corresponding to our problem (at $\zeta=0$ the end of the plate is fastened, at $\zeta=1$ the free end is loaded by a concentrated mass m and a bending moment M_σ), have the form

$$\left.\begin{aligned}&v(0,t)=0,\quad \left.\frac{\partial v(\zeta,t)}{\partial\zeta}\right|_{\zeta=0}=0,\\&\left.\frac{\partial^2 v(\zeta,t)}{\partial\zeta^2}\right|_{\zeta=1}=p\left.\frac{\partial v(\zeta,t)}{\partial\zeta}\right|_{\zeta=1},\\&\left.\frac{\partial^3 v(\zeta,t)}{\partial\zeta^3}\right|_{\zeta=1}=d_0\left.\frac{\partial v(\zeta,t)}{\partial\zeta}\right|_{\zeta=1},\end{aligned}\right\} \quad (6.113)$$

where

$$p=\frac{kl}{EI},\quad d_0=\frac{l^2 m\omega^2(r_0+l)}{EI},$$

k being a fixed coefficient.

Putting $\varepsilon=0$ in the right-hand side of equation (6.111), we obtain the unperturbed equation

$$\frac{\partial^4 v}{\partial\zeta^4}-\frac{l^2}{EI}\left(p_0-\alpha r_0 l\zeta-\frac{\alpha l^2}{2}\zeta^2\right)\frac{\partial^2 v}{\partial\zeta^2}+\frac{\alpha l^3}{EI}(r_0+l\zeta)\frac{\partial v}{\partial\zeta}+\frac{\mu l^4}{EI}\frac{\partial^2 v}{\partial t^2}=0, \quad (6.114)$$

describing free transversal vibrations of constant cross section without taking into account the energy dispersion in the material.

Unlike the equations of unperturbed motion considered in preceding sections of the present chapter, equation (6.114) is an equation with variable coefficients.

Integrating this equation, we have to take into account the boundary conditions (6.113). Applying the Fourier method, we are looking for the solution of equation (6.114) in the form of a product

$$v(\zeta,t)=\varphi(\zeta)\,a\cos(\omega_c t+\psi), \quad (6.115)$$

* For a detailed derivation of equation (6.110) see /132/.

where ω_c is the natural frequency of transversal vibrations of the plate.

The separation of variables leads to the following ordinary differential equation for the deformation function $\varphi(\zeta)$:

$$\frac{d^4\varphi}{d\zeta^4}-\frac{l^2}{EI}\left(p_0-ar_0l\zeta-\alpha\frac{l^2}{2}\zeta^2\right)\frac{d^2\varphi}{d\zeta^2}+\frac{\alpha l^3}{EI}(r_0+l\zeta)\frac{d\varphi}{d\zeta}-\frac{\mu l^4\omega_c^2}{EI}\varphi=0, \tag{6.116}$$

with the boundary conditions

$$\left.\begin{aligned}&\varphi(0)=0,\quad \left.\frac{d\varphi(\zeta)}{d\zeta}\right|_{\zeta=0}=0,\\ &\left.\frac{d^2\varphi(\zeta)}{d\zeta^2}\right|_{\zeta=1}=p\left.\frac{d\varphi(\zeta)}{d\zeta}\right|_{\zeta=1},\quad \left.\frac{d^3\varphi(\zeta)}{d\zeta^3}\right|_{\zeta=1}=d_0\left.\frac{d\varphi(\zeta)}{d\zeta}\right|_{\zeta=1}.\end{aligned}\right\} \tag{6.117}$$

The most convenient solution of the linear homogeneous equation with polynomial coefficients (6.116) is in the form of a power series

$$\varphi(\zeta)=\sum_{i=0}^{\infty}c_i\zeta^i. \tag{6.118}$$

The coefficients are obtained as recurrence formulas after substituting (6.118) into equation (6.116). We shall not perform the calculations; the interested reader is referred to /121, 132/. The arbitrary constants and the frequency equation for determining the natural frequencies ω_c are found with the help of the boundary conditions (6.117) and the additional condition

$$\varphi(1)=1, \tag{6.119}$$

corresponding to the fact that the quantity a is chosen as the total vibration amplitude of the plate's end.

We find the lowest frequency of normal vibrations given by

$$\omega_c^{(1)}=\sqrt{\frac{rEI}{\mu l^4}}, \tag{6.120}$$

and the series

$$\varphi^{(1)}(\zeta)=\sum_{i=2}^{\infty}c_i\zeta^i, \tag{6.121}$$

expressing the deformation function corresponding to the lowest frequency $\omega_c^{(1)}$. This series converges for any $0\leqslant\zeta\leqslant 1$.

We can now construct an approximate solution (in the first approximation) of the equation of perturbed motion (6.111), close to the vibrations of the basic mode of the unperturbed system.

Assuming that a transition through the principal resonance may occur in the system under investigation, we look for an approximate solution for the relative deflection $v(\zeta, t)$ in the form

$$v^{(1)}(\zeta, t)=\varphi^{(1)}(\zeta)\,a\cos(\theta+\psi), \tag{6.122}$$

where $\varphi^{(1)}(\zeta)$ is the deformation function corresponding to the first (lowest) frequency of free vibrations $\omega_c^{(1)}$, and a and ψ are to be determined from the equations of the first approximation, which we construct by means of the energy method.

We have the following expression for the perturbing generalized force acting on an element $d\zeta$ of the plate:

$$\varepsilon Q\left(\zeta, \theta, \frac{\partial^2 v}{\partial\zeta^2}\right)=\varepsilon\left\{\frac{q_0 l^3}{EI}\sin\theta-\frac{l}{EI}\frac{\partial^2}{\partial\zeta^2}\left[\overleftarrow{\Phi}\left(\frac{\partial^2 v}{\partial\zeta^2}\right)\right]\right\}d\zeta, \tag{6.123}$$

and, consequently, the perturbing force acting on an element of the plate in a mode of sinusoidal vibrations (6.122) is, up to terms of the first order of smallness,

$$\varepsilon Q_0^{(1)}\left(\zeta, \theta, \frac{\partial^2 v^{(1)}}{\partial \zeta^2}\right) = \frac{\varepsilon l}{EI}\left\{q_0 l^2 \sin\theta - \frac{\partial^2}{\partial \zeta^2}\left[\overleftrightarrow{\Phi}\left(\frac{d^2\varphi^{(1)}(\zeta)}{d\zeta^2} a\cos(\theta+\psi)\right)\right]\right\} d\zeta. \quad (6.124)$$

We can now find the virtual work dW performed by the perturbing forces (6.124) along the virtual displacements

$$\delta v^{(1)}(\zeta, t) = \varphi^{(1)}(\zeta)\cos(\theta+\psi)\,\delta a - \varphi^{(1)}(\zeta)\, a\sin(\theta+\psi)\,\delta\psi, \quad (6.125)$$

corresponding to amplitude and phase variations of the basic mode of normal vibrations. We have

$$\delta W = \varepsilon \int_0^1 Q_0^{(1)}\left(\zeta, \theta, \frac{\partial^2 v^{(1)}(\zeta)}{\partial \zeta^2}\right)\delta v^{(1)}(\zeta) \quad (6.126)$$

or, explicitly,

$$\delta W = \frac{\varepsilon l}{EI}\left[\cos(\theta+\psi)\,\delta a - a\sin(\theta+\psi)\,\delta\psi\right]\left\{q_0 l^2 \sin\theta \int_0^1 \varphi^{(1)}(\zeta)\, d\zeta - \right.$$

$$\left. - \int_0^1 \frac{\partial^2}{\partial \zeta^2}\left[\overleftrightarrow{\Phi}\left(\frac{d^2\varphi^{(1)}(\zeta)}{d\zeta^2} a\cos(\theta+\psi)\right)\right]\varphi^{(1)}(\zeta)\, d\zeta\right\}. \quad (6.127)$$

Let us now derive the average value of this work over a full vibration period. We have

$$\overline{\delta W} = \frac{1}{2\pi}\int_0^{2\pi} \overline{\delta W}\, d(\theta+\psi). \quad (6.128)$$

Substituting (6.127) into the integrand and performing elementary manipulations, we obtain the expression

$$\overline{\delta W} = -\frac{\varepsilon l}{2EI}\left\{\left[\frac{1}{\pi}\Phi_c(a) + q_0\beta l^2 \sin\psi\right]\delta a + a\left[q_0\beta l^2\cos\psi - \frac{1}{\pi}\Phi_s(a)\right]\delta\psi\right\}, \quad (6.129)$$

where we have used the notation

$$\beta = \int_0^1 \varphi^{(1)}(\zeta)\, d\zeta, \quad (6.130)$$

$$\Phi_c(a) = \oint \int_0^1 \frac{\partial^2}{\partial \zeta^2}\left[\overleftrightarrow{\Phi}\left(\frac{d^2\varphi^{(1)}(\zeta)}{d\zeta^2} a\cos(\theta+\psi)\right)\right]\varphi^{(1)}(\zeta)\cos(\theta+\psi)\, d\zeta\, d(\theta+\psi), \quad (6.131)$$

$$\Phi_s(a) = \oint \int_0^1 \frac{\partial^2}{\partial \zeta^2}\left[\overleftrightarrow{\Phi}\left(\frac{d^2\varphi^{(1)}(\zeta)}{d\zeta^2} a\cos(\theta+\psi)\right)\right]\varphi^{(1)}(\zeta)\sin(\theta+\psi)\, d\zeta\, d(\theta+\psi). \quad (6.132)$$

In a mode of sinusoidal vibrations, the explicit expression of the function $\varepsilon\overleftrightarrow{\Phi}$, which takes into account the losses due to internal friction, has according to the hypothesis of Davidenkov-Pisarenko /43, 132/ the form

$$\varepsilon\overleftrightarrow{\Phi}\left[\frac{d^2\varphi^{(1)}(\zeta)}{d\zeta^2} a\cos(\theta+\psi)\right] =$$

$$= \mp\frac{E\nu_1}{nl^n} a^n \left(\frac{d^2\varphi^{(1)}(\zeta)}{d\zeta^2}\right)^n \left[(1\pm\cos(\theta+\psi))^n - 2^{n-1}\right]\int_{(F)} z^{n+1}\, dF, \quad (6.133)$$

where n and ν_1 are constants determined by experiment, z is the distance of the surface element dF of the plate cross section from the neutral axis of the plate, F, the area of the plate.

For m_1 we find (according to (4.61)) the expression

$$m_1 = \frac{\mu l^4}{EI} \int_0^1 [\varphi^{(1)}(\zeta)]^2 d\zeta. \tag{6.134}$$

Deriving now the nonvanishing σ-th "partial derivatives" of the average virtual work (6.129) and substituting them and the values of m_1 given by

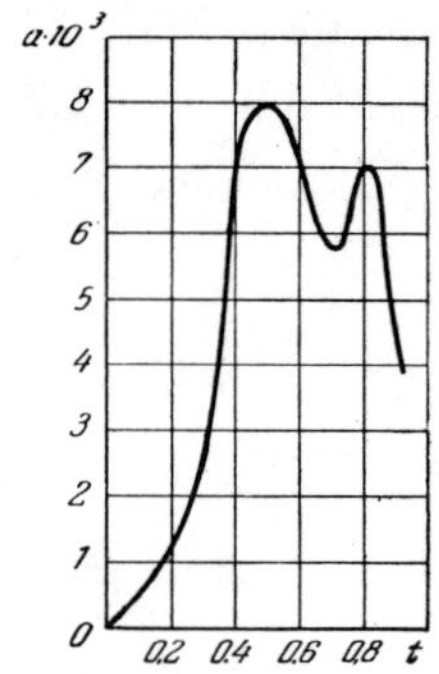

FIGURE 112

(6.134) and of $\omega_c^{(1)}$ given by (6.120) into the right-hand side of the system (6.7) (putting herewith $p = q = 1$), we obtain finally the following system of equations of the first approximation:

$$\left.\begin{aligned} \frac{da}{dt} &= -\frac{\varepsilon\Phi_s(a)}{\alpha_1\omega_c^{(1)}\pi} + \frac{2\varepsilon q_0 l^2\beta}{\alpha_1[\omega_c^{(1)}+\nu(\tau)]}\cos\psi, \\ \frac{d\psi}{dt} &= \omega_c^{(1)} - \nu(\tau) - \frac{\varepsilon\Phi_c(a)}{\alpha_1 a\omega_c^{(1)}\pi} - \frac{2\varepsilon q_0 l^2\beta}{\alpha_1 a[\omega_c^{(1)}+\nu(\tau)]}\sin\psi, \end{aligned}\right\} \tag{6.135}$$

where

$$\alpha_1 = -2\mu l^3 \int_0^1 [\varphi^{(1)}(\zeta)]^2 d\zeta. \tag{6.136}$$

Taking actual numerical data, as in /132/, and integrating the system (6.135) numerically, we obtain the resonance curve describing a transition through resonance (cf. Figure 112).

§ 6. Transversal-torsional vibrations of a turbine plate during a nonstationary mode

We give another example of an investigation of nonstationary vibrations in a system with distributed parameters, discussed in detail in /23, 61/, — the transversal-torsional vibrations of a turbine plate during a transition through resonance.

We introduce the following notation: EI, the rigidity of the plate under bending; GI_d, the rigidity under torsion; m, the mass of unit length of the plate; I_m, the moment of inertia of unit length of the plate; x_c, the distance between the center of gravity and the center of rigidity of a cross section; $y(z, t)$, the deflection; $\theta(z, t)$, the torsion angle; $q(z)$, the load intensity along the axis of elasticity; $m_z(z)$, the intensity of the moment of this load with respect to the axis of torsion; $\frac{d\theta}{dt} = \nu(\tau)$, the frequency of the external force; l, the length of the plate; $\tau = \varepsilon t$, the "slowing" time; ε, a small positive parameter.

G. S. Pisarenko, N. V. Vasilenko and Yu. A. Klikh /23, 61/ have obtained under the assumption that the forces of internal friction during transversal $f_1(z, t)$ and torsional $f_2(z, t)$ vibrations as well as the perturbing loads are proportional to the small parameter ε, the system of differential equations

$$\left.\begin{aligned} &\frac{\partial^2}{\partial z^2}\left(EI\frac{\partial^2 y}{\partial z^2}\right) + m\frac{\partial^2 y}{\partial t^2} + mx_c\frac{\partial^2\theta}{\partial t^2} = \varepsilon\,[q(z)\cos\nu t - f_1(z, t)],\\ &-\frac{\partial}{\partial z}\left(GI_d\frac{\partial\theta}{\partial z}\right) + I_m\frac{\partial^2\theta}{\partial t^2} + mx_c\frac{\partial^2 y}{\partial t^2} = \varepsilon\,[-m_z(z)\cos\nu t + f_2(z, t)],\end{aligned}\right\} \tag{6.137}$$

describing the perturbed transversal-torsional vibrations and neglecting insignificant displacements in the direction of the largest rigidity of the plate cross section.

We assume for simplicity that the plate has a constant cross section.

Then, putting $\varepsilon = 0$ in the system (6.137), we obtain the corresponding system of equations for the unperturbed motion

$$\left.\begin{aligned} &EI\frac{\partial^4 y}{\partial z^4} + m\frac{\partial^2 y}{\partial t^2} + mx_c\frac{\partial^2\theta}{\partial t^2} = 0,\\ &-GI_d\frac{\partial^2\theta}{\partial z^2} + I_m\frac{\partial^2\theta}{\partial t^2} + mx_c\frac{\partial^2 y}{\partial t^2} = 0,\end{aligned}\right\} \tag{6.138}$$

which describes the free transversal-torsional vibrations without taking into account the forces of internal friction.

As usual, we look for the solution of the system (6.138) in the form

$$\left.\begin{aligned} &y(z, t) = Y(z)\,a\cos(\omega t + \psi),\\ &\theta(z, t) = \Phi(z)\,b\cos(\omega t + \psi),\end{aligned}\right\} \tag{6.139}$$

where ω is the natural frequency of the transversal-torsional vibrations; $Y(z)$ and $\Phi(z)$, functions describing the shape of the transversal-torsional vibrations of the plate, corresponding to the natural frequency ω; a, b and ψ, constants; a and b, amplitude coefficients; ψ, phase of the harmonic transversal-torsional vibrations.

Substituting (6.139) into equations (6.138) and dividing by $\cos(\omega t + \psi)$, we obtain the following system of ordinary differential equations for $Y(z)$ and $\Phi(z)$:

$$\left.\begin{aligned} &EIa\frac{d^4Y}{dz^4} - m\omega^2 aY - mx_c\omega^2 b\Phi = 0,\\ &GI_d b\frac{d^2\Phi}{dz^2} + I_m\omega^2 b\Phi + mx_c\omega^2 aY = 0.\end{aligned}\right\} \tag{6.140}$$

We find from this system of equations

$$\left.\begin{aligned} Y(z) = {} & C_1\,\mathrm{sh}\,k_1z + C_2\,\mathrm{ch}\,k_1z + C_3\sin k_2z + C_4\cos k_2z +\\ & + C_5\sin k_3z + C_6\cos k_3z,\\ \Phi(z) = {} & \frac{1}{\beta}\,[\alpha_1(C_1\,\mathrm{sh}\,k_1z + C_2\,\mathrm{ch}\,k_1z) + \alpha_2(C_3\sin k_2z +\\ & + C_4\cos k_2z) + \alpha_3(C_5\sin k_3z + C_6\cos k_3z)],\end{aligned}\right\} \tag{6.141}$$

where

$$\left.\begin{aligned} &\alpha_i = \frac{EIk_i^4}{m\omega^2 x_c} - \frac{1}{x_c} \quad (i = 1, 2, 3),\\ &\beta = \frac{b}{a},\end{aligned}\right\} \tag{6.142}$$

and the arbitrary constants C_k ($k = 1, 2, \ldots, 6$) are derived with the help of the boundary conditions.

In the given case of a turbine plate it is natural to consider it as a cantilever: one end is rigidly fastened ($z = 0$), and the other ($z = 1$) is free. The boundary conditions are then written in the form

$$\left.\begin{aligned} Y(0) = Y'(0) = Y''(1) = Y'''(1) = 0, \\ \Phi(0) = \Phi'(0) = 0, \\ Y^{\mathrm{IV}}(0) = 0, \ Y'(1) - \frac{EI}{m\omega^2} Y^{\mathrm{V}}(1) = 0. \end{aligned}\right\} \quad (6.143)$$

Satisfying these conditions, we obtain a system of equations for the arbitrary constants C_k ($k = 1, 2, \ldots, 6$), and also the frequency equation

$$\Delta(\omega) = 0, \quad (6.144)$$

from which we derive the frequencies of normal vibrations ω_i ($i = 1, 2, \ldots$).

Determining the lowest frequency $\omega = \omega_1$ for the equations (6.140), i.e., the frequency of the basic tone of free transversal-torsional vibrations, we obtain the corresponding values of the functions $Y^{(1)}(z)$ and $\Phi^{(1)}(z)$, which are determined up to the factor β and an arbitrary factor N.

Additional conditions are necessary to determine β and N. As such conditions we can take

$$Y^{(1)}(1) = 1, \quad \Phi^{(1)}(1) = 1, \quad (6.145)$$

which may be expressed by means of the deformation function $y(z, t)$ and the torsion angle $\theta(z, t)$ as

$$y^{(1)}(z, t)\Big|_{\substack{z=1 \\ \omega_1 t + \psi = 0}} = a, \qquad \theta^{(1)}(z, t)\Big|_{\substack{z=1 \\ \omega_1 t + \psi = 0}} = b. \quad (6.146)$$

The amplitude coefficients a and b are in this case, correspondingly, equal to the amplitudes of the transversal and torsional vibrations of the plate's free end, and the relation between them

$$\frac{b}{a} = \beta$$

is determined from condition (6.145).

After this preliminary discussion on solving the unperturbed system (6.138) we now study the system (6.137).

We look for a solution of this system of equations in the first approximation, corresponding to perturbed vibrations close to the unperturbed vibrations at the frequency ω_1, in the form

$$\left.\begin{aligned} y^{(1)}(z, t) = Y^{(1)}(z)\, a \cos(\theta + \psi), \\ \theta^{(1)}(z, t) = \Phi^{(1)}(z)\, \beta a \cos(\theta + \psi), \end{aligned}\right\} \quad (6.147)$$

where the amplitude a and the phase ψ are functions of the time t and are to be determined from the system of equations of the first approximation, which we construct with the help of our energy method.

In the specified case of perturbed transversal-torsional vibrations described by the system of equations (6.137), the generalized coordinates are the deflection $y(z, t)$ and the torsion angle $\theta(z, t)$, and the corresponding generalized perturbing forces acting on an element of length dz of the plate are

$$\left.\begin{aligned} \varepsilon Q_y\left(z, t, \frac{\partial^2 y}{\partial z^2}\right) = \varepsilon\,[q(z) \cos \nu t - f_1(z, t)]\, dz, \\ \varepsilon Q_\theta\left(z, t, \frac{\partial \theta}{\partial z}\right) = \varepsilon\,[-m_z(z) \cos \nu t + f_2(z, t)]\, dz. \end{aligned}\right\} \quad (6.148)$$

The generalized perturbing forces in the mode of sinusoidal vibrations (6.147) are written, with an accuracy of first order terms, in the form

$$\left.\begin{aligned}\varepsilon Q_{y_0}^{(1)}\left(z,\ t,\ \frac{\partial^2 y^{(1)}}{\partial z^2}\right)&=\varepsilon\left[q(z)\cos\nu t-f_{10}^{(1)}(z,\ t)\right]dz,\\ \varepsilon Q_{\theta_0}^{(1)}\left(z,\ t,\ \frac{\partial\theta^{(1)}}{\partial z}\right)&=\varepsilon\left[-m_z(z)\cos\nu t+f_{20}^{(1)}(z,\ t)\right]dz,\end{aligned}\right\}\tag{6.149}$$

where $f_{10}^{(1)}(z,\ t)$ and $f_{20}^{(1)}(z,\ t)$ are determined by formulas (6.156) (cf. below), and the virtual work of these forces along the virtual displacements of the generalized coordinates

$$\left.\begin{aligned}\delta y^{(1)}&=Y^{(1)}(z)[\cos(\nu t+\psi)\,\delta a-a\sin(\nu t+\psi)\,\delta\psi],\\ \delta\theta^{(1)}&=\Phi^{(1)}(z)\,\beta\,[\cos(\nu t+\psi)\,\delta a-a\sin(\nu t+\psi)\,\delta\psi],\end{aligned}\right\}\tag{6.150}$$

corresponding to variations of the amplitude and phase of transversal-torsional vibrations, is expressed by the integral

$$\delta W=\varepsilon\int_0^1\left[Q_{y_0}^{(1)}\left(z,\ t,\ \frac{\partial^2 y^{(1)}}{\partial z^2}\right)\delta y^{(1)}+Q_{\theta_0}^{(1)}\left(z,\ t,\ \frac{\partial\theta^{(1)}}{\partial z}\right)\delta\theta^{(1)}\right]\tag{6.151}$$

or, by virtue of (6.149) and (6.150),

$$\begin{aligned}\delta W=[\cos(\nu t+\psi)\,\delta a-a\sin(\nu t+\psi)\,\delta\psi]\Big\{A\cos\nu t+\\ +\int_0^1\left[-\varepsilon f_{10}^{(1)}(z,\ t)\,Y^{(1)}(z)+\beta\varepsilon f_{20}^{(1)}(z,\ t)\,\Phi^{(1)}(z)\right]dz\Big\}.\end{aligned}\tag{6.152}$$

Here

$$A=\varepsilon\int_0^1 q(z)\,Y^{(1)}(z)\,dz+\varepsilon\beta\int_0^1 m_z(z)\,\Phi^{(1)}(z)\,dz.\tag{6.153}$$

We shall now calculate the average value of the virtual work over a complete vibration period.

We have

$$\begin{aligned}\overline{\delta W}=\frac{1}{2}A(\cos\psi\,\delta a-a\sin\psi\,\delta\psi)+\frac{\varepsilon}{2\pi}\oint\int_0^1\{[-f_{10}^{(1)}(z,\ t)\,Y^{(1)}(z)+\\ +\beta f_{20}^{(1)}(z,\ t)\,\Phi^{(1)}(z)]\,dz\,[\cos(\nu t+\psi)\delta a-a\sin(\nu t+\psi)\,\delta\psi]\}\,d(\nu t+\psi),\end{aligned}\tag{6.154}$$

where, as usual, $\oint$ denotes integration over the closed cycle connected with the circumvention of the hysteresis loop.

We introduce the notation

$$\left.\begin{aligned}\chi_1(a)=\varepsilon\oint\int_0^1\left[-f_{10}^{(1)}(z,\ t)\,Y^{(1)}(z)+\beta f_{20}^{(1)}(z,\ t)\,\Phi^{(1)}(z)\right]\times\\ \times\cos(\nu t+\psi)\,dz\,d(\nu t+\psi),\\ \chi_2(a)=\varepsilon\oint\int_0^1\left[-f_{10}^{(1)}(z,\ t)\,Y^{(1)}(z)+\beta f_{20}^{(1)}(z,\ t)\,\Phi^{(1)}(z)\right]\times\\ \times\sin(\nu t+\psi)\,dz\,d(\nu t+\psi).\end{aligned}\right|\tag{6.155}$$

The functions of energy dispersion in the mode of perturbed transversal-torsional vibrations have, according to G. S. Pisarenko /132/, the form

$$\begin{aligned}\varepsilon f_{10}^{(1)}(z,\ t)=\mp\frac{\nu_1 E a^n}{n l^{2(n+1)}}\frac{d^2}{dz^2}\left[\left(\frac{d^2Y^{(1)}(z)}{dz^2}\right)^n\right]\times\\ \times[(1\pm\cos(\nu t+\psi))^n-2^{n-1}]\int_{(F)}\xi^{n+1}\,dF,\end{aligned}\tag{6.156}$$

$$\varepsilon f_{20}^{(1)}(z,\ t) = \mp \frac{\eta G a^k \beta^k}{k l^{k+1}} \frac{d}{dz} \left[\left(\frac{d^2\Phi^{(1)}(z)}{dz^2} \right)^k \right] \times$$

$$\times [(1 \pm \cos(\nu t + \psi))^k - 2^{k-1}] \int\limits_{(F)} \varrho \, dF, \qquad (6.156)$$

where ν_1, n, η and k are geometric parameters of the hysteresis loop, ξ, the distance from the neutral axis to the surface element of the rod cross section dF, and ϱ, the distance from the axis of torsion to the same surface element.

The expression (6.154) may hence be written in the form

$$\overline{\delta W} = \frac{1}{2} A \left(\cos\psi \, \delta a - a \sin\psi \, \delta\psi \right) + \frac{1}{2\pi} [\chi_1(a) \, \delta a - a \chi_2(a) \, \delta\psi] \qquad (6.157)$$

or

$$\overline{\delta W} = \frac{1}{2} \left\{ \left[A \frac{e^{i\psi} + e^{-i\psi}}{2} + \frac{1}{\pi} \chi_1(a) \right] \delta a - a \left[A \frac{e^{i\psi} - e^{-i\psi}}{2i} + \frac{1}{\pi} \chi_2(a) \right] \delta\psi \right\}. \qquad (6.158)$$

We now derive the generalized mass m_1 from the formula

$$m_1 = 2T(\varphi^{(1)}). \qquad (6.159)$$

We must first find the expression for the kinetic energy of the unperturbed motion in the case of combined transversal-torsional vibrations of the plate. In the given case of a constant cross section we have the following expression for the kinetic energy:

$$T = \frac{1}{2} \int\limits_0^1 \left[m \left(\frac{\partial y}{\partial t} \right)^2 + 2 m x_c \frac{\partial y}{\partial t} \frac{\partial \theta}{\partial t} + I_m \left(\frac{\partial \theta}{\partial t} \right)^2 \right] dz. \qquad (6.160)$$

Replacing $\frac{\partial y}{\partial t}$ and $\frac{\partial \theta}{\partial t}$ in (6.160) by the normal functions $Y^{(1)}(z)$ and $\beta\Phi^{(1)}(z)$, respectively, we obtain for m_1 the expression

$$m_1 = m \int\limits_0^1 \{[Y^{(1)}(z)]^2 + 2\beta x_c Y^{(1)}(z) \Phi^{(1)}(z) + \beta^2 I_m [\Phi^{(1)}(z)]^2\} \, dz. \qquad (6.161)$$

It only remains to derive the "partial derivatives" from (6.158) and substitute them into equations (6.7). We then obtain the equations of the first approximation in the form

$$\left.\begin{aligned} \frac{da}{dt} &= -\frac{\chi_2(a)}{2\pi m_1 \omega_1} - \frac{A}{m_1(\omega_1 + \nu)} \sin\psi, \\ \frac{d\psi}{dt} &= \omega_1 - \nu - \frac{\chi_1(a)}{2\pi m_1 \omega_1 a} - \frac{A}{a m_1 (\omega_1 + \nu)} \cos\psi. \end{aligned}\right\} \qquad (6.162)$$

With the help of this system of equations it is not difficult to construct the amplitude curves of the coupled transversal and torsional vibrations of the free end of the turbine plate (the curves describing the change of a and b) and of any cross section of the plate (to do this we must multiply the values of a and b by the functions $Y^{(1)}(z)$ and $\Phi^{(1)}(z)$, respectively) for the stationary mode as well as for the nonstationary case — transition through resonance.

As an actual example we quote the results obtained in /23/ for the following numerical values:

$$\left.\begin{aligned} &E = 2.2 \cdot 10^6 \ \text{kg/cm}^2, \ G = 0.859 \cdot 10^6 \ \text{kg/cm}^2, \ I_m = 9.21 \cdot 10^{-5} \ \text{kg} \cdot \text{sec}^2, \\ &m = 3.33 \cdot 10^{-5} \ \text{kg} \cdot \text{sec}^2/\text{cm}^2, \ l = 28 \ \text{cm}, \ I = 1.37 \ \text{cm}^4, \\ &I_d = 0.896 \ \text{cm}^4, \ q = 0.065 \ \text{kg/cm}, \ m_z = 0.0128 \ \text{kg} \cdot \text{cm/cm}, \\ &x_c = 0.28 \ \text{cm}, \ n = 2.0, \ \gamma = 4.18, \ k = 2.0, \ \eta = 26.0. \end{aligned}\right\} \qquad (6.163)$$

For the unperturbed system (according to the above procedure), we find that the lowest frequency ω_1 and β are

$$\omega_1 = 1343 \text{ sec}^{-1}, \ \beta = 0.0318 \text{ rad/cm}. \tag{6.164}$$

Substituting the values of (6.163) and (6.164) into the right-hand sides of (6.161) and (6.162), we obtain (after elementary calculations) the equations of the first approximation in the form

$$\left.\begin{aligned} \frac{da}{dt} &= -13.18\,a^2 - \frac{3088}{1343+\nu}\sin\psi,\\ \frac{d\psi}{dt} &= 1343 - \nu - 31.04\,a - \frac{3088}{a\,(1343+\nu)}\cos\psi. \end{aligned}\right\} \tag{6.165}$$

Equating the right-hand sides of the system (6.165) to zero, we obtain a relation for constructing the stationary resonance curves.

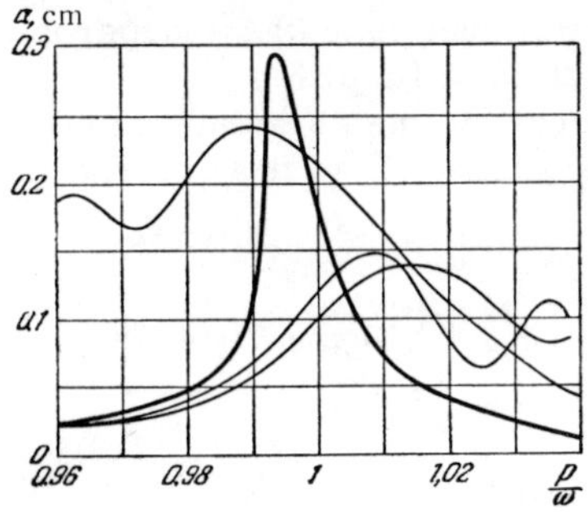

FIGURE 113

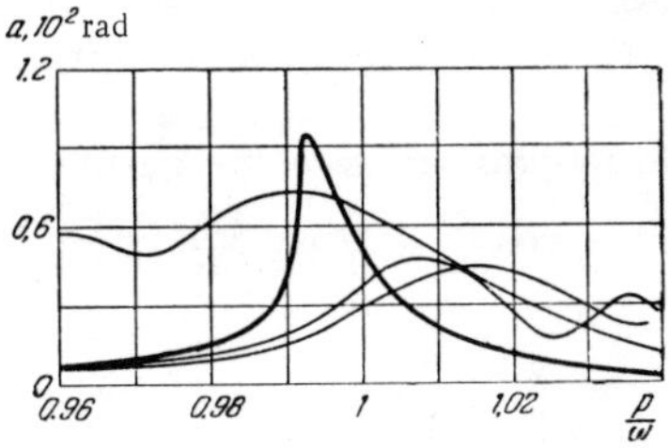

FIGURE 114

Assuming in (6.165)

$$\nu = \nu_0 \pm \alpha t,$$

where $\alpha = 230.6 \text{ sec}^{-2}$ in one case, and $\alpha = 115.3 \text{ sec}^{-2}$ in the other, and integrating numerically with the help of an analogue computer, as was done by Yu. A. Klikh /62/, we obtain the curves of transition through resonance shown in Figures 113 and 114. The curve in Figure 113 is the resonance curve for transversal vibrations, and in Figure 114— for torsional vibrations.

§ 7. Asymptotic expansions for nonlinear partial differential equations, close to hyperbolic equations

In the preceding sections we have illustrated by means of numerous examples the effectiveness of the energy method when constructing approximate solutions corresponding to monofrequency vibrational processes in so-called systems with distributed parameters. The motion of these systems is described, as known, by partial differential equations of the hyperbolic type.

The essential advantage of this method is that it allows us to construct approximate solutions proceeding directly from the expressions of the kinetic and potential energies of the system and the expressions of the external perturbing forces, avoiding the construction of the exact differential equations of the problem. This is especially convenient when under certain boundary conditions it is impossible to separate the variables in the perturbed partial differential equations and thereby to reduce the problem of a partial equation to the study of an infinite system of ordinary differential equations containing a small parameter.

The elaborated formalism of asymptotic expansions is, as was demonstrated by the quoted examples, suitable for practical calculations and gives a good qualitative, and also quantitative agreement with experiment.

This method is to some extent formal, and in the general case requires a mathematical justification and a number of criteria determining the limits of the effective applicability of asymptotic expansions corresponding to a monofrequency vibrational process.

Such a justification is given in detail in Chapter VIII for the case of vibrating systems with a finite number of degrees of freedom. In this book we shall not consider the mathematical justification of the method for constructing asymptotic expansions for systems with infinitely many degrees of freedom described by partial differential equations of the hyperbolic type. This problem may be solved without essential difficulties.

We discuss briefly some general concepts in the construction of asymptotic solutions for partial differential equations, close to a hyperbolic equation.

Let this equation have the quite general form

$$A(\tau)\frac{\partial^2 u}{\partial t^2}+2B(\tau)\frac{\partial^2 u}{\partial t\,\partial x}+C(\tau)\frac{\partial^2 u}{\partial x^2}+D(\tau)\frac{\partial u}{\partial t}+E(\tau)\frac{\partial u}{\partial x}+F(\tau)u= \\ =\varepsilon\Phi\left(\tau,\ \theta,\ x,\ u,\ \frac{\partial u}{\partial t},\ \frac{\partial u}{\partial x}\right) \qquad (6.166)$$

where ε is a small positive parameter; $\tau=\varepsilon t$, the "slowing" time; $\frac{d\theta}{dt}=\nu(\tau)$; the functions $A(\tau), B(\tau),\ C(\tau),\ D(\tau),\ E(\tau), F(\tau), \nu(\tau)$ and $\Phi\left(\tau,\ \theta,\ x,\ u,\ \frac{\partial u}{\partial t}\cdot\frac{\partial u}{\partial x}\right)$ satisfy all the desired conditions; and $\Phi\left(\tau,\ \theta,\ x,\ u,\ \frac{\partial u}{\partial t},\ \frac{\partial u}{\partial x}\right)$ are periodic in θ with period 2π.

In addition, the coefficients $A(\tau),\ B(\tau)$, and $C(\tau)$ satisfy for all τ in the interval $0\leqslant\tau\leqslant L$ the inequality

$$B^2(\tau)-A(\tau)C(\tau)>0. \qquad (6.167)$$

To apply our method, it is first convenient to transform equation (6.166) to another form. If condition (6.167) is satisfied, this equation may be reduced with the help of already known substitutions of the dependent and independent variables to the form (we retain the notation $u,\ x$, and t in the new variables)

$$\frac{\partial^2 u}{\partial t^2}-a^2(\tau)\frac{\partial^2 u}{\partial x^2}=\lambda(\tau)u+\varepsilon F\left(\tau,\ \theta,\ x,\ u,\ \frac{\partial u}{\partial t},\ \frac{\partial u}{\partial x}\right). \qquad (6.168)$$

Let it be required to solve this equation for the following initial and boundary conditions:

$$u(0,\ t)=u(l,\ t)=0, \qquad (6.169)$$

$$\left.\begin{aligned} u|_{t=0}&=f(x),\\ \left.\frac{\partial u}{\partial t}\right|_{t=0}&=F(x), \end{aligned}\right\} \qquad (6.170)$$

where $f(x)$ and $F(x)$ are continuous functions satisfying all the desired conditions.

Before solving equation (6.168) approximately, we consider the unperturbed equation which is obtained by putting $\varepsilon=0$ in (6.168) and regarding τ

as a constant parameter

$$\frac{\partial^2 u}{\partial t^2} - a^2(\tau)\frac{\partial^2 u}{\partial x^2} = \lambda(\tau)\,u \tag{6.171}$$

with the same initial and boundary conditions

$$u(0,\ t) = u(l,\ t) = 0, \qquad u|_{t=0} = f(x), \qquad \left.\frac{\partial u}{\partial t}\right|_{t=0} = F(x).$$

Assuming that the condition

$$\left(\frac{\pi a(\tau)\,n}{l}\right)^2 - \lambda(\tau) > 0 \qquad (n = 1,\ 2,\ 3,\ \ldots), \tag{6.172}$$

is satisfied for any constant τ in the interval $0 \leqslant \tau \leqslant L$, we find the solution of equation (6.171) by means of the Fourier method in the form of the series

$$u(x,\ t) = \sum_{n=1}^{\infty}\{A_n \cos\omega_n t + B_n \sin\omega_n t\}\sin\frac{\pi n}{l}x, \tag{6.173}$$

where $\omega_n = \sqrt{\left(\frac{\pi a(\tau)\,n}{l}\right)^2 - \lambda(\tau)}$ $(n = 1,\ 2,\ 3,\ \ldots)$ are the frequencies of the normal vibrations, and A_n and B_n $(n = 1,\ 2,\ 3,\ \ldots)$ are constants which are to be determined from the initial conditions.

Proceeding from the solution (6.173) of the unperturbed equations (6.171) and assuming that the shape of vibrations of the normal modes is, because of the smallness of the parameter ε, determined in the presence of a perturbation with a sufficient accuracy by the same function $\sin\frac{n\pi}{l}x$, we look for the solution of the perturbed equation (6.168) in the form

$$u(x,\ t,\ \varepsilon) = \sum_{n=1}^{\infty} z_n(t)\sin\frac{\pi n}{l}x. \tag{6.174}$$

This expression may be considered as a change of variables transforming (6.168) to quasi-normal coordinates. We recall that the solution of equation (6.168) by virtue of the vanishing boundary conditions (6.169), may be sought for as the sum of products (6.174).

We discussed the above point when we showed that the energy method enables us to construct the equations of the first and higher approximations in those cases when the boundary conditions do not permit us to apply the method of separation of variables directly.

Substituting (6.174) into (6.168), multiplying by $\sin\frac{\pi m}{l}x$ $(m = 1,\ 2,\ 3,\ \ldots)$ and integrating from 0 to l, we obtain the following infinite system of equations for the determination of $z_n(t)$ $(n = 1,\ 2,\ 3,\ \ldots)$:

$$\frac{d^2 z_n}{dt^2} + \omega_n^2(\tau)\,z_n = \varepsilon F_n(\tau,\ \theta,\ z_1,\ z_2,\ \ldots,\ \dot{z}_1,\ \dot{z}_2,\ \ldots) \qquad (n = 1,\ 2,\ 3,\ \ldots). \tag{6.175}$$

To construct asymptotic approximate solutions for the system (6.175) we can either formally apply the results of the u-method presented above, or, performing a change of variables according to the formulas

$$\left.\begin{aligned} z_n &= x_n e^{i\int^t \omega_n(\tau)\,dt} + x_{-n} e^{-i\int^t \omega_n(\tau)\,dt}, \\ \frac{dz_n}{dt} &= i\omega_n x_n e^{i\int^t \omega_n(\tau)\,dt} - i\omega_n x_{-n} e^{-\int^t \omega_n(\tau)\,dt} \qquad (n = 1,\ 2,\ 3,\ \ldots), \end{aligned}\right\} \tag{6.176}$$

where x_n and x_{-n} ($n = 1, 2, 3, \ldots$) are new unknown complex conjugate functions, reduce the system of equations (6.175) to the standard form

$$\frac{dx}{dt} = \varepsilon X(\tau, \theta, x, \varepsilon), \qquad (6.177)$$

where x is a vector in Hilbert space, and $X(\tau, \theta, x, \varepsilon)$, a vector function in Hilbert space.

It is not difficult to show (using Chapter VIII and the results of /13/) that under certain conditions imposed on the right-hand side of (6.177) it is possible to construct with the help of the method of averaging asymptotic solutions in the first and higher approximations with any previously specified accuracy in a sufficiently large, but finite, interval $0 \leqslant t \leqslant \frac{L}{\varepsilon}$.

The construction of asymptotic solutions for the system (6.175) or (6.177) is in practice possible when there exists in the system a stable vibration mode depending on a finite number of arbitrary constants (with the desired accuracy).

Chapter VII

METHODS OF CONSTRUCTING ASYMPTOTIC SOLUTIONS FOR SYSTEMS OF DIFFERENTIAL EQUATIONS CONTAINING SLOWLY VARYING PARAMETERS

§ 1. Relaxation systems with slowly varying parameters

In the present chapter we shall generalize and extend our studies to differential equations of a more general type, those containing slowly varying parameters.

We first of all consider a relaxation vibrating system with slowly varying parameters, in which the "free" vibrations are described by the system of differential equations

$$\frac{dx_k}{dt} = X_k(\tau, x_1, \ldots, x_n) \qquad (k = 1, 2, \ldots, n), \tag{7.1}$$

where $\tau = \varepsilon t$ is the "slowing" time, and ε is a small positive parameter.

For simplicity we denote the set of n quantities $x_1, x_2, \ldots, x_n$ by the single letter x, and, correspondingly, the set of functions $X_1(\tau, x_1, \ldots, x_n), \ldots, X_n(\tau, x_1, \ldots, x_n)$ by $X(\tau, x)$. Hence, x is regarded in the sequel as a vector in the n-dimensional Euclidean space E_n with the components $x_1, x_2, \ldots, x_n$, and, correspondingly, $X(\tau, x)$— as an n-dimensional vector function.

Equations (7.1) may then be written in the abbreviated form

$$\frac{dx}{dt} = X(\tau, x). \tag{7.2}$$

We assume that for any constant τ belonging to the interval $0 \leqslant \tau \leqslant L$ we know a family of stable periodic solutions for the system of equations (7.2)

$$x = x^0(\tau, \omega t + \varphi) \tag{7.3}$$

with period 2π with respect to ψ $(\psi = \omega t + \varphi)$

$$x^0(\tau, \psi + 2\pi) = x^0(\tau, \psi), \tag{7.4}$$

depending on one arbitrary constant φ, and also on τ as a parameter. Generally

$$\omega = \omega(\tau). \tag{7.5}$$

It is evident that if one puts $\tau = \varepsilon t$ in the solution (7.3) and in the equation (7.2), then the solution (7.3) satisfies equations (7.2) only with an accuracy of the first order in ε.

We set up variational equations for the system (7.2) corresponding to the solution (7.3) (τ is herewith regarded as a constant parameter). We have

$$\frac{d\delta x}{dt} = X'_x[\tau, x^0(\tau, \omega t + \varphi)]\,\delta x. \tag{7.6}$$

The system of equations (7.6) is a system of homogeneous linear equations with periodic coefficients, the coefficients and their periods depending on the parameter τ.

According to our assumption that the solution (7.3) of the system of equations (7.2) is stable for any τ $(0 \leqslant \tau \leqslant L)$ it follows that the $n-1$ characteristic exponents of the system of the variational equations (7.6)

$$\lambda_1(\tau),\ \lambda_2(\tau),\ \ldots,\ \lambda_{n-1}(\tau) \tag{7.7}$$

have for any constant τ from the interval $0 \leqslant \tau \leqslant L$ negative real parts (these are quite large in their absolute magnitude, since the basic property of relaxation vibrational systems consists in a rapid approach to a stationary mode). The n-th characteristic exponent is, as known, equal to zero, as the derivative of the solution (7.3) with respect to φ, i.e., $\frac{\partial x^0(\tau, \omega t + \varphi)}{\partial \varphi}$, represents the solution of the system of variational equations (7.6), and is a periodic function of ψ with period 2π.

It is now assumed that a small external perturbation is acting from a certain moment of time, which we choose as the initial time $t=0$, on the given relaxation vibrational system. Until that moment the system was in the stationary state (7.3) ($\tau = \mathrm{const}$). This perturbation has the form

$$\varepsilon X^*(\tau,\ \theta,\ x,\ \varepsilon), \tag{7.8}$$

where ε is a small positive parameter, $\tau = \varepsilon t$, $\frac{d\theta}{dt} = \nu(\tau) > 0$ for any τ in the interval $0 \leqslant \tau \leqslant L$. The functions

$$X^*(\tau,\ \theta,\ x,\ \varepsilon)\ [X_1^*(\tau,\ \theta,\ x_1,\ \ldots,\ x_n,\ \varepsilon),\ \ldots,\ X_n^*(\tau,\ \theta,\ x_1,\ \ldots,\ x_n,\ \varepsilon)]$$

are periodic in θ with period 2π, and may be represented as Fourier sums

$$X^*(\tau,\ \theta,\ x,\ \varepsilon) = \sum_{k=-N}^{N} e^{ik\theta} X_{(k)}^*(\tau,\ x,\ \varepsilon), \tag{7.9}$$

where the coefficients $X_{(k)}^*(\tau,\ x,\ \varepsilon)$ are polynomials in x possessing the desired number of derivatives with respect to τ for any θ in some neighborhood of the curve

$$x = x^0(\tau,\ \omega t + \varphi) \qquad (\tau = \mathrm{const})$$

in an n-dimensional space.

The perturbed equation of relaxation vibrations takes the form

$$\frac{dx}{dt} = X(\tau,\ x) + \varepsilon X^*(\tau,\ \theta,\ x,\ \varepsilon) \qquad (\tau = \varepsilon t). \tag{7.10}$$

Since the formulas are valid at the moment $t=0$, the initial conditions for the relaxation vibrations are

$$x(0) = x^0(\tau,\ \varphi) \qquad (\tau = \mathrm{const}). \tag{7.11}$$

In order to construct an approximate one-parametric family of solutions of the system (7.10), close to the family of solutions (7.3), we transform the system (7.10) to new variables.

We shall first consider a simpler case when all the $n-1$ characteristic exponents (7.7) have negative real parts for any τ belonging to the interval $0 \leqslant \tau \leqslant L$, and when the following conditions are satisfied:

$$\left.\begin{aligned} &m_1\gamma_1(\tau) + m_2\gamma_2(\tau) + \ldots + m_{n-1}\gamma_{n-1}(\tau) \neq \gamma_q(\tau), \\ &m_1 + m_2 + \ldots + m_{n-1} \geqslant 2, \end{aligned}\right\} \tag{7.12}$$

where $\gamma_q(\tau)$ $(q = 1, 2, \ldots, n-1)$ are the real parts of the characteristic exponents (7.7), and $m_1, m_2, \ldots, m_{n-1}$ are non-negative integers.

We also assume that the functions

$$X(\tau, x) + \varepsilon X^*(\tau, \theta, x, \varepsilon) \qquad (\tau = \text{const}) \tag{7.13}$$

are analytic functions of x and ε in the neighborhood of the curve

$$x = x^0(\tau, \psi), \qquad \varepsilon = 0 \tag{7.14}$$

in an $(n+1)$-dimensional space.

On the grounds of A. Poincaré's well-known theorems, the general solution of the unperturbed equations (7.2) has the form

$$x = f\left(\tau, \omega t + \varphi, Ce^{\int_0^t \lambda(\tau)dt}\right), \tag{7.15}$$

in the neighborhood of the periodic solution (7.3), where φ, C $(C_2, C_3, \ldots, C_n)$ are integration constants, $\lambda(\tau)$ denotes the set $(\lambda_2(\tau), \lambda_3(\tau), \ldots, \lambda_n(\tau))$,

$$\left.\begin{aligned} f(\tau, \psi + 2\pi, h) &\equiv f(\tau, \psi, h), \\ f(\tau, \psi, 0) &\equiv x^0(\tau, \psi) \end{aligned}\right\} \tag{7.16}$$

while the $f(\tau, \psi, h)$ are analytic regular functions for sufficiently small values of h.

It is actually not difficult to derive explicit expressions for the functions $f(\tau, \psi, h)$. We denote by $v_k^q(\tau, \omega t + \varphi)$ $(q, k = 1, 2, \ldots, n)$ a basic system of solutions of the variational equations (7.6)

$$\left.\begin{aligned} v_k^{(1)}(\tau, \omega t + \varphi) &= \frac{\partial x_k^{(0)}}{\partial \varphi} = f_{k,1}(\tau, \omega t + \varphi), \\ v_k^{(q)}(\tau, \omega t + \varphi) &= f_{k,q}(\tau, \omega t + \varphi)\, e^{\int_0^t \lambda_q(\tau)dt} \\ & (k, q = 1, 2, \ldots, n), \end{aligned}\right\} \tag{7.17}$$

where $\tau = \text{const}$, $f_{k,q}(\tau, \psi + 2\pi) = f_{k,q}(\tau, \psi)$ $(k, q = 1, 2, \ldots, n)$.

We denote by $w_k^{(q)}(\tau, \omega t + \varphi)$ $(k, q = 1, 2, \ldots, n)$ a basic system of solutions of the system of equations conjugate with (7.6), i.e., the system

$$\frac{d\delta x}{dt} + \overline{X}'_x(\tau, x^0)\,\delta x = 0, \tag{7.18}$$

where $\overline{X}'_x(\tau, x^0)$ is the transposed matrix of the matrix $X'_x(\tau, x^0)$.

Then, obviously,

$$w_k^{(q)}(\tau, \omega t + \varphi) = (-1)^{k+q} \frac{M_{kq}[v_r^{(s)}(\tau, \omega t + \varphi)]}{\Delta[v_r^{(s)}(\tau, \omega t + \varphi)]}, \tag{7.19}$$

where $\Delta[v_r^{(s)}(\tau, \omega t + \varphi)]$ is the Wronskian of the system of solutions (7.17), and $M_{kq}[v_r^{(s)}(\tau, \omega t + \varphi)]$ the cofactor of the element $v_k^{(q)}(\tau, \omega t + \varphi)$.

Performing simple operations, we obtain

$$\left.\begin{aligned} w_k^{(1)}(\tau, \omega t + \varphi) &= f_{k,1}^*(\tau, \omega t + \varphi), \\ w_k^{(q)}(\tau, \omega t + \varphi) &= f_{k,q}^*(\tau, \omega t + \varphi)\, e^{-\int_0^t \lambda_q(\tau)dt} \\ & (k, q = 1, 2, \ldots, n), \end{aligned}\right\} \tag{7.20}$$

where $f_{k,q}^*(\tau, \omega t + \varphi)$ are the conjugate functions of $f_{k,q}(\tau, \omega t + \varphi)$, and $f_{k,q}^*(\tau, \psi + 2\pi) = f_{k,q}^*(\tau, \psi)$ $(k, q = 1, 2, \ldots, n)$.

We now look for the general solution of the unperturbed equations (7.2) in the neighborhood of the periodic solution (7.3), in the form of a power expansion of some parameter μ

$$x = x^{(0)} + \mu x^{(1)} + \mu^2 x^{(2)} + \ldots + \mu^m x^{(m)} + \ldots. \tag{7.21}$$

Substituting (7.21) into (7.2), expanding the right-hand side in powers of μ, and equating the coefficients of equal powers of μ, we obtain

$$\frac{dx^{(0)}}{dt} = X(\tau, x^0), \tag{7.22}$$

$$\frac{dx^{(1)}}{dt} - X'_x(\tau, x^0)\, x^{(1)} = 0, \tag{7.23}$$

$$\frac{dx^{(2)}}{dt} - X'_x(\tau, x^0)\, x^{(2)} = X''_{xx}(\tau, x^0)\, x^{(1)} x^{(1)}, \tag{7.24}$$

$$\ldots\ldots\ldots\ldots\ldots\ldots\ldots\ldots\ldots\ldots$$

From (7.22) we find the already known solution (7.3) depending on a single arbitrary constant φ (τ is everywhere regarded as a constant parameter)

$$x = x^0(\tau, \omega t + \varphi). \tag{7.25}$$

We can derive the solution of (7.23) depending on the $n-1$ arbitrary constants $C_2, C_3, \ldots, C_n$. With the notation of (7.17) we have

$$x_k^{(1)} = \sum_{q=2}^{n} C_q e^{\int\limits_0^t \lambda_q(\tau)dt} f_{kq}(\tau, \omega t + \varphi) \qquad (k = 1, 2, \ldots, n). \tag{7.26}$$

Having found $x^{(1)}$, we already know the right-hand side of (7.24) for $x^{(2)}$. We obtain

$$\frac{dx^{(2)}}{dt} - X'_x(\tau, x^0)\, x^{(2)} = f^{(2)}, \tag{7.27}$$

where $f^{(2)}$ is a vector function with the components $(f_1^{(2)}, f_2^{(2)}, \ldots, f_n^{(2)})$, and

$$\left.\begin{aligned} f_k^{(2)} &= \sum_{q=2}^{n} \sum_{q_1=2}^{n} C_q C_{q_1} e^{\int\limits_0^t [\lambda_q(\tau)+\lambda_{q_1}(\tau)]dt} F_{k,q,q_1}(\tau, \omega t + \varphi), \\ F_{k,q,q_1}(\tau, \psi + 2\pi) &= F_{k,q,q_1}(\tau, \psi) \qquad (k = 1, 2, \ldots, n), \end{aligned}\right\} \tag{7.28}$$

where we have used the notation

$$F_{k,q,q_1}(\tau, \omega t + \varphi) = \sum_{s=1}^{n} \sum_{r=1}^{n} X''_{kx_s x_r}(\tau, x_1^0, \ldots, x_n^0)\, f_{s,q}(\tau, \omega t + \varphi)\, f_{r,q}(\tau, \omega t + \varphi)$$
$$(k, q, q_1 = 1, 2, \ldots, n). \tag{7.29}$$

Deriving the particular solution corresponding to its right-hand side from (7.27), we have, with the notation of (7.17) and (7.20),

$$x_k^{(2)} = \sum_{i=1}^{n} \sum_{j=1}^{n} v_k^{(i)}(\tau, \omega t + \varphi) \int\limits_0^t f_j^{(2)} w_j^{(i)}(\tau, \omega t + \varphi)\, dt \qquad (k = 1, 2, \ldots, n), \tag{7.30}$$

or, using (7.28), (7.17), and (7.20), we finally obtain for $x_k^{(2)}$:

$$x_k^{(2)} = \sum_{q=2}^{n} \sum_{q_1=2}^{n} \sum_{i=1}^{n} \sum_{j=1}^{n} C_q C_{q_1} e^{\int\limits_0^t \lambda_i(\tau)dt} f_{k,i}(\tau, \omega t + \varphi) \times$$
$$\times \int\limits^t \{ e^{\int\limits_0^t [\lambda_q(\tau)+\lambda_{q_1}(\tau)-\lambda_i(\tau)]dt} f^*_{j,i}(\tau, \omega t + \varphi)\, F_{j,q,q_1}(\tau, \omega t + \varphi) \}\, dt \tag{7.31}$$
$$(k = 1, 2, \ldots, n)$$

or

$$x_k^{(2)} = \sum_{q=2}^{n} \sum_{q_1=2}^{n} C_q C_{q_1} e^{\int^t [\lambda_q(\tau)+\lambda_{q_1}(\tau)]dt} \Phi^{(2)}_{k,q,q_1}(\tau, \omega t + \varphi) \tag{7.32}$$

$$(k = 1, 2, \ldots, n),$$

where the functions $\Phi^{(2)}_{k,q,q_1}(\tau, \psi)$ are periodic in ψ with period 2π.

Continuing this procedure, we find

$$x_k^{(m)} = \sum_{q=2}^{n} \sum_{q_1=2}^{n} \ldots \sum_{q_{m-1}=2}^{n} C_q C_{q_1} \ldots C_{q_{m-1}} e^{\int^t [\lambda_q(\tau)+\lambda_{q_1}(\tau)+\ldots+\lambda_{q_{m-1}}(\tau)]dt} \Phi^{(m)}_{k,q,q_1,\ldots,q_{m-1}} \tag{7.33}$$

$$(k = 1, 2, \ldots, n),$$

where

$$\Phi^{(m)}_{k,q,q_1,\ldots,q_{m-1}}(\tau, \psi + 2\pi) = \Phi^{(m)}_{k,q,q_1,\ldots,q_{m-1}}(\tau, \psi).$$

Substituting the values obtained for x^0, $x^{(1)}$, $x^{(2)}$, ..., $x^{(m)}$, ... into the right-hand side of (7.21) and putting $\mu = 1$, we obtain the solution of the unperturbed equation (7.2) in the form

$$x_k = x_k^{(0)}(\tau, \omega t + \varphi) + \sum_{m=1}^{\infty} \Big\{ \sum_{q=2}^{n} \ldots \sum_{q_{m-1}=2}^{n} C_q \ldots C_{q_{m-1}} \times$$

$$\times e^{\int^t [\lambda_q(\tau)+\ldots+\lambda_{q_{m-1}}(\tau)]dt} \Phi^{(m)}_{k,q,\ldots,q_{m-1}}(\tau, \omega t + \varphi) \Big\} \quad (k = 1, 2, \ldots, n), \tag{7.34}$$

where

$$q_0 = q, \quad \Phi^{(1)}_{k,q}(\tau, \omega t + \varphi) = f_{k,q}(\tau, \omega t + \varphi).$$

The series (7.34) may obviously be written in the form (7.15), and it converges quickly when condition (7.12) is satisfied.

After finding the general solution of the system (7.2), i.e., the functions (7.15), where the right-hand sides have the form (7.34), we introduce in equations (7.10) the new variables

$$\psi, \; h(h_1, h_2, \ldots, h_{n-1}) \tag{7.35}$$

defined by the formulas

$$x = f(\tau, \psi, h), \tag{7.36}$$

where $\tau = \varepsilon t$.

Substituting (7.36) into equations (7.10), we obtain

$$\frac{\partial f}{\partial \psi} \frac{d\psi}{dt} + \frac{\partial f}{\partial h} \frac{dh}{dt} + \varepsilon \frac{\partial f}{\partial \tau} = X[\tau, f(\tau, \psi, h)] + \varepsilon X^*[\tau, \theta, f(\tau, \psi, h), \varepsilon]. \tag{7.37}$$

Next, substituting (7.17) into equation (7.2), we obtain the identities

$$\frac{\partial f}{\partial \psi} \omega(\tau) + \frac{\partial f}{\partial h} \lambda(\tau) h = X[\tau, f(\tau, \psi, h)]. \tag{7.38}$$

Subtracting (7.38) from (7.37), we find

$$\frac{\partial f}{\partial \psi} \Big[\frac{d\psi}{dt} - \omega(\tau) \Big] + \frac{\partial f}{\partial h} \Big[\frac{dh}{dt} - \lambda(\tau) h \Big] =$$

$$= \varepsilon \Big\{ X^*[\tau, \theta, f(\tau, \psi, h), \varepsilon] - \frac{\partial f}{\partial \tau} \Big\}. \tag{7.39}$$

Solving the system (7.39) for the variables

$$\frac{d\psi}{dt} - \omega(\tau), \quad \frac{dh}{dt} - \lambda(\tau) h, \tag{7.40}$$

which can always be done, since the determinant of the system (7.39)

$$\mathrm{Det} \left\| \frac{\partial f}{\partial \psi} \ \frac{\partial f}{\partial h} \right\| \tag{7.41}$$

is, by construction of the functions $f(\tau, \psi, h)$, different from zero for $h = 0$, we obtain the following system of equations in the variables (7.35), equivalent to equations (7.10):

$$\left.\begin{aligned} \frac{d\psi}{dt} &= \omega(\tau) + \varepsilon P(\tau, \theta, \psi, h, \varepsilon), \\ \frac{dh}{dt} &= \lambda(\tau) h + \varepsilon R(\tau, \theta, \psi, h, \varepsilon). \end{aligned}\right\} \tag{7.42}$$

Here the function $P(\tau, \theta, \psi, h, \varepsilon)$ and the $(n-1)$-dimensional vector function $R(\tau, \theta, \psi, h, \varepsilon)$ are, on the grounds of the conditions mentioned above, analytic and regular for sufficiently small h and ε, periodic in the angular variables θ and ψ with period 2π, and have the desired number of derivatives with respect to τ for any τ in the interval $0 \leqslant \tau \leqslant L$.

Let us now consider a more complicated case, when all the $n-1$ characteristic exponents (7.7) of equations (7.6) have negative real parts for any τ belonging to the interval $0 \leqslant \tau \leqslant L$, but condition (7.12) is not satisfied.

In that case, on the grounds of well-known results of Floquet-Lyapunov on the properties of linear differential equations with periodic coefficients, it is obvious that the transformation

$$\delta x = \frac{\partial x^0}{\partial \psi} \delta u_0 + A(\tau, \psi) \delta u, \tag{7.43}$$

where $A(\tau, \psi)$ is a matrix of n rows and $n-1$ columns, the elements of which are periodic functions of ψ with period 2π and have continuous derivatives with respect to ψ and the parameter τ, and δu is a vector with components $\delta u_1, \delta u_2, \ldots, \delta u_{n-1}$, may be used to reduce the equations with periodic coefficients (7.6) to the following system of differential equations with constant coefficients (depending on the parameter τ):

$$\frac{d\delta u_0}{dt} = 0, \quad \frac{d\delta u}{dt} = H(\tau) \delta u. \tag{7.44}$$

The system (7.44) is such that the roots of the equation

$$\mathrm{Det} \left| I_{n-1} p - H(\tau) \right| = 0, \tag{7.45}$$

where $H(\tau)$ is a square matrix of the $n-1$-th order with elements depending on the parameter τ, are the characteristic exponents (7.7) of the equations with periodic coefficients (7.6).

The matrices $A(\tau, \psi)$, $H(\tau)$, and the vector δu are generally complex quantities.

Substituting (7.43) into equations (7.6), and taking into account that $\frac{\partial x^0}{\partial \psi} \delta u_0$, where δu_0 is a constant, is a solution of the system (7.6), we obtain identically

$$\frac{\partial A(\tau, \psi)}{\partial \psi} \omega(\tau) + A(\tau, \psi) H(\tau) = X'_x(\tau, x^0) A(\tau, \psi), \tag{7.46}$$

and also

$$\frac{\partial \bar{A}(\tau, \psi)}{\partial \psi} \omega(\tau) + \bar{A}(\tau, \psi) \bar{H}(\tau) = X'_x(\tau, x^0) \bar{A}(\tau, \psi), \tag{7.47}$$

where $\bar{A}(\tau, \psi)$ and $\bar{H}(\tau)$ are complex conjugate to $A(\tau, \psi)$ and $H(\tau)$.

After these preliminary remarks we turn to the transformation of the system (7.10).

We introduce into this system new variables ψ, $h(h_1, h_2, \ldots, h_{n-1})$ defined by the formulas

$$x = x^0(\tau, \psi) + \frac{1}{2}\{A(\tau, \psi) h + \bar{A}(\tau, \psi) \bar{h}\}, \tag{7.48}$$

where $\bar{h}$ is the complex conjugate of h, and $\tau = \varepsilon t$.

Substituting (7.48) into (7.10), we obtain

$$\begin{gathered}
\frac{\partial x^0}{\partial \psi}\frac{d\psi}{dt} + \frac{1}{2}\left\{\frac{\partial A(\tau, \psi)}{\partial \psi} h + \frac{\partial \bar{A}(\tau, \psi)}{\partial \psi} \bar{h}\right\}\frac{d\psi}{dt} + \\
+ \frac{1}{2}\left\{A(\tau, \psi)\frac{dh}{dt} + \bar{A}(\tau, \psi)\frac{d\bar{h}}{dt}\right\} = \\
= X\left(\tau, x^0 + \frac{1}{2}(Ah + \bar{A}\bar{h})\right) + \varepsilon X^*\left(\tau, \theta, x^0 + \frac{1}{2}(Ah + \bar{A}\bar{h}), \varepsilon\right) - \\
- \varepsilon\left\{\frac{\partial x^0}{\partial \tau} + \frac{1}{2}\left[\frac{\partial A(\tau, \psi)}{\partial \tau} h + \frac{\partial \bar{A}(\tau, \psi)}{\partial \tau} \bar{h}\right]\right\}.
\end{gathered} \tag{7.49}$$

The right-hand side of this system may obviously be expressed in the form

$$\begin{gathered}
X\left(\tau, x^0 + \frac{1}{2}(Ah + \bar{A}\bar{h})\right) + \varepsilon X^*\left(\tau, \theta, x^0 + \frac{1}{2}(Ah + \bar{A}\bar{h}), \varepsilon\right) - \\
- \varepsilon\left\{\frac{\partial x^0}{\partial \tau} + \frac{1}{2}[A(\tau, \psi) h + \bar{A}(\tau, \psi)\bar{h}]\right\} = \\
= X(\tau, x^0) - X(\tau, x^0) + \frac{1}{2} X'_x(\tau, x^0)[A(\tau, \psi) h + \bar{A}(\tau, \psi)\bar{h}] - \\
- \frac{1}{2} X'_x(\tau, x^0)[A(\tau, \psi) h + \bar{A}(\tau, \psi)\bar{h}] + X\left(\tau, x^0 + \frac{1}{2}(Ah + \bar{A}\bar{h})\right) + \\
+ \varepsilon X^*\left(\tau, \theta, x^0 + \frac{1}{2}(Ah + \bar{A}\bar{h}), \varepsilon\right) - \\
- \varepsilon\left\{\frac{\partial x^0}{\partial \tau} + \frac{1}{2}\frac{d}{d\tau}[A(\tau, \psi) h + \bar{A}(\tau, \psi)\bar{h}]\right\} = X(\tau, x^0) + \\
+ \frac{1}{2} X'_x(\tau, x^0)[A(\tau, \psi) h + \bar{A}(\tau, \psi)\bar{h}] + Y(\tau, \theta, \psi, h, \varepsilon),
\end{gathered} \tag{7.50}$$

where we have used the notation*

$$\begin{aligned}
Y(\tau, \theta, \psi, h, \varepsilon) = &- X(\tau, x^0) - \frac{1}{2} X'_x(\tau, x^0)[A(\tau, \psi) h + \\
&+ \bar{A}(\tau, \psi)\bar{h}] + X\left(\tau, x^0 + \frac{1}{2}(Ah + \bar{A}\bar{h})\right) + \\
&+ \varepsilon X^*\left(\tau, \theta, x^0 + \frac{1}{2}(Ah + \bar{A}\bar{h}), \varepsilon\right) - \\
&- \varepsilon\left\{\frac{\partial x^0}{\partial \tau} + \frac{1}{2}\frac{d}{d\tau}[A(\tau, \psi) h + \bar{A}(\tau, \psi)\bar{h}]\right\}.
\end{aligned} \tag{7.51}$$

Evidently, when $h = 0$

$$Y(\tau, \theta, \psi, h, \varepsilon)|_{h=0} = \varepsilon X^*(\tau, \theta, x^0, \varepsilon) - \varepsilon\frac{\partial x^0}{\partial \tau} \to 0 \text{ for } \varepsilon \to 0. \tag{7.52}$$

On the grounds of (7.2) and (7.3) we have identically, for any constant τ

* For simplicity, we do not indicate here or in the sequel the dependence of the functions on $\bar{h}$.

in the interval $0 \leqslant \tau \leqslant L$,

$$\frac{\partial x^0}{\partial \psi} \omega(\tau) = X(\tau, x^0). \tag{7.53}$$

Therefore, subtracting (7.53) from (7.49), and taking into account (7.50) and the identities (7.46) and (7.47), we obtain the system

$$\left\{\frac{\partial x^0}{\partial \psi} + \frac{1}{2}\left[\frac{\partial A(\tau, \psi)}{\partial \psi} h + \frac{\partial \bar{A}(\tau, \psi)}{\partial \psi} \bar{h}\right]\right\}\left(\frac{d\psi}{dt} - \omega(\tau)\right) + \\ + \frac{1}{2} A(\tau, \psi)\left[\frac{dh}{dt} - H(\tau) h\right] + \frac{1}{2}\bar{A}(\tau, \psi)\left[\frac{d\bar{h}}{dt} - \bar{H}(\tau)\bar{h}\right] = \\ = Y(\tau, \theta, \psi, h, \varepsilon), \tag{7.54}$$

where the real n-dimensional vector function $Y(\tau, \theta, \psi, h, \varepsilon)$ is periodic in θ and ψ with period 2π, has the desired number of derivatives with respect to τ for any τ in the interval $0 \leqslant \tau \leqslant L$, is analytic and regular in h and ε for sufficiently small h and ε, and tends to zero as $h \to 0$ and $\varepsilon \to 0$.

As in /9/, we determine from the system (7.54) the variables

$$\frac{d\psi}{dt} - \omega(\tau), \quad \frac{dh}{dt} - H(\tau) h = R \tag{7.55}$$

so as to satisfy the condition

$$R = \bar{R} \tag{7.56}$$

and, consequently, the condition

$$\frac{d\bar{h}}{dt} - \bar{H}(\tau)\bar{h} = R. \tag{7.57}$$

Substituting the expressions (7.55) into equations (7.54), and taking (7.56) into account, we obtain

$$\left\{\frac{\partial x^0}{\partial \psi} + \frac{1}{2}\left[\frac{\partial A(\tau, \psi)}{\partial \psi} h + \frac{\partial \bar{A}(\tau, \psi)}{\partial \psi} \bar{h}\right]\right\}\left(\frac{d\psi}{dt} - \omega(\tau)\right) + \\ + \frac{1}{2}[A(\tau, \psi) + \bar{A}(\tau, \psi)] R = Y(\tau, \theta, \psi, h, \varepsilon). \tag{7.58}$$

We assume that the determinant of this system

$$\operatorname{Det}(\tau, \psi, h) = \left|\frac{\partial x^0}{\partial \psi} + \frac{1}{2}\left[\frac{\partial A(\tau, \psi)}{\partial \psi} h + \frac{\partial \bar{A}(\tau, \psi)}{\partial \psi} \bar{h}\right], \ \frac{1}{2}[A(\tau, \psi) + \bar{A}(\tau, \psi)]\right| \tag{7.59}$$

does not vanish for $h = 0$ and any τ in the interval $0 \leqslant \tau \leqslant L$. Then, by virtue of continuity, it does not vanish also in some δ-neighborhood of the point $h = 0$. We can always find such a small δ that x, defined by formula (7.48), would lie in D_ρ — a ρ-neighborhood of the curve (7.14).

Solving the system (7.58) for the variables (7.55) in the domain

$$\psi, \theta \in \Omega, \quad h \in D_\delta, \quad 0 \leqslant \varepsilon \leqslant \varepsilon_0, \tag{7.60}$$

where Ω is a circle, we find the following system of equations, equivalent to the system (7.10):

$$\left.\begin{aligned} \frac{d\psi}{dt} &= \omega(\tau) + P_1(\tau, \theta, \psi, h, \varepsilon), \\ \frac{dh}{dt} &= H(\tau) + R_1(\tau, \theta, \psi, h, \varepsilon). \end{aligned}\right\} \tag{7.61}$$

Here the function $P_1(\tau, \theta, \psi, h, \varepsilon)$ and the $(n-1)$-dimensional vector function $R_1(\tau, \theta, \psi, h, \varepsilon)$ possess the same properties as the corresponding terms in the right-hand sides of the system of equations (7.42).

Having reduced the system of equations (7.10) to equations of the type (7.42) or (7.61), we proceed to construct an algorithm for deriving an approximate solution.

We apply the method developed in /81/, according to which we eliminate the variables h from the system (7.42), thus reducing the solution of (7.42) to the solution of a single equation. To do this we derive from the system of equations

$$\frac{dh}{dt}=\lambda(\tau)h+\varepsilon R(\tau,\theta,\psi,h,\varepsilon) \tag{7.62}$$

the particular solutions h, corresponding to the perturbing forces $\varepsilon R(\tau,\theta,\psi,h,\varepsilon)$, periodic in θ and ψ with period 2π, and depending on the parameter τ (the solutions corresponding to the homogeneous system — the system (7.62) with $\varepsilon=0$, attenuate rapidly since the exponents $\lambda(\tau)$ have quite large negative real parts).

We are looking for h in the form of the series

$$h(\tau,\theta,\psi,\varepsilon)=\varepsilon h_1(\tau,\theta,\psi)+\varepsilon^2 h_2(\tau,\theta,\psi)+\ldots \tag{7.63}$$

Substituting the expression for h given by (7.63) into equation (7.62) (as usual, we put $\tau=\varepsilon t$ in the series (7.63) during differentiation and regard τ as a constant parameter during integration), and equating the coefficients of equal powers of ε, we obtain the equations

$$\frac{\partial h_1}{\partial\theta}\nu(\tau)+\frac{\partial h_1}{\partial\psi}\omega(\tau)=\lambda(\tau)h_1+R(\tau,\theta,\psi,0,0), \tag{7.64}$$

$$\begin{aligned}\frac{\partial h_2}{\partial\theta}\nu(\tau)+\frac{\partial h_2}{\partial\psi}\omega(\tau)&=\lambda(\tau)h_2+R'_h(\tau,\theta,\psi,0,0)h_1+\\&+R'_\varepsilon(\tau,\theta,\psi,0,0)-\frac{\partial h_1}{\partial\psi}P(\tau,\theta,\psi,0,0)-\frac{\partial h_1}{\partial\tau},\end{aligned} \tag{7.65}$$

. . . .

We derive from equation (7.64), with an accuracy of the order of ε, the expression

$$h_1(\tau,\theta,\psi)=\sum_{n,m}\frac{r_{n,m}(\tau)e^{i(n\theta+m\psi)}}{i[n\nu(\tau)+m\omega(\tau)]-\lambda(\tau)}, \tag{7.66}$$

where

$$r_{n,m}(\tau)=\frac{1}{4\pi^2}\int_0^{2\pi}\int_0^{2\pi}R(\tau,\theta,\psi,0,0)e^{-i(n\theta+m\psi)}d\theta\,d\psi. \tag{7.67}$$

The denominator on the right-hand side of the expressions (7.66) cannot vanish for any values of n and m, as the $\lambda(\tau)$ have negative real parts for any τ in the interval $0\leqslant\tau\leqslant L$.

Having found $h_1(\tau,\theta,\psi)$, the right-hand side of equations (7.65) becomes known, and we can derive an expression for $h_2(\tau,\theta,\psi)$.

We obtain with the same degree of accuracy

$$h_2(\tau,\theta,\psi)=\sum_{n,m}\frac{s_{n,m}(\tau)e^{i(n\theta+m\psi)}}{i[n\nu(\tau)+m\omega(\tau)]-\lambda(\tau)}, \tag{7.68}$$

where

$$\begin{aligned}s_{n,m}(\tau)=\frac{1}{4\pi^2}\int_0^{2\pi}\int_0^{2\pi}\Big\{&R'_h(\tau,\theta,\psi,0,0)h_1(\tau,\theta,\psi)+\\&+R'_\varepsilon(\tau,\theta,\psi,0,0)-\frac{\partial h_1}{\partial\psi}P(\tau,\theta,\psi,0,0)-\frac{\partial h_1}{\partial\tau}\Big\}e^{-i(n\theta+m\psi)}\,d\theta\,d\psi.\end{aligned} \tag{7.69}$$

Continuing this procedure, we derive $h_3(\tau, \theta, \psi), \ldots, h_m(\tau, \theta, \psi), \ldots$, and substitute the value of $h(\tau, \theta, \psi, \varepsilon)$ given by (7.63) into the right-hand side of the first equation of the system (7.42). We obtain

$$\frac{d\psi}{dt} = \omega(\tau) + \varepsilon P[\tau, \theta, \psi, h(\tau, \theta, \psi, \varepsilon), \varepsilon], \tag{7.70}$$

where $\tau = \varepsilon t$ and $\frac{d\theta}{dt} = \nu(\tau)$.

Deriving the value of ψ from equation (7.70) and inserting it into the right-hand side of (7.15), we obtain a family of particular solutions of the system of equations (7.10), depending on a single arbitrary constant, and lying close (by orbit) to the periodic solution (7.3), in the form

$$x = f[\tau, \theta, \psi(\tau, t), h(\tau, \theta, \psi(\tau, t), \varepsilon)]. \tag{7.71}$$

In the case when the system of equations (7.10) reduces to equations of the type (7.61), we find by an analogous procedure the following expression for $h_1(\tau, \theta, \psi)$:

$$h_1(\tau, \theta, \psi) = \sum_{n, m} \frac{r_{1n, m}(\tau)\, e^{i(n\theta + m\psi)}}{\operatorname{Det} | Ii[n\nu(\tau) + m\omega(\tau)] - H(\tau)|}, \tag{7.72}$$

where

$$r_{1n, m}(\tau) = \frac{1}{4\pi^2} \int_0^{2\pi}\int_0^{2\pi} R_1(\tau, \theta, \psi, 0, 0)\, e^{-i(n\theta + m\psi)}\, d\theta\, d\psi, \tag{7.73}$$

and I is the $(n-1)$-dimensional unit matrix.

Deriving $h_2(\tau, \theta, \psi), \ldots, h_m(\tau, \theta, \psi), \ldots$ and substituting $h(\tau, \theta, \omega)$ given by (7.63) into the right-hand side of the first equation of the system (7.61), we obtain the equation

$$\frac{d\psi}{dt} = \omega(\tau) + P_1[\tau, \theta, \psi, h(\tau, \theta, \psi, \varepsilon), \varepsilon]. \tag{7.74}$$

Solving this equation, we obtain $\psi = \psi(\tau, t, \varepsilon)$.

Then the family of solutions of the system of equations (7.10), depending on a single arbitrary parameter and lying close (by orbit) to the periodic solution (7.3), has the form

$$x = x^0(\tau, \psi(\tau, t, \varepsilon)) + \frac{1}{2}\{A(\tau, \psi(\tau, t, \varepsilon))\, h(\tau, \theta, \psi(\tau, t, \varepsilon), \varepsilon) + \\ + \bar{A}(\tau, \psi(\tau, t, \varepsilon))\, \bar{h}(\tau, \theta, \psi(\tau, t, \varepsilon), \varepsilon)\}, \tag{7.75}$$

in the case when condition (7.12) is not satisfied, where $\bar{h}(\tau, \theta, \psi, \varepsilon)$ is defined by the formula

$$\bar{h}(\tau, \theta, \psi, \varepsilon) = \varepsilon \bar{h}_1(\tau, \theta, \psi) + \varepsilon^2 \bar{h}_2(\tau, \theta, \psi) + \ldots, \tag{7.76}$$

where $\bar{h}_1(\tau, \theta, \psi), \bar{h}_2(\tau, \theta, \psi), \ldots$ are the conjugate functions of $h_1(\tau, \theta, \psi)$, $h_2(\tau, \theta, \psi), \ldots$ and may be expressed by formulas of the type

$$\bar{h}_1(\tau, \theta, \psi) = \sum_{n, m} \frac{r_{1n, m}(\tau)\, e^{i(n\theta + m\psi)}}{| Ii[n\nu(\tau) + m\omega(\tau)] - \bar{H}(\tau)|}. \tag{7.77}$$

§ 2. Asymptotic representations for two-parametric families of solutions

We now consider some results obtained by O. B. Lykova /82/ for the case when the unperturbed system (7.2) admits a stable periodic family of solutions depending on two arbitrary constants.

Let us again consider the system of equations (7.10),

$$\frac{dx}{dt} = X(\tau, x) + \varepsilon X^*(\tau, \theta, x, \varepsilon), \tag{7.78}$$

where, as in the preceding section, x is a vector in the n-dimensional Euclidean space E_n; $X(\tau, x)$, $X^*(\tau, \theta, x, \varepsilon)$ are n-dimensional vector functions; $\tau = \varepsilon t$ is the "slowing" time; ε is a small positive parameter; $\frac{d\theta}{dt} = \nu(\tau)$.

We assume that for the system of the unperturbed equations

$$\frac{dx}{dt} = X(\tau, x) \qquad (\tau = \text{const}), \tag{7.79}$$

a family of stable periodic solutions

$$x = x^0(\tau, \omega t + \varphi, a), \tag{7.80}$$

depending on two arbitrary constants φ and a, and on τ as a parameter is known for any τ in the interval $0 \leqslant \tau \leqslant L$. Generally, ω is a function of a and depends on τ as a parameter

$$\omega = \omega(\tau, a). \tag{7.81}$$

We construct the variational equations for the system (7.79) corresponding to the family of periodic solutions (7.80) (τ is, evidently, regarded as a constant parameter). We have

$$\frac{d\delta x}{dt} = X'_x[\tau, x^0(\tau, \omega t + \varphi, a)]\,\delta x. \tag{7.82}$$

The system of equations (7.82) is a system of linear homogeneous differential equations with periodic coefficients. These coefficients and also their periods depend in the general case on the parameter τ and the arbitrary constant a.

It follows from our assumption that the family of solutions (7.80) is stable for any τ in the interval $0 \leqslant \tau \leqslant L$, that the $n - 2$ characteristic exponents of the system of variational equations (7.82)

$$\lambda_3(\tau, a),\ \lambda_4(\tau, a),\ \ldots,\ \lambda_n(\tau, a) \tag{7.83}$$

have for any τ ($0 \leqslant \tau \leqslant L$) and a, negative real parts (these exponents obviously depend on the parameter τ and on the arbitrary constant a), and that two of the characteristic exponents of the variational equations (7.82) vanish by the assumption that the family of solutions of the system (7.79) depends on two arbitrary constants.

We assume that the functions

$$X(\tau, x) + \varepsilon X^*(\tau, \theta, x, \varepsilon) \tag{7.84}$$

have the desired number of derivatives with respect to τ, x, ε in the domain

$$0 \leqslant \tau \leqslant L,\ \theta \in \Omega,\ \ x \in U_\rho,\ \ 0 \leqslant \varepsilon \leqslant \varepsilon_0, \tag{7.85}$$

where U_ρ is a ϱ-neighborhood of the family of the closed orbits

$$x = x^0(\tau, \omega t + \varphi, a)$$

in the n-dimensional Euclidean space E_n, that they are periodic in θ with period 2π, and that the condition (7.9) is satisfied.

Under these conditions let us first transform the system (7.78).

The system of equations (7.82) may, as above, be reduced with the help of a transformation of the type

$$\delta x = \frac{\partial x^0}{\partial \psi}\delta u_1 + \frac{\partial x^0}{\partial a}\delta u_2 + A(\tau, \psi, a)\,\delta u, \tag{7.86}$$

where $\frac{\partial x^0}{\partial \psi}\delta u_1$, $\frac{\partial x^0}{\partial a}\delta u_2$ are solutions of the variational equations, to a system of differential equations with constant coefficients (depending on the parameter τ and the arbitrary constant a)

$$\frac{d\delta u_1}{dt} = \omega_a'(\tau, a)\,\delta u_2, \quad \frac{d\delta u_2}{dt} = 0, \quad \frac{d\delta u}{dt} = H(\tau, a)\,\delta u. \tag{7.87}$$

This system is such that the roots of the equation

$$\operatorname{Det}|I_{n-2}\,p - H(\tau, a)| = 0, \tag{7.88}$$

where $H(\tau, a)$ is a square matrix of the $(n-2)$-th order, the elements of which depend on τ and a, are the characteristic exponents of the system of variational equations (7.82).

The matrix $A(\tau, \psi, a)$ consists of n rows and $n-2$ columns, its elements being periodic functions of ψ with period 2π. The matrix $A(\tau, \psi, a)$ and also the matrix $H(\tau, a)$ are, generally speaking, complex quantities; δu is a vector with components $\delta u_3, \delta u_4, \ldots, \delta u_n$.

We obviously have the following identity:

$$\frac{\partial A(\tau, \psi, a)}{\partial \psi}\,\omega(\tau, a) + A(\tau, \psi, a)\,H(\tau, a) = X_x'(x^0)\,A(\tau, \psi, a), \tag{7.89}$$

from which it follows that

$$\frac{\partial \overline{A}(\tau, \psi, a)}{\partial \psi}\,\omega(\tau, a) + \overline{A}(\tau, \psi, a)\,\overline{H}(\tau, a) = X_x'(x^0)\,\overline{A}(\tau, \psi, a), \tag{7.90}$$

where $\overline{A}(\tau, \psi, a)$ and $\overline{H}(\tau, a)$ are the complex conjugates of $A(\tau, \psi, a)$ and $H(\tau, a)$.

We shall now introduce into the equations (7.78) the new variables ψ, a, and $h\,(h_3, h_4, \ldots, h_n)$ by means of the transformation

$$x = x^0(\tau, \psi, a) + \frac{1}{2}\left(A(\tau, \psi, a)\,h + \overline{A}(\tau, \psi, a)\,\overline{h}\right) \qquad (\tau = \varepsilon t). \tag{7.91}$$

Substituting (7.91) into (7.78), we obtain

$$\begin{gathered}\frac{\partial x^0}{\partial \psi}\frac{d\psi}{dt} + \frac{\partial x^0}{\partial a}\frac{da}{dt} + \frac{1}{2}\left(\frac{\partial A}{\partial \psi}h + \frac{\partial \overline{A}}{\partial \psi}\overline{h}\right)\frac{d\psi}{dt} + \frac{1}{2}\left(\frac{\partial A}{\partial a}h + \frac{\partial \overline{A}}{\partial a}\overline{h}\right)\frac{da}{dt} + \\ + \frac{1}{2}\left(A\frac{dh}{dt} + \overline{A}\frac{d\overline{h}}{dt}\right) = X\left(\tau, x^0 + \frac{1}{2}(Ah + \overline{A}\overline{h})\right) + \\ + \varepsilon X^*\left(\tau, \theta, x^0 + \frac{1}{2}(Ah + \overline{A}\overline{h}), \varepsilon\right) - \varepsilon\left[\frac{\partial x^0}{\partial \tau} + \frac{1}{2}\frac{d}{d\tau}(Ah + \overline{A}\overline{h})\right].\end{gathered} \tag{7.92}$$

The right-hand side of the system just obtained may be expressed in the form

$$\begin{gathered}X\left(\tau, x^0 + \frac{1}{2}(Ah + \overline{A}\overline{h})\right) + \varepsilon X^*\left(\tau, \theta, x^0 + \frac{1}{2}(Ah + \overline{A}\overline{h}), \varepsilon\right) - \\ - \varepsilon\left[\frac{\partial x^0}{\partial \tau} + \frac{1}{2}\frac{d}{d\tau}(Ah + \overline{A}\overline{h})\right] = X(\tau, x^0) - X(\tau, x^0) + \\ + \frac{1}{2}X_x'(\tau, x^0)(Ah + \overline{A}\overline{h}) - \frac{1}{2}X_x'(\tau, x^0)(Ah + \overline{A}\overline{h}) + \\ + X\left(\tau, x^0 + \frac{1}{2}(Ah + \overline{A}\overline{h})\right) + \varepsilon X^*\left(\tau, \theta, x^0 + \frac{1}{2}(Ah + \overline{A}\overline{h}), \varepsilon\right) - \\ - \varepsilon\left[\frac{\partial x^0}{\partial \tau} + \frac{1}{2}\frac{d}{d\tau}(Ah + \overline{A}\overline{h})\right] = X(\tau, x^0) + \\ + \frac{1}{2}X_x'(\tau, x^0)(Ah + \overline{A}\overline{h}) + Y(\tau, \theta, \psi, a, h, \varepsilon),\end{gathered} \tag{7.93}$$

where

$$Y(\tau, \theta, \psi, a, h, \varepsilon) = -X(\tau, x^0) - \frac{1}{2} X'_x(\tau, x^0)(Ah + \overline{A}\overline{h}) +$$
$$+ X\left(\tau, x^0 + \frac{1}{2}(Ah + \overline{A}\overline{h})\right) + \varepsilon X^*\left(\tau, \theta, x^0 + \frac{1}{2}(Ah + \overline{A}\overline{h}), \varepsilon\right) -$$
$$-\varepsilon\left[\frac{\partial x^0}{\partial \tau} + \frac{1}{2}\frac{d}{d\tau}(Ah + \overline{A}\overline{h})\right]. \qquad (7.94)$$

The right-hand sides of the system (7.93) are real, as the equations are considered in the real domain and the substitution (7.91) is real.

Evidently, when $h=0$

$$Y(\tau, \theta, \psi, a, h, \varepsilon) = \varepsilon X^*(\tau, \theta, x^0, \varepsilon) - \varepsilon \frac{\partial x^0}{\partial \tau} \to 0 \text{ for } \varepsilon \to 0. \qquad (7.95)$$

Since on the grounds of (7.2)

$$\frac{\partial x^0}{\partial \psi} \omega(\tau, a) = X(\tau, x^0)$$

is valid for any constant τ belonging to the interval $0 \leqslant \tau \leqslant L$, the system of equations (7.92) may, by virtue of the identities (7.89) and (7.90), be written in the form

$$\left[\frac{\partial x^0}{\partial \psi} + \frac{1}{2}\left(\frac{\partial A}{\partial \psi} h + \frac{\partial \overline{A}}{\partial \psi}\overline{h}\right)\right]\left(\frac{d\psi}{dt} - \omega(\tau, a)\right) + \left[\frac{\partial x^0}{\partial a} + \frac{1}{2}\left(\frac{\partial A}{\partial a} h + \frac{\partial \overline{A}}{\partial a}\overline{h}\right)\right]\frac{da}{dt} +$$
$$+ \frac{1}{2} A\left(\frac{dh}{dt} - H(\tau, a)h\right) + \frac{1}{2}\overline{A}\left(\frac{d\overline{h}}{dt} - \overline{H}(\tau, a)\overline{h}\right) = Y(\tau, \theta, \psi, a, h, \varepsilon), \qquad (7.96)$$

where the real functions $Y(\tau, \theta, \psi, a, h, \varepsilon)$ are defined in the domain

$$0 \leqslant \tau \leqslant L, \quad \theta \in \Omega, \quad \psi \in \Omega, \quad h \in U_\delta, \quad 0 \leqslant \varepsilon \leqslant \varepsilon_0, \qquad (7.97)$$

have the desired number of derivatives with respect to τ, θ, ψ, a, h, ε, are periodic in θ and ψ with period 2π, and are quantities of the order of ε when $h = 0$

As in the preceding section, we derive from the system (7.96)

$$\frac{da}{dt}, \quad \frac{d\psi}{dt} - \omega(\tau, a), \quad \frac{dh}{dt} - H(\tau, a)\mathbf{h} = R \qquad (7.98)$$

so that the condition

$$R = \overline{R} \qquad (7.99)$$

be satisfied.

The system of equations (7.96) may on account of (7.98) and (7.99) be written in the form

$$\left[\frac{\partial x^0}{\partial \psi} + \frac{1}{2}\left(\frac{\partial A}{\partial \psi} h + \frac{\partial \overline{A}}{\partial \psi}\overline{h}\right)\right]\left(\frac{d\psi}{dt} - \omega(\tau)\right) + \left[\frac{\partial x^0}{\partial a} + \frac{1}{2}\left(\frac{\partial A}{\partial a} h + \frac{\partial \overline{A}}{\partial a}\overline{h}\right)\right]\frac{da}{dt} +$$
$$+ \frac{1}{2}[A + \overline{A}]R = Y(\tau, \theta, \psi, a, h, \varepsilon). \qquad (7.100)$$

We suppose that

$$\Delta(\tau, a) \neq 0 \quad \text{for} \quad a \in \mathfrak{A} \qquad (\mathfrak{A} = [a_0, a_1]), \qquad (7.101)$$

where

$$\Delta(\tau, a) = \min_{\psi} |\operatorname{Det}(\tau, \psi, a, h)|\Big|_{h=0} \qquad (\tau \in [0, L]).$$

$\operatorname{Det}(\tau, \psi, a, h)$ is the determinant of the system (7.100). It is then possible to find some nonvanishing positive constant which would be smaller than the modulus of this determinant

$$|\operatorname{Det}(\tau, \psi, a, 0)| \geqslant \alpha,$$

and, consequently, for all ψ and τ $(0 \leqslant \tau \leqslant L)$:

$$\left|\frac{1}{\operatorname{Det}(\tau, \psi, a, 0)}\right| \leqslant A, \text{ where } A = \frac{1}{\alpha}.$$

Solving the system (7.100) in the domain

$$0 \leqslant \tau \leqslant L, \quad \psi \in \Omega, \quad \theta \in \Omega, \quad h \in U_\delta, \quad a \in \mathfrak{A}, \quad 0 \leqslant \varepsilon \leqslant \varepsilon_0, \tag{7.102}$$

where U_δ is a δ-neighborhood of the point $h = 0$ within which the determinant $\operatorname{Det}(\tau, \psi, a, h)$ of the system (7.100) does not vanish in view of the continuity and condition (7.101), we obtain for the quantities (7.98)

$$\left.\begin{aligned}
\frac{d\psi}{dt} &= \omega(\tau, a) + P(\tau, \theta, \psi, a, h, \varepsilon), \\
\frac{da}{dt} &= Q(\tau, \theta, \psi, a, h, \varepsilon), \\
\frac{dh}{dt} &= H(\tau, a) h + R(\tau, \theta, \psi, a, h, \varepsilon),
\end{aligned}\right\} \tag{7.103}$$

where the functions $P(\tau, \theta, \psi, a, h, \varepsilon)$, $Q(\tau, \theta, \psi, a, h, \varepsilon)$, and the $(n-2)$-dimensional vector function $R(\tau, \theta, \psi, a, h, \varepsilon)$ possess the same properties as the functions (7.84) and reduce to quantities of the order of ε when $h = 0$.

Let us now construct the two-parametric family of approximate solutions of the system (7.78), close for sufficiently small ε to the family of periodic solutions (7.80).

To do this we derive from the system of $n-2$ equations

$$\frac{dh}{dt} = H(\tau, a) h + R(\tau, \theta, \psi, a, h, \varepsilon) \tag{7.104}$$

the particular solution h, which corresponds to the external perturbing forces $R(\tau, \theta, \psi, a, h, \varepsilon)$, and depends on τ as a parameter /82/.

We look for h in the form of a series

$$h(\tau, \theta, \psi, a, \varepsilon) = \varepsilon h_1(\tau, \theta, \psi, a) + \varepsilon^2 h_2(\tau, \theta, \psi, a) + \ldots \tag{7.105}$$

Substituting (7.105) into the left-hand side of the system (7.104), we obtain

$$\begin{aligned}
&\varepsilon\left[\frac{\partial h_1}{\partial\theta}\frac{d\theta}{dt} + \frac{\partial h_1}{\partial\psi}\frac{d\psi}{dt} + \frac{\partial h_1}{\partial a}\frac{da}{dt} + \varepsilon\frac{\partial h_1}{\partial\tau}\right] + \\
&\quad + \varepsilon^2\left[\frac{\partial h_2}{\partial\theta}\frac{d\theta}{dt} + \frac{\partial h_2}{\partial\psi}\frac{d\psi}{dt} + \frac{\partial h_2}{\partial a}\frac{da}{dt} + \varepsilon\frac{\partial h_2}{\partial\tau}\right] + \varepsilon^3[\ldots] + \ldots = \\
&\quad = \varepsilon\left\{\frac{\partial h_1}{\partial\theta}\nu(\tau) + \frac{\partial h_1}{\partial\psi}[\omega(\tau, a) + P(\tau, \theta, \psi, a, h, \varepsilon)] + \right. \\
&\quad \left. + \frac{\partial h_1}{\partial a} Q(\tau, \theta, \psi, a, h, \varepsilon) + \varepsilon\frac{\partial h_1}{\partial\tau}\right\} + \\
&\quad + \varepsilon^2\left\{\frac{\partial h_2}{\partial\theta}\nu(\tau) + \frac{\partial h_2}{\partial\psi}[\omega(\tau, a) + P(\tau, \theta, \psi, a, h, \varepsilon)] + \right. \\
&\quad \left. + \frac{\partial h_2}{\partial a} Q(\tau, \theta, \psi, a, h, \varepsilon) + \varepsilon\frac{\partial h_2}{\partial\tau}\right\} + \varepsilon^3[\ldots] + \ldots = \\
&\quad = \varepsilon\left\{\frac{\partial h_1}{d\theta}\nu(\tau) + \frac{\partial h_1}{\partial\psi}[\omega(\tau, a) + \varepsilon P(\tau, \theta, \psi, a, 0, 0) + \varepsilon^2(\ldots) + \ldots] + \right. \\
&\quad \left. + \frac{\partial h_1}{\partial a}[\varepsilon Q(\tau, \theta, \psi, a, 0, 0) + \varepsilon^2(\ldots) + \ldots]\right\} + \\
&\quad + \varepsilon^2\left\{\frac{\partial h_2}{\partial\theta}\nu(\tau) + \frac{\partial h_2}{\partial\psi}[\omega(\tau, a) + \varepsilon P(\tau, \theta, \psi, a, 0, 0) + \varepsilon^2(\ldots) + \ldots] + \right. \\
&\quad \left. + \frac{\partial h_2}{\partial a}[\varepsilon Q(\tau, \theta, \psi, a, 0, 0) + \varepsilon^2(\ldots) + \ldots]\right\} + \varepsilon^3\{\ldots\} + \ldots =
\end{aligned} \tag{7.106}$$

$$= \varepsilon\left\{\frac{\partial h_1}{\partial \theta}\nu(\tau) + \frac{\partial h_1}{\partial \psi}\omega(\tau, a)\right\} + \varepsilon^2\left\{\frac{\partial h}{\partial \psi}P(\tau, 0, \psi, a, 0, 0) + \right.$$

$$\left. + \frac{\partial h_1}{\partial a}Q(\tau, 0, \psi, a, 0, 0) + \frac{\partial h_1}{\partial \tau} + \frac{\partial h_2}{\partial \theta}\nu(\tau) + \frac{\partial h_2}{\partial \psi}\omega(\tau, a)\right\} + \varepsilon^3\{\ldots\} + \ldots \quad (7.106)$$

Expanding the right-hand side of the system (7.104) in a series in the neighborhood of the point ψ, a, $h = 0$, $\varepsilon = 0$, and equating the coefficients of equal powers of ε in the expression thus obtained and in the right-hand side of (7.106), we obtain the following system of equations for the determination of h_1, h_2, ...:

$$\varepsilon\frac{\partial h_1}{\partial \theta}\nu(\tau) + \varepsilon\frac{\partial h_1}{\partial \psi}\omega(\tau, a) = H(\tau, a)\,\varepsilon h_1 + R(\tau, \theta, \psi, a, 0, 0), \quad (7.107)$$

$$\varepsilon^2\frac{\partial h_2}{\partial \theta}\nu(\tau) + \varepsilon^2\frac{\partial h_2}{\partial \psi}\omega(\tau, a) = H(\tau, a)\,\varepsilon^2 h_2 + R'_h(\tau, \theta, \psi, a, 0, 0)\varepsilon h_1 +$$

$$+ \varepsilon R'_\varepsilon(\tau, \theta, \psi, a, 0, 0) - \varepsilon\frac{\partial h_1}{\partial \psi}P(\tau, \theta, \psi, a, 0, 0) -$$

$$- \varepsilon\frac{\partial h_1}{\partial a}Q(\tau, \theta, \psi, a, 0, 0) - \varepsilon^2\frac{\partial h_1}{\partial \tau}. \quad (7.108)$$

We derive $h_1(\tau, \theta, \psi, a)$ from the system (7.107) as the particular solution corresponding to the perturbing force $R(\tau, \theta, \psi, a, 0, 0)$:

$$\varepsilon h_1(\tau, \theta, \psi, a) = \sum_{n, m}\frac{F^{(1)}_{n, m}(\tau, a)\,e^{i(n\theta + m\psi)}}{|Ii\,[n\nu(\tau) + m\omega(\tau, a)] - H(\tau, a)|}, \quad (7.109)$$

where $F^{(1)}_{n, m}(\tau, a)$ are the Fourier coefficients of the function appearing on the right-hand side of (7.107)

$$F^{(1)}_{n, m}(\tau, a) = \frac{1}{4\pi^2}\int_0^{2\pi}\int_0^{2\pi} R(\tau, \theta, \psi, a, 0, 0)\,e^{-i(n\theta + m\psi)}\,d\theta\,d\psi.$$

The denominator in (7.109) is different from zero, as the roots of equation (7.88) have negative real parts.

Having found the expression for $h_1(\tau, \theta, \psi, a)$, we know the right-hand side of equations (7.108), and we can derive from them an expression for $h_2(\tau, \theta, \psi, a)$:

$$\varepsilon^2 h_2(\tau, \theta, \psi, a) = \sum_{n, m}\frac{F^{(2)}_{n, m}(\tau, a)\,e^{i(n\theta + m\psi)}}{|Ii\,[n\nu(\tau) + m\omega(\tau, a)] - H(\tau, a)|}, \quad (7.110)$$

where $F^{(2)}_{n, m}(\tau, a)$ are the Fourier coefficients of the functions appearing on the right-hand side of equations (7.108)

$$F^{(2)}_{n, m}(\tau, a) = \frac{1}{4\pi^2}\int_0^{2\pi}\int_0^{2\pi}\left\{\varepsilon R'_h(\tau, \theta, \psi, a, 0, 0)h_1 + \right.$$

$$+ \varepsilon R'_\varepsilon(\tau, \theta, \psi, a, 0, 0) - \varepsilon\frac{\partial h_1}{\partial \psi}P(\tau, \theta, \psi, a, 0, 0) -$$

$$\left. - \varepsilon\frac{\partial h_1}{\partial a}Q(\tau, \theta, \psi, a, 0, 0) - \varepsilon^2\frac{\partial h_1}{\partial \tau}\right\}e^{-i(n\theta + m\psi)}\,d\theta\,d\psi. \quad (7.111)$$

We derive analogously $h_3(\tau, \theta, \psi, a)$, $h_4(\tau, \theta, \psi, a)$, ...

After deriving an expression for $h(\tau, \theta, \psi, a, \varepsilon)$, the first two equations of the considered system (7.103) take the form

$$\left.\begin{aligned} \frac{d\psi}{dt} &= \omega(\tau, a) + P[\tau, \theta, \psi, a, h(\tau, \theta, \psi, a, \varepsilon), \varepsilon], \\ \frac{da}{dt} &= Q[\tau, \theta, \psi, a, h(\tau, \theta, \psi, a, \varepsilon), \varepsilon], \end{aligned}\right\} \quad (7.112)$$

where $\frac{d\theta}{dt} = \nu(\tau)$.

Using our method to derive a and ψ from the system (7.112) (cf. Chapter II, § 5), and substituting them together with $h(\tau, \theta, \psi, a)$ given by (7.105) into (7.91), we obtain the following family of solutions of the system (7.78), depending on two arbitrary constants:

$$x = x^0[\tau, \psi(\tau, t, \varepsilon), a(\tau, t, \varepsilon)] + \\ + \frac{1}{2}\{A[\tau, \psi(\tau, t, \varepsilon), a(\tau, t, \varepsilon)]\, h[\tau, \theta, \psi(\tau, t, \varepsilon), a(\tau, t, \varepsilon), \varepsilon] + \\ + \overline{A}[\tau, \psi(\tau, t, \varepsilon), a(\tau, t, \varepsilon)]\, \overline{h}[\tau, \theta, \psi(\tau, t, \varepsilon), a(\tau, t, \varepsilon), \varepsilon]\}, \tag{7.113}$$

where $\overline{h}(\tau, \theta, \psi, a, \varepsilon)$ is representable in the form

$$\overline{h}(\tau, \theta, \psi, a, \varepsilon) = \varepsilon\overline{h}_1(\tau, \theta, \psi, a) + \varepsilon^2\overline{h}_2(\tau, \theta, \psi, a) + \ldots,$$

and $\overline{h}_1(\tau, \theta, \psi, a)$, $\overline{h}_2(\tau, \theta, \psi, a)$, ... are conjugate to $h_1(\tau, \theta, \psi, a)$, $h_2(\tau, \theta, \psi, a)$, ... and may be defined by formulas of the type

$$\varepsilon\overline{h}_1(\tau, \theta, \psi, a) = \sum_{n, m} \frac{F^{(1)}_{n, m}(\tau, a)\, e^{i(n\theta + m\psi)}}{|Ii[n\nu(\tau) + m\omega(\tau, a)] - \overline{H}(\tau, a)|}. \tag{7.114}$$

We shall present another method of constructing two-parametric families of solutions of the system of equations (7.78).

As in the preceding section, besides the condition requiring that the real parts of the characteristic exponents of the system of variational equations (7.82) be negative, let also the following inequalities be satisfied for any value of τ $(0 \leqslant \tau \leqslant L)$ and a:

$$\left.\begin{array}{c} m_3\gamma_3(\tau, a) + m_4\gamma_4(\tau, a) + \ldots + m_n\gamma_n(\tau, a) \neq \gamma_q(\tau, a), \\ m_3 + m_4 + \ldots + m_n \geqslant 2, \end{array}\right\} \tag{7.115}$$

where $\gamma_q(\tau, a)$ $(q = 3, 4, \ldots, n)$ are the real parts of the characteristic exponents, and $m_3, m_4, \ldots, m_n$ are non-negative integers.

It is in this case possible to construct, as in the preceding section, the general solution of the unperturbed system (7.79) in the form

$$x = f(\tau, \omega t + \varphi, a, Ce^{\int \lambda(\tau)\, dt}), \tag{7.116}$$

thereafter to reduce the system of equations (7.78) to the form (7.112), and then to apply the method developed in Chapter II, § 5 for solving the system obtained (for deriving approximate solutions in the first, second, etc., approximations).

Without going into details, we show below that if conditions (7.115) are satisfied, it is possible to construct the first, second, etc., approximations by applying the u-method directly, and to obtain a two-parametric family of solutions for the system (7.78) in the form of an asymptotic series.

We denote by $v_k^{(q)}(\tau, \psi, a)$ $(q, k = 1, 2, \ldots, n)$ a basic system of solutions of the variational equations (7.82)

$$\left.\begin{array}{l} v_k^{(1)}(\tau, \psi, a) = \dfrac{\partial x_k^{(0)}}{\partial \varphi} = f_{k1}(\tau, \psi, a), \\ v_k^{(2)}(\tau, \psi, a) = \dfrac{\partial x_k^{(0)}}{\partial a} = f_{k2}(\tau, \psi, a) + t\,\dfrac{d\omega}{da}\, f_{k1}(\tau, \psi, a), \\ v_k^{(q)}(\tau, \psi, a) = f_{kq}(\tau, \psi, a)\, e^{\int^t \lambda_q(\tau, a)\, dt} \\ \qquad (q = 3, 4, \ldots, n;\ k = 1, 2, \ldots, n), \end{array}\right\} \tag{7.117}$$

where

$$f_{kq}(\tau, \psi + 2\pi, a) = f_{kq}(\tau, \psi, a) \qquad (k, q = 1, 2, \ldots, n). \tag{7.118}$$

We denote by $w_k^{(q)}(\tau, \psi, a)$ $(q, k = 1, 2, \ldots, n)$ a basic system of solutions of the system of equations conjugate to (7.82), i.e., of the system

$$\frac{d\delta x}{dt}+\overline{X}'_x(\tau, x^0)\,\delta x=0, \qquad (7.119)$$

where $\overline{X}'_x(\tau, x^0)$ is the transposed matrix of $X'_x(\tau, x^0)$.

We then have, as known,

$$w_k^{(q)}(\tau, \psi, a)=(-1)^{k+q}\frac{M_{kq}[v_r^{(s)}(\tau, \psi, a)]}{\Delta[v_r^{(s)}(\tau, \psi, a)]} \qquad (k, q=1, 2, \ldots, n), \qquad (7.120)$$

where $\Delta[v_r^{(s)}(\tau, \psi, a)]$ is the Wronskian of the system of solutions (7.117), and $M_{kq}[v_r^{(s)}(\tau, \psi, a)]$ is the cofactor of the element $v_k^{(q)}(\tau, \psi, a)$.

One easily verifies by elementary manipulations that

$$\left.\begin{aligned} &w_k^{(1)}(\tau, \psi, a)=f_{k1}^*(\tau, \psi, a)-t\frac{d\omega}{da}f_{k2}^*(\tau, \psi, a),\\ &w_k^{(2)}(\tau, \psi, a)=f_{k2}^*(\tau, \psi, a),\\ &w_k^{(q)}(\tau, \psi, a)=f_{kq}^*(\tau, \psi, a)\,e^{-\int^t \lambda_q(\tau, a)\,dt}\\ &\qquad (q=3, 4, \ldots, n;\ k=1, 2, \ldots, n); \end{aligned}\right\} \qquad (7.121)$$

where

$$f_{kq}^*(\tau, \psi+2\pi, a)=f_{kq}^*(\tau, \psi, a) \qquad (k, q=1, 2, \ldots, n) \qquad (7.122)$$

and the functions $f_{kq}^*(\tau, \psi, a)$ are conjugate to $f_{kq}(\tau, \psi, a)$.

The solution of the system (7.78), which depends on two arbitrary constants and is close to the two-parametric family (7.80), is sought for in the form of the asymptotic series

$$x=x^0(\tau, p\varphi+\psi, a)+\varepsilon u^{(1)}(\tau, \theta, p\varphi+\psi, a)+\varepsilon^2 u^{(2)}(\tau, \theta, p\varphi+\psi, a)+\ldots, \qquad (7.123)$$

where $u^{(i)}(\tau, \theta, p\varphi+\psi, a)$ $(i=1, 2, 3, \ldots)$ is an n-dimensional vector function, a and ψ are functions of time which are to be determined from the system of equations

$$\left.\begin{aligned} &\frac{da}{dt}=\varepsilon A_1(\tau, a, \psi)+\varepsilon^2 A_2(\tau, a, \psi)+\ldots,\\ &\frac{d\psi}{dt}=\omega(\tau, a)-\frac{p}{q}\nu(\tau)+\varepsilon B_1(\tau, a, \psi)+\varepsilon^2 B_2(\tau, a, \psi)+\ldots, \end{aligned}\right\} \qquad (7.124)$$

p and q are mutually prime integers whose choice depends on the particular resonance under investigation, $\varphi=\frac{1}{q}\theta$.

We substitute the series (7.123) into equations (7.78), assume $\tau=\varepsilon t$, differentiate a and ψ, and use equations (7.124), the variational equations (7.82), their conjugate equations (7.119), and the basic systems of solutions (7.117) and (7.121). Then, equating the coefficients of equal powers of ε and imposing the usual conditions requiring the absence of secular terms in the right-hand sides of the series (7.123), we obtain a series of equations for determining the functions appearing on the right-hand sides of (7.123) and of the equations (7.124).

Thus the solution of the system (7.78), close to the two-parametric family of particular solutions of the unperturbed system, is in the first approximation

$$x=x^0(\tau, p\varphi+\psi, a), \qquad (7.125)$$

where a and ψ are to be determined from the system of equations of the first approximation

$$\left.\begin{aligned}\frac{da}{dt} &= \varepsilon A_1(\tau, a, \psi),\\ \frac{d\psi}{dt} &= \omega(\tau, a) - \frac{p}{q}\nu(\tau) + \varepsilon B_1(\tau, a, \psi).\end{aligned}\right\} \tag{7.126}$$

Here $A_1(\tau, a, \psi)$ and $B_1(\tau, a, \psi)$, periodic in ψ with period 2π, are particular solutions of the system

$$\left.\begin{aligned}&\left[\omega(\tau, a) - \frac{p}{q}\nu(\tau)\right]\frac{\partial A_1}{\partial\psi} - 2a\omega(\tau, a) B_1 =\\ &\qquad = \underset{t}{M}\left\{\sum_{k=1}^{n} X_k^*(\tau, \theta, x_1^0, \ldots, x_n^0, 0) f_{k2}^*(\tau, \psi, a)\right\},\\ &\left[\omega(\tau, a) - \frac{p}{q}\nu(\tau)\right] a\frac{\partial B_1}{\partial\psi} + 2\omega(\tau, a) A_1 =\\ &\qquad = -\frac{d\omega(\tau, a)}{d\tau} - \underset{t}{M}\left\{\sum_{k=1}^{n} X_k^*(\tau, \theta, x_1^0, \ldots, x_n^0, 0) f_{k1}^*(\tau, \psi, a)\right\},\end{aligned}\right\} \tag{7.127}$$

where $\underset{t}{M}$ denotes averaging over the explicit time.

It is not difficult to see that the same equations (7.126) with (7.127) for a and ψ are obtained in the first approximation from the system of equations (7.112) according to the method given in Chapter II, § 5.

If conditions (7.115) are satisfied, then, as follows from (7.125) and (7.127), it is sufficient to know only the explicit expression of the family of solutions (7.80) for constructing the first approximation. The basic system of solutions (7.117) is required only for constructing the improved first approximation and higher approximations.

§ 3. Application of the method of averaging to the study of vibrating systems with slowly varying parameters

In the present section we shall briefly discuss some aspects of the application of the method of averaging to the study of general vibrating systems with many degrees of freedom, containing slowly varying parameters.

The differential equations describing vibrational processes and containing a "small" parameter may, as known, be reduced in many cases to a common form in which the right-hand sides are proportional to a small parameter

$$\frac{dx}{dt} = \varepsilon X(t, x), \tag{7.128}$$

where x is an n-dimensional vector, $X(t, x)$, an n-dimensional vector function, and ε, a small positive parameter.

For sufficiently small values of ε, equation (7.128), according to the method of averaging, is replaced by the averaged equation

$$\frac{dx_0}{dt} = \varepsilon X_0(x_0), \tag{7.129}$$

in a finite time interval, where

$$X_0(x_0) = \lim_{T\to\infty}\frac{1}{T}\int_0^T X(t, x_0)\, dt. \tag{7.130}$$

The application of the method of averaging for constructing approximate solutions of the system (7.128) in the case when the right-hand side of this equation is periodic in t, was studied by Fatou, L. I. Mandel'shtam, N. D. Papaleksi, and others. N. M. Krylov and N. N. Bogolyubov generalized this method to the case when only condition (7.130) must be satisfied by the right-hand sides of equations (7.128). They proved a fundamental theorem which states that under certain conditions the solutions of the system (7.128) are as close as desired to the solutions of the averaged system (7.129) in as large as desired an interval $t \sim \frac{1}{\varepsilon}$ for sufficiently small values of ε, and worked out a method of constructing higher approximations. This method is based on a successive change of variables accompanied by the averaging of the right-hand sides of the equations obtained for the new variables in each step.

The method of averaging may in many cases be successfully applied also to the construction of approximate solutions for equations with slowly varying parameters (obviously with appropriate corrections; thus, for example, these parameters are regarded as constants during the averaging, and as slowly varying functions of time in all operations connected with differentiation). This was already shown by us in several places in the preceding chapters, where we have applied the method of averaging for obtaining the first approximation, in addition to the u-method.

The idea of applying the method of averaging to the study of systems with slowly varying parameters is already found in the work of N. N. Bogolyubov and D. N. Zubarev /8/, which deals with a vibrating system with a rapidly rotating phase. The study of this system reduces to the examination of differential equations of the type

$$\left.\begin{aligned} \frac{d\psi}{dt} &= \lambda\omega(x) + A(\psi, x), \\ \frac{dx}{dt} &= X(\psi, x), \end{aligned}\right\} \tag{7.131}$$

where x is an n-dimensional vector; X, an n-dimensional vector function; ψ, a scalar, the functions $X(\psi, x)$ and $A(\psi, x)$ are periodic in ψ with period 2π; λ, a large parameter.

Changing the independent variable in (7.131) according to the formula $\lambda t = \tau$, and introducing the small parameter $\varepsilon = \frac{1}{\lambda}$, we obtain in place of (7.131) the system

$$\left.\begin{aligned} \frac{d\psi}{d\tau} &= \omega(x) + \varepsilon A(\psi, x), \\ \frac{dx}{d\tau} &= \varepsilon X(\psi, x), \end{aligned}\right\} \tag{7.132}$$

in which the variable ψ represents (according to the widely accepted terminology) rapid motion, and the vector x, slow motion.

One finds in /8/ a formal method, based on the grounds of the ideas pertaining to the method of averaging, for eliminating the variable ψ from the right-hand sides of the system (7.132) with any degree of accuracy in powers of $\varepsilon = \frac{1}{\lambda}$. This elimination is carried out by means of a substitution of variables of the form

$$\left.\begin{aligned} \psi &= \bar{\psi} + \sum_{n=1}^{\infty} \varepsilon^n U_n(\bar{\psi}, \bar{x}), \\ x &= \bar{x} + \sum_{n=1}^{\infty} X^{(n)}(\bar{\psi}, \bar{x}), \end{aligned}\right\} \tag{7.133}$$

after which we obtain the following system of equations for the new variables $\bar{x}$, $\bar{\psi}$:

$$\left.\begin{aligned} \frac{d\bar{\psi}}{dt} &= \lambda\omega(\bar{x}) + \sum_{n=1}^{\infty} \frac{1}{\lambda^n}\Omega_n(\bar{x}), \\ \frac{d\bar{x}}{dt} &= \sum_{n=0}^{\infty} X^{(n)}(\bar{x}), \end{aligned}\right\} \tag{7.134}$$

where the right-hand sides do not depend any more on the rapid angular variable ψ.

The physical meaning of the transformation (7.133) lies in the separation of the actual motion described by the coordinates x, ψ into an average slow motion with coordinates $\bar{x}$, and a rapid vibration described by the angle $\bar{\psi}$ and the functions

$$U_n(\bar{\psi}, \bar{x}), \quad X^{(n)}(\bar{\psi}, \bar{x}).$$

We already applied this method to the study of the system (3.301) in Chapter III, § 14.

We proceed to consider more general systems of differential equations containing slowly varying parameters (1.12)

$$\left.\begin{aligned} \frac{dx}{dt} &= X(t, x, y, \varepsilon), \\ \frac{dy}{dt} &= \varepsilon Y(t, x, y, \varepsilon), \end{aligned}\right\} \tag{7.135}$$

where x, y are vectors of dimensions k and m, respectively, $(k+m=n)$ of the n-dimensional Euclidean space E_n, and $X(t, x, y, \varepsilon)$ and $Y(t, x, y, \varepsilon)$ are k- and m-dimensional vector functions, respectively.

V. M. Volosov developed for such systems a method of averaging /32, 33/ based on /6, 8/.

We shall discuss the formalism of this method which has a few advantages when deriving asymptotic approximations for the system of equations (7.135).

If $Y(t, x, y, \varepsilon) \equiv 1$ in (7.135), then $y = \varepsilon t = \tau$, and we are led to a system of equations containing the slowing time which was considered in the preceding sections.

Expanding the right-hand sides of the system (7.135) in powers of the small parameter ε, we write it in the form

$$\left.\begin{aligned} \frac{dx}{dt} &= X_0(t, x, y) + \varepsilon X_1(t, x, y) + \ldots, \\ \frac{dy}{dt} &= \varepsilon Y_1(t, x, y) + \varepsilon^2 Y_2(t, x, y) + \ldots \end{aligned}\right\} \tag{7.136}$$

Together with this system we consider the unperturbed system

$$\left.\begin{aligned} \frac{dx}{dt} &= X_0(t, x, y), \\ \frac{dy}{dt} &= 0, \quad y = \text{const}, \end{aligned}\right\} \tag{7.137}$$

which is obtained by putting $\varepsilon = 0$ in (7.136).

We assume that the limit

$$\bar{Y}_1 = \lim_{T\to\infty} \int_0^T Y[t, x(t, y), y, 0]\, dt, \tag{7.138}$$

exists along any integral curve $x = x(t, y)$ of the system (7.137).

In the method of averaging, averaging along an integral curve usually means simply an averaging over the explicit time. In Chapter III, § 13, while considering the system (3.301), we averaged the second equation of this system along an integral curve of the equation (3.302), i. e., along $x = a \cos(\theta + \psi)$.

The problem of averaging may for the system (7.136) be approached in two ways.

First approach. Since the variables x vary slowly in (7.136), while the variables y vary rapidly, the affect of the variables x on y consists in a rapid influence of the variables x on the slow velocities εY of the variables y because $Y = Y(t, x, y, \varepsilon)$. Averaging this influence of the rapid motions over a long time interval, we are led to consider in place of the second equation of the system (7.135) the averaged equation

$$\frac{d\bar{y}}{dt} = \varepsilon \bar{Y}_1. \tag{7.139}$$

If the right-hand sides of the system (7.139) depend only on y and, consequently, do not depend on the choice of the trajectory $x = x(t, y)$ of the unperturbed system (7.137), then such an averaging simplifies the problem considerably since in that case the system (7.136) separates into two independent systems of the k-th and m-th order.

The asymptotic approximation of the solution of the system (7.136) reduces in that case to the problem of approximation by means of solutions of the averaged system (7.139). The solutions of the averaged system (7.139) give the first approximations for the solutions (for y) of the original system (7.136).

An essential restriction, inherent in this approach, is the requirement that the right-hand side of equation (7.139) does not depend on the choice of the trajectory of the unperturbed system (7.137). However, in many cases of practical importance it is possible to work out methods consisting in a special change of variables, which permit us to reduce the general case, when the average (7.138) depends on the trajectory of the unperturbed system (7.137), to the case considered above. We shall not discuss this problem, sending the interested reader to the special literature /33, 118, 135/.

Second approach. The second, more general, approach consists in constructing higher approximations for y, and approximations for the rapid motions x.

It reduces to constructing directly, for the entire system of equations (7.136), an averaged system of the form

$$\left.\begin{aligned} \frac{d\bar{x}}{dt} &= X_0(t, \bar{x}, \bar{y}) + \varepsilon A_1(\bar{y}) + \varepsilon^2 A_2(\bar{y}) + \dots, \\ \frac{d\bar{y}}{dt} &= \varepsilon \bar{Y}_1(\bar{y}) + \varepsilon^2 B_2(\bar{y}) + \dots, \end{aligned}\right\} \tag{7.140}$$

the solutions of which should approximate to the solutions of the system of equations (7.136) with any degree of accuracy in a time interval of the order of $\frac{1}{\varepsilon}$, and the so far unknown functions $A_1(\bar{y})$, $A_2(\bar{y})$, ..., $B_2(\bar{y})$, ...should be derived by averaging along a trajectory of the unperturbed system (7.137).

If we confine ourselves to a finite number of terms, then the system of equations (7.140) is integrated easier than the original system (7.136) because the rapid and slow variables x and y are here separated.

Let us derive a formal change of variables which, as usual in the method of averaging, reduces the system (7.136) to the averaged system of equations (7.140).

While considering in the preceding sections systems with slowing time, we have assumed that there exists for the unperturbed system a family of particular solutions depending on one, two, etc., arbitrary constants. We then constructed an integral manifold, and reduced the problem of solving the original system to the study of a system of a lower order in that manifold, the solutions of which could be obtained by means of the u-method, the method of averaging, etc.

In the present case we make a more general assumption that the general solution of the unperturbed system (7.137) of the form

$$x=\varphi(t, y, x_0, t_0), \tag{7.141}$$

exists and is known in a certain domain. Here x_0 is a constant k-dimensional vector, and $\varphi(t_0, y, x_0, t_0) \equiv x_0$. We shall also assume that the vector function $\varphi(t, y, x_0, t_0)$ has the desired number of derivatives, and that all the conditions necessary for the following arguments are satisfied for this function and for the right-hand sides of the system (7.136).

With these assumptions we are looking for a change of variables of the form

$$\left.\begin{aligned} x&=\bar{x}+\varepsilon u_1(t, \bar{x}, \bar{y})+\varepsilon^2 u_2(t, \bar{x}, \bar{y})+\ldots, \\ y&=\bar{y}+\varepsilon v_1(t, \bar{x}, \bar{y})+\varepsilon^2 v_2(t, \bar{x}, \bar{y})+\ldots, \end{aligned}\right\} \tag{7.142}$$

where $u_1(t, \bar{x}, \bar{y}), u_2(t, \bar{x}, \bar{y}), \ldots$ are k-dimensional vector functions, and $v_1(t, \bar{x}, \bar{y}), v_2(t, \bar{x}, \bar{y}), \ldots$ are m-dimensional vector functions, which are to be determined in such a manner that the substitution (7.142) would transform the system (7.136) to the system of averaged equations (7.140). It is herewith clear that when $\varepsilon=0$, the systems (7.136) and (7.140) degenerate to the unperturbed system (7.137), and the substitution (7.142) gives at the same time $x=\bar{x}$, $y=\bar{y}$ when $\varepsilon=0$.

Thus, to solve the problem stated, it is necessary to find relations defining the vector functions

$$\left.\begin{aligned} &u_1(t, \bar{x}, \bar{y}), \quad u_2(t, \bar{x}, \bar{y}), \ldots, v_1(t, \bar{x}, \bar{y}), \quad v_2(t, \bar{x}, \bar{y}), \ldots, \\ &\qquad A_1(\bar{y}), \quad A_2(\bar{y}), \ldots, \quad B_2(\bar{y}), \ldots \end{aligned}\right\} \tag{7.143}$$

Substituting (7.142) into equations (7.136), taking (7.140) into account, and expanding the result in powers of the small parameter ε, we obtain, after equating the coefficients of equal powers of ε, the expressions

$$\frac{\partial u_1}{\partial t}+\frac{\partial u_1}{\partial \bar{x}} X_0(t, \bar{x}, \bar{y})=X_1(t, \bar{x}, \bar{y})+\frac{\partial X_0}{\partial \bar{x}} u_1+\frac{\partial X_0}{\partial \bar{y}} v_1-A_1(\bar{y}), \tag{7.144}$$

$$\begin{aligned} \frac{\partial u_2}{\partial t}+\frac{\partial u_2}{\partial \bar{x}} X_0(t, \bar{x}, \bar{y})&=X_2(t, \bar{x}, \bar{y})+\frac{1}{2}\left\{\frac{\partial^2 X_0}{\partial \bar{x}^2} u_1^2+\frac{\partial^2 X_0}{\partial \bar{y}^2} v_1^2+2 \frac{\partial^2 X_0}{\partial \bar{x}\, \partial \bar{y}} u_1 v_1\right\}+ \\ &+\frac{\partial X_0}{\partial \bar{x}} u_2+\frac{\partial X_0}{\partial \bar{y}} v_2+\frac{\partial X_1}{\partial \bar{x}} u_1+\frac{\partial X_1}{\partial \bar{y}} v_1-\frac{\partial u_1}{\partial \bar{x}} A_1(\bar{y})-\frac{\partial u_1}{\partial \bar{y}} Y_1(t, \bar{x}, \bar{y})-A_2(\bar{y}), \end{aligned} \tag{7.145}$$

. . . .

$$\frac{\partial v_1}{\partial t}+\frac{\partial v_2}{\partial \bar{x}} X_0(t, \bar{x}, \bar{y})=Y_2(t, \bar{x}, \bar{y})-\bar{Y}_1(\bar{y}), \tag{7.146}$$

$$\begin{aligned} &\frac{\partial v_2}{\partial t}+\frac{\partial v_2}{\partial \bar{x}} X_0(t, \bar{x}, \bar{y})= \\ &=Y_2(t, \bar{x}, \bar{y})-B_2(\bar{y})+\frac{\partial Y_1}{\partial \bar{x}} u_1+\frac{\partial Y_1}{\partial \bar{y}} v_1-\frac{\partial v_1}{\partial \bar{x}} A_1(\bar{y})-\frac{\partial v_1}{\partial \bar{y}} \bar{Y}_1(\bar{y}), \end{aligned} \tag{7.147}$$

. . . .

from which the unknown functions (7.143) are to be determined.

Evidently, the structure of these equations allows us to derive successively the functions (7.143), however, this determination is not unique and, consequently, some additional conditions are necessary. Such an additional condition for constructing the first approximation may, for example, be the requirement that the average (7.138) be independent of the initial values x_0, t_0.

As regards the accuracy with which we can find x and y from the equations of the first, second, and higher approximations, we may use here the same arguments as those applied in Chapters II and III during the study of systems close to exactly integrable ones.

Proceeding from the arguments given in Chapter III, § 10 with reference to the system (3.233), we consider in the first approximation the system

$$\left.\begin{aligned} \frac{d\bar{x}}{dt} &= X_0(t, \bar{x}, \bar{y}), \\ \frac{d\bar{y}}{dt} &= \varepsilon \bar{Y}_1(\bar{y}), \end{aligned}\right\} \tag{7.148}$$

which approximates the solution y of the system (7.136) with an error of the order of $O(\varepsilon)$, and gives in general no approximations for the variables x.

In the second approximation we take the system

$$\left.\begin{aligned} \frac{d\bar{x}}{dt} &= X_0(t, \bar{x}, \bar{y}) + \varepsilon A_1(\bar{y}), \\ \frac{d\bar{y}}{dt} &= \varepsilon \bar{Y}_1(\bar{y}) + \varepsilon^2 B_2(\bar{y}), \end{aligned}\right\} \tag{7.149}$$

which approximates the solution y of the system (7.136) up to terms of the order of $O(\varepsilon^2)$, and the variables x– up to terms of the order of $O(\varepsilon)$.

V. M. Volosov has proved /33/ for this method of constructing the averaged system (7.140), which yields an asymptotic approximate solution for the original system (7.136), theorems which are analogous to and extensions of well-known theorems of N. N. Bogolyubov. We shall, however, not discuss them here, sending the interested readers to the special literature.

In addition to numerous problems of vibrations of systems with slowly varying parameters, also an extended class of equations with a small parameter preceding the highest derivative (cf., for example, /118, 135/) may be reduced to systems of the type (7.135). Such equations play an important role in the theory of relaxation vibrations, when small so-called "parasitic" parameters (small masses, small inductances, etc.) are taken into account. The study of such systems leads in the general case to systems of differential equations of the type

$$\left.\begin{aligned} \varepsilon \frac{dx}{dt_1} &= X(t_1, x, y, \varepsilon), \\ \frac{dy}{dt_1} &= Y(t_1, x, y, \varepsilon), \end{aligned}\right\} \tag{7.150}$$

where x, y are k- and m-dimensional vectors respectively; $X(t_1, x, y, \varepsilon)$ and $Y(t_1, x, y, \varepsilon)$ are k- and m-dimensional vector functions; ε is a small positive parameter; t_1 is the time. Usually, x is called the rapid motion, and y— the slow.

We have already indicated in the first chapter than an equation with a small parameter preceding the highest derivative may be reduced by introducing a new time scale to an equation with slowly varying parameters. In the example given there, the equation (1.3) is reduced to equation (1.5) with slowly varying time.

Introducing a new time scale in the system (7.150) according to the relation $t=\frac{t_1}{\varepsilon}$, we obtain in place of (7.150) the system of equations

$$\left.\begin{aligned} &\frac{dx}{dt}=X(\tau, x, y, \varepsilon),\\ &\frac{dy}{dt}=\varepsilon Y(\tau, x, y, \varepsilon) \qquad (\tau=\varepsilon t), \end{aligned}\right\} \tag{7.151}$$

which coincides with the system (7.135), the right-hand sides depending now on the slowing time $\tau = \varepsilon t$. The system (7.151) may therefore be reduced to the system

$$\left.\begin{aligned} &\frac{dx}{dt}=X(x, z, \varepsilon),\\ &\frac{dz}{dt}=\varepsilon Z(x, z, \varepsilon), \end{aligned}\right\} \tag{7.152}$$

where z is an $(m+1)$-dimensional vector with the components $y_1, y_2, \ldots, y_m, \tau$, and $Z(x, z, \varepsilon)$ is an $(m+1)$-dimensional vector function with the components $Y_1(x, z, \varepsilon), Y_2(x, z, \varepsilon), \ldots, Y_m(x, z, \varepsilon), 1$.

The system (7.152) is a particular case of the system (7.135) for which the right-hand sides do not depend on t (but of an order greater by one—$n+1$).

Chapter VIII

THE MATHEMATICAL FOUNDATION OF THE ASYMPTOTIC METHOD

§ 1. Asymptotic convergence of approximate solutions. Estimation of the error of the *m*-th approximation

In the preceding chapters we have derived approximate solutions of equations (2.1), (3.1), and of the systems of equations (4.3), (5.4), and (7.10). As the convergence of the expansions (2.5), (3.4), and also (4.9), (5.10), and (7.123) cannot be demonstrated in the general case*, we are obliged to consider them as formal expansions. Retaining in these expansions, for example in (4.9), the first m terms, we obtain an approximate solution satisfying the original equations up to terms of the order of ε^{m+1}. This approximate solution is of practical value if, over a sufficiently long time interval, it represents the exact solution with the required accuracy.

In order to justify the asymptotic method developed above, a theorem must first be proved. This states that under certain quite general conditions the difference between a particular exact solution of the system of equations (4.3)

$$\frac{d}{dt}\left\{\sum_{i=1}^{N} a_{ij}(\tau)\dot{q}_i\right\} + \sum_{i=1}^{N} b_{ij}(\tau) q_i =$$
$$= \varepsilon Q_j(\tau, \theta, q_1, \ldots, q_N, \dot{q}_1, \ldots, \dot{q}_N, \varepsilon) \qquad (j = 1, 2, \ldots, N) \tag{8.1}$$

and our particular solution of the m-th approximation,

$$q_i^{(m-1)} = \varphi_i^{(1)}(\tau)\, a\cos(p\varphi+\psi) + \varepsilon u_i^{(1)}(\tau, a, \theta, p\varphi+\psi) +$$
$$+ \varepsilon^2 u_i^{(2)}(\tau, a, \theta, p\varphi+\psi) + \ldots + \varepsilon^{m-1} u_i^{(m-1)}(\tau, a, \theta, p\varphi+\psi) \qquad (i = 1, 2, \ldots, N) \tag{8.2}$$

is of the order of ε^m and, consequently, may be made as small as desired for a sufficiently small ε in some finite interval $0 \leqslant t \leqslant T$. T should be taken as L/ε, since the solutions depend on t through the product εt; L can be made as large as desired for a sufficiently small ε. The same may also be said concerning the approximate asymptotic solutions obtained in the other chapters of this book.

We have shown in the preceding chapter that by introducing new quasi-normal variables $x_1, x_2, \ldots, x_N$ by means of the formulas

$$q_i = \sum_{k=1}^{N} \varphi_i^{(k)}(\tau)\, x_k \qquad (\tau = \varepsilon t) \qquad (i = 1, 2, \ldots, N), \tag{8.3}$$

* In order to establish the convergence, for example, of the series in (4.9) and those appearing in the right-hand sides of equations (4.10), such strict conditions must be imposed on the functions (4.2) that the problem acquires the character of an "exceptional case". The same may be said about the series representing the solutions in other chapters.

where $\varphi^{(k)}(\tau)$ $(i,\ k=1,\ 2,\ \ldots,\ N)$ are normal functions depending on the parameter τ and satisfying the orthogonality condition, the system of equations (8.1) may be reduced to the system of differential equations

$$\frac{d^2x}{dt^2}+\omega_k^2(\tau)\,x_k=\frac{\varepsilon}{m_k(\tau)}X_k(\tau,\theta,x_1,\ldots,x_N,\dot{x}_1,\ldots,\dot{x}_N,\varepsilon)\qquad (k=1,2,\ldots,N).\quad (8.4)$$

Reversing equations (8.3), we find

$$x_k=\frac{1}{m_k(\tau)}\sum_{i,\,j=1}^{N}a_{ij}(\tau)\,\varphi_i^{(k)}(\tau)\,q_j\qquad (k=1,\ 2,\ \ldots,\ N).\quad (8.5)$$

Replacing the q_j in (8.5) by their m-th approximations (8.2), we obtain the m-th approximation for x_k:

$$x_k^{(m-1)}=\frac{1}{m_k(\tau)}\sum_{i,\,j=1}^{N}a_{ij}(\tau)\,\varphi_i^{(k)}(\tau)\,[\varphi_j^{(1)}(\tau)\,a\cos(p\varphi+\psi)+\varepsilon u_j^{(1)}(\tau,a,\theta,p\varphi+\psi)+$$
$$+\varepsilon^2u_j^{(2)}(\tau,a,\theta,p\varphi+\psi)+\ldots+\varepsilon^{m-1}u_j^{(m-1)}(\tau,a,\theta,p\varphi+\psi)]\qquad (k=1,2,\ldots,N),\quad (8.6)$$

where a and ψ are to be determined from the system of equations of the m-th approximation

$$\left.\begin{aligned}\frac{da}{dt}&=\varepsilon A_1(\tau,a,\psi)+\varepsilon^2A_2(\tau,a,\psi)+\ldots+\varepsilon^mA_m(\tau,a,\psi),\\ \frac{d\psi}{dt}&=\omega_1(\tau)-\frac{p}{q}\nu(\tau)+\varepsilon B_1(\tau,a,\psi)+\ldots+\varepsilon^mB_m(\tau,a,\psi).\end{aligned}\right\}\quad (8.7)$$

We shall thus consider instead of the system (8.1) and the approximate solutions (8.2), the system (8.4) and its m-th particular solutions (8.6).

Before formulating and proving the above-mentioned theorem, it is appropriate to present a well-known theorem of N. N. Bogolyubov /6/ concerning the averaging of standard systems of the form

$$\frac{dx}{dt}=\varepsilon X(t,x)\quad (8.8)$$

over the interval $t\sim\frac{1}{\varepsilon}$. We consider it advisable to recall this theorem since in the preceding chapters we have frequently used the method of averaging for constructing approximate solutions.

We shall only formulate the theorem of averaging.

Theorem 1. *Let the functions $X(t,x)$ in the system of equations (8.8) satisfy the conditions:*

1) for some domain D positive constants M and λ exist such that the inequalities

$$|X(t,x)|\leqslant M,\quad |X(t,x')-X(t,x'')|\leqslant\lambda|x'-x''|$$

are satisfied for all real values of $t\geqslant 0$ and for any points x, x', x'' from this domain;

2) there exists, uniformly with respect to x in the domain D, the limit

$$\lim_{T\to\infty}\frac{1}{T}\int_0^T X(t,x)\,dt=X_0(x).$$

Then, corresponding to any positive, and as small as desired, ϱ and η, and for as large a value of L as required, we can find a positive ε_0 such that if $\xi=\xi(t)$ is a solution of the equation

$$\frac{d\xi}{dt}=\varepsilon X_0(\xi),$$

defined in the interval $0 \leqslant t < \infty$ *and lying in the domain* D *along with its* ϱ*-neighborhood, then the inequality*

$$|x(t) - \xi(t)| < \eta,$$

is valid for $0 \leqslant \varepsilon < \varepsilon_0$ *in the interval* $0 \leqslant t < \frac{L}{\varepsilon}$; *in this inequality* $x(t)$ *represents the solution of (8.8) coinciding with* $\xi(t)$ *at* $t = 0$.

This theorem formulates a statement about the smallness of the error of $x(t) - \xi(t)$ of the first approximation.

We pass now to the problem stated at the beginning of this section.

We shall formulate a theorem about the smallness of the error $x_k(t) - x_k^{(m-1)}(t)$ of the m-th approximation in the interval $0 \leqslant t \leqslant T$; here $x_k(t)$ $(k = 1, 2, 3, \ldots, N)$ are exact particular solutions of the system (8.4) $x_k^{(m-1)}(t)$ $(k = 1, 2, \ldots, N)$ are the m-th approximate particular solutions defined by (8.6). Analogous theorems may also be formulated for more complicated systems of differential equations, discussed by us before, for example for systems of the type (5.4), while similar theorems for equations (2.1) and (3.1) may be obtained as particular cases of the theorem quoted below.

Theorem 2. Let the following conditions be satisfied:

1) the quadratic forms

$$T(\dot{q}) = \frac{1}{2}\sum_{i,\,j=1}^{N} a_{ij}(\tau)\,\dot{q}_i\dot{q}_j, \quad V(q) = \frac{1}{2}\sum_{i,\,j=1}^{N} b_{ij}(\tau)\,q_i q_j$$

be positive definite in any finite interval

$$0 \leqslant \tau \leqslant L;$$

2) the functions $a_{ij}(\tau)$, $b_{ij}(\tau)$, $\nu(\tau)$, $X_k(\tau, \theta, x_1, \ldots, x_N, \dot{x}_1, \ldots, \dot{x}_N, \varepsilon)$ $(i, j, k = 1, 2, \ldots, N)$ *be differentiable to any desired order for all finite values of their arguments and sufficiently small* ε*;

3) the expressions $\left(\frac{\partial^m X_k}{\partial \varepsilon^m}\right)_{\varepsilon=0}$ *be finite trigonometric polynomials of the angle* θ;

4) the inequality

$$A_1(\tau, a, \psi) \leqslant Ca + C_1, \tag{8.9}$$

where C *and* C_1 *are some constants, be valid in the entire interval* $0 \leqslant \tau \leqslant L$.

Then, corresponding to any L *as large as desired and for any constants* M, S, *we may find positive* ε_0 *and* K_m, *such that the inequalities*

$$\left.\begin{aligned} &|x_k^{(m-1)} - x_k| < K_m \varepsilon^m, \\ &|\dot{x}_k^{(m-1)} - \dot{x}_k| < K_m \varepsilon^m \qquad (k = 1, 2, \ldots, N), \end{aligned}\right\} \tag{8.10}$$

are valid for all $\varepsilon (0 \leqslant \varepsilon < \varepsilon_0)$ *in the interval* $0 \leqslant t \leqslant \frac{L}{\varepsilon}$; *in these inequalities* $x_k^{(m-1)}$ *are the* m*-th approximations of particular solutions of the system (8.4), and* x_k *— the particular solutions of the system (8.4) which satisfy the initial conditions*

$$\left.\begin{aligned} &|x_k^{(m)}(0) - x_k(0)| \leqslant S\varepsilon^m, \quad |x_k^{(m)}(0)| \leqslant M, \\ &|\dot{x}_k^{(m)}(0) - \dot{x}_k(0)| \leqslant S\varepsilon^m, \quad |\dot{x}_k(0)| \leqslant M \qquad (k = 1, 2, \ldots, N), \end{aligned}\right\} \tag{8.11}$$

where $x_k^{(m)}$ *denote the* m*-th improved approximations.*

* Even if $X_k(\tau, \theta\, x_1, \ldots, x_N, \dot{x}_1, \ldots, \dot{x}_N, \varepsilon)$ $(k = 1, 2, \ldots, N)$ are not differentiable with respect to $x_1, \ldots, x_N, \dot{x}_1, \ldots, \dot{x}_N$, it is still possible in a number of cases to give a rigorous mathematical justification.

Note. By the conditions of the theorem:

1) The eigenfrequencies $\omega_k(\tau)$ $(k = 1, 2, \ldots, N)$ and the basic functions $\varphi_i^{(k)}(\tau)$ $(i, k = 1, 2, \ldots, N)$ are bounded in the interval $0 \leqslant \tau \leqslant L$, consequently, it is always possible to find positive nonvanishing constants ω_k^*, ω_k^0, m_k^*, m_k^0 $(k = 1, 2, \ldots, N)$ such that the inequalities

$$\omega_k^* \leqslant \omega_k(\tau) \leqslant \omega_k^0, \quad m_k^* \leqslant m_k(\tau) \leqslant m_k^0 \qquad (k = 1, 2, \ldots, N) \tag{8.12}$$

hold in the entire interval $0 \leqslant \tau \leqslant L$.

2) The $\omega_k(\tau)$, $\varphi_i^{(k)}(\tau)$ $(i, k = 1, 2, \ldots, N)$ are, according to /161/, unlimitedly differentiable for all finite values of τ.

3) For any finite x_i, x_i^*, $\dot{x}_i$, $\dot{x}_i^*$ $(i = 1, 2, \ldots, N)$, the Lifshitz condition

$$\begin{aligned}&|X_k(\tau, \theta, x_1, \ldots, x_N, \dot{x}_1, \ldots, \dot{x}_N, \varepsilon) - \\ &\qquad - X_k(\tau, \theta, x_1^*, \ldots, x_N^*, \dot{x}_1^*, \ldots, \dot{x}_N^*, \varepsilon)| \leqslant \\ &\leqslant \lambda\{|x_1 - x_1^*| + |x_2 - x_2^*| + \ldots + |\dot{x}_N - \dot{x}_N^*|\} \quad (k = 1, 2, \ldots, N),\end{aligned} \tag{8.13}$$

is satisfied, where λ is a positive constant.

The conditions imposed are not necessary, and may be replaced by other, in some cases even less severe, conditions. Thus, for example, for obtaining asymptotic approximations (especially the first approximation) and for estimating the error, differentiability is not always necessary— this condition may be replaced by a less severe one.

Before proving the above theorem, we shall demonstrate the validity of two auxiliary propositions.

Lemma 1. If the conditions of Theorem 1 are satisfied, then the m-th improved approximations of the solutions of the system of equations (8.4) (and also their first derivatives) are uniformly bounded in the interval $0 \leqslant \tau \leqslant L$.

Proof. When the conditions of the lemma are satisfied, $x_k^{(m)}$ (and also $\dot{x}_k^{(m)}$) are bounded in the interval $0 \leqslant \tau \leqslant L$, if $a^{(m)}$, determined from the system of equations of the m-th approximation (8.7)*, is bounded.

We now prove that $a^{(m)}$ is bounded.

Let $a^{(m)}(0) \leqslant \varrho$, where $\varrho \neq 0$ is an arbitrary constant; it is then possible to choose a constant

$$M_1 = \varrho e^{CL} + \frac{C_1}{C}(e^{CL} - 1) \tag{8.14}$$

and a small positive ε_0, such that

$$\varrho e^{CL} + \frac{C_1 + \varepsilon P}{C}(e^{CL} - 1) = 2M_1 \qquad (\varepsilon < \varepsilon_0) \tag{8.15}$$

for all

$$\varrho \leqslant a \leqslant 2M_1.$$

Here P denotes the maximum of the expression

$$|A_2| + \varepsilon|A_3| + \ldots + \varepsilon^{m-2}|A_m|. \tag{8.16}$$

If $a^{(m)}(t) \leqslant \varrho$ for all t in the interval $0 \leqslant t \leqslant \frac{L}{\varepsilon}$, then the lemma is valid, since $a^{(m)}$ is bounded. We show that if $a^{(m)}(t) \geqslant \varrho$ after a certain moment $t = t^*$, then $a^{(m)}(t) \leqslant 2M_1$ in the entire interval $0 \leqslant t \leqslant \frac{L}{\varepsilon}$ for any $\varepsilon \leqslant \varepsilon_0$.

* The functions $u_i^{(l)}(\tau, a, \theta, p\varphi + \psi)$ $(i = 1, 2, \ldots, N;\ l = 1, 2, \ldots, m)$, appear in the expression of $x_k^{(m)}$; the convergence of the series representing these functions is ensured by the fulfillment of the first three conditions of the theorem.

Indeed, according to (8.7), (8.9), and (8.16), the inequality

$$\frac{da^{(m)}}{dt} \leqslant \varepsilon C a^{(m)} + \varepsilon C_1 + \varepsilon^2 P \qquad (a^{(m)}(t^*) = \varrho), \tag{8.17}$$

is valid for $\varrho \leqslant a \leqslant 2M_1$.

Subtracting the equation

$$\frac{d\alpha}{dt} = \varepsilon C\alpha + \varepsilon C_1 + \varepsilon^2 P \qquad (\alpha(t^*) = \varrho) \tag{8.18}$$

from inequality (8.17), we obtain

$$\frac{d}{dt}(a^{(m)} - \alpha) \leqslant \varepsilon C (a^{(m)} - \alpha), \tag{8.19}$$

or

$$\frac{d}{dt}(a^{(m)} - \alpha) - \varepsilon C (a^{(m)} - \alpha) \leqslant 0. \tag{8.20}$$

Multiplying this inequality by $e^{-\varepsilon Ct}$, we find that

$$\frac{d}{dt}[e^{-\varepsilon Ct}(a^{(m)} - \alpha)] \leqslant 0, \tag{8.21}$$

from which

$$e^{-\varepsilon Ct}(a^{(m)} - \alpha) \leqslant 0, \tag{8.22}$$

consequently,

$$a^{(m)} \leqslant \alpha. \tag{8.23}$$

From equation (8.18) we obtain

$$\alpha = \varrho e^{\varepsilon C(t-t^*)} + \frac{C_1 + \varepsilon P}{C}(e^{\varepsilon C(t-t^*)} - 1). \tag{8.24}$$

Substituting (8.24) into (8.23), and taking (8.15) into account, we have

$$a^{(m)} \leqslant \varrho e^{\varepsilon C(t-t^*)} + \frac{C_1 + \varepsilon P}{C}(e^{\varepsilon C(t-t^*)} - 1) \leqslant 2M_1 \tag{8.25}$$

for any t in the interval $t^* \leqslant t \leqslant \frac{L}{\varepsilon}$, Q. E. D.

Lemma 2. *If the conditions of Theorem 1 are satisfied, it is always possible to find, for any positive number M, positive numbers ε_0 and $R_m^{(M)}$ for which the m-th improved approximations satisfy in the interval $0 \leqslant \tau \leqslant L$ the inequality*

$$\left| \frac{d^2 x_k^{(m)}}{dt^2} + \omega_k^2(\tau) x_k^{(m)} - \frac{\varepsilon}{m_k(\tau)} X_k(\tau, \theta, x_1^{(m)}, \ldots, x_N^{(m)}, \dot{x}_1^{(m)}, \ldots, \dot{x}_N^{(m)}, \varepsilon) \right| \leqslant$$
$$\leqslant R_m^{(M)} \varepsilon^{m+1} \qquad (k = 1, 2, \ldots, N), \tag{8.26}$$

where

$$0 < \varepsilon < \varepsilon_0, \quad |x_k^{(m)}(0)| \leqslant M, \quad |\dot{x}_k^{(m)}(0)| \leqslant M,$$

i.e., satisfy the original system (8.4) with an accuracy of the order of ε^{m+1}.

Actually, the right-hand side of the inequality (8.26) consists of a product of ε^{m+1} and a sum of terms of the type

$$u_i^{(l)}(\tau, a, \theta, p\varphi + \psi), \ldots, \quad A_l(\tau, a, \psi), \ldots, \quad B_l(\tau, a, \psi), \ldots$$
$$\ldots, \frac{\partial u_i^{(l)}}{\partial a}, \quad \frac{\partial u_i^{(l)}}{\partial \theta}, \ldots, \quad \frac{\partial^2 u_i^{(l)}}{\partial a\, \partial \theta}, \quad \frac{\partial^2 u_i^{(l)}}{\partial a^2}, \ldots$$
$$\ldots, \frac{\partial A_l(\tau, a, \psi)}{\partial a}, \ldots, \frac{\partial B_l(\tau, a, \psi)}{\partial \psi}, \ldots, \frac{\partial^l X_k}{\partial x_p \ldots \partial x_q}, \ldots, \frac{\partial^l X_k}{\partial \dot{x}_p \ldots \partial \dot{x}_q}$$
$$(i, k = 1, 2, \ldots, N; \; l = 1, 2, \ldots, m)$$

and various products of these terms. Denoting this sum by $R_{km}^{(M)}/m_k$, we conclude that under the conditions of the lemma

$$|R_{km}^{(M)}| \leqslant R_m^{(M)}, \tag{8.27}$$

where $R_m^{(M)}$ is some constant.

We can now prove the theorem formulated above.

Proof of Theorem 2. Substituting the expressions of the m-th improved approximation into equations (8.4), we obtain

$$\frac{d^2 x_k^{(m)}}{dt^2} + \omega_k^2(\tau) x_k^{(m)} = \frac{\varepsilon}{m_k(\tau)} X_k(\tau, \theta, x_1^{(m)}, \ldots, x_N^{(m)}, \dot{x}_1^{(m)}, \ldots, \dot{x}_N^{(m)}, \varepsilon) + \\ + \frac{\varepsilon^{m+1}}{m_k(\tau)} R_{km}^{(M)} \qquad (k = 1, 2, \ldots, N). \tag{8.28}$$

Subtracting (8.4) from (8.28), we obtain

$$\frac{d^2(x_k^{(m)} - x_k)}{dt^2} + \omega_k^2(\tau)(x_k^{(m)} - x_k) = \frac{\varepsilon}{m_k(\tau)} \{X_k(\tau, \theta, x_1^{(m)}, \ldots, x_N^{(m)}, \ldots \\ \ldots, \dot{x}_1^{(m)}, \ldots, \dot{x}_N^{(m)}, \varepsilon) - X_k(\tau, \theta, x_1, \ldots, x_N, \dot{x}_1, \ldots, \dot{x}_N, \varepsilon)\} + \varepsilon^m R_{km}^{(M)} \\ (k = 1, 2, \ldots, N). \tag{8.29}$$

Using Lemma 1 and the initial conditions (8.11), we can choose $\tau_0 \leqslant L$, such that x_k, $\dot{x}_k$ be bounded in the interval $0 \leqslant \tau \leqslant \tau_0$. In fact initially, by virtue of conditions (8.11),

$$|x_k(0)| \leqslant M + \varepsilon^m S, \quad |\dot{x}_k(0)| \leqslant M + \varepsilon^m S \qquad (k = 1, 2, \ldots, N).$$

Being continuous, $x_k(\tau)$, $\dot{x}_k(\tau)$ cannot change considerably for some small $\tau \leqslant \tau_0$.

At the moment $\tau = \tau_0$ let

$$|x_k(\tau_0)| \leqslant M + \varepsilon^m S_1, \quad |\dot{x}_k(\tau_0)| \leqslant M + \varepsilon^m S_1 \qquad (k = 1, 2, \ldots, N).$$

Choosing ε_0 so small that $S_1 \varepsilon^m \leqslant 1$, we find that

$$|x_k| \leqslant M + 1, \quad |\dot{x}_k| \leqslant M + 1 \qquad (k = 1, 2, \ldots, N)$$

in the entire interval $0 \leqslant \tau \leqslant \tau_0$.

Let us first prove the theorem in the interval $0 \leqslant \tau \leqslant \tau_0$.

We perform the following change of variables in equations (8.29):

$$\left.\begin{aligned} x_k^{(m)} - x_k &= u_k \cos\varphi_k + v_k \sin\varphi_k, \\ \dot{x}_k^{(m)} - \dot{x}_k &= -u_k \omega_k \sin\varphi_k + v_k \omega_k \cos\varphi_k \\ &\left(\varphi_k = \int_0^t \omega_k(\varepsilon t)\, dt, \quad k = 1, 2, \ldots, N\right). \end{aligned}\right\} \tag{8.30}$$

Then instead of (8.29) we obtain the equations

$$\left.\begin{aligned} \frac{du_k}{dt} &= -\frac{F_k(t)\sin\varphi_k}{\omega_k} - \frac{\varepsilon}{\omega_k}(u_k \omega_k' \sin\varphi_k - v_k \omega_k' \cos\varphi_k)\sin\varphi_k, \\ \frac{dv_k}{dt} &= \frac{F_k(t)\cos\varphi_k}{\omega_k} + \frac{\varepsilon}{\omega_k}(u_k \omega_k' \sin\varphi_k - v_k \omega_k' \cos\varphi_k)\cos\varphi_k \\ &\qquad (k = 1, 2, \ldots, N). \end{aligned}\right\} \tag{8.31}$$

Integrating these, we obtain

$$\left.\begin{aligned} u_k &= u_k(0) - \int_0^t \left\{\frac{F_k(t)\sin\varphi_k}{\omega_k} + \frac{\varepsilon}{\omega_k}(u_k \omega_k' \sin\varphi_k - v_k \omega_k' \cos\varphi_k)\sin\varphi_k\right\} dt, \\ v_k &= v_k(0) + \int_0^t \left\{\frac{F_k(t)\cos\varphi_k}{\omega_k} + \frac{\varepsilon}{\omega_k}(u_k \omega_k' \sin\varphi_k - v_k \omega_k' \cos\varphi_k)\cos\varphi_k\right\} dt \\ &\qquad (k = 1, 2, \ldots, N). \end{aligned}\right\} \tag{8.32}$$

Here $F_k(t)$ denotes the right-hand side of equation (8.29).
The inequalities

$$|u_k(0)|\leqslant S\varepsilon^m, \quad |v_k(0)|\leqslant \frac{S\varepsilon^m}{\omega_k^*} \qquad (k=1, 2, \ldots, N) \tag{8.33}$$

hold on account of (8.11).

Maximizing the right-hand sides of the expressions (8.32), and taking into account (8.12), (8.33), we obtain the inequalities

$$\left.\begin{aligned} |u_k| &\leqslant \frac{1}{\omega_k^*}\int_0^t |F_k(t)|\,dt + \frac{\varepsilon\omega_k^{0\prime}}{\omega_k^*}\int_0^t [|u_k|+|v_k|]\,dt + S\varepsilon^m, \\ |v_k| &\leqslant \frac{1}{\omega_k^*}\int_0^t |F_k(t)|\,dt + \frac{\varepsilon\omega_k^{0\prime}}{\omega_k^*}\int_0^t [|u_k|+|v_k|]\,dt + \frac{S\varepsilon^m}{\omega_k^*} \\ &\qquad (k=1, 2, \ldots, N) \end{aligned}\right\} \tag{8.34}$$

($\omega_k^{0\prime}$ is the highest value of ω_k' in the interval $0\leqslant\tau\leqslant L$), which may, according to Lemma 1, (8.13) and (8.30), be expressed in the form

$$\left.\begin{aligned} |u_k| \leqslant \frac{\varepsilon\lambda(1+\omega_*)}{m_k^*\omega_k^*}\sum_{r=1}^N\int_0^t [|u_r|+|v_r|]\,dt &+ \\ +\varepsilon\frac{\omega_k^{0\prime}}{\omega_k^*}\int_0^t [|u_k|+|v_k|]\,dt &+ \frac{\varepsilon^{m+1}}{m_k^*\omega_k^*}R_m^{(M)}t + \varepsilon^m S_*, \\ |v_k| \leqslant \frac{\varepsilon\lambda(1+\omega_*)}{m_k^*\omega_k^*}\sum_{r=1}^N\int_0^t [|u_r|+|v_r|]\,dt &+ \\ +\varepsilon\frac{\omega_k^{0\prime}}{\omega_k^*}\int_0^t [|u_k|+|v_k|]\,dt &+ \frac{\varepsilon^{m+1}}{m_k^*\omega_k^*}R_m^{(M)}t + \varepsilon^m S_*. \end{aligned}\right\} \tag{8.35}$$

Here S_* is the greater of the numbers S, $\frac{S}{\omega_k^*}$, and ω_* is the largest of the ω_k^0 $(k=1, 2, \ldots, N)$. For simplicity we use the notation

$$\sum_{k=1}^N \{|u_k|+|v_k|\} = z_m. \tag{8.36}$$

Then, adding inequalities (8.35) to each other and summing the result over k, we obtain

$$z_m \leqslant \varepsilon g_1\int_0^t z_m(t)\,dt + \varepsilon^{m+1}g_2R_m^{(M)}t + 2\varepsilon^m S_*, \tag{8.37}$$

where

$$g_1 = 2\lambda(1+\omega_*)\sum_{k=1}^N\frac{1}{m_k^*\omega_k^*} + 2\sum_{k=1}^N\frac{\omega_k^{0\prime}}{\omega_k^*}, \quad g_2 = 2\sum_{k=1}^N\frac{1}{m_k^*\omega_k^*}.$$

We consider together with the inequality (8.37) the equation

$$\xi_m = \varepsilon g_1\int_0^t \xi_m(t)\,dt + \varepsilon^{m+1}g_2R_m^{(M)}t + 2\varepsilon^m S_*. \tag{8.38}$$

Subtracting it from inequality (8.37), we obtain

$$z_m - \xi_m \leqslant \varepsilon g_1\int_0^t (z_m(t)-\xi_m(t))\,dt, \tag{8.39}$$

or

$$z_m - \xi_m - \varepsilon g_1 \int_0^t (z_m(t) - \xi_m(t))\, dt \leqslant 0. \tag{8.40}$$

Multiplying this inequality by $e^{-\varepsilon g_1 t}$, we obtain

$$(z_m - \xi_m) e^{-\varepsilon g_1 t} - \varepsilon g_1 e^{-\varepsilon g_1 t} \int_0^t (z_m(t) - \xi_m(t))\, dt \leqslant 0, \tag{8.41}$$

or

$$\frac{d}{dt}\left[e^{-\varepsilon g_1 t} \int_0^t (z_m(t) - \xi_m(t))\, dt \right] \leqslant 0, \tag{8.42}$$

which immediately implies the inequality

$$\int_0^t (z_m(t) - \xi_m(t))\, dt \leqslant 0. \tag{8.43}$$

Comparing (8.40) and (8.43), we find

$$z_m - \xi_m \leqslant 0,$$

or

$$z_m \leqslant \xi_m. \tag{8.44}$$

From equation (8.38) we have

$$\xi_m = \varepsilon^m \left\{ \frac{g_2}{g_1} R_m^{(M)} (e^{\varepsilon g_1 t} - 1) + 2S_* e^{\varepsilon g_1 t} \right\}. \tag{8.45}$$

Replacing εt by τ_0 in the right-hand side of the last expression and taking into account (8.44), we obtain the inequality

$$z_m \leqslant \varepsilon^m \left\{ \frac{g_2}{g_1} R_m^{(M)} (e^{g_1 \tau_0} - 1) + 2S_* e^{g_1 \tau_0} \right\}, \tag{8.46}$$

which implies, according to the notation introduced above,

$$|x_k^{(m)} - x_k| \leqslant S_m \varepsilon^m, \quad |\dot{x}_k^{(m)} - \dot{x}_k| \leqslant S_m \varepsilon^m \qquad (k = 1, 2, \ldots, N), \tag{8.47}$$

where S_m is some constant.

We have thus obtained an estimate for the m-th improved approximation which, as known, differs from the m-th approximation by a quantity of the order of ε^m, i. e.,

$$|x_k^{(m)} - x_k^{(m-1)}| \leqslant \varepsilon^m r_m, \quad |\dot{x}_k^{(m)} - \dot{x}_k| \leqslant \varepsilon^m r_m \qquad (k = 1, 2, \ldots, N), \tag{8.48}$$

where r_m is a constant.

Consequently, choosing k_m such that $r_m < \frac{k_m}{2}$, $S_m < \frac{k_m}{2}$, we obtain the inequalities

$$\left.\begin{aligned} |x_k^{(m-1)} - x_k| &\leqslant \varepsilon^m S_m + \varepsilon^m r_m < \varepsilon^m k_m, \\ |\dot{x}_k^{(m-1)} - \dot{x}_k| &\leqslant \varepsilon^m S_m + \varepsilon^m r_m < \varepsilon^m k_m. \end{aligned}\right\} \tag{8.49}$$

We have thus proved the theorem in the interval $0 \leqslant \tau \leqslant \tau_0$. We shall now show that τ_0 may be taken equal to L.

If this were not so, the inequalities (8.49) could not be satisfied in the entire interval $0 \leqslant \tau \leqslant L$, since then x_k, $\dot{x}_k$ would not be bounded there. Since these inequalities are satisfied for sufficiently small τ, there obviously exists such a τ_1 that they are satisfied for $\tau < \tau_1$, and for $\tau = \tau_1$

$$\left.\begin{aligned} &|x_k^{(m-1)}(\tau_1) - x_k(\tau_1)| > \varepsilon^m k_m - \varrho_1, \\ &|\dot{x}_k^{(m-1)}(\tau_1) - \dot{x}_k(\tau_1)| > \varepsilon^m k_m - \varrho_1 \qquad (k = 1, 2, \ldots, N), \end{aligned}\right\} \tag{8.50}$$

where ϱ_1 may be made as small as desired.

Let $\tau_0 = \tau_1$ (this is possible since x_k, $\dot{x}_k$ are bounded for $\tau = \tau_1$), and let ϱ_1 be chosen as

$$\varrho_1 = \frac{\varepsilon^m}{2}\left\{\frac{k_m}{2} - r_m\right\}. \tag{8.51}$$

Substituting r_m from (8.51) and $\tau = \tau_1$ into the inequalities (8.49), we obtain

$$\left.\begin{aligned} &|x_k^{(m-1)}(\tau_1) - x_k(\tau_1)| \leqslant \varepsilon^m S_m + \frac{\varepsilon^m k_m}{2} - 2\varrho_1 < \varepsilon^m k_m - 2\varrho_1, \\ &|\dot{x}_k^{(m-1)}(\tau_1) - \dot{x}_k(\tau_1)| \leqslant \varepsilon^m S_m + \frac{\varepsilon^m k_m}{2} - 2\varrho_1 < \varepsilon^m k_m - 2\varrho_1, \end{aligned}\right\} \tag{8.52}$$

which contradicts (8.50). Hence our assumption is wrong and we can take $\tau_0 = L$. This implies that the inequalities (8.49) are satisfied in the entire interval $0 \leqslant \tau \leqslant L$, Q. E. D.

Analogous theorems can be proved, considering instead of the system (8.4) systems of equations of the type

$$\frac{dx}{dt} = \varepsilon X(\tau, \theta, x, \varepsilon), \tag{8.53}$$

where x, X are points of the n-dimensional Euclidean space E_n, $\frac{d\theta}{dt} = \nu(\tau)$, $\tau = \varepsilon t$, or of the type

$$\left.\begin{aligned} &\frac{dx}{dt} = X(y, \theta, x, \varepsilon), \\ &\frac{dy}{dt} = \varepsilon Y(y, \theta, x, \varepsilon), \end{aligned}\right\} \tag{8.54}$$

to which most of the equations discussed in this book may be reduced, as shown in the preceding chapter.

The methods of constructing asymptotic solutions for systems with slowly varying parameters, developed in the preceding chapters, may be reduced to a modified version of the well-known method of averaging (evidently taking into account several correction terms connected with the presence of the "slowing time" τ or the slowly changing variables y) applied to systems of equations of the type (8.53) or, in more general cases, to systems of the type (8.54), discussed in detail by V. M. Volosov /32/. It is thus possible to formulate and prove the theorems about the error estimation in a more general form. We shall not consider this here, but proceed to justify our method of constructing families of particular solutions corresponding to monofrequency vibrations.

§ 2. Some stability criteria of a monofrequency mode in vibrating systems with slowly varying parameters

It has already been indicated that the expansions (8.2) do not give an approximate expression for the general solution of the system of equations

(8.1), but only for a certain two-parametric family of particular solutions. Any solution of equations (8.1) starting from the initial point $(q_{10}, \ldots, q_{N0})$ which lies sufficiently close to the initial point $(q_{10}^{(m)}, \ldots, q_{N0}^{(m)})$ of the two-parametric family (8.2), remains, according to the above proved Theorem 1, sufficiently close to the obtained approximate particular solution in the entire interval $0 \leqslant \tau \leqslant L$.

However, the family of solutions (8.2) has in some important cases a special property of strong stability, consisting in the fact that any solution of the system of equations (8.1), whose initial values may even not be close to the initial values of the two-parametric family of solutions (8.2), tends to the solutions of (8.2) (to the family of solutions as a whole) when t increases.

In the present section we consider some particular cases of the system (8.1) for which the two-parametric family (8.2) possesses the property of strong stability. These cases are of importance for application.

Let the perturbing forces appearing on the right-hand side of the system (8.1) have the form

$$Q_j(\tau, \theta, q_1, \ldots, q_N, \dot{q}_1, \ldots, \dot{q}_N, \varepsilon) = \\ = Q_j(q_1, \ldots, q_N, \dot{q}_1, \ldots, \dot{q}_N, \varepsilon) + E_j(\tau)\cos\theta \qquad (j = 1, 2, \ldots, N), \tag{8.55}$$

and the coefficients $a_{ij}(\tau)$ and $b_{ij}(\tau)$ in (8.1) be independent of τ. Then, transforming by means of the change of variables

$$q_i = \sum_{k=1}^{N} \varphi_i^{(k)} x_k \qquad (i = 1, 2, \ldots, N), \tag{8.56}$$

where $\varphi_i^{(k)}$ do not depend any more on τ, to quasi-normal coordinates, we obtain the system of differential equations

$$\frac{d^2x_k}{dt^2} + \omega_k^2 x_k = \varepsilon X_k(x_1, \ldots, x_N, \dot{x}_1, \ldots, \dot{x}_N, \varepsilon) + \varepsilon F_k(t) \qquad (k = 1, 2, \ldots, N), \tag{8.57}$$

where

$$F_k(t) = U_k(\tau)\cos\theta, \quad \omega_k = \text{const}, \quad \tau = \varepsilon t, \quad \frac{d\theta}{dt} = \nu(\tau). \tag{8.58}$$

For simplicity we shall study the fundamental resonance.

The following theorem is valid:

***Theorem 3**. Let the following conditions be satisfied:*

1) positive constants λ_k exist for which the functions $X_k(x_1, \ldots, x_N, \dot{x}_1, \ldots, \dot{x}_N, \varepsilon)$ $(k = 1, 2, \ldots, N)$ satisfy the conditions

$$|X_k(x_1^{(1)}, \ldots, x_N^{(1)}, \dot{x}_1^{(1)}, \ldots, \dot{x}_N^{(1)}, \varepsilon) - X_k(x_1^{(2)}, \ldots, x_N^{(2)}, \dot{x}_1^{(2)}, \ldots, \dot{x}_N^{(2)}, \varepsilon)| \leqslant \\ \leqslant \lambda_k \sum_{m=1}^{N} \{|x_m^{(1)} - x_m^{(2)}|\,\omega_m + |\dot{x}_m^{(1)} - \dot{x}_m^{(2)}|\} \qquad (k = 1, 2, \ldots, N); \tag{8.59}$$

2) the functions $F_k(t)$ are bounded in any interval $0 \leqslant t \leqslant \frac{L}{\varepsilon}$:

$$|F_k(t)| \leqslant I_k \qquad (k = 1, 2, \ldots, N); \tag{8.60}$$

3) for any $T \geqslant T_0$ the inequality

$$\frac{1}{T}\int_0^T \sum_{k=1}^{N} X_k(x_1^{(0)}, \ldots, x_N^{(0)}, \dot{x}_1^{(0)}, \ldots, \dot{x}_N^{(0)}, \varepsilon)\,\dot{x}_k^0\,dt \leqslant -\beta \sum_{k=1}^{N} a_k^2, \tag{8.61}$$

is valid, where

$$x_k^{(0)} = a_k \cos(\omega_k t + \varphi_k) \quad (k = 1, 2, \ldots, N),$$

and the constants a_k *and* φ_k *are defined by the expressions*

$$a_k = \sqrt{x_k^2(t_0) + \frac{\dot{x}_k^2(t_0)}{\omega_k^2}}, \quad \operatorname{tg} \varphi_k = \frac{x_k(t_0)}{\omega_k x_k(t_0)} \quad (k = 1, 2, \ldots, N). \tag{8.62}$$

Then positive ε_0, M_k, N_k, *and* T^* *exist, such that the inequalities*

$$|x_k(t)| \leqslant M_k, \quad |\dot{x}_k(t)| \leqslant N_k \quad (k = 1, 2, \ldots, N) \tag{8.63}$$

are satisfied for any solution of the system of differential equations (8.57) for all $t \geqslant T^*$.

Before proving this theorem, we prove the following lemma.

Lemma 3. If the first two conditions of Theorem 3 are satisfied, then corresponding to any positive L *as large as desired we can always find constants* C, C_1, C_2 *and a small positive* ε_0, *such that the inequalities*

$$\left.\begin{aligned} &|x_k(t) - a_k \cos(\omega_k t + \varphi_k)| \leqslant C\sqrt{\varepsilon}\Big(\sum_{m=1}^{N} |a_m| + C_1\Big), \\ &|\dot{x}_k(t) - a_k \omega_k \sin(\omega_k t + \varphi_k)| \leqslant C\omega_k \sqrt{\varepsilon}\Big(\sum_{m=1}^{N} |a_m| + C_2\Big) \\ &\qquad (k = 1, 2, \ldots, N), \end{aligned}\right\} \tag{8.64}$$

in which the constants a_k *and* φ_k *are defined by (8.62), are satisfied in the interval*

$$|t - t_0| < \frac{L}{\sqrt{\varepsilon}} \tag{8.65}$$

for all $\varepsilon \leqslant \varepsilon_0^2$.

Proof. Integrating equations (8.57) between the limits t_0 and t, we obtain

$$\left.\begin{aligned} x_k(t) &= x_k(t_0)\cos\omega_k(t - t_0) + \frac{\dot{x}_k(t_0)}{\omega_k}\sin\omega_k(t - t_0) + \\ &\quad + \frac{\varepsilon}{\omega_k}\int_{t_0}^{t} F_k(t_1)\sin\omega_k(t - t_1)\,dt_1 + \\ &\quad + \frac{\varepsilon}{\omega_k}\int_{t_0}^{t} X_k(x_1, \ldots, x_N, \dot{x}_1, \ldots, \dot{x}_N, \varepsilon)\sin\omega_k(t - t_1)\,dt_1, \\ \dot{x}_k(t) &= -x_k(t_0)\,\omega_k\sin\omega_k(t - t_0) + \dot{x}_k(t_0)\cos\omega_k(t - t_0) + \\ &\quad + \varepsilon\int_{t_0}^{t} X_k(x_1, \ldots, x_N, \dot{x}_1, \ldots, \dot{x}_N, \varepsilon)\cos\omega_k(t - t_1)\,dt_1 + \\ &\quad + \varepsilon\int_{t_0}^{t} F_k(t_1)\cos\omega_k(t - t_1)\,dt_1 \quad (k = 1, 2, \ldots, N). \end{aligned}\right\} \tag{8.66}$$

We use the notation

$$\left.\begin{aligned} &x_k(t_0)\cos\omega_k(t - t_0) + \frac{\dot{x}_k(t_0)}{\omega_k}\sin\omega_k(t - t_0) = x_k^{(0)}(t), \\ &-x_k(t_0)\,\omega_k\sin\omega_k(t - t_0) + \dot{x}_k(t_0)\cos\omega_k(t - t_0) = \dot{x}_k^{(0)}(t) \\ &\qquad (k = 1, 2, \ldots, N). \end{aligned}\right\} \tag{8.67}$$

Next, maximizing the right-hand sides of (8.66), we obtain the inequalities

$$|x_k(t) - x_k^{(0)}(t)| \leqslant \frac{\varepsilon}{\omega_k}\int_{t_0}^{t} |X_k(x_1, \ldots, x_N, \dot{x}_1, \ldots, \dot{x}_N, \varepsilon)|\,dt_1 + \frac{\varepsilon}{\omega_k} I_k(t - t_0), \tag{8.68}$$

$$|\dot{x}_k(t)-\dot{x}_k^{(0)}(t)|\leqslant \varepsilon\int_{t_0}^{t}|X_k(x_1,\ \dots,\ x_N,\ \dot{x}_1,\ \dots,\ \dot{x}_N,\ \varepsilon)|\,dt_1+ \\ +\varepsilon I_k(t-t_0)\qquad (k=1,\ 2,\ \dots,\ N). \tag{8.68}$$

With the help of (8.59) we can express these inequalities in the form

$$\left.\begin{array}{l} |x_k(t)-x_k^{(0)}(t)|\leqslant \frac{\varepsilon\lambda_k}{\omega_k}\sum_{m=1}^{N}\int_{t_0}^{t}\{|x_m-x_m^{(0)}|\,\omega_m+|\dot{x}_m-\dot{x}_m^{(0)}|\}\,dt_1+ \\ +\frac{\varepsilon}{\omega_k}\int_{t_0}^{t}|X_k(x_1^{(0)},\ \dots,\ x_N^{(0)},\ \dot{x}_1^{(0)},\ \dots,\ \dot{x}_N^{(0)},\ \varepsilon)|\,dt_1+\frac{\varepsilon}{\omega_k}I_k(t-t_0), \\ |\dot{x}_k(t)-\dot{x}_k^{(0)}(t)|\leqslant \varepsilon\lambda_k\sum_{m=1}^{N}\int_{t_0}^{t}\{|x_m-x_m^{(0)}|\,\omega_m+|\dot{x}_m-\dot{x}_m^{(0)}|\}\,dt_1+ \\ +\varepsilon\int_{t_0}^{t}|X_k(x_1^{(0)},\ \dots,\ x_N^{(0)},\ \dot{x}_1^{(0)},\ \dots,\ \dot{x}_N^{(0)},\ \varepsilon)|\,dt_1+\varepsilon I_k(t-t_0) \\ (k=1,\ 2,\ \dots,\ N). \end{array}\right\} \tag{8.69}$$

Adding these inequalities to each other, and summing the result over k, we obtain

$$z(t)\leqslant 2\varepsilon\sum_{k=1}^{N}\lambda_k\int_{t_0}^{t}z(t_1)\,dt_1+ \\ +2\varepsilon\sum_{k=1}^{N}\int_{t_0}^{t}|X_k(x_1^{(0)},\ \dots,\ x_N^{(0)},\ \dot{x}_1^{(0)},\ \dots,\ \dot{x}_N^{(0)},\ \varepsilon)|\,dt_1+2\varepsilon\sum_{k=1}^{N}I_k(t-t_0), \tag{8.70}$$

where

$$z(t)=\sum_{m=1}^{N}\{|x_m(t)-x_m^{(0)}(t)|\,\omega_m+|\dot{x}_m(t)-\dot{x}_m^{(0)}(t)|\}. \tag{8.71}$$

We estimate the second term of the right-hand side of (8.70). By the condition of the lemma and using (8.67) we have

$$|X_k(x_1^{(0)},\ \dots,\ x_N^{(0)},\ \dot{x}_1^{(0)},\ \dots,\ \dot{x}_N^{(0)},\ \varepsilon)|\leqslant|X_k^{(0)}|+\lambda_k\sum_{m=1}^{N}\{|x_m^{(0)}|\,\omega_m+|\dot{x}_m^{(0)}|\}\leqslant \\ \leqslant|X_k^{(0)}|+2\lambda_k\sum_{m=1}^{N}|a_m|\,\omega_m\qquad (k=1,\ 2,\ \dots,\ N). \tag{8.72}$$

Here $X_k^{(0)}=X_k(0,\ 0,\ \dots,\ 0,\ \varepsilon)$. Thus, the inequality (8.70) may be expressed in the form

$$z(t)\leqslant 2\varepsilon\sum_{k=1}^{N}\lambda_k\int_{t_0}^{t}z(t_1)\,dt_1+\ 2\varepsilon\left\{\sum_{k=1}^{N}\lambda_k|X_k^{(0)}|+\sum_{k=1}^{N}I_k+2\sum_{k=1}^{N}\lambda_k\sum_{m=1}^{N}|a_m|\,\omega_m\right\}(t-t_0). \tag{8.73}$$

Denoting the right-hand side of (8.73) by $\zeta(t)$, we obtain the inequality

$$z(t)\leqslant\zeta(t), \tag{8.74}$$

where $\zeta(t)$ is the solution of the equation

$$\zeta(t)=2\varepsilon\sum_{k=1}^{N}\lambda_k\int_{t_0}^{t}\zeta(t_1)\,dt_1+\ 2\varepsilon\left\{\sum_{k=1}^{N}\lambda_k|X_k^{0}|+\sum_{k=1}^{N}I_k+2\sum_{k=1}^{N}\lambda_k\sum_{m=1}^{N}|a_m|\,\omega_m\right\}(t-t_0), \tag{8.75}$$

solving which, we obtain

$$\zeta(t)=\Big[\sum_{k=1}^{N}\lambda_k|X_k^{(0)}|+\sum_{k=1}^{N}I_k+2\sum_{k=1}^{N}\lambda_k\sum_{m=1}^{N}|a_m|\,\omega_m\Big]\times \\ \times\{\exp\,[2\varepsilon\sum_{k=1}^{N}\lambda_k(t-t_0)]-1\}\,(\sum_{k=1}^{N}\lambda_k)^{-1}. \tag{8.76}$$

We now put

$$\frac{1}{L}=2\sum_{k=1}^{N}\lambda_k.$$

Then, substituting $L/\sqrt{\varepsilon}$ into (8.76) instead of $t-t_0$, and taking into account (8.73), we deduce that

$$z(t)\leqslant(e^{\sqrt{\varepsilon}}-1)\{2\sum_{m=1}^{N}|a_m|\omega_m+S\}, \tag{8.77}$$

where

$$S=\sum_{k=1}^{N}|X_k^{(0)}|+\sum_{k=1}^{N}I_k|\sum_{k=1}^{N}\lambda_k|. \tag{8.78}$$

Neglecting terms of the first and higher order in ε in the right-hand side of (8.77) we obtain (on account of (8.70)) for the interval $|t-t_0|\leqslant L/\sqrt{\varepsilon}$ the inequalities (8.64), Q. E. D.

We now prove the theorem.

Proof of Theorem 3. We multiply equations (8.37) by $\dot{x}_k$ and integrate them from t_0 to t_0+T ($T=L/\varepsilon$, ε being such that $L/\sqrt{\varepsilon}\geqslant T$. Summing the result over k, we obtain

$$\sum_{k=1}^{N}\frac{1}{2}[\dot{x}_k^2+\omega_k^2x_k^2]\Big|_{t_0}^{t_0+T}=$$
$$=\sum_{k=1}^{N}\varepsilon\int_{t_0}^{t_0+T}X_k(x_1,\ldots,x_N,\dot{x}_1,\ldots,\dot{x}_N,\varepsilon)\dot{x}_k\,dt+\sum_{k=1}^{N}\varepsilon\int_{t_0}^{t_0+T}F_k(t)\dot{x}_k\,dt. \tag{8.79}$$

Noting that $\dot{x}_k^2+\omega_k^2x_k^2=a_k^2(t)\omega_k^2$, $x_k^{(0)}=a_k(t_0)\cos(\omega_kt+\varphi_k(t_0))$ $(k=1,2,\ldots,N)$ we can write (8.79) in the form

$$\omega_k^2\sum_{k=1}^{N}\frac{1}{2}[a_k^2(t_0+T)-a_k^2(t_0)]=$$
$$=\varepsilon\sum_{k=1}^{N}\int_{t_0}^{t_0+T}X_k(x_1^{(0)},\ldots,x_N^{0},\dot{x}_1^{(0)},\ldots,\dot{x}_N^{(0)},\varepsilon)\dot{x}_k^{(0)}\,dt+$$
$$+\varepsilon\sum_{k=1}^{N}\int_{t_0}^{t_0+T}F_k(t)\dot{x}_k^{(0)}\,dt+\sum_{k=1}^{N}D_k, \tag{8.80}$$

where the D_k satisfy (by the 1-st and 2-nd conditions of the theorem) the inequalities

$$|D_k|\leqslant\varepsilon I_k\int_{t_0}^{t_0+T}|\dot{x}_k-\dot{x}_k^{(0)}|\,dt+$$
$$+\varepsilon\int_{t_0}^{t_0+T}|X_k(x_1^{(0)},\ldots,x_N^{(0)},\dot{x}_1^{(0)},\ldots,\dot{x}_N^{(0)},\varepsilon)||\dot{x}_k-\dot{x}_k^{(0)}|\,dt+$$
$$+\varepsilon\int_{t_0}^{t_0+T}|\dot{x}_k^{(0)}|\lambda_k\sum_{m=1}^{N}\{|x_m^{(0)}-x_m|\omega_m+|\dot{x}_m^{(0)}-\dot{x}_m|\}\,dt+$$
$$+\int_{t_0}^{t_0+T}|\dot{x}_k-\dot{x}_k^{(0)}|\lambda_k\sum_{m=1}^{N}\{|x_m^{(0)}-x_m|\omega_m+|\dot{x}_m^{(0)}-\dot{x}_m|\}\,dt\qquad(k=1,2,\ldots,N). \tag{8.81}$$

Summing these inequalities over k, we can verify, on the grounds of the lemma proved above and the expressions (8.70), (8.76), and (8.77), that

$$\sum_{k=1}^{N}|D_k|\leqslant\varepsilon\sqrt{\varepsilon}\,[S_1(\sum_{k=1}^{N}|a_k|)^2+S_2\sum_{k=1}^{N}|a_k|+S_3]\,T, \tag{8.82}$$

where

$$\left.\begin{aligned} S_1 &= 2C\sum_{k=1}^{N}\left[\lambda_k\omega_k+\sqrt{\varepsilon}\,C\lambda_k\omega_k\sum_{m=1}^{N}\omega_m+2C\lambda_k\sum_{m=1}^{N}\omega_m\right],\\ S_2 &= \sum_{k=1}^{N}\left[CI_k\omega_k+2C_1C_2\lambda_k\omega_k+C\,|X_k^{(0)}|\,\omega_k+\right.\\ &\left.\quad+\sqrt{\varepsilon}\,C^2(C_1+3C_2)\,\omega_k\lambda_k\sum_{m=1}^{N}\omega_m+C(C_1+C_2)\,\lambda_k\sum_{m=1}^{N}\omega_m\right],\\ S_3 &= \sum_{k=1}^{N}\left[CC_2I_k\omega_k+CC_2\omega_k\,|X_k^{(0)}|+C^2C_2(C_1+C_2)\sqrt{\varepsilon}\,\omega_k\lambda_k\sum_{m=1}^{N}\omega_m\right].\end{aligned}\right\}\qquad(8.83)$$

Applying the Bunyakovskii-Schwarz inequality

$$\Big(\sum_{k=1}^{N}|a_k|\Big)^2\leqslant N\sum_{k=1}^{N}a_k^2,\qquad(8.84)$$

we can write (8.82) in the form

$$\sum_{k=1}^{N}|D_k|\leqslant\varepsilon\sqrt{\varepsilon}\,\Big\{S_1N\sum_{k=1}^{N}a_k^2+S_2\sum_{k=1}^{N}|a_k|+S_3\Big\}\,T.\qquad(8.85)$$

We choose ε_0 so that

$$\sqrt{\varepsilon}\,S_1N\leqslant\frac{\beta}{2}\qquad(8.86)$$

for $\varepsilon\leqslant\varepsilon_0^2$.

Maximizing the right-hand side of (8.80), we then obtain, on account of the inequalities (8.85), (8.86), and (8.77)

$$\sum_{k=1}^{N}\frac{1}{2}\left[a_k^2(t_0+T)-a_k^2(t_0)\right]\leqslant-\frac{\beta\varepsilon}{2}\sum_{k=1}^{N}a_k^2(t_0)\,T+$$
$$+\varepsilon\sqrt{\varepsilon}\,S_2\sqrt{\sum_{k=1}^{N}a_k^2(t_0)\,T}+\varepsilon\sqrt{\varepsilon}\,S_3T+\varepsilon I_1\sqrt{\sum_{k=1}^{N}a_k^2(t_0)\,T},\qquad(8.87)$$

where $I_1=\sqrt{N}\,I_{k\,\max}$. Using the notation

$$\sum_{k=1}^{N}\omega_k^2a_k^2(t_0)=\eta^2(t_0),\qquad t_s=sT,\qquad(8.88)$$

we find

$$\eta^2(t_{s+1})\leqslant\eta^2(t_s)\,(1-\beta\varepsilon T)+2\varepsilon I_1\eta(t_s)\,T+2\varepsilon\sqrt{\varepsilon}\,S_2\eta(t_s)\,T+2\varepsilon\sqrt{\varepsilon}\,S_3T.\qquad(8.89)$$

Let us consider the inequality

$$-\frac{\beta\varepsilon}{2}\,\eta^2(t_s)+2\varepsilon I_1\eta(t_s)+2\varepsilon\sqrt{\varepsilon}\,S_2\eta(t_s)+2\varepsilon\sqrt{\varepsilon}\,S_3\leqslant0,\qquad(8.90)$$

which will be satisfied if

$$\eta(t_s)\geqslant\frac{2}{\beta}\left\{I_1+S_2\sqrt{\varepsilon}+\sqrt{(I_1+S_2\sqrt{\varepsilon})^2+S_3\beta\sqrt{\varepsilon}}\right\}=M.\qquad(8.91)$$

We use the notation

$$M^2(1-\beta\varepsilon T)+2\varepsilon T\left[I_1M+\sqrt{\varepsilon}\,(S_2M+S_3)\right]=M_1^2.\qquad(8.92)$$

If $\eta(t_s)\geqslant M$, then

$$\eta^2(t_{s+1})\leqslant M^2\left(1-\frac{\beta\varepsilon}{2}T\right),\qquad(8.93)$$

by virtue of the inequalities (8.89)–(8.91).

If, however, $\eta(t_s) \leqslant M$, then according to (8.89)

$$\eta^2(t_{s+1}) \leqslant M_1^2. \tag{8.94}$$

It follows from the inequalities (8.93), (8.94) that the inequality

$$\eta(t_s) \leqslant \sup(M, M_1), \tag{8.95}$$

is satisfied from a certain $s \geqslant s_0$ on.

Indeed, if $\eta(t_s) \geqslant M$, and $M \geqslant M_1$, then by virtue of (8.93) we can always find some s_0, such that $\eta(t_{s_0}) \leqslant M$, hence, according to (8.94) $\eta(t_{s_0+1}) \leqslant M_1$, i.e., (8.95) is valid for all $s \geqslant s_0$

If $\eta(t_s) \geqslant M$, but $M_1 \geqslant M$, then for a certain s_0 we have $\eta(t_{s_0}) \leqslant M \leqslant M_1$, $\eta(t_{s_0+1}) \leqslant M$, and the inequality (8.95) is valid.

Now let $\eta(t_s) \leqslant M$, and $M \geqslant M_1$; then, according to (8.94), we have $\eta(t_{s_0}) \leqslant M_1 \leqslant M$, $\eta(t_{s_0+1}) \leqslant M_1 \leqslant M$ and the inequality (8.95) holds for all $s \geqslant s_0$. If $\eta(t_s) \leqslant M$, but $M \leqslant M_1$, then $\eta(t_{s_0}) \leqslant M_1$ for a certain s_0, and we obtain one of the cases considered above.

Hence the inequality (8.95) is always satisfied beginning from some $s \geqslant s_0$, which implies the validity of the inequalities (8.63), Q. E. D.

We shall now prove the theorem about the boundedness of the solutions of the system (8.57) under different conditions imposed on the functions $X_k(x_1, \ldots, x_N, \dot{x}_1, \ldots, \dot{x}_N, \varepsilon)$.

We assume that the generalized forces acting on the system may be separated into potential and dissipative forces

$$\varepsilon X_k(x_1, \ldots, x_N, \dot{x}_1, \ldots, \dot{x}_N, \varepsilon) = -\varepsilon \frac{\partial V(x_1, \ldots, x_N, \varepsilon)}{\partial x_k} + \varepsilon R_k(\dot{x}_1, \ldots, \dot{x}_N, \varepsilon). \tag{8.96}$$

Equations (8.57) become in this case

$$\frac{d^2x}{dt^2} + \omega_k^2 x_k = -\varepsilon \frac{\partial V}{\partial x_k} + \varepsilon R_k + \varepsilon F_k(t) \qquad (k = 1, 2, \ldots, N), \tag{8.97}$$

and the theorem about the boundedness may be formulated as follows.

Theorem 4. *If*

$$\left.\begin{aligned} &\sum_{k=1}^{N} \omega_k^2 x_k^2 + \varepsilon V \geqslant \omega^2 \sum_{k=1}^{N} x_k^2, \\ &\sum_{k=1}^{N} \omega_k^2 x_k^2 + \varepsilon \sum_{k=1}^{N} \frac{\partial V}{\partial x_k} x_k \geqslant \omega^2 \sum_{k=1}^{N} x_k^2, \\ &\varepsilon \sum_{k=1}^{N} R_k \dot{x}_k \leqslant -\beta \sum_{k=1}^{N} \dot{x}_k^2, \quad |R_k| \leqslant S' \sum_{k=1}^{N} |\dot{x}_k|, \quad |F_k(t)| \leqslant I, \end{aligned}\right\} \tag{8.98}$$

where ω, β, S', *and* I *are positive constants, then corresponding to any* ε_0 *as small as desired, we can always find positive constants* M_k, N_k, *and* T^*, *for any solution of the system of equations (8.97) such that the inequalities*

$$|x_k(t)| \leqslant M_k, \quad |\dot{x}_k(t)| \leqslant N_k \qquad (k = 1, 2, \ldots, N) \tag{8.99}$$

hold for all $\varepsilon \leqslant \varepsilon_0$ *and* $t \geqslant T^*$*.

Proof of Theorem 4. Multiplying both sides of equations (8.97) by the corresponding $\dot{x}_k$, and then by x_k, and summing over k, we obtain the two

* The constants N_k and M_k may be fixed independently of the choice of the solution.

equations

$$\frac{d}{dt}\left\{\frac{1}{2}\sum_{k=1}^{N}(\dot{x}_k^2+\omega_k^2x_k^2)+\varepsilon V\right\}=\varepsilon\sum_{k=1}^{N}R_k\dot{x}_k+\varepsilon\sum_{k=1}^{N}F_k(t)\dot{x}_k, \tag{8.100}$$

$$\frac{d}{dt}\left\{\sum_{k=1}^{N}\dot{x}_kx_k\right\}-\sum_{k=1}^{N}\dot{x}_k^2+\sum_{k=1}^{N}\omega_k^2x_k^2=$$

$$=-\varepsilon\sum_{k=1}^{N}\frac{\partial V}{\partial x_k}x_k+\varepsilon\sum_{k=1}^{N}R_kx_k+\varepsilon\sum_{k=1}^{N}F_k(t)\,x_k. \tag{8.101}$$

Multiplying equation (8.101) by some positive constant α and adding to (8.100), we find

$$\frac{d}{dt}E=\varepsilon\sum_{k=1}^{N}R_k\dot{x}_k+\varepsilon\sum_{k=1}^{N}F_k(t)\,\dot{x}_k+\alpha\sum_{k=1}^{N}\dot{x}_k^2-$$

$$-\left[\sum_{k=1}^{N}\omega_k^2x_k^2+\varepsilon\sum_{k=1}^{N}\frac{\partial V}{\partial x_k}x_k\right]+\alpha\varepsilon\sum_{k=1}^{N}R_kx_k+\alpha\varepsilon\sum_{k=1}^{N}F_k(t)\,x_k, \tag{8.102}$$

where

$$E=\frac{1}{2}\sum_{k=1}^{N}[\dot{x}_k^2+\omega_k^2x_k^2]+\varepsilon V+\alpha\sum_{k=1}^{N}\dot{x}_kx_k. \tag{8.103}$$

Maximizing the right-hand side of the expression (8.102), we obtain (in view of the conditions of the theorem) the inequality

$$\frac{d}{dt}E\leqslant-\beta\sum_{k=1}^{N}\dot{x}_k^2+\varepsilon I\sum_{k=1}^{N}|\dot{x}_k|+\alpha\sum_{k=1}^{N}\dot{x}_k^2-\omega^2\alpha\sum_{k=1}^{N}x_k^2+$$

$$+\alpha\varepsilon S'\sum_{k=1}^{N}|x_k|\sum_{k=1}^{N}|\dot{x}_k|+\alpha\varepsilon I\sum_{k=1}^{N}|x_k|, \tag{8.104}$$

which, according to the Bunyakovskii-Schwarz inequality, may be written in the form

$$\frac{d}{dt}E\leqslant-\{(\beta-\alpha)X^2+\omega^2\alpha Y^2-\alpha\varepsilon S'NXY\}+\varepsilon I\sqrt{N}X+\alpha\varepsilon I\sqrt{N}Y, \tag{8.105}$$

where we have used the notation

$$X^2=\sum_{k=1}^{N}\dot{x}_k^2,\quad Y^2=\sum_{k=1}^{N}x_k^2. \tag{8.106}$$

We choose now α, β, and ε_0 such that

$$\alpha\leqslant\frac{\omega}{2},\quad \alpha=\frac{\beta}{2},\quad \varepsilon_0S'N<2\omega. \tag{8.107}$$

Maximizing the right-hand side of (8.103), we have

$$E\geqslant\frac{1}{2}(X^2+\omega^2Y^2)-\alpha XY\geqslant\gamma^*(X^2+Y^2), \tag{8.108}$$

where γ^* is some positive constant. On the other hand, we obtain from the inequality (8.105)

$$\frac{d}{dt}E\leqslant-\frac{\beta}{4}(X^2+\omega^2Y^2)+\varepsilon M^*\sqrt{X^2+Y^2}, \tag{8.109}$$

where $M^*=I\sqrt{N(1+\alpha^2)}$.

We can conclude by virtue of (8.108) and (8.109) that E, while remaining positive, will decrease until

$$X^2+\omega^2Y^2>\left(\frac{4\varepsilon M^*}{\beta}\right)^2. \tag{8.110}$$

The quantities X^2 and ω^2Y^2 will decrease together with E.

Thus, whatever the initial values of coordinates and velocities, we obtain from a certain moment $t=T^*$ the inequality

$$X^2+\omega^2Y^2\leqslant\left(\frac{4\varepsilon M^*}{\beta}\right)^2, \tag{8.111}$$

which is satisfied for all $t\geqslant T^*$.

Indeed, a violation of the inequality (8.111) immediately implies a decrease of E so that the inequality remains valid.

Using (8.111) and (8.106) we deduce the validity of the theorem.

We now formulate the theorem about the strong stability of a two-parametric family of particular solutions

$$\begin{aligned} q_i^{(m-1)} &= \varphi_i^{(1)}a\cos(\theta+\psi)+\varepsilon u_i^{(1)}(\tau, a, \theta, \theta+\psi)+\\ &+\varepsilon^2 u_i^{(2)}(\tau, a, \theta, \theta+\psi)+\ldots+\varepsilon^{m-1}u_i^{(m-1)}(\tau, a, \theta, \theta+\psi) \quad (i=1, 2, \ldots, N). \end{aligned} \tag{8.112}$$

Theorem 5. *Let the following conditions be satisfied:*

1) the functions $U_k(\tau)$ *and* $U_k'(\tau)$ *be bounded in the interval* $0\leqslant\tau\leqslant L$, *i.e.,*

$$|U_k(\tau)|\leqslant I, \quad |U_k'(\tau)|\leqslant I \qquad (k=1, 2, \ldots, N);$$

2) the inequalities

$$|\nu(\tau)-\omega_k|\geqslant\alpha \qquad (k=2, 3, \ldots, N), \quad \left|\frac{d\nu(\tau)}{d\tau}\right|\leqslant\lambda, \tag{8.113}$$

where α *and* λ *are positive constants, be valid in any interval* $0\leqslant\tau\leqslant L$;

3) for any A_k $(k=1, 2, \ldots, N)$ $\beta>0$ *can be found such that*

$$\lim_{T\to\infty}\frac{1}{T}\int_0^T\sum_{k=2}^N X_k(x_1^{(0)}, \ldots, x_N^{(0)}, \dot{x}_1^{(0)}, \ldots, \dot{x}_N^{(0)}, \varepsilon)\,\dot{x}_k^{(0)}\,dt\leqslant-\beta\sum_{k=2}^N a_k^2 \tag{8.114}$$

for $|a_k|\leqslant A_k$.

Then ε_0, K_1, K_2 *exist such that the inequalities*

$$|x_k|\leqslant\sqrt{\varepsilon}K_1, \quad |\dot{x}_k|\leqslant\sqrt{\varepsilon}K_2 \quad (\varepsilon\leqslant\varepsilon_0^2, \ k=2, 3, \ldots, N) \tag{8.115}$$

hold for any solution of equations (8.57), beginning at some T^*.

We first consider two lemmas.

Lemma 4. *If the second condition of Theorem 4 is satisfied, then the inequality*

$$\left|\frac{1}{T}\int_t^{t+T}e^{i(\theta-\psi)}\,dt\right|\leqslant\frac{2}{T\alpha}+\frac{\lambda\varepsilon}{\alpha^2}, \tag{8.116}$$

where

$$\frac{d\theta}{dt}-\nu(\tau), \quad \frac{d\psi}{dt}=\omega_k=\text{const}, \qquad \tau=\varepsilon t,$$

is valid for any T *and* ε.

Proof. Integrating the function $e^{i(\theta-\psi)}$ between the limits t and $t+T$, we obtain

$$\int_t^{t+T}e^{i(\theta-\psi)}\,dt=\left.\frac{e^{i(\theta-\psi)}}{\nu(\tau)-\omega_k}\right|_t^{t+T}-i\varepsilon\int_t^{t+T}\frac{1}{[\nu(\tau)-\omega_k]^2}\frac{d\nu(\tau)}{d\tau}e^{i(\theta-\psi)}\,dt. \tag{8.117}$$

Maximizing the right-hand sides by means of the inequalities (8.113), we obtain (8.116), Q. E. D.

Lemma 5. *If the first and second conditions of Theorem 4 are satisfied, then corresponding to any* T *and* ε *we can find positive constants* M_0 *and* M_1 *such that the inequality*

$$\left|\frac{\varepsilon}{T}\int_t^{t+T} U_k(\tau)\cos\theta\sin\psi\,dt\right| \leqslant \frac{M_0\varepsilon}{T}+M_1\varepsilon^2 \quad \left(\frac{d\theta}{dt}=\nu(\tau),\ \frac{d\psi}{dt}=\omega_k\right). \tag{8.118}$$

is valid.

Proof. Integrating the expression $U_k(\tau)\cos\theta\sin\psi$ by parts between the limits t and $t+T$, we obtain

$$\int_t^{t+T} U_k(\tau)\cos\theta\sin\psi\,dt = -\frac{U_k(\tau)\cos(\theta+\psi)}{2[\nu(\tau)+\omega_k]}\Bigg|_t^{t+T} - \frac{U_k(\tau)\cos(\theta-\psi)}{2[\omega_k-\nu(\tau)]}\Bigg|_t^{t+T} +$$
$$+\int_t^{t+T}\frac{U_k'(\tau)[\omega_k+\nu(\tau)]+U_k(\tau)\nu'(\tau)}{2[\omega_k+\nu(\tau)]^2}\cos(\theta+\psi)\,dt +$$
$$+\int_t^{t+T}\frac{U_k'(\tau)[\omega_k-\nu(\tau)]+U_k\nu'(\tau)}{2[\omega_k-\nu(\tau)]^2}\cos(\theta-\psi)\,dt. \tag{8.119}$$

Maximizing the right-hand side of this equality we obtain by Lemma 4 the inequality (8.118) in which

$$M_0=\frac{I}{\alpha^3}(2\alpha^2+\alpha+\lambda), \quad M_1=\frac{I\lambda}{2\alpha^4}(1+\alpha), \tag{8.120}$$

Q. E. D.

It is not difficult now to prove the theorem about the strong stability.

Proof of Theorem 5. We multiply equations (8.57) (except the first) by the corresponding $\dot{x}_k$, and integrate between the limits t and $t+T$; summing the result over k, we obtain

$$\sum_{k=2}^{N}\frac{1}{2}[a_k^2(t+T)-a_k^2(t)] = \sum_{k=2}^{N} D_k +$$
$$+\varepsilon\int_t^{t+T}\sum_{k=2}^{N} X_k(x_1^{(0)},\ldots,x_N^{(0)},\dot{x}_1^{(0)},\ldots,\dot{x}_N^{(0)},\varepsilon)\,\dot{x}_k^{(0)}\,dt +$$
$$+\varepsilon\int_t^{t+T}\sum_{k=2}^{N} F_k(t)\,\dot{x}_k^{(0)}\,dt, \tag{8.121}$$

where D_k satisfies the inequalities (8.81), and

$$\sum_{k=2}^{N}|D_k| \leqslant \varepsilon\sqrt{\varepsilon}\,\{S_1(\sum_{k=2}^{N}a_k)^2+S_2\sum_{k=2}^{N}a_k+S_3\}\,T \tag{8.122}$$

for $\varepsilon\leqslant\varepsilon_0^2$.

Maximizing the right-hand side of (8.121) by means of (4.122), (8.86), the third condition of the theorem, and the results of Lemma 5, and taking $t=nT$, $T=\frac{L}{\sqrt{\varepsilon}}$ we obtain for all $t\geqslant T^*$, with an accuracy of the first order, the inequality

$$\sum_{k=2}^{N} a_k^2(nT) \leqslant K(1-\beta\varepsilon T)^n+\varepsilon C_1(1-\beta\varepsilon T)^{n-1}+\varepsilon C_1\left(1-\frac{\beta\varepsilon T}{2}\right)^{n-1}+\varepsilon C_2, \tag{8.123}$$

where K, C_1, and C_2 are some positive constants. The first three terms of the right-hand side of this inequality tend to zero as $n \to \infty$, consequently,

$$\sum_{n=2}^{N} a_k^2 (nT) \leqslant \varepsilon C_2. \tag{8.124}$$

Using the notation of (8.62), we obtain the inequalities (8.115), Q. E. D.

Further, by the substitution of variables (8.56) and the transformation formulas (8.6), it is not difficult to see that any solution of the system (8.1) becomes in our case, from a certain time, close to our family of particular solutions.

The generalization of these theorems concerning the stability of a monofrequency mode to the case of a vibrating system described by more complicated differential equations, for example of the type (8.53) or (8.54), presents no essential difficulty. It is also possible to improve the estimates considerably. We consider these questions in the following section.

§ 3. The existence and stability of integral manifolds for nonlinear systems with slowly varying parameters

N. N. Bogolyubov suggested that the problem of justifying the developed method of constructing approximate solutions corresponding to nonstationary monofrequency vibrations in systems with slowly varying coefficients is closely connected with the theory of integral manifolds /6/. This theory has recently been extensively applied to the solution of various problems encountered in the theory of differential equations containing a small parameter.

The special feature of the ideas developed in the method of integral manifolds is a new approach in the qualitative theory of differential equations. The particular solutions of these equations are very sensitive to small changes in the right-hand sides of the equations.

The theory of integral manifolds deals, not with particular solutions, but with integral manifolds (not with curves, but with hyper-surfaces), which are more stable under small changes of the right-hand sides of the equations than the individual solutions. We can thus prove a number of theorems concerning integral manifolds which could be obtained for individual solutions only under sufficiently strict conditions on the right-hand sides of the equations.

The question of existence and stability of integral manifolds is also very important in studying the individual solutions, since in the case of existence of a stable integral manifold, to which all the solutions of the system tend in course of time, it is sufficient to consider instead of the entire phase space only the solutions belonging to the integral manifold— on the hypersurface.

While constructing approximate two-parametric solutions corresponding to a monofrequency mode in a nonlinear vibrating system with slowly varying parameters, we are led directly to such manifolds, since the particular solutions we derived depend on two arbitrary constants and constitute a parametric representation of some two-dimensional integral manifold. Such an integral manifold is of practical value only when for any t in the interval $0 \leqslant t \leqslant T$ it lies in an asymptotic narrow neighborhood of the exact integral manifold of the system of equations, and if it possesses the property of strong stability, i. e., all solutions of the system of equations tend to this manifold.

We cannot discuss in full the extremely interesting and complicated theory of integral manifolds, not even its application to the class of differential equations with slowly varying parameters. This problem requires a special treatise which is much beyond the scope of this monograph.

However, in view of the importance and effectiveness of this method, we quote a few simple theorems in order to illustrate the applicability of this method to the study of the behavior of a family of solutions of a system of differential equations with slowly varying parameters.

We first consider the following system of equations, to which many problems of the theory of nonstationary vibrational processes may be reduced:

$$\left.\begin{aligned} \frac{dh}{dt} &= H(t)\,h + Q(t, g, h, \varepsilon), \\ \frac{dg}{dt} &= \omega(t) + P(t, g, h, \varepsilon), \end{aligned}\right\} \tag{8.125}$$

where h is an $(n-1)$-dimensional vector, $H(t)$, a bounded square matrix of the $(n-1)$-th order, and $\omega(t)$, a function bounded for all real t.

Systems of the type (8.125), where H is a constant matrix, ω, a constant, and P, Q are almost periodic functions of t and periodic functions of g, were investigated by N. N. Bogolyubov /6/. The existence of an integral manifold representable by a relation of the type

$$h = f(t, g, \varepsilon),$$

where $f(t, g, \varepsilon)$ is a periodic function of g with period 2π and an almost periodic function of t, was proved for such systems, and its properties were established.

We prove the existence and establish some properties of the integral manifold of the system (8.125), when H and ω are not constant but depend on t, and P, Q are arbitrary functions of t /112/.

For the system under consideration, we assume that there exist positive ε_0, ϱ_0, such that the following conditions are satisfied:

1) the function $P(t, g, h, \varepsilon)$ and the $(n-1)$-dimensional vector function $Q(t, g, h, \varepsilon)$ are defined in the domain

$$-\infty < t < \infty, \quad -\infty < g < \infty, \quad h \in U_{\varrho_0}, \quad 0 \leqslant \varepsilon \leqslant \varepsilon_0, \tag{8.126}$$

where U_{ϱ_0} is a ϱ_0-neighborhood of the point $h=0$;

2) in the domain

$$-\infty < t < \infty, \quad -\infty < g < \infty, \quad 0 \leqslant \varepsilon \leqslant \varepsilon_0, \tag{8.127}$$

the inequalities

$$|P(t, g, 0, \varepsilon)| \leqslant M(\varepsilon), \quad |Q(t, g, 0, \varepsilon)| \leqslant M(\varepsilon), \tag{8.128}$$

are valid, where $M(\varepsilon) \to 0$ as $\varepsilon \to 0$;

3) the inequalities

$$\left.\begin{aligned} |P(t, g', h', \varepsilon) - P(t, g'', h'', \varepsilon)| &\leqslant \lambda(\varepsilon, \varrho)\{|g' - g''| + |h' - h''|\}, \\ |Q(t, g', h', \varepsilon) - Q(t, g'', h'', \varepsilon)| &\leqslant \lambda(\varepsilon, \varrho)\{|g' - g''| + |h' - h''|\}, \end{aligned}\right\} \tag{8.129}$$

where $\lambda(\varepsilon, \varrho) \to 0$ as $\varepsilon \to 0$, $\varrho \to 0$, are valid for any positive $\varrho < \varrho_0$ in the domain

$$-\infty < t < \infty, \quad -\infty < g' < \infty, \quad -\infty < g'' < \infty, \quad h' \in U_\varrho, \quad h'' \in U_\varrho, \quad 0 \leqslant \varepsilon \leqslant \varepsilon_0;$$

4) the $(n-1)$-dimensional square matrix $H(t)$ is bounded and uniformly continuous in t on the entire real axis, and the roots $p_i(t)$ $(i=1, 2, \ldots, n-1)$ of the corresponding characteristic equation

$$\operatorname{Det} \| pE - H(t) \| = 0 \tag{8.130}$$

are such that

$$R\{p_i(t)\} < -2\gamma, \quad \gamma > 0, \tag{8.131}$$

where E is the unit $(n-1)$-dimensional matrix. Under these conditions the matrix $U(t, \tau)$ satisfying the equation

$$\frac{dU(t, \tau)}{dt} = H(t) U(t, \tau), \quad U(t, \tau)|_{t=\tau} = E, \tag{8.132}$$

satisfies, according to Hale's lemma /51/, the estimate

$$|U(t, \tau)| \leqslant K e^{-\gamma(t-\tau)}, \qquad t \geqslant \tau, \tag{8.133}$$

where K and γ are positive constants.

With these assumptions we prove by means of a known method /6/ the existence and stability of the integral manifold for the system of equations (8.125) depending on one parameter.

As in /9/, we fix some positive numbers D, Δ $(\Delta < \varrho)$ and consider the class $C(D, \Delta)$ of $(n-1)$-dimensional vector functions $F(t, g)$ defined in the domain

$$-\infty < t < \infty, \quad -\infty < g < \infty \tag{8.134}$$

and satisfying in this domain the inequalities

$$|F(t, g)| \leqslant D, \quad |F(t, g') - F(t, g'')| \leqslant \Delta |g' - g''|. \tag{8.135}$$

We examine an equation of the form

$$\frac{dg}{dt} = \omega(t) + P(t, g, F(t, g), \varepsilon), \tag{8.136}$$

for some function $F(t, g)$ belonging to the class $C(D, \Delta)$.

In the domain (8.127), by virtue of the conditions (8.128) and (8.129), we have the inequalities

$$\left.\begin{aligned} &|P(t, g', F(t, g'), \varepsilon)| \leqslant M(\varepsilon) + \lambda(\varepsilon, D) D, \\ &|P(t, g', F(t, g'), \varepsilon) - P(t, g'', F(t, g''), \varepsilon)| \leqslant \\ &\qquad \leqslant \lambda(\varepsilon, D)(1+\Delta)|g' - g''|. \end{aligned}\right\} \tag{8.137}$$

With arbitrary initial conditions $t=t_0$, $g=g_0$, and with the help of Cauchy's theorem, we can construct the unique solution of equation (8.136). We write this solution symbolically in the form

$$g_t = T^F_{z, t_0}(g_0), \quad \text{where } z = t - t_0. \tag{8.138}$$

Let $F(t, g)$ and $F^*(t, g)$ be functions belonging to the class $C(D, \Delta)$.

Then, on the grounds of (8.136) and (8.137), we have

$$\left| \frac{d(g_t^* - g_t)}{dt} \right| \leqslant \lambda(\varepsilon, D) \| F^* - F \| + \lambda(\varepsilon, D)(1+\Delta)|g_t^* - g_t|, \tag{8.139}$$

where

$$\| F \| = \sup_{t, g} |F(t, g)|.$$

Putting $g_t^* - g_t = g_0^* - g_0$ at $t = t_0$, we find from (8.139)

$$|T^{F^*}_{z, t_0}(g_0^*) - T^F_{z, t_0}(g_0)| \leqslant |g_0^* - g_0| \exp\{\lambda(\varepsilon, D)(1+\Delta)|z|\} + \\ + \frac{\|F^* - F\|}{1+\Delta}\{\exp[\lambda(\varepsilon, D)(1+\Delta)|z|] - 1\}. \tag{8.140}$$

After these known results we examine the transformation

$$SF = \int_{-\infty}^{0} U(t, t+z) Q(t+z;\ T^F_{z,t}(g);\ F(t+z; T^F_{z,t}(g));\ \varepsilon)\, dz, \tag{8.141}$$

which transforms the function F belonging to the class $C(D, \Delta)$ to the function SF.

From the inequality (8.133) we have

$$|U(t, t+z)| \leqslant Ke^{-\gamma|z|}. \tag{8.142}$$

Maximizing the right-hand side of (8.141) and taking account of the inequalities (8.128), (8.129), and (8.142), we obtain

$$|SF| \leqslant \{M(\varepsilon) + \lambda(\varepsilon, D)D\} K \int_{-\infty}^{0} e^{-\gamma|z|}\, dz = \frac{K}{\gamma}\{M(\varepsilon) + \lambda(\varepsilon, D)D\}. \tag{8.143}$$

Taking account of the inequalities (8.128), (8.129), (8.140), and (8.142), we similarly obtain

$$|SF^* - SF| \leqslant \frac{K\lambda(\varepsilon, D)(1+\Delta)|g_0^* - g_0|}{|-\gamma + \lambda(\varepsilon, D)(1+\Delta)|} + \frac{K\lambda(\varepsilon, D)\|F^* - F\|}{|-\gamma + \lambda(\varepsilon, D)(1+\Delta)|}. \tag{8.144}$$

We now choose D and Δ as functions of the parameter ε: $D = D(\varepsilon)$, $\Delta = \Delta(\varepsilon)$, so that the inequalities

$$\left.\begin{aligned} &\frac{K}{\gamma}\{M(\varepsilon) + \lambda(\varepsilon, D)D\} < D, \quad \frac{K\lambda(\varepsilon, D)(1+\Delta)}{|-\gamma + \lambda(\varepsilon, D)(1+\Delta)|} < \Delta, \\ &\frac{K\lambda(\varepsilon, D)}{|-\gamma + \lambda(\varepsilon, D)(1+\Delta)|} < \frac{1}{2}, \end{aligned}\right\} \tag{8.145}$$

hold for all positive $\varepsilon < \varepsilon_1$; this is possible since $M(\varepsilon) \to 0$, $\lambda(\varepsilon, D) \to 0$ as $\varepsilon \to 0$, $D \to 0$.

The inequalities (8.143) and (8.144) now become

$$\left.\begin{aligned} &|SF| < D(\varepsilon), \\ &|SF^* - SF| \leqslant \Delta(\varepsilon)|g_0^* - g_0| + \tfrac{1}{2}\|F^* - F\|, \end{aligned}\right\} \tag{8.146}$$

which immediately implies that the transformation S, for $\varepsilon < \varepsilon_1$, maps the class of functions $C(D, \Delta)$ onto itself. Since

$$|SF^* - SF| \leqslant \tfrac{1}{2}\|F^* - F\|, \tag{8.147}$$

the equation

$$F = SF, \tag{8.148}$$

(according to Tikhonov's theorem concerning a fixed point in the class of functions $C(D, \Delta)$), has a unique solution F, which we denote by $f(t, g, \varepsilon)$:

$$F = f(t, g, \varepsilon). \tag{8.149}$$

We shall now show that the functions (8.149)

$$f(t, g, \varepsilon) = \int_{-\infty}^{0} U(t, t+z) Q(t+z;\ T^f_{z,t}(g);\ f(t+z; T^f_{z,t}(g); \varepsilon);\ \varepsilon)\, dz \tag{8.150}$$

define an integral manifold for the system of equations (8.125).

To do this we replace g by $T^f_{t-t_0,\,t_0}(g)$ in (8.150), noting that

$$T^f_{z,\,t}[T^f_{t-t_0,\,t_0}(g)]=T^f_{z+t-t_0,\,t_0}(g).$$

Introducing the notation

$$h_t=\int\limits_{-\infty}^{t}U(t,\,\tau)\,Q(\tau,\,g_\tau,\,h_\tau,\,\varepsilon)\,d\tau \tag{8.151}$$

and taking in place of z the new integration variable $\tau=z+t$, we obtain instead of (8.150)

$$h_t=\int\limits_{-\infty}^{t}U(t,\,\tau)\,Q(t,\,g_\tau,\,h_\tau,\,\varepsilon)\,d\tau. \tag{8.152}$$

Differentiating (8.152) with respect to the parameter t and taking (8.132) into account, we obtain

$$\frac{dh_t}{dt}=H(t)\,h_t+Q(t,\,g_t,\,h_t,\,\varepsilon). \tag{8.153}$$

On the other hand, according to the definition of the operator $T^f_{t-t_0,\,t_0}(g)$, g_t and h_t satisfy the second equation of the system (8.125) and, consequently, g_t and h_t, defined by the formulas (8.151), represent the solution of the system (8.125), and (8.150) is an integral manifold of this system.

We shall now show that this integral manifold possesses the property of stability, i. e., the trajectories of any solutions of the system (8.125), whose initial values lie in the domain of definition of the manifold, approach the manifold exponentially with time.

To do this we consider the integrodifferential system, equivalent to the system of differential equations (8.125),

$$\left.\begin{aligned} h_t&=\int\limits_{t_0}^{t}U(t,\,\tau)\,Q(\tau,\,g_\tau,\,h_\tau,\,\varepsilon)\,d\tau+U(t,\,t_0)\,A,\\ \frac{dg_t}{dt}&=\omega(t)+P(t,\,g_t,\,h_t,\,\varepsilon),\quad t>t_0,\quad g_t=g_0\ \text{for}\ t=t_0,\end{aligned}\right\} \tag{8.154}$$

where A is an arbitrary fixed $(n-1)$-dimensional vector.

To solve the system (8.154) we introduce the auxiliary system of equations

$$\left.\begin{aligned} h&=-\int\limits_{0}^{t_0-t}U(t,\;t+z)\,Q(t+z;\;g,\;h,\;\varepsilon)\,dz+U(t,\,t_0)\,A,\\ \frac{dg}{dt}&=\omega(t)+P(t,\,g,\,h,\,\varepsilon),\ t>t_0,\ g_t=g_0\ \text{for}\ t=t_0,\end{aligned}\right\} \tag{8.155}$$

which we solve by the method of successive approximations, and then prove that the functions to which these successive approximations converge satisfy (8.154).

We take an arbitrary function $\bar{f}_0(t,\,g)$ satisfying the Lipschitz condition in the domain $-\infty<t<\infty$, $-\infty<g<\infty$,

$$|\bar{f}_0(t,\,g')-\bar{f}_0(t,\,g'')|\leqslant\Delta\,|g'-g''|, \tag{8.156}$$

where Δ is a constant independent of g, and a vector A such that the inequality

$$|\bar{f}_0(t,\,g)|+K\,|A|\leqslant D(\varepsilon)<\varrho, \tag{8.157}$$

where K is the constant from (8.133), holds.

We put

$$f_0(t,\,g,\,A)=\bar{f}_0(t,\,g)+U(t,\,t_0)\,A. \tag{8.158}$$

It is obvious (from (8.157) and (8.133)), that

$$|f_0(t, g, A)| \leqslant D(\varepsilon) < \varrho, \tag{8.159}$$

and also that $f_0(t, g, A)$ satisfies, according to (8.156) and (8.133), a Lipschitz inequality of the form

$$|f_0(t, g', A') - f_0(t, g'', A'')| \leqslant \Delta |g' - g''| + Ke^{-\gamma|t-t_0|} |A' - A''|. \tag{8.160}$$

We now consider the equation

$$\frac{dg}{dt} = \omega(t) + P(t, g, f_0(t, g, A), \varepsilon). \tag{8.161}$$

As the right-hand side of this equation satisfies the Cauchy conditions, we can find the solution for the initial values $t = t_0$, $g = g_0$, which we represent in the form*

$$g_t^{f_0} = T_{t, t_0}^{f_0}(g_0, A). \tag{8.162}$$

To obtain the values of h in the first approximation we replace g and h in the right-hand sides of equations (8.155) by the functions

$$T_{z, t}^{f_0}(g, A), \quad f_0(t+z;\ T_{z, t}^{f_0}(g, A);\ A). \tag{8.163}$$

We obtain for h a first approximation in the form

$$f_1(t, g, A, \varepsilon) = -\int_0^{t_0 - t} U(t, t+z)\, Q(t+z;\ T_{z, t}^{f_0}(g, A);$$
$$f_0(t+z;\ T_{z, t}^{f_0}(g, A);\ A);\ \varepsilon)\, dz + U(t, t_0) A. \tag{8.164}$$

According to (8.161) and the inequalities (8.137), (8.133), we have

$$\left| \frac{d[T_{t, t_0}^{f_0}(g', A') - T_{t, t_0}^{f_0}(g'', A'')]}{dt} \right| \leqslant$$
$$\leqslant \lambda(\varepsilon, D)(1+\Delta) |T_{t, t_0}^{f_0}(g', A') - T_{t, t_0}^{f_0}(g'', A'')| + \lambda(\varepsilon, D) Ke^{-\gamma(t-t_0)} |A' - A''|,$$

from which we find

$$|T_{t, t_0}^{f_0}(g', A') - T_{t, t_0}^{f_0}(g'', A'')| \leqslant |g' - g''|\, e^{\lambda(\varepsilon, D)(1+\Delta)(t-t_0)} +$$
$$+ \frac{\lambda(\varepsilon, D) K}{|-\gamma + \lambda(\varepsilon, D)(1+\Delta)|} |A' - A''| \{e^{-\gamma(t-t_0)} + e^{\lambda(\varepsilon, D)(1+\Delta)(t-t_0)}\}. \tag{8.165}$$

It is then not difficult to verify that $f_1(t, g, A, \varepsilon)$ satisfies an inequality of the form

$$|f_1(t, g', A', \varepsilon) - f_1(t, g'', A'', \varepsilon)| \leqslant$$
$$\leqslant \nu_1(\varepsilon, D) |g' - g''| + K_1(\varepsilon, D) e^{-\gamma(t-t_0)} |A - A''|, \tag{8.166}$$

where $\nu_1(\varepsilon, D) \to 0$, $K_1(\varepsilon, D) \to K$ when $\varepsilon \to 0$, $D \to 0$. According to (8.164), and the inequalities (8.128), (8.133), and (8.159), we also have

$$|f_1(t, g, A, \varepsilon)| \leqslant \frac{K}{\gamma}\{M(\varepsilon) + \lambda(\varepsilon, D) D\} + K|A|. \tag{8.167}$$

We choose a positive $\varepsilon_2 \leqslant \varepsilon_1$, such that the inequality

$$\frac{K}{\gamma}\{M(\varepsilon) + \lambda(\varepsilon, D) D\} + K|A| \leqslant D(\varepsilon), \tag{8.168}$$

holds for all positive $\varepsilon < \varepsilon_2$. This is always possible since $M(\varepsilon) \to 0$, $\lambda(\varepsilon, D) \to 0$ when $\varepsilon \to 0$, $D \to 0$. We then obtain

$$|f_1(t, g, A, \varepsilon)| \leqslant D(\varepsilon). \tag{8.169}$$

* For simplicity, we do not indicate the dependence of T on ε either here or in the sequel.

Having derived the first approximation $h=f_1(t, g, A, \varepsilon)$, we can find from the equation

$$\frac{dg}{dt}=\omega(t)+P(t, g, f_1(t, g, A, \varepsilon), \varepsilon),\ t=t_0,\ g=g_0, \tag{8.170}$$

the value of $g_t^{f_1}$ corresponding to $h=f_1(t, g, A, \varepsilon)$,

$$g_t^{f_1}=T_{t, t_0}^{f_1}(g_0, A). \tag{8.171}$$

To determine h in the second approximation we also obtain the expression

$$\begin{aligned} f_2(t, g, A, \varepsilon)=-\int_0^{t_0-t} & U(t, t+z)\, Q(t+z;\ T_{z, t}^{f_1}(g, A); \\ & f_1(t+z;\ T_{z, t}^{f_1}(g, A);\ A;\ \varepsilon);\ \varepsilon)\, dz+U(t, t_0)\, A. \end{aligned} \tag{8.172}$$

The functions $f_2(t, g, A, \varepsilon)$ satisfy the inequalities

$$|f_2(t, g, A, \varepsilon)|\leqslant D(\varepsilon), \tag{8.173}$$

$$\begin{aligned} |f_2(t, g', A', \varepsilon)-f_2(t, g'', A'', \varepsilon)| & \leqslant \nu_2(\varepsilon, D)|g'-g''|+ \\ & +K_2(\varepsilon, D)\, e^{-\gamma(t-t_0)}|A'-A''|, \end{aligned} \tag{8.174}$$

where $\nu_2(\varepsilon, D)\rightarrow 0$, $K_2(\varepsilon, D)\rightarrow K$ as $\varepsilon\rightarrow 0$, $D\rightarrow 0$.

This procedure yields a sequence of functions

$$f_0(t, g, A),\ f_1(t, g, A, \varepsilon),\ f_2(t, g, A, \varepsilon),\ \ldots,\ f_n(t, g, A, \varepsilon),\ \ldots, \tag{8.175}$$

satisfying the conditions

$$\left.\begin{aligned} |f_n(t, g, A, \varepsilon)| & \leqslant D(\varepsilon), \\ |f_n(t, g', A', \varepsilon)-f_n(t, g'', A'', \varepsilon)| & \leqslant \nu_n(\varepsilon, D)|g'-g''|+ \\ & +K_n(\varepsilon, D)\, e^{-\gamma(t-t_0)}|A'-A''|, \end{aligned}\right\} \tag{8.176}$$

where

$$\nu_n(\varepsilon, D)\rightarrow 0,\ K_n(\varepsilon, D)\rightarrow K \text{ for } \varepsilon\rightarrow 0,\ D\rightarrow 0. \tag{8.177}$$

We shall now show that the series (8.175) converges to a certain function $f(t, g, A, \varepsilon)$ belonging to the domain U_D and satisfying the integral equation in the system (8.154).

To do this we estimate the difference $f_2(t, g, A, \varepsilon)-f_1(t, g, A, \varepsilon)$:

$$\begin{aligned} |f_2(t, g, A, \varepsilon)&-f_1(t, g, A, \varepsilon)|= \\ =\Bigg|\int_0^{t_0-t} & U(t, t+z)\{Q(t+z;\ T_{z, t}^{f_1}(g, A);\ f_1(t+z;\ T_{z, t}^{f_1}(g, A), A, \varepsilon);\ \varepsilon)- \\ & -Q(t+z;\ T_{z, t}^{f_0}(g, A);\ f_0(t+z;\ T_{z, t}^{f_0}(g, A);\ A);\ \varepsilon)\}\, dz\Bigg|. \end{aligned} \tag{8.178}$$

From (8.129) we have

$$\begin{aligned} |Q(t+z;\ T_{z, t}^{f_1}(g, A);\ & f_1(t+z;\ T_{z, t}^{f_1}(g, A);\ A;\ \varepsilon);\ \varepsilon)- \\ -Q(t+z;\ & T_{z, t}^{f_0}(g, A);\ f_0(t+z;\ T_{z, t}^{f_0}(g, A);\ A);\ \varepsilon)|\leqslant \\ & \leqslant\lambda(\varepsilon, D)\{|T_{z, t}^{f_1}(g, A)-T_{z, t}^{f_0}(g, A)|+ \\ & +|f_1(t+z;\ T_{z, t}^{f_1}(g, A);\ A;\ \varepsilon)-f_0(t+z;\ T_{z, t_0}^{f_0}(g, A); A)|\}. \end{aligned} \tag{8.179}$$

As the sequence of functions (8.175) satisfies conditions (8.177), we can write

$$\begin{aligned} |f_1(t+z;\ T_{z, t}^{f_1}(g, A);\ A;\ \varepsilon)&-f_0(t+z;\ T_{z, t}^{f_0}(g, A);\ A)|\leqslant \\ \leqslant|f_1(t+z;\ T_{z, t}^{f_1}(g, A);\ A;\ \varepsilon)&-f_0(t+z;\ T_{z, t}^{f_1}(g, A);\ A)|+ \\ & +\Delta|T_{z, t}^{f_1}(g, A)-T_{z, t}^{f_0}(g, A)|. \end{aligned} \tag{8.180}$$

At the same time we have, according to the inequality (8.140),

$$|T^{f_1}_{z,\,t}(g,\,A)-T^{f_0}_{z,\,t}(g,\,A)|\leqslant\frac{\|f_1-f_0\|}{1+\Delta}\,[e^{\lambda(\varepsilon,\,D)(1+\Delta)|z|}-1], \tag{8.181}$$

where

$$\|f_1\|=\sup_{t,\,g}|f_1(t,\,g,\,A,\,\varepsilon)|.$$

Taking into account (8.179), (8.180), and (8.181), we obtain for the left-hand side of (8.178)

$$|f_2(t,\,g,\,A,\,\varepsilon)-f_1(t,\,g,\,A,\,\varepsilon)|\leqslant$$
$$\leqslant K\lambda(\varepsilon,\,D)\|f_1-f_0\|\left|\int_0^{t-t_0}e^{\gamma z+\lambda(\varepsilon,\,D)(1+\Delta)|z|}\,dz\right|\leqslant$$
$$\leqslant K\lambda(\varepsilon,\,D)\|f_1-f_0\|\frac{|e^{[-\gamma+\lambda(\varepsilon,\,D)](1+\Delta)(t-t_0)}-1|}{|-\gamma+\lambda(\varepsilon,\,D)(1+\Delta)|}\,. \tag{8.182}$$

We now choose a positive $\varepsilon^*\leqslant\varepsilon_2$, such that the inequalities

$$\left.\begin{aligned}&\lambda(\varepsilon,\,D)(1+\Delta)<\frac{\gamma}{2}\,,\\&\frac{K\lambda(\varepsilon,\,D)}{|-\gamma+\lambda(\varepsilon,\,D)(1+\Delta)|}<1,\end{aligned}\right\} \tag{8.183}$$

hold for any positive $\varepsilon<\varepsilon^*$. This is possible, since $\lambda(\varepsilon,\,D)\to 0$, $\Delta(\varepsilon)\to 0$ as $D\to 0$, $\varepsilon\to 0$, and also $D(\varepsilon)\to 0$ as $\varepsilon\to 0$. We finally obtain

$$|f_2(t,\,g,\,A,\,\varepsilon)-f_1(t,\,g,\,A,\,\varepsilon)|\leqslant m\|f_1-f_0\|,$$

where $m<1$.

Performing similar estimations, we find

$$|f_3(t,\,g,\,A,\,\varepsilon)-f_2(t,\,g,\,A,\,\varepsilon)|\leqslant m^2\|f_1-f_0\|$$

and generally

$$|f_n(t,\,g,\,A,\,\varepsilon)-f_{n-1}(t,\,g,\,A,\,\varepsilon)|\leqslant m^{n-1}\|f_1-f_0\|,$$

where $m<1$.

Thus, the terms of the series

$$f_0+(f_1-f_0)+(f_2-f_1)+\ldots+(f_n-f_{n-1})+\ldots$$

do not exceed, correspondingly, the terms of the converging numerical series

$$M_0+M+mM+m^2M+\ldots m^{n-1}M+\ldots$$

where $m<1$, $M_0=\|f_0\|$, $M=\|f_1-f_0\|$ and, consequently, the sequence of functions (8.175) converges uniformly with respect to t and g (for any $0\leqslant\varepsilon<\varepsilon^*$) to some function $f(t,\,g,\,A,\,\varepsilon)$, which satisfies Lipschitz inequalities of the form

$$|f(t,\,g',\,A',\,\varepsilon)-f(t,\,g'',\,A'',\,\varepsilon)|\leqslant\nu(\varepsilon,\,D)\,|g'-g''|+$$
$$+K(\varepsilon,\,D)\,e^{-\gamma(t-t_0)}\,|A'-A''|, \tag{8.184}$$

where $\nu(\varepsilon,\,D)\to 0$, $K(\varepsilon,\,D)\to K$ as $\varepsilon\to 0$, $D\to 0$.

It is not difficult to show, with the help of usual methods, that the obtained function $f(t,\,g,\,A,\,\varepsilon)$ satisfies the first equation of the system (8.155), being also a unique solution of this equation.

We shall now show that $f(t,\,g,\,A,\,\varepsilon)$ is a solution of the integrodifferential system (8.154).

We have the identity

$$f(t,\ g,\ A,\ \varepsilon) = -\int_0^{t_0-t} U(t,\ t+z)\, Q(t+z;\ T^f_{z,t}(g,\ A);$$
$$f(t+z;\ T^f_{z,t}(g,\ A);\ A;\ \varepsilon);\ \varepsilon)\, dz + U(t,\ t_0)\, A. \qquad (8.185)$$

As above, we replace g by $T^f_{t,t_0}(g_0,\ A)$, noting that

$$T^f_{z,t}[T^f_{t,t_0}(g_0, A),\ A] = T^f_{z+t,t_0}(g_0,\ A).$$

Then, using the notation

$$h_t = f(t;\ T^f_{t,t_0}(g_0,\ A);\ A;\ \varepsilon), \quad g_t = T^f_{t,t_0}(g_0;\ A) \qquad (8.186)$$

and introducing in (8.185) the new integration variable $\tau = z+t$ in place of z, we obtain

$$h_t = \int_{t_0}^{t} U(t,\ \tau)\, Q(\tau,\ g_\tau,\ h_\tau,\ \varepsilon)\, d\tau + U(t,\ t_0)\, A,$$

where g_t satisfies by construction the second equation of the system (8.154). Thus, h_t and g_t, defined by the expressions (8.186), represent a solution of the integrodifferential system (8.154) converging for $t = t_0$ to A and g_0, and satisfying the inequality (8.184).

It is also clear on the grounds of (8.132) that the solutions of the integrodifferential system (8.154) are also solutions of the differential equations.

On the other hand, let g_τ, h_τ be any solutions of the system (8.125), for which $g = g_0$, $h = h_0 \in U_D$ when $\tau = t_0$. Then, substituting this solution into equation (8.125) and multiplying on the left by $U(t,\ \tau)$, we obtain the identity

$$U(t,\ \tau)\frac{dh_\tau}{d\tau} = U(t,\ \tau)\, H(\tau)\, h_\tau + U(t,\ \tau)\, Q(\tau,\ g_\tau,\ h_\tau,\ \varepsilon). \qquad (8.187)$$

Integrating this identity between t_0 and t, we find

$$\int_{t_0}^{t} U(t,\tau)\frac{dh_\tau}{d\tau}\, d\tau = \int_{t_0}^{t} U(t,\tau)\, H(\tau)\, h_\tau\, d\tau + \int_{t_0}^{t} U(t,\tau)\, Q(\tau, g_\tau, h_\tau, \varepsilon)\, d\tau. \qquad (8.188)$$

Integrating the left-hand side of (8.188) by parts, we obtain

$$\int_{t_0}^{t} U(t,\tau)\frac{dh}{d\tau}\, d\tau = U(t,\ t)\, h_t - U(t,\ t_0)\, h_0 - \int_{t_0}^{t} \frac{dU(t,\tau)}{d\tau}\, h_\tau\, d\tau. \qquad (8.189)$$

Noting that according to condition (4) (p. 355)

$$\frac{dU(t,\ \tau)}{d\tau} = -U(t,\ \tau)\, H(\tau),$$

we finally obtain

$$h_t = \int_{t_0}^{t} U(t,\ \tau)\, Q(\tau,\ g_\tau,\ h_\tau,\ \varepsilon)\, d\tau + U(t,\ t_0)\, h_0. \qquad (8.190)$$

Thus, any solution of the system (8.125) satisfying the initial conditions ($g = g_0$, $h = h_0 \in U_D$ for $t = t_0$) is a solution of the integrodifferential system (8.154) for $A = h_0$.

Hence, any solution of the differential system (8.125) for which $h_0 < D$, is a solution of the integrodifferential system for $A = h_0$. Any solution lying in the integral manifold

$$h = f(t,\ g,\ \varepsilon), \qquad (8.191)$$

satisfies the condition $|f(t_0, g_0, \varepsilon)| < D$ and, consequently, is a solution of the integrodifferential system for some $A = A'$.

Therefore, as in /6/, we can write according to (8.184),

$$|f(t, g, \varepsilon) - f(t, g, A, \varepsilon)| \leqslant K(\varepsilon, D) e^{-\gamma(t-t_0)} |A' - A| \tag{8.192}$$

or, replacing the arbitrary g by $g_t = T^f_{t\ t_0}(g_0, A)$,

$$|f(t, g_t, \varepsilon) - h_t| \leqslant K(\varepsilon, D) e^{-\gamma(t-t_0)} |f(t_0, g_0, \varepsilon) - h_0|. \tag{8.193}$$

According to the inequality (8.193) any solution h_t of the system (8.125), for which $h_0 < D$, tends with time to the manifold (8.191), but the inequality (8.193) in no case implies an approach of any solution to a particular curve belonging to this manifold. It is also clear from (8.184) and (8.165) that any solution h_t, for which the initial value of g coincides with the initial value of g in the manifold, will not approach with time the integral curves lying in the manifold (for $A' \neq A$).

Proceeding from these results, we can formulate the following fundamental theorem which has already been proved in the foregoing argument.

Theorem 6. *If the conditions 1)-4) are satisfied for the system of differential equations (8.125), then it is always possible to find positive constants* ϱ^* *and* ε^**, such that the system of equations (8.125) has a unique one-parametric integral manifold for all positive* $\varepsilon < \varepsilon^*$*, explicitly representable by the relations*

$$h = f(t, g, \varepsilon), \tag{8.194}$$

where $f(t, g, \varepsilon)$ *is defined as a function of* t *and* g *for* $-\infty < t < \infty$, $-\infty < g < \infty$, *and satisfies the inequalities*

$$|f(t, g, \varepsilon)| \leqslant D(\varepsilon) < \varrho^*, \tag{8.195}$$

$$|f(t, g', \varepsilon) - f(t, g'', \varepsilon)| \leqslant \Delta(\varepsilon) |g' - g''|, \tag{8.196}$$

where $\Delta(\varepsilon) \to 0$, $D(\varepsilon) \to 0$ *as* $\varepsilon \to 0$.

This integral manifold (8.194) possesses the property of strong stability, i.e., any solution of the system (8.125) $h = h_t$, *the initial values of which belong to the domain* U_{ρ^*}, *tends with time to the manifold (8.194) according to the law*

$$|h_t - f(t, g, \varepsilon)| \leqslant K(\varepsilon, D) e^{-\gamma(t-t_0)}, \tag{8.197}$$

where γ *is a certain positive constant.*

§ 4. Theorems on the stability of one- and two-parametric families of solutions in general form

The fundamental theorem obtained in the preceding section allows us to prove in general form a number of interesting theorems about the stability of one- and two-parametric families of solutions of systems of differential equations containing the slowing time.

Omitting the intermediate operations, we shall formulate in this section a few theorems for systems of differential equations discussed in Chapter VII.

First of all we consider the system of differential equations (7.10)

$$\frac{dx}{dt} = X(\tau, x) + \varepsilon X^*(\tau, \theta, x, \varepsilon), \tag{8.198}$$

where x is an n-dimensional vector belonging to the Euclidean space E_n; X, X^* are n-dimensional vector functions; t is the time; ε is a small positive parameter; $\tau=\varepsilon t$; and $\frac{d\theta}{dt}=\nu(\tau)>0$.

From Theorem 6 and the transformations performed in the preceding chapter, we can formulate for the system of equations (8.198), and prove with the help of a method similar to the one developed in the preceding section, the following theorems /113/.

Theorem 7. *Let the following conditions be satisfied for the system (8.198):*

1) For the unperturbed equations

$$\frac{dx}{dt}=X(\tau,\ x), \tag{8.199}$$

in which τ is regarded as a constant parameter, one knows for any τ a family of stable periodic solutions

$$x=x^0(\tau,\ \omega t+\varphi) \tag{8.200}$$

with period 2π in ψ ($\psi=\omega t+\varphi$, $\omega=\omega(\tau)$) depending on one arbitrary constant φ (and also on τ as a parameter). Consequently, the $n-1$ characteristic exponents $\alpha_1(\tau)$, $\alpha_2(\tau)$, ..., $\alpha_{n-1}(\tau)$ of the system of variational equations

$$\frac{d\delta x}{dt}=X'_x(\tau,\ x^0)\,\delta x, \tag{8.201}$$

set up for the equations (8.199) and the solutions (8.200), have negative real parts

$$\gamma_q(\tau)<0 \qquad (q=1,\ 2.\ \ldots,\ n-1). \tag{8.202}$$

2) Let the real parts of the characteristic exponents of the system of equations (8.201) satisfy, in addition to (8.202), also the conditions

$$\left.\begin{array}{c} m_1\gamma_1(\tau)+m_2\gamma_2(\tau)+\ldots+m_{n-1}\gamma_{n-1}(\tau)\neq\gamma_q(\tau)\\ (q=1,\ 2,\ \ldots,\ n-1),\\ m_1+m_2+\ldots+m_{n-1}\geqslant 2, \end{array}\right\} \tag{8.203}$$

where m_1, m_2, ..., m_{n-1} are non-negative integers.

3) There exist positive constants ϱ and ε_0, such that in the domain

$$-\infty<t<\infty,\ \theta\in\Omega,\quad x\in D_\rho,\quad 0<\varepsilon<\varepsilon_0, \tag{8.204}$$

the functions

$$X(\tau,\ x)+\varepsilon X^*(\tau,\ \theta,\ x,\ \varepsilon) \tag{8.205}$$

$\left(\tau=\varepsilon t,\ \frac{d\theta}{dt}=\nu(\tau)\right)$ *are periodic in θ with period 2π, are bounded, and have the desired number of bounded derivatives with respect to x, t, ε, where D_ρ is a ϱ-neighborhood of the orbit (8.200) in the n-dimensional space E_n, and Ω is a circle.*

It is then possible to find positive constants ϱ^*, ε^* ($\varrho^*<\varrho$, $\varepsilon^*\leqslant\varepsilon_0$) such that for all positive $\varepsilon<\varepsilon^*$:

a) The system of equations (8.198) has a unique one-parametric family of solutions of the form

$$x(\varepsilon t,\ \theta,\ \psi)=f[\varepsilon t,\ \psi,\ h(\varepsilon t,\ \theta,\ \psi,\ \varepsilon)], \tag{8.206}$$

where $f[\varepsilon t, \psi, h(\varepsilon t, \theta, \psi, \varepsilon)]$ *is defined for all real* t, ψ, θ, *and is periodic in* θ *and* ψ *with period* 2π. *Herewith, there exists a positive function* $\delta(\varepsilon)$ ($\delta(\varepsilon) \to 0$ *when* $\varepsilon \to 0$) *such that*

$$|f[\varepsilon t, \psi, h(\varepsilon t, \psi, \theta, \varepsilon)] - x^0(\varepsilon t, \psi)| \leqslant \delta(\varepsilon). \tag{8.207}$$

Here $f(\tau, \psi, h)$ *are analytic, regular for sufficiently small values of* h, *and satisfy the condition*

$$f(\tau, \psi, 0) \equiv x^0(\tau, \psi), \tag{8.208}$$

and $h(\varepsilon t, \theta, \psi, \varepsilon)$ *is a parametric representation of an integral manifold depending on a single parameter, for the system of equations*

$$\left.\begin{aligned} \frac{d\psi}{dt} &= \omega(\tau) + \varepsilon P_1(\tau, \theta, \psi, h, \varepsilon), \\ \frac{dh}{dt} &= \lambda(\tau) h + \varepsilon R_1(\tau, \theta, \psi, h, \varepsilon), \end{aligned}\right\} \tag{8.209}$$

where $\tau = \varepsilon t$, $\lambda(\tau)$ *is an* $(n-1)$*-dimensional vector, and the function* $P_1(\tau, \theta, \psi, h, \varepsilon)$ *and the* $(n-1)$*-dimensional vector function* $R_1(\tau, \theta, \psi, h, \varepsilon)$ *are periodic in* θ *and* ψ *with period* 2π.

b) The system of equations (8.198), for the solutions belonging to the two-parametric family (8.206), is equivalent to a single differential equation of the form

$$\frac{d\psi}{dt} = \omega(\tau) + \varepsilon P_1[\tau, \theta, \psi, n(\tau, \theta, \psi, \varepsilon), \varepsilon] \qquad (\tau = \varepsilon t), \tag{8.210}$$

where $P_1[\tau, \theta, \psi, h(\tau, \theta, \psi, \varepsilon), \varepsilon]$ *defined for all real* t, θ, ψ, *is periodic in* θ *and* ψ *with period* 2π, *and possesses the same differentiability properties as the functions (8.205).*

c) The one-parametric family of solutions (8.206) possesses the property of stability, i.e., any solution of the system of equations (8.198), the initial values of which lie in a sufficiently close neighborhood of the family of solutions (8.206), approach with time the family of solutions (8.206) according to the law

$$|x(\varepsilon\tau, \theta, \psi) - x(t)| \leqslant Ce^{-\gamma(t-t_0)}, \tag{8.211}$$

where C *and* γ *are positive constants, and* $x(t)$ *is any solution of the system of equations (8.198).*

This theorem, as is clear from the results of the preceding chapter, may also be formulated with less rigid conditions on the characteristic exponents of the system of equations for the variations (8.201).

Let the conditions (8.203) not be satisfied. Then the following theorem is valid /113/.

Theorem 8. Let conditions (1)–(3) of Theorem 7 be satisfied for the system of differential equations (8.198).

It is then possible to find positive constants ϱ^*, ε^* ($\varrho^* < \varrho$, $\varepsilon^* \leqslant \varepsilon_0$) *such that for all positive* $\varepsilon < \varepsilon^*$ *the following holds:*

1) The system of equations (8.198) has a unique one-parametric family of solutions, representable by relations of the type

$$x(\varepsilon t, \theta, \psi) = x^0(\varepsilon t, \psi) + \frac{1}{2}\{A(\varepsilon t, \psi) h(\varepsilon t, \theta, \psi, \varepsilon) + \\ + \bar{A}(\varepsilon t, \psi) \bar{h}(\varepsilon t, \theta, \psi, \varepsilon)\}, \tag{8.212}$$

where $A(\tau, \psi)$ *is a matrix of* n *rows and* $n-1$ *columns, the elements of which are periodic functions in* ψ *with period* 2π; $\bar{A}(\tau, \psi)$ *and* $\bar{h}$ *are conjugate to* $A(\tau, \psi)$ *and* h; $h(\varepsilon t, \theta, \psi, \varepsilon)$ *is a parametric representation of an integral manifold, depending on a single parameter, of the system of equations*

$$\left.\begin{aligned}\frac{d\psi}{dt} &= \omega(\tau) + P(\tau, \theta, \psi, h, \varepsilon),\\ \frac{dh}{dt} &= H(\tau)h + R(\tau, \theta, \psi, h, \varepsilon).\end{aligned}\right\} \tag{8.213}$$

The $(n-1)$*-dimensional square matrix* $H(\tau)$ *is such that, according to condition (1) and Hale's lemma mentioned in the preceding section, the solutions of the system of equations*

$$\frac{dU}{dt} = H(\tau)U \tag{8.214}$$

satisfy the inequality

$$|U(t, t_0)| \leqslant Ke^{-\gamma(t-t_0)}, \tag{8.215}$$

where K *and* γ *are positive constants, and the initial conditions are* $U(t, t_0)_{t=t_0} = E_{n-1}$ *where* E_{n-1} *is the* $(n-1)$*-dimensional unit matrix.*

The solution (8.212) is periodic in θ *and* ψ *with period* 2π; *there exists a positive function* $\delta(\varepsilon)$ $(\delta(\varepsilon) \to 0$ *when* $\varepsilon \to 0)$, *such that*

$$|x(\varepsilon t, \theta, \psi) - x^0(\varepsilon t, \psi)| \leqslant \delta(\varepsilon). \tag{8.216}$$

2) The system of equations (8.198), for solutions belonging to the one-parametric family of solutions (8.212), is equivalent to a single differential equation of the form

$$\frac{d\psi}{dt} = \omega(\tau) + P[\tau, \theta, \psi, h(\tau, \theta, \psi, \varepsilon), \varepsilon] \qquad (\tau = \varepsilon t), \tag{8.217}$$

where $P[\tau, \theta, \psi, h(\tau, \theta, \psi, \varepsilon), \varepsilon]$ *defined for all real* t, θ, ψ, *is periodic in* θ *and* ψ *with period* 2π.

3) The one-parametric family of solutions possesses the property of stability, i.e., any solutions of the system of equations (8.198), whose initial values lie in a sufficiently close neighborhood of the family of solutions (8.212), tend to this family with time according to the law

$$|x(\varepsilon t, \theta, \psi) - x(t)| \leqslant Ce^{-\gamma(t-t_0)}, \tag{8.218}$$

where, as in the preceding theorem, C *and* γ *are positive constants, and* $x(t)$ *is any solution of the system (8.198).*

Theorems, analogous to those quoted above, may be formulated and proved also in a more general case.

In the preceding chapter we considered the case when the unperturbed system of equations admits a stable periodic solution depending on two arbitrary constants, and suggested an algorithm for deriving a two-parametric family of solutions. O. B. Lykova /85/ proved for this case the following theorem.

Theorem 9. *Let the following conditions be satisfied for the system of equations (8.198):*

1) For the unperturbed equations

$$\frac{dx}{dt} = X(\tau, x), \tag{8.219}$$

in which τ *is regarded as a constant parameter, one knows for any* $\tau \in [0, L]$

a family of stable periodic solutions

$$x = x^0(\tau,\ \omega t + \varphi,\ a), \tag{8.220}$$

depending on two arbitrary constants φ, a *and on* τ *as a parameter, while* ω *is in the general case a function of* a *and depends on* τ *as a parameter*

$$\omega = \omega(\tau,\ a). \tag{8.221}$$

Let ω *be a bounded function with respect to* τ, *and let it satisfy the Lipschitz condition with respect to* a.

As the family of solutions (8.220) is stable, the $n-2$ *characteristic exponents* $\alpha_3(\tau),\ \alpha_4(\tau),\ \ldots,\ \alpha_n(\tau)$ *of the system of the variational equations*

$$\frac{d\delta x}{dt} = X'_x(x^0)\,\delta x, \tag{8.222}$$

set up for the solutions (8.220), have for any τ *negative real parts, while* $\alpha_1(\tau),\ \alpha_2(\tau)$ *vanish by virtue of our assumption that (8.220) is a periodic solution depending on two arbitrary constants. The characteristic exponents* α_i $(i = 1, 2, \ldots, n)$, *as mentioned in the preceding chapter, depend in general also on the parameter* a: $\alpha_i = \alpha_i(\tau, a)$ $(i = 1, 2, \ldots, n)$ *and are continuous functions of* τ.

2) In the domain

$$-\infty < t < \infty, \quad \theta \in \Omega, \quad x \in U_{\rho_0}, \quad 0 \leqslant \varepsilon \leqslant \varepsilon_0, \tag{8.223}$$

positive constants ρ_0 *and* ε_0 *exist such that all the arguments of the functions*

$$X(\tau,\ x) + \varepsilon X^*(\tau,\ \theta,\ x,\ \varepsilon) \tag{8.224}$$

are continuous and possess the desired number of derivatives with respect to ε *and* x, *where* U_{ρ_0} *is a* ρ_0*-neighborhood of the family of solutions (8.220) in the space* E_n. *In addition,* $X^*(\tau, \theta, x, \varepsilon)$ *are periodic functions of* θ *with period* 2π.

It is then possible to find positive constants ρ^*, ε^* $(\rho^* < \rho_0,\ \varepsilon^* \leqslant \varepsilon_0)$ *such that the following is valid for any positive* $\varepsilon < \varepsilon^*$:

a) The system of equations (8.198) has a unique two-parametric family of solutions, representable (in the general case of arbitrary negative real parts of the characteristic exponents $\alpha_i(\tau)$ $(i = 3, 4, \ldots, n)$*) by relations of the form*

$$x(\varepsilon t,\ \theta,\ \psi,\ a) = x^0(\varepsilon t,\ \psi,\ a) + \frac{1}{2}\{A(\varepsilon t,\ \psi,\ a)\,h(\varepsilon t,\ \theta,\ \psi,\ a,\ \varepsilon) + \overline{A}(\varepsilon t,\ \psi,\ a)\,\overline{h}(\varepsilon t,\ \theta,\ \psi,\ a,\ \varepsilon)\}, \tag{8.225}$$

where $A(\tau, \psi, a)$ *is a matrix of* n *rows and* $n-2$ *columns, the elements of which are periodic functions of* ψ *with period* 2π, *and depend on the parameter* τ *and the arbitrary constant* a; $\overline{A}(\tau, \psi, a)$ *and* $\overline{h}$, *are conjugate to* $A(\tau, \psi, a)$ *and* h; $h(\tau, \theta, \psi, a, \varepsilon)$ *is a parametric representation of an integral manifold, depending on two parameters, of the system of equations*

$$\left.\begin{aligned} \frac{d\psi}{dt} &= \omega(\tau,\ a) + P(\tau,\ \theta,\ \psi,\ a,\ h,\ \varepsilon),\\ \frac{da}{dt} &= Q(\tau,\ \theta,\ \psi,\ a,\ h,\ \varepsilon),\\ \frac{dh}{dt} &= H(\tau,\ a)\,h + R(\tau,\ \theta,\ \psi,\ a,\ h,\ \varepsilon); \end{aligned}\right\} \tag{8.226}$$

and the $(n-2)$*-dimensional matrix* $H(\tau, a)$ *is such that the solutions of* *the system*

$$\frac{dU}{dt} = H(\tau, a) U \tag{8.227}$$

satisfy, according to condition (1) and Hale's lemma mentioned in the preceding chapter, the inequality

$$|U(t, t_0)| \leqslant K e^{-\gamma(t-t_0)} \tag{8.228}$$

for all values of a, *for which the family of solutions (8.220) is stable. Here* K *and* γ *are positive constants; the initial conditions are* $U(t, t_0)|_{t=t_0} = E_{n-2}$; E_{n-2} *is the* $(n-2)$*-dimensional unit matrix.*

b) The system of equations (8.198) is, for the solutions belonging to the two-parametric family (8.225), equivalent to two differential equations of the form

$$\left.\begin{aligned} \frac{d\psi}{dt} &= \omega(\tau, a) + P[\tau, \theta, \psi, a, h(\tau\ \theta, \psi, a, \varepsilon), \varepsilon], \\ \frac{da}{dt} &= Q[\tau, \theta, \psi; a, h(\tau, \theta, \psi, a, \varepsilon), \varepsilon], \end{aligned}\right\} \tag{8.229}$$

where $\tau = \varepsilon t$, $P[\tau, \theta, \psi, a, h(\tau, \theta, \psi, a, \varepsilon), \varepsilon]$ *and* $Q[\tau, \theta, \psi, a, h(\tau, \theta, \psi, a, \varepsilon), \varepsilon]$ *defined for all real values of* t, θ, ψ, *and* $a \in \mathfrak{A}$ *are periodic in* θ *and* ψ *with period* 2π *and possess certain properties depending upon the properties of the functions (8.224).*

c) The two-parametric family of solutions possesses the property of stability, i.e., any solutions $x = x(t)$ *of the system of equations (8.198), with initial conditions lying in a sufficiently close neighborhood of the family of solutions (8.225), tend with time to this family of solutions according to the law*

$$|x(\varepsilon t, \theta, \psi, a) - x(t)| \leqslant C e^{-\gamma(t-t_0)}, \tag{8.230}$$

where C *and* γ *are positive constants.*

As shown in /86/, an analogous theorem may also be formulated for the case when the vibrating system with slowly varying parameters is described by a more complicated system of differential equations of the form

$$\left.\begin{aligned} \frac{dx}{dt} &= X(y, x) + \varepsilon X^*(y, \theta, x, \varepsilon), \\ \frac{dy}{dt} &= \varepsilon Y(y, \theta, x, \varepsilon), \end{aligned}\right\} \tag{8.231}$$

where $\frac{d\theta}{dt} = \nu(y)$; x is an n-dimensional vector; y, an m-dimensional vector; X and X^*, n-dimensional vector functions; Y, an m-dimensional vector function; and ε, a small positive parameter.

APPENDIX*

The study of nonstationary processes in nonlinear vibrating systems with a single as well as with many degrees of freedom reduces in a number of cases (with the help of the asymptotic method expounded in the present monograph), to the study of a system of two equations of the type

$$\left.\begin{aligned}\frac{da}{dt}&=f_1(\tau,\ a,\ \psi),\\ \frac{d\psi}{dt}&=f_2(\tau,\ a,\ \psi),\end{aligned}\right\} \tag{1}$$

whose solutions (in many practical applications) are slowly varying functions of time.

The system (1) can generally not be integrated in a closed form, and the dependence of a and ψ on t has to be derived by numerical integration.

One of the main features and advantages of the asymptotic method exhibited in this monograph lies in the fact that the study of a vibrational process described by a differential equation of the type (2.1), (3.1), or (in the case of a monofrequency process) by systems of differential equations of the type (4.3), (5.4), etc., reduces to the integration of two first order equations for the amplitude and phase of vibration.

Since during the investigation of a vibrational process we are in most cases not so interested in the vibrational process itself (i. e., the "sinusoid" describing the dependence of the vibrating quantity on time) as in its envelope, it is quite sufficient for practical purposes to determine only the dependence of a and ψ on t; this characterizes sufficiently well the specific features of the vibrating process under investigation.

At the same time, as was already mentioned, the numerical integration of systems of equations of the first (and higher) approximation, i. e., systems of the type (1), is far simpler than the numerical integration of the equations of motion (3.1), (4.3), etc. (especially in the case of systems with many degrees of freedom). This is because when integrating equations (1) we find in general the vibration amplitude i. e., the envelope of the vibrational process which varies smoothly, and to obtain a complete description of the vibrational process it is, therefore, sufficient to compute a relatively small number of points (as is evident from the example in Chapter III, § 7). If to obtain a complete description of the vibrational process we directly integrate the equations of motion (3.1), (4.3), etc. numerically, we would have to compute the "sinusoid"; this requires calculations in a few tens and even a few hundred times as many points. Practically, it is impossible to solve this problem, not only because it requires much more time, but also because of the accumulation of a large systematic error leading to absolutely inadmissible results.

* The appendix was written by the author in collaboration with Yu. A. Klikh.

It should be mentioned, however, that even the numerical integration of systems of the type (1), describing smoothly varying quantities a and ψ, is quite cumbersome for many variants of parameter values of the system.

A very easy method for solving equations of the type (1) is to apply electronic analogues of various constructions. Because of the slow variation of the quantities a and ψ, not many nonlinear units are required in actual cases for the set of nonlinear functions appearing on the right-hand sides of equations (1) and hence the structure of the analogue becomes very simple.

The application of continuously operating electronic computers permits an easy change of the system parameters in the course of simulation. This makes the investigation much easier and widens considerably the range of technical problems which may be solved by asymptotic methods.

The study of equations of the type (1) may be performed with the help of one of the widely extended analogue computers MN-2, MN-7, MN-8, etc.

The variable quantities are represented during the operation of an analogue computer by direct current voltages varying between the limits ± 100 v, the interaction between which is expressed in the computing elements of the analogue by the system of differential equations under investigation.

An analogue computer for the system (1) contains several computing elements; units of summation, integration, multiplication, multiplication by constant coefficients, and nonlinear function generating units for single variable functions. The number of units depends on the order of the system and the number of operations on the variable quantities appearing in the equations of the system. The summation and integration units consist of direct current amplifiers with a corresponding feedback. The multiplication of variable voltages by constant coefficients is performed with the help of voltage dividing units (BDN) or input resistors (BVS). These units allow for a variation of the constant coefficients from 0.001 to 10.

The multiplication of two variable quantities X_1 and X_2 is performed in a multiplication unit (BP) by a realization of the formula

$$Y = -c\left[\frac{X_1+X_2}{2}\right]^2 + c\left[\frac{X_1-X_2}{2}\right]^2 = -cX_1X_2.$$

The squares of the half-sum and half-difference of the input quantities are obtained by a piecewise linear approximation of parabolas with the help of functional systems— square law generators.

The nonlinear units allow us to generate single-valued functions by their approximation. The diode units BN-3A are capable of generating a function of a single variable by a piecewise linear approximation of 12-14 sections. The electromechanical unit BN-5 allows us to generate a nonlinear function of a single variable $\alpha(X_1)$ with a simultaneous multiplication by X_2 by means of a step-approximation of 50 or 100 steps.

The equations of the original system are transformed to a form convenient for simulation, and the block diagram of the computing elements is designed in correspondence with the analogue system of equations. In the block diagram the transmission coefficients of the separate computing elements must be accounted for according to the coefficients of the analogous system and the chosen scales M_{x_i} for representing the variable quantities x_i by the corresponding voltages X_i. To determine M_{x_i} the limits of variation of the corresponding variables must be known.

The initial conditions in the form of voltages according to the scale chosen for the corresponding variables are inserted at the outputs of the integrators.

The result of simulation is fixed on the screen of the cathode-ray indicator I-5M in the form of curves characterizing the amplitude and phase variation according to the chosen scale.

It is possible to investigate equations of the type (1) successfully with the help of the analogue computers described above, using Yu. A. Klikh's method /62/.

We describe this method as applied to an actual example of the transition through resonance of transversal-torsional vibrations of a turbine plate (cf. Chapter VI, § 6).

In this case system (1) has the form

$$\left.\begin{aligned} \frac{da}{dt} &= -\frac{\chi_2(a)}{2\pi m_1\omega_1} - \frac{A}{m_1(\omega_1+\nu)}\sin\psi, \\ \frac{d\psi}{dt} &= \omega_1 - \nu - \frac{\chi_1(a)}{2\pi m_1 a\omega_1} - \frac{A}{m_1 a(\omega_1+\nu)}\cos\psi, \end{aligned}\right\} \qquad (2)$$

where a is the amplitude; ψ, the phase shift; ω_1, the frequency of normal vibrations; ν, perturbation frequency;

$$\chi_1(a) = -a^n P 2\int_0^{\pi}(1-\cos\tau)^n\cos\tau\,d\tau + a^k Q 2\int_0^{\pi}(1-\cos\tau)^k\cos\tau\,d\tau;$$

$$\chi_2(a) = a^n P\frac{n-1}{n+1}2^{n+1} - a^k Q\frac{k-1}{k+1}2^{k+1};$$

$\tau = \omega t + \psi$ is the full phase of vibrations; and n, k are experimentally determined parameters of the hysteresis loop. The constants A, m_1, P, and Q are determined by the physical parameters of the mechanical system and the solution of the corresponding linear unperturbed system of differential equations /23/.

Let us consider the case of transition through resonance for a linear variation of the perturbation frequency, i. e.,

$$\nu = \nu_0 \pm \beta t,$$

where β is a parameter (the velocity of transition through resonance).

The resonance zone, the parameter values, the initial conditions, and the variation limits of the amplitude are found with the help of the resonance curve of the stationary mode $\left(\text{i. e., when } \frac{da}{dt} = 0 \text{ and } \frac{d\psi}{dt} = 0\right)$.

Equating the right-hand sides of equations (2) to zero, we obtain, up to quantities of the first order of smallness, relations for the stationary amplitude and phase values

$$\left.\begin{aligned} \left(\frac{\nu}{\omega_1}\right)^2 &= 1 - \frac{A}{m_1\omega_1^2 a}\left(\cos\psi - \frac{\chi_1(a)}{\chi_2(a)}\sin\psi\right), \\ \sin\psi &= -\frac{\chi_2(a)}{\pi A}. \end{aligned}\right\} \qquad (3)$$

We next proceed as follows:

1) We choose the resonance zone $\nu_{0(+\beta)} \leqslant \nu \leqslant \nu_{0(-\beta)}$ so that the resonance curve for the stationary mode be close to a horizontal line outside it.

2) For a given time of transition through resonance t_{tran}, we determine the corresponding transition velocity from the relation

$$\beta = \frac{\nu_{0(-\beta)} - \nu_{0(+\beta)}}{t_{\text{tran}}}. \qquad (4)$$

The initial conditions are obtained by equating a and ψ outside the resonance zone to their values in the stationary mode. Such a choice of the initial conditions is justified since the curve of transition through resonance is outside the resonance zone close to the resonance curve for the stationary mode.

The terms $\chi_1(a)$ and $\chi_2(a)$ may be neglected for small values of a (outside the resonance zone under consideration) as quantities of a higher order of smallness.

Hence, we have for the determination of the initial conditions

$$\sin\psi = 0, \quad \frac{\cos\psi}{a} = \frac{(\nu_{0(\pm\beta)}^2 - \omega_1^2) m_1}{A}. \tag{5}$$

4) The limits of the amplitude variation are obviously

$$0 < a < a_{\max},$$

where $a_{\max}$ is the maximum amplitude value for the stationary mode which may be obtained from the second equation of (3) by putting $\sin\psi = -1$.

If the argument of the functions $\sin\psi$ and $\cos\psi$ varies within sufficiently small limits, then these functions may be approximated by means of nonlinear units. However, we then need to know the variation interval of the phase.

It is difficult to simulate system (1) directly, since it is impossible to indicate the limits of variation of ψ, and in some problems also because of the large variation interval of the phase.

At the beginning of the simulation $\sin\psi$ should remain equal to zero a long time (approximately one third of the time required for the solution of the given problem); this does not happen in practice, causing a considerable increase in the error.

We therefore rewrite the system (2) in the form

$$\left.\begin{aligned} \frac{da}{dt} &= -\frac{\chi_2(a)}{2\pi m_1\omega_1} - \frac{A}{\sqrt{2}\,m_1}\frac{1}{\omega_1+\nu}\cos\left(\psi - \frac{\pi}{4}\right) - \\ &\qquad - \frac{A}{\sqrt{2}\,m_1}\frac{1}{\omega_1+\nu}\sin\left(\psi - \frac{\pi}{4}\right), \\ \frac{d\psi}{dt} &= \omega_1 - \nu - \frac{\chi_1(a)}{2\pi m_1 a\omega_1} - \frac{1}{a}\left[\frac{A}{\sqrt{2}\,m_1}\frac{1}{\omega_1+\nu}\cos\left(\psi - \frac{\pi}{4}\right) - \right. \\ &\qquad \left. - \frac{A}{\sqrt{2}\,m_1}\frac{1}{\omega_1+\nu}\sin\left(\psi - \frac{\pi}{4}\right)\right], \end{aligned}\right\} \tag{6}$$

and replace the obtained system of two equations (6) by the equivalent system of five equations

$$\left.\begin{aligned} \frac{da}{dt} &= -\frac{\chi_2(a)}{2\pi m_1\omega_1} - \frac{A}{\sqrt{2}\,m_1}u\frac{1}{2\omega_1 - y} - \frac{A}{\sqrt{2}\,m_1}v\frac{1}{2\omega_1 - y}, \\ \frac{dy}{dt} &= \mp\beta, \\ \frac{du}{dt} &= vf, \\ \frac{dv}{dt} &= -uf, \\ \frac{d\psi}{dt} &= f, \end{aligned}\right\} \tag{7}$$

where

$$f = y - \frac{\chi_1(a)}{2\pi m_1 a\omega_1} - \frac{1}{a}\left[\frac{A}{\sqrt{2}\,m_1}v\frac{1}{2\omega_1 - y} - \frac{A}{\sqrt{2}\,m_1}u\frac{1}{2\omega_1 - y}\right],$$

$$y=\omega_1-\nu, \quad u=\sin\left(\psi-\frac{\pi}{4}\right), \quad v=\cos\left(\psi-\frac{\pi}{4}\right).$$

After approximating the nonlinear functions $\frac{1}{a}$, $\chi_2(a)$, and $\frac{1}{2\omega_1-y}$ in the corresponding units and choosing the scales of the variables, the system (7) becomes, in terms of the computer variables (for the case $n=k=2$),

$$\left.\begin{aligned}
&\frac{dX_1}{d\tau}=-m_0X_1^2-m_2X_4\alpha(X_3)-m_3X_5\alpha(X_3),\\
&\frac{dX_3}{d\tau}=\mp\beta m_4,\\
&\frac{dX_4}{d\tau}=m_5X_5F,\\
&\frac{dX_5}{d\tau}=-m_6X_4F,\\
&\frac{dX_2}{d\tau}=m_{12}F,
\end{aligned}\right\} \tag{8}$$

where

$$\alpha(X_3)=\frac{74112}{2\omega_1M_y-X_3},$$

$$F=m_7X_3-m_8X_1-m_{11}\frac{50}{X_1}\left[m_9X_5\alpha(X_3)-m_{10}X_4\alpha(X_3)\right],$$

$$m_0=\frac{4(P-Q)M_{\text{BN-3A No. 1}}}{3m_1\omega_1M_aM_t}, \quad m_6=\frac{M_vM_f}{M_uM_t},$$

$$m_2=\frac{AM_aM_yM_{\text{BN-5}}}{74112\sqrt{2}\,m_1M_uM_t}, \quad m_7=\frac{M_uM_{\text{BP}}}{10M_yM_vM_t},$$

$$m_3=\frac{AM_aM_yM_{\text{BN-5}}}{74112\sqrt{2}\,m_1M_vM_t}, \quad m_8=\frac{(\dot{P}-Q)M_uM_{\text{BP}}}{10m_1\omega_1M_aM_vM_t},$$

$$m_4=\frac{M_y}{M_t}, \quad m_9=m_{10}=0.5,$$

$$m_5=\frac{M_uM_f}{M_vM_t}, \quad m_{11}=\frac{2AM_aM_yM_uM_{\text{BP}}M_{\text{BN-5}}M_{\text{BN-3A No. 2}}}{741120M_v^2M_t}, \quad m_{12}=\frac{10M_\psi}{M_{\text{BP}}}.$$

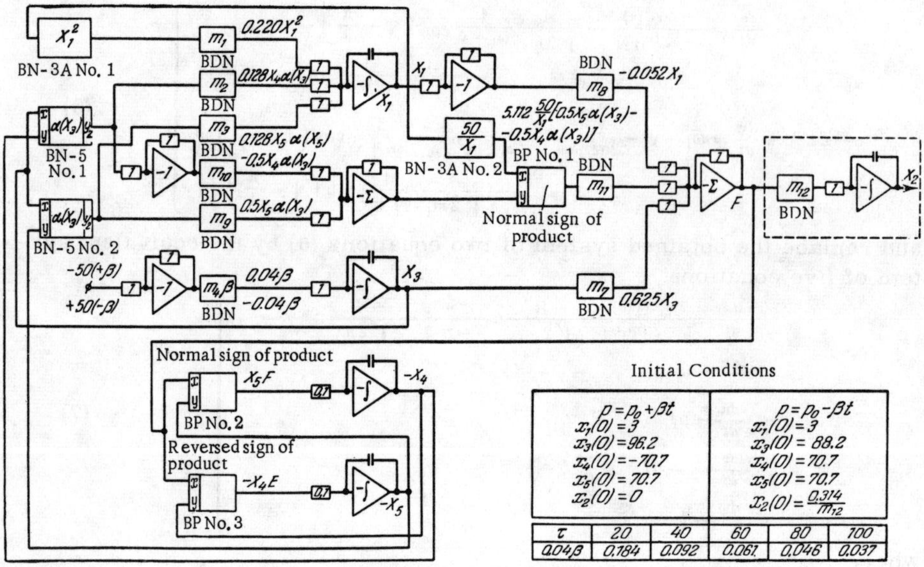

FIGURE 115

M_{BP} is the scale factor introduced by the multiplication unit; for the units BP-2, $M_{BP} = 100$. $M_{BN\text{-}5}$ is the scale factor introduced by the nonlinear unit BN-5. The function $\alpha(X_3)$ is generated by a step-approximation in the unit BN-5 with a simultaneous multiplication by X_4 or X_5. In the considered example $M_{BN\text{-}5} = 36.17$. $M_{BN\text{-}3A\,No.\,1}$, and $M_{BN\text{-}3A\,No.\,2}$ are the scale factors introduced by the nonlinear units BN-3A, where the functions $\chi_2(a)$ and $\frac{1}{a}$ are generated by means of a step-approximation. In our example $M_{BN\text{-}3A\,No.\,1} = 100$, $M_{BN\text{-}3A\,No.\,2} = 0.02$. M_a, M_ψ, M_y, M_u, M_v, M_t are the scales of $a, \psi, y, u, v,$ and the time t. In our example $M_a = 3\cdot10^2$ v/cm, $M_y = 0.8$ v·sec/rad, $M_u = M_v = 100$ v, and $M_t = 20$.

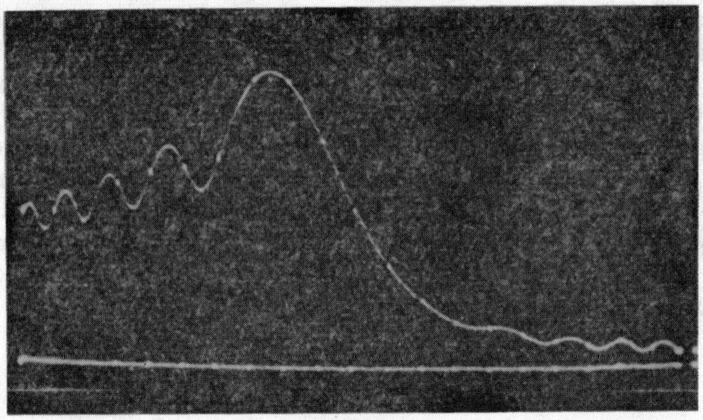

FIGURE 116

The block diagram of the analogous circuit for solving the system (8) is shown in Figure 115.

The first four equations of the system (8) do not contain X_2 explicitly, which allows us to simulate them independently of the last equation.

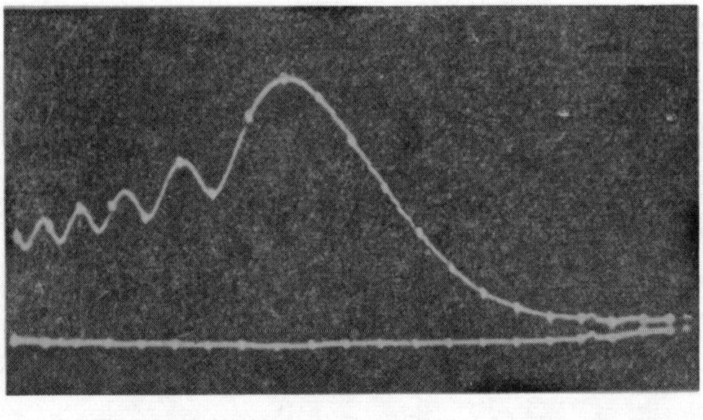

FIGURE 117

The unknown scale of the variable ψ is found by choosing in the voltage dividers (BDN) a scale factor m_{12} such that X_2 does not exceed the limits ± 100 v during the simulation of the last equation.

To pass from the voltages to the original variables we must take into account their scales

$$a = \frac{1}{M_a} X_1, \quad \psi = \frac{10}{m_{12} M_{\text{BDN}}} X_2, \quad t = \frac{\tau}{M_t}.$$

Samples of the results of simulation photographed from the screen of the cathode-ray indicator I-5M for actual numerical values of the parameters of a turbine plate (cf. Chapter VI, § 6) are shown in Figure 116 (for $\beta > 0$) and Figure 117 (for $\beta < 0$).

For other values of the parameters n and k of the hysteresis loop we must approximate the function $\chi_1(a)$ in the unit BN-3A, and introduce the corresponding change in the block diagram of the analogue circuit.

If the perturbation frequency varies according to another known law $\nu = \nu(t)$, then only the second equation of the system (8) is affected.

Systems of equations of the type (1) may be solved by means of digital computing machines as illustrated in /63/. However, their application is not advisable in the considered example. The application of continuously operating mathematical machines to the study of systems of the type (1) give sufficient accuracy for practical purposes; they are more convenient for obtaining a large number of variants of the solution, simpler to operate, and more economical. Besides, it is possible to solve successfully, with the help of continuously operating machines, the inverse problem — the derivation of the parameters of a given vibrational system satisfying a specified nonstationary mode.

BIBLIOGRAPHY*

[Note: References are listed in Cyrillic alphabetic order.]

1. ANDRONOV, A. A., A. A. VITT, and S. E. KHAIKIN. Teoriya kolebanii (The Theory of Oscillations). — Fizmatgiz. 1959.

2. Actes du colloque international des vibrations nonlinéaires (Ile de Porquerolles, 1951). — Publications Scientifiques et Techniques du Ministère de l'Air, Paris. 1953.

3. BABAKOV, I. M. Teoriya kolebanii (The Theory of Oscillations). — Gostekhizdat. 1958.

4. BARBER, N. F. and F. URSELL. The Response of a Resonant System to a Gliding Tone. — Phil. Mag., Vol. 39: 345. 1948.

5. BELKIN, M. K. K teorii otrazhatel'nogo klistrona (The Theory of the Reflex Klystron). — Radiotekhnika, Vol. 6, No. 5. 1951.

6. BOGOLYUBOV, N. N. O nekotorykh statisticheskikh metodakh v matematicheskoi fizike (Some Statistical Methods in Mathematical Physics). — Izdatel'stvo AN SSSR. 1945.

7. BOGOLYUBOV, N. N. Odnochastotnye svobodnye kolebaniya v nelineinykh sistemakh so mnogimi stepenyami svobody (Monofrequency Vibrations in Nonlinear Systems with Many Degrees of Freedom). -- In: Sbornik trudov Instituta stroitel'noi mekhaniki AN USSR, No. 10. 1949.

8. BOGOLYUBOV, N. N. and D. N. ZUBAREV. Metod asimptoticheskogo priblizheniya dlya sistem s vrashchayushcheisya fazoi i ego primenenie k dvizheniyu zaryazhennykh chastits v magnitnom pole (The Method of Asymptotic Approximation for Systems with Rotating Phase and its Application to the Motion of Charged Particles in a Magnetic Field). — UMZh, Vol. 7, No. 1. 1955.

9. BOGOLYUBOV, N. N and Yu. A. MITROPOL'SKII. Asimptoticheskie metody v teorii nelineinykh kolebanii (Asymptotic Methods in the Theory of Nonlinear Oscillations). — Fizmatgiz. 1958.

10. [Japanese translation of reference 9.]—Shuppan Kyoritsu, Tokyo. 1961.

11. BOGOLYUBOV, N. N. and Yu. A. MITROPOL'SKII. Asymptotic Methods in the Theory of Nonlinear Oscillations. — Gordon and Breach, Science Publishers, New York; Hindustan Publishing Corpn., Delhi, 6. 1961. [English translation of reference 9.]

12. BOGOLYUBOV, N. N. and Yu. A. MITROPOL'SKII. Analiticheskie metody v teorii nelineinykh kolebanii (Analytical Methods in the Theory of Nonlinear Oscillations). — In: Trudy I s"ezda po teoreticheskoi i prikladnoi mekhanike, Izdatel'stvo AN SSSR. 1961.

13. BOGOLYUBOV, N. N. and Yu. A. MITROPOL'SKII. Metod integral'nykh mnogoobrazii v nelineinoi mekhanike (The Method of Integral Manifolds in Nonlinear Mechanics). — In: Trudy Mezhdunarodnogo simpoziuma po nelineinym kolebaniyam, Vol. I, Izdatel'stvo AN USSR. 1963.

14. BOLOTIN, V. V. Dinamicheskaya ustoichivost' uprugikh sistem (Dynamic Stability of Elastic Systems). — Gostekhizdat. 1956.

15. BOLOTIN, V. V. O vozdeistvii podvizhnoi nagruzki na mosty (The Action of a Moving Load on Bridges). — Trudy MIIT, No. 74. 1950.

16. BREUS, K. A. O privodimosti kanonicheskoi sistemy differentsial'nykh uravnenii s periodicheskimi koeffitsientami (The Reducibility of a Canonical System of Differential Equations with Periodic Coefficients). — DAN SSSR, Vol. 123, No. 1. 1958.

17. BREUS, K. A. Ob odnom klasse lineinykh differentsial'nykh uravnenii s periodicheskimi koeffitsientami (A Class of Linear Differential Equations with Periodic Coefficients). — UMZh, Vol. 12, No. 4. 1960.

* [For list of Russian abbreviations appearing in this Bibliography see p. 385.]

18. BULGAKOV, B. V. O normal'nykh koordinatakh (Normal Coordinates). — Prikladnaya Matematika i Mekhanika, Vol. 10, No. 2. 1946.

19. BULGAKOV, B. V. Kolebaniya (Vibrations). — Gostekhizdat. 1954.

20. BYKOVA, N. O. Vozdeistvie napryazhenii menyayushcheisya chastoty na rezonansnye sistemy (The Action of Stresses of Varying Frequency on Resonance Systems). — Trudy MAP, No. 28. 1948.

21. LEFSCHETZ, S. (ed.). Contributions to the Theory of Nonlinear Oscillations, Vol. I, 1950; Vol. II, 1952; Vol. III, 1956; Vol. IV, 1958; Vol. V, 1960. — Princeton, Princeton University Press.

22. VASILEVSKAYA, D. P., A. A. GLAZOV, V. I. DANILOV, Yu. N. DENISOV, V. P. DZHELEPOV, V. P. DMITRIEVSKII, B. I. ZAMOLODCHIKOV, N. L. ZAPLATIN, V. V. KOL'GA, A. A. KRONIN, LIU NIEH-CHWAN, V. S. RYBALKO, A. L. SAVENKOV, and L. A. SARKISYAN. Tsiklotron s prostranstvennoi variatsiei napryazhennosti magnitnogo polya (Cyclotron with Space Variable Magnetic Field Intensity). — Atomnaya Energiya, Vol. 8, No. 3. 1960.

23. VASILENKO, M. V. and G. S. PISARENKO. Zmusheni poperechno-krutyl'ni kolyvannya sterzhniv z urakhuvannyam vnutrishn'oho tertya (Forced Transverse Torsional Oscillations of Beams under Consideration of Internal Friction). — Dopovidi AN URSR, No. 8. 1959. [In Ukrainian.]

24. VASIL'EVA, A. B. O differentsirovanii reshenii differentsial'nykh uravnenii, soderzhashchikh malye parametry (The Differentiation of Solutions of Differential Equations Containing Small Parameters). — Matematicheskii Sbornik, 31(79): 587-664. 1952.

25. VEKSLER, V. I., A. A. KOLOMENSKII, V. A. PETUKHOV, and M. S. RABINOVICH. Fizicheskie osnovy sooruzheniya sinkhrofazotrona na 10 Bev (Physical Bases for the Construction of a 10 Bev Synchrophasotron). — In: Uskoriteli elementarnykh chastits, Atomizdat. 1957.

26. VOLOSOV, V. M. Nelineinye differentsial'nye uravneniya vtorogo poryadka s malym parametrom pri starshei proizvodnoi (Nonlinear Second Order Differential Equations with a Small Parameter Attached to the Highest Derivative). — Matematicheskii Sbornik, 30(2): 245-270. 1952.

27. VOLOSOV, V. M. O reshenii nekotorykh differentsial'nykh uravnenii vtorogo poryadka, zavisyashchikh ot parametra (The Solution of Second Order Differential Equations Depending on a Parameter). — Matematicheskii Sbornik, 31(3): 675-686. 1952.

28. VOLOSOV, V. M. Differential'nye uravneniya dvizheniya, soderzhashchie parametr medlennosti (Differential Equations of Motion Containing a Slowness Parameter). — DAN SSSR, Vol. 106, No. 1. 1956.

29. VOLOSOV, V. M. O reshenii nelineinykh differentsial'nykh uravnenii vtorogo poryadka s medlenno izmenyayushchimisya koeffitsientami (Solutions of Second Order Nonlinear Differential Equations with Slowly Varying Coefficients). — DAN SSSR, Vol. 114, No. 6. 1957.

30. VOLOSOV, V. M. O nelineinykh kolebaniyakh s odnoi stepen'yu svobody sistemy s medlenno izmenyayushchimisya parametrami (Nonlinear Vibrations with a Single Degree of Freedom in a System with Slowly Varying Parameters). — DAN SSSR, Vol. 117, No. 6. 1957.

31. VOLOSOV, V. M. Uravneniya kolebanii s medlenno izmenyayushchimisya parametrami (Vibration Equations with Slowly Varying Parameters). — DAN SSSR, Vol. 121, No. 1. 1958.

32. VOLOSOV, V. M. Usrednenie nekotorykh vozmushchennykh sistem (Averaging of Some Perturbed Motions). — DAN SSSR, Vol. 133, No. 2. 1960. [Soviet Mathematics — Doklady, Vol. 1: 825.]

33. VOLOSOV, V. M. O metode usredneniya (The Method of Averaging). — DAN SSSR, Vol. 137, No. 1. 1961. [Soviet Mathematics — Doklady, Vol. 2: 221.]

34. VOLOSOV, V. M. Metod usredneniya i nekotorye zadachi teorii nelineinykh kolebanii (The Method of Averaging and some Problems of the Theory of Nonlinear Vibrations). — Doctoral Thesis. Kiev, Institut Matematiki AN USSR. 1961.

35. GORCHAKOV, N. G. Krutil'nye kolebaniya kolenchatykh valov dvigatelei vnutrennego sgoraniya pri prokhozhdenii cherez rezonans (Torsional Vibrations of Crankshafts of Internal Combustion Motors during a Transition through Resonance). — In: Sbornik trudov laboratorii problem bystrokhodnykh mashin i mekhanizmov AN USSR. 1946.

36. GRADSHTEIN, I. S. O resheniyakh na vremennoi polupryamoi differentsial'nykh uravnenii s malymi mnozhitelyami pri proizvodnykh (Solutions on the Time Semiaxis of Differential Equations with Small Coefficients of the Derivatives). — Matematicheskii Sbornik, Vol. 32: 533-544. 1953.

37. GROBOV, V. A. O poperechnykh kolebaniyakh vala pri peremennoi skorosti vrashcheniya (Transversal Vibrations of a Shaft with a Variable Rotation Velocity). — Izvestiya AN LatvSSR, No. 5. 1955.

38. GROBOV, V. A. Poperechnye kolebaniya rotora s raspredelennoi po dline massoi pri peremennoi skorosti vrashcheniya (Transversal Vibrations of a Rotor with a Longitudinally Distributed Mass and Variable Rotation Frequency). — Izvestiya AN LatvSSR, No. 5. 1955.

39. GROBOV, V. A. Nestatsionarnye kolebaniya vala turbiny v oblasti kriticheskikh chisel oborotov (Nonstationary Vibrations of a Turbine Shaft in the Domain of Critical Numbers of Revolutions). — Izvestiya AN LatvSSR, No. 8. 1957.

40. GROBOV, V. A. Nestatsionarnye kolebaniya uprugo opertykh rotorov (Nonstationary Vibrations of Elastically Supported Rotors). — In: Sbornik "Problemy prochnosti v mashinostroenii", No. 1, Izdatel'stvo AN SSSR. 1958.

41. GROBOV, V. A. O nestatsionarnykh kolebaniyakh nelineinykh sistem s giroskopicheskimi chlenami pri uchete svyazi s istochnikom energii maloi moshchnosti (Nonstationary Vibrations of Nonlinear Systems with Gyroscopic Terms with an Account of Coupling to a Low Power Energy Source). — UMZh, Vol. 12, No. 2. 1961.

42. GROBOV, V. A. Asimptoticheskie metody rascheta izgibnykh kolebanii valov turbomashin (Asymptotic Methods for Calculating Deformation Vibrations of Turbomachine Shafts). — Izdatel'stvo AN SSSR. 1961.

43. DAVIDENKOV, N. N. O rasseyanii energii pri vibratsiyakh (Energy Scattering during Vibrations). — ZhTF, Vol. 7, No. 6. 1938.

44. DEN-HARTOG, J. P. Mechanical Vibrations. New York-Toronto-London. 1956.

45. [Russian translation of reference 44.]— Fizmatgiz. 1960.

46. DOLGINOV, A. I. Reznonans v elektricheskikh tsepyakh i sistemakh (Resonance in Electrical Circuits and Systems). — Gosenergoizdat. 1957.

47. ZADIRAKA, K. V. O periodicheskikh resheniyakh sistemy nelineinykh differentsial'nykh uravnenii s malym parametrom pri proizvodnykh (Periodic Solutions of a System of Nonlinear Differential Equations with a Small Parameter Attached to the Derivatives). — DAN SSSR, Vol. 121, No. 2. 1958.

48. ZADIRAKA, K. V. Issledovanie reshenii sistemy nelineinykh differentsial'nykh uravnenii s malym parametrom pri nekotorykh proizvodnykh (Investigation of the Solutions of a System of Nonlinear Differential Equations with a Small Parameter Attached to Some of the Derivatives). — UMZh, Vol. 10, No. 2. 1958.

49. ZUBOV, V. I. Kolebaniya v nelineinykh i upravlyaemykh sistemakh (Oscillations in Nonlinear and Controlled Systems). — Sudpromgiz. 1962.

50. HALEJACK, K. and A. P. STOKES. Behavior of Solutions Near Integral Manifolds. — Archive for Rational Mechanics and Analysis, Vol. 6, No. 2. 1960.

51. HALEJACK, K. and G. SEIFERT. Bounded and Almost Periodic Solutions of Singularly Perturbed Equations. — Journal of Mathematical Analysis and Applications, Vol. 3, No. 1. 1960.

52. HALEJACK, K. Integral Manifolds of Perturbed Differential Systems. — Annals of Mathematics, Vol. 73, No. 3. 1961.

53. HASSELGRUBER und SCHWINGERS. Ein Naherungsverfahren zur Ermittlung der Wechselbeanspruchung in den Wellen und Kupplungen beim Durchlaufen torsionkritischen Drehzahlen. — Forschungsgebeite des Ingenieurwesens, Bd. 21, No. 4. 1955.

54. HOK, G. Response of Linear Resonant Systems to Excitation of a Frequency Varying Linearly with Time. — Journ. Appl. Phys., Vol. 19 : 242. 1948.

55. ISHLINSKII, A. Yu. Ob odnom integro-differentsial'nom sootnoshenii v teorii uprugoi niti (kanata) peremennoi dliny (An Integrodifferential Relation in the Theory of Elastic Thread (Rope) of Variable Length). — UMZh, Vol. 5, No. 4. 1953.

56. KAZAK, S. A. Kolebaniya kovsha draglaina (Vibration of a Drag-line Scoop). — In: Voprosy teorii i raboty pod"emno-transportnykh mashin, Ural'skii Politekhnicheskii Institut, No. 17, Mashgiz. 1953.

57. KAPITSA, P. L. Ustoichivost' i perekhod cherez kriticheskie oboroty bystro vrashchayushchikhsya rotorov pri nalichii treniya (Stability and Transition through Critical Numbers of Revolutions of Rapidly Rotating Rotors in the Presence of Friction). — Zhurnal Tekhnicheskoi Fiziki, Vol. 9, No. 2. 1939.

58. KAUDERER, H. Nichtlineare Mechanik. — Springer-Verlag, Berlin, Göttingen-Heidelberg. 1958.

59. [Russian translation of reference 58.]— IL. 1961.

60. KATS, A. M. Vynuzhdennye kolebaniya pri prokhozhdenii cherez rezonans (Forced Oscillations During Transition through Resonance). —Inzhenernyi Sbornik, Vol. 3, No. 2, Izdatel'stvo AN SSSR. 1947.

61. KLIKH, Yu. O. Doslidzhennya kolyvan' mekhanichnoyi systemy z dvoma stepenyamy vil'nosti z urakhuvannyam rozsiyannya enerhiyi v upruhykh elementakh (Investigation of Oscillations of a Mechanical System with Two Degrees of Freedom under Consideration of Energy Dispersion in the Elastic Elements). —Prykladna mekhanika, Vol. 7, No. 3. 1961. [In Ukranian.]

62. KLIKH, Yu. O. Doslizhdennya nestazionarnykh protsesiv v neliniinykh kolyvnykh systemakh z dopomohoyu elektroonnoyi moleyuyuchoyi ustanovky (Investigation of Nonstationary Processes in Nonlinear Oscillation Systems by Means of Electronic Equipment). — Dopovidi AN URSR, No. 5. 1962. [In Ukrainian.]

63. KLYACHKO, M. D. and L. E. YAKIMCHUK. Fazovye sootnosheniya pri perekhode lineinoi sistemy cherez rezonans (Phase Relations during the Transition of a Linear System through Resonance). — Izvestiya AN SSSR, Otdelenie Tekhnicheskikh Nauk, Mekhanika i Mashinostroenie, No. 6. 1961.

64. KONONENKO, V. O. O nelineinykh kolebaniyakh v sistemakh s izmenyayushchimisya parametrami (Nonlinear Vibrations in Systems with Varying Parameters). — DAN SSSR, Vol. 105, No. 2. 1955.

65. KONONENKO, V. O. O kolebaniyakh v nelineinykh sistemakh so mnogimi stepenyami svobody (Vibrations of Nonlinear Systems with Many Degrees of Freedom). — DAN SSSR, Vol. 105, No. 4. 1955.

66. KONONENKO, V. O. and S.S. KORABL'OV. Pro kolyvannya vala z dyskamy pry vrakhuvanni oberrenoho zv'yazku z kolyval'noyu systemoyu (Oscillations of a Shaft with Discs in [Feed-back ?] Connection with an Oscillating System). — Prykladna Mekhanika, Vol. 6, No. 2. 1960. [In Ukrainian.]

67. KONONENKO, V. O. Nekotorye avtonomnye zadachi teorii nelineinykh kolebanii (Autonomous Problems of Nonlinear Vibrations). — In: Trudy Mezhdunarodnogo simpoziuma po nelineinym kolebaniyam, Izdatel'stvo AN USSR, Vol. III. 1963.

68. KRYLOV, A. N. Lektsii o priblizhennykh vycnisleniyakh (Lectures on Approximate Calculations). Leningrad. 1933.

69. KRYLOV, N. M. and N. N. BOGOLYUBOV. Novye metody nelineinoi mekhaniki v ikh primenenii k izucheniyu raboty elektronnykh generatorov (New Methods of Nonlinear Mechanics Applied to Studying the Operation of Electronic Oscillators), Part I. — ONTI. 1934.

70. KRYLOV, N. M. and N. N. BOGOLYUBOV. Prilozhenie metodov nelineinoi mekhaniki k teorii statsionarnykh kolebanii (Application of the Methods of Nonlinear Mechanics in the Theory of Stationary Vibrations). — Izdatel'stvo AN USSR. 1934.

71. KRYLOV, N. M. and N. N. BOGOLYUBOV. Raschet vibratsii ramnykh konstruktsii s uchetom normal'nykh sil pri pomoshchi metodov nelineinoi mekhaniki (Calculation of Frame Construction Vibrations Taking Account of Normal Forces Using the Methods of Nonlinear Mechanics). — In: Ukrainskii kompleksnyi nauchno-issledovatel'skii institut sooruzhenii, Issledovaniya kolebanii konstruktsii, sbornik statei, Gostekhizdat USSR. 1935.

72. KRYLOV, N. M. and N. N. BOGOLYUBOV. Vvedenie v nelineinuyu mekhaniku (Introduction to Nonlinear Mechanics). — Izdatel'stvo AN USSR. 1937. [For a free translation of this book see: Lefschetz, S. Introduction to Nonlinear Mechanics. — Princeton University Press. 1943.]

73. KUZMAK, G. E. Asimptoticheskie resheniya uravneniya dvizheniya s odnoi stepen'yu svobody i medlenno izmenyayushchimisya parametrami (Asymptotic Solutions of the Equation of Motion with One Degree of Freedom and Slowly Varying Parameters). — DAN SSSR, Vol. 121, No. 1. 1958.

74. KUZMAK, G. E. Asimptoticheskoe reshenie nelineinykh differentsial'nykh uravnenii vtorogo poryadka s peremennymi koeffitsientami (Asymptotic Solution of Second Order Nonlinear Differential Equations with Varying Coefficients). — Prikladnaya Matematika i Mekhanika, Vol. 23, No. 3. 1959.

75. LEFSCHETZ, S. Lecture on Differential Equations. — Princeton University Press. 1946.

76. LEFSCHETZ, S. Differential Equations: Geometric Theory. London. 1957.

77. [Russian translation of reference 76.]—IL. 1961.

78. LEWIS, F. M. Vibration during Acceleration through a Critical Speed. — Transactions of the A. S. M. E., 54(23): 253. 1932.

79. LOITSYANSKII, L. G. and A. I. LUR'E. Teoreticheskaya mekhanika (Theoretical Mechanics), Part III. — GTTI. 1933.

80. LUR'E, A. I. and A. I. CHEKMAREV. Vynuzhdennye kolebaniya v nelineinoi sisteme s kharakteristikoi iz dvukh pryamolineinykh otrezkov (Forced Vibrations in a Nonlinear System with a Characteristic Composed of Two Rectilinear Segments). — Prikladnaya Matematika i Mekhanika, Vol. 1, No. 3. 1938.

81. LYKOVA, O. B. Pro odnochastotni kolyvannya v systemakh z bahat'ma stepenyamy vil'nosti, bliz'kykh do tochno intehrovnykh (Single-Frequency Oscillations in Systems with Many Degrees of Freedom. Approximation and Direct Integration.). — Dopovidi AN URSR, No. 1. 1957. [In Ukrainian.]

82. LYKOVA, O. B. Ob odnochastotnykh kolebaniyakh v sistemakh s medlenno menyayushchimisya parametrami (Monofrequency Vibrations in Systems with Slowly Varying Parameters). — UMZh, Vol. 9, No. 2. 1957.

83. LYKOVA, O. B. and Yu.O. MITROPOL'S'KYI. Do pytannya pro neliniini rivnyannya z periodychnymy koefitsientamy (Investigation of Nonlinear Equations with Periodic Coefficients). — Visnyk Kyyivs'koho Universytetu, No. 2, seriya astronomiyi, matematyky ta mekhaniky, vyp. 2. 1959. [In Ukrainian.]

84. LYKOVA, O. B. and Yu. O. MITROPOL'S'KYI. Pro neliniini dyferentsial'ni rivnyannya z periodychnymy koefitsientamy ta povil'no zminnymy parametramy (Nonlinear Differential Equations with Periodic Coefficients and Slowly Varying Parameters). — Visnyk Kyyivs'koho Universytetu, No. 3, seriya matematyky ta mekhaniky, vyp. 2. 1960. [In Ukrainian.]

85. LYKOVA, O. B. O nekotorykh svoistvakh reshenii sistem nelineinykh differentsial'nykh uravnenii s medlenno menyayushchimisya parametrami (Properties of the Solutions of Nonlinear Differential Equations with Slowly Varying Parameters). —UMZh, Vol. 12, No. 3. 1960.

86. LYKOVA, O. B. Do pytannya isnuvannya intehral'nykh mnohoobrazii (Investigation of the Existence of Integral Manifolds). — Dopovidi AN URSR, No. 12. 1962. [In Ukrainian.]

87. MALKIN, I. G. Nekotorye zadachi teorii nelineinykh kolebanii (Problems of Nonlinear Vibrations). — Gostekhizdat. 1956.

88. MANDEL'SHTAM, L. I. and N. D. PAPALEKSI. O yavleniyakh rezonansa n-go roda (The Phenomena of n-th Type Resonance). — In: Mandel'shtam, L. I. Sobranie sochinenii, Vol. II, Izdatel'stvo AN SSSR. 1947.

89. MANDEL'SHTAM, L. I. Sistemy s periodicheskimi koeffitsientami so mnogimi stepenyami svobody i s maloi nelineinost'yu (Systems with Periodic Coefficients with Many Degrees of Freedom and a Small Nonlinearity). — In: Sobranie sochinenii, Vol. II, Izdatel'stvo AN SSSR. 1947.

90. MESHCHERSKII, I. V. Rabota po mekhanike tel peremennoi massy (A Treatise on the Mechanics of Bodies with Varying Mass). — Gostekhizdat. 1949.

91. MINORSKY, N. Introduction to Nonlinear Mechanics. — Ann Arbor, Michigan. 1947.

92. MINORSKY, N. Nonlinear Oscillations. — D. Van Nostrand Company, Inc., Princeton, New York-London-Toronto. 1962.

93. MITROPOL'SKII, Yu. A. Sobstvennye kolebaniya nelineinoi sistemy s medlenno menyayushchimisya parametrami (Natural Vibrations of a Nonlinear System with Slowly Varying Parameters). — In: Sbornik trudov Instituta stroitel'noi mekhaniki AN USSR, No. 11. 1948.

94. MITROPOL'SKII, Yu. A. Issledovanie sobstvennykh kolebanii nelineinoi sistemy, blizkoi k tochno integriruyushcheisya (Investigation of the Natural Vibrations of a Nonlinear System Close to an Exactly Integrable One). — Ibid.

95. MITROPOL'SKII, Yu. A. Issledovanie kolebanii v nelineinykh sistemakh so mnogimi stepenyami svobody medlenno menyayushchimisya parametrami (Investigation of Vibrations in Nonlinear Systems with Many Degrees of Freedom and Slowly Varying Parameters). — UMZh, Vol. 1, No. 2. 1949.

96. MITROPOL'SKII, Yu. A. Primenenie simbolicheskikh metodov k issledovaniyu nelineinykh sistem s medlenno menyayushchimisya parametrami (The Application of Symbolic Methods to the Study of Nonlinear Systems with Slowly Varying Parameters). — In: Sbornik trudov Instituta stroitel'noi mekhaniki AN USSR, No. 13. 1949.

97. MITROPOL'SKII, Yu. A. Issledovanie kolebanii nelineinoi sistemy s medlenno menyayushchimisya parametrami (Investigation of Vibrations in a Nonlinear System with Slowly Varying Parameters). — Ibid., No. 14. 1950.

98. MITROPOL'SKII, Yu. A. Medlennye protsessy v nelineinykh kolebatel'nykh sistemakh s mnogimi stepenyami svobody (Slow Processes in Nonlinear Systems with Many Degrees of Freedom). — Prikladnaya Matematika i Mekhanika, Vol. 14, No. 2. 1950.

99. MITROPOL'SKII, Yu. A. O prokhozhdenii cherez rezonans v nelineinykh sistemakh so mnogimi stepenyami svobody (The Transition through Resonance in Nonlinear Systems with Many Degrees of Freedom). — In: Sbornik trudov Instituta stroitel'noi mekhaniki AN USSR, No. 17. 1952.

100. MITROPOL'SKII, Yu. A. Vynuzhdennye kolebaniya v nelineinykh sistemakh pri prokhozhdenii cherez rezonans (Forced Vibrations in Nonlinear Systems during the Transition through Resonance). — Inzhenernyi Sbornik, Vol. XV. 1953.

101. MITROPOL'SKII, Yu. A. O kolebaniyakh v giroskopicheskikh sistemakh pri prokhozhdenii cherez rezonans (Vibrations in Gyroscopic Systems during the Transition through Resonance). — UMZh, Vol. 5, No. 3. 1953.

102. MITROPOL'SKII, Yu. A. O nestatsionarnykh kolebaniyakh v sistemakh so mnogimi stepenyami svobody (Nonstationary Vibrations in Systems with Many Degrees of Freedom). — UMZh, Vol. 6, No. 2. 1954.

103. MITROPOL'SKII, Yu. A. O vozdeistvii na nelineinyi vibrator sinusoidal'noi sily s modulirovannoi chastotoi (The Action of a Sinusoidal Force with Modulated Frequency on a Nonlinear Vibrator). — UMZh, Vol. 6, No. 4. 1954.

104. MITROPOL'SKII, Yu. A. Nestatsionarnye protsessy v nelineinykh kolebatel'nykh sistemakh (Nonstationary Processes in Nonlinear Oscillatory Systems). — Izdatel'stvo AN USSR. 1955.

105. [Chinese translation of reference 104.] — Published by "Nauka", Peking. 1958.

106. [Japanese translation of reference 104.]

107. MITROPOLSKY, Yu. A. Nonstationary Processes in Nonlinear Oscillatory Systems. — Air Technical Intelligence Translation, ATIC—270579, F-TS-9085/v. [English translation of reference 104.]

108. MITROPOL'SKII, Yu. A. K voprosu o prokhozhdenii cherez rezonans vtorogo roda (The Problem of Transition through a Resonance of the Second Kind). — UMZh, Vol. 7, No. 1. 1955.

109. MITROPOL'SKII, Yu. A. K voprosu o vnutrennem rezonanse v nelineinykh kolebatel'nykh sistemakh (The Problem of Internal Resonance in Nonlinear Oscillatory Systems). — Naukovi Zapysky KDU, Vol. 16, issue II, Matematychnyi zbirnyk, No. 9. 1957.

110. MITROPOL'S'KYI, Yu.O. Pro neustaleni protsesy v deyakykh relaksatsiinykh kolyvnykh systemakh (Nonsteady Processes in Some Relaxation Oscillation Systems). — Naukovi Zapysky KDU, Vol. 16, issue XVI, Matematychnyi zbirnyk, No. 10. 1957. [In Ukrainian.]

111. MITROPOL'S'KYI, Yu. O. Pro asymptotychne predstavlennya rozv'yazkiv systemy neliniinykh rivnyan' zi zminnymy kofitsiyentamy (Asymptotic Representation of the Solutions of a System of Nonlinear Equations with Variable Coefficients). — Naukovyi shchorichnyk KDU za 1956, 1957.

112. MITRIPOL'SKII, Yu. A. Ob issledovanii integral'nogo mnogoobraziya dlya sistemy nelineinykh uravnenii s peremennymi koeffitsientami (The Investigation of Integral Manifolds for a System of Nonlinear Equations with Variable Coefficients). — UMZh, Vol. 10, No. 3. 1958.

113. MITROPOL'SKII, Yu. A. Ob ustoichivosti odnoparametricheskogo semeistva reshenii sistemy uravnenii s peremennymi koeffitsientami (The Stability of a One-Parametric Family of Solutions of a System of Equations with Variable Coefficients). — UMZh, Vol. 10, No. 4. 1958.

114. MITROPOL'SKII, Yu. A. O nekotorykh uravneniyakh, blizkikh k tochno integriruyushchimsya (Equations Close to Exactly Integrable Ones). — Avtomaticheskoe Upravlenie i Vychislitel'naya Tekhnika, No. 2: 221-248. 1959.

115. MITROPOL'S'KYI, Yu. O. and B. I. MOSYEYENKOV. Doslidzhennya nestatsionarnykh kolyval'nykh rezhymiv v systemakh z rozpodilenymy parametramy (Nonstationary Oscillation Regimes in Systems with Unstable Parameters). — Visnyk KDU, seriya astronomiyi, matematyky ta mekhaniky, No. 2, issue 1. 1959. [In Ukrainian.]

116. MITROPOL'S'KYI, Yu. O. and B. I. MOSYEYENKOV. Doslidzhennya kolyvan' v systemakh z rozpodi lenymy parametramy (asymptotychni metody) (Investigation of Oscillations in Systems with Unstable Parameters (Asymptotic Method)). — Vydavnytstvo KDU. 1962. [In Ukrainian.]

117. MISHCHENKO, E. F. and L. S. PONTRYAGIN. Periodicheskie resheniya sistem differentsial'nykh uravnenii, bliskikh k razryvnym (Periodic Solutions of Systems of Differential Equations Close to Discontinuous Ones). — DAN SSSR, Vol. 102, No. 5. 1955.

118. MISHCHENKO, E. F. Asimptoticheskoe vychislenie periodicheskikh reshenii sistem differentsial'nykh uravnenii, soderzhashchikh malye parametry pri proizvodnykh (Asymptotic Derivation of Periodic Solutions for Systems of Differential Equations Containing Small Parameters Attached to the Derivatives). — Izvestiya AN SSSR, seriya matematicheskaya, Vol. 21, No. 5. 1957.

119. MOISEYEV, N. N. Methods of Nonlinear Mechanics in the Problems of Satellites. (Presented at the conference on nonlinear vibrations, Warsaw, 17-21 September, 1962).

120. MOSEENKOV, B. I. O kolebaniyakh sistem s raspredelennymi parametrami pri prokhozhdenii cherez rezonans (The Vibration of Systems with Distributed Parameters during the Transition through Resonance). — In: Students'ki naukovi pratsi, Zbirnyk XVI, Matematyka, Vidavnytstvo KDU. 1955.

121. MOSYEYENKOV, B. I. Poperechni kolyvannya sterzhnya dvoyakoyi zhorstkosti v perekhidnomu rezhymi (Transverse Oscillations of a Beam Rigid with Respect to Two Axes in a Transient Regime). — Prykladna Mekhanika, Vol. 3, No. 2. Vydavnytstvo AN URSR. 1957. [In Ukrainian.]

122. MOSYEYENKOV, B. I. Poperechni kolyvannya sterzhnya dvoyakoyi zhorstkosti v statsionarnomu rezhymi obertannya (Transverse Oscillations of a Beam Rigid with Respect to Two Axes in a Stationary Rotation Regime). —Prykladna Mekhanika, Vol. 4, No. 2. Vydavnytstvo AN URSR. 1958. [In Ukrainian.]

123. MOSYEYENKOV, B. I. Pro yavyshche rezonansu pid diyeyu syl vlasnoyi vahy obertovoho sterzhnya dvoyakoyi zhorstkosti (Resonance Phenomena of a Rotating Beam Rigid along Two Axes under the Action of the Forces of Its Own Weight). — Dopovidi AN URSR, No. 9. 1959. [In Ukrainian.]

124. NATANZON, B. Krutil'nye kolebaniya kolenchatykh valov (Torsional Vibrations of Crankshafts). — Trudy TsIAM, No. 40. 1943.

125. NEIMAN, I. Sh. Krutil'nye kolebaniya mnogomassovoi nelineinoi sistemy (Torsional Vibrations of a Multi-Mass Nonlinear System). — Oborongiz. 1947.

126. NEMYTSKII, V. V. and V. V. STEPANOV. Kachestvennaya teoriya differentsial'nykh uravnenii (Qualitative Theory of Differential Equations). — Gostekhizdat. 1950.

127. NIKOLAI, E. L. Teoriya giroskopov (The Theory of Gyroscopes). — Gostekhizdat. 1948.

128. PANOVKO, Ya. G. Istoricheskii ocherk razvitiya teorii dinamicheskogo deistviya podvizhnoi nagruzki (Historical Survey of the Development of the Theory of Dynamic Action of a Moving Load). — Trudy Leningradskoi Voenno-Vozdushnoi Inzhenernoi Akademii, No. 17. 1948.

129. PANOVKO, Ya. G. Dinamicheskii raschet sooruzhenii (Dynamical Calculation of Constructions). — In: Sbornik "Stroitel'naya mekhanika v SSSR. 1917-1957", Gosstroiizdat. 1957.

130. PAPALEKSI, N. D. Sobranie trudov (Collected Articles). — Izdatel'stvo AN SSSR. 1948.

131. PISARENKO, G. S. Vynuzhdennye poperechnye kolebaniya sterzhnei postoyannogo secheniya s uchetom gisterezisnykh poter' (Forced Transversal Vibrations of Rods of Constant Cross Section Taking Account of Hysteresis Losses). — Inzhenernyi Sbornik, Vol. 5, No. 1. Izdatel'stvo AN SSSR. 1947.

132. PISARENKO, G. S. Kolebaniya uprugikh sistem s uchetom rasseyaniya energii v materiale (Vibrations of Elastic Systems Taking Account of Energy Scattering in the Substance). — Izdatel'stvo AN USSR. 1955.

133. PISARENKO, G. S. Rasseyanie energii pri mekhanicheskikh kolebaniyakh (Energy Scattering during Mechanical Vibrations). — Izdatel'stvo AN USSR. 1962.

134. PONTRYAGIN, L. S. Sistemy obyknovennykh differentsial'nykh uravnenii s malym parametrom pri vysshikh proizvodnykh (Systems of Ordinary Differential Equations with a Small Parameter Attached to the Highest Derivatives). — In: Trudy III Vsesoyuznogo matematicheskogo s"ezda, Vol. 2. 1956.

135. PONTRYAGIN, L. S. Asimptoticheskoe povedenie reshenii sistem differentsial'nykh uravnenii s malym parametrom pri vysshikh proizvodnykh (Asymptotic Behavior of the Solutions of Systems of Differential Equations with a Small Parameter Attached to the Highest Derivatives). — Izvestiya AN SSSR, seriya matematicheskaya, Vol. 21, No. 5. 1957.

136. POPOV, E. P. Priblizhennoe issledovanie perekhodnykh protsessov v nelineinykh avtomaticheskikh sistemakh (Approximate Study of Transient Processes in Nonlinear Automatic Systems). — Izvestiya AN SSSR, OTN, No. 9. 1956.

137. POPOV, E. P. and I. P. PAL'TOV. Priblizhennye metody issledovaniya nelineinykh avtomaticheskikh sistem (Approximate Methods of Investigating Nonlinear Automatic Systems). — Fizmatgiz. 1960.

138. PÖSCHL, T. Das Anlaufen eines einfachen Schwingers. — Ing. Arch., Vol. 4: 98. 1933.

139. PROSKURYAKOV, A. P. Postroenie periodicheskikh reshenii avtonomnykh sistem s odnoi stepen'yu svobody v sluchae proizvol'nykh veshchestvennykh kornei uravneniya osnovnykh amplitud (Construction of Periodic Solutions for Autonomous Systems with a Single Degree of Freedom in the Case of Arbitrary Real Roots of the Equation of the Basic Amplitudes). — Prikladnaya Matematika i Mekhanika, Vol. 22, No. 4. 1958.

140. PROSKURYAKOV, A. P. K postroeniyu periodicheskikh reshenii avtonomnykh sistem s odnoi stepen'yu svobody (The Construction of Periodic Solutions for Autonomous Systems with a Single Degree of Freedom). — Ibid., Vol. 21, No. 4. 1957.

141. POINCARÉ, H. Les méthodes nouvelles de la mécanique céleste. — Paris, Gauthier-Villars, Vol. I, 1892; Vol. II, 1893; Vol. III, 1899.

142. POINCARÉ, H. Sur les courbes définies par une equation differentialle. — Oeuvres, Vol. I, Paris, Gauthier-Villars. 1950. [Russian translation. 1947.]

143. PUST, L. Vliyanie svoistv istochnika peremennoi sily na kolebaniya mekhanicheskoi sistemy (The Influence of the Properties of a Source of Varying Power on the Vibrations of a Mechanical System). — Aplikace matematiky, Vol. 3, No. 6. 1958.

144. PUST, L. and A. TONDL. Úvod do theorie nelinéarnich a quasiharmonických kmitů mechanickýkh soustav. — Nakladatelsvi Československé Akademie Věd., Praha. 1956.

145. RABINOVICH, M. S. Osnovy teorii sinkhrofazotrona (Bases of the Theory of the Synchrophasotron). — Trudy Fizicheskogo Instituta imeni P. N. Lebedeva, Vol. 10, Izdatel'stvo AN SSSR. 1958.

146. RUBANIK, V. P. Vzaimnoe vliyanie garmonik pri prokhozhdenii cherez rezonans (Mutual Influence of Harmonics during a Transition through Resonance). — Naukovi Students'kii Pratsi Kyyivs'kogo Derzhuniversytetu, Matematyka, No. 16. 1954.

147. RUBANIK, V. P. Prokhozhdenie cherez rezonans v sisteme s dvumya stepenyami svobody (Transition through Resonance in a System with Two Degrees of Freedom). — Ibid.

148. SAVIN, G. N. and O. A. GOROSHKO. Dinamika niti peremennoi dliny (The Dynamics of a Thread of Varying Length). p. 331. — Kiev, Izdatel'stvo AN USSR. 1962.

149. SANSONE, G. and R. CONTI. Equazioni differenziali non-lineari. Roma. 1956.

150. SERENSEN, S. V. Dinamicheskaya prochnost' v mashinostroenii (Dynamical Strength in Mechanical Engineering). — Mashgiz. 1945.

151. SLUTSKA, O. B. Doslidzhennya nestatsionarnykh avtokolyvan' systemy dvyhun-transmisiya traktora pry yiyi roz"honi (Nonstationary Auto-Oscillations, under Acceleration of an Engine-and-Transmission System of a Tractor). — Prykladna Mekhanika, Vol. 6, No. 4. 1960. [In Ukrainian.]

152. SMELKOV, A. A. O kolebaniyakh v giroskopicheskikh sistemakh pri bol'shikh skorostyakh vrashcheniya (Vibrations in Gyroscopic Systems during Large Rotation Velocities). — Inzhenernyi Sbornik, Vol. 8, Izdatel'stvo AN SSSR. 1950.

153. STOKER, J. J. Nonlinear Vibrations in Mechanical and Electrical Systems. — Interscience Publishers, Inc. 1950. [Translated into Russian. 1953.]

154. TIMOSHENKO, S. P. Vibration Problems in Engineering. — D. Van Nostrand Co., New York. 1937. [Translated into Russian. 1959.]

155. TIKHONOV, A. N. O zavisimosti reshenii differentsial'nykh uravnenii ot malogo parametra (The Dependence of the Solutions of Differential Equations on a Small Parameter). — Matematicheskii Sbornik, Vol. 22, No. 2. 1948.

156. TIKHONOV, A. N. Sistemy differentsial'nykh uravnenii, soderzhashchie malye parametry pri proizvodnykh (Systems of Differential Equations with Small Parameters Attached to the Derivatives). — Matematicheskii Sbornik, Vol. 31, No. 5. 1952.

157. Trudy Mezhdunarodnogo simpoziuma po nelineinoi mekhanike (Proceedings of the International Symposium on Nonlinear Mechanics), Vols. I–III. — Izdatel'stvo AN USSR. 1963.

158. TURBOVICH, I. T. K voprosu o sistemakh s peremennymi parametrami (Systems with Variable Parameters). — Izvestiya AN SSSR, OTN, No. 2. 1948.

159. WHITTAKER, E. T. A Treatise on the Analytical Dynamics of Particles and Rigid Bodies. — Dover Publ., New York, 4th Ed. 1937. [Translated into Russian. 1937.]

160. FEDORCHENKO, A. M. O dvizhenii tyazhelogo nesimmetrichnogo giroskopa s vibriruyushchei tochkoi podvesa (The Motion of a Heavy Asymmetric Gyroscope with a Vibrating Suspension Point). — UMZh, Vol. 10, No. 2. 1958.

161. FESHCHENKO, S. F. Pytannya teoriyi zvychainykh liniinykh dyferentsial'nykh rivnyan' druhoho poryadku z povil'no-zminnymy koefitsiyentamy ta deyaki zastosuvannya (Contribution to the Theory of Ordinary Linear Differential Equations of the Second Order with Slowly Varying Coefficients, and Some Applications). — Naukovi Zapysky Kyyivs'kogo Derzhavnoho Pedahohichnoho Institutu, Vol. 6, No. 3. 1948. [In Ukrainian.]

162. FESHCHENKO, S. V. Asimptoticheskoe predstavlenie integralov lineinogo neodnorodnogo differentsial'nogo uravneniya vtorogo poryadka s medlenno menyayushchimisya koeffitsientami (Asymptotic Representation of the Integrals of a Linear Inhomogeneous Second Order Differential Equation with Slowly Varying Coefficients). — Doklady AN SSSR, No. 2. 1947.

163. FILATOV, A. N. O primenenii ryadov Li k zadacham nelineinoi mekhaniki (The Application of Lie Series to the Problems of Nonlinear Mechanics). — In: Trudy Mezhdunarodnogo simpoziuma po nelineinym kolebaniyam, Izdatel'stvo AN USSR, Vol. I. 1963.

164. FILIPPOV, A. P. Kolyvannya balky pid diyeyu rukhomoho vantazhu (Oscillations of a Beam under the Action of a Moving Load). — Prykladna Mekhanika, Vol. 1, No. 3. 1955. [In Ukrainian.]

165. FILIPPOV, A. P. Kolebaniya uprugikh sistem (Vibrations of Elastic Systems). — Izdatel'stvo AN USSR. 1956.

166. FILIPPOV, A. P. Vynuzhdennye kolebaniya lineinoi sistemy pri perekhode cherez rezonans (Forced Vibrations of a Linear System during a Transition through Resonance). — In: Sbornik "Kolebeniya v turbomashinakh" (Institut mashinovedeniya AN SSSR), Izdatel'stvo AN SSSR. 1956.

167. CHIHIRO HAYASHI. Vynuzhdennye kolebaniya v nelineinykh sistemakh (Forced Oscillations in Nonlinear Systems). — IL. 1957. [Translated into English. — Nippon Printing and Publishing Company, Ltd. 1953.]

168. CHEKMAREV, A. I. K voprosu o raschete krutil'nykh kolebanii sistem s mayatnikovymi antivibratorami (The Calculation of Torsional Vibrations of Systems with Floating Shock Absorbers). — In: Dinamika i prochnost' kolenchatykh valov, Izdatel'stvo AN SSSR. 1948.

169. CHEKMAREV, A. I. Vzaimnoe vliyanie garmonik v nelineinykh sistemakh (Mutual Influence of Harmonics in Nonlinear Systems). — Ibid., Sbornik (Collection) II. 1950.

170. CHELOMEI, V. N. O kolebaniyakh sterzhnei, podverzhennykh deistviyu periodicheski menyayushchikhsya prodol'nykh sil (The Vibrations of Rods Subjected to the Action of Periodically Varying Longitudinal Forces). — Trudy KAI, Issue 8. 1937.

171. CHELOMEI, V. N. O vozmozhnosti povysheniya ustoichivosti uprugikh sistem pri pomoshchi vibratsii (The Possibility of Increasing the Stability of Elastic Systems by Means of Vibrations). — DAN SSSR, Vol. 110, No. 3. 1956.

172. SHATALOV, K. T. Voprosy eksperimental'nykh issledovanii krutil'nykh kolebanii valov dvigatelei (Problems in the Experimental Study of Torsional Vibrations of Motor Shafts). — In: Dinamika i prochnost' kolenchatykh valov, Sbornik (Collection) II, Izdatel'stvo AN SSSR. 1950.

173. SHIMANOV, S. N. K teorii kolebanii kvazilineinykh sistem (The Theory of Vibrations of Quasi-Linear Systems). — Prikladnaya Matematika i Mekhanika, Vol. 18, No. 2. 1954.

174. SHTOKALO, I. Z. Lineinye differential'nye uravneniya s peremennymi koeffitsientami (Linear Differential Equations with Variable Coefficients). — Izdatel'stvo AN USSR. 1960. [Translated into English. Delhi, Hindustan Russian Corpn. 1961.]

175. SHTOKALO, I. Z. Operatsionnye metody i ikh razvitie v teorii lineinykh differentsial'nykh uravnenii s peremennymi koeffitsientami (Operational Methods and their Development in the Theory of Linear Differential Equations with Variable Coefficients). — Izdatel'stvo AN USSR. 1961.

EXPLANATORY LIST OF ABBREVIATED NAMES OF USSR INSTITUTIONS, ORGANIZATIONS, PERIODICALS, ETC., APPEARING IN THE BIBLIOGRAPHY

Abbreviation	Full name (transliterated)	Translation
AN SSSR	Akademiya Nauk SSSR	Academy of Sciences of the USSR
DAN	Doklady Akademii Nauk	Reports of the Academy of Sciences
KAI	Kuibyshevskii Aviatsionnyi Institut	Kuibyshev Aviation Institute
	Kazanskii Aviatsionnyi Institut	Kazan Aviation Institute
KDU	Kiivs'kii Derzhavnii Universitet	Kiev State University
MAP	Ministerstvo Aviatsionnoi Promyshlennosti	Ministry of Aviation Industry of the USSR
MIIT	Moskovskii Institut Inzhenerov Zheleznodorozhnogo Transporta	Moscow Institute of Railroad Transportation Engineers
OTN	Otdelenie Tekhnicheskikh Nauk	Technical Sciences Section [of the Academy of Sciences of the USSR]
TsIAM	Tsentral'nyi Nauchno-Issledovatel'skii Institut Aviatsionnogo Motorostroeniya	Central Scientific-Research Institute of Aviation Engine Construction
UMZH	Ukrainskii Matematicheskii Zhurnal	Ukrainian Journal of Mathematics
ZhTF	Zhurnal Tekhnicheskoi Fiziki	Journal of Technical Physics